Air Pollution

Air Pollution

KL DOREN

CBS Publishers & Distributors Pvt Ltd

New Delhi • Bengaluru • Chennai • Kochi • Kolkata • Mumbai • Pune
Hyderabad • Nagpur • Patna • Pune • Vijayawada

Air Pollution

ISBN: 978-81-239-2901-9 (Soft Cover)

ISBN: 978-81-239-2924-8 (Hard Cover)

Copyright © Publisher

First Edition: 2016
Reprint: 2017

Published by Satish Kumar Jain and produced by Varun Jain for

CBS Publishers & Distributors Pvt Ltd

4819/XI Prahlad Street, 24 Ansari Road, Daryaganj, New Delhi 110 002, India.
Ph: 23289259, 23266861, 23266867   Website: www.cbspd.com
Fax: 011-23243014   e-mail: delhi@cbspd.com; cbspubs@airtelmail.in.
Corporate Office: 204 FIE, Industrial Area, Patparganj, Delhi 110 092
Ph: 4934 4934   Fax: 4934 4935   e-mail: publishing@cbspd.com; publicity@cbspd.com

Branches

- **Bengaluru:** Seema House 2975, 17th Cross, K.R. Road,
  Banasankari 2nd Stage, Bengaluru 560 070, Karnataka
  Ph: +91-80-26771678/79   Fax: +91-80-26771680   e-mail: bangalore@cbspd.com
- **Chennai:** 7, Subbaraya Street, Shenoy Nagar, Chennai 600 030, Tamil Nadu
  Ph: +91-44-26680620. 26681266   Fax: +91-44-42032115   e-mail: chennai@cbspd.com
- **Kochi:** Ashana House, No. 39/1904, AM Thomas Road, Valanjambalam,
  Ernakulam 682 016, Kochi, Kerala
  Ph: +91-484-4059061-65   Fax: +91-484-4059065   e-mail: kochi@cbspd.com
- **Kolkata:** 6/B, Ground Floor, Rameswar Shaw Road, Kolkata-700 014, West Bengal
  Ph: +91-33-22891126. 22891127, 22891128   e-mail: kolkata@cbspd.com
- **Mumbai:** 83-C, Dr E Moses Road, Worli, Mumbai-400018, Maharashtra
  Ph: +91-22-24902340/41   Fax: +91-22-24902342   e-mail: mumbai@cbspd.com

Representatives

- **Hyderabad** 0-9885175004
- **Jharkhand** 0-9811541605
- **Nagpur** 0-9021734563
- **Patna** 0-9334159340
- **Pune** 0-9623451994
- **Uttarakhand** 0-9716462459

Printed at India Binding House, Noida, UP, India

Preface

Air pollution is the introduction of chemicals, particulate matter, or biological materials that cause harm or discomfort to humans or other living organisms, or cause damage to the natural environment or built environment, into the atmosphere. The atmosphere is a complex dynamic natural gaseous system that is essential to support life on planet earth. Stratospheric ozone depletion due to air pollution has long been recognised as a threat to human health as well as to the earth's ecosystems. Indoor air pollution and urban air quality are listed as two of the world's worst pollution problems. Pollutants can be in the form of solid particles, liquid droplets, or gases. In addition, they may be natural or man-made.

This reference textbook on air pollution is divided into five sections. Each chapter covers an important aspects of air pollution and its control. Section I is devoted to general considerations of air pollution. Chapter 1 is focused on introduction to environment which is constituted by the interacting systems of physical, biological and cultural elements which are inter-linked individually as well as collectively in many ways. Chapter 2 deals with particulate matter which are fine particles and so-called soot, are tiny subdivision of solid matter suspended in a gas or liquid. Chapter 3 concentrates on gaseous pollutants which are caused by burning of fuels. Chapter 4 discusses odour pollution and its control. Controlling odour is an important consideration for protecting the environment, because there are neither indicative compounds within odour plumes nor electronic or other devices for measuring odour emissions. Chapter 5 deals with indoor air pollution which leads to severe health hazards like irritation to the eyes, nose, throat, headache and most importantly the respiratory problems. Chapter 6 focuses on meteorological aspects of air pollution.

Section II discusses global warming which is driven primarily by CO_2 produced by fossil fuel energy usage and by industry. Chapter 7 is devoted to impact of air pollution on local to global scale. Chapter 8 concentrates on hurricanes which is a very powerful, sometimes violent storm with strong winds and heavy rains. Chapter 9 focuses on volcanoes which are opening or rupture, in a planet's surface or crust, which allows hot magma, volcanic ash and gases to escape below the surface. Volcanoes are generally found where tectonic plates are diverging or converging. Chapter 10 discusses tornado which is a violent, dangerous, rotating column of air that is in contact with both the surface of the earth and a cumulonimbus cloud or in rare cases, the base of a cumulus cloud. Chapter 11 focuses on tsunami or tidal wave is a series of water waves (called tsunami wave train) caused by displacement of a large volume of a body of water, usually on ocean, but can occur in large lakes also. Chapter 12 concentrates on El Nino and La Nina which can be described as a type of abnormal warming that occurs on the surface of ocean waters in the part of the eastern tropical pacific that is known as the southern oscillation. Chapter 13 deals with Kyoto Protocol which is an international agreement linked to United Nations framework convention on climate change. The major feature of the Kyoto protocol is that it sets binding targets for 37 industrialised countries and the European community for reducing greenhouse gas (GHG) emissions. These amounts to an average of five per cent against 1990 levels over the five year period 2008–2012.

Section III focuses on air pollution from chemical, metallurgical and miscellaneous industries. Chapter 14 is devoted to metallurgical industries. Metallurgical industries being the largest manpower employers and the largest material handling processing industries, contribute significantly to air, water and land pollution. Moreover, most metallurgical industries of today came into existence at a time when concept of pollution and implementation of pollution laws were not considered very seriously. Some polluting industries discussed are iron and steel, aluminium, lead, zinc, copper and foundries, which are the first among industrial plants producing air pollutants. Chapter 15 discusses chemical and allied industries which over the years have developed several methodologies aimed at early detection of hazards and attenuate residual risks. The polluting industries discussed are—petroleum refinery, pesticides, fertilisers, paints and dyes, textiles, drugs and pharmaceutical, leather and tannery, caustic soda, sodium carbonate. Various food processing industries such as sugar, dairy, meat and seafood are also discussed. Chapter 16 deals with miscellaneous industries which are hazardous and polluting are—nuclear and radioactive, thermal power plants and cement.

Section IV focuses on special topics. Chapter 17 deals with instrumental techniques in environmental chemical analysis. Chapter 18 concentrates on fugitive emissions from chemical and cement industries. Fugitive emissions are leaks or released that occur whenever there are discontinuities and solid barrier that maintains containment. Fugitive emissions include pumps and compressors, storage and processing vessels, loading facilities, flow control and pressure relief valves and leakage from pipelines carrying materials from one process to another. Chapter 19 is devoted to chemical toxicology which is a branch of biology and medicine concerned with the study of the adverse effects of chemicals on living organisms. Chapter 20 discusses about environmental implications of nanotechnology and the possible effects of the use of nanotechnological materials and devices.

Section V discusses various case studies related to air pollution. Chapter 21 therefore concentrates on important case studies such as Bhopal gas tragedy, Chernobyl explosion–Russia and London smog, etc.

Glossary and index have been provided at the end for quick reference. Diagrams, figures and tables supplement the text. All the topics have been covered in a cogent and lucid style to help the reader grasp the information quickly and easily.

It may not be wrong to hold that the present reference textbook on *Air Pollution* is a complete treatise on environment and air pollution. It is essential reading for all students and teachers of environment, engineering and life sciences. In addition, researchers in environmental and allied fields will also find it highly useful and informative.

The reference textbook also caters to the requirement of the syllabus prescribed by various Indian universities for undergraduate student pursuing engineering, life sciences, environment and allied courses. It has been prepared with meticulous care, aiming at making the book error-free. Constructive suggestions are always welcome from users of this book.

K. L. Doren

Contents at a Glance

SECTION IV

SECTION V

Contents

SECTION I

General Considerations

Chapter 1

Introduction to Environment

INTRODUCTION

Environment is viewed in different ways and angles by different groups of people and disciplines. It may be safely argued that environment is an inseparable whole and is constituted by the interacting systems of physical, biological and cultural elements which are inter-linked individually as well as collectively in many ways.

Physical elements (space, landform, water bodies, climate, soils, rocks and minerals) determine the variable character of the human habitat, its opportunities as well as its limitations. Biological elements (plants, animal, micro-organisms and man) constitute the biosphere; cultural elements (economic, social and political) are essentially man-made feature, which go into the making of cultural milieu.

TYPES OF ENVIRONMENT

The environment is both physical and biological concept, it includes both the non-living (abiotic) and living (biotic) components of the planet earth. Thus based on basic structure, the environment can be divided into two basic types:

1. Physical or abiotic environment.
2. Biological or biotic environment.

Based on physical characteristics and state, abiotic or physical environment is subdivided into three broad categories:

1. Lithosphere (solid earth).
2. Hydrosphere (water component).
3. Atmosphere (gas).

The biotic components of the environment consist of plant (Flora) and animal (Fauna) including man as an important component and micro-organisms.

Relationship Between Man and Environment

The study of relationship between man and environment has always attracted attention. The relationship of man and environment has also influenced the development of human society. It may be noted that, of all the organisms, man is the most skilled and civilised and therefore, it is significant to note the following three aspects of man:

1. Physical man is a component of the biological community and as such, requires basic elements of physical environment such as air, water, food and habitat, etc. like other biological population and release wastes in the eco-system.

2. Social man establishes the social institutions, forms the social organisations and formulates laws and policies to safeguard his existence, interests and welfare.
3. Economic man derives and utilises resource from the physical and biological.

Changing relationship of man with environment

The changing relationship of man with the environment from pre-historic to modern times can be divided into the following four periods:

1. Hunting and food gathering.
2. Animal domestication and agriculture.
3. Plant domestication and agriculture.
4. Science, technology and industrialisation.

Period of hunting and food gathering

This period is related to the most primitive man when he had been basically a part of natural environment and was functionally as a 'biological man' or 'physical man' because his basic requirements were limited to food and shelter. The natural environment satisfied all his wants. The relationship between man and the environment was very friendly. Man was leading a nomadic life. Then a stage came when he learnt to hunt animals. The discovery of 'fire', which was accidental, taught man to cook animal flesh before eating. So we can say that the discovery of fire and subsequently, invention of tools and weapons made man capable of exploiting natural resources for his benefit. Some would also conclude by saying that 'fire' was the first major ecological tool used by man to change the environment for his own benefit.

Period of animal domestication and pastoralism

With the passage of time, primitive man learnt to domesticate animals for his benefits. In the beginning, he might have domesticated some milch/cattle and animals for meat and slowly his herd of domesticated animals must have increased. Domestication of animals might have given birth to group or community life among early people in order to protect their flock and themselves from wild animals. They still stuck to nomadic way of life, as they had to move from one place to other in search of water, food for themselves and fodder for animals.

Period of plant domestication and agriculture

Domestication of plants for food became a hallmark in the development of human skills of timing and controlling the biotic component of the natural environment system. Domestication of plant initiated primitive type of agriculture and sedentary settled life of people who were nomads. Cultivation of food crops resulted in the formation of social groups and organisation. Now man started settling down in the river valleys due to availability of water and fertile land which came to be known as 'river-valley civilisation'. From here onwards started the journey of man to transform the natural environmental resources around him through improved farming practices resulting in gradual increase in human population. This led to clearing of forests to have access more agricultural land. With the passage of time, man developed his own cultural environment by building houses and creating towns and cities, constructing roads and bridges.

Period of science, technology and industrialisation

The advancement of industrial revolution in late nineteenth century and emergence of science and development of sophisticated technology, embittered the friendly relationship between man and his natural environment. This impact of modern technology on natural environment is highly complex and controversial. Highly advanced technologies and scientific techniques led to indiscriminate exploitation of natural environment which have created most of the present day environmental problems.

From the days of early primitive man till today, man has venerated nature in different forms (trees, plants, animals, rivers, mountains, etc.). Nature worship commands great sanctity in most of the communities. In the Indian tradition, nature and man form an inseparable, part of the life support system. The five elements, i.e. air, water, land, flora and fauna are inter-related and inter-dependent.

NATURAL ENVIRONMENT

The natural environment, encompasses all living and non-living things occurring naturally on earth or some region thereof. It is an environment that encompasses the interaction of all living species. The concept of the natural environment can be distinguished by the following components:

1. Complete ecological units that function as natural systems without massive human intervention, including all vegetation, micro-organisms, soil, rocks, atmosphere and natural phenomena that occur within their boundaries.
2. Universal natural resources and physical phenomena that lack clear-cut boundaries, such as air, water, and climate, as well as energy, radiation, electric charge, and magnetism, not originating from human activity.

The natural environment is contrasted with the built environment, which comprises the areas and components that are strongly influenced by humans. A geographical area is regarded as a natural environment, if the human impact on it is kept under a certain limited level.

Composition

Earth science generally recognises four spheres, the lithosphere, the hydrosphere, the atmosphere, and the biosphere as correspondent to rocks, water, air, and life. Some scientists include, as part of the spheres of the earth, the cryosphere (corresponding to ice) as a distinct portion of the hydrosphere, as well as the pedosphere (corresponding to soil) as an active and intermixed sphere. Earth science (also known as geoscience, the geosciences or the earth sciences), is an all-embracing term for the sciences related to the planet earth. There are four major disciplines in earth sciences, namely geography, geology, geophysics and geodesy. These major disciplines use physics, chemistry, biology, chronology and mathematics to build a qualitative and quantitative understanding of the principal areas or spheres of the earth system.

Geological Activity

The earth's crust, or lithosphere, is the outermost solid surface of the planet and is chemically and mechanically different from underlying mantle. It has been generated largely by igneous processes in which magma (molten rock) cools and solidifies to form solid rock. Beneath the lithosphere lies the mantle which is heated by the decay of radioactive elements. The mantle though solid is in a state of rheic convection. This convection process causes the lithospheric plates to move, albeit slowly. The resulting process is known as plate tectonics. Volcanoes result primarily from the melting of subducted crust material or of rising mantle at mid-ocean ridges and mantle plumes.

Water on Earth

Oceans

An ocean is a major body of saline water, and a component of the hydrosphere. Approximately 71 per cent of the earth's surface (an area of some 361 million square kilometers) is covered by ocean, a continuous body of water that is customarily divided into several principal oceans and smaller seas. The concept of a global ocean as a continuous body of water with relatively free interchange among its parts is of fundamental importance to oceanography. The major oceanic divisions are defined in part by the continents, various archipelagos, and other criteria: these divisions are (in descending order of size) the Pacific Ocean, the Atlantic Ocean, the Indian Ocean, the Southern Ocean and the Arctic Ocean.

Rivers

A river is a natural watercourse, usually freshwater, flowing toward an ocean, a lake, a sea or another river. In a few cases, a river simply flows into the ground or dries up completely before reaching another body of water. Small rivers may also be termed by several other names, including stream, creek and brook. Rivers are a part of the hydrological cycle. Water within a river is generally collected from precipitation through surface runoff, groundwater recharge, springs, and the release of water stored in glaciers and snowpacks.

Streams

A stream is a flowing body of water with a current, confined within a bed and stream banks. Streams play an important corridor role in connecting fragmented habitats and thus in conserving biodiversity. The study of streams and waterways in general is known as surface hydrology. Types of streams include creeks, tributaries, which do not reach an ocean and connect with another stream or river, brooks, which are typically small streams and sometimes sourced from a spring or seep and tidal inlets.

Lakes

A lake is a terrain feature, a body of water that is localised to the bottom of basin. A body of water is considered a lake when it is inland, is not part of a ocean, is larger and deeper than a pond, and is fed by a river.

Natural lakes on earth are generally found in mountainous areas, rift zones, and areas with ongoing or recent glaciation. Other lakes are found in endorheic basins or along the courses of mature rivers. In some parts of the world, there are many lakes because of chaotic drainage patterns left over from the last Ice Age. All lakes are temporary over geologic time scales, as they will slowly fill in with sediments or spill out of the basin containing them.

Ponds

A pond is a body of standing water, either natural or man-made, that is usually smaller than a lake. A wide variety of man-made bodies of water are classified as ponds, including water gardens designed for aesthetic ornamentation, fish ponds designed for commercial fish breeding, and solar ponds designed to store thermal energy. Ponds and lakes are distinguished from streams via current speed. While currents in streams are easily observed, ponds and lakes possess thermally driven micro-currents and moderate wind driven currents. These features distinguish a pond from many other aquatic terrain features, such as stream pools and tide pools.

Atmosphere, Climate and Weather

The atmosphere of the earth serves as a key factor in sustaining the planetary ecosystem. The thin layer of gases that envelops the earth is held in place by the planet's gravity. Dry air consists of 78 per cent nitrogen, 21 per cent oxygen, 1 per cent argon and other inert gases, such as carbon dioxide. The remaining gases are often referred to as trace gases, among which are the greenhouse gases such as water vapour, carbon dioxide, methane, nitrous oxide, and ozone. Filtered air includes trace amounts of many other chemical compounds. Air also contains a variable amount of water vapour and suspensions of water droplets and ice crystals seen as clouds. Many natural substances may be present in tiny amounts in an unfiltered air sample, including dust, pollen and spores, sea spray, volcanic ash, and meteoroids. Various industrial pollutants also may be present, such as chlorine (elementary or in compounds), fluorine compounds, elemental mercury, and sulphur compounds such as sulphur dioxide.

The ozone layer of the earth's atmosphere plays an important role in depleting the amount of ultraviolet (UV) radiation that reaches the surface. As DNA is readily damaged by UV light, this serves to protect life at the surface. The atmosphere also retains heat during the night, thereby reducing the daily temperature extremes.

Atmospheric layers

Earth's atmosphere can be divided into five main layers. These layers are mainly determined by whether temperature increases or decreases with altitude. From highest to lowest, these layers are: (i) Exosphere, (ii) Thermosphere, (iii) Mesosphere, (iv) Stratosphere, and (v) Troposphere.

Other layers
Within the five principal layers determined by temperature are several layers determined by other properties. There are: (i) Ozone layer, (ii) Ionosphere, (iii) Homosphere, and (iv) Planetary boundary layer.

Effects of global warming
The potential dangers of global warming are being increasingly studied by a wide global consortium of scientists, who are increasingly concerned about the potential long-term effects of global warming on our natural environment and on the planet. Of particular concern is how climate change and global warming caused by anthropogenic, or human-made releases of greenhouse gases, most notably carbon dioxide, can act interactively, and have adverse effects upon the planet, its natural environment and humans' existence. Efforts have been increasingly focused on the mitigation of greenhouse gases that are causing climatic changes, on developing adaptative strategies to global warming, to assist humans, animal and plant species, ecosystems, regions and nations in adjusting to the effects of global warming. Some examples of recent collaboration to address climate change and global warming include.

Climate
Climate encompasses the statistics of temperature, humidity, atmospheric pressure, wind, rainfall, atmospheric particle count and numerous other meteorological elements in a given region over long periods of time. Climate can be contrasted to weather, which is the present condition of these same elements over periods up to two weeks.

Weather
Weather is a set of all the phenomena occurring in a given atmospheric area at a given time. Most weather phenomena occur in the troposphere, just below the stratosphere. Weather refers, generally, to day-to-day temperature and precipitation activity, whereas climate is the term for the average atmospheric conditions over longer periods of time. When used without qualification, 'weather' is understood to be the weather of earth.

Weather occurs due to density (temperature and moisture) differences between one place and another. These differences can occur due to the sun angle at any particular spot, which varies by latitude from the tropics. The strong temperature contrast between polar and tropical air gives rise to the jet stream.

Surface temperature differences in turn cause pressure differences. Higher altitudes are cooler than lower altitudes due to differences in compressional heating. Weather forecasting is the application of science and technology to predict the state of the atmosphere for a future time and a given location. The atmosphere is a chaotic system, and small changes to one part of the system can grow to have large effects on the system as a whole. Human attempts to control the weather have occurred throughout human history, and there is evidence that human activity such as agriculture and industry has inadvertently modified weather patterns.

Life

Evidence suggests that life on earth has existed for about 3.7 billion years. All known life forms share fundamental molecular mechanisms, and based on these observations, theories on the origin of life attempt to find a mechanism explaining the formation of a primordial single cell organism from which all life originates. There are many different hypotheses regarding the path that might have been taken from simple organic molecules via pre-cellular life to protocells and metabolism.

Ecosystems

An ecosystem is a natural unit consisting of all plants, animals and micro-organisms (biotic factors) in an area functioning together with all of the non-living physical (abiotic) factors of the environment.

Central to the ecosystem concept is the idea that living organisms are continually engaged in a highly interrelated set of relationships with every other element constituting the environment in which they exist.

A greater number or variety of species or biological diversity of an ecosystem may contribute to greater resilience of an ecosystem, because there are more species present at a location to respond to change and thus 'absorb' or reduce its effects. This reduces the effect before the ecosystem's structure is fundamentally changed to a different state. This is not universally the case and there is no proven relationship between the species diversity of an ecosystem and its ability to provide goods and services on a sustainable level.

The term ecosystem can also pertain to human-made environments, such as human ecosystems and human-influenced ecosystems, and can describe any situation where there is relationship between living organisms and their environment. Fewer areas on the surface of the earth today exist free from human contact, although some genuine wilderness areas continue to exist without any forms of human intervention.

Biomes

Biomes are terminologically similar to the concept of ecosystems, and are climatically and geographically defined areas of ecologically similar climatic conditions on the earth, such as communities of plants, animals, and soil organisms, often referred to as ecosystems. Biomes are defined on the basis of factors such as plant structures (such as trees, shrubs, and grasses), leaf types (such as broadleaf and needleleaf), plant spacing (forest, woodland, savanna), and climate. Unlike ecozones, biomes are not defined by genetic, taxonomic, or historical similarities. Biomes are often identified with particular patterns of ecological succession and climax vegetation.

Biogeochemical Cycles

Global biogeochemical cycles are critical to life, most notably those of water, oxygen, carbon, nitrogen and phosphorus.

1. The nitrogen cycle is the transformation of nitrogen and nitrogen-containing compounds in nature. It is a cycle which includes gaseous components.
2. The water cycle, is the continuous movement of water on, above, and below the surface of the earth. Water can change states among liquid, vapour, and ice at various places in the water cycle.
3. The carbon cycle is the biogeochemical cycle by which carbon is exchanged among the biosphere, pedosphere, geosphere, hydrosphere, and atmosphere of the earth.
4. The oxygen cycle is the movement of oxygen within and between its three main reservoirs: the atmosphere, the biosphere, and the lithosphere. The main driving factor of the oxygen cycle is photosynthesis, which is responsible for the modern earth's atmospheric composition and life.
5. The phosphorus cycle is the movement of phosphorus through the lithosphere, hydrosphere, and biosphere.

ENVIRONMENTAL POLLUTION

Presence of a pollutant in environment which affects our living and non-living resources adversely, is called environmental pollution. Under the Environment (Protection) Act, 1986, the environmental pollution has been defined as under:

1. Environmental pollution means the presence in the environment of any environmental pollutant.
2. Environment pollution can be defined as any adverse change in the environment caused directly or indirectly by man, in the process of his living.

With the increasing accent on industrialisation, man has become both the cause and victim of pollution. Industrialisation involves progressively greater exploitation of natural resources, with the consequent loading of the environment with harmful by-products. The result is that pollution has assumed giant proportion. The environmental pollution can mainly occur in the following forms:

Air Pollution

There are many pollutants present in the environment. Air is obviously the first and most susceptible component of our environment towards pollution. It being our fundamental need to keep ourselves alive, any change in its composition will create serious problem. Air pollution may be defined as the occurence or release into the atmosphere of any foreign materials or gases which are harmful to man, vegetation, animals or property. Burning of coal, or other fossils fuels releases considerable amount of sulphur compounds like sulphur oxides into the air.

Some of main causes of air pollution are domestic fuels like coal and firewood, combustion engines such as automobiles, plants, and industries. These cause an accummulation of solid particles, poisonous liquid particles and gases in clear air. Air pollution is a serious threat for those living in urban, congested, industrialised cities with heavy vehicular traffic not only in India but also all over the world.

Water Pollution

In this modern era, the problems like population increase, sewage disposal, industrial wastes, radioactive wastes, agriculture wastes, etc. have polluted our water resources so much that more than 70 per cent rivers and streams contain polluted water. The water is said to be polluted when the presence of some foreign organic, inorganic, biological, radiological, and physical substance or property in the water tends to deteriorate its quality and either constitutes a health hazard or otherwise decreases the utility of water for consumption purposes.

Water pollution is caused due to the presence of physical, chemical and bacteriological impurities in the water. Though, some pollution occurs in its natural form of soil erosion, disposing of animal wastes, etc. but most of the water pollution is the direct result of human activity.

The water pollution can be:
1. Surface water pollution.
2. Underground water pollution.
3. Marine water pollution.

Land Pollution

There is intimate relationship between the land and other components of the environment. Soil is the most important component of the earth's crust. Like air and water, soil is also subject to pollution. Any substance that is common or foreign to the soil system which adversely effects the productivity of the soil is termed as a soil pollutant. In addition to the natural causes of land pollution, man has been mainly responsible for disturbing the natural environment of the soil in different ways, through his domestic, agricultural and industrial activities which directly or indirectly pollute the soil. Landscape is one of the important part of our environment which causes some form of pollution. Loss of soil productivity due to erosion, unplanned irrigation, overgrazing, deforestration lead to formation of deserts. With huge pressure of population on the land and the intensive and extensive employment of technology and industrial inputs, the land degradation has become major problem in the developing countries.

Noise Pollution

A recent addition to the galaxy of pollutants is noise. Noise is neither tangible nor measureable except scientifically. The amount of harm it does to a person depends on his senstivity to it. Scientifically speaking, the intensity of the pitch and the frequency of the sound determine the quality of the noise. Noise pollution is a serious threat to the quality of man's environment.

Noise pollution is an unwanted sound dumped into atmosphere which adversely affects the surroundings. Noise can disturb air, affect our hearing and interefere with our conversation. This may result in lower efficiency in offices, increase in cough, hoarseness, throat pain, and headaches due to strain of talking loudly in a noisy atmosphere. The maximum comfortable level for the human ear is assesed to be about 55 decibles (db). The continued exposure to levels above 100 db is likely to affect hearing ability and may damage the inner ear. The modern machines and gadgets like loud speakers, radios, stereos, television, printing presses, industries, transport vehicles, autos, motorcycles, aeroplanes, etc. are the major sources of noise pollution.

There is another serious type of pollution caused by radioactive substances known as radiation pollution. Radiation has been responsible for hazards in recent years which may be caused by ultraviolet rays, X-rays, alpha, beta, gamma rays, and neutrons, etc. These radiations act as molecular and nuclear level and are at times responsible for mutations. Carcinogenic properties of these rays are now a widely accepted scientific truth.

SOURCES OF POLLUTION

The pollution can occur in a natural process as well as due to unmindful activities of a man. Man as a biological organism has developed in an environment which is changing more rapidly than ever before.

Certain environmental conditions are necessary to sustain life. Man has controlled vast forces of nature and subdued them for his material gain. But he has not yet succeeded in creating an environment which is favourable to life's full growth. He has failed to maintain a balance between himself and nature. Thus, the sources of environmental pollution can be divided mainly into two parts:

1. Natural sources of pollution.
2. Man-made sources of pollution.

Natural Sources of Pollution

Nature plays a very important role in the growth and decline of human beings. With the ever noticed changes, there is ecological imbalance. The change in climate, soil erosion, earthquakes, landscapes, industrial disastors and biological hazards directly or indirectly affect the living and non-living environment.

The main threats to the environmental pollution due to natural sources are:

1. Floods, glaciers, droughts, earthquakes.
2. Critically low population of species.
3. Diseases.
4. Inadequacy of existing regulative mechanism.
5. Lack of pollinators.
6. Invasion by exotics or other aggressive spices.
7. Air and water pollution (which can also be man made).

Man-made Sources of Pollution

Some of the environmental hazards are natural, but increasingly environmental pollution result from the man's activities. Some time man made hazards have their impact directly on other men and resources, but they may be indirect in their influence, acting through other biological systems or over burdening the capacity of natural systems for renewal, dispersion or assimilation. The man-made threats to the environment are mainly summarised as under:

1. Destruction or modification of habitats.
2. Over-exploitation for commercial, scientific and educational purposes.
3. Overgrasing by domesticated animals.
4. Regeneration of scrub.
5. Change in arable farming.
6. Forestry.
7. Traditional rural practices.
8. Industrialisation, urbanisation, building townships, roads and dams.
9. Tourism and tourist development that is coastal and inland.
10. Mining and quarrying.
11. Pressure from introduced plants.
12. Population pressure.
13. Destruction of ecological balance because of food habits, game and commercial exploitation.
14. Use of drugs and chemicals.

The sources of pollution can also be classified as:

1. Controllable sources.
2. Un-controllable sources.

The man has no control on the natural catastrophies. The untimely changes in nature are beyond the control of man and it will affect the environment. The pollution created through the unmindful activities of a man can however, easily be controlled. But the speed and nature of man induced environmental changes have brought increasing disharmony between him and nature. The degradation of environment has been caused mostly by his activities such as burning of wood fuel, smelting of ores, tanning of leather, primitive of sewage disposal, etc.

MAJOR SOURCES OF ENVIRONMENTAL POLLUTION

Some of the major sources of environmental pollution can be discussed as follows:

Population Density

Increase in population leads to greater demands on our natural resources with limited land resource. As the population density of a country increases, the availability of land, wood, minerals, fuels and various other resources decreases and simultaneously creates serious economic, socio-political and environmental problems. Further, because people are forced to live in increasing proximity, pollution too closes to them. More job needs arise, leading to more commuters whereby more traffic pressure is inevitable. All these affects are cummulative.

Standard of Living

With the fast scientific development, urbanisation and economic growth the standard of living of people has gone up in the society. The people spend most of their income for buying luxrious goods like four wheelers, etc. At the same time, it is contributing to pollution by way of wastes and ignorance on the part of individuals. Hence, environmental pollution further increases as the standard of living goes up.

Degree of Recycling

Low degree of recycling of agricultural and non-agricultural wastes is aggravating the problem of pollution, though, the pollution is inversely related to the degree of recycling. In other words, if the waste product is effectively recycled and reused, the pollution level will decrease. But the Waste utilisation in India is neglected by majority of the people which affect pollution.

Technological Innovations

With increasing awareness of energy and environmental problem, several efforts have been made by the central government and state governments in the recent years to promote new technology, to meet the requirements of the society, e.g. bio-gas plants, fuel wood plantations, fuel conserving smokeless chulahas, laterines and hand pumps and also in the fields of automobiles, transportation, communication, industrialisation, etc. This fast changing development in the field of technology has failed to keep pace with the developments envisaged in the society. Technological advancements have also negative impact on the environment. The increase in air pollution, water pollution and noise pollution is mainly the result of new technological innovations in the process of economic development.

Industries

The fast development of industrialisation in the country poses various problems of environmental pollution. Some of the major pollutants contributed by the industrialisation are sulphur dioxide, suspended particles like flyash, mineral dusts from cement, waste from glasses and asbestos plants, poisonous gases from chemicals industries, radioactive fallouts, etc. Prolonged inhalation of sulphur dioxide is known to damage the respiratory system causing asthama, tuberculosis, etc. These oxides are readily absorbed by plants and cause damage to them. In the chambur area of Mumbai, the air was found to have the second highest level of sulphur dioxide in the city. These gases when found in abundance in the rain came down with precipitation causing acid rains, and increase the acidity of the soil in addition to having their deleterious effects on the existing flora and fauna.

The water of rivers, oceans and lakes, etc. is being polluted on a large scale by the effluents and wastes from the industries. The air pollution like smoke from, furnances, injurious chemical fumes from chemical industries, metallurgical plants, iron and steel plants have increased to such an extent that stringent steps are required immediately to reduce it.

Thermal Power Station

Thermal or heat pollution is common in the vicinity of power plants generating electricity. The water used for cooling the working steam is led back to the source at an appriciable higher temperature, affecting the acquatic organisms. The use of nuclear power for the industries opens up the possibility of radio-active wastes leaking out, which might cause enormous damage to the bio-sphere around.

The fuels like coal and oil used in the power plants prove very harmful to the environment as they release very heavy amounts of pollutants in the atmosphere causing air pollution. Thermal power station though primarily encouraged for the development has become a major source of pollution in the environment.

Transport Vehicles

The pollution from the increasing numbers of automobiles have been proportionally more. It is estimated that the annual pollutant load of the world atmosphere exceeds 100 million tons. Most of our vehicles of transport function with combustion engines, using petrol or diesel as fuel. These fuels are hydrocarbons. While functioning these engines burn the fuels in the presence of a limited supply of air resulting in the release of carbon monoxide, unburnt hydrocarbons, and oxides of nitrogen.

The problem of unburnt hydrocarbons is of greatest concern for two-stroke engines which are used in scooters, auto-rikshaws whose number is increasing day by day. Where as carbon monoxide is a problem confined to gasoline engines. Thus, the problem of pollution is equally shared between automobiles and all other sources put together in the environment.

Modern Agricultural Methods

The modern agricultural methods employed in cultivation are also contributing to pollution in the forms of smoke, noise, chemical residues, etc. though mechanisation in agriculture is yet to bring the desired level of achievement in Indian economy. The use of agricultural implements, tractors, spraying pumps, etc. are contributing their bit to pollution.

Urbanisation

Although industrialisation has greatly benefitted mankind, it has led to urbanisation. Urbanisation is not undesireable in itself but the haphazard growth of modern cities and towns has created a very unhealthy environment. The major consequences of increasing urbanisation are:
1. Overcrowding, leading to problem of sanitation and sewage disposal.
2. Transportation and associated traffic problem.
3. Environmental pollution generated by industrial activities and automobiles.

4. Noise pollution.
5. Socio-economic and cultural changes.

If the people cannot breath fresh air and live in open space, amidst noise and excitement and lead hectic life full of stress and strain, they cannot be healthy and strong and cannot lead peaceful lives.

Insecticides and Pesticides

In rural areas, especially in agricultural country like ours, insecticides and pesticides have proved a definite health hazard. Insecticides are not only harmful to insects but are harmful to man and animals when used in higher doses.

Pesticides play a major role in controlling a wide variety of agricultural pests. The bulk of their use is for spraying and dusting crops, with a small percentage being used for seed and soil treatment, grain storage, etc. Modern techniques of air spraying, dusting of particles over large areas on standing crops is still in the process of adoption on large scale. These agricultural chemicals are known to cause mutations in micro-organisms and act variously to molecular level. Such non-judicious use of these insecticides and pesticides fertilisers and other chemicals is contributing in a big way towards pollution especially in a fragile agricultural eco-system of India.

Ecological Changes

The environment is affected with the ecological changes at national as well as at international levels. The frequent ecological changes takes place due to mismanagement of natural resources, large scale deforestration, unplanned discharge of residues and wastes, handling of toxic chemicals, construction and expansion of settlement activities. These changes with adverse affects pose threat to the environment. Hence, the issues arising from environmental pollution have to be dealt on the local, national and internation levels.

SOCIETY AND ENVIRONMENT

Environment can be described as the natural world of land, water, air, plants and animals the exist around us. It forms the basis of our existence and development. As already discussed environment therefore, refers to the sum total of conditions, which surround man at a given point of time. As of today, environmental issues are no danger confined to geography and allied disciplines but have also drawn the attention of the common man.

Effects of Environment on Human Society

Effect of environment on human society has been emerging as a major challenge for quite sometime. Development was for long associated with under exploitation of natural resources. It was little realised that obsession with under exploitation may result in over-exploitation. We seemed to have believed that natural resources are inexhaustible. Environmental process includes those physical processes, which operate on the surface of the earth both internally and externally. Though man began to interfere with the natural processes right from the beginning of sedentary life, it assumed greater in proportion after the industrial revolution.

The impact of modern technology on environment is varied and highly complex as the transformation or modification of our natural conditions and process leads to as series of changes in the biotic and abiotic components of natural environment process.

We have seen that man, equipped with modern technologies and advanced scientific knowledge, has become an important factor in changing the environmental processes. It has to be realised that disturbances in one of the elements of nature (i.e. air, water, land, flora and fauna) gives rise to an imbalance in others. Natural processes or human factors some times aggravate natural environmental process to cause disaster for human society like (earthquakes, volcanic eruptions, floods, cyclones, etc.). They result in heavy loss of life and property. Environmental hazards for human health are as follows:

1. Air pollution causes respiratory diseases.
2. Water pollution causes enteric diseases.
3. Solid waste pollution causes vector-borne diseases.
4. Toxic waste causes cancer and neurological disorders.

VITAL ASPECTS OF ENVIRONMENTAL PROTECTION

Environmental protection cannot be looked at only from the regulatory angle; it has to be part of an integrated multi-dimensional holistic strategy. There are six basic aspects or tools for effective protection of environment. Thus we have to pay for the environment we want, otherwise we will get the environment we deserve.

Protection of the environment and the prevention and control of pollution require the use of a multipronged strategy and a variety of tools. Legislation (Regulation) is only one of them. The basic tools are:

1. Policy.
2. Planning.
3. Legislation.
4. Technology.
5. Economics.
6. Awareness.

To enable the regulatory mechanism to function and to translate regulation into compliance, it is necessary that all these tools are used. Only through the use of all these tools in a sustained and planned manner, the overall objective of environmental protection is possible.

Policy

It is essential that an overall policy is framed that permits development without environmental degradation. This requires a judicious trade off between economic growth and environmental protection.

As an example, significant pollution is caused by small scale industries in sectors like bulk drugs, dyes and intermediates, electroplating, etc. Should these sectors be open to small scale industry when we know that it is not realistic to expect them to meet the standards of pollution control? The result is that in the absence of a 'negative list' for the small scale industries at the policy level, the courts are closing them down, thus perhaps performing the function of the executive. Just as there is a minimum size based on economy of scale there is also an environmentally acceptable size.

Similarly, some of the chemicals which have high export potential are the ones that also have high environmental costs. In fact, this is the reason why countries importing them shy away from their manufacture. To control pollution from such units is extremely difficult technically, and unviable financially.

Planning

A large number of our problems with industrial pollution control today stem from improper planning of industrial estates. Many of these estates are land locked, have industrial units that are highly polluting and have no ultimate disposal outlet even if they meet standards of discharge. We have learnt some lessons from these past mistakes but the application of this knowledge at the planning level is slow and sporadic. Environmental impact assessment (EIA) is an excellent planning tool providing reasonably objective assessment of the impact of siting an industry at a specific location, not only on the environment, but also on the socioeconomic structure of the area. This tool is still not made serious use of by industries which tend to consider it a necessary evil. As large industries under take such exercise, industrial estates also should undergo such EIAs.

Similarly, internal planning of an industrial estate/zone into polluting and non-polluting areas, clustering of similar or complementary industries, etc. would reduce costs of environmental protection and result in more effective systems of pollution control.

Legislation

As mentioned above, legislation is only one of the tools and excessive dependence on this alone is bound to fail. There are, however, issues directly related to Regulation which have affected its implementation and consequent impact on the environment.

One of them is the 'standards' set for the discharges of waste-waters and emissions of air pollutants. To be effective these standards are necessary, should be implementable and enforceable. Although most of these standards in our country meet the above criteria there is a need to review them.

It is also necessary to have ranges for some of the parameters which have a high standard deviation in their analytical methods. For example, the measurement for biochemical oxygen demand (BOD) has a standard deviation of almost 20 per cent. This means that when the result of analysis shows 30 milligrams per litre (mg/l), the usual standard for BOD, it could be anywhere from 24 to 36 mg/l. This statistical range is, hence, necessary.

Today, as the regulation stands, the compliance has to be 100 per cent, meaning any sample taken at any time must comply with the standard. This is almost impossible as the raw waste-water characteristics coming into the treatment plant continuously vary. All over the world compliance is on a statistical basis. The other problems in enforcement is related to the Pollution Control Boards which have grown from Water Pollution Control Boards to Environmental Pollution Control Boards but a corresponding increase in their budgets has not materialised. This has left the Boards with a comprehensive mandate to enforce the regulation with meagre resources in terms of manpower, equipment and other infrastructure.

Industries also have continued to complain about standards rather than doing what can be done and taking up issues of difficult standards separately. The attitude of 'all or nothing' has been used to postpone pollution control measures as long as possible.

Technology

Many of the problems associated with protection of the environment stem from the unjustified belief that all problems have a technological solution. As mentioned above some problems are created by policy and planning deficiencies. Not all of these problems have a technological solution.

For example, the first step in the control of pollution from the small scale industries is the common effluent treatment plant (CETP). For a cluster of industrial units in the same sub-sector, say tanneries, a CETP has a good possibility of controlling pollution. For a group of diverse industries, however, the problem is much more complex and to comply with the regulations is almost impossible. There are certain pollution parameters such as total dissolved solids (TDS) and chemical oxygen demand (COD) which have almost no technology of control or technologies which are not economically viable. In other countries these problems are less pronounced as the infrastructure of treating industrial waste-water with municipal sewage exists. A number of problems associated with industrial waste-water treatment would be easier to handle with such combined effluent treatment plants as sewage provides dilution which improves the treatability of industrial waste-water. It is, hence, necessary to look at the pollution problem in its totality and not as industrial and municipal and integrate this concept in the planning process. Technology of manufacturing process also requires attention. Most of our existing industrial units use, technologies of production which are economically and environmentally inefficient. Atleast the newer projects should move towards low waste or so-called cleaner technologies. The attempt all over the world now is towards pollution prevention with the use of inplant control measures. These generally, consist of reduction in the quantity and quality of pollutants being discharged through conservation, process modification/control, equipment modification, recycle and reuse, by-product recovery, etc. Environmental audit or statement is another excellent tool to provide insight into potential of in-plant control. Once again this tool is not used effectively by the industries.

Economics

A number of economic incentives for pollution control are provided for by the government of India. These, however, need to be integrated and focussed towards desired goals, The incentives, for example, could be specifically directed towards adoption of in-plant control measures by the industries.

There is also need for a disincentive which would make the cost of not controlling pollution significantly higher than that of pollution control. This could be in the form of pollution charge or tax. The present cess provided for is much too low to be effective.

Awareness

Industry should become more sensitive towards environmental issues. One of the important messages to propagate is the fact that environmental protection makes eminent economic sense. Environment impact assessments (EIAs) and environmental audits results in more efficient use of resources and less waste generation. It is important that such awareness programs be undertaken by industry associations. All around the world, environmental protection has been directly related to public awareness and opinion building. This is evident in India also with significant environmental movement. Active role of the judiciary has also helped the cause in India.

To continue to be effective, however, the public environmental has to move from the 'fault finding and problem identification' stage to the 'problem solving' stage. Once again this has been the case the world over in the evolution of environmental movements.

Public awareness has to grow into public understanding. This understanding would include not only identification of a problem but the options available, the problems with each of the options and cost involved (both economic and environmental). It should also include the understanding that there are no absolutes in the field of environmental protection, only trade offs and 'least cost' solutions.

The basic principles of understanding are very well enunciated by the following statements (very probably by barry commoner):

1. Everything is connected in the environment.
2. Everything has to go someplace.
3. There is no free lunch.

That there is no free lunch is the most important of all the above. We have to learn to pay for the environment we want, because we always get the environment that we deserve.

ENVIRONMENTAL ETHICS

Environmental ethics is the part of environmental philosophy which considers extending the traditional boundaries of ethics from solely including humans to including the non-human world. It exerts influence on a large range of disciplines including law, sociology, theology, economics, ecology and geography.

There are many ethical decisions that human beings make with respect to the environment. For example:

1. Should we continue to clear cut forests for the sake of human consumption?
2. Should we continue to propagate?
3. Should we continue to make gasoline powered vehicles?
4. What environmental obligations do we need to keep for future generations?
5. Is it right for humans to knowingly cause the extinction of a species for the convenience of humanity?

Environmental ethics believes in the ethical relationship between human beings and the natural environment. Human beings are a part of the society and so are the other living beings. When we talk about the philosophical principle that guides our life, we often ignore the fact that even plants and animals are a part of our lives. They are an integral part of the environment and hence have a right to be considered a part of the human life. On these lines, it is clear that they should also be associated with our guiding principles as well as our moral and ethical values. We are cutting down forests for making our homes. We are continuing with an excessive consumption of natural resources. Their excessive use is resulting in their depletion, risking the life of our future generations. Is this ethical? This is the issue that environmental ethics takes up. Scientists like Rachel Carson and the environmentalists who led philosophers to consider the philosophical aspect of environmental problems, pioneered in the development of environmental ethics as a branch of environmental philosophy.

The Earth Day celebration of 1970 was also one of the factors, which led to the development of environmental ethics as a separate field of study. This field received impetus when it was first discussed in the academic journals in North America and Canada. Around the same time, this field also emerged in Australia and Norway. Today, environmental ethics is one of the major concerns of mankind.

When industrial processes lead to destruction of resources, is it not the industry's responsibility to restore the depleted resources? Moreover, can a restored environment make up for the originally natural one? Mining processes hamper the ecology of certain areas; they may result in the disruption of plant and animal life in those areas. Slash and burn techniques are used for clearing the land for agriculture.

Most of the human activities lead to environmental pollution. The overly increasing human population is increasing the human demand for resources like food and shelter. As the population is exceeding the carrying capacity of our planet, natural environments are being used for human inhabitation.

Thus human beings are disturbing the balance in the nature. The harm we, as human beings, are causing to the nature, is coming back to us by resulting in a polluted environment. The depletion of natural resources is endangering our future generations. The imbalance in nature that we have caused is going to disrupt our life as well. But environmental ethics brings about the fact that all the life forms on earth have a right to live. By destroying the nature, we are depriving these life forms of their right to live. We are going against the true ethical and moral values by disturbing the balance in nature. We are being unethical in treating the plant and animal life forms, which coexist in society.

Human beings have certain duties towards their fellow beings. On similar lines, we have a set of duties towards our environment. Environmental ethics says that we should base our behaviour on a set of ethical values that guide our approach towards the other living beings in nature. Environmental ethics is about including the rights of non-human animals in our ethical and moral values. Even if the human race is considered the primary concern of society, animals and plants are in no way less important. They have a right to get their fair share of existence.

We, the human beings, along with the other forms of life make up our society. We all are a part of the food chain and thus closely associated with each other. We, together form our environment. The conservation of natural resources is not only the need of the day but also our prime duty.

Chapter 2

Particulate Matter

INTRODUCTION

Particulates, alternatively known to as particulate matter (PM) or fine particles and also called soot, are tiny subdivisions of solid matter suspended in a gas or liquid. In contrast, aerosol refers to particles and/or liquid droplets and the gas together. Sources of particulate matter can be man made or natural. Air pollution and water pollution can take the form of solid particulate matter, or be dissolved. Salt is an example of a dissolved contaminant in water, while sand is generally a solid particulate.

To improve water quality, solid particulates can be removed by water filters or settling, and is referred to as insoluble particulate matter. Dissolved contaminants in water are often collected by distilling— allowing the water to evaporate and the contaminants to return to particle form and precipitate.

Some particulates occur naturally, originating from volcanoes, dust storms, forest and grassland fires, living vegetation, and sea spray. Human activities, such as the burning of fossil fuels in vehicles, power plants and various industrial processes also generate significant amounts of aerosols. Averaged over the globe, anthropogenic aerosols—those made by human activities—currently account for about 10 per cent of the total amount of aerosols in our atmosphere. Increased levels of fine particles in the air are linked to health hazards such as heart disease, altered lung function and lung cancer.

Particulate matter is released from natural sources, such as volcanic eruptions, seasalt, or soils. Anthropogenic sources, such as car exhausts, industry, shipping and coal and ore fabrication, cause increases in airborne particluate matter concentrations on many locations. Power plants add to emissions of particulate matter. Emissions from private houses stem from fire places and barbecues. Dutch epidemiologists state that emissions from traffic, particularly from diesel engines, are most damaging to human health.

TYPES OF PARTICULATES

Smoke

Smoke particulates consist of solid and liquid particles ranging in size from 0.05 to 1.0 micron, which are formed during incomplete combustion of carbonaceous materials. This includes smoke of gaseous pollutants like oxides of sulphur, oxides of nitrogen, carbon monoxide and hydrocarbons, etc. Smoke can come from fires, cigarettes, cigars, and cooking. It can burn your eyes, make you sick, give you cancer, breathing problems, and it can make you dizzy if you are around it too much. It can give you second-hand smoke. It can release toxins into the air, do damage to your home, and many other things.

Smoke can stain just about any surface. Smoke can make your eyes really red and make your skin smell, and it can also stain your teeth and your skin yellow if you smoke.

Incineration does not get rid of unwanted material, but only releases the material into the air as smoke. Smoke is made up of particulate matter and gases, and carries many different types of chemicals. Some are more harmful than others, but even seemingly benign materials can cause problems if they are present in high concentrations. The contents of smoke depend on what material is being burned. The most worrisome sources of smoke-related air pollution are power plants (particularly coal), vehicle emissions, waste incinerators, and burning forests or agricultural land. Cigarette smoke also releases toxins into the air.

Smoke from coal and fossil fuels: One of the largest sources of air pollution is smoke from coal and vehicles or other engines run on fossil fuels. Combustion of coal and petroleum releases sulphur dioxide, nitric oxide, carbon dioxide and carbon monoxide. These emissions contribute to acid rain and the greenhouse effect, exacerbating global climate change and damaging sensitive habitats. Although nitric oxide and carbon dioxide are natural components of the atmosphere, in large amounts they can be harmful. Smoke from coal and fossil fuels also contains high levels of fine particles, which cause health problems when inhaled and may exacerbate heart and lung conditions. Some of the particulate matter may also contain toxins.

Other types of smoke: Incineration of waste also generates particularly dangerous smoke. Burning plastic or foam (such as styrofoam) releases chlorofluorocarbons (CFCs), which break down atmospheric ozone and contribute to the greenhouse effect. Burning biomass, such as forest or grassland, adds carbon dioxide and nitric oxide to the atmosphere, so even 'natural' fires, with only plant material as fuel, contribute to overall air pollution. Cigarettes pack a heavy dose of pollution for something so small— cigarette smoke contains 4000 different chemicals, many of them toxic. The EPA regulates air quality, but emissions standards are difficult to meet and there is still a long way to go before smoke is no longer a major environmental concern.

Whether you are concerned with particulate matter, carbon monoxide, carbon dioxide or hydrocarbons, all smoke components from wildland fires are generated from the incomplete combustion of fuel. The amount of smoke produced can be derived from knowledge of area burned, fuel loading (tons/acre), fuel consumption (tons/acre), and pollutant specific emission factors. Multiplying a pollutant specific emission factor (lbs/ton) by the fuel consumed, and adding the time variable to the emission production and fuel consumption equations results in emission and heat release rates that allow the use of smoke dispersion models.

Fuel characteristics: Fuel consumption and smoke production are influenced by preburn fuel loading categories such as grasses, shrubs, woody fuels, litter, moss, duff, and live vegetation; condition of the fuel (live, dead, sound, rotten); fuel moisture; arrangement; and continuity. These characteristics can vary widely across fuelbed types and within the same fuelbed type.

Dust

Dust is a general name for solid particles with diameters less than 20 thou (500 micrometers). Particles in the atmosphere arise from various sources such as soil dust lifted up by wind, volcanic eruptions, and pollution. Dust in homes, offices, and other human environments contains small amounts of plant pollen, human and animal hairs, textile fibres, paper fibres, minerals from outdoor soil, and many other materials which may be found in the local environment.

Domestic dust and humans: Insects and other small fauna found in houses subtly interact with dust and may have adverse impact on the health of humans. Dust may worsen hay fever. Circulating outdoor air through a house by keeping doors and windows open, or at least slightly ajar, may reduce the risk of hay fever-causing dust. In colder climates, occupants seal even the smallest air gaps, and eliminate outside fresh air circulating inside the house. So it is essential to manage dust and airflow.

House dust mites exist on all indoor surfaces and even suspended in the air. They feed on minute particles of organic matter, the main constituent of house dust. Dust mites flourish in the fibres of bedding, furniture, and carpets. They excrete enzymes to digest the organic particles, and excrete feces, that together become part of the house dust, and may irritate allergies.

Alternately, the hygiene hypothesis posits that the modern obsession with cleanliness is as much a problem as house dust mites. The hygiene hypothesis argues that our lack of prior pathogenic exposure may in fact encourage the development of ailments including hay fever and asthma.

Atmospheric dust: Airborne dust is considered an aerosol and can have a strong local radiative forcing on the atmosphere and significant effects on climate. In addition, if enough minute particles are dispersed within the air in a given area (such as flour or coal dust), under certain circumstances can cause an explosion hazard. Coal dust is responsible for the lung disease known as Pneumoconiosis, including black lung disease, that occurs among coal miners. The danger of coal dust resulted in environmental legislation regulating work place air quality in some jurisdictions.

Road dust: Dust kicked up by vehicles travelling on roads may make up 33 per cent of air pollution Road dust consists of deposition of vehicle exhausts and industrial exhausts, tyre and brake wear, dust from paved roads or potholes, and dust from construction sites. Road dust represents a significant source contributing to the generation and release of particulate matter into the atmosphere. Control of road dust is a significant challenge in urban areas, and also in other spheres with high levels of vehicular traffic upon unsealed roads such as mines and garbage dumps. Road dust may be suppressed by mechanical methods like sweeping vehicles, with vegetable oils, or with water sprayers.

Source of dust

The pulverisation or crushing which fragments a solid material by mechanical force and makes fine particles is the oldest mechanical operation of humankind. By this operation, the handling of solid becomes easier and the rate of the reaction and dissolution and the catalyst reactivity are improved because the surface area of solid is increased.

These effects are readily seen in such processes as coal pulverisation in the pulverised coal combustion facility and ore processing in the pudding. Also, the polishing is an operation to smooth out the surface of solid materials. The paritculate material that is generated and scattered during smashing or other mechanical operations, or associated with the accumulation of the materials obtained through these processes, is called the dust and is attracting attention as a source of air pollution.

The major source of dust includes the pulverising equipment, polishing machine, sieve, particulate transport facility, pile of ore or soil, and cokes oven. Generally, the size of dust generated by the mechanical operation such as pulverisation or crashing is relatively large and is about a few μm and its range is rather broad, unlike the fine fume generated by the chemical reaction.

Pulverisation or crushing process: There are various types of pulverisation process including the mining of materials of cement, limestone, coal, etc. crushing of clinker in the cement factory, production of pulverised coal in the pulverised coal burning power station, production of materials for glass, material preparation for china and porcelain, production of aggregate to use construction of building, waste treatment, and so on. Also, the type of pulverisation or crushing equipment is diverse.

Pulverisation or crushing equipment: The pulverisation or crushing equipment is categorised into three types according to the type of force that the machine exerts on the solid to crush it finely. That is, the equipment that uses pressure by sandwiching the material to be crushed between two surfaces, the one that uses impact by hitting the material with a hammer so that it is crushed instantly by a high speed collision of a hard body, and the one that uses shear force exerted perpendicular to the direction of sear force.

Polishing process: Polishing is an operation that finishes the surface of material by using the polishing machine such as the belt sander, drum sander and wide belt sander. The belt sander polishes the surface of material by moving the polishing paper or cloth with 2–4 pulleys. The drum sander polishes the material by rotating a cylinder, wrapped around with the polishing paper or cloth, at high speed.

The wide belt sander is a machine to polish the material by upper and lower drums wrapped with an endless polishing paper. These machines are used in the factories for paper and pulp manufacturing, furniture manufacturing, and plywood manufacturing, and produce dusts.

Sieving of particles: The sieve discriminates between smaller particles, which pass through small holes, and larger particles, which do not. It is used to classify the dust that corresponds to a mesh of the sieve by moving the dust on it. The mesh and thickness of the sieve are specified by the Japanese Industrial Standard (JIS). If the dust is classified by stacking a sieve with a larger mesh on another sieve with a smaller mesh, its particle size distribution can be obtained.

For industrial purposes, the rotation sieve which has a sieve of cylindrical wall and the flat sieve which uses a moving flat sieve are used. The dust is generated at the particle feed portion, on the mesh of a sieve, and at the particle discharge portion. The protection measures such as cover and a dust collecting equipment are installed.

Transportation of particles: For the transportation of particles, belt conveyers and bucket conveyers are used, and they also scatter dusts. The belt conveyer is a transportation facility which has an endless belt stretched between two pulleys on both ends of a frame and transfers loads on the belt by moving it continuously. This facility is quite economical for transporting loose materials such as coal, ore, or gravel, but sometimes accompanied by dust scattering. Also, the bucket conveyer is a facility to transport loose materials from the lower to higher positions vertically or at a steep angle. Buckets are attached at a constant distance to a chain or a belt stretched upward, and loose materials thrown into the bucket are continuously transported upward. When it is used for transporting dusts, they are scattered around. The protection cover is installed or the dust is collected by a dust collecting equipment while arranging an enclosure hood.

Pile of particles: Industrial raw materials such as coal and ore are often stored in the open air as a pile, but some are lost through the scattering of dust. For instance, it is reported that about 6.4 mg per 1 kg per year will be lost from the pile of coal. Also, about 13.2 pounds (1 pound = 0.4536 kg) per acre of pile area (1 area = 4046 m^2) per day of rock and gravel is reported to be lost from the rock pile operated 24 hours a day. Measures such as the water sprinkling or the chemical sprinkling by a sprinkler are used.

Mist

Mist is a phenomenon of small droplets suspended in air. It can occur as part of natural weather or volcanic activity, and is common in cold air above warmer water, in exhaled air in the cold, and in a steam room of a sauna. It can also be created artificially with aerosol canisters if the humidity conditions are right. The only difference between mist and fog is visibility. This phenomenon is called fog if the visibility is one kilometre (1100 yards) or less (in the UK for driving purposes the definition of fog is visibility less than 200 metres, for pilots the distance is 1 kilometre). Otherwise it is known as mist. Seen from a distance, mist is bluish, and haze is more brownish. Religious connotations are associated with mist in some cultures; it is used as a metaphor in 2 Peter 2:17. Mist makes a beam of light visible from the side via refraction and reflection on the suspended water droplets. 'Scotch mist' is a light steady drizzle, the name being typical of the Scottish penchant for understatement (and of Scottish weather). Mist usually occurs near the shores, and is often associated with fog. Mist can be as high as mountain tops when extreme temperatures are low.

Freezing mist: Freezing mist is similar to freezing fog, only the density is less and the visibility greater. When mist falls below 0 degrees Celsius in temperature it becomes known as freezing mist.

Spray

Spray constitutes liquid particles obtained from the parent liquid by the process of mechanical disintegration like atomisation.

Fumes

Visible or invisible vapours given off by inorganic or organic liquids, mixed gases resulting from incomplete combustion such as in a gasoline engine, one or more gases produced by a chemical reaction, submicroscopic particulate-matter emitted by heated metals or metallic compounds. Fumes are generally obtained by the condensation of vapours by sublimation, distillation, boiling, calcination and by several other chemical reactions. Generally organic solvents, metals and metallic oxides form fume particles having a size less than 1 micron.

SCALE CLASSIFICATION OF PARTICULATES

Among the most common categorisations imposed on particulates are those with respect to size, referred to as fractions. As particles are often non-spherical (for example, asbestos fibres), there are many definitions of particle size. The most widely used definition is the aerodynamic diameter. A particle with an aerodynamic diameter of 10 micrometers moves in a gas like a sphere of unit density (1 gram per cubic centimetre) with a diameter of 10 micrometers. PM diameters range from less than 10 nanometers to more than 10 micrometers. These dimensions represent the continuum from a few molecules up to the size where particles can no longer be carried by a gas. The notation PM_{10} is used to describe particles of 10 micrometers or less and $PM_{2.5}$ represents particles less than 2.5 micrometers in aerodynamic diameter.

But because no sampler is perfect in the sense that no particle larger than its cutoff diameter passes the inlet, all reference methods allow a high margin of error. These are also sometimes referred to with other equivalent numeric values. Everything below 100 nm, down to the size of individual molecules is classified as ultrafine particles.

Fraction	Size range
PM_{10} (thoracic fraction)	$<= 10 \ \mu m$
$PM_{2.5}$ (respirable fraction)	$<= 2.5 \ \mu m$
PM_1	$<= 1 \ \mu m$
Ultrafine (UFP or UP)	$<= 0.1 \ \mu m$
PM_{10}-$PM_{2.5}$ (coarse fraction)	$2.5 \ \mu m - 10 \ \mu m$

Note that PM_{10}–$PM_{2.5}$ is the difference of PM_{10} and $PM_{2.5}$, so that it only includes the coarse fraction of PM_{10}.

These are the formal definitions. Depending on the context, alternative definitions may be applied. In some specialised settings, each fraction may exclude the fractions of lesser scale, so that PM_{10} excludes particles in a smaller size range, e.g. $PM_{2.5}$, usually reported separately in the same work. Such a case is sometimes emphasised with the difference notation, e.g. PM_{10}–$PM_{2.5}$. Other exceptions may be similarly specified. This is useful when not only the upper bound of a fraction is relevant to a discussion. The facts that some particle size ranges require greater filter strength and the smallest ones can outstrip the body's ability to keep them out of cells both serve to guide understanding of related public policy, environment, and health topics.

Composition

The composition of aerosol particles depends on their source. Windblown mineral dust tends to be made of mineral oxides and other material blown from the Earth's crust; this aerosol is light-absorbing. Sea salt is considered the second-largest contributor in the global aerosol budget, and consists mainly of sodium chloride originated from sea spray; other constituents of atmospheric sea salt reflect the composition of sea water, and thus include magnesium, sulphate, calcium, potassium, etc. In addition, sea spray aerosols may contain organic compounds, which influence their chemistry.

Secondary particles derive from the oxidation of primary gases such as sulphur and nitrogen oxides into sulphuric acid (liquid) and nitric acid (gaseous). The precursors for these aerosols, i.e. the gases from which they originate—may have an anthropogenic origin (from fossil fuel combustion) and a natural biogenic origin.

In the presence of ammonia, secondary aerosols often take the form of ammonium salts, i.e. ammonium sulphate and ammonium nitrate (both can be dry or in aqueous solution); in the absence of ammonia, secondary compounds take an acidic form as sulphuric acid (liquid aerosol droplets) and nitric acid (atmospheric gas). Secondary sulphate and nitrate aerosols are strong light-scatterers. This is mainly because the presence of sulphate and nitrate causes the aerosols to increase to a size that scatters light effectively.

Organic matter (OM) can be either primary or secondary, the latter part deriving from the oxidation of VOCs; organic material in the atmosphere may either be biogenic or anthropogenic. Organic matter influences the atmospheric radiation field by both scattering and absorption. Another important aerosol type is constitute of elemental carbon (EC, also known as black carbon, BC): this aerosol type includes strongly light-absorbing material and is thought to yield large positive radiative forcing. Organic matter and elemental carbon together constitute the carbonaceous fraction of aerosols.

The chemical composition of the aerosol directly affects how it interacts with solar radiation. The chemical constituents within the aerosol change the overall refractive index. The refractive index will determine how much light is scattered and absorbed.

Removal Processes

In general, the smaller and lighter a particle is, the longer it will stay in the air. Larger particles (greater than 10 micrometers in diameter) tend to settle to the ground by gravity in a matter of hours whereas the smallest particles (less than 1 micrometer) can stay in the atmosphere for weeks and are mostly removed by precipitation. Diesel particulate matter (DPM) is highest near the source of emission. Any info regarding DPM and the atmosphere, flora, height, and distance from major sources would be useful to determine health effects.

Effects of Aerosols on Electromagnetic Radiation

All aerosols both absorb and scatter solar and terrestrial radiation. This is quantified in the Single Scattering Albedo (SSA), the ratio of scattering alone to scattering plus absorption (extinction) of radiation by a particle. The SSA tends to unity if scattering dominates, with relatively little absorption, and decreases as absorption increases, becoming zero for infinite absorption. For example, sea-salt aerosol has an SSA of 1, as a sea-salt particle only scatters, whereas soot has an SSA of 0.23, showing that it is a major atmospheric aerosol absorber. Aerosols, natural and anthropogenic, can affect the climate by changing the way radiation is transmitted through the atmosphere. Direct observations of the effects of aerosols are quite limited so any attempt to estimate their global effect necessarily involves the use of computer models. The Intergovernmental Panel on Climate Change (IPCC), says: While the radiative forcing due to greenhouse gases may be determined to a reasonably high degree of accuracy, the uncertainties relating to aerosol radiative forcing remain large, and rely to a large extent on the estimates from global modelling studies that are difficult to verify at the present time.

Sulphate aerosol

Sulphate aerosol has two main effects, direct and indirect. The direct effect, via albedo, is to cool the planet: the IPCC's best estimate of the radiative forcing is -0.4 watts per square meter with a range of -0.2 to -0.8 W/m^2 but there are substantial uncertainties. The effect varies strongly geographically, with most cooling believed to be at and downwind of major industrial centres. Modern climate models attempting to deal with the attribution of recent climate change need to include sulphate forcing, which appears to account (at least partly) for the slight drop in global temperature in the middle of the 20th century. The indirect effect (via the aerosol acting as cloud condensation nuclei, CCN, and thereby modifying the cloud properties 'albedo and lifetime') is more uncertain but is believed to be a cooling.

Physical Methods Involved in Particulate Formation

Some of the important physical methods involved in particulate formation are:

1. Dispersion process mainly yields dusts which are solid dispersion aerosols. Particulates of nearly 1 μ in size are formed by the disintegration of larger particles.
2. Adhesion of smaller particles by chemical process yields particulates of size ranging from 10 μ to 20 μ.
3. Natural sources also produce dispersed aerosols as from sea sprays, wind blown dust during cultivation and volcanic dust, etc.
4. By coagulation, extremely small particles form larger aggregates; sorption, absorption, adsorption and chemisorption also results in the formation of particulate matter.

Formation of Inorganic Particulate Matter

Inorganic particulates generally originate from metallic oxides, sulphides, carbonates, etc. These are produced when fuels containing metals are burned. For instance:

1. The particulate Fe_3O_4 is formed during the combustion of pyrites containing coal.
$$3FeS_2 + 8O_2 \rightarrow Fe_3O_4 + 6SO_2$$
2. A part of calcium carbonate in the ash fraction of coal gets converted to calcium oxide, which is emitted into the atmosphere through stack.
$$CaCO_3 \xrightarrow{\text{Heat}} CaO + CO_2$$

3. Organic vanadium in residual fuel oil is converted to particulate vanadium oxide (V_2O_3).
4. Lead halides are generated by the combustion of leaded gasoline. Tetra ethyl lead, $Pb(C_2H_5)_4$ in leaded gasoline combines with oxygen and halogenated scavengers, dibromo ethane, dichloro-ethane, etc. These lead halides emerge through the exhaust system and condense to form particulates, entering the atmosphere.

$$Pb(C_2H_5)_4 + O_2 + C_2H_4Cl_2 + C_2H_4Br_2 \rightarrow CO_2 + H_2O + PbCl_2 + PbBr_2 + PbBrCl.$$

This resultant, lead pollution can be minimised by using unleaded gasoline.
5. Aerosol mists are formed by the oxidation of atmospheric sulphur dioxide to sulphuric acid:

$$2SO_2 + O_2 + 2H_2O \rightarrow 2H_2SO_4$$

This forms salts with basic air pollutants such as ammonia or calcium oxide:

$$H_2SO_4 + 2NH_3 \rightarrow (NH_4)_2SO_4 \quad H_2SO_4 + CaO \rightarrow CaSO_4 + H_2O$$
$$\text{(droplet) (gas)} \qquad \text{(droplet)} \qquad \text{(droplet) (particle)} \qquad \text{(droplet)}$$

Organic particulate matter (OPM)

Organic particulate matter originates mainly from combustion of fuels, automobiles and vegetation. Polycyclic aromatic hydrocarbons (PAH) like chrysene, benzofluoranthene, benzo α-pyrene, benzidine, etc. are some of the constituents of organic particulate matter of carcinogenic nature. PAH compounds mostly occur in urban atmosphere at levels of ~20 $\mu g/m^3$. Aldehydes, ketones, peroxides, epoxides, esters, quinones and lactones are found among the oxygenated neutral organic compounds. Organic acids present in OPM include lauric, palmitic, stearic, myristic, oleic, linoleic, behemic acids, etc.

Particulate paraffins are pyrolysed to yield $C_{10}H_{22}$ which again disintegrates into fine particles. Oxidised polymerised hydrocarbons and nitrogenous azoheterocyclic compounds are released into the atmosphere from automobile exhaust.

PAH compounds remain absorbed in soot particles. Soot itself is a highly condensed product of these compounds. A soot particle is composed of several thousand inter-connected crystallites, i.e. graphitic platelets, each having about 100 condensed aromatic rings. It consists of 1–3 per cent hydrogen and 5–10 per cent oxygen due to partial oxidation. Soot particles act as a disease carrier containing toxic trace metals like Be, Cr, Mn, Ni, Cd, Fe and poisonous organics such as benzo-(α-) pyrene.

Effects of particulate pollutants

Some of the effects of particulate pollutants are:

Effects on plants

1. Plants are adversely affected by gaseous pollutants and deposition of particulates on soil. This deposition of toxic metals on the soil makes it unsuitable for plant growth.
2. Several particulate pollutants fall on the soil by acid rain which tend to lower its pH, making it more acidic and infertile.
3. Particulates such as dust, fog, soot, deposited on plant leaves block the stomata of plants, thus inhibiting the rate of transpiration of minerals from the soil.
4. Deposited particulates restrict the absorption of CO_2, thereby reducing the rate of photosynthesis, retarding plant growth and crop production.
5. Dust mixed with mist or light rain forms a thick crust on the upper leaf surfaces, which shields the bright sunlight necessary for carbon assimilation.

6. Some plants are very sensitive to the traces of toxic metals, where particulates inhibit the action of their enzyme system.

7. Arsenic is a cumulative, potent, protoplasmic poison which inhibits SH-group in enzymes. This is present in almost all types of soils in minute quantities and affects plant growth.

Effects on humans

The effects of particulate pollutants are largely dependent on the particle size. Airborne particles, i.e. dust, soot, fumes and mists, are potentially dangerous for human health.

1. Particulate pollutants have a bearing on the penetration of particles beyond the respiratory passages into the lungs. Nasal passage prevents coarser particulates bigger than 5 microns from entering into the respiratory system. Particles with a size of about 1 micron enter into lungs easily and rapidly. Actually, aerosols less than 1 micron may reach the alveoli of lungs and damage lung tissues.

2. Soluble aerosols will be absorbed into the blood from the alveoli while insoluble aerosols are carried to the lymphatic stream and get deposited in pulmonary lymphatic depot points or in the lymph glands, where they create toxicity on the respiratory system.

3. Workers exposed to pollutant asbestos mostly develop cancer called mesothelioma, which occurs in the tissue lining the abdomen. They also have a greatly elevated risk of lung cancer.

4. Insoluble particulates which cannot be phagocytised by white blood corpuscles (WBCs) pass through the alveolar walls into lymph channels. They also accumulate in various specific organs and their increased concentration exerts actions on lungs.

5. Lead, the most serious pollutant released from automobile exhaust, is reported to have detrimental effect on children's brains.

6. Lead interferes with the development and maturation of red blood cells.

7. Workers exposed to lead excrete porphyrins, the precursors of haemoglobin, in their urine.

8. It has been reported that smokers can more easily develop symptoms of asthma, which is also due to excess concentration of lead, than non-smokers.

9. Silcosis, a chronic disease of lungs, is caused by inhalation of dust containing free silica, SiO_2.

10. Acid particulates and aldehydes cause eye, nose and throat irritation.

11. Formaldehyde is extremely toxic to human health. Acrolein causes irritation to mucus membranes and also poses bronchioconstriction.

12. Lead and asbestos act as cumulative poison and are dangerous to children, causing brain damage and cancer. Small fibres of asbestos irritate lung tissues causing asbestosis, a condition characterised by lung fibrosis.

13. Black lung disease is common among coal miners, while white lung disease occurs frequently among textile workers.

14. Carbon, in the form of soot, deposits in the nose, throat and respiratory tract.

15. Particulates such as silica, asbestos and different forms of carbon are capable of exerting a noxious or fibrotic local action in the interstitial areas of the lungs and in the lymphatic tissues.

16. Beryllium compounds like $BeSO_4$ and $BeCl_2$ cause acute inflammation of the lungs.

17. Small exposure of cadmium causes cardiovascular diseases and hypertension. It also interferes with copper and zinc metabolism.

18. Traces of mercury cause nerve damage and death, while arsenic creates acute and chronic cancer.

19. Fine particulates of less than 2 μ size are the worst causes of lung damage while the larger particles (3 μ) are trapped in the nose and throat. These particles create various breathing troubles due to nasal tract blockage and irritation of lung capillaries.

20. Arsenic is absorbed through the lungs and skin, and causes diarrhoea, peripheral neuritis, conjunctivitis, hyperketosis, lung and skin cancer. Chronic exposure to arsenic leads to the so-called 'black foot' disease.

Effects on materials

1. Particulates affect a variety of materials in various ways. They damage buildings, paints, furniture, etc. Painted surfaces are very susceptible to damage in wet conditions. Particulate fumes and mists react directly with any painted surfaces and cause cracks in them.

2. Particulates accelerate corrosion of metals, mainly in urban and industrial areas.

3. Particles, including fumes, dust, soot, mists and aerosols, can cause severe damage to soil, buildings, sculpture and monuments.

4. As a result of the 1984 Bhopal disaster, soil within 160 km, radius was coated with thick dust due to methyl isocyanate leakage and soil fertility was lost for at least the next ten years.

5. Particulates cause extensive damage when they themselves are corrosive and carry toxic harmful chemicals along with them.

6. Particulate pollutants are responsible for polluted cloud formation, rain and snow in which the particles act as the nuclei on which water droplets condense.

7. Particulates accumulate on soil surfaces causing erosion. The particles are generally sticky, tarry and acidic, so they adhere to surfaces and act as acid reservoirs.

8. Precipitation effect over localities or suburbs is due to particulate pollutants. They also influence the formation of rain, dew and snow.

Effects on solar radiation

1. Particulates reduce visibility by absorption and scattering of solar radiation. Decreased visibility creates chronic and annoying problems.

2. They cause illumination problems by reducing sunlight by nearly one-third.

3. In winter, increased heating requires more power which releases more particulate particles in the atmosphere.

4. They also upset the delicate heat balance of the earth's atmosphere. Increased content of CO_2 (10 per cent) causes global warming but the particulates (i.e. aerosols) have the ability to reject more solar radiations thus compensating the climatic effect of increased carbon dioxide concentration.

CONTROL OF PARTICULATES

There are three broad approaches to the control of particulates: (i) dilution in the atmosphere, (ii) control at source, and (iii) control by using pollution control equipment.

Dilution in the Atmosphere

Dilution of particulates and gases can be accomplished by the use of tall stacks. Pollutants released from taller stacks disperse easily and hence low ground level concentrations are observed. Tall stacks penetrate the inversion layer and disperse the contaminants easily so that the ground level concentrations are less harmful. However, dilution is only a short-term control measure and has highly undesirable long-range effects. In India, industries have to ensure a minimum stack height of 30 metres. Theoretically, the minimum height of stack, H, required for effective dispersion of particulates is given by $H = 74\ Q^{0.27}$ where, Q is particulate emission rate in tonnes per hour. This often demands stack heights in excess of 400 m, especially in cement industries and thermal power plants, and therefore may prove more uneconomical when compared to particulate control by treatment.

Control at Source

The most effective means of dealing with the problem of air pollution is to prevent emission at the source itself. In the case of industrial pollutants, this can often be achieved by investigating various approaches at the early stage of process design and development, and selecting those methods which do not contribute to air pollution, (or) have the minimum air pollution potential. It may be difficult to implement these methods in existing plants, but still some of the methods can be applied without upsetting the economy of the operation. Control of particulates at the source can be accomplished in several ways through raw material changes, operational changes, modification or replacement of process equipment and by more effective operation of existing equipment. Source locations through allocation of land usage, i.e. proper planning and zoning of industrial areas, is also an important means of control of pollutants. Some of the source correction methods are discussed briefly as follows:

Raw material changes

Some raw materials are primarily responsible for causing air pollution. Use of pure grade of raw material is often beneficial and may reduce the formation of undesirable impurities and by-products, and may even eliminate troublesome effluents. Ore handling operations usually result in the emission of large quantities of dust into the atmosphere. In the steel industry, replacement of raw ore with briquetted or pelleted sintered ore has greatly reduced dust emissions during ore handling and also helped reduce blast furnace 'slips', which result in emission of enormous amounts of uncleaned, blast furnace gas when the safety dampers open, by-passing the dust collectors.

Process changes

Changing the processing methods is still another important method of controlling emissions at their source. For example, petroleum/chemical industries have undergone radical changes in processing methods which emphasise continuous automatic operations, often computer controlled and completely enclosed systems that minimise the release of materials into the atmosphere. The volatile substances from storage tanks, etc. are recovered by condensation and the non-condensable gases are recycled for additional reactions such as polymerisation and alkylation of gaseous hydrocarbons to produce gasoline. Hydrogen sulphide, which was once flared in refineries, is now recycled using the Clauss process to recover elemental sulphur.

Replacing open hearth furnaces with controlled basic oxygen furnaces or electric furnaces can reduce smoke, carbon monoxide and metal fumes while at the same time conserve energy. Such changes coupled with various combinations of gas cleaning devices can be very effective in reducing air pollution.

In cement plants, rotary kilns are the major sources of dust generation. This can be reduced to some degree by adjusting operating conditions like reduction of the gas velocities within the kiln, modification of the rate and location of feed, introduction and employment of a dense curtain of light weight chain at the discharge end of the kiln, etc.

Emissions of sulphurous materials from smelting and paper industries are highly objectionable. These emissions are being curtailed by major process changes such as hydro-metallurgical separations of ores and avoiding the use of sulphides in paper making. In the steel industry, a radically different process has been proposed to lower sulphurous emissions during combustion. In this process, the sulphur bearing fuel, limestone and air are injected into a molten bath. The combustibles in the fuel are partially oxidised to carbon monoxide within the molten iron bath. Here the gaseous CO comes off at the top of the molten iron and is burnt efficiently in a conventional manner, the sulphur is retained in the iron bath and forms a slag with the limestone, which is removed.

Other examples involving process changes

Other examples involving process changes include:

1. Reduction of the formation of nitric oxides in combustion chambers by low excess air combustion in two stages, fuel gas recirculation and water injection.
2. Washing the coal before pulverisation to reduce fly-ash emissions.
3. Substitution of bauxite flux for fluorine-containing fluorspar in the open hearth furnace method.
4. The use of liquid and gaseous fertiliser chemicals like anhydrous ammonia, applied by injection into the earth instead of being spread across the surface as finely divided powder.
5. Reduction in oxidation of SO_2 to SO_3, by reducing excess air from 20 per cent to less than 1 per cent when burning fossil fuels, has eliminated sulphuric acid emissions. However, care should be taken as absence of excess air tends to result in greater soot production.

Equipment modification/replacement

Another method of control of pollutants at the source involves the proper use of existing equipment, and modification and replacement of equipment. For example, the unburnt carbon monoxide and hydrocarbons in the cylinders of an automobile engine, which are otherwise emitted into the atmosphere through the tail pipe, can be burnt by injecting air into the hot exhaust manifold of the engine. Similar results can be obtained by suitable modification in the carburetion and ignition systems.

Hydrocarbon vapours from petroleum refineries are released into the atmosphere from storage tanks due to temperature changes, direct evaporation and displacement during filling. These losses can be minimised by designing the tanks with floating roof covers or by pressurising them.

In addition to the above mentioned source correction methods, air pollutants emitted from industrial operations can be reduced by proper equipment maintenance, handkeeping, and cleanliness in the facilities and premises. Often, changes in the design of local exhaust hood and proper installation can minimise the emission of pollutants. Chemical process plants often have excessive leakage around ducts, piping, valves and pumps. Many such leaks can be prevented by routine checking of the seals and gaskets. Floors, decks, storage bins and silos, loading areas and material transfer conveyors must also be kept clean to reduce dust pollution.

Particulate Control by Using Equipment

The most effective methods of particulate control are reduction at the source by the application of control equipment and process control. If air pollution problems are properly considered in an industry prior to its design, real economy can be effected. But unfortunately in most cases, air pollution control is an after-thought and the ways and means to control it are derived and designed hurriedly at the eleventh hour.

To remove particulate matter from gas streams, various types of control equipment are available. But to select the required equipment, certain basic data must be available.

The required data are:

1. Quantity of gas to be treated and its variation with time.
2. Nature and concentration of the particulate matter to be removed.
3. Temperature and pressure of the gas stream.
4. Nature of the gas phase (for solubility and corrosive effects).
5. Desired quality of the treated effluent, i.e. efficiency of removal of particulates required.

Objectives of using control equipment

Objectives of using control equipment includes:

1. Prevention of nuisance.
2. Prevention of physical damage to property.
3. Elimination of health hazards to plant personnel and the general population.
4. Recovery of valuable waste products.
5. Minimisation of economic losses through reduction of plant maintenance.
6. Improvement of product quality.

As pollutants, originating from a variety of sources, primarily from industrial processes, airborne particulates exert a significant influence on atmospheric phenomena, plants, property and on animals, including man. Many techniques for the control of particulate emission have been developed.

Gravity Settling Chamber

It consists of a simple horizontal rectangular gravity apparatus for the removal of particulates from gas streams. Effluent gases are led into this chamber which allows larger particles (250 μ) to settle down. However, the method is unsuitable for finer particles which have longer settling time.

Cyclone Collector

A gas containing of particulates flowing in a tight circular spiral creates a centrifugal force on suspended particles forcing them to move away from the gas stream to a wall where they are collected. This inertial separator due to centrifugal force exerts greater inertial effects on the dispersoids to effect particulate and gas separation. It is one of the least expensive and efficient techniques by which 95 per cent of particulates in the range of 5 to 20 μ can be separated.

Cyclonic Separators and Trajectory Separators

These have been extensively employed for industrial dust collection. They are also used for the control of gas-borne dispersoids in industrial operations, e.g. manufacture of cement, paper and pulp industries, mineral processing, feed and grain processing, and wood working factories.

Filters

Filters are now widely used to collect extremely fine particulates. Solid dispersoids can be removed from the carrier gas by filtration of the effluents through porous cloth or fibre. Next to cyclonic separators, filters constitute the most effective devices of controlling particulate pollutants.

Scrubbers

Scrubbers may be classified as wet washers, cyclonic scrubbers, gravity spray towers, impingement scrubbers, disintegrator scrubbers and spray de-dusters.

Wet scrubbers

These utilise water to remove solid, liquid or gaseous contaminants. The effectiveness of the device depends upon the degree of contact and interaction between the liquid phase and the contaminants to be removed. Spray chambers or towers are used, where the liquid is introduced into the gas stream as a fine spray. Another recent technique makes the liquid percolate downward through a bed of materials, while the gas stream moves upward and gets separated.

Electrostatic Precipitators (ESP)

The electrostatic precipitator is one of the most widely used devices for controlling particulate emissions from industrial installations manufacturing household appliances to power plants, cement and paper mills and oil refineries. In most cases, the particulates to be collected are by-products of combustion. In others, they are dust fibres or other small particles such as acid mists from process industries. The electrostatic precipitators are particulate collection devices that utilise electrical energy directly to assist in the removal of particulate matter (Fig. 2.1). These are successfully used for removal of fine dust from all kinds of waste gases with very high efficiency. Particles as small as a tenth of a micron also can be removed efficiently. The principle on which this equipment operates is that when a gas containing aerosols is passed between two electrodes that are electrically insulated from each other and between which there is a considerable difference in electric potential, aerosol particles precipitate and which are subsequently removed.

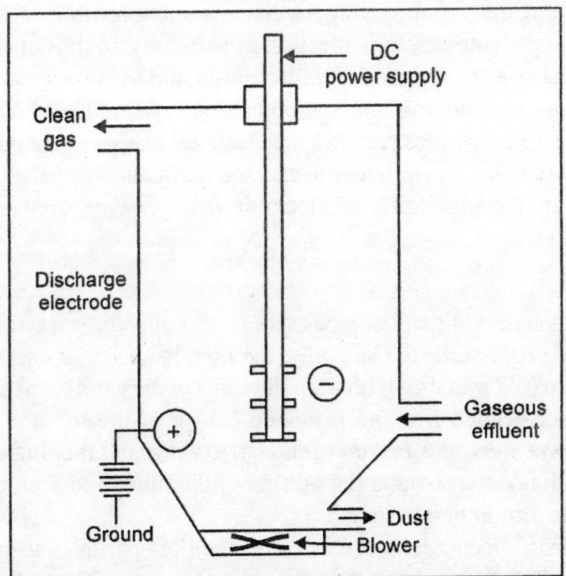

Fig. 2.1. Schematic diagram of an electrostatic precipitator.

There are various types of electrostatic precipitators used for different industrial purposes. They can also be used for air cleaning in public buildings, theatres, railways, cars, etc. An electrostatic precipitator consists of six major components, viz.

1. A source of high voltage.

2. Discharge electrodes and collecting electrodes.
3. Inlet and outlet for the gas.
4. An electronic cleaning system.
5. 'Hopper' for collection and disposal of particulates.
6. An outer casing (shell) to form an enclosure around the electrodes.

Principle of ESP

Electrostatic precipitation is a physical process by which particles (solid or liquid) can be removed from gaseous streams. The gas stream is passed between a pair of electrodes, across which a high potential difference is maintained. One of them is a discharge at a high potential due to which a powerful ionising field is formed. Under the action of electrical field, gas ions formed in the corona move rapidly towards the collecting electrode and transfer their charge to the particles by colliding with them. The electrical field interacting with the charge on the particles then causes them to drift towards and deposit on the collecting electrode. These particles lose their charge and are then removed mechanically by rapping or vibration to a hopper below the electrical treatment zone, and are collected for ultimate disposal. When the particles are liquid droplets, they coalesce on the collecting electrode and drip off the bottom of that electrode into a collecting sump.

Types of precipitators

The main electrical mechanisms for precipitation of particles are: (i) supplying an electrical charge to the particles for gas ionisation, and (ii) supplying the electrostatic force that causes the charged particles to drift towards the collecting electrode (for collection of particles). In the usual industrial electrostatic precipitators, both these charges are supplied simultaneously and the precipitator acts as a single stage unit. In some cases like air conditioning applications and a few industrial applications, a two-stage precipitator is used. One set of electrodes supplies the electrical charge to the particles and a second set supplies the electrostatic force that precipitates the charged particles. Almost all industrial precipitators are of the single stage design. Depending on the electrode arrangements, they may be classified as pipe type or plate type precipitators.

Pipe-type precipitator

In the pipe-type precipitator, a nest of parallel pipes acts as the collecting electrode. The pipes may be of round, square or octagonal cross-section. Generally, the pipe is about 30 cm or less in diameter. The discharge electrode is a wire (2.8 mm dia) with a small radius of curvature, suspended along the axis of each pipe. The wires are suspended from an insulated hanger at the top and kept under tension by weights attached to their lower ends, and are strong enough to withstand rapping or vibrating for cleaning purposes. The gas flow path is down around the outside of the tubes and then up through their inside. The pipe electrodes are 2 to 5 m in height/length.

As the gas flows upwards, electrostatic forces cause the dust particles to migrate to the collector electrode where they stick. The cleaned gas then emerges at the top. The collected dust (aerosols) is removed periodically from the collector electrodes by rapping it; this dust then falls to the dust hopper and accumulates there for periodic removal. A power supply furnishes a large DC voltage of the order of 50 kV, which is either steady or pulsed; modern usage favours the pulsed DC voltage. Generally, the pipe-type precipitators are used for the removal of liquid particles, in which case, no cleaning mechanism is required.

Plate-type precipitators

In the plate-type precipitators, the collecting electrodes consist of parallel plates of size 1–2 m wide and 3-6 m high. These parallel plates are spaced at 20–30 cm. The number of plates in the precipitators depend upon the inlet flow, so that the inlet gas velocities are 1–2 m/s in each channel. The discharge electrodes are similar to those used in pipe-type precipitator. Sometimes electrodes of square rods (4 to 5 mm) and twisted square rods (3.2 to 6.4 mm) are used. These discharge electrodes (i.e. wires), made from non-corrosive materials like tungsten, alloys of steel and copper, are suspended from the top and hang free with a weight attached at the bottom to keep them straight. The collection of the aerosols takes place on the inner sides of the parallel plates. The dust material collected can be removed by periodic rapping and vibrating. The plate-type precipitators are generally employed for the collection of solid particulates. The collection efficiency of the precipitator depends upon the collection surface, bulking resistance and resistance to corrosion of the collection electrode.

Collection efficiency of the ESP

The electrostatic precipitator is a high efficiency collector. The objective of visually clear stacks has made it necessary for many installations to operate at efficiencies of 98–99 per cent and, in some cases, in the 99.5–99.9 per cent range. Some materials ionise more readily than others and are thus more suitable for removal by electrostatic precipitation. Acid mists and catalyst recovery units often have efficiencies in excess of 99 per cent. Over 60 per cent of the installed capacity of precipitators is used for fly ash recovery and the remaining capacity is used mainly for the detarring of fuel gases and for carbon black recovery. Carbon black cannot be removed efficiently because of its agglomerating tendency. However, by proper combination of an ESP with a cyclonic collector, high efficiencies may be obtained in collecting carbon black.

The gas flow in a precipitator is normally turbulent because of 'electric wind' effect. The electric wind is the motion of the gas at right angles to the main gas flow, induced by the motion of the gas ions streaming from the discharge electrode to the collecting electrode. As a result of the turbulence, the collection efficiency takes on an exponential form. Scientists, W. K. Rose and M. S. Wood, modified the collection efficiency equation with the assumption that turbulent mixing effect is intense enough so that the particle concentration is constant over any cross-sectional area at right angles to the main gas flow. The given equation for efficiency of ESP is a function of gas flow rate and precipitator size, and is applicable to both cylindrical and parallel plate type precipitation. The collection efficiency equation is:

$$\text{Efficiency} = 1 - \exp\left(\frac{-A_c V_p}{Q}\right)$$

where,

V_p = particle migration velocity, m/sec
A_c = collector surface area, m^2
Q = volumetric flow rate of the gas, m^3/sec

Dust removal

The dusts particles are deposited on the collector electrode during the operation of ESP. After some time, the thickness of the dust collected increases to 1 to 2 cms, which decreases the efficiency of the collection electrode. When the dust resistivity becomes more than 2×10^{10} ohm-cm, the system is stopped. A dust resistivity of 10^9 to 10^{10} ohm-cm is desired for ESPs. If resistivity increases, it becomes difficult to maintain corona discharge and hence back-ionisation occurs decreasing the efficiency. Particles of high resistivity may be conditioned with moisture to bring them into the acceptable range. This can be achieved by spraying water or steam into the gas stream at the inlet. If the resistivity is less, particles are charged easily, but dissipate it so quickly that the particles are re-entrained in the gas stream, again decreasing the efficiency. 'Dust resistivity' is a function of the composition of the dust, the continuity of the dust layer, operating temperature and the voltage gradient in the dust layer. Thus, when dust resistivity is high, the collecting electrodes should be cleaned to 0.2 to 0.3 cm of dust thickness, by rapping or vibrating the collecting electrode. The dislodged particles are collected in the dust hopper. If the particulates are 'liquid droplets', they are automatically drained into the hoppered bottom by gravity. Both collecting and discharge electrodes must be cleaned of dust to reduce electrical resistance of the dust layers and permit continued operation. Dust build-up on wires is difficult to be removed and occasionally, the deposited dust resembling 'dough nuts' or 'grape fruits' must be removed by hand cleaning or electrode rapping. The frequency and intensity of the rapping cycle have an important effect on the collection efficiency of the precipitator. A high collection efficiency requires that the dust, when rapped loose from the collecting plate, should fall as coarse aggregates, so that it is not redispersed into the gas stream. Cylindrical ESPs are preferred to collect liquid particles with high efficiency.

Re-entrainment

In the derivation of collection efficiency equations, re-entrainment of the deposited dust is usually neglected. The re-entrainment has no effect on liquid droplets but for dry particles, re-entrainment losses can markedly reduce efficiency. There are a number of different causes of re-entrainment. Gas flow through the hoppers can sweep the collected dust back into the gas stream. This can be minimised by providing baffles in the hoppers to reduce gas circulation.

Advantages and disadvantages of electrostatic precipitators

Advantages

1. High collection efficiency.
2. Particles as small as 0.1 micron can be removed.
3. Low maintenance and operating costs.
4. Low pressure drop (0.25 to 1.25 cm of water).
5. Satisfactory handling of large quantities of high temperature gas.
6. Treatment time is negligible (0.1 to 10 seconds).
7. Cleaning is easy by removing units of the precipitator from operation.
8. There is no limit to solid, liquid or corrosive chemical usage.

Disadvantages

1. High initial cost.
2. Space requirement is more because of the large size of the equipment.
3. Possible explosion hazards during collection of combustible gases or particulates. Well trained personnel are needed.
4. The poisonous gas, ozone is produced by the negatively charged discharge electrodes during gas ionisation.
5. Precautions are necessary to maintain safety during operation (i.e. proper gas flow distribution, gas resistivity, particulate conductivity, etc.).
6. Gases cannot be removed by ESPs.

Applications of industrial precipitators

The important applications of electrostatic precipitators in industries are given in Table 2.1.

Table 2.1. Important applications of electrostatic precipitators.

Industry	Application
Cement factories	Cleaning of flue gas from cement kilns
	Recovery of cement dust from kilns
Pulp and paper	Soda fume recovery in kraft pulp mills
Steel plants	Cleaning blast furnace gas
	Removing tars from coke oven gases
	Cleaning open hearth and electric furnace gases
Chemical industries	Collection of SO_x, phosphoric acid mist
	Cleaning various types of gases, i.e. hydrogen, CO_2, SO_2.
	Removing dust from elemental phosphorus in the vapour state
Petroleum industry	Recovery of catalyst dust
Carbon black industry	Agglomeration and collection of carbon black
Thermal power plants	Collecting fly ash from coal fired boilers

AIRBORNE PARTICULATE MATTER: POLLUTION PREVENTION AND CONTROL

Airborne particulate matter (PM) emissions can be minimised by pollution prevention and emission control measures. Prevention, which is frequently more cost-effective than control, should be emphasised. Special attention should be given to pollution abatement measures in areas where toxics associated with particulate emissions may pose a significant environmental risk.

Approaches to Pollution Prevention

Management

Measures such as improved process design, operation, maintenance, housekeeping, and other management practices can reduce emissions. By improving combustion efficiency, the amount of products of incomplete combustion (PICs), a component of particulate matter, can be significantly reduced. Proper fuel-firing practices and combustion zone configuration, along with an adequate amount of excess air, can achieve lower PICs.

Choice of fuel

Atmospheric particulate emissions can be reduced by choosing cleaner fuels. Natural gas used as fuel emits negligible amounts of particulate matter. Oil-based processes also emit significantly fewer particulates than coal-fired combustion processes.

Low-ash fossil fuels contain less noncombustible, ash-forming mineral matter and thus generate lower levels of particulate emissions. Lighter distillate oil-based combustion results in lower levels of particulate emissions than heavier residual oils. However, the choice of fuel is usually influenced by economic as well as environmental considerations.

Fuel cleaning

Reduction of ash by fuel cleaning reduces the generation of PM emissions. Physical cleaning of coal through washing and beneficiation can reduce its ash and sulphur content, provided that care is taken in handling the large quantities of solid and liquid wastes that are generated by the cleaning process. An alternative to coal cleaning is the co-firing of coal with higher and lower ash content. In addition to reduced particulate emissions, low-ash coal also contributes to better boiler performance and reduced boiler maintenance costs and downtime, thereby recovering some of the coal cleaning costs. For example, for a project in East Asia, investment in coal cleaning had an internal rate of return of 26 per cent.

Choice of technology and processes

The use of more efficient technologies or process changes can reduce PIC emissions. Advanced coal combustion technologies such as coal gasification and fluidised-bed combustion are examples of cleaner processes that may lower PICs by approximately 10 per cent. Enclosed coal crushers and grinders emit lower PM.

Approaches to Emission Control

A variety of particulate removal technologies, with different physical and economic characteristics, are available. Inertial or impingement separators rely on the inertial properties of the particles to separate them from the carrier gas stream. Inertial separators are primarily used for the collection of medium-size and coarse particles. They include settling chambers and centrifugal cyclones (straight-through, or the more frequently used reverse-flow cyclones). Cyclones are low-cost, low-maintenance centrifugal collectors that are typically used to remove particulates in the size range of 10–100 microns (mm). The fine-dust-removal efficiency of cyclones is typically below 70 per cent, whereas electrostatic precipitators (ESPs) and baghouses can have removal efficiencies of 99.9 per cent or more. Cyclones are, therefore, often used as a primary stage before other PM removal mechanisms. They typically cost about US $35 per cubic metre/minute flow rate (m^3/min.), or US $1 per cubic foot/minute (cu. ft/min.).

Electrostatic precipitators (ESPs) remove particles by using an electrostatic field to attract the particles onto the electrodes. Collection efficiencies for well-designed, well-operated, and well maintained systems are typically in the order of 99.9 per cent or more of the inlet dust loading. ESPs are especially efficient in collecting fine particulates and can also capture trace emissions of some toxic metals with an efficiency of 99 per cent. They are less sensitive to maximum temperatures than are fabric filters, and they operate with a very low pressure drop.

Their consumption of electricity is similar to that of fabric filters (Table 2.1). ESP performance is affected by fly-ash loading, the resistance of fly-ash, and the sulphur content of the fuel. Lower sulphur concentrations in the flue gas can lead to a decrease in collection efficiency. ESPs have been used for the recovery of process materials such as cement, as well as for pollution control. They typically add 1–2 per cent to the capital cost of a new industrial plant.

Filters and dust collectors (baghouses) collect dust by passing flue gases through a fabric that acts as a filter. The most commonly used is the bag filter, or baghouse. The various types of filter media include woven fabric, needled felt, plastic, ceramic, and metal. The operating temperature of the baghouse gas influences the choice of fabric.

Accumulated particles are removed by mechanical shaking, reversal of the gas flow, or a stream of high-pressure air. Fabric filters are efficient (99.9 per cent removal) for both high and low concentrations of particles but are suitable only for dry and free-flowing particles.

Their efficiency in removing toxic metals such as arsenic, cadmium, chromium, lead, and nickel is greater than 99 per cent. They also have the potential to enhance the capture of sulphur dioxide (SO_2) in installations downstream of sorbent injection and dry-scrubbing systems. They typically add 1–2 per cent to the capital cost of new power plants.

Wet scrubbers rely on a liquid spray to remove dust particles from a gas stream. They are primarily used to remove gaseous emissions, with particulate control a secondary function. The major types are venturi scrubbers, jet (fume) scrubbers, and spray towers or chambers. Venturi scrubbers consume large quantities of scrubbing liquid (such as water) and electric power and incur high pressure drops. Jet or fume scrubbers rely on the kinetic energy of the liquid stream.

The typical removal efficiency of a jet or fume scrubber (for particles 10 mm or less) is lower than that of a venturi scrubber. Spray towers can handle larger gas flows with minimal pressure drop and are therefore often used as precoolers. Because wet scrubbers may contribute to corrosion, removal of water from the effluent gas of the scrubbers may be necessary.

Another consideration is that wet scrubbing results in a liquid effluent. Wet-scrubbing technology is used where the contaminant cannot be removed easily in a dry form, soluble gases and wettable particles are present, and the contaminant will undergo some subsequent wet process (such as recovery, wet separation or settling or neutralisation). Gas flow rates range from 20 to 3000 m^3/min. Gas flow rates of approximately 2000 m^3/min. may have a corresponding pressure drop of 25 cm water column.

Equipment Selection

The selection of PM emissions control equipment is influenced by environmental, economic, and engineering factors:

1. Environmental factors include: (i) the impact of control technology on ambient air quality, (ii) the contribution of the pollution control system to the volume and characteristics of waste-water and solid waste generation, and (iii) maximum allowable emissions requirements.
2. Economic factors include: (i) the capital cost of the control technology, (ii) the operating and maintenance costs of the technology, and (iii) the expected lifetime and salvage value of the equipment.
3. Engineering factors include: (i) contaminant characteristics such as physical and chemical properties—concentration, particulate shape, size distribution, chemical reactivity, corrosivity, abrasiveness, and toxicity, (ii) gas stream characteristics such as volume flow rate, dust loading, temperature, pressure, humidity, composition, viscosity, density, reactivity, combustibility, corrosivity, and toxicity, and (iii) design and performance characteristics of the control system such as pressure drop, reliability, dependability, compliance with utility and maintenance requirements, and temperature limitations, as well as size, weight, and fractional efficiency curves for particulates and mass transfer or contaminant destruction capability for gases or vapours.

Table 2.2 presents the principal advantages and disadvantages of the particulate control technologies discussed here. ESPs can handle very large volumetric flow rates at low pressure drops and can achieve very high efficiencies (99.9 per cent). They are roughly equivalent in costs to fabric filters and are relatively inflexible to changes in process operating conditions. Wet scrubbers can also achieve high efficiencies and have the major advantage that some gaseous pollutants can be removed simultaneously with the particulates.

Table 2.2. Advantages and disadvantages of particulate control technologies.

Advantages	Disadvantages
Inertial or impingement (cyclone) separators	
Low capital cost (approximately US $1/cu. ft/min. flow rate)	Relatively low overall particulate collection efficiencies, especially for particulate sizes below 10 mm
Relative simplicity and few maintenance problems	Inability to handle sticky materials
Relatively low operating pressure drop (for the degree of particulate removal obtained) in the range of approximately 5–15 cm (2–6 inches) water column	
Temperature and pressure limitations imposed only by the materials of construction used	
Dry collection and disposal	
Relatively small space requirements	
Wet scrubbers	
No secondary dust sources	Potential water disposal/effluent treatment problem
Relatively small space requirement	Corrosion problems (more severe than with dry systems)
Ability to collect gases, as well as particulates (especially 'sticky' ones)	Potentially objectionable steam plume opacity or droplet entrainment
Ability to handle high-temperature, high-humidity gas streams	Potentially high pressure drop—approximately 25 cm (10 inches) water column and horsepower requirements
Low capital cost (if waste-water treatment system is not required)	Potential problem of solid buildup at the wet-dry interface
Insignificant pressure-drop concerns for processes where the gas stream is already at high pressure	Relatively high maintenance costs
High collection efficiency of fine particulates (albeit at the expense of pressure drop)	
Electrostatic precipitators	
Collection efficiencies of 99.9% or greater for coarse and fine particulates at relatively low energy consumption	High capital cost
	High sensitivity to fluctuations in gas stream conditions (flow rates, temperature, particulate and gas composition, and particulate loadings)
Dry collection and disposal of dust	Difficulties with the collection of particles with extremely high or low resistivity
Low pressure drop—typically less than 1–2 cm (0.5 inch) water column	Relatively large space requirement for installation
Continuous operation with minimum maintenance	Explosion hazard when dealing with combustible gases or particulates
Relatively low operation costs	
Operation capability at high temperatures (up to 700°C, or (1300°F) and high pressure (up to 10 atmospheres, or 150 pounds per square inch, psi) or under vacuum	Special precautionary requirements for safeguarding personnel from high voltage during ESP maintenance by deenergising equipment before work commencement
Capability to handle relatively large gas-flow rates (on the order of 50,000 m³/min	Production of ozone by the negatively charged electrodes during gas ionisation
	Highly trained maintenance personnel required

(Contd ...)

Advantages	Disadvantages
Fabric filter systems (baghouses)	
Very high collection efficiency (99.9%) for both coarse and fine particulates	Requirement of costly refractory mineral or metallic fabric at temperatures in excess of 290°C (550°F)
Relative insensitivity to gas stream fluctuations and large changes in inlet dust loadings (for continuously cleaned filters)	Need for fabric treatment to remove collected dust and reduce seepage of certain dusts
	Relatively high maintenance requirements
Recirculation of filter outlet air	Explosion and fire hazard of certain dusts at concentration
Dry recovery of collected material for subsequent processing and disposal	($\sim$50 g/m^3) in the presence of accidental spark or flame, and fabric fire hazard in case of readily oxidisable dust
No corrosion problems	collection
Simple maintenance, flammable dust collection in the absence of high voltage	Shortened fabric life at elevated temperatures and in the presence of acid or alkaline particulate or gas constituents.
High collection efficiency of submicron smoke and gaseous contaminants through the use of selected fibrous or granular filter aids	Potential crusty caking or plugging of the fabric or need for special additives due to hygroscopic materials, moisture condensation, or tarry adhesive components
Various configurations and dimensions of filter collectors	Respiratory protection requirement for fabric replacement
Relatively simple operation	Medium pressure-drop requirements—typically in the range of 10–25 centimetres (4–10 inches) in water column

However, they can only handle smaller gas flows (up to 3000 m^3/min.), can be very costly to operate (owing to a high pressure drop), and produce a wet sludge that can present disposal problems. For a higher flue gas flow rate and greater than 99 per cent removal of PM, ESPs and fabric filters are the equipment of choice, with very little difference in costs.

Recommendations

For effective PM$_{10}$ control in industrial application, the use of ESPs or baghouses are recommended. They should be operated at their design efficiencies. In the absence of a specific emissions requirement, a maximum level of 50 milligrams per normal cubic meter (mg/Nm^3) should be achieved.

For gases containing soluble toxics and where the gas flow rate is less than 3000 m^3/min., wet scrubbers may be used. Cyclones and mechanical separators should be used only as precleaning devices upstream of a baghouse or an ESP.

Key Issues for Pollution Prevention and Control Planning

The principal methods for controlling the release of particulate matter are summarised here:
1. Identify measures for improving operating and management practices.
2. Consider alternative fuels such as gas instead of coal.
3. Consider fuel-cleaning options such as coal washing, which can reduce ash content by up to 40 per cent.
4. Consider alternative production processes and technologies, such as fluidised bed combustion, that result in reduced PM emissions.
5. Select optimal particulate removal devices such as ESPs and baghouses.

Note: However, controlling emissions of many heavy metals, such as cadmium, lead, and mercury, that are present as trace elements in fuels is a difficult and largely unsolved problem.

Chapter 3

Gaseous Pollutants and Their Control

INTRODUCTION

Pollutants such as sulphur dioxide, hydrogen sulphide, and nitrogen dioxide combine with moisture in the air to form acids that attack and damage library material. Generally, gaseous pollution is caused by the burning of fuels. Pollutants such as sulphur dioxide, hydrogen sulphide, and nitrogen dioxide combine with moisture in the air to form acids that attack and damage library material. Ozone is a powerful oxidant which severely damages all organic materials. It is a product of the combination of sunlight and nitrogen dioxide from car exhaust; it may also be produced by electrostatic filtering systems used in some air conditioners, as well as by electrostatic photocopy machines.

Smoking, cooking, and off-gassing from unstable materials (cellulose nitrate film, paint finishes, fire-retardant coatings, and adhesives) may also produce harmful gaseous pollutants. Wood, particularly oak, birch and beech, emit acetic and other acids, and vulcanised rubber releases volatile sulphides that are especially damaging to photographs. The composition of all equipment, materials, and finishes used for the storage, transport, and display of objects should be tested by recognised methods to ascertain whether they are likely to produce harmful emissions.

CONTROL OF GASEOUS POLLUTANTS

Emission of gaseous pollutants can be controlled by the following methods: (i) absorption, (ii) adsorption, and (iii) combustion.

Absorption

Absorption involves the transfer of pollutants from the gas phase to the liquid phase across the interface in response to a concentration gradient, with the concentration decreasing in the direction of mass transfer. Absorption of a gaseous contaminant by a liquid occurs because the latter is not saturated with the contaminant at the conditions existing in the absorber. The difference between the concentration of the contaminant in the liquid if it was saturated and the actual concentration provides the driving force for absorption. Hence, the more soluble a contaminant is in the liquid phase, the greater is the overall efficiency that can be attained. In the absorption process, effluent gases are passed through absorbers (scrubbers), which contain liquid absorbents, that remove one or more pollutants from the gas stream. Absorbents are being used to remove SO_2, H_2S, SO_3, F and oxides of nitrogen. The absorbents may be either reactive or non-reactive with the pollutants removed by them. The reactive type absorbents may be either regenerative or non-generative. Various equipment which use the principle of absorptions are:

(i) spray towers, (ii) packed towers, (iii) plate towers, (iv) bubble plate towers, and (v) ventury scrubbers. Of them, spray towers, packed power and ventury scrubbers can be simultaneously used for removing particulate pollutants. Figure 3.1 shows a plate tower while Fig. 3.2 shows a bubble plate tower. A plate tower consists of a vertical shell in which are mounted a large number of equally spaced circular perforated plates. Gases and vapours bubble upward through the liquid seal above each plate. A bubble plate tower consists of a vertical shell in which are mounted a large number of equally spaced circular bubble plates. The dispersion of gas phase within the liquid phase is accomplished by breaking the gas phase into bubbles within the liquid phase by supplying the gas under pressure to bubble plates.

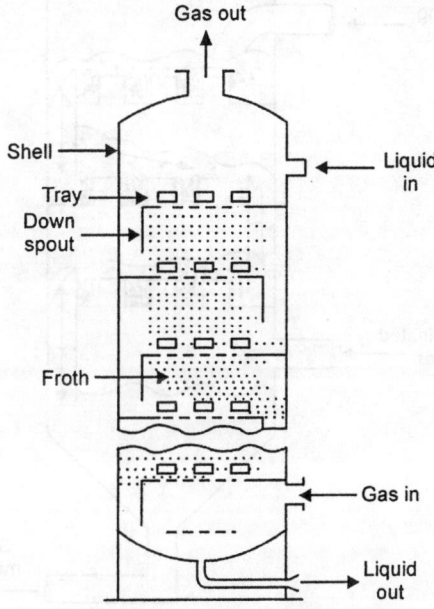

Fig. 3.1. Plate tower.

Adsorption

An alternative to absorption by liquids is the adsorption of air pollutants on solids. The commonly used adsorbers include activated carbon, molecular sieves such as dehydrated zeolites, silica gel, activated alumina, lithium, chloride, bauxite, etc. (Fig. 3.3). Adsorption is a surface phenomenon by which gas or liquid molecules are captured by and adhere to the surface of the solid adsorbent. The attractive forces holding the molecules on the surface may be either physical (physical adsorption) or chemical (chemisorption) in nature. The steps necessary for the effective removal of gaseous pollutants by adsorbents are: (i) intimate contact between gaseous pollutants and the solid adsorbent, (ii) separation of unadsorbed gases from the adsorbent and adsorbate, and for final disposal, and (iii) separation (or desorption) of the adsorbed gaseous pollutant from the solid adsorbent by regeneration or replacement of the adsorbent.

The first two steps are carried out in the adsorption equipment, which may take the form of a stationary or fixed bed of adsorbent or a continuous flow of solid phase through a series of fluidised beds. The most important characteristics of an adsorbent are its large surface-to-volume ratio and its preferential affinity to specific substances. However, almost all the adsorbents are subject to destruction at moderately high temperatures (such as 150°C for activated carbon, 400°C for silica gel, 500°C for activated aluminium and 600°C for molecular sieves). Hence, they are very inefficient for purifying industrial gases at such high temperatures.

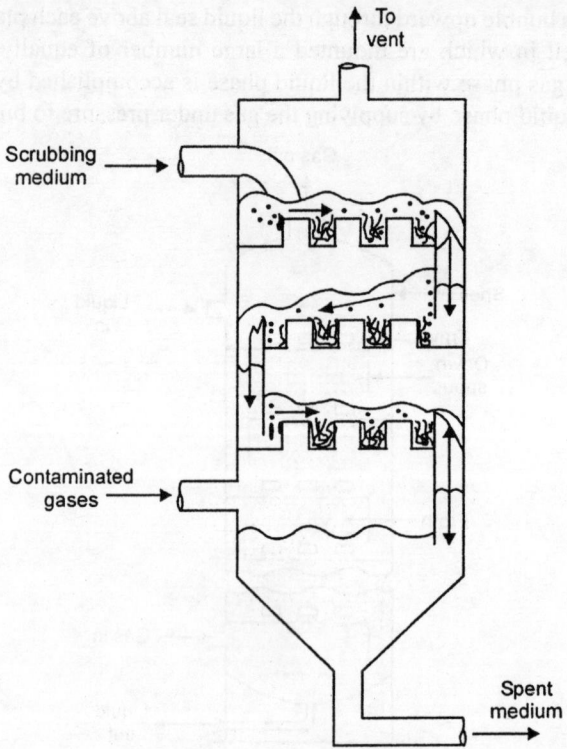

Fig. 3.2. Bubble plate tower.

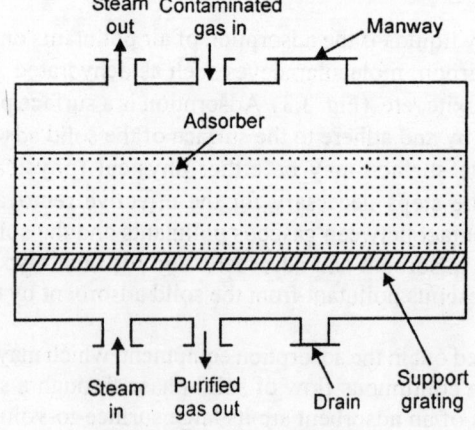

Fig. 3.3. Fixed-bed adsorber.

Combustion or Incineration

This is used when the pollutants in the gas stream are oxidisable to an inert gas. Pollutants like hydrocarbons, and carbon monoxide can be easily burned, oxidised and removed from the combustion equipment. This is achieved by: (i) direct flame incineration, and (ii) catalytic incineration.

Direct flame incineration

Direct flame incineration is a control technique for combustible organic air pollutants (Fig. 3.4). This is accomplished in the presence of a flame and sufficient oxygen by raising the temperature of the gases above their ignition temperature and then maintaining this temperature until the oxidation reactions are complete.

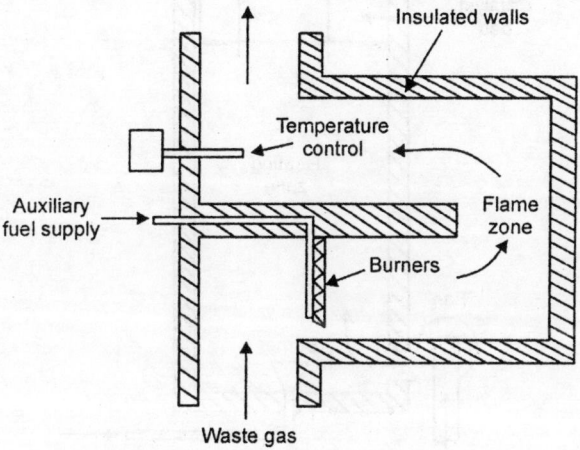

Fig. 3.4. Direct flame incinerator.

Catalytic Incineration

In catalytic incineration (Fig. 3.5), a mixture of dilute organic gases and oxygen is exposed to a catalytic surface at a temperature which is high enough for oxidation to occur and for a length of time sufficient for the oxidation to be completed. Catalysts are usually solids that are neither reactants nor products of a reaction, yet they alter the rate of chemical reactions. The effect of the catalyst is to reduce the temperature required to oxidise the organic compounds, and hence the inlet gases need not be heated to ignition temperature. This requires less fuel than would be needed for direct flame incineration. The catalytic oxidation reaction occurs in a very rapid sequence at the catalytic surface. Hence, the overall *residence time* required is very much less than that in a direct flame incinerator. Due to this, the size of the unit is reduced.

SULPHUR OXIDES

Sulphur oxides (SO_x) are compounds of sulphur and oxygen molecules. Sulphur dioxide (SO_2) is the predominant form found in the lower atmosphere. It is a colourless gas that can be detected by taste and smell in the range of 1,000 to 3,000 micrograms per cubic meter ($\mu g/m^3$). At concentrations of 10,000 $\mu g/m^3$, it has a pungent, unpleasant odour. Sulphur dioxide dissolves readily in water present in the atmosphere to form sulphurous acid (H_2SO_3). About 30 per cent of the sulphur dioxide in the atmosphere is converted to sulphate aerosol (acid aerosol), which is removed through wet or dry deposition processes. Sulphur trioxide (SO_3), another oxide of sulphur, is either emitted directly into the atmosphere or produced from sulphur dioxide and is rapidly converted to sulphuric acid (H_2SO_4).

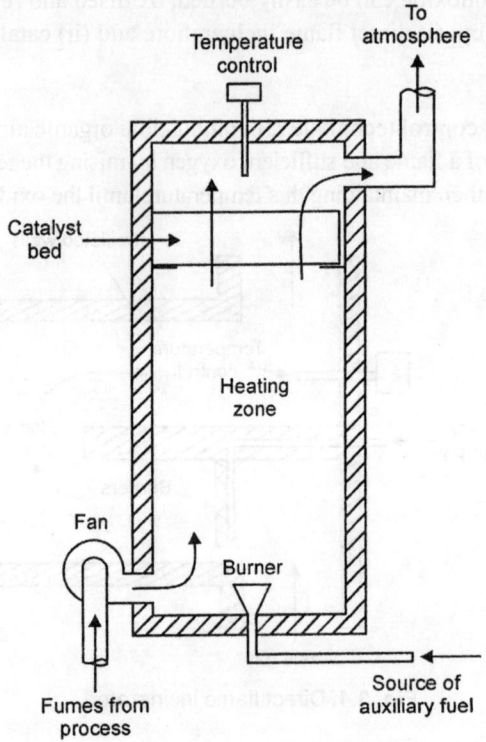

Fig. 3.5. Catalytic incinerator.

Major Sources

Most sulphur dioxide is produced by burning fuels containing sulphur or by roasting metal sulphide ores, although there are natural sources of sulphur dioxide (accounting for 35–65 per cent of total sulphur dioxide emissions) such as volcanoes. Thermal power plants burning high-sulphur coal or heating oil are generally the main sources of anthropogenic sulphur dioxide emissions worldwide, followed by industrial boilers and nonferrous metal smelters. Emissions from domestic coal burning and from vehicles can also contribute to high local ambient concentrations of sulphur dioxide.

SULPHUR COMPOUNDS (SO_x) AS POLLUTANTS

Oxides of sulphur, i.e. sulphur dioxide (SO_2) and sulphur trioxide (SO_3), represented as SO_x, hydrogen sulphide (H_2S), carbonyl sulphide (COS), carbon disulphide (CS_2), dimethyl sulphide [$(CH_3)_2S$] and sulphates (SO_4) are the most serious air pollutants.

Oxides of Sulphur (SO_x)

SO_2 is the second most important contribution of all air pollutants as it accounts for about 29 per cent of the total weight of all pollutants. Sulphur in low concentration is essential for the metabolism of both animals and plants, but it becomes injurious when its concentration increases. There are two sources of SO_2: (i) natural, and (ii) anthropogenic or man-made.

Natural sources (e.g. volcanoes) provide about 67 per cent of the SO_x pollution all over the globe, while man-made sources contribute about 33 per cent of SO_x pollution, which is mainly localised in some urban areas. Among man-made sources, fossil fuel combustion (coal, etc.) accounts for 74 per cent, industries 22 per cent and transportation 2 per cent of the total SO_x emission. This clearly indicates that coal fired power stations are mainly responsible for the SO_x pollution, followed by industrial plants.

The burning of fossil fuels in thermal power plants, manufacture of sulphuric acid and fertilisers, smelting industries and other processes like electric power plants accounts for 75 per cent of total SO_2 emission, while automobiles and refineries contribute to the remaining 25 per cent.

Reaction of SO_2 in the Atmosphere

SO_2 undergoes several chemical reactions in air, forming particulate matter and aerosols, etc. which are scavenged from the atmosphere. SO_2 and SO_3 gases are washed down from the atmosphere in the form of sulphuric acid (H_2SO_4). The presence of increasing concentrations of H_2SO_4 in the troposphere is evident from the increasing and more widespread occurrence of acid rain. A number of factors, such as temperature, light intensity, humidity, air traffic, suspended particulate matter (SPM) may enter into these reactions. However, SO_2 reacts through several ways in the atmosphere. These are:

1. Photochemical reactions.
2. Chemical reactions in the presence of NO_x and hydrocarbons.
3. Chemical reactions in water droplets and solid particles.

Effects of SO_2

SO_2 is a highly irritating gas, which adversely affects men, animals, plants and materials. It is perhaps the most damaging among the various gaseous air pollutants.

Health and Environmental Impacts

Periodic episodes of very high concentrations of sulphur dioxide are believed to cause most of the health and vegetation damage attributable to sulphur emissions. Depending on wind, temperature, humidity, and topography, sulphur dioxide can concentrate close to ground level. During the London fog of 1952, levels reached 3,500 µg/m³ (averaged over 48 hours) in the center of the city and remained high for a period of 5 days. High levels have been recorded during temperature inversions in Central and Eastern Europe, in China, and in other localities.

Health

Exposure to sulphur dioxide in the ambient air has been associated with reduced lung function, increased incidence of respiratory symptoms and diseases, irritation of the eyes, nose, and throat, and premature mortality. Children, the elderly, and those already suffering from respiratory ailments, such as asthmatics, are especially at risk. Health impacts appear to be linked especially to brief exposures to ambient concentrations above 1,000 µg/m³ (acute exposures measured over 10 minutes). Some epidemiologic studies, however, have shown an association between relatively low annual mean levels and excess mortality. It is not clear whether long-term effects are related simply to annual mean values or to repeated exposures to peak values.

Health effects attributed to sulphur oxides are due to exposure to sulphur dioxide, sulphate aerosols, and sulphur dioxide adsorbed onto particulate matter. Alone, sulphur dioxide will dissolve in the watery fluids of the upper respiratory system and be absorbed into the bloodstream. Sulphur dioxide reacts with other substances in the atmosphere to form sulphate aerosols. Since most sulphate aerosols are part of $PM_{2.5}$ (fine particulate matter, with an aerodynamic diameter of less than 2.5 microns), they may have an important role in the health impacts associated with fine particulates. However, sulphate aerosols can be transported long distances through the atmosphere before deposition occurs. Average sulphate aerosol concentrations are about 40 per cent of average fine particulate levels in regions where fuels with high sulphur content are commonly used. Sulphur dioxide adsorbed on particles can be carried deep into the pulmonary system. Therefore, reducing concentrations of particulate matter may also reduce the health impacts of sulphur dioxide. Acid aerosols affect respiratory and sensory functions.

Environment

Sulphur oxide emissions cause adverse impacts to vegetation, including forests and agricultural crops. Studies in the United States and elsewhere have shown that plants exposed to high ambient concentrations of sulphur dioxide may lose their foliage, become less productive, or die prematurely. Some species are much more sensitive to exposure than others. Plants in the immediate vicinity of emissions sources are more vulnerable. Studies have shown that the most sensitive species of plants begin to demonstrate visible signs of injury at concentrations of about 1,850 $\mu g/m^3$ for 1 hour, 500 $\mu g/m^3$ for 8 hours, and 40 $\mu g/m^3$ for the growing season. In studies carried out in Canada, chronic effects on pine forest growth were prominent where concentrations of sulphur dioxide in air averaged 44 $\mu g/m^3$, the arithmetic mean for the total 10 year measurement period; the chronic effects were slight where annual concentrations of sulphur dioxide averaged 21 $\mu g/m^3$.

Trees and other plants exposed to wet and dry acid depositions at some distance from the source of emissions may also be injured. Impacts on forest ecosystems vary greatly according to soil type, plant species, atmospheric conditions, insect populations, and other factors that are not well understood.

Agricultural crops may also be injured by exposure to depositions. Alfalfa and rye grass are especially sensitive. It appears that leaf damage must be extensive before exposure affects the yields of most crops. It is possible that over the long-term, sulphur input to soils will affect yields. However, sulphur dioxide may not be the primary cause of plant injury, and other pollutants such as ozone may have a greater impact.

Acid depositions can damage freshwater lake and stream ecosystems by lowering the pH of the water. Lakes with low buffering capacity, which could help neutralise acid rain, are especially at risk. Few fish species can survive large shifts in pH, and affected lakes could become completely devoid of fish life. Acidification also decreases the species variety and abundance of other animal and plant life.

Sulphate aerosols, converted from sulphur dioxide in the atmosphere, can reduce visibility by scattering light. In combination with warm temperatures, abundant sunlight, high humidity, and reduced vertical mixing, such aerosols can contribute to haziness extending over large areas.

Materials

Sulphur dioxide emissions may affect building stone and ferrous and nonferrous metals. Sulphurous acid, formed from the reaction of sulphur dioxide with moisture, accelerates the corrosion of iron, steel, and zinc. Sulphur oxides react with copper to produce the green patina of copper sulphate on the surface of the copper. Acids in the form of gases, aerosols, or precipitation may chemically erode building materials such as marble, limestone, and dolomite. Of particular concern is the chemical erosion of historical monuments and works of art. Sulphurous and sulphuric acids formed from sulphur dioxide and sulphur trioxide when they react with moisture may also damage paper and leather. Details are given below.

Effects of SO₂ on materials

1. SO_2 also rapidly attacks marble, limestone, roofing, slate, electrical contacts, paper, textiles and buildings. It can even dissolve nylon. Some textile fibres obtained from vegetable sources lose strength when exposed to H_2SO_4. However, wool is somewhat more resistant to SO_2.
2. Paper also absorbs SO_2 which is oxidised to H_2SO_4, causing it to become brittle and fragile.
3. Leather too has much affinity towards SO_2, which affects its strength and causes it to disintegrate.
4. SO_2 is also involved in erosion of building materials such as marble, mortar and in the deterioration of statues.
5. Pollutant emissions of SO_2 from the nearby railway marshalling yard, thermal power stations and about 300 foundries are accelerating the deterioration of world famous Taj Mahal in Agra. Pollutants from Mathura refinery may also cause serious damage to the Taj Mahal.
6. Petroleum refineries, craft paper mills, smelters and industries liberating SO_2 adversely affect historic monuments.
7. Acid rain produced by the oxidation of SO_2 corrodes metals, attacks fibres and washes out basic materials like lime from the soil. The rapid attack of H_2SO_4 on marble is known as 'stone leprosy'.
8. Long exposure to SO_2 increases the drying and hardening time of paints. It affects durability in paint films. For example, exposure of linseed oil paint films 1–2 ppm SO_2 increased drying times by 50–100 per cent.
9. SO_2 polluted air accelerates corrosion rates of metals such as Fe, steel, Zn and Cu. High humidity, particulate matter and temperature also enhance the corrosion of metals.

Effects of SO₂ on man

SO_2 affects human health in various ways:

1. It causes intense irritation to eyes and respiratory tract even at 2.5 ppm levels.
2. SO_2 is absorbed by the nasal system, leading to swelling and stimulated mucus secretion. It severely affects the aged and chronically ill persons.
3. Lung cancer is known to result from increased levels of SO_2 in the atmosphere.
4. SO_2 inhalation causes bronchitis, emphysema and other lung diseases. The intensity increases with the increased atmospheric concentration of SO_2.
5. London or sulphurous acid smog, called the killer smog of January 1956, was formed when SO_2 concentration rose to 0.40 ppm. As a result, the mortality rate among the aged increased from 130 to 180 per day.
6. Moisture and fog enhance SO_2 danger due to the formation of H_2SO_3 and H_2SO_4, and also sulphates. H_2SO_4 is nearly 5 to 20 times as irritant as SO_2.
7. SO_2 is considered as the most serious single air pollutant causing health hazard, obstructing breathing.
8. Oxides of sulphur are the major contributors to lung diseases, cough and choking. Their increased concentration causes acute and chronic asthma.
9. SO_2 is a severe allergenic agent.

Effects of SO₂ on plants

Effects of SO$_2$ on plants

Plants are relatively more sensitive to SO_2 than humans and animals. The threshold levels of SO_2 injury in plants are quite low as compared to man and animals. Some of the effects of SO_2 on plants are:

1. SO_2 damages vegetable crops and affects plant growth, and nutrient quality of plant products.
2. Acute exposure to high levels of SO_2 kills leaf tissues causing leaf necrosis. The edges and area between leaf veins are severely damaged.
3. Chronic SO_2 exposure to plants causes bleaching of leaf pigments due to conversion of chlorophylls to phecophytin— reducing plant productivity.
4. Concentration of SO_2 even as much as 1.00 ppm is injurious to trees causing chlorosis and dwarfing.
5. Susceptible species like cucumber, oats and spinach may be damaged by exposure to air containing 0.05 to 0.5 ppm of SO_2 for 8 hours. Cotton, wheat, barley and apple are most sensitive to SO_2.
6. Plants are mostly injured from SO_2 when their stomata are open during the day time.

CONTROL/REMOVAL OF SOₓ POLLUTANTS

There are six procedures for controlling SO_x emissions. They are either in-plant control measures or effluent treatment methods.

1. Natural dispersion by dilution.
2. Using alternate fuels.
3. Removal of sulphur from fuels (desulphurisation).
4. Process modifications.
5. Control of SO_x in the combustion process.
6. Treatment of flue gas emissions.

These different control methods are described further.

Natural Dispersion by Dilution

The control method is based on natural dispersion at high elevation so that the ground-level concentrations are acceptable at all times. In India, minimum stack heights of 30 m are recommended.

It is also a common practice to stop discharging the effluents into the atmosphere during adverse meteorological conditions. However, this technique is not possible in large-scale power plants, etc.

Using Alternate Fuels

A switch to natural gas from the conventional high-sulphur fuels like coal and petroleum to lessen SO_x emissions is an available alternative. Liquefied natural gas, LNG, also is an effective alternative. However, for utility use, its cost will be much higher than that of other alternatives. Low sulphur coal is another alternative, but obtaining low sulphur coal from the ground is neither quick nor cheap.

Removal of Sulphur from Fuels

The process of removing sulphur prior to combustion is theoretically attractive but practically ineffective. Coal consists of sulphur in both organic and inorganic forms. The inorganic form of sulphur is iron disulphide (FeS_2) mostly available in the forms of pyrites and marcasites. Apparently, washing seems to be an effective process with more than 30 per cent of sulphur being removed. But this results in a loss of combustible material and may increase the requirement of coal and thus the cost. Organic sulphur is present in the form of crystine, thiols, sulphides and some other cyclic compounds, which can only be removed by chemical processing.

Hydro-desulphurisation of coal using a solvent extraction process can remove both organic sulphur as well as inorganic sulphur. In this process finely ground coal is mixed with anthracene oil to form slurry, is heated at a very high temperature of 450°C. The ash residue consisting of both organic and inorganic forms of sulphur is eliminated by pressure filteration. To avoid repolymerisation a small amount of hydrogen is also added. The coal solution filterate is sent into a chamber to remove lighter fractions. The hot liquid residue is cooled to a brittle solid fuel which can be pulverised. The product will be liquefied at about 450°C. It has a higher heating value than raw coal, and contains less than 1 per cent sulphur.

Coal gasification is another process widely used in the India. There are several processes to convert coal into the gaseous form. The earlier processes like water-gas and producer-gas processes produced gas of lower quality. The new processes give a better quality, besides removing both organic and inorganic types of sulphur. *Catalytic hydrogenation* of coal suspended in tar at 100–250 bars at 450°C can achieve 75 per cent desulphurisation with the consumption of 20 kg hydrogen per ton of feed coal. About 5 per cent of this hydrogen is transformed to hydrogen sulphide, an unwanted by-product. Similarly, impure natural gas can be freed of hydrogen sulphide by scrubbing with monoethanolamine or other amines. The sulphur can be recovered by dry catalytic conversion when molten sulphur is obtained. Usually 0.5 to 5 per cent of sulphur is present in heavy fuel oils. The hydro-desulphurisation process is used for desulphurisation of fuel oil. In this process, temperatures are maintained between 400°C to 550°C. The fuel oil is treated with hydrogen in the presence of a catalyst. The residue is eliminated by pressure filtration, with pressures ranging from 35 to 70 atmospheres (atm). The sulphur is recovered by sending it into a flash evaporator where lighter fractions are removed. Almost all coals dissolve in solvent oils at high temperatures (550°C) and pressures (75 bars) in the presence of hydrogen which prevents polymerisation and helps in the removal of organic sulphur. Coal gasification appears to be promising for the abundant coal reserves in India. Oil gas can be scrubbed free of sulphur and can be used in gas-burning devices.

Process changes

Process changes involve new or modified techniques lowering atmospheric pollutant emissions. The process is known as Double Catalysis Double Absorption (DCDA), (Fig. 3.6). In this process, sulphur is burnt with air in a horizontal-spray type combustion chamber.

The emitted SO_2 gas from the sulphur burner has a very high temperature. It is cooled in the waste heat boiler, which recovers surplus heat as the by-product, steam. From the waste heat boiler, the gas flows to the converter system in different stages. The gases from the converter, after 90 per cent of SO_2 has been converted to SO_3, are interrupted and passed at an intermediate stage to an absorber to remove SO_3. The gases are then reheated and returned to the converter for further conversion. They then pass through the additional catalyst, are cooled, pass through a second absorber and then to the atmosphere. The unconverted gases, after being heated by the gases entering the absorber, are returned to the next stage of the converter. As a result, the overall SO_2 conversion efficiency increases. It is also possible to use, higher inlet concentrations of SO_2 (10–12 per cent) as against the usually employed 8 per cent concentration.

The thermal efficiency of the system can further be improved with suitable waste heat recovery methods like utilising heat from sulphur burners or heat from oxidation of SO_2 to SO_3. At high pressures of about 22 atm and 9 per cent SO_2, conversion efficiencies of up to 99.7 per cent with SO_2 concentration as low as 40 ppm can be achieved. The use of oxygen instead of air further decreases the plant size and volume of gases. Thus, the DCDA process requires just one extra absorption tower than the conventional plant. It has been established that any plant of capacity 50 ton of acid per day and above can be converted into a double absorption system economically. Another advantage of the DCDA process is that it can be adapted to a wide range of SO_2 concentrations. This process has been proved to be economical and may be strongly recommended for almost all the H_2SO_4 manufacturing units.

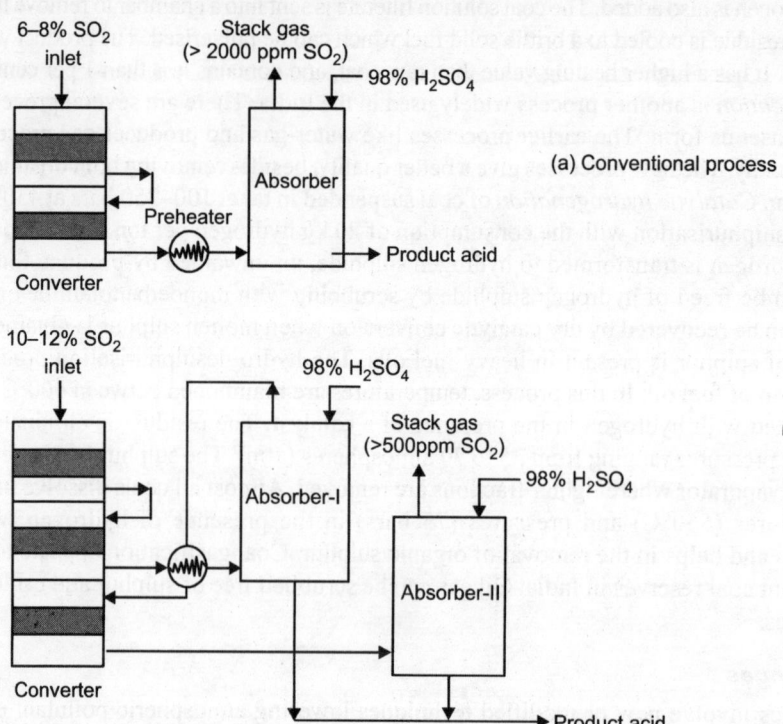

Fig. 3.6. Manufacturing of H_2SO_4 by DCDA process.

Control of SO$_x$ in the combustion process

Here, finely powdered limestone is injected directly into the conventional combustion chamber. The limestone is calcinated to CaO and it reacts with SO_2 contained in the flue gas to form sulphites and sulphates. The unreacted materials and flyash are removed by dry collectors.

$$CaCO_3 \rightarrow CaO + CO_2$$

$$CaO + SO_2 \rightarrow CaSO_3$$

$$CaO + SO_2 + 1/2O_2 \rightarrow CaSO_4$$

The formation of $CaSO_4$ is most favoured at temperatures above 1000°C. Due to smaller residence time in combustion zone, limestone does not completely react with SO_2 to produce a stoichiometric yield of $CaSO_4$. At temperatures above 1200°C, $CaSO_3$ is unstable and SO_2 removal by sulphite formation is inhibited. As a result, the process removes less than half of the sulphur oxides. Due to this, the dry limestone injection process, although relatively simple and easy to operate, is not very attractive from the point of view of control technology. In the widely used *fluid bed combustion process*, limestone and crushed coal together form the fluidised bed and oil is used as the fluidising medium. The operating temperatures are of the order of 700°–1000°C, since the fluidised beds are capable of transferring high heat. This process proves to be quite effective as it removes more than 90 per cent of sulphur. In addition to high SO_2 removal efficiencies, the fluidised bed combustion process prevents the onset of ash fusion and as a result, the fouling and corrosion of boiler tubes associated with the molten slag are considerably reduced. Also, the formation of nitrogen oxides by the nitrogen fixation reaction are reduced. The method, however, has some drawbacks. It requires design modification for boilers and also additional installations for the preparation of limestone. If the limestone is ground to the same size as coal, then a practically inseparable mixture of ash and lime is produced. The fine ash can be elutriated and partially sulphated lime is regenerated.

$$2CaSO_4 + C \rightarrow 2CaO + 2SO_2 + CO_2$$

This regeneration, which requires temperatures in excess of 1000°C will substantially suppress the consumption of limestone.

Treatment of sulphur from flue gas emissions

It seems unlikely that a single desulphurisation method will be developed that is capable of controlling effluents from all types of sources. The control techniques to be employed depend upon such factors as boiler size, configuration, load pattern, geographical location and the like. Nearly 50 flue gas desulphurisation processes have been proposed, but as such no ideal process exists. The general classification of these processes may be wet and dry.

Dry methods

The dry methods can be divided, broadly, into two types:
1. Oxidation/reduction methods.
2. Metal oxides usage.

Oxidation/reduction

CatOx process: This method of oxidation produces sulphuric acid. In this process, fly ash is first removed from the flue gas by a high temperature electrostatic precipitator (Fig. 3.7). SO_2 is then catalytically oxidised to SO_3 and recovered as sulphuric acid. If the exit gases are at a lower temperature, they may be heated. V_2O_5 at 400–500°C is used as a catalyst for good conversion efficiency. In a modified oxidation process, SO_2 and oxygen present in the stack gas are absorbed on the surface of an active carbon catalyst, which catalyses the oxidation of SO_2 to SO_3. SO_3 reacts with the moisture present to form H_2SO_4 in the pores of active carbon. The combined effect of absorption and catalysis by the active carbon leads to a complete conversion of SO_2.

Westvaco process: This process also involves oxidation/reduction. This process is unique in that it utilises fluidised beds of high efficiency activated carbon and converts H_2SO_4 to sulphur. Flue gas is contacted with activated carbon in the absorber unit and SO_2 is oxidised to SO_3.

$$SO_2 \rightarrow O_2, H_2O \rightarrow H_2SO_4$$
$$\text{Activated carbon}$$

The carbon which is used as a catalyst is fed to a sulphur generator where it is contacted with H_2S to form sulphur.

$$H_2SO_4 + 3H_2S \rightarrow 4S + 4H_2$$

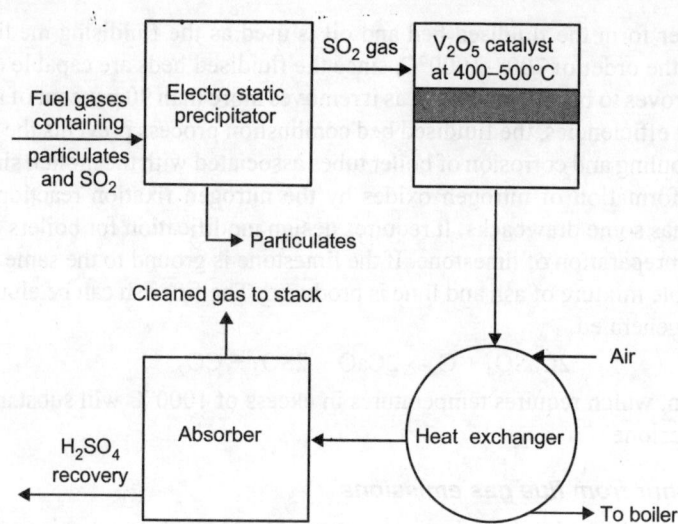

Fig. 3.7. CatOx process of removal of SO_x.

A fraction of the sulphur is recovered by vapourisation and is recondensed as a molten product. The remaining fraction of sulphur reacts with hydrogen in a hydrogen sulphide generator to form H_2S.

SCOT process: In the SCOT (Shell Clauss Off-gas Treatment) process, developed in the Netherlands, almost complete removal of sulphur or its compounds is possible. The gases are reduced while sulphur and its compounds are converted to H_2S. In this process, SO_2 reacts with copper oxide at a temperature of about 400°C to form copper sulphate.

$$SO_2 + 1/2O_2 + CuO \rightarrow CuSO_4$$

Copper sulphate is then reduced at the same temperature in a hydrogen rich gas

$$CuSO_4 + 2H_2 \rightarrow Cu + SO_2 + 2H_2O$$

The concentrated SO_2 is then sent to a Clauss sulphur recovery plant. The Clauss off-gas is heated with the addition of a reducing gas (H_2,CO) and passed through a reactor containing a cobalt-molybdenum catalyst. The absorber used for removal of H_2S is often a diisopropanolamine (DIPA) solution. SO_2 concentrations of less than 250 ppm can be obtained with this process.

Use of metal oxides

In this process, sodium aluminate ($Na_2O \cdot Al_2O_3$) is used to remove SO_2 in a fluidised bed. The dust free flue gas is fed to a reactor wherein the absorbent, a porous form of sodium aluminate ($Na_2O \cdot Al_2O_3$), adsorbs SO_2 at a temperature of 300°C.

$$Na_2O \cdot Al_2O_3 + SO_2 + 1/2O_2 \rightarrow Na_2SO_4 + Al_2O_3$$

The product from the above reaction is contacted with a reducing gas such as hydrogen in a regenerator at about 680°C to form hydrogen sulphide.

$$Na_2SO_4 + Al_2O_3 + 4H_2 \rightarrow Na_2O \cdot Al_2O_3 + H_2S + 3H_2O$$

From this, the sodium aluminate is recycled and hence the process is called a cyclic adsorption process. Sulphur can also be produced as a by-product from the H_2S gas by sending it into a Clauss unit.

$$2H_2S + SO_2 \rightarrow 3S + 2H_2O$$

However, maintenance of the granular strength of sorbent is the main problem relating to this process. The rigorous temperature and chemical cycling to which the sorbent is subjected, deteriorate the sorbent by causing crystalline growth and loss of surface area.

Wet processes

The main unit responsible in a wet process is the scrubber, a spray tower, a cyclone scrubber, a venturi scrubber or a packed tower. Most of the currently available wet flue gas desulphurisation methods use slurries of compounds of calcium, magnesium and sodium. In the wet process, the treated gases are kept at low temperatures in the range of 25°–50°C. This creates a problem in the dispersion of the flue gas. Hence, reheating is required which consumes heat in the range of 3 per cent of the heat of fuel used in the combustion chamber. Some of the wet processes are described below:

Calsox Process (Lime and Limestone Scrubbing): Here, the flue gas is scrubbed with a 5 to 15 per cent slurry of calcium sulphite/sulphate salts which also contains some amounts of lime (CaO) and limestone ($CaCO_3$). The SO_2 reacts with the slurry to form additional sulphite and sulphate salts. The solids are continuously separated from the slurry and discharged into a settling pond. The remaining liquor, at a pH of 6 to 8, is recycled to the scrubbing tower after fresh lime or limestone has been added. A schematic representation of the process is shown in Fig. 3.8.

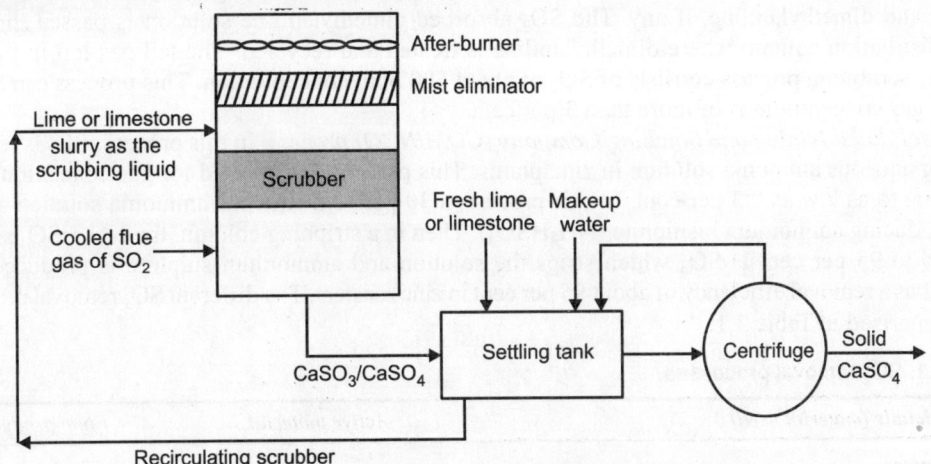

Fig. 3.8 SO_x controlled by lime/limestone scrubbing

The problems in this process include scaling, corrosion, erosion and solid waste disposal. A sizeable disposal area adjacent to plant is required. Another drawback of the process is the necessity for reheating the cleaned flue gas. Reheating is accomplished by installing a gas cooler before the scrubber and a gas stack heater after it. Thus, two additional units must be provided to the gas flow system.

Chemico process (magnesium oxide scrubbing): In this process, known as the 'Chemico Process', magnesium oxide acts as a venturi scrubber by absorbing SO_2 and generating magnesium sulphite and sulphate. The mixed sulphite/sulphate slurry along with unreacted MgO is separated from the liquid phase in a centrifuge. This process is a regenerative one as the mother liquor is regenerated. Concentrated SO_2 (10 to 15 per cent) removed from the flue gas is used in making elemental sulphur or reprocessed to manufacture H_2SO_4.

Welman Lord (Single Alkali) process: This process removes sulphur dioxide by washing fuel gases with an aqueous solution of sodium sulphite. It is quite a common practice in chemical industries, in which 90 per cent of desulphurisation is possible. In this process, sulphite is converted to bisulphite as the SO_2 from flue gases is absorbed by a saturated solution of sodium sulphite. The sodium bisulphite formed is led to a double effect evaporator cum crystalliser where it decomposes into sodium sulphite at a temperature of around 100°C. This results in the ejection of SO_2 and steam.

Clear solutions of either sodium or ammonia are excellent absorbers of SO_2. The regeneration step can be carried out at a relatively low temperature in a liquid system. The one advantage that sodium scrubbing has over ammonia is that the cation is non-volatile. Fume development is another problem in almost all ammonia scrubbers. Both processes produce an unavoidable side product, sodium sulphate in one case and ammonium sulphate in the other.

CuO/CuSO₄ process: This process removes NO_x and SO_x simultaneously by using copper oxide (CuO) supported on stabilised alumina placed in two or more parallel passage reactors.

Flue gas is introduced at about 400°C into one of the reactors where SO_2 reacts with CuO to form copper sulphate ($CuSO_4$). The $CuSO_4$ and to a lesser extent, the CuO act as catalysts in the reduction of NO_x with NH_3. When the reactor is saturated with $CuSO_4$ flue gas is switched to a fresh reactor for acceptance of the flue gases and the spent reactor is regenerated. In the regeneration cycle, hydrogen (H_2) is used to reduce $CuSO_4$ to copper (Cu), yielding an SO_2 stream of sufficient concentration for conversion to sulphur or sulphuric acid.

The American Smelting and Refining Company (ASARCO) process: In this process, emissions of SO_x are controlled by absorption of flue gases in dimethylaniline. The desulphurised gas is further sent through sodium carbonate solution and then to diluted H_2SO_4, where the gas is scrubbed for traces of sulphur and dimethylaniline, if any. The SO_2 absorbed dimethylaniline solution is passed through a steam distillation column where dimethylaniline is treated and recycled. The tail gas left in the stack after this scrubbing process consists of SO_2 of about 500 ppm concentration. This process can be used for SO_2 gas concentrations of more than 3 per cent.

Consolidated Mining and Smelting Company (COMINCO) process. In this process, SO_2 is removed by using aqueous ammonia solution in zinc plants. This process can be used for gas concentrations of 5 per cent to as low as 0.3 per cent. In this process, a 30 per cent aqueous ammonia solution absorbs SO_2, producing ammonium bisulphite (NH_4HSO_4). Then in a stripping column, the NH_4HSO_4 solution is added to 95 per cent H_2SO_4 which strips the solution and ammonium sulphite is produced. This process has a removal efficiency of about 95 per cent in zinc roasters. The different SO_x removal processes are summarised in Table 3.1.

Table 3.1. SO_2 removal processes.

Process details (material used)	Active material	Final product
Dry processes		
Carbon adsorption at 125°C reaction with H_2S to S and H_2 to H_2S	Activated carbon/H_2	Sulphur
Spray dryer, absorption by sodium carbonate or slaked lime solution	Na_2CO_3/Ca $(OH)_2$	Na_2SO_3/Na_2SO_4 Na_2SO_3/$CaSO_4$
Throwaway scrubbing processes		
Limestone or lime slurry	CaO/$CaCO_3$	$CaSO_3$/$CaSO_4$
Na_2SO_3 solution	Na_2CO_3	Na_2SO_4

(Contd...)

Process details (material used)	Active material	Final product
Double alkali: Na₂SO₃ solution regenerated by CaO or CaCO₃	$CaCO_3/Na_2SO_3$ $CaO/NaOH$	$CaSO_3/CaSO_4$
MgSO₃ solution regenerated by lime/limestone	$MgO/MgSO_4$	$CaSO_3/CaSO_4$
Regenerative scrubbing processes		
Mg (OH)₂ slurry	MgO	15% SO_2
Na₂ SO₃ solution	Na_2SO_3	90% SO_2
Sodium citrate solution	H_2S	Sulphur
Ammonia solution	NH_4OH	Sulphur

HYDROGEN SULPHIDE

1. Hydrogen sulphide gas is a naturally occuring chemical (chemical formula H_2S).
2. The gas has a characteristic rotten egg odour at low concentrations. About half of the population can smell it at concentrations as low as 8 parts per billion (ppb) in air, and more than 90 per cent can smell it at levels of 50 ppb. At higher concentrations, hydrogen sulphide rapidly deadens the sense of smell. For most people, this occurs at approximately 150 ppm.
3. Hydrogen sulphide is heavier than air, and it often settles in low-lying areas where it can accumulate in concentrations that can injure or kill livestock, wildlife, and human beings. Additionally, hydrogen sulphide has been found to migrate into surface soils and groundwater.

Sources of H₂S: Hydrogen sulphide occurs naturally in the environment (e.g. in volcanic gases, marshes, swamps, sulphur springs, decaying organic matter). It is produced by living organisms, including human beings, through the digestion and metabolism of sulphur-containing materials. Hydrogen sulphide is also a by-product of many industrial processes, such as paper manufacturing, sewage treatment, landfills, or concentrated animal feed operations (CAFOs).

Hydrogen sulphide gas also is found in petroleum and natural gas. Oil or natural gas is considered sour if it has a high percentage of hydrogen sulphide. Natural gas can contain up to 28 per cent hydrogen sulphide gas, consequently, it may be an air pollutant near petroleum refineries and in oil and gas extraction areas. The principal source of anthropogenic hydrogen sulphide is as a by-product in the purification of natural gas and refinement of crude oil. Atmospheric releases of hydrogen sulphide represent the most significant public health concern for the geothermal energy industry. The chief industrial sources of H_2S are users of sulphur containing fuels.

Sour gas: As mentioned above, oil or natural gas is considered sour if it has a high percentage of hydrogen sulphide. It has been estimated that 15 to 25 per cent of natural gas in the US may contain hydrogen sulphide. Worldwide, the percentage could be as high as 30 per cent. It has been reported, as well, that new drilling is increasingly being focused on deep gas formations that tend to be sour. Although the exact number of sour wells are not available, the EPA has reported that in the US 'the potential for routine H₂S emissions [at oil and gas wells] is significant'.

Releases of hydrogen sulphide from sour gas wells or facilities may occur in a number of ways. US EPA has collected documentation of sour gas well blowouts, line releases, extinguished flares, collection of sour gas in low-lying areas, and leakage from idle or abandoned wells that have impacted the public near oil and gas extraction sites. Other possible sources of hydrogen sulphide emissions at sour oil and gas operations are gas venting, and fugitive emissions (leaks) from well head equipment and compressors.

Environmental Concentrations of H₂S

Typically, areas that are not exposed to industrial releases of hydrogen sulphide have airborne concentrations of less than 1 part per billion (ppb) hydrogen sulphide. Some non-industrial areas, however, do have higher ambient levels than 1 ppb, because hydrogen sulphide is a natural by-product of decomposing organic matter than contains sulphur. For example, Cooper reported short-term hydrogen sulphide levels as high as 62 ppb in Florida wetlands.

In areas with industrial sources of hydrogen sulphide, average concentrations of hydrogen sulphide in nearby neighbourhoods may be present in the low parts per billion range, although maximum concentrations can be in the 100s of parts per billion range or higher (e.g. if there has been a large, industrial release). Also, spills, leaks, malfunctions or build-up of hydrogen sulphide in enclosed or low-lying areas can result in much higher, and sometimes lethal levels.

Effects of Human Exposure

Although numerous case studies of acutely toxic effects of H_2S exist, there is inadequate occupational or epidemiological information for specific chronic effects in humans exposed to H_2S.

Smith and collen showed that 16 healthy subjects exposed for short durations to 5 ppm (7 mg/m^3) H_2S under conditions of moderate exercise exhibited impaired lactate and oxygen uptake in the blood. Smith and Collen also reported that exposure of 42 individuals to 2.5 to 5 ppm (3.5 to 7 mg/m^3) H_2S caused coughing and throat irritation after 15 minutes.

In another study, ten asthmatic volunteer subjects were exposed to 2 ppm H_2S for 30 minutes and pulmonary function was tested. All subjects reported detecting 'very unpleasant' odour but 'rapidly became accustomed to it'. Three subjects reported headache following exposure. No significant changes in mean FVC or FEV$_1$ were reported. Although individual values for specific airway resistance (SR$_{aw}$) were not reported, the difference following exposure ranged from –5.95 per cent to +137.78 per cent. The decrease in specific airway conductance, SG$_{aw}$, ranged from –57.7 per cent to +28.9 per cent. The increase in mean SR$_{aw}$ and decrease in mean SG$_{aw}$ were not statistically significant.

Kilburn and Warshaw investigated whether people exposed to sulphide gases, including H_2S, as a result of working at or living downwind from the processing of 'sour' crude oil demonstrated persistent neurobehavioral dysfunction. They studied thirteen former workers and 22 neighbours (of a California coastal oil refinery) who complained of headaches, nausea, vomiting, depression, personality changes, nosebleeds, and breathing difficulties. Their neurobehavioural functions and a profile of mood states were compared to 32 controls (matched for age and educational level). The exposed subjects' mean values were statistically significantly different (abnormal) compared to controls for several tests [two-choice reaction time; balance (as speed of sway); colour discrimination; digit symbol; trail-making A and B; immediate recall of a story]. Their profile of mood states scores were much higher than those of controls. Visual recall was significantly impaired in neighbours, but not in the former workers. The authors concluded that neurophysiological abnormalities were associated with exposure to reduced sulphur gases, including H_2S from crude oil desulphurisation.

Linger conducted a retrospective epidemiological study in a large petrochemical complex in Beijing, China in order to assess the possible association between petrochemical exposure and spontaneous abortion. The facility consisted of 17 major production plants divided into separate workshops, which allow for the assessment of exposure to specific chemicals. Married women (n = 2853), who were 20–44 years of age, had never smoked, and who reported at least one pregnancy during employment at the plant, participated in the study. According to their employment record, about 57 per cent of these workers reported occupational exposure to petrochemicals during the first trimester of their pregnancy. There was a significantly increased risk of spontaneous abortion for women working in all of the production plants with frequent exposure to petrochemicals compared with those working in non-chemical plants. Also, when a comparison was made between exposed and non-exposed groups within each plant, exposure to petrochemicals was consistently associated with an increased risk of spontaneous abortion (overall odds ratio (OR) = 2.7 [95 per cent confidence interval (CI) = 1.8 to 3.9] after adjusting for potential confounders. When the analysis was performed with the exposure information obtained from interview responses for (self reported) exposures, the estimated OR for spontaneous abortions was 2.9 (95 per cent CI = 2.0 to 4.0). When the analysis was repeated by excluding those 452 women who provided inconsistent reports between recalled exposure and work history, a comparable risk of spontaneous abortion (OR 2.9; 95 per cent CI = 2.0 to 4.4) was found. In analyses for exposure to specific chemicals, an increased risk of spontaneous abortion was found with exposure to most chemicals. There were 106 women (3.7 per cent of the study population) exposed only to hydrogen sulphide, and the results for hydrogen sulphide (OR 2.3; 95 per cent CI = 1.2 to 4.4) were significant. No hydrogen sulphide exposure concentration was reported.

Four workers were exposed for several minutes to concentrations of hydrogen sulphide sufficient to cause unconciousness. Four other workers were exposed chronically to H_2S and developed lacrimation, eye irritation, nausea, vomiting, headache, sore throat, and skin irritation but retained conciousness as the result of a 150 minute release. Both groups were subjected to olfactory testing 2 to 3 years later. Six of eight workers showed deficits in odour detection and identification, with the workers who had experienced unconciousness most severely affected in the followup tests.

Three patients exposed acutely to unknown concentrations of hydrogen sulphide developed persistent cognitive impairment. While standard neurological and physical examinations were unremarkable, all three subjects had prolonged P-300 latencies and persistent neurological and neurobehavioral deficits.

Effects of Animal Exposure

Rats (Fischer and Sprague-Dawley, 15 per group) were exposed to 0, 10.1, 30.5, or 80 ppm (0, 14.1, 42.7, or 112 mg/m^3, respectively) H_2S for 6 hrs/day, 5 days/week for 90 days. Measurements of neurological and hematological function revealed no abnormalities due to H_2S exposure. A histological examination of the nasal turbinates also revealed no significant exposure-related changes. A significant decrease in body weight was observed in both strains of rats exposed to 80 ppm (112 mg/m^3).

In a companion study, the Chemical Industry Institute of Toxicology conducted a 90 day inhalation study in mice (10 or 12 mice per group) exposed to 0, 10.1, 30.5, or 80 ppm (0, 14.1, 42.7, or 112 mg/m^3, respectively) H_2S for 6 hrs/day, 5 days/week. Neurological function was measured by tests for posture, gait, facial muscle tone, and reflexes. Ophthalmological and hematological examinations were also performed, and a detailed necropsy was included at the end of the experiment. The only exposure-related histological lesion was inflammation of the nasal mucosa of the anterior segment of the noses of mice exposed to 80 ppm (112 mg/m^3) H_2S. Weight loss was also observed in the mice exposed to 80 ppm. Neurological and hematological tests revealed no abnormalities. The 30.5 ppm (42.5 mg/m^3) level was considered the No Observed Adverse Effect Level (NOAEL) for histological changes in the nasal mucosa. (Adjustments were made by US EPA to this value to calculate an RfC of 0.9 mg/m^3.)

Fischer F344 rats inhaled 0, 1, 10, or 100 ppm hydrogen sulphide for 8 hrs/day for 5 weeks. No effects were noted on baseline measurements of airway resistance, dynamic compliance, tidal volume, minute volume, or heart rate. Two findings were noted more frequently in exposed rats: (i) proliferation of ciliated cells in the tracheal and bronchiolar epithelium, and (ii) lymphocyte infiltration of the bronchial submucosa. Some exposed animals responded similarly to controls to aerosol methacholine challenge, whereas a subgroup of exposed rats were hyperreactive to concentrations as low as 1 ppm.

Male rats were exposed to 0, 10, 200, or 400 ppm H_2S for 4 hours. Samples of bronchoalveolar and nasal lavage fluid contained increased inflammatory cells, protein, and lactate dehydrogenase in rats treated with 400 ppm. Lopez and associates later showed that exposure to 83 ppm (116 mg/m^3) for 4 hrs resulted in mild perivascular edema.

A study by Saillenfait investigated the developmental toxicity of H_2S in rats. Rats were exposed 6 hrs/day on days 6 through 20 of gestation to 100 ppm hydrogen sulphide. No maternal toxicity or developmental defects were observed.

Hayden exposed gravid Sprague-Dawley rat dams continuously to 0, 20, 50, and 75 ppm H_2S from day 6 of gestation until day 21 postpartum. The animals demonstrated normal reproductive parameters until parturition when delivery time was extended in a dose dependent manner (with a maximum increase of 42 per cent at 75 ppm). Pups which were exposed in utero and neonatally to day 21 postpartum developed with a subtle decrease in time of ear detachment and hair development and with no other observed change in growth and development through day 21 postpartum.

The adverse effects reported in chronic animal studies occur at higher concentrations than effects seen in acute human exposures. For example, human irritation was reported at concentrations of 2.5–5 ppm for 15 minutes, yet no effects on laboratory animals were observed at concentrations up to 80 ppm for 90 days. This suggests either that humans are more sensitive to H_2S, or that the measurements in laboratory animals are too crude to detect subtle measures of irritation. However, the uncertainty factor and HEC attempt to account for these interspecies differences.

Data Strengths and Limitations for Development of the REL

Hydrogen sulphide is the leading chemical agent causing human fatalities following inhalation exposures. Although lower concentration acute exposures have been quantitatively studied with human volunteers, the dose-response relationship for human toxicity due to hydrogen sulphide exposure is not known. Thus, a major area of uncertainty is the lack of adequate long-term human exposure data. Subchronic (but not chronic) studies have been conducted with several animal species and strains, and these studies offer an adequate basis for quantitative risk assessment.

The strengths of the inhalation REL include the availability of controlled exposure inhalation studies in multiple species at multiple exposure concentrations, adequate histopathogical analysis, and the observation of a NOAEL.

Effects of Hydrogen Sulphide (H$_2$S) and Organic Sulphides

1. Mostly sulphides cause odour nuisance when present even in minute concentrations.
2. H_2S causes headache, nausea, collapse, coma and death even at 1–3 ppm.
3. H_2S at 5 ppm affects the digestive system destroying appetite.
4. An exposure to 150 ppm of H_2S gives rise to conjunctivitis and irritation of the mucos membrane.
5. H_2S gas rapidly passes through the alveolar membrane of lungs and penetrates blood. It causes death due to respiratory failure.
6. Even a short exposure, for 10–30 minutes at 500 ppm, of H_2S causes colic diarrhoea and bronchial pneumonia.
7. H_2S reacts with lead paints to form lead sulphide, thereby producing brown to black discolouration.

Hydrogen sulphide has a strong unpleasant odour. The threshold for detection of this odour is low, but shows wide variation among individuals. A level of 7 mg/m^3, based on a 30 minute averaging time, was estimated by a Task Force of the International Program on Chemical Safety (IPCS) to not produce odour nuisance in most situations. On the other hand, the current California ambient air quality standard for hydrogen sulphide, based on a 1 hour averaging time, is 42 mg/m^3 (30 ppb).

Amoore analysed a large number of reports from the scientific literature and found that reported thresholds for detection were log-normally distributed, with a geometric mean of 10 mg/m^3 (8 ppb). Detection thresholds for individuals were reported to be log-normally distributed in the general population, with a geometric standard deviation of 4.0, i.e. 68 per cent of the general population would be expected to have a detection threshold for hydrogen sulphide between 2.5 and 40 mg/m^3 (2 and 32 ppb). Sources of variation included age, sex, medical conditions, and smoking. Training and alertness of the subject in performing the test also affected the results.

Health Effects from H_2S Exposure

Common symptoms of exposure to long-term, low levels of hydrogen sulphide include headache, skin complications, respiratory and mucous membrane irritation, respiratory soft tissue damage and degeneration, confusion, impairment of verbal recall, memory loss, and prolonged reaction time. Exposure to high concentrations can cause unconsciousness, known as 'knockdown,' and can be lethal.

Exposure to hydrogen sulphide is one of the leading causes of sudden death in the workplace. At concentrations greater than 500 parts per million, inhalation of hydrogen sulphide can lead to immediate collapse and unconsciousness. A single breath at 1000 ppm results in immediate loss of consciousness, cardiac arrest and death unless the unconscious victim is successfully revived. Unconsciousness and death have occurred in situations of prolonged exposure to hydrogen sulphide at concentrations of 50 ppm. Many occupational and community studies have documented the adverse health effects of exposure to relatively high levels of H_2S.

Almost all organ systems are affected by hydrogen sulphide, but the most susceptible are those with exposed mucous membranes (e.g. eyes, noise and throat) and those with high oxygen demands (e.g. lungs, brain). Neurotoxicity of the central nervous system (causing nausea, dizziness, confusion, headaches and sleeping problems) and pulmonary edema (build-up of fluid in the lungs) are other well-documented effects of hydrogen sulphide poisoning. Cardiovascular and gastrointestinal toxicity are also associated with H_2S exposure.

Research conducted by Kaye Kilburn, a medical doctor and professor of medicine at the University of Southern California, suggests that exposure to hydrogen sulphide may cause long-term, irreversible human health effects. Kilburn performed physiologic and psychological measurements on nineteen exposed individuals, and compared results with 202 unexposed subjects. Of the 19 exposed subjects, 10 were exposed at work sites, which included four oil and gas operations, and nine were exposed in their residences, which were near various sources of hydrogen sulphide. The concentrations to which the subjects were exposed are not known. Kilburn found that depression, anger, fatigue, tension, confusion and respiratory ailments were significantly higher in exposed subjects than the control group.

Increasingly, scientific research is revealing that even low concentrations of hydrogen sulphide (in the low parts per million or even the parts per billion range) can affect human health, especially when exposure occurs over an extended period of time. For example, data collected in a study of sewer workers indicated that low-level exposure to hydrogen sulphide may be associated with reduced lung function.

The following studies provide more information on the potential association between low-level exposures to hydrogen sulphide and health effects:

1. A study of hydrogen sulphide in the workplace found that workers complained of eye pain at a level of 6.4 ppm.
2. Clinical studies suggest that short-term exposure to hydrogen sulphide at concentrations of 2 ppm may induce bronchial obstruction. In a study investigating the effects of hydrogen sulphide on asthmatics, two out of ten subjects exhibited a pronounced response when exposed to 2 ppm hydrogen sulphide. Airway resistance and conductivity were affected by more than 30 per cent, suggesting significant bronchial obstruction.
3. Former workers and residents living downwind of a crude oil processing plant had neurophysiological abnormalities. Residents in this study were exposed to hydrogen sulphide at 10 ppb, although concentrations occasionally reached 100 ppb.
4. Residents near pulp and paper mills in Finland have reported an excess of health symptoms compared to residents living in a community without any industrial hydrogen sulphide sources. The annual mean concentrations of hydrogen sulphide in the affected community was 8 $\mu g/m^3$ (5.7 ppb). Symptoms included respiratory, eye and nasal problems.

Recommendations

In the long-term, countries should seek to ensure that ambient exposure to sulphur dioxide does not exceed the guidelines recommended by WHO. In the interim, countries should set ambient standards or sulphur dioxide that take into account the benefits to human health and sensitive ecosystems of reducing exposure to sulphur dioxide; the concentration levels achievable by pollution prevention and control measures; and the costs involved in meeting the standards. In adopting new ambient air quality standards or guidelines, countries should set appropriate phase in periods. Where large differences exist between the costs and the benefits of meeting air quality standards and guidelines, it may be appropriate to establish area-specific ambient standards case by case.

Prior to carrying out an environmental assessment (EA), a trigger value for the annual average concentrations of sulphur dioxide should be agreed on by the country and the World Bank. Countries may wish to adopt EU, USEPA, or WHO guidelines or standards as their trigger values. The trigger value should be equal to or lower than the country's ambient standard. The trigger value is not an ambient air quality standard but simply a threshold. If, as a result of the project, the trigger value is predicted to be exceeded in the area affected by the project, the EA should seek mitigation alternatives on a regional or sectoral basis. In the absence of an agreed value, the world bank group will classify airsheds as moderately degraded if concentration levels are above 80 $\mu g/m^3$ annual average or if the 98th percentile of 24 hours mean values over a period of one year is estimated to exceed 150 $\mu g/m^3$. Airsheds will be classified as having poor air quality with respect to sulphur dioxide if either the annual mean value of sulphur dioxide is greater than 100 $\mu g/m^3$ or the 95th percentile of 24 hours mean value for sulphur dioxide for the airshed over a period of one year is estimated to exceed 150 $\mu g/m^3$.

In addition, good practice in airshed management should encompass the establishment of an emergency response plan during industrial plant operation. It is recommended that this plan be put into effect when levels of air pollution exceed one or more of the emergency trigger values (determined for short-term concentrations of sulphur dioxide, nitrogen oxides, particulates, and ozone). The recommended emergency trigger value for sulphur dioxide is 150 $\mu g/m^3$ for the 24 hours average concentrations.

SULPHUR OXIDES: POLLUTION PREVENTION AND CONTROL

Traditionally, measures designed to reduce localised ground-level concentrations of sulphur oxides (SO_x) used high-level dispersion. Although these measures reduced localised health impacts, it is now realised that sulphur compounds travel long distances in the upper atmosphere and can cause damage far from the original source. Therefore the objective must be to reduce total emissions.

The extent to which SO_x emissions harm human health depends primarily on ground-level ambient concentrations, the number of people exposed, and the duration of exposure. Source location can affect these parameters; thus, plant siting is a critical factor in any SO_x management strategy.

The human health impacts of concern are short-term exposure to sulphur dioxide (SO_2) concentrations above 1000 micrograms per cubic meter, measured as a 10 minute average. Priority therefore must be given to limiting exposures to peak concentrations. Industrial sources of sulphur oxides should have emergency management plans that can be implemented when concentrations reach predetermined levels. Emergency management plans may include actions such as using alternative low-sulphur fuels.

Traditionally, ground-level ambient concentrations of sulphur dioxide were reduced by emitting gases through tall stacks. Since this method does not address the problem of long-range transport and deposition of sulphur and merely disperses the pollutant, reliance on this strategy is no longer recommended. Stack height should be designed in accordance with good engineering practice.

Approaches for Limiting Emissions

The principal approaches to controlling SO_x emissions include use of low-sulphur fuel; reduction or removal of sulphur in the feed; use of appropriate combustion technologies; and emissions control technologies such as sorbent injection and flue gas desulphurisation (FGD).

Choice of fuel

Since sulphur emissions are proportional to the sulphur content of the fuel, an effective means of reducing SO_x emissions is to burn low-sulphur fuel such as natural gas, low-sulphur oil, or low-sulphur coal. Natural gas has the added advantage of emitting no particulate matter when burned.

Fuel cleaning

The most significant option for reducing the sulphur content of fuel is called beneficiation. Up to 70 per cent of the sulphur in high-sulphur coal is in pyritic or mineral sulphate form, not chemically bonded to the coal. Coal beneficiation can remove 50 per cent of pyritic sulphur and 20–30 per cent of total sulphur. (It is not effective in removing organic sulphur.) Beneficiation also removes ash responsible for particulate emissions. This approach may in some cases be cost-effective in controlling emissions of sulphur oxides, but it may generate large quantities of solid waste and acid waste-waters that must be properly treated and disposed of. Sulphur in oil can be removed through chemical desulphurisation processes, but this is not a widely used commercial technology outside the petroleum industry.

Selection of technology and modifications

Processes using fluidised-bed combustion (FBC) reduce air emissions of sulphur oxides. A lime or dolomite bed in the combustion chamber absorbs the sulphur oxides that are generated.

Emissions control technologies

The two major emissions control methods are sorbent injection and flue gas desulphurisation:

1. Sorbent injection involves adding an alkali compound to the coal combustion gases for reaction with the sulphur dioxide. Typical calcium sorbents include lime and variants of lime. Sodium-

based compounds are also used. Sorbent injection processes remove 30–60 per cent of sulphur oxide emissions.

2. Flue gas desulphurisation may be carried out using either of two basic FGD systems: regenerable and throwaway. Both methods may include wet or dry processes. Currently, more than 90 per cent of utility FGD systems use a wet throwaway system process.

Throwaway systems use inexpensive scrubbing mediums that are cheaper to replace than to regenerate. Regenerable systems use expensive sorbents that are recovered by stripping sulphur oxides from the scrubbing medium. These produce useful by-products, including sulphur, sulphuric acid, and gypsum. Regenerable FGDs generally have higher capital costs than throwaway systems but lower waste disposal requirements and costs.

In wet FGD processes, flue gases are scrubbed in a liquid or liquid/solid slurry of lime or limestone. Wet processes are highly efficient and can achieve SO$_x$ removal of 90 per cent or more. With dry scrubbing, solid sorbents capture the sulphur oxides. Dry systems have 70–90 per cent sulphur oxide removal efficiencies and often have lower capital and operating costs, lower energy and water requirements, and lower maintenance requirements, in addition to which there is no need to handle sludge. However, the economics of the wet and dry (including 'semidry' spray absorber) FGD processes vary considerably from site to site. Wet processes are available for producing gypsum as a by-product. Table 3.2 compares removal efficiencies and capital costs of systems for controlling SO$_x$ emissions.

Table 3.2. Comparison of SO$_x$ emissions control systems.

System	Per cent SO$_x$ reduction	Capital cost ($/kilowatt)
Sorbent injection	30–70	50–100
Dry flue gas desulphurisation	70–90	80–170
Wet flue gas sulphurisation	>90	80–150

Monitoring

The three types of SO$_x$ monitoring systems are continuous stack monitoring, spot sampling, and surrogate monitoring. Continuous stack monitoring (CSM) involves sophisticated equipment that requires trained operators and careful maintenance. Spot sampling is performed by drawing gas samples from the stack at regular intervals. Surrogate monitoring uses operating parameters such as fuel sulphur content.

Recommendations

The traditional method of SO$_x$ dispersion through high stacks is not recommended, since it does not reduce total SO$_x$ loads in the environment. Natural gas is the preferred fuel in areas where it is readily available and economical to use. Methods of reducing SO$_x$ generation, such as fuel cleaning systems and combustion modifications, should be examined. Implementation of these methods may avoid the need for FGD systems. Where possible and commercially feasible, preference should be given to dry SO$_x$ removal systems over wet systems.

Ambient Standards and Guidelines

The main goal of almost all the major national and international standards and guidelines produced over the last two decades has been to protect human ealth. Early research appeared to indicate a threshold or 'no-effects' level below which health impacts were negligible for even the most vulnerable groups, such as asthmatics and smokers. Standards were then set below this level to provide a margin of safety. The EU standards recognize the possibility that exposure to both sulphur dioxide and particulate matter may have an additive or synergistic effect on health. (This is also recognised by WHO.) The EU limit value for ambient sulphur dioxide therefore varies depending on the concentration of particulate matter in the ambient air. Table 3.3 summarises key reference standards and guidelines for ambient SO_2 concentrations.

Table 3.3. Reference standards and guidelines for ambient sulphur dioxide concentrations.

Standard or guideline	Annual average		Winter		24 hours		1-hour, sulphur dioxide
	Sulphur dioxide	Associated particulate levels	Sulphur dioxide	Associated particulate levels	Sulphur dioxide	Associated particulate levels	
EU limit values	80[a]	>40[b]	130[c]	>60[b]	250[d]	>150[b]	
	120[a]	≤40[b]	180[c]	≤60[b]	350[d]	≤150[b]	
	80[a]	>150[e]	130[c]	>200[e]	250[d]	>350[e]	
	120[a]	≤150[e]	180[c]	≤200[e]	350[d]	≤350[e]	
USEPA standards	80[f]				365[g]		
WHO guidelines	40–60[f]				100–150[d]		
WHO guidelines for Europe	50[f]				125[d]		350
ECE critical value	10/20/30		20/30				

[a] Median of daily values taken throughout the year.
[b] Black smoke method.
[c] Median of daily values taken throughout the winter.
[d] 98th percentile of all daily values taken throughout the year; should not be exceeded more than 7 days a year.
[e] Gravimetric method.
[f] Arithmetic mean.
[g] Not to be exceeded more than once a year.

CARBON MONOXIDE (CO)

Carbon monoxide is a colourless, odourless, tasteless and toxic gas produced as a by-product of combustion. Any fuel burning appliance, vehicle, tool or other device has the potential to produce dangerous levels of carbon monoxide gas. Examples of carbon monoxide producing devices commonly in use around the home include:

1. Fuel fired furnaces (non-electric).
2. Gas water heaters.
3. Fireplaces and woodstoves.
4. Gas stoves.
5. Gas dryers.
6. Charcoal grills.
7. Lawnmowers, snowblowers and other yard equipment.
8. Automobiles.

Oxides of Carbon as Pollutants

CO, a significant contaminant is produced in the atmosphere by natural processes, i.e. forest fires, natural gas emission, marsh gas production and volcanic actions contribute to form CO. However, human activities, mainly automobile exhausts, contribute significantly (about 80 per cent) to CO emission. Its concentration varies depending upon the density of vehicular traffic.

Sinks of carbon monoxide (CO)

1. The possible sinks which oxidise CO into CO_2 in the atmosphere are atomic oxygen, hydroxyl radical, NO_2, N_2O, O_3 and excited oxygen molecules.
2. The major CO sinks are some soil micro-organisms.
3. Plants are world's natural pollutant sink of CO and CO_2. They fix and metabolise CO also with the help of chlorophyll in the presence of light. This is known as photosynthesis and even in the dark, plants can carry on the fixation process non-photosynthetically.

CO is removed by converting it into CO_2 which then metabolises. Thus, the fixation of CO by plants is of immense importance as green plants are a major global sink of CO. This CO absorption by plants increases linearly with the increase of CO concentration. Therefore, in cities where CO concentration is higher, the rate of CO absorption by plants may be greater by a factor of 10 to 100. Thus, plants play a vital role in the global CO sink.

Distribution and concentration of CO

Although CO sinks absorb five times the annual discharge of CO into the atmosphere, considerable concentration of CO still exists in the atmosphere. The reason is that neither CO nor the soil sinks are distributed uniformly. Even the largest CO producing areas have the least amount of soil sink available. Higher CO levels are generally observed in areas of heavy vehicular traffic, as the automobiles constitute the single largest (about 59.2 per cent) CO producing source. According to an estimate, CO concentration in heavy traffic areas is nearly three times that in commercial areas and five times that in industrial areas.

Effects of CO Pollutants

1. On plants: CO has some detrimental effects on plants, when they are exposed to CO for long duration. For example, it inhibits the nitrogen fixation ability of bacteria when they are exposed to CO levels of 2000 ppm for 33–38 hours. Similarly, nitrogen fixing ability of bacteria living in clover roots is inhibited when exposed to a 100 ppm CO concentration for 35 days. Since the CO levels in the atmosphere normally does not reach 100 ppm it creates no significant effects on plants and vegetation. However, CO concentration from 100–10,000 ppm affects leaf drop, leaf curling, reduction in leaf size, synthesis and chlorophyll with premature ageing, etc.
2. On humans: All gaseous pollutants cause severe damage to the respiratory system. But the adverse effects of CO in the human body are unique. On inhalation, it passes through the lungs into the blood stream. It actually reacts with the haemoglobin (Hb) of the red blood corpuscles (RBC) forming a stable co-ordinated complex, called carboxy-haemoglobin, which restricts transport of oxygen from lungs to cells. A relatively low concentration of CO, because of its high affinity to haemoglobin, is able to displace considerable amount of oxygen forming carboxy haemoglobin complex (COHb).

Carbon Monoxide Poisoning

Carbon monoxide poisoning occurs after enough inhalation of carbon monoxide (CO). Carbon monoxide is a toxic gas, but, being colourless, odourless, tasteless, and non-irritating, it is very difficult for people to detect. Carbon monoxide is a product of incomplete combustion of organic matter with insufficient oxygen supply to enable complete oxidation to carbon dioxide (CO_2) and is often produced in domestic or industrial settings by older motor vehicles and other gasoline-powered tools, heaters, and cooking equipment. Exposures at 100 ppm or greater can be dangerous to human health.

Symptoms of mild acute poisoning include headaches, vertigo, and flu-like effects; larger exposures can lead to significant toxicity of the central nervous system and heart, and even death. Following acute poisoning, long-term sequelae often occur. Carbon monoxide can also have severe effects on the fetus of a pregnant woman. Chronic exposure to low levels of carbon monoxide can lead to depression, confusion, and memory loss. Carbon monoxide mainly causes adverse effects in humans by combining with haemoglobin to form carboxyhemoglobin (HbCO) in the blood. This prevents oxygen binding to haemoglobin, reducing the oxygen-carrying capacity of the blood, leading to hypoxia. Additionally, myoglobin and mitochondrial cytochrome oxidase are thought to be adversely affected. Carboxy haemoglobin can revert to hemoglobin, but the recovery takes time because the HbCO complex is fairly stable.

Treatment of poisoning largely consists of administering 100 per cent oxygen or providing hyperbaric oxygen therapy, although the optimum treatment remains controversial. Oxygen works as an antidote as it increases the removal of carbon monoxide from haemoglobin, in turn providing the body with normal levels of oxygen. The prevention of poisoning is a significant public health issue. Domestic carbon monoxide poisoning can be prevented by early detection with the use of household carbon monoxide detectors. Carbon monoxide poisoning is the most common type of fatal poisoning in many countries. Historically, it was also commonly used as a method to commit suicide, usually by deliberately inhaling the exhaust fumes of a running car engine. Modern cars with electronically controlled combustion and catalytic converters produce so little carbon monoxide that this is much less viable. Carbon monoxide poisoning has also been implicated as the cause of apparent haunted houses. Symptoms such as delirium and hallucinations have led people suffering poisoning to think they have seen ghosts or to believe their house is haunted.

Signs and symptoms

Carbon monoxide is toxic to all aerobic forms of life. It is easily absorbed through the lungs. Carbon monoxide is colourless, odourless, tasteless, and non-irritating, which makes it difficult for humans to detect. Inhaling even relatively small amounts of the gas can lead to hypoxic injury, neurological damage, and even death. Different people and populations may have a different carbon monoxide tolerance level. On average, exposures at 100 ppm or greater is dangerous to human health. In the United States, the OSHA limits long-term workplace exposure levels to less than 50 ppm averaged over an 8 hours period; in addition, employees are to be removed from any confined space if a upper limit ('ceiling') of 100 ppm is reached. Carbon monoxide exposure may lead to a significantly shorter life span due to heart damage.

The carbon monoxide tolerance level for any person is altered by several factors, including activity level, rate of ventilation, a pre-existing cerebral or cardiovascular disease, cardiac output, anemia, sickle cell disease and other hematological disorders, barometric pressure, and metabolic rate.

The acute effects produced by carbon monoxide in relation to ambient concentration in parts per million are listed below:

Concentration	Symptoms
35 ppm (0.0035%)	Headache and dizziness within six to eight hours of constant exposure
100 ppm (0.01%)	Slight headache in two to three hours
200 ppm (0.02%)	Slight headache within two to three hours; loss of judgment
400 ppm (0.04%)	Frontal headache within one to two hours
800 ppm (0.08%)	Dizziness, nausea, and convulsions within 45 min; insensible within 2 hours
1,600 ppm (0.16%)	Headache, tachycardia, dizziness, and nausea within 20 min; death in less than 2 hours
3,200 ppm (0.32%)	Headache, dizziness and nausea in five to ten minutes. Death within 30 minutes.
6,400 ppm (0.64%)	Headache and dizziness in one to two minutes. Convulsions, respiratory arrest, and death in less than 20 minutes.
12,800 ppm (1.28%)	Unconsciousness after 2–3 breaths. Death in less than three minutes

Acute poisoning

The main manifestations of poisoning develop in the organ systems most dependent on oxygen use, the central nervous system and the heart. The initial symptoms of acute carbon monoxide poisoning include headache, nausea, malaise, and fatigue. These symptoms are often mistaken for a virus such as influenza or other illnesses such as food poisoning or gastroenteritis. Headache is the most common symptom of acute carbon monoxide poisoning; it is often described as dull, frontal, and continuous. Increasing exposure produces cardiac abnormalities include fast heart rate, low blood pressure, and cardiac arrhythmia; central nervous system symptoms include delirium, hallucinations, dizziness, unsteady gait, confusion, seizures, central nervous system depression, unconsciousness, respiratory arrest, and even death. Less common symptoms of acute carbon monoxide poisoning include myocardial ischemia, atrial fibrillation, pneumonia, pulmonary edema, high blood sugar, lactic acidosis, muscle necrosis, acute kidney failure, skin lesions, and visual and auditory problems.

One of the major concerns following acute carbon monoxide poisoning is the severe delayed neurological manifestations that may occur. Problems may include difficulty with higher intellectual functions, short-term memory loss, dementia, amnesia, psychosis, irritability, a strange gait, speech disturbances, Parkinson's disease-like syndromes, cortical blindness, and a depressed mood. Depression may even occur in those who did not have pre-existing depression. These delayed neurological sequelae may occur in up to 50 per cent of poisoned patients after 2 to 40 days. It is difficult to predict who will develop delayed sequelae; however, advancing age, loss of consciousness while poisoned, and initial neurological abnormalities may increase the chance of developing delayed symptoms.

The poisonous effect of CO is due to the fact that it prevents the red cells, saturated with CO from absorbing oxygen and carrying it to different parts of the body. Death is caused by asphyxiation. The normal level of blood COHb is 0.5 per cent. COHb is formed in the body as a result of CO produced by the body during the destructive metabolism of heme, which is a component of haemoglobin. A little of it comes from low levels of CO in the ambient air. Smoke contains CO concentration of about 20,000 ppm, which is diluted to 400-500 ppm during inhalation. So it leads to fertility problems, premature births, spontaneous abortions and deformed babies in pregnant women. In chain smokers, CO may cause an adaptive response, even producing as much as 8 per cent more haemoglobin. At 10 per cent COHb in blood due to smoking, there may be lowered tolerance to CO. Cigar smokers have increased haematocrit (per cent volume of red blood corpuscles). The levels of COHb in the blood is directly related to the CO concentration of inhaled air. The inhaled CO combines with blood haemoglobin to form COHb nearly 210–240 times faster than oxygen does. However, COHb level slowly changes as the CO concentration of ambient air decreases.

CO poisoning remedies

CO is the most widespread human poison on record. In poisoning by CO, asphyxiation takes place because it has 200–300 times as great affinity than oxygen for the haemoglobin of the blood and death may result instantaneously. In case of inhalation of CO, carry the patient to fresh air immediately and not allow him to walk. Loosen all tight clothings. Apply artificial respiration if breathing has stopped or is irregular. Wrap the patient in blankets to prevent chilling. If the patient is convulsing, keep him in bed in a semi-dark room, avoid jarring or noise. Life can also be saved by blood transfusion, by injecting methylene blue or by making the person inhale carbogen (a mixture of 95 per cent O_2 and 5 per cent CO_2).

Control of CO pollution through law

To check air pollution, some standards through Motor Vehicles Act and other Acts for design of engine, etc. have been enforced recently, to check and control the amount of CO emitted by automobiles. The CO level in vehicles is tested by experts of the Directorate of Transport. The carburettor is also adjusted to correct air fuel ratio at service stations, and properly tuned for CO emissions. These efforts will certainly control CO pollution by vehicular exhaust and will increase the mileage per litre. Successful abatement scheme for CO pollution will save the life of living biota from its hazardous effects and will protect the natural ecosystems.

Effects of Carbon Dioxide (CO_2)

Carbon dioxide is non-toxic, therefore it is not harmful to human health, unlike CO. CO_2 is utilised by green plants to prepare starch during photosynthesis. But today its increasing concentration in the atmosphere (10 per cent) has long-term effects. It produces adverse physiological effects only at very high levels. The increased amount of CO_2 in air is mainly responsible for global warming. CO_2 molecules absorb heat energy and tend to prevent the long wave infrared heat radiation from earth from escaping into space and it deflects these radiations back to earth. The phenomena called 'atmospheric effect', 'green house effect' or 'global warming' has become a serious threat to global food production. The increased temperature may also lead to melting of polar ice caps and hence flooding of low lakes a caged bird into lying areas.

A person may die in an atmosphere of CO_2, not because it is poisonous, but because of lack of oxygen. Birds and animals are more sensitive to CO_2. If a person takes a caged bird into a long deep cave, which is suspected to contain CO_2 in dangerous proportion, the bird will die first. This serves as a warning to the person entering the cave. Mixed with oxides of sulphur and carbon monoxide, CO_2 gas becomes poisonous, pungent, suffocating and causes diseases like chronic asthma and bronchitis. Mixed with CO, it causes headache, dizziness, vomitting and difficulty in breathing.

Safe disposal of carbon dioxide

Depleting forests, growing population and increasing industrialisation are serious causes of the imbalance of carbon dioxide concentrations in the earth's atmosphere. Carbon dioxide — the number one greenhouse gas—has much more power to affect the earth's temperature. Increasing carbon dioxide concentrations are linked to dangerous environmental hazards like global warming. For about 25 years, data have been telling us that the earth is getting warmer and humans are causing it. Now, the immense danger posed by greenhouse gas emissions can no longer be denied.

In just the past 150 years, mankind has boosted carbon dioxide concentrations by 32 per cent. Scientist say that if we continue to increase greenhouse gas emissions at this rate, temperatures will rise between 2°C and 3°C this century, making the Earth as warm as it was 3 mn years ago, when the seas were between 15 m and 35 m higher than they are today. Scientists scrutinising the atmosphere, ice, earth and seas, say global warming is approaching a tipping point. Some see global warming as the greatest threat civilisation has ever faced.

But we still have time to keep it from reaching catastrophic levels. As worries over the impact of carbon dioxide emissions on global climate change soar, researchers are increasingly searching for ways to rid the atmosphere of greenhouse gas. But, so far, industrial-scale projects have been limited. Notable among them are a project to build power plants that derive hydrogen from fossil fuels and sequester the carbon dioxide by-product.

Sea-floor injections

Sea-floor injections, a safe, high-capacity method, could make carbon sequestration more practical. A better way to store carbon dioxide is to pump it into the sea floor in liquid form. There high pressure and cold temperatures make it more dense than water in the surrounding rock, preventing it from rising to the surface. Researchers at Harvard and Columbia Universities have proposed a new method for trapping nearly limitless amounts of carbon dioxide—a technique they say will be secure, as well as a practical option for areas located far from underground reservoirs.

The researchers, propose that carbon dioxide be pumped into the porous sediment a few hundred meters into the sea floor in deep parts of the ocean (more than 3000 m deep), in what one of the researchers calls 'a fairly simple, permanent solution'. The key was finding a 'sweet spot', where the pressure and temperature of the surrounding environment make carbon dioxide denser than surrounding fluids, thereby trapping it in place. This situation occurs at the bottom of the ocean because of a combination of high pressure and low temperatures—a fact others have also noted in proposals to store carbon dioxide in deep parts of the ocean.

The carbon dioxide, in liquid form, would be brought to the sequestration site by ship or pipeline, and piped into the sea floor with equipment like that used by the oil industry for drilling deepsea wells. Once beneath the sea floor, the carbon dioxide would interact with the surrounding fluids and produce hydrate ice crystals, which would plug the rock pores, serving as a secondary cap on the carbon dioxide. Over hundreds of years, the carbon dioxide would dissolve in the surrounding water, and then would only have the potential of leaking out by diffusion, a slow process that would take millions of years. Within the next five years the scientists hope to run a large-scale field test of this new approach.

The potential

If all the known geologic reservoirs for conventional storage were useable, they could store all the carbon dioxide currently produced each year, and continue doing so for 80 years at current emission rates. Indeed, the costs for the new seafloor method will vary, but will probably be slightly more than for land-based storage. It could, however, be more economical for areas near the ocean, especially those far from a known geological reservoir.

The costs

The cost for any method of largescale sequestration is still unclear. Moreover, such injections would kill ocean life, and, unless sequestered in deep trenches, the carbon dioxide could be carried by currents to shallow areas, where it could re-enter the atmosphere.

The researchers' insight was that injections into the sea floor could take advantage of the pressure and temperature of the ocean, while avoiding the negative side-effects of earlier proposals. The need for robust, potentially inexpensive carbon sequestration schemes is enormous; while it still requires more experimental validation, it is potentially very important and should be considered very seriously.

Along with this, efforts should be made to identify which plants (land- or ocean-based) are most efficient at removing carbon dioxide through photosynthesis and start growing them. At current levels of depletion of rain forest, and increasing production of carbon dioxide, we are burning (literally) the candle at both ends. Storage of CO_2 will just defer the problem to future generations, while we continue our greenhouse gas emitting lifestyles.

Unfortunately, we've already delayed long enough that even when we finish converting to sustainable energy we're going to have to take out at least as much CO_2 as we've already put in. Moreover, the oceans are already becoming too acidic from CO_2 for many important organisms. Enough of the 'trapped' CO_2 will likely diffuse into the ocean to make the problem worse. The scientific community should think towards engineering primitive photosynthetic organisms, which would harvest CO_2, trap some of the sunlight reaching the earth's surface and then concentrate these organisms into a thick sludge. This could potentially be a source of biofuels, at best, or at worst, the sludge could be pumped back into depleted oil wells and then could turn back to petroleum in several tens of millions of years.

Converting carbon dioxide into plastics and fuels

Every one is worried about the increasing concentration of carbon dioxide (CO_2) in the atmosphere and its impact on global climate. Global warming has become a topic of everybody's concern nowadays, but without giving any thought towards adopting the lifestyle compatible with controlling the alarming emission of greenhouse gases, in general, and CO_2, in particular.

Due to its large emissions from various man-made combustion sources, CO_2 has been identified as the main cause of global warming. Currently, between 80–85 per cent of our energy comes from fossil fuels. However, fossil fuel resources are finite and distributed unevenly beneath the earth's surface. When fossil fuel is turned into useful energy through combustion, it often produces environmental pollutants that are harmful to human health and greenhouse gases that threaten global climate.

In contrast, solar resources are widely available and have a benign effect on the environment and climate; making it an appealing alternative energy source. Sunlight is not only the most plentiful energy resource on earth, but also one of the most versatile, converting readily to electricity, fuel and heat. The challenge is to raise its conversion efficiency by factors of five or ten. That requires understanding the fundamental conversion phenomena at the nanoscale.

Various methods are being advocated for control of CO_2 emissions at pre- and post-combustion stages of fossil fuels, but with little significant success so far. Researchers have known for a long time that theoretically it might be possible to recycle CO_2 and convert it into some useful end-products, such as biofuels or plastics, but many thought it could not be practical, technically or economically.

Scientists are now working towards making use of solar energy in recycling CO_2. In this way, it will be possible to make use of cheaply and abundantly available solar energy not only to get rid of CO_2, but to convert it into some useful end-product like fuel.

CO_2 splitting produces carbon monoxide (CO), an important industrial chemical, normally produced from natural gas. So with CO_2 splitting we can save fuel, produce a useful chemical and reduce a greenhouse gas. Although CO is poisonous, it is highly sought after. Millions of pounds of it are used each year to manufacture chemicals, including detergents and plastics. It can also be converted into liquid fuel using technology that has been around a long time. Rising environmental concern and fuel prices make it economically competitive to convert CO into fuel.

Two methods are touted for converting CO_2 to CO:

1. Using direct solar energy through a solar furnace.
2. First converting solar energy into electricity using semiconductors.

Each of these methods for converting CO_2 into CO has its advantages and disadvantages.

Using direct solar energy through a solar furnace

A research team from Sandia National Laboratories (USA) is working to develop a method of getting a liquid fuel–methanol, gasoline or other liquid fuels–from water and CO produced using solar energy.

Using concentrated solar energy to reverse combustion, scientists at the laboratory are building a prototype device intended to chemically 'reenergise' CO_2 into CO using concentrated solar power. The CO could then be used to make hydrogen or serve as a building block to synthesise a liquid combustible fuel, such as methanol or even gasoline, diesel and jet fuel.

The prototype device will break a carbon-oxygen bond in CO_2 to form CO and oxygen in two distinct steps. The Sandia researchers, over the past year, have shown proof of concept and are completing a prototype device that will use concentrated solar energy to re-energise CO_2 or water, the products of combustion. This will form carbon monoxide, hydrogen, and oxygen, which ultimately could be used to synthesise liquid fuels. Researchers think the prototype will successfully break down CO_2 in a clever and viable two-step process. The invention, though probably a good 15–20 years away from coming on the market, holds promise of being able to reduce CO_2 emissions, while preserving options to keep using existing fuels. Recycling CO_2 into fuels provides an attractive alternative to burying it.

What's exciting about this invention is that it will result in fossil fuels being used at least twice, meaning less CO_2 being put into the atmosphere and a reduction of the rate that fossil fuels are pulled out of the ground. The CO_2 from the burning of the coal would be captured and reduced to carbon monoxide in the device. The carbon monoxide would then be the starting point for making gasoline, jet fuel, methanol, or almost any type of liquid fuel. The prospect of a liquid fuel is significant because it fits in with current gasoline and oil infrastructure. After the synthesised fuel is made from the carbon monoxide, it could be transported through a pipeline or put in a truck and hauled to a gas station, just like gasoline refined from petroleum is now. Plus it would work in ordinary gasoline and diesel engine vehicles. Researchers say that while the first step would be to capture the CO_2 from sources where it is concentrated, like power plants, smokestacks and breweries, the ultimate goal would be to snatch it out of the atmosphere/air.

Converting solar energy into electricity

Chemists at the University of California, San Diego have demonstrated the feasibility of exploiting sunlight to transform a greenhouse gas into a useful product. Prof. Clifford Kubiak and his team members have developed a prototype device that can capture energy from the sun, convert it to electrical energy and 'split' CO_2 into CO and oxygen. The device designed by Kubiak and Sathrum utilises a semiconductor and two thin layers of catalysts. It splits CO_2 to generate CO and oxygen in a three-step process.

1. The first step is the capture of solar energy photons by the semiconductor.
2. The second step is the conversion of optical energy into electrical energy by the semiconductor.

3. The third step is the deployment of electrical energy to the catalysts. The catalysts convert CO_2 to CO on one side of the device and to oxygen on the other side. Because electrons are passed around in these reactions, a special type of catalyst that can convert electrical energy to chemical energy is required. Researchers have created a large molecule with three nickel atoms at its heart that has proven to be an effective catalyst for this process.

Choosing the right semiconductor is critical to making CO_2 splitting practical, say the researchers. Semiconductors have bands of energy to which electrons are confined. Sunlight causes the electrons to leap from one band to the next, creating an electrical energy potential. The energy difference between the bands—the band gap—determines how much solar energy will be absorbed and how much electrical energy is generated.

Researchers initially used a silicon semiconductor to test the merits of their device because silicon is wellstudied. However, silicon absorbs in the infrared range and the researchers say it is 'too wimpy' to supply enough energy. The conversion of sunlight by silicon supplied about half of the energy needed to split CO_2, and the reaction worked if the researchers supplied the other half of the energy needed.

They are now building the device using a gallium-phosphide semiconductor. It has twice the band gap of silicon and absorbs more energetic visible light. Therefore, they predict that it will absorb the optimal amount of energy from the sun to drive the catalytic splitting of CO_2.

Converting CO_2 into plastics

Researchers first found a way to make biodegradable plastics called aliphatic polycarbonates from CO_2 in 1969. They used CO_2 and a class of compounds called epoxides. But the process requires expensive catalysts, high temperatures and pressure. The plastic costs more than US $100 a pound and is used only in specialty products such as biomedical and electronic devices.

Prof. C.J. Li, from McGill University (Canada), is working on a process that will enable us to use waste CO_2 to create polymers. This has multiple commercial and environmental benefits, not the least of which is the elimination of petrochemicals and toxic solvents from the plastics-making process. It also has huge implications for global warming. Instead of burying waste CO_2, to reduce its greenhouse effects, it is far more efficient to turn it into something useful. This new plastic has extremely good properties and biodegradable. If people throw it out, it decomposes and becomes CO_2 again.

A Cornell University (USA) spinoff technology centres on a catalyst that converts CO_2 into a polymer that could be used to make everyday items such as packaging, cups and forks. The plastic, which was originally created by Cornell chemist, Geoffrey Coates, is also safe and strong enough to be used in medical implants and devices.

Novomer, a company based in Ithaca (NY, USA), and co-founded by Coates is working on commercialising a process to convert CO_2 into different forms of biodegradable polymers, including a honeylike liquid and a powder. The plastic is being made on a pilot scale, and Novomer declines to give details of its commercial-scale manufacturing plans. At present, while it is hard to predict the product's final cost, it is expected to be cost competitive with traditional petroleum-based plastics as CO_2 is a cheap feedstock.

Novomer uses the same raw materials—CO_2 and epoxides—but its product is distinguished by a metallic catalyst developed by Coates. The zinc-based catalyst works at room temperature and low pressure, and faster. The polymer has different properties—it can be hard, soft, transparent, or opaque —based on the type of epoxide used. It is also biodegradable, since the carbon-oxygen bonds in Novomer's polymer are relatively easy for bacteria to break down. Though the company has not tested the degradability of the polymer, aliphatic polycarbonates, in general, have been shown to degrade in six months in composts under ideal conditions. Though, in terms of biodegradability, the CO_2 plastic will have to compete with several other plant-based plastics now on the market, the use of CO_2 and CO as inexpensive feedstocks, instead of the corn-based feedstocks used by other biodegradable plastics, means the CO_2 plastic formation won't compete with food production.

HYDROCARBONS AS POLLUTANTS

Among a variety of hydrocarbons involved in air pollution, 56 have been clearly identified by making use of the technique of gas chromatography. Natural sources, particularly trees, emit huge quantities of hydrocarbons in air.

$$2CH_2O \xrightarrow{\text{Bacteria}} CH_4 + CO_2$$

Methane has mean residence time of 3 to 7 years in the atmosphere. Human activities contribute nearly 20 per cent of the hydrocarbons emitted to the atmosphere every year and animals contribute about 80–85 million tons of methane in the atmosphere every year. About 45 per cent of reactive hydrocarbons in the atmosphere originate from the combustion of gasoline while 15 per cent are produced by other pollutant sources. Emissions are reported to occur from the crank-case of an automobile also. Automobile exhausts emit maximum hydrocarbons in the atmosphere.

Industrial processes like processing, storage and transfer of products, etc. constitute the next largest source of hydrocarbons. The evaporation of organic solvents leads to about 10 per cent hydrocarbons in the air during industrial operation. These solvents are ingredients of paints, varnishes, lacquers, undercoatings and other products. During their manufacture, a large quantity of reactive hydrocarbons are emitted into the atmosphere. Gasoline hydrocarbons evaporate and get mixed with air, increasing their concentration in the atmosphere. Hydrocarbons in the presence of nitrogen oxides (NO_x) are mainly responsible for the formation of photochemical smog. Photochemical reactions initiate most of the chemical processes, culminating in the removal of hydrocarbons from the atmosphere. The chief culprit in photochemical smog is NO_2.

Effects of Hydrocarbons

Effects on human beings

1. Hydrocarbons at high concentration (500–1000 ppm) have carcinogenic effects on lungs, mainly causing swelling when they enter the lungs.
2. Aromatic hydrocarbons like benzene, toluene; etc. are more dangerous than acyclic and alicyclic hydrocarbons. The inhalation of their vapours causes much irritation to the mucous membrane. However, their different levels create different types symptoms in the body (Table 3.4).
3. Secondary pollutants (PAN) produced by hydrocarbon and NO_x results in the formation of photochemical smog which causes irritation of eyes, nose, throat and respiratory distress.
4. Excess of hydrocarbon increases mucous secretion as a result of which respiratory tracts are blocked and a person coughs continuously, exerting pressure on the trachea of lungs due to which the lining membrane of alveoli bursts. So very little area is left for exchange of oxygen and carbon dioxide.
5. Benzpyrene, which is present as trace amounts in tobacco, charcoal, boiler stacks and gasoline exhausts, etc. is a dangerous cancer inducing hydrocarbon pollutant.

6. Methane (marsh gas) is a severe gaseous pollutant and occurs in air by volume to the extent of 0.0002 per cent. Its higher levels in the absence of oxygen create narcotic effects on human beings.
7. A group of hydrocarbons, especially carcinogenic hydrocarbons, cause cancer in man and animals affecting DNA and cell growth. The effects of carcinogens on animals have been studied by US Occupational Safety and Health Administration (Table 3.5).

Table 3.4 Effects of toxic hydrocarbons

Hydrocarbons	Content ppm	Adverse effect
Benzene	100	Mucous membrane irritation
	3000	Injures sensitive parts of the body; respiratory irritant
	7500	Lung cancer, dangerous for health
	20,000	Fatal, causes death
Benzpyrene	100	Induces cancer
Toluene	200	Headache, weakness, fatigue
	600	Affects nervous system

Effects on plants

1. Hydrocarbons and photochemical oxidants are injurious to plants. Exposure to high levels of ozone to plants causes chlorosis, i.e. yellowing of green portion of leaves.
2. Ozone enhances plant injury creating light flecks or stipples (clusters of dead cells) on the upper leaf surface inhibiting photosynthetic activity.
3. Ethylene even at 1 ppm concentration shows adverse effects on vegetation.
4. Acetylene and propylene at 50–500 ppm show extreme toxicity towards plants damaging their growth.
5. Ethylene hydrocarbons inhibits plant growth and damages leaf tissues and flowering plants.

Table 3.5. Hazards of carcinogenic hydrocarbons.

Compound	Hazards	Compound	Hazards
Benzidine	Causes bladder cancer	Ethyleneimine	Causes cancer
β-Naphthylamine	Causes cancer in urinary bladder	β-propiolactone	Potential carcinogen
Bis-chloromethyl	Creates lung cancer	α-Naphthylamine	Causes bladder cancer
Ethylene dichloride	Causes stomach, spleen and lung cancer	Nitrophenol	Causes bladder cancer
Vinyl chloride	Causes liver cancer	3–3′ dichlorobenzidine	Causes cancer

Effects on materials

1. Even low levels of ozone induces chemical alteration in natural synthetic textiles, paper, rubber and polymers. Higher the number of carbon-to-carbon double bonds in the material, greater is the susceptibility of their attack.
2. Hydrocarbon pollutants damage long chains of carbon atoms in a polymer, which loses its tensile strength.
3. Ozone forms new carbon chain links between parallel carbon chains so that the material becomes less elastic and more brittle.

Control of hydrocarbons

Hydrocarbons and NO_x produce PAN and O_3, etc. which are chronic secondary pollutants. So their control ultimately depends on the control of the primary precursors, i.e. hydrocarbons and NO_x which are the main culprit of air pollution. (Refer to NO_x control). Hydrocarbons from auto-exhaust emissions can be controlled by applying the techniques like incineration, absorption, adsorption and condensation, etc. By adopting these methods, all the three pollutants (hydrocarbons, NO_x and CO) can be converted to less harmful end products, i.e.

$$\text{Hydrocarbons} \xrightarrow{\text{Combustion}} CO + H_2O \quad CO \xrightarrow{\text{Combustion}} CO_2$$

POLYCYCLIC AROMATIC HYDROCARBON (PAH)

PAHs are a group of chemicals that are formed during the incomplete burning of coal, oil, gas, wood, garbage, or other organic substances, such as tobacco and charbroiled meat. There are more than 100 different PAHs. PAHs generally occur as complex mixtures (for example, as part of combustion products such as soot), not as single compounds. PAHs usually occur naturally, but they can be manufactured as individual compounds for research purposes; however, not as the mixtures found in combustion products. As pure chemicals, PAHs generally exist as colourless, white, or pale yellow-green solids. They can have a faint, pleasant odour. A few PAHs are used in medicines and to make dyes, plastics, and pesticides. Others are contained in asphalt used in road construction. They can also be found in substances such as crude oil, coal, coal tar pitch, creosote, and roofing tar. They are found throughout the environment in the air, water, and soil. They can occur in the air, either attached to dust particles or as solids in soil or sediment.

Although the health effects of individual PAHs are not exactly alike, the following 19 PAHs are considered toxic:

1. Acenaphthene.
2. Acenaphthylene.
3. Acenaphthene.
4. Acenaphthylene.
5. Anthracene.
6. Benz[a]anthracene.
7. Benzo[a]pyrene.
8. Benzo[e]pyrene.
9. Benzo[b]fluoranthene.
10. Benzo[g,h,i]perylene.
11. Benzo[j]fluoranthene.
12. Benzo[k]fluoranthene.
13. Chrysene.
14. Dibenz[a,h]anthracene.
15. Fluoranthene.
16. Fluorene.
17. Indeno[1,2,3-c,d]pyrene.
18. Phenanthrene.
19. Pyrene.

These 19 PAHs were chosen to be included in this profile because (i) more information is available on these than on the others, (ii) they are suspected to be more harmful than some of the others, and they exhibit harmful effects that are representative of the PAHs, (iii) there is a greater chance that you will be exposed to these PAHs than to the others, and (iv) of all the PAHs analyzed, these were the PAHs identified at the highest concentrations at NPL hazardous waste sites.

Exposure to PAHs

PAHs are present throughout the environment, and you may be exposed to these substances at home, outside, or at the workplace. Typically, you will not be exposed to an individual PAH, but to a mixture of PAHs. In the environment, you are most likely to be exposed to PAH vapours or PAHs that are attached to dust and other particles in the air. Sources include cigarette smoke, vehicle exhausts, asphalt roads, coal, coal tar, wildfires, agricultural burning, residential wood burning, municipal and industrial waste incineration, and hazardous waste sites. Background levels of some representative PAHs in the air are reported to be $0.02-1.2$ nanograms per cubic meter (ng/m^3; a nanogram is one-millionth of a milligram) in rural areas and $0.15-19.3$ ng/m^3 in urban areas. You may be exposed to PAHs in soil near areas where coal, wood, gasoline, or other products have been burned. You may be exposed to PAHs in the soil at or near hazardous waste sites, such as former manufactured-gas factory sites and wood-preserving facilities. PAHs have been found in some drinking water supplies in the United States. Background levels of PAHs in drinking water range from 4 to 24 nanograms per liter (ng/L; a liter is slightly more than a quart).

In the home, PAHs are present in tobacco smoke, smoke from wood fires, creosote-treated wood products, cereals, grains, flour, bread, vegetables, fruits, meat, processed or pickled foods, and contaminated cow's milk or human breast milk. Food grown in contaminated soil or air may also contain PAHs. Cooking meat or other food at high temperatures, which happens during grilling or charring, increases the amount of PAHs in the food. The level of PAHs in the typical US diet is less than 2 parts of total PAHs per billion parts of food (ppb) or less than 2 micrograms per kilogram of food ($\mu g/kg$; a microgram is one-thousandth of a milligram).

The primary sources of exposure to PAHs for most of the US population are inhalation of the compounds in tobacco smoke, wood smoke, and ambient air, and consumption of PAHs in foods. For some people, the primary exposure to PAHs occurs in the workplace. PAHs have been found in coal tar production plants, coking plants, bitumen and asphalt production plants, coal-gasification sites, smoke houses, aluminum production plants, coal tarring facilities, and municipal trash incinerators. Workers may be exposed to PAHs by inhaling engine exhaust and by using products that contain PAHs in a variety of industries such as mining, oil refining, metalworking, chemical production, transportation, and the electrical industry. PAHs have also been found in other facilities where petroleum, petroleum products, or coal are used or where wood, cellulose, corn, or oil are burned. People living near waste sites containing PAHs may be exposed through contact with contaminated air, water, and soil.

Pathways for PAHs in the Environment

PAHs enter the environment mostly as releases to air from volcanoes, forest fires, residential wood burning, and exhaust from automobiles and trucks. They can also enter surface water through discharges from industrial plants and waste water treatment plants, and they can be released to soils at hazardous waste sites if they escape from storage containers. The movement of PAHs in the environment depends on properties such as how easily they dissolve in water, and how easily they evaporate into the air. PAHs in general do not easily dissolve in water. They are present in air as vapours or stuck to the surfaces of small solid particles. They can travel long distances before they return to earth in rainfall or particle settling. Some PAHs evaporate into the atmosphere from surface waters, but most stick to solid particles and settle to the bottoms of rivers or lakes. In soils, PAHs are most likely to stick tightly to particles. Some PAHs evaporate from surface soils to air. Certain PAHs in soils also contaminate underground water. The PAH content of plants and animals living on the land or in water can be many times higher than the content of PAHs in soil or water. PAHs can break down to longer-lasting products by reacting with sunlight and other chemicals in the air, generally over a period of days to weeks. Breakdown in soil and water generally takes weeks to months and is caused primarily by the actions of micro-organisms.

Pathways for PAHs in the Body

PAHs can enter your body through your lungs when you breathe air that contains them (usually stuck to particles or dust). Cigarette smoke, wood smoke, coal smoke, and smoke from many industrial sites may contain PAHs. People living near hazardous waste sites can also be exposed by breathing air containing PAHs. However, it is not known how rapidly or completely your lungs absorb PAHs. Drinking water and swallowing food, soil, or dust particles that contain PAHs are other routes for these chemicals to enter your body, but absorption is generally slow when PAHs are swallowed. Under normal conditions of environmental exposure, PAHs could enter your body if your skin comes into contact with soil that contains high levels of PAHs (this could occur near a hazardous waste site) or with used crankcase oil or other products (such as creosote) that contain PAHs. The rate at which PAHs enter your body by eating, drinking, or through the skin can be influenced by the presence of other compounds that you may be exposed to at the same time with PAHs. PAHs can enter all the tissues of your body that contain fat. They tend to be stored mostly in your kidneys, liver, and fat. Smaller amounts are stored in your spleen, adrenal glands, and ovaries. PAHs are changed by all tissues in the body into many different substances. Some of these substances are more harmful and some are less harmful than the original PAHs. Results from animal studies show that PAHs do not tend to be stored in your body for a long time. Most PAHs that enter the body leave within a few days, primarily in the feces and urine.

Health Effects of PAHs

PAHs can be harmful to your health under some circumstances. Several of the PAHs, including benz[a]anthracene, benzo[a]pyrene, benzo[b]fluoranthene, benzo[j]fluoranthene, benzo[k]fluoranthene, chrysene, dibenz[a,h]anthracene, and indeno[1,2,3-c,d]pyrene, have caused tumors in laboratory animals when they breathed these substances in the air, when they ate them, or when they had long periods of skin contact with them. Studies of people show that individuals exposed by breathing or skin contact for long periods to mixtures that contain PAHs and other compounds can also develop cancer.

Mice fed high levels of benzo[a]pyrene during pregnancy had difficulty reproducing and so did their offspring. The offspring of pregnant mice fed benzo[a]pyrene also showed other harmful effects, such as birth defects and decreased body weight. Similar effects could occur in people, but we have no information to show that these effects do occur. Studies in animals have also shown that PAHs can cause harmful effects on skin, body fluids, and the body's system for fighting disease after both short- and long-term exposure. These effects have not been reported in people.

The Department of Health and Human Services (DHHS) has determined that benz[a]anthracene, benzo[b]fluoranthene, benzo[j]fluoranthene, benzo[k]fluoranthene, benzo[a]pyrene, dibenz[a,h]anthracene, and indeno[1,2,3-c,d]pyrene are known animal carcinogens. The International Agency for Research on Cancer (IARC) has determined the following: benz[a]anthracene and benzo[a] pyrene are probably carcinogenic to humans; benzo[b]fluoranthene, benzo[a]fluoranthene, benzo[k]fluoranthene, and indeno[1,2,3-c,d]pyrene are possibly carcinogenic to humans; and anthracene, benzo[g,h,i]perylene, benzo[e]pyrene, chrysene, fluoranthene, fluorene, phenanthrene, and pyrene are not classifiable as to their carcinogenicity to humans. EPA has determined that benz[a]anthracene, benzo[a]pyrene, benzo[b]fluoranthene, benzo[k]fluoranthene, chrysene, dibenz[a,h]anthracene, and indeno[1,2,3,-c,d]pyrene are probable human carcinogens and that acenaphthylene, anthracene, benzo[g,h,i] perylene, fluoranthene, fluorene, phenanthrene, and pyrene are not classifiable as to human carcinogenicity. Acenaphthene has not been classified for carcinogenic effects by the DHHS, IARC, or EPA.

Medical Tests for PAHs

In your body, PAHs are changed into chemicals that can attach to substances within the body. The presence of PAHs attached to these substances can then be measured in body tissues or blood after exposure to PAHs. PAHs or their metabolites can also be measured in urine, blood, or body tissues. Although these tests can show that you have been exposed to PAHs, these tests cannot be used to predict whether any health effects will occur or to determine the extent or source of your exposure to the PAHs. It is not known how effective or informative the tests are after exposure is discontinued. These tests to identify PAHs or their products are not routinely available at a doctor's office because special equipment is required to detect these chemicals.

OXIDES OF NITROGEN AS POLLUTANTS

Of the six or seven oxides of nitrogen known, only three — nitrous oxide (N_2O), nitric oxide (NO), and nitrogen oxide (NO_2) — are formed in any appreciable quantities in the atmosphere. Often NO and NO_2 are analysed together in air and are referred to as NO_x. The various reasons for the increase of oxides of nitrogen in the ambient atmosphere may be the increased emission of NO_x from industries and automobiles, and various other sources. Well over 90 per cent of all man-made nitrogen oxides that enter our atmosphere are produced by the combustion of various fuels.

The real danger posed by NO_x at the concentration found in metropolitan areas lies in its role in photochemical reactions leading to smog formation. These atmospheric reactions lead to the formation of chemical compounds that have a direct adverse effect on human beings and plants. In some situations, NO_x may be present in a high enough concentration, yet not react to form smog because other necessary conditions for the reaction are absent. However, nearly every major city in India experiences the effects induced by the presence of NO_x.

The actual quantities of NO_x produced by any given industry can be quite large. For example, a 750 MW gas or coal-fired power plant produces around 75 to 100 tons of NO_x per day. The type of fuel used can change the amount of NO_x released significantly.

About 60 per cent of NO_x is contributed by fuel combustion in the stationary sources and about 40 per cent by transportation. The stable gaseous oxides of nitrogen include N_2O (nitrous oxide), NO (nitric oxide), N_2O_3 (nitrogen trioxide), NO_2 (nitrogen dioxide) and N_2O_5 (nitrogen pentoxide). An unstable form, NO_3 also exists. Of all these oxides of nitrogen, the only ones present in the atmosphere in any significant amount are N_2O, NO and NO_2. These three are the potential contributors to air pollution. The main chemical reactions in the formation of nitrogen oxides are:

$$1/2N_2 + 1/2O_2 \rightarrow NO \text{ and } NO + 1/2O_2 \rightarrow NO_2$$

Nitric oxide (NO) is a colourless gas and its ambient concentration is usually far less than 0.5 ppm. At this concentration, its biological toxicity in terms of human health is insignificant. N_2O is an inert gas with an anaesthetic characteristic. Its ambient air concentration is 5 ppm. It has a balanced environmental cycle which is independent of other oxides of nitrogen.

However, nitric oxide is a precursor to the formation of nitrogen dioxide and is an active compound in photochemical reactions leading to air pollution. NO_2 (nitrogen dioxide) is a reddish brown gas and is quite visible in sufficient amounts. NO_2 directly affects human health only if it is present in extremely high concentrations. Exposure to 15 ppm of NO_2 causes eye and nose irritation, and pulmonary, discomfort is noted at 25 ppm for exposure of less than 1 hour. NO_2 concentration of 1 ppm can be detected by the eye. The threshold limit in ambient air is considered to be 5 ppm for daily exposure. In general, the concentrations of oxides of nitrogen are expressed under the formula NO_x (or on an 'equivalent NO_2' basis). The various NO_x emission sources are:

Coal
    Household and commercial
    Industry and utilities
Fuel Oil
    Household and commercial
    Industry
    Utility
Natural Gas
    Household and commercial
    Industry
    Utility
    Gas turbines
Waste disposal
    Conical incinerator
    Municipal incinerator
Mobile source combustion
    Gasoline–powered vehicle
    Diesel–powered vehicle
    Aircraft–conventional; fan-type jet

NO_2 from Passive Filters

Nitrogen dioxide (NO_2) emissions

Diesel engines emit oxides of nitrogen (NO_x), consisting of nitrogen oxide (NO) and nitrogen dioxide (NO_2). The percentage of each does vary, but typically NO_2 might make up 10 per cent of the total NO_x coming out of the engine. Reports have linked continuously regenerating filter systems to increased emissions of NO_2 from the tailpipe. These do not always contain sufficient information to enable the reader to make an informed judgement on the issues.

Some key points are therefore set out below:

1. NO_2 produced on the catalyst is the key component used to burn the soot collected in the filter, and so passive filter systems using platinum group metals can lead to an increase in tailpipe NO_2 over engine-out.

2. The amount of excess NO_2 produced varies greatly according to factors such as the actual amount of particulate and NO_x produced by each engine as well as the driving pattern of the vehicle and is therefore difficult to design out by optimising the filter system.

3. Continuously regenerating filter systems do not increase total NO_x but change the balance of NO to NO_2 at the tailpipe. NO reacts more slowly in the atmosphere to form NO_2 in any case so direct NO_2 (that emitted from the tailpipe) mostly affects air quality close to busy roads.
4. All of the NO emitted by an engine would ultimately be converted to NO_2 in the atmosphere, irrespective of the presence of aftertreatment systems.
5. The health benefits from reduced PM emissions (plus HC and CO) are widely considered to greatly outweigh any disadvantage from increased NO_2 formation. The World Health Organisation and California Air Resources Board amongst others have published the statements below about the health impacts of the two emissions.
6. More than 120,000 CRT® systems have brought significant air quality improvement in many cities around the world, and will continue to do so.
7. Direct NO_2 emissions from these catalysed filter systems are only part of picture—NO_2 formation in the atmosphere and from other sources needs to be considered.

NO_2—The few long-term studies have not shown evidence for association between NO_2 and mortality.

Occurrence of Nitrogen Oxides in the Atmosphere

These oxides occur in atmosphere as follows:
1. NO: It is the main product of combustion of nitrogen, and automobile exhaust produced by the combustion of gasoline. It is slowly oxidised to NO_2 by O_2 but rapidly by ozone.
2. N_2O: Present in air at concentration level of 0.25 ppm. Maximum level is 0.5 ppm. It is not a product of combustion.
3. NO_2: In atmosphere NO_2 levels are about 0.001 ppm. It is a strong absorber of ultraviolet light and is the chief constituent of photochemical smog. It initiates photochemical reactions in the troposphere.
4. N_2O_3: It reacts with water vapour to form HNO_3, which combines with ammonia to form ammonium nitrate.
5. N_2O_5: Forms HNO_3 with water and thus reduces the pH of rain water.

Reactions of NO_x in the atmosphere

The average residence life of nitric oxide (NO) is four days and that of NO_2 is three days in the atmosphere. The residence time of NO_x may decrease in a highly polluted atmosphere because of the formation of photochemical smog. Nitric oxide (NO) reacts with O_2 to form various NO_x, the end product being nitric acid (HNO_3), which is precipitated as nitrate salts with ammonia. HNO_3 reacts rapidly to form particulate substances in the atmosphere and causes air pollution. Reactions are given below.

Nitric oxide (NO)

In the thermosphere, NO is formed by the reaction of O_2 with atomic nitrogen.

$$N_2 \xrightarrow{hv} N+N \qquad N+O_2 \rightarrow NO+O$$

High temperature favours the formation of NO, hence automobiles such as motor, and heavy vehicle engines which operate at higher temperatures are the major sources of NO.

Oxidation of atmospheric NO is a very slow process which may generally take place over a period of six days.

$$2NO+O_2 \rightarrow 2NO_2 \qquad NO+O+M \rightarrow NO_2+M \ (M=O_2 \text{ or } N_2)$$

When NO and O_3 are formed in equal amounts, they combine together to produce NO_2 and O_2.

$$O_2 + O + M \leftrightarrows O_3 + M \ (M = O_2 \text{ or } N_2) \qquad NO + O_3 \leftrightarrows O_2 + NO_2$$

Nitrogen oxide (NO_2)

It is a highly reactive significant gas occurring in atmosphere. It penetrates the troposphere and absorbs ultraviolet light as well as visible light, thereby forming excited molecules in the wavelength range of 380 nm to 600 nm.

$$NO_2 \xrightarrow[380–600 \text{ nm}]{hv} NO_2$$

NO_2 undergoes photodissociation using photo energy below 380 nm.

$$NO_2 \xrightarrow{hv} NO + O$$

Effectively, NO and NO_2 concentrations remain constant in the atmosphere.

$$N_2O \xrightarrow{hv} N_2 + O \qquad\qquad\qquad (\lambda = 337 \text{ nm})$$

$$N_2O \xrightarrow{hv} NO + N \qquad\qquad\qquad (\lambda = 250 \text{ nm})$$

Micro-organisms reduce N_2O under anaerobic conditions producing the pollutant NO. NO_2 mainly exists as N_2O_4 (10 per cent) in solid state. In gaseous state, there is 90 per cent NO_2 and 10 per cent N_2O_4 at about 100°C. NO_2 is produced from NO.

$$2NO + O_2 \rightleftarrows 2NO_2 \rightleftarrows N_2O_4$$

Photodissociation of NO_2 results in several significant reactions

$$NO_2 + O \rightarrow NO + O_2$$

$$O + NO_2 + M \rightarrow NO_3 + M$$

$$NO_3 + NO \rightarrow 2NO_2$$

$$NO_2 + O_3 \rightarrow NO_3 + O_2$$

$$NO_3 + NO_2 \rightarrow N_2O_5$$

$$O + NO + M \rightarrow NO_2 + M$$

NO_2 reacts with water forming nitric acid, nitrates and photochemical smog.

$$2NO_2 + H_2O \rightarrow HNO_2 + HNO_3 \text{ or } N_2O_5 + H_2O \rightarrow 2HNO_3$$

NO_2 also reacts with hydroxyl ions in the stratosphere:

$$HO\cdot + NO_2 \rightarrow HNO_3$$

Nitric acid, in this region, is reduced to NO_2 by the attack of hydroxyl radical

$$HO\cdot + HNO_3 \rightarrow H_2O + NO_3 \text{ or } HNO_3 \xrightarrow{hv} HO\cdot + NO_2$$

Actually, HNO_3 acts as a temporary sink for NO_2 in the stratosphere. It rapidly reacts with NH_3 and particulate lime, etc. forming particulate nitrates.

Sources of NO_x Pollution

Natural stratospheric NO_x are also produced by the action of cosmic rays in the upper atmosphere. Man-made sources of NO_x varies depending upon global areas. NO_x are 10 to 100 times greater in urban atmosphere as compared to rural areas. Major man-made activities include combustion of coal, oil, natural gas and gasoline, etc. which produce up to 50 ppm. of nitrogen.

NO_x are also produced as by-products of some chemical industries like nitric acid and sulphuric acid industry. They are also formed during the manufacture of nylon intermediates. Another potential serious source of man-made stratospheric NO_x involves rising fire balls associated with atmospheric nuclear explosion. Abundant fluxes of nuclear produced NO_x are immediately transported to high altitudes by aerodynamic forces created by the explosion itself. The average residence time of NO and NO_2 in the atmosphere are four days and three days, respectively. This indicates that natural processes, including photochemical reactions, take care of NO_x, the product being HNO_3. The latter is precipitated as nitrate salts in either rain or as dust.

$$O_3 + NO_2 \rightarrow NO_3 + O_2$$
$$NO_3 + NO_2 \rightarrow N_2O_5$$
$$N_2O_5 + H_2O \rightarrow 2HNO_3$$

The end product of NO_x, however, is HNO_3. The latter reacts rapidly to form various particulate nitrates.

Effects of nitrogen oxides (NO_x)

Mostly the oxides of nitrogen are not so dangerous, but the role they play in the formation of photochemical oxidants, etc. constitute the most harmful effect. These NO_x affect plants, human health, and the atmosphere.

Effects of NO_x on plants

1. Higher concentrations of NO_2 damage leaves of plants, retard the photosynthetic activity and cause chlorosis.
2. Plants exposed to 100 ppm of NO_2 suffer from leaf spotting and breakdown of tissues.
3. Exposure to 10 ppm of NO checks the metabolic activities in plant tissues, e.g. bean and tomato plants on fumigation showed decreased activity of CO_2 absorption and photosynthetic rate.
4. Damage to vegetation probably results from the production of secondary pollutants such as O_3, PAN and smog, etc.
5. NO_2 is highly injurious to plants.

Effects of NO_x on human health

1. Acute toxicity has, however, not been found in man due to NO_x, NO is biochemically inert and not extremely toxic.
2. The health effects of NO_2 vary with the degree of exposure. An exposure to 50 to 100 ppm of NO_2 for 30 to 50 minutes for a period of 5 to 8 weeks causes inflammation of lung tissues.
3. NO_2 has irritating effects on mucous membrane.
4. Higher doses of NO_2 cause bronchitis and respiratory problems.
5. An exposure to 500–600 ppm of NO_2 for 2 to 10 days results in death of the victim.
6. Death often results due to irritation caused by gases containing NO_x from burning celluloid and nitro cellulose films.
7. Cigarette smokers readily develop lung diseases as cigarettes contain 330–1500 ppm of NO_x.
8. NO_2 lowers the resistance to influenza; irritates eyes.
9. Nitric, nitrous acid and several nitrates cause respiratory, digestive and nervous ailments.
10. NO gets attached to haemoglobin and reduces oxygen transport efficiency of blood.
11. Higher levels of NO_x cause gum inflammtion, internal bleeding, pneumonia, oxygen deficiency and lung cancer, etc.

12. Both NO and NO_2 are fairly toxic at low concentrations.
13. Oxides of nitrogen are the second-most abundant atmospheric pollutants. These are extremely dangerous to human health. Their acute effects are sometimes more severe than that of carbon monoxide.

Effects of NO_x on materials

1. NO_x fades away a number of textile materials like cotton, rayon, and acetates. It is found that NO_x level reaches to 1 to 2 ppm, during the combustion of natural gas which is used to heat dryers.
2. However, damaging effects of NO_x on various textile fibres is not much pronounced. Higher levels of NO_x cause 10 per cent loss of fibre strength in cotton and rayon.
3. NO_2 produces aerosols which damage nylon fibres.
4. Abnormally high levels of NO_2 present in ambient air, during dynamite blast operations in New York had adverse effects on nylon stockings and cotton.
5. NO_2 along with hydrocarbons produces peroxides, which combine with ozone and cause cracks in rubber.
6. Oxides of nitrogen are involved in the generation of serious air pollutants such as ozone, PAN and peroxy organic compounds, which are toxic at 0.1 ppm levels and responsible for photochemical smog. Hydrocarbons and NO_x are the key factors in multiplying photochemical smog problems.
7. Higher concentrations of particulate nitrates in airborne dust accumulate adjacent to cracked areas of metals like nickel and brass.

NO_x and acid rain

Oxides of nitrogen (NO, NO_2 and N_2O) through photochemical chain reactions, produce irritating gases which are toxic and corrosive. These gases produce nitirc acid which is washed down and contributes to acid rain. Acidity of rain water increases when the amount of oxides of nitrogen and sulphur increases in the atmosphere to produce more nitric and sulphuric acid. Acid rain reduces fertility of the soil. It is also affecting the Taj Mahal in Agra. NO_2 is a part of the photolytic NO_2 cycle.

Effect of NO_x on the stratospheric atmosphere

In the stratosphere, SST (super sonic transport), airplanes produce NO_x which catalyses the stratospheric ozone layer depletion. Ozone absorbs harmful ultraviolet (uv) radiations near 250–300 nm and disintegrates into oxygen and atomic oxygen:

$$O_3 \xrightarrow{hv} O_2 + O \qquad O_2 \xrightarrow{hv} O + O$$

atomic oxygen reacts with O_2 in presence of energy absorbing bodies (M):

$$O_2 + O + M \rightarrow O_3 + M$$

Ozone reacts with NO_x in the stratosphere

$$NO + O_3 \rightarrow NO_2 + O_2 \qquad NO_2 + O \rightarrow NO + O_2$$

Thus, a chain reaction decreases the concentration of O_3 allowing the chronic UV radiation to reach the earth's surface.

$$O + O_3 \rightarrow O_2 + O_2$$

The cycle, alongwith other air pollutants, is responsible for several adverse effects in the atmosphere.

Control of NO$_x$ Pollution

NO$_x$ emissions from flue gases can be reduced by: (i) dilution in the atmosphere by increasing stack height or by discharging effluents into the atmosphere only during favourable meteorological conditions, (ii) modification of operating and design conditions, and (iii) treatment of flue gases. The method of control of NO$_x$ by dilution is similar to that of SO$_x$, particulates and other pollutants.

NO$_x$ control by modification of operating and design conditions

On the basis of the thermodynamics and kinetics involved in the formation of NO$_x$, the following in-plant measures can be adopted to reduce NO$_x$ emissions. These are based on the principles that NO$_x$ emissions can be reduced by: (i) reducing the peak temperature, (ii) reducing the residence time at the peak temperature, and (iii) reducing the availability of oxygen, i.e. by maintaining higher N/O ratios. The in-plant control measures are:

Low excess air combustion

The presence of excess air affects both the temperature and oxygen concentration of gases in the post-combustion zone. Normal boiler units operate in the presence of 10–20 per cent excess air to ensure complete combustion of the fuel, but this excess air provides enough oxygen for reacting with nitrogen.

Decreasing combustion air temperature

In many industries, waste heat is available to help preheat the air entering a combustion process. Although this process leads to appreciable energy savings, the added energy increases the flame temperature. Thus, NO$_x$ emissions increase. Data from full-sized boiler tests indicate a three-fold increase in NO$_x$ emissions when combustion air is preheated from 25° to 300°C. Since temperature has a major influence on nitric oxide formation, an effective method of control is cooling of primary flame zone by heat transfer to surrounding surfaces.

Two stage combustion

This is one of the most effective control methods for the suppression of the formation of nitrogen oxides. In the first stage, the fuel and air are burned at near — stoichiometric conditions. All the fuel is fired with only 85–95 per cent of total air requirements at the bottom of the furnace followed by secondary air injection higher up in the furnace to complete the combustion. Thus, in the primary zone, incomplete combustion of the fuel takes place resulting in reduced concentrations of NO$_x$ in furnace gases. Heat removal and gas dilution between zones cause the temperature of the gases after the primary zone to decrease so that the final stage of combustion process occurs at a lower temperature. Reduction in the emission of NO$_x$ by 38 per cent for coal and oil firing and by 50 per cent for natural gas combustion has been observed under these conditions.

Flue gas recirculation

The recirculation of coal flue gas into the combustion chamber is found to be quite effective in reducing NO$_x$ formation from stationary sources. A portion of the cooled flue-gas is injected back into the combustion zone. This additional gas acts as a thermal sink and reduces the overall combustion temperature. In addition, the oxygen concentration is lowered as the flue gases going out have a higher N/O ratio, i.e. lower oxygen availability. Both these effects favour a reduction in NO$_x$ emissions. One disadvantage is the increased cost of duct work, since large volumes of gas are mixed with the primary air prior to combustion.

Injection of water and steam

Injection of water or steam into the combustion zone is also effective in reducing emissions of nitrogen oxides. This method is similar to that in flue gas circulation. Injecting water to the extent of 10–15 per cent of the total combustion air can reduce the temperature level in the furnace to some extent.

Modification of furnace burner configuration

The two basic furnace designs are the tangential and the horizontal methods of firing. In the tangential method of firing, the flame and the combustion products rotate in an upward spiral around the walls of the furnace, and the furnace itself is used as the burner. This results in low peak flame temperature and consequently reduction in NO_x emissions. In the horizontal firing, the flame is at right angles to the walls of the fire box. This tends to concentrate the hot gases, leading to higher flame temperatures and more NO_x emissions.

NO_x control by treatment

The modification of design and combustion techniques, in general, is to prevent or to suppress the emission of nitrogen oxides in the combustion chambers. However, if the measures are not enough or too costly, stack gas treatment is necessary. Stack gas treatment is an efficient method for the control of NO_x. The methods are (i) absorption by liquids, (ii) adsorption by solids, (iii) catalytic reduction— selective and non-selective, and (iv) electron beam irradiation.

Absorption by liquids

The oxides of nitrogen can be absorbed by water, hydroxide and carbonate solutions, sulphuric acid, organic solutions and molten alkali carbonates and hydroxides. Several scrubbing techniques were developed which initially used solutions of sodium and calcium hydroxide. Some of the commonly adopted methods are:

1. Absorption by alkaline solutions: NO_x absorption by using aqueous alkaline solutions like NaOH and MgOH also yields good results. This is a common method adopted during desulphurisation of power plant emissions by such alkaline solutions. In the desulphurisation process, about 10 per cent of NO is oxidised to NO_2 before the flue gas reaches the scrubber. The scrubber then removes about 20 per cent of the total NO_x in equal parts of NO and NO_2.
2. Absorption by lime: Aqueous suspension of calcium hydroxide can be used as the scrubber to reduce NO_x levels to 200 ppm. The calcium nitrite in the solution can further be converted to more valuable calcium nitrate by treating with sulphuric acid.

$$2H_2SO_4 + 3Ca\,(NO_2)_2 \rightarrow 2CaSO_4 + 4NO + Ca(NO_3)_2 + 2H_2O$$

   The NO evolved may be recycled to the nitric acid plant and calcium nitrate can be used as a fertiliser. Thus, the process reduces NO_x, recovers NO and gives a valuable fertiliser simultaneously.
3. NO_x absorption by magnesium hydroxide: In this process, the oxides of nitrogen are absorbed by hydroxide liquor in an absorption tower. The magnesium nitrite solution leaving the absorber is taken to a pressure reactor where the nitrite is converted to nitrate. The by-product NO is oxidised to NO_2. The liquid leaving the pressure reactor, consisting of Mg $(NO_3)_2$/Mg $(OH)_2$ is sent to a settling chamber where nitrate is separated from the hydroxide. Part of the NO_2 from the oxidiser is sent to the absorber to maintain equimolar concentrations of NO and NO_2, while the rest of NO_2 is used for nitric acid production.

A continuous catalytic absorption process using stripped nitric acid as the absorbing medium has been reported. The advantage of this is that it not only reduces NO_x in the tail gas to a tolerable level, but also recovers NO as nitric acid.

Some of the scrubbing techniques, correctly used are:
1. Two-stage absorption, first in water and then in sodium hydroxide yielding nitrite and nitrate salts.
2. Absorption in various types of ammoniacal solutions such as waste caustic ammonia liquor, ammonium bicarbonate and ammonium bisulphate.
3. Absorption with an aqueous suspension of lime, where calcium nitrite and nitrate can be recovered for use as fertilisers.

Adsorption by solids

The adsorbents that show some capacity for oxidising NO and NO_2, and for adsorbing nitrogen dioxide are activated carbon, silica gel, molecular sieves, ion-exchange resins, and certain metal oxides, particularly manganese and alkalised ferric oxides. The use of activated carbon to adsorb oxides of nitrogen has been studied extensively. Activated carbon has a high adsorption rate and capacity compared to other materials. However, regeneration may be a problem.

A potential fire and explosion hazard may be another difficulty with this material since O_2 is usually present in most stack gases. Thus, the efficiency of activated carbon decreases with quantity of O_2 present in flue gas. Manganese oxides and alkalised ferric oxides show technical potential. However, sorbent attrition is a major technical stumbling block. The most suitable adsorbent for NO_x is the one which can be regenerated, and at the same time, which does not preferentially react with water vapour or with CO_2 in the flue gas. The most promising adsorbent is ferrous salt. Molecular sieves also can be used for NO_x control, particularly for NO_x from nitric acid plants. In this process, two beds operate batch wise — one adsorbs NO_x from the tail gas while the other is regenerated.

Catalytic reduction

Selective catalytic reduction (SCR): Selective catalytic reduction refers to a process that chemically reduces NO_x with NH_3 over a heterogeneous catalyst in the presence of O_2. The process is termed selective because the reducing agent NH_3 preferentially attacks NO_x rather than O_2. However, the O_2 enhances the reaction and is indeed a necessary part of the reaction scheme. Non-selective catalytic reduction: Reducing agents other than NH_3, such as CO, H_2 or CH_4, completely reduce NO at temperatures of 300–400°C. Non-noble metal catalysts such as supported copper oxides may also be used for NO_x control. These reductants act non-selectively, reacting with O_2 and SO_x present in the flue gas, in addition to NO_x and thus huge amounts of reductants must be added. For non-selective catalytic reduction of NO_x in stationary combustion sources, stoichiometric or sub-stoichiometric air must be used in the primary combustion zone. This reduces the cost of the reducing agent. NO_x removal by CH_4 is as follows: When hydrocarbons like methane are used for removing NO_x, higher oxides of nitrogen are converted to NO in the first stage. In the second stage, NO is reduced to nitrogen. Oxygen in the tail gases react simultaneously with the hydrocarbons to produce water vapour and carbon dioxide.

Electron beam irradiation

This process simultaneously removes 90 per cent of NO_x and SO_x with NH_3 activated by electron beam irradiation, without the need for catalysts. The process first requires removal of fly ash from the flue gas. Then ammonia is added and the gas enters the electron beam reactors. There, ammonia in the presence of electrons converts NO_x and SO_x into a dry powder of ammonium sulphate $\{(NH_4)_2SO_4\}$ and ammonium nitrate sulphate $\{(NH_4)_2SO_4, 2\,NH_4NO_3)\}$. After the dry powder is removed by a second particulate collection device, the treated flue gas leaves the system hot enough to be exhausted through the stack without reheat. The waste generated also is a potential fertiliser feedstock.

CONTROLLING NO$_x$ EMISSIONS IN PROCESS INDUSTRY

Unlike its polluting cousins, NO$_x$ has not yet received the attention it deserves in many countries. In spite of regulations, NO$_x$ levels continue to rise having a serious impact on public health and environment whereas 'most of the other major air pollutants like CO, SO$_2$, lead, ozone, etc. have seen a decline in most of the countries. NO$_x$ continues to be a serious health and environment concern around the world ever since its role in generation of smog and acid rain came to light in 1986. NO$_x$ emissions are directly responsible for ground level ozone (smog) which causes serious respiratory ailments, acid rain and indirectly for many other problems. The section deals with the various aspects of NO$_x$ emissions; their formation, available technologies, current trends and ways to control and reduce these emissions in process industries.

Nitrogen oxides or NO$_x$ is the generic term for a group of highly reactive gases, all of which contain nitrogen and oxygen in varying amounts. As already discussed NO$_x$ are formed when fuel is burned at high temperatures as in a combustion process. Emissions of NO$_x$ from combustion are primarily in the form of NO and NO$_2$ at temperatures normally above 760°C. NO$_x$ generation increases rapidly at temperatures above 1,300°C and is generated to the limit of available oxygen in combustion air. Combustion NO$_x$ is generated as a function of air to fuel ratio and more pronounced when the mixture is on the fuel lean side of the stoichiometric ratio. The primary sources of NO$_x$ are motor vehicles, electric utilities, power plants, boilers, process plants, refineries and other industrial, commercial, and residential sources that burn fuel.

Why Should We Control NO$_x$?

The role of NO$_x$ in the production of smog and acid rain came to light in 1986 and since then it has been recognised as a serious health and environment concern around the world. NO$_x$ emissions are directly responsible for ground level ozone (smog) which causes serious respiratory ailments, acid rain, global warming and indirectly for eutrophication of water bodies such as lakes and reservoirs.

Human health impacts appear to be related to the peak exposures to NO$_x$. In addition to the potentially damaging human health, nitrogen oxides are precursors to ozone (O$_3$) formation (found in the ambient air which we breathe), which is harmful to human health and vegetation. The dispersion of nitrogen oxides may contribute to ozone formation and acid rain tar away from the source; hence the long-term objective must be to reduce total emissions and not to rely on plant location factor.

Emission Standards

NO$_x$ emissions cause acid rain, photochemical smog and tropospheric ozone destruction. This has led to establishment of regulatory measures and to the development of technologies to reduce NO$_x$ emissions from both existing and new plants. United states environment protection agency (EPA) mainly regulates nitrogen dioxide (NO$_2$), because it is the most prevalent form of NO$_x$ in the atmosphere that is generated by anthropogenic (human) activities. The four main regulatory agencies that influence emission norms in India are:

1. Ministry of Environment and Forests.
2. Central Pollution Control Board (CPCB).
3. State Pollution Control Boards (SPCBs)—in respective states.
4. Pollution Control Committees—in respective territories.

NO$_x$ limits for a plant in India are based on the Environmental Impact Assessment (EIA) study for that facility. The central pollution control board of India is executing a program of ambient air quality monitoring known as National Air Quality Monitoring Program (NAMP) and has included NO$_x$ limits in the national ambient air quality standards. Some states have fixed the limit whereas some other states are yet to do so. However, the tropospheric ozone continues to be a significant air pollution problem in India and is the primary constituent of smog. Large portions of the country do not meet the ozone National Ambient Air Quality Standards (NAAQS) and thereby expose large segments of the population to unhealthy levels of ozone in the air.

Types of NO_x

In all combustion processes, the three main sources for NO_x formation are:

Thermal NO_x: are formed due to the oxidation of free nitrogen in the combustion air or fuel at high temperatures. Its production increases directly with air to fuel ratio, and exponentially with combustor inlet temperature, and decreases exponentially with increasing water or steam injection or increasing specific humidity.

Fuel (organic) NO_x: are formed due to the oxidation of organically bound nitrogen in the fuel (fuel bound nitrogen). Oxidation of fuel bound nitrogen to NO_x is very efficient and could be up to 100 per cent at low FBN (fuel bound nitrogen) concentrations (0.01–0.04 per cent). At higher levels of FBN (=1.0 per cent), the efficiency falls off to approximately 20 per cent. Reductions of flame temperatures to abate thermal NO_x have an adverse effect on organic NO_x formation.

Prompt NO_x: is formed from molecular nitrogen in the air, combining with fuel in fuel-rich conditions. The abundance of prompt NO_x in combustion products is however disputed by some authorities in this field. In a process industry, there are various sources of NO_x depending on the type of firing equipment and fuels that are used. NO_x generation is much higher in plants employing older technologies and plants using coal, naphtha or fuel oil as fuel compared to modern plants with latest technologies (e.g. low NO_x burners) and using natural gas as fuel. In refineries and other process industries, NO_x is normally generated from the four major sources namely: furnaces/fired heaters, fired boilers, gas turbines and HRSG (heat recovery and steam generation).

NO_x Abatement and Control Methods

To reduce NO_x emission in an existing installation, each source of NO_x requires a separate strategy depending on the type of firing equipment, type of fuel and operating conditions. No common method is applicable for all cases. Also every method of NO_x abatement, in some way or the other adds some cost to the existing plant costs. A simple block diagram of NO_x control methods is shown in Fig. 3.9.

There are different methods to control NO_x emissions depending on the system and the regulatory requirements. One way is to use low nitrogen fuels and the other is to modify combustion conditions or equipment (e.g. low NO_x burners) to generate less NO_x at source. Other methods convert NO_x in the flue gases by using various gas treating technologies such as 'selective catalytic reduction' (SCR) or 'selective non-catalytic reduction' (SNCR). The methods used to abate NO_x can be divided into precombution and post-combution methods. Figure 3.10 shows both pre-combustion and post-combustion NO_x control options. The LHS of the figure shows technologies used in the combustor, i.e. low-NO_x burners, over fire air and reburning. The top and RHS of the figure shows technologies used in post combustion-SNCR, SCR (hot-side and post-FGD), and combined NO_x/SO_x technologies.

Pre-combustion Technologies

The method involves the following strategies:
1. Control of combustion air preheat temperature.
2. Control of excess air.
3. Use of diluent injection (steam, water).

4. Use of low NO_x burners and flue gas recirculation.
5. Control of fuel bound nitrogen.

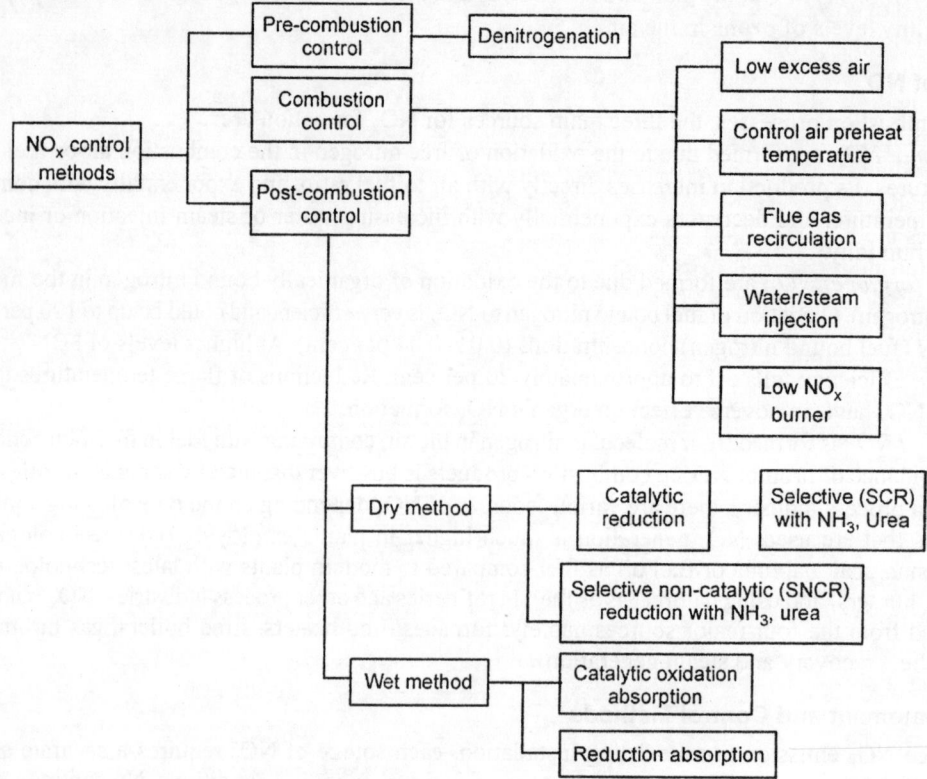

Fig. 3.9. Block diagram of NO_x control methods.

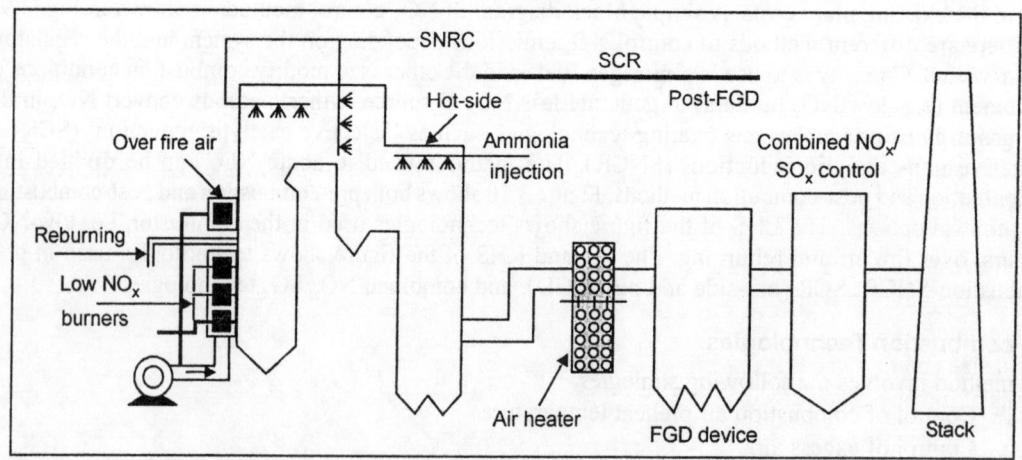

Fig. 3.10. Pre-combustion and post-combustion NO_x control options.

Combustion Control

The method of combustion control will depend on the type of combustion equipment and the method of firing fuel. It involves any of the three strategies: reducing peak temperatures in the combustion zone, reducing the gas residence time in the high temperature zone, and reducing oxygen concentrations in the combustion zone. These changes in the combustion process can be achieved either through process modifications or by modifying operating conditions on existing furnaces. Increase in atomising steam quantity reduces NO_x formation due to its quenching effects but higher temperature of atomising steam increases NO_x formation. Pre-heating combustion air also increases NO_x formation.

Process modifications include: use of specially designed low NOx burners, re-burning, combustion staging, flue gas recirculation, burners out of service (BOOS), over fire air (OFA), catalytic combustion, reduced air preheat/firing rates, water or steam injection and low excess air (LEA) firing. These modifications are capable of reducing NO_x emissions by 50 to 80 per cent. In addition to these techniques, some software are available claiming to achieve NO_x reductions up to 30 per cent by optimising combustion process conditions. Basically software make use of 'advance process control' techniques to control NO_x levels by optimising condition over a range of excess air levels and air preheat temperatures. Extensive research has been carried out in developing simple and accurate kinetic models for NO_x formation. With the availability of information on NO_x kinetics, a kinetic based model can be developed for the predictive emissions monitoring system (PEMS) to control NO_x.

Choice of Fuel

According to some estimates, coals and residual fuel oils containing organically bound nitrogen contribute to over 50 per cent of total overall emissions of NO_x. Purge gases from ammonia plant synthesis loop containing ammonia, if used as fuel in primary reformer can also reduce NO_x emission. In many circumstances, the most cost-effective way of reducing NO_x emission's will be to use low nitrogen fuels such as natural gas.

Low NO_x Burners

Low NO_x burners (LNB) are effective in reducing NO_x emissions from both new plants and existing plants as a retrofit. It limits the formation of NO_x by controlling the mixing of fuel and air, in effect by low excess air firing or staged combustion (Fig. 3.11). LNB drives the major fraction of the nitrogen compounds into the fuel rich conditions. A secondary zone provides the necessary oxygen for complete combustion in a temperature window designed to minimise the formation of thermal NO_x.

Compared with conventional burners, LNB reduces NO_x emissions by 20–60 per cent depending on the type of equipment and fuel. It can be combined with other primary measures such as overfire air, reburning or flue gas recirculation. The combination of low NOx burners with other primary measures can achieve up to 74 per cent NO_x removal efficiency. LNB can generally achieve NO_x emissions below 30 ppmv. Because LNB are relatively less expensive and introduce least changes in the existing system, they are natural choice for reducing NO_x emissions in process industries especially where burners are large in size and few in numbers. In fact, LNB are now a standard feature of all latest designs. Detailed analysis of furnace geometry and dimensions are very important for any LNB retrofit in existing furnaces.

Ultra-low NO_x Burners

The ultra-low NO_x burner with flue gas recirculation may control NO_x emissions to less than 5 ppmv without any significant efficiency reduction. The burner achieves ultra-low NO_x emissions by merging technology advancements in advanced lean pre-mixed burners and fuel pretreatment. The resulting burner system combines a low-swirl flame stabilisation method with internal flue gas recirculation (IFGR) benefits to produce very low NO_x emissions.

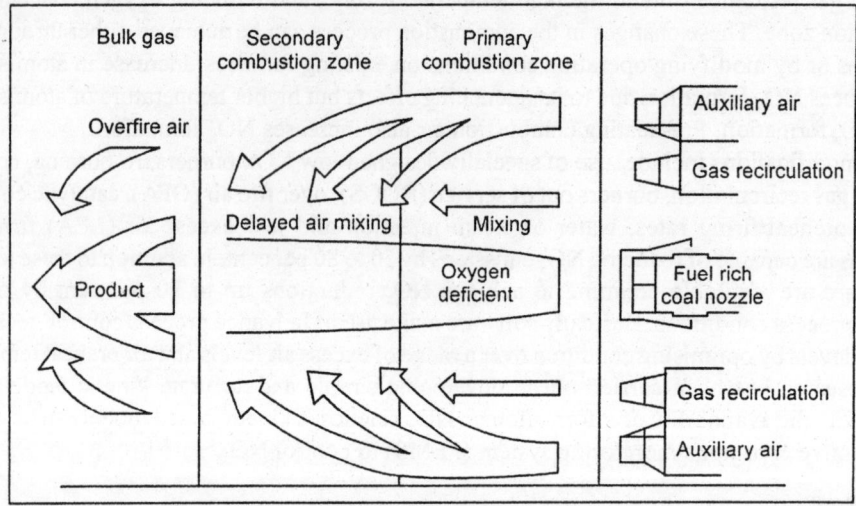

Fig. 3.11. Staged combustion in a low NO_x burner.

Increased system efficiency, with operation at less than 10 per cent excess air over the entire turndown range makes burner design suitable for new or retrofit applications to a wide range of combustion chamber configurations.

Post Combustion Technologies

Post combustion methods are based on flue gas treatment; hence these technologies are applicable to a broad range of sources and fuels, and have become important where interference with the existing firing equipment design is not desirable. These techniques are more effective in reducing NO_x emissions than combustion controls, although at a higher cost. These techniques utilise a chemically reducing substance to remove oxygen from nitrogen oxides. SCR uses ammonia and SNCR uses ammonia or urea. Nonthermal plasma, an emerging technology, when used with a reducing agent, chemically reduces NO_x. All of these technologies attempt to chemically reduce the valence level of nitrogen to zero after the valence has become higher.

Selective Catalytic Reduction (SCR)

One of the most-effective NO_x control techniques is selective catalytic reduction. SCR is currently the most developed and widely applied FGT (flue gas treatment) technology. SCR involves catalytic reduction of NO_x in the flue gas by ammonia in a controlled temperature reactor. It can achieve up to a 94 per cent DRE (destruction or recovery efficiency) and is one of the most effective NO_x abatement techniques. Precious metal catalysts were used earlier in SCR but now base metal and zeolite catalysts are more cost effective yielding better results. SCR can be installed in any furnace depending on space availability in the desired temperature range and is independent of the fuel or firing equipment. A simple representation of the SCR process is shown in Fig. 3.12.

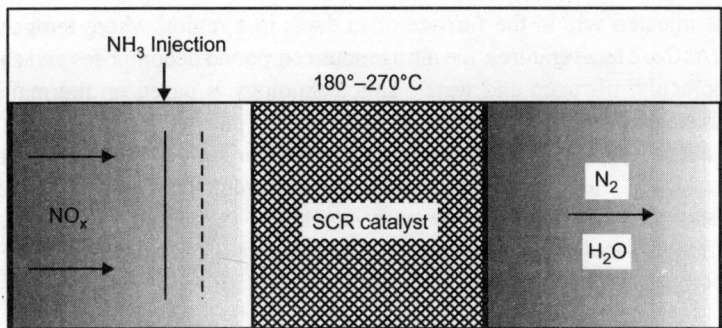

Fig. 3.12. A simple representation of SCR process.

The principal reactions involved in SCR include:

$$4NH_3 + 6NO \quad = \quad 5N_2 + 6H_2O$$
$$8NH_3 + 6NO_2 \quad = \quad 7N_2 + 12H_2O$$
$$4NH_3 + 4NO + O_2 \quad = \quad 4N_2 + 6H_2O$$
$$4NH_3 + 2NO_2 + O_2 \quad = \quad 3N_2 + 6H_2O$$

Anhydrous or aqueous ammonia is injected into the flue gas upstream of a catalyst bed. The NO_x and NH_3 combine at the catalyst surface forming an ammonium salt intermediate that subsequently decomposes to produce elemental nitrogen and water. The catalyst lowers the activation energy of the NO_x-decomposition reaction, thereby enabling the use of this technology at lower flue gas temperatures. In SCR, the NO_x-reduction reactions occur only in a specified temperature range, generally 180°–270°C, depending on the type of catalyst.

SCR Process

The catalyst location within the overall process is important and is one of the main factors that ultimately determine the technical feasibility of SCR. Besides operating temperature, several factors like the catalyst, system configuration, sulphur and metals content of the fuel, and the design of the ammonia injection system influence the SCR process. Catalysts are available in a wide variety of shapes (honeycomb plates, parallel ridged plates, rings and pellets) and materials (base metals, such as titanium, vanadium, platinum, zeolites and ceramics).

Each combination presents various advantages and disadvantages in terms of allowable operating temperatures, catalyst fouling and pressure drop. Fuel oil firing may result in deposition on the catalyst surface causing catalyst deactivation and poisoning. SCR technology is generally not very suitable for fuel oil fired in many refineries. SCR is one of the most effective NO_x reduction methods and achieves maximum reduction (up to 94 per cent) and may be the right choice for NO_x abatement, keeping a futuristic view when NO_x standards are going to be very stringent. Capital and operating costs of SCR design have dropped rapidly over the past decade as a result of technological innovations, increased manufacturing expertise, cheaper and long life catalysts and competition among suppliers.

Selective Non-catalytic Reduction (SNCR)

SNCR is probably one of the most cost-effective post combustion NO_x abatement method due to its low cost compared to SCR. Current trends indicate that SNCR systems can reduce NO_x emissions by 30–70 per cent. In SNCR (also called as Thermal DeNO$_x$ process), a nitrogenous compound typically ammonia or urea is injected within the furnace or in ducts in a region where temperature is between 850°C and 1100°C. At these temperatures, the nitrogenous compound decomposes and chemically reduces the NO_x to form molecular nitrogen and water. This technology is based on thermal ionisation of the ammonia or urea instead of using a catalyst or non-thermal plasma. This temperature 'window' is important because outside of it, either more ammonia 'slips' through or more NO_x is generated than is being chemically reduced. The temperature 'window' is different for urea and ammonia. The capital expenditure for installation for SNCR is normally much less compared to SCR system. Gas phase reaction between ammonia and NO_x uses either aqueous or anhydrous ammonia.

The two basic competing reactions are:

$$NO_x + NH_3 + O_2 \rightarrow N_2 + H_2O$$
(Desired DeNO$_x$ reaction)

$$NH_3 + O_2 \rightarrow NO_x + H_2O$$
(Undesired reaction)

The efficiency of the chemical reaction depends on factors that include flue gas temperature, residence time at that temperature, amount and time of nitrogenous reagent injected, mixing effectiveness and uncontrolled NO_x levels. It would be very important to monitor NO_x in the flue gas, as the amount of ammonia or urea injected has to be in stoichiometric proportion to the amount of NO_x. Most of the excess reagent will degrade to nitrogen and carbon dioxide. However, some quantities of ammonia may slip and trace quantity of carbon monoxide may also form. A comparison of SNCR and SCR systems is given in Table 3.6.

Table 3.6. Comparison of typical SNCR and SCR systems.

Design criteria	SNCR	SCR
NO_x reduction efficiency, %	40–70	60–90
Temperature window, °C	850–11 00	180–270
Operating cost	Low	High
Capital investment	Low	High
Plot requirement	Minor	Major
Retrofit	Easy	Difficult
Ammonia slip	5–20 ppmv	5–10 ppmv
Effect of SOx	No effect	Low-medium
Dust	No effect	Catalyst
Soot	No effect	Poisoning

Thus though there are many techniques available to reduce NO_x in various applications, Low NO_x burners, SCR, SNCR and combustion optimisation are the most effective and suitable for conditions normally present in process industries. A trade-off has to be made between level of reduction, capital cost, operating cost and impact on existing operations. Also, it is very important to choose the best option available for total NO_x reduction as required by the law with optimum capital expenditure. In USA, this is called 'Emission averaging' which allows plants to reduce NO_x emissions at units where it is technically easier and less expensive to the least possible extent and leave the places where it is technically difficult or more costly so that the average overall NO_x levels are controlled as desired by the regulation.

Chapter 4

Odour Pollution and Its Control

INTRODUCTION

Odour can be defined as the 'perception of smell' or in scientific terms as 'a sensation resulting from the reception of stimulus by the olfactory sensory system'. Whether pleasant or unpleasant, odours are induced by inhaling airborne volatile organics or inorganics. With growing population, industrialisation and urbanisation, the odour problem has been assuming objectionable proportion. Urbanisation without proper sanitation facilities is a major cause of odour problems. Rapidly growing industrialisation has aggravated the problem through odourous industrial operations. Undesirable odours contribute to air quality concerns and affect human lifestyles. Odour is undoubtedly the most complex of all the air pollution problems. Unlike conventional air pollutants, odour has distinctly different characteristics, which, to an extent, can be comparable with noise pollution. Similar to noise, nuisance is the primary effect of odour on people. Some of such characteristics are:

Substances of similar or dissimilar chemical constitution may have similar odours. Nature and strength of odour may change on dilution.

1. Weak odours are not perceived in presence of strong odours.
2. Odours of same strength blend to produce a combination in, which one or both may be unrecognisable.
3. Constant intensity of odours causes an individual to quickly loose awareness of the sensation and only noticed when it varies in intensity.
4. Fatigue for one odour may not affect the perception of dissimilar odours but will interfere with the perception of similar odours.
5. An unfamiliar odour is more likely to cause complaint than a familiar one.
6. Two or more odourous substances may cancel the smell of each other.
7. Odours travel downwind.
8. Person can smell at a distance.
9. Many animals have keener sense of olfaction than man.

Likes and dislikes often depend on association of the scent with pleasant or unpleasant experiences.

Controlling odour is an important consideration for protecting the environment and our community amenity. Since there are neither indicative compounds within odour plumes nor electronic or other devices for measuring odour emissions, it is difficult to develop meaningful regulatory threshold limits for odour. However, based on the international experience on odour control, our country also needs to adopt effective odour control policies.

EFFECTS OF ODOUR

Odour affects human beings in a number of ways. Strong, unpleasant or offensive smells can interfere with a person's enjoyment of life especially if they are frequent and/or persistent. Major factors relevant to perceived odour nuisance are:

1. Offensiveness.
2. Duration of exposure to odour.
3. Frequency of odour occurrence.
4. Tolerance and expectation of the receptor.

Though foul odour may not cause direct damage to health, toxic stimulants of odour may cause ill health or respiratory symptoms. Secondary effects, in some, may be nausea, insomnia and discomfort. Very strong odour can result in nasal irritation, trigger symptoms in individuals with breathing problems or asthma. On the economic front, loss of property value near odour causing operations/ industries and odourous environment is partly a consequence of offensive odour.

SOURCES OF ODOUR

Most commonly reported odour-producing compounds are hydrogen sulphide (rotten egg odour) and ammonia (sharp pungent odour). Carbon disulphide, mercaptans, product of decomposition of proteins (especially of animal origin) phenols and some petroleum hydrocarbons are other common odourants. Most offensive odour are created by the anaerobic decay of wet organic matter such as flesh, manure, feed or silage. For example, odour originating from livestock manure are a result of a broad range of over 168 odour-producing compounds. Warm temperatures enhance anaerobic decay and foul odour production, as represented in Fig. 4.1.

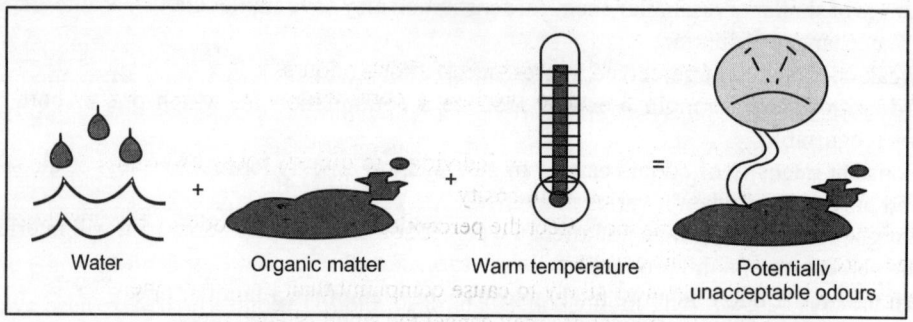

Water   +   Organic matter   +   Warm temperature   =   Potentially unacceptable odours

Fig. 4.1. Conditions for potential foul odour.

Odour sources can be classified as:

1. Point sources: Point sources are confined to emissions from vents, stacks and exhausts.
2. Area sources: Area sources may be unconfined like swine operations, sewage treatment plant, waste water treatment plant, solid waste landfill, composting, household manure spreading, settling lagoons or a cattle feedlot, etc.
3. Building sources: Building sources of odour may like from hog confinement chicken and pig sheds.
4. Fugitive sources: In this source of odour, emissions are of fugitive nature like odour emissions from soil bed or biofilter surface.

Odour can arise from many sources. Most of the sources are man-made. Garbage/improper dumping in vacant land is a common phenomenon. It leads to foul smell due to putrefaction of dumped garbage, which lies uncollected for days together. Unscientific design of landfill, increased sewage production and improper sewage treatment practices produce unpleasant odour.

Large livestock operations, poultry farms, tanneries, slaughterhouses, food and meat processing industries, and bone mills are among major contributors to odour pollution. Agricultural activities like decaying of vegetation, production and application of compost etc. also contribute to odour pollution.

In urban areas, improper handling of public amenities like toilets of cinema hall, bus/railway stations, hospitals, shopping complex, etc. generate pungent odour, which affects the users as well as neighbourhood residents. Congested markets do not allow the escape of odour from markets products, thus causing problems to shop-owners as well as to customers.

Vehicular sector also has its share in odour pollution. Rapidly growing vehicular population as well as harmful pollutants emitted by them generate very harmful and pungent odour that have marked effects on pedestrians as well as nearby residents. Table 4.1 indicates the various odourous chemicals emitted from industrial operations.

Table 4.1. Sources of odour.

Industry	Odourous material
Pulp and paper	Mercaptans, hydrogen sulphide
Tanneries	Hides, flesh
Fertilisers	Ammonia, nitrogen compounds
Petroleum	Sulphur compounds from crude oil, mercaptans
Chemical	Ammonia, phenols, mercaptans, hydrogen sulphide, chlorine, organic products
Foundries	Quenching oils
Pharmaceuticals	Biological extracts and wastes, spent fermentation liquors
Food	Cannery waste, dairy waste, meat products, packing house wastes, fish cooking odours, coffee roaster effluents
Detergent	Animal fats
General	Burning rubber, solvents, incinerator, smoke
Swine operations	Hydrogen sulphide and ammonia
Waste-water treatment plant	Hydrogen sulphide
Municipal solid waste landfill	Hydrogen sulphide

MEASUREMENT AND MONITORING OF ODOUR

Terms Associated with Odour Measurement

For better understanding of the methods of measurement of odour, definition of following few terms is required:

1. Odour intensity: Odour intensity is the strength of the perceived odour sensation. It is related to the odourant concentration. The odour intensity is usually stated according to a predetermined rating system. Widely used scale for odour intensity is the following:
    (a) 0 = No odour.
    (b) 1 = Threshold level.

(c) 2 = Definite odour.

(d) 3 = Strong odour.

(e) 4 = Overpowering odour.

(Half score is used when the observer is undecided).

2. Odour detectability or threshold: Odour detectability or threshold is a sensory property referring to the minimum concentration that produces an olfactory response or sensation. With odour intensity at or just above 'threshold' odour become difficult to perceive.

3. Odour character: Odour character or quality is the property to identify an odour and to differentiate it from another odour of equal intensity.

4. Hedonic tone: Hedonic tone is a property of an odour relating to its pleasantness. When an odour is evaluated for its hedonic tone in the neutral context of an olfactometric presentation, the panellist is exposed to a controlled stimulus in terms of intensity and duration. The degree of pleasantness or unpleasantness is determined by each panellist experience and emotional associations.

Sampling of Odours

Odours are measured adopting olfactometric testing methods, which are psycho-physical methods. Olfactometery employs a panel of human noses as sensors. In these methods, the olfactory responses of individuals sniffing diluted odour presented by an olfactometer to determine odour strength or odour concentration. Odour measurement requires representative samples of the air to be drawn into a sample bag and rapidly transported to an odour laboratory for olfactometric testing. Sampling strategies and techniques depends on emission sources characteristics. Each type of source has special requirements for sampling and sample collection.

1. Point sources: Typically, a point source will be a stack with a known flow rate. Odour samples are taken into Tedlar sampling bags loaded in a vacuum drum through Teflon tube inserted into the stack at different points.

2. Area sources: Area source will be water or a solid surface. A portable wind tunnel system can be used to determine the specific emission odour rate (SEOR). The specific emission odour rate may be defined as the quantity of odour emitted per unit time from a unit surface area. The quantity of odour is not determined directly by olfactometry but is calculated from the concentration of odour (as measured by olfactometry), which is then multiplied by the volume of air passing through the hood per unit time.

3. Building sources: Building sources, such as chicken and pig sheds, have a number of openings. For animal sheds odour samples are normally taken from several points within a shed. Experience indicates that one composite sample is sufficient to represent a single shed at a particular time.

4. Fugitive sources: Typically fugitive sources include odour emissions from bed or biofilter surface. The emission normally has an outgoing or upward gas flow. Odourants in the atmosphere or gas stream can be collected by passing known volume of air or gas through a column of activated carbon or by condensing techniques.

Odour Measurement

The olfactometric methods of odour measurement falls into two categories.

Determination of the threshold concentration of odouriferous gases

For determining threshold concentration by the olfactometery testing procedure, a diluted odourous mixture and an odour free gas (as a reference) are presented separately from two sniffing ports at 20 lpm to a group of eight panellists in succession. In comparing the gases emitted from each port, the panellists are asked to report the presence of odour together with confidence level such as guessing, inkling or certainty. The gas dilution ratio is then decreased by a factor of two. The panellists are asked to repeat their judgment. This continues for six different dilution levels, resulting a total of $8 \times 6 \times 2 = 96$ judgments (sniffing) from eight panellists. Using panellist responses over a range of dilution settings, odour concentration expressed as odour unit per cubic meter (ou/m^3) can be calculated from individual threshold estimates. European threshold concentration ranges for some unpleasant odours are presented in Table 4.2.

Table 4.2. European threshold concentration ranges.

Compound	Detection threshold (mg/m^3)	Compound	Detection threshold (mg/m^3)
Acetic acid	25–10,000	Indole	0.6
Propanic acid	3–890	3-Methyl indole	0.4–0.8
Butanoic	4–3000	Methanethiol	0.5
3-Methyl butanoic acid	5	Dimethyl sulphide	2–30
Pentanoic acid	0.8–70	Dimethyl disulphide	3–14
Phenol	22–4000	Dimethyl trisulphide	7.3
4-Methly phenol	0.22–35	Hydrogen sulphide	0.1–180

Determination of the type and intensity of odour

For odour intensity measurement, generally a panel of 6 to 12 persons of normal health are employed. The panel members sniff the air at a given location at the same time and report individually the nature and intensity of the odour. By averaging the values recorded by members of the panel, a single value can be assigned to the odour intensity at a given location.

Generally odour intensity increases with the odourant concentration. The relationship between intensity and concentration can be expressed as:

$$P = K \log S$$

where, P = Odour intensity.
    K = Constant.
    S = Odour concentration.

Currently, the preferred and internationally standardised methods of measuring odour are the Dutch Standard Method (NVN 2820) and the more recent European Standard Method (CEN TC 264). A joint Australia New Zealand standard based on the draft CEN standard is in the process of preparation.

Critical Factors in Measurement of Odour Concentration

A dynamic olfactometer is a gas diluting apparatus and an interface between a panel of human observers and an odourous gas sample diluted at various concentrations. Olfactometery requires very high standards of testing conditions. These include following requirements:

1. Odour free testing: An odour free testing environment is an important element in the olfactometery testing process.

2. Olfactometer calibration: The olfactometer must be calibrated against a tracer gas to check the dilution setting of the olfactometer.
3. Panellist management: Panellist should be trained and screened using reference air incorporating certified butanol at a concentration of 60 ppm.

Limitations in Odour Measurement

Odour concentration is only one of the four dimensions that are used to express odour sensation as experience by humans. The odour concentration is determined in an odour free environment and do not reflects the actual perception of the odour. Common standardised instrument calibration and panel selection procedure are a prerequisite for comparison of odour concentration data reported in the literature. In the absence of standardised procedures, odour concentration levels reported might simply reflect the experience of the operator, the design of the olfactometer, its operational mode, its mixing method, the flow rate presented to panellist and the number of panellist employed.

ODOUR IMPACT ASSESSMENT

Wind movement in the atmosphere carries away odourous gases emitted from a source. Odour annoyance occurs when a person exposed to an odour perceives the odour as unwanted. Using an air dispersion models, it is possible to predict the downwind odour concentration on the basis of odour emission rates, topography and meteorological data. The results can be used against odour impact criteria to derive an odour impact area. Odour impact assessment is an effective tool for the following purposes:

1. Preparation of environmental management plans.
2. Development of appropriate regional and local planning and development control instruments.
3. Odour regulation.

In essence, odour impact assessment uses inputs of source odour concentration, ventilation rate and odour emission rate, topography information together with meteorological data (one-year data), dispersion model to model odour dispersion about the source. Odour impact area is defined by plotting isopleths of odour concentration corresponding to same value for odour impact criteria. The approach to odour impact assessment is illustrated in Fig. 4.2. Within an odour impact area, typical receptors (e.g. residents) may be expected to experience a certain degree of odour nuisance. Odour impact criteria are not ambient air quality standards but rather provide a scientifically derived benchmark for making of informed decisions in planning, design, environmental management and regulation.

ODOUR CONTROL TECHNOLOGIES

Odour control depends on type of sources and are discussed below.

Odour Control from Area Sources

For large area sources following methods can be used to reduce odour complaints.

1. Excluding development close to the site: Development close to the site is to be excluded. A reasonable 'buffer zone' around the area sources has to be determined. The actual size of this zone will depend upon a number of factors, including the size of the area from which odours emanate, the intensity of the odours being emitted, the duration and frequency of the odour emissions, the actual process being undertaken, the topography of the site, the weather conditions that prevails at the site. Green belt development in the buffer zone may help at least partially to obfuscate the odour.

2. Ensuring that the operation is carried out under the best management practice: Best management practices (BMP) will vary according to the industry producing the odour. However, for all new developments, BMPs will start with the site selection and the building of the facilities.

3. Nozzles, sprayers and atomisers that spray ultra-fine particles of water or chemicals can be used along the boundary lines of area sources to suppress odours. Rotary atomiser is one such technique widely recommended for adoption for effective control of odour in case of area sources. The atomiser uses centrifugal action by a spinning inner mesh to force droplets on to an outer mesh which 'cuts' the water into atoms (Fig. 4.3). The rotary atomiser produces millions of microscopic droplets of water—up to 238 billion from a single litre droplets that are thinner than a human hair and a fine spray which covers up to 30 linear metres. This creates a fine mist, which is more effective with minimal use of water and electricity.

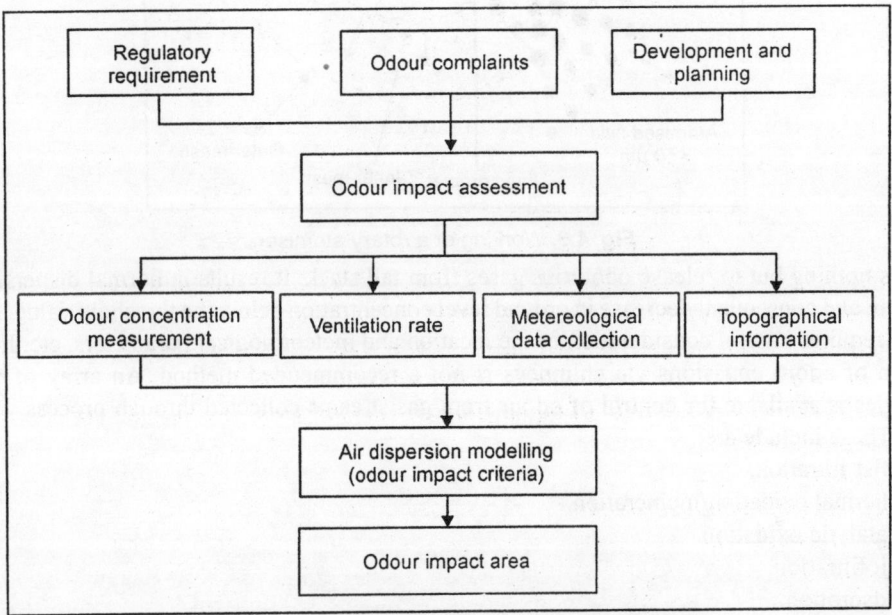

Fig. 4.2. Odour impact assessment flow chart.

There are a large number of chemicals and proprietary products that claim to reduce odour when they are applied to area sources. Atmospheric odours that are contained in a restricted area can be oxidised by atomisation of the chlorine dioxide. Odour from sources such as holding ponds, lagoons, and sewage pre- or post-treatment effluent can be controlled by atomised spray of chlorine dioxide. To reduce odour, chemicals have to be applied over very large area, the cost of materials and labours would be very high. The large quantity of these compounds required could themselves cause pollution. The spray/atomiser techniques are used to conceal odours also from building and fugitive sources.

Odour Control from Point Sources

In case of point sources such as that of industries, the odour-causing gas stream can be collected through piping and ventilation system and made available for treatment. Dispersion method is the simplest of the methods that can be adopted for odour abatement.

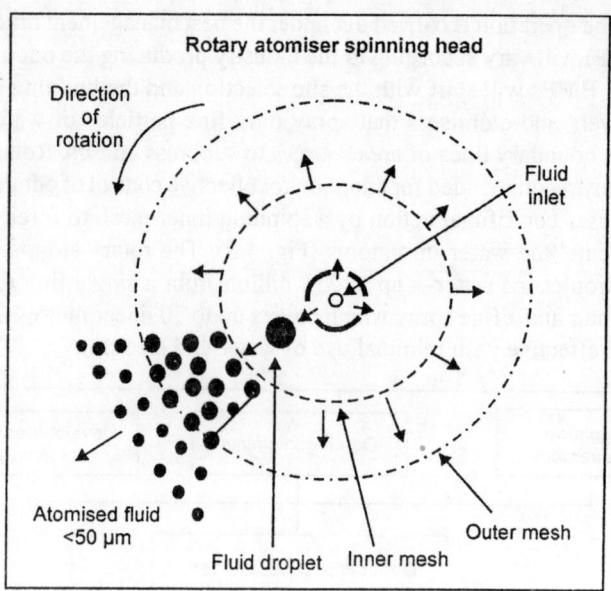

Fig. 4.3. Working of a rotary atomiser.

This is nothing but to release odourous gases from tall stack. It results in normal dispersion in the atmosphere and consequent decrease in ground level concentration below the threshold value. Dispersal by stacks requires careful consideration of the location and meteorological parameters, etc. In general, dispersion of odour emissions via chimneys is not a recommended method. An array of treatment technologies is available for control of odour from gas streams collected through process ventilation systems. These include are:

1. Mist filtration.
2. Thermal oxidation/incineration.
3. Catalytic oxidation.
4. Biofiltration.
5. Adsorption.
6. Wet scrubbing/absorption.
7. Chemical treatment.
8. Irradiation.

The choice of the technology is often influenced by the following factors:

1. The volume of gas (or vapour) being produced and its flow rate.
2. The chemical composition of the mixture causing the odour.
3. The temperature.
4. The water content of the stream.

Mist filtration

While gases cause most odour, problems may also result from aerosols in the fumes. Odourous air streams frequently contain high concentration of moisture. If these vapour discharge can be cooled to less than 40°C, a substantial quantity of the water vapour will be condensed and so reduce the volume of gases to be incinerated. Mist filters can be used for this purpose. Mist filters can also remove solids and liquids from gas stream; if the odour is caused by these particles, then it will result in odour reduction.

Thermal oxidation/incineration

Thermal oxidation/incineration is the oxidation of the odour into carbon dioxide and water by the combustion of the odour with fuel and air. The reaction takes place at temperatures ranging from 750° to 850°C. This is generally above the auto-ignition temperature of most solvents and other VOCs and is a reflection of the heat required to maintain the reaction at dilute concentrations with additional process heat losses. In this regime, the destruction efficiency is almost 100 per cent, assuming adequate oxygen supply. In some cases, other compounds may be formed depending on the mixture of fuel and air used, the flame temperature and the composition of the odour. These compounds may include carbon monoxide, oxide of nitrogen and sulphur oxides. Thermal oxidiser is a refractory-lined furnace fitted with one or more burners. The furnace consists of two chambers-mixing chamber and combustion chamber. There are three types of thermal oxidiser:

1. Direct-fired thermal oxidiser.
2. Recuperative thermal oxidiser.
3. Regenerative thermal oxidiser.

Direct-fired thermal oxidiser

Direct-fired thermal oxidiser is effectively a combustion chamber with a burner and the appropriate control system. The exhaust from a direct-fired unit is typically at the combustion temperature with no primary or secondary heat recovery. This is used where heat recovery is not required (e.g. when fuel for the burner is free or very cheap). In many cases the fuel cost of heating a process stream to the combustion temperature leads to the inclusion of some sort of heat recovery mechanism. Where the level of VOC is significant, then the heat release from the VOC can be recovered to improve the cost effectiveness of the system. Both recuperative and regenerative thermal oxidiser technologies include heat recovery systems to recover heat as a utility for other energy requirements.

Recuperative systems

Recuperative systems are basic thermal oxidisers with built-in primary shell and tube heat exchangers. A primary heat exchanger can recover up to 70 per cent of the heat input by the burner or released during the oxidation process by heating up the inlet stream thus reducing the required burner load to maintain the required oxidation temperature (typically 750°–800°C). These are a simple, cost effective, means of destroying VOC where the inlet concentration is relatively high or particularly where heat can be usefully recovered for other processes.

Regenerative thermal oxidiser (RTO)

Regenerative thermal oxidiser (RTO) is the most often used type of thermal oxidiser because of its robust performance and its ability to operate at high thermal efficiency. The RTO utilises beds of ceramic media to provide the thermal efficiency. Two or more beds are used in a controlled cycle and alternatively operate to heat incoming air and to cool exit air. The unit can operate at thermal efficiencies of between 80 and 98 per cent and can handle most types of fume. This means that where an exhaust stream contains a significant level of VOC, then auto-thermal burning (without the use of burners) is possible. At lower concentrations also, the RTO often provides cost-effective operation because of its very high thermal efficiency.

Catalytic oxidation

Catalytic oxidation reaction can be forced to proceed at much lower temperatures (e.g. 200°C) in the presence of a catalyst. Thus, the advantage of this process over thermal oxidation is the reduction in required energy input. Catalytic systems are therefore more favourable where auto-thermal operation is not practical and heat cannot be economically used elsewhere.

· A number of transition and precious metal catalysts can be used in catalytic oxidiser to destroy various VOCs over a wide range of process conditions.

Biofiltration

This method is becoming an acceptable and successful way of reducing odours from biological process. Biofiltration is a natural process that occurs in the soil that has been adopted for commercial use. Biofilters contain micro-organisms that break down VOC's and oxidise inorganic gases and vapours into non-malodourous compounds such as water and CO_2. The bacteria grow on inert supports, allowing intimate contact between the odourous gases and the bacteria. The process is self-sustaining. Biofilters constructed of various materials including compost, straw, wood chips, peat, soil, and other inexpensive biologically active materials.

Another type of biofilter is the soil-bed filter. Here the odourous gas stream is allowed to flow through a porous soil with a typical depth of 60 cm. The bacteria in the soil are responsible for the destruction of the odourous compounds.

Adsorption

A method that is suitable for controlling odourous substances, even at low concentrations, is adsorption on to activated carbon. For effectivity, the contaminated air stream must be free from dusts and particulates that might clog the carbon particles. Regeneration of carbon for reuse will produce either waste water, which will require further treatment before disposal, or a concentrated vapour stream, which can be incinerated more cheaply than the original air stream. There are also systems that use activated alumina impregnated with potassium permanganate for adsorption. The alumina absorbs the odourous substances so that the permanganate can oxidise them, usually to carbon dioxide, water, nitrogen and sulphur dioxide, depending on their composition. The alumina bed is replaced progressively as the permanganate is exhausted. This has an advantage over carbon because no further treatment is needed; this may offset the cost of alumina.

Wet scrubbing/absorption

Wet scrubbing of gases to remove odour involve either absorption in a suitable solvent or chemical treatment with a suitable reagent. It is important that hot, moist streams are cooled before they contact scrubbing solutions. If this is not done the scrubbing solution will be heated and become less efficient, the scrubbing medium will become diluted from condensation of water vapour.

Wet scrubbing or absorption systems can be either ventury systems or packed tower systems. Venturi systems are co-current scrubbers that accelerate the gas stream into a high density liquor spray. The aqueous droplets then impinge or impact at high relative velocity with solids in the gas stream. The resulting conglomerated particle is then separated from the gas stream in a disengagement tower by virtue of inertial forces. The high density spray also provides reasonable mass transfer to the absorption of gaseous contaminants. Packed Towers are typically counter current scrubbers that utilise high surface area media as a contact zone for the gas stream with suitable scrubbing liquor. The media facilitates high efficiency mass transfer to provide >99.9 per cent removal of gaseous contaminants. When the odour is caused by the presence of unsaturated organic compounds, it may be necessary to use an oxidising agent such as chlorine, diluted sulphuric acid and sodium hydroxide to treat odour.

Absorption is applicable when the odourous gases are soluble or emulsifiable in a liquid or react chemically in solution. Wet scrubbing is a useful process to handle acid gas streams, ammonia or streams with solids that might foul other equipment. It has been suggested that liquid scrubbing becomes economically attractive compared to incineration and adsorption on activated carbon when the volume of odourous gas to be treated is greater than 5000 cubic meters per hour.

Chemical treatment

Injecting controlled quantities of chemicals such as chlorine or ozone into process-gas stream can control odour. Similarly, unlike various other 'odour control' treatments, chlorine dioxide will destroy the odour at source. Chlorine dioxide is several times more effective than chlorine and other commonly used treatments, and will not form hazardous by-products, such as chlorinated organic, which can cause more problems than the original odour itself.

Odours arising from water bodies can generally be eliminated by adding the chlorine dioxide solution directly to the odouriferous fluid. The first action of chlorine dioxide is to rapidly oxidise the vapour gases dissolved in the fluid to their oxide form. As the dissolved gases are oxidised and the amount of chlorine dioxide will increase, next action of chlorine dioxide is the oxidation of small molecular material (micro-organisms), and, as the amount of chlorine dioxide will further increase, the larger molecules and compounds are oxidised. Due to this versatility, chlorine dioxide can be used in all aspects of the odour control process, from air scrubbers and waste-water treatment with stabilised chlorine dioxide solutions.

Irradiation

Ultraviolet irradiation can be used to control of odour. Here, the action is probably due to ozone formation or bactericidal effect.

BIOTECHNOLOGY FOR AIR POLLUTION ABATEMENT AND ODOUR CONTROL

Sulphur oxides, nitrogen oxides, carbon monoxides, hydrogen sulphides, hydrocarbons and particulate matter are the major components of air pollution and are responsible for health and environmental hazards. Equally important are the substances that cause unpleasant/offensive odour. A range of malodorous substances, including phenol, styrene, trichloroethane (TCE), volatile organic compounds (VOCs), amines, hydrogen sulphide, mercaptans and ammonia, are present in gaseous effluents of various industries, treatment plants, animal rendering activities, etc. Deodourisation technology makes use of physical, chemical and biological means for odour removal. Biodeodorisation, though applied since 1923, is relatively less studied and less applied. Its importance and application is, however, increasing.

The Problem

Increase in environmental awareness has resulted in an increasing attention of the people to pollution problems. Offensive odour from pollution is easily sensed and regulations obviously target this aspect, leading to the development of deodourisation technology. A number of industries produce offensive waste gases. These include pesticide, petrochemical, explosive, mining, meat processing, paints and

varnishes, textile, chemical, pharmaceutical, animal rendering, and fermentation industries. The sources of malodorous substances are many and they may originate:
1. During the production process.
2. From storage areas.
3. From pumps and compressors which leak.
4. During transfer of material.
5. From open waste-water treatment plants and garbage composting plants.

Abatement of these odours is difficult because:
1. A number of different compounds are often involved.
2. Odour producing compounds may be present in very low concentrations—some mercaptans, for example, have an odour threshold of below 1 ppb.
3. Sources are often complex, multi-point and difficult to accurately trace.
4. Odours that escape from transfer, filtration and drying operations are more difficult to contain.

Deodourisation Processes

Preventive as well as corrective methods are useful for control of odour. In the first category, process modification and equipment modification can be included, while processes of corrective nature are roughly classified into physical, chemical and biological methods. Dispersion, water washing, adsorption, thermal incineration and catalytic incineration are amongst the dominant physical methods, while chemical methods include catalytic oxidation. Physical and chemical methods, in general, are not flexible for volume, concentrations, or composition of gas changes that may occur. This can be overcome by biological methods. Biological processes earlier required skilled control and large space, but recent developments have overcome some of these restrictions. These processes are now characterised by low running costs (one-third of other processes), easy operation/maintenance and control, and low energy intensity. There are five types of biological waste gas purification systems in operation—bioscrubbers (Fig. 4.4), biofilters (Fig. 4.5), biobeds (Fig. 4.6), trickling filters (Fig. 4.7), biotrickling filters (Fig. 4.8), and the differences in them are shown in the following table:

	Aqueous phase	
Microbial flora	Mobile	Stationary
Dispersed	Bioscrubber	—
Immobilised	Biotrickling filters	Biofilters (biobeds)

Applications of biological processes depend on physical and microbiological phenomena. While the former include: mass transfer between gas and liquid phase; mass transfer to micro-organisms; and average residence time of mobile phase; microbiological phenomena are dependent on rate of degradation; substrate/product inhibition; diauxy, etc. There is a lot in literature on laboratory experiments and successful field applications. Although biological deodourisation is considered to be an effective tool, applications are relatively limited. At present, biological deodourisation systems are treating odours from other treatment units; however, it may be possible some day to seed reactors with specially-cultured micro-organisms so that odourous substances/gas will not be produced. Biologically active materials, like peat, compost, humus, woody heather, bushwood carrying micro-organisms, activated sludge of effluent treatment units or mixture of organisms, or a single organism immobilised as a biofilm on an inert material or in suspended form, are used in biological oxidation of gases. In the biological deodourisation process, target molecules are decomposed by micro-organisms, but the deodorisation mechanism is not clear in many cases.

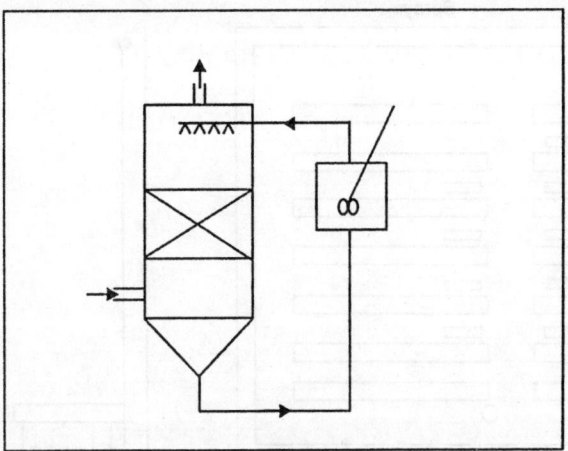

Fig. 4.4. Bioscrubber.

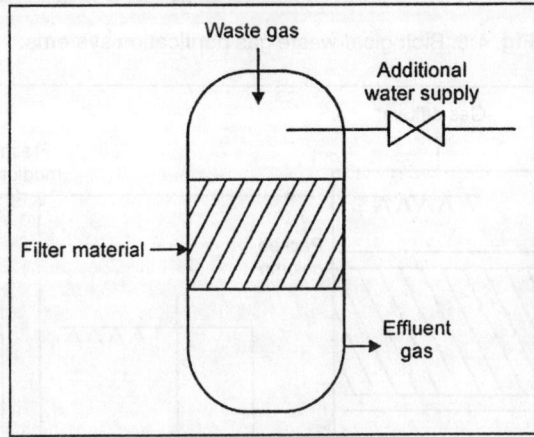

Fig. 4.5. Biofilters.

Bioscrubbers

A typical bioscrubber consists of an absorption column and one or more bioreactors. Biological oxidation takes place in these bioreactors. The reaction tanks are aerated and supplied with a nutrient solution. The microbial mass mainly remains in the circulating liquor, which passes through the absorption column. Circulation rate is fast and not much of the biofilm will develop in the absorption column. If any biofilm develops in the packing of the column, it has to be removed from time to time. Waste air to be treated is first brought to a temperature range (10°–43°C) suitable for micro-organisms. Dust in the air, if any, should be removed by line filters. Construction of the bioscrubber is such that air velocity will be 0.8 m/s, residence time in packing is 1.8 seconds, liquor circulation rate is 5–6 kg/(hr)·(m²) and residence time of liquor in reaction tank is 50 minutes.

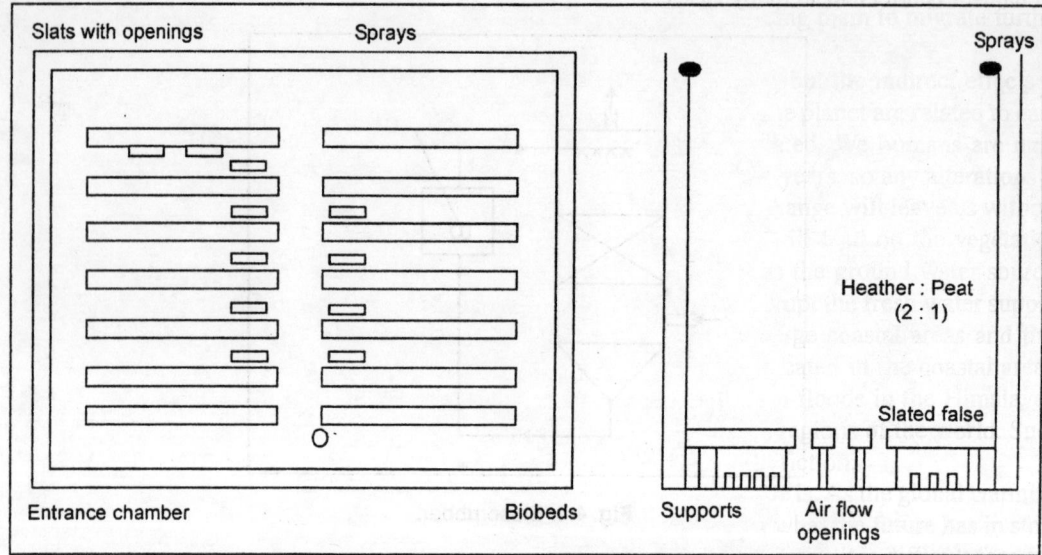

Fig. 4.6. Biological-waste gas purification systems.

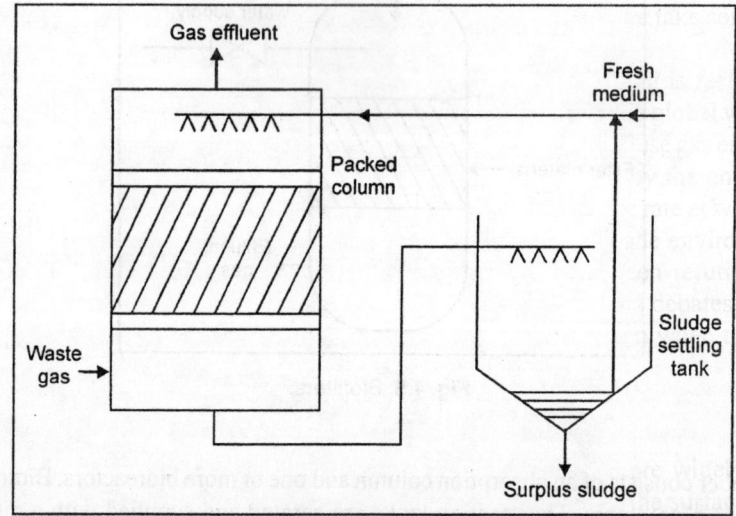

Fig. 4.7. Trickling filters (activated carbon columns).

Bioscrubbers require a lot of skilled attention and are reported to be successful in experimental works. They are used in the food industry, livestock farming and in foundries.

Bioscrubbers are more suitable for water-soluble hydrocarbons, where the concentration of biodegradable compound is <100–500 mg/m³ air.

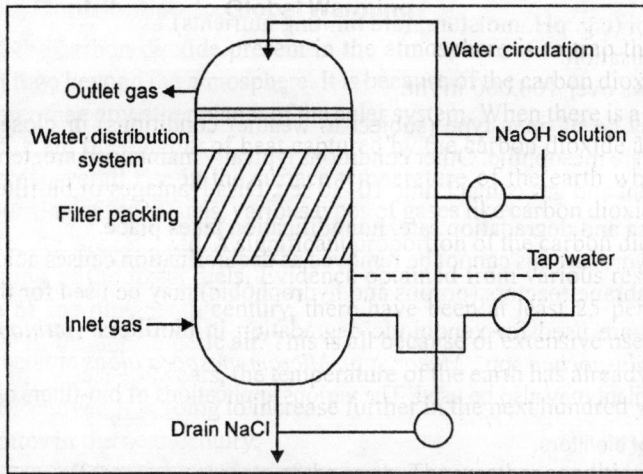

Fig. 4.8. Biotrickling filter.

The use of activated carbon in the absorber improves mass transfer, buffer capacity and immobilisation of micro-organisms. The venturi scrubber has 0.2 to 1 kg TSS biomass /m³ and gas flow 0.5–1 m/s and gives 90 per cent conversions. Emission of micro-organisms is considered to be a risk, especially by the food and pharmaceutical industries. Typically, 10^3–10^4 organisms per m³ are present in treated emission, but this is the same as that in normal air.

Biofilters (Biobeds)

Biofiltration is the most accepted technique and Germany and the Netherlands are leaders in this technique. Soil, compost, peat, heather, bark, etc. which contain micro-organisms with ability to oxidise VOCs and other odour causing compounds—are used in combination in biobeds. Moisture content and weed growth are the main problems of biobeds, but weedicides cannot be used.

Uniformity and permeability are important factors for proper gas treatment, and 'by-passing' or choking should not occur. Proper drainage is essential at the bottom. Beds require a lot of space (7.5 m³/hm³), and are turned two-three times in a year, but in Europe, designs requiring much less space have been developed, based on specially-prepared media with a high percentage of void space and very large microbiologically active material surface. Distribution of gas should be proper throughout the bed. The residence time depends on the substrate and is about 28–56s. Height of packing bed is 1m and flow rate of gases is 130 m³/hm³.

Peat and heather (in a 1:2 ratio) are commonly used for biobeds in Britain and Ireland, while municipal waste compost is popular in Germany. Bark may be used instead of heather, but it disintegrates quickly and requires repeated replacements. Acidic pH is maintained (pH 3) in the case of peat for better functioning. After an initial acclimatisation period of three months, most VOC components are reduced by 99 per cent with residence time of 51s. Acetone, 2-butanol, n-butanol, butyl acetate, butyl benzene, ethyl benzene, n-heptane, MEK, 2-propanol, styrene, TCE, toluene, xylene show 99 per cent reduction, while n-octane and n-pentane show 70 per cent and 20 per cent reduction respectively. A medium-size biofilter typically treats 40,000 m³ of air per hour. Future developments in biofilters are:

1. The use of specific micro-organisms.
2. Reduction in cost.

3. Process control (e.g. pH, moisture, rate limiting nutrients).
4. More standardisation.
5. Use of air flows over 100,000 m³/hr.

Biofilters used may be of open-type (subject to weather conditions) or closed-type (costly), and micro-organisms used are mesophilic. Other conditions typically maintained are: temperature 15°–40°C, moisture 40–60 per cent and gas contact time 10–30 sec. Disadvantages of biofilters are:

1. At high loading and degradation rate, humidification takes place.
2. Chlorinated hydrocarbons cannot be removed as dechlorination causes acidification of packing material. Membrane reactors (porous and hydrophobic) may be used for this purpose.

Various organisms are used for xenobiotic degradation in biofilters. *Actinomyces globisporous, Penicilium* spp. *Cephalosporium* spp., *Mucor* spp., *Micromonospora albus* are useful. Organisms from waste-water treatment plant may also be used. The various applications of bio-filters are given in Table 4.3.

Table 4.3. Application of biofilters.

Industry	Application	% Efficiency
Gelatine production	Elimination of odour	70–93
Cocoa and Chocolate processing	Elimination of odour	99
Fishmeal factory	Elimination of odour	50–90
Tobacco industry	Elimination of nicotine	95
Waste-water treatment	Elimination of odour, H_2S, acetone	90–95
Flavour and fragrance	Removal of odour, H_2S, acetone	98
Paint production	Organic solvent	75
Pharmaceuticals	Aromatic, aliphatic, chlorinated compounds	80
Photofilm production	Organic solvent	75
Food processing	Elimination of oil, odour	93
Ceramics processing	Ethanol	98
Metal foundry	Benzene	80

Biotrickling Filters

These have limited applications. Degradation of halogenated hydrocarbons, NH_3, H_2S, etc. encounters acid production, which has to be neutralised, as it can have an inhibitory effect on the microbiological process. Trickling filters can be used to solve this problem.

$$CH_2Cl_2 + O_2 \xrightarrow{\text{\textit{Hypomicrobium} spp.}} CO_2 + 2HCl$$

Applications

The use of biological deodourisation of air has been increasing, with improved knowledge of process conditions, improvements in biofilter characteristics, composition of filters, etc. Envirogen Inc. has developed a biocatalytic route for degradation of trichloroethylene (TCE).

Here, instead of naturally occurring microbes, a pure culture of *Pseudomonas* strain is used. In the first field trial carried out in New York, 90 per cent of TCE in contaminated air from air stripper treating groundwater was successfully degraded.

The culture is kept alive on a subsistence diet of phenol and toluene, and careful control of temperature, pH and exclusions of microbial predators is required. The company also has a process where genetically engineered *E. coli* is used as an efficient TCE degrader. *E. coli* can be fed on glucose and no competitive substrate (as phenol and toluene) is required.

EG and G. Rotron (New York), EG and G. Idaho and US Department of Energy's Idaho National Engineering Laboratory have developed a process of aerobic biofiltration. The method is called 'Biocube' and employs naturally occurring micro-organisms, mostly *actinomycetes* and *Pseudomonas*, to remove more than 90 per cent of aliphatic and aromatic substances and their derivatives from gas streams. Biocube's filter beds are modular trays filled with soil compost mixture containing micro-organisms. Beds are kept moist and at proper temperature so that biofilm develops on the surface. The volatile organic compounds (VOCs) are degraded and end products are CO_2, water, biomass and inorganic salts. The process is cheaper than thermal and catalytic oxidation alternatives. Biocube can also handle malodorous gases such as H_2S.

An exhaust gas treatment system for H_2S and SO_2, based on bacteria, has been mooted in Japan. The Dowa Mining Company uses *Chilobacillus ferroxidans*, found in the natural environment of mines, to oxidise Fe^{+2} to Fe^{+3} for energy and give solid sulphur from H_2S. Bacteria are circulated in intimate contact with ferrosulphate and ferric sulphide. The system consists of H_2S reaction tank, bacteria tank, a sulphur recovery unit and a simple closed circuit. Compared to conventional process, which uses caustic soda for neutralisation, the biological system works at only one-third of the cost. Energy conservation, compact system, room temperature treatment and easy operational control are its other advantages. The biological system has potential applications in petroleum and chemical processing plants. A biofiltration unit can remove offensive odours produced at animal rendering plants using 25 per cent peat and 75 per cent heather. The odorous air is sprayed with water to lower its temperature and increase its humidity before pumping through the filter. Efficiencies of such plants in Germany are 95 per cent. There are over 70 biofilters in operation to treat sewage treatment plants' gaseous emission. Animal rendering units collect and process animal bodies, slaughterhouse offal, blood, etc. which produces odourous emissions. C_2–C_{11} straight chain alkanals, methyl propanal, 2,3-methylbutanal, C_2–C_6 straight chain organic acids, furans, sulphur compounds, thiophene, H_2S, NH_3, etc. are present. Degradation of aldehydes and ketones is better, while that of sulphur compounds, thiophene, H_2S, and NH_3 is poor. Biological elimination of ammonia gas in the exhaust air from livestock production has also been reported. It is a two-step process, converting NH_3 to O_2 and to NO_3. Packing material is gradually acidified due to the process and buffers keep pH constant.

Diffusion and reaction rate of biofilter decide the overall elimination capacity. Maximum elimination capacity at laboratory scale is 9 gram/m^3 packing material per hour, while at pilot plant scale, it is 2 gram/m^3 of packing material per hour. Rapid microbial deodourisation of agricultural and animal wastes has been reported, in which animal house faeces of pigs, cows, sludge, domestic garbage is mixed with seed culture (5:1 w/w) and is blended with rice hulls or rice straw or saw dust. It is then left in a wooden box. Temperature rises on its own to 70°C and deodourisation occurs. Sulphites, hydrogen sulphite, mercaptans, low molecular weight fatty acids, etc. are metabolised. Organisms active are *Streptomyces griseus*, *Streptomyces antibioticus*, *Thermoactinomyces* spp. There are many advantages of biological deodorisation of this kind:

1. Simple and rapid process.
2. No aeration, no heating.
3. Low cost.

4. Application at small or large-scale.
5. Deodourised wastes are recycled as seed.
6. Coliforms decrease.
7. No flies.
8. Drying in sun.
9. Deodourised wastes are useful as fertiliser and fodder.

An efficient and economical bioscrubber system to remove styrene and volatile organic compounds (VOC) from industrial waste gases has also been developed. The process uses water to strip styrene and VOCs from industrial waste gases in a packed column scrubber. The styrene-laden waste-water is pumped to a fermenter, where selected strains of naturally occurring bacteria decompose styrene to CO_2 and water. Clean wash water is then recycled to the scrubber. Since 1994, the first commercial unit has run continuously on 20,000 m^3/hr waste gas stream at a German automotive parts manufacturer, cutting the styrene concentration from 400 ppm to 5 ppm. The operating costs are reported to be only about 20 per cent of those of comparable biofilters, while the capital costs are at least 40 per cent less.

Envirogen Inc. is scaling up its biofiltration systems to permit handling of 2000–2,00,000 Scfm of airflow. Styrene is a hazardous air pollutant and manufacturers of polystyrene are required to achieve 90 per cent reduction in its release by the year 2002 AD. In Envirogen's system, naturally occurring microbes are immobilised on a porous filter substrate such as compost or peat. Concentrated vapour stream passes through the filter bed, pollutants from vapour phase are transferred to the immobilised biofilm and are oxidised to CO_2 and water. The biological route has 30–70 per cent lower operating costs than other physical/chemical methods.

TNO was one of the first organisations to appreciate the true potential of biological treatment. They have developed a special low-cost biofilter containing compost and wood bark, which treats VOCs. They have also developed a fast-acting biofiltration system, which removes toluene, xylene, propene and styrene. Fungi well dispersed on ceramic carrier are used in this system. TNO has also developed a two-stage system in which a photoreactor with UV radiation improves the biodegradability of off-gases containing hydro-phobic pollutants, which are then biotreated. Styrenes, NO_x and alkanes are being treated this way. The last compound is tried by improved bioscrubber.

Biotechnology can clean up sulphur from gas streams and produce elemental sulphur. A flue gas desulphurisation process called Biostar has been developed by Paques BV (Netherlands) and Hoogovens Technical Services Energy and Environment BV. First sulphur dioxide is absorbed and converted to sulphite by reaction with sodium hydroxide, then sulphate reducing bacteria converts it to H_2S, which in turn, is converted to elemental sulphur by *Thiobacilli*. Another Paque bioprocess has been installed in Dutch paper mills and is producing 0.2 mt/d of sulphur from a gas stream, reducing the H_2S content from 12,000 ppm to 40 ppm.

NKK Japan uses *Thiobacillus ferroxidans* bacteria in its Bio-SR process. A ferric sulphate solution absorbs H_2S from a gas stream, producing elemental sulphur and ferrous sulphate solution. After sulphur is filtered, the solution is regenerated to ferric form by the *T. ferroxidans*. Sulphate reducing bacteria (anaerobic) may also offer a way to deal with the mountains of gypsum accumulated from wet scrubbing of SO_2 from stack gases. In a process developed by Idaho National Engineering Laboratory, bacteria and slurried gypsum are mixed in a stirred tank along with a nutrient of starch from potato wastes. The bacteria produce H_2S, which can be converted to elemental sulphur.

Idaho National Engineering Laboratory is also developing a way to treat stack gas directly by sparging it into a tank containing water, bacteria and starch. The SO_2 dissolves to form sulphurous acid, from which the bacteria produce H_2S. The process is yet to be scaled up. At present, the technology to remove sulphur from process streams and effluents are fast developing. Recovered sulphur may one day exceed the total demand and there will be no native sulphur production from mining. INEL is reported to be experimenting with a bioreactor for removing nitrogen oxides from flue gas. Tests using a gas stream containing 250 ppm. nitric oxide showed that bacteria can remove up to 99 per cent of the NO, leaving a residual concentration of only 2.5 ppm. Flue gas from coal typically contains 100–400 ppm. of NO. The flue gas is passed through a column of 100 mm. diameter and 1 metre long. Compost inside the column immobilises the *Pseudomonas denitrificans* bacteria and also serves as source of nutrients. A sugar solution dropped over the bed every few days provides a food supplement. With flow rates of 1–2 l/min., the residence time is of the order of 1 min. Researchers attribute the impressive performance by bacteria to the fact that in the gas phase, the mass transfer is better than that in a similar liquid system. The bacteria grow best at 30°–45°C, so the system has to be in the coolest part of the flue gas duct.

Odours from food processing and sewage treatment plants that are caused by mercaptans, alkyl sulphides and hydrogen sulphide are removed by sulphur eating bacteria in a technique developed by Obayashi Corp. (Tokyo) and Hitachi Zosen Corp. (Osaka). The equipment costs almost the same as conventional activated carbon systems, but operating costs are lower, because regeneration is not needed. The bacteria grow on 2–30 mm diameter ceramic pellets that are packed in a tower. Exhaust gas passes through the tower from bottom in about 20 seconds, which is sufficient for complete odour removal. Water is sprinkled from top periodically to keep the bacteria alive. Until now, there is no single bacterial treatment method for all three types of compounds. Odours from mercaptans and alkyl sulphides have been treated by bacteria that work at neutral pH and those from hydrogen sulphide by acid loving bacteria. The companies solved this problem by means of bacteria that work in neutral environment. The main factor in maintaining that environment is the development of ceramic pellets made by mixing process sludge with an undisclosed component. Phenol derivatives are common constituents of gaseous effluents of resin production, petrochemicals, pharmaceuticals, pesticide, explosive, textile, colour and coffee industry. Phenol is toxic and malodourous, and its removal from waste gases with biological filters by *Pseudomonas putida, Candida tropicalis, Fusarium flocciferium, Trichosoporan cutaneum* and heterogenous population is reported.

The tobacco industry emits odourous air during manufacturing. The biofilter system of Clean Air TechniQ Pty Ltd., Australia, consists of one or more filter housings where the compounds in question are absorbed and oxidised by the selected bacterial species. The filter volume is calculated based on previous experience and using a computer aided model. The polluted air is delivered to the biofiltration system by a centrifugal fan. The air is then passed through an air washer where the particulates are removed and the gas is conditioned to the correct temperature and humidity (20°–40°C, 100 per cent RH). After leaving the air washer, it travels to the inlet plenum where it is distributed prior to moving through the biological filter. After passing through the filter, the air is discharged into the atmosphere via a stack. Table 4.4 highlights various air pollutants degradation for various source industries.

The biofilter is 300 m^3 in volume, and the housing is normally 18 m long by 16 m wide and 5 m high. The biological bed is located on a grid approximately 1 metre from the ground and covers the entire housing to a depth of 1 metre. The entire bed is moistened by PLC operated spray system located above the bed. The spray system is controlled by the moisture content of the bed. Optimal performance is obtained between 50–60 per cent moisture (w/w), 90 per cent odour removal and 98 per cent particulate matter removal is reported. Air volume treated is 1,00,000 m^3/hr and inlet air odour concentration is 5000–8000 OU/m^3.

Table 4.4. Air pollutants degradation for various source industries.

Industry	Pollutants	Systems/Micro-organisms	Technology
Food processing, Sewage treatment	Mercaptans, alkyl sulphides and H_2S		Obayashi Corp. and Hitachi Zosen Corp.
Coal industry	NO_x from flue gas	*Pseudomonas denitrificans*	Idaho National Engineering Laboratory
	SO_2 from flue gas	Sulphate reducing bacteria, *Thiobacillus ferroxidans*	Paques BV and Hoogovens Technical Services Energy and Environment BV
	Trichloroethylene	Pseudomonas or genetically engineered *E. coli*	Envirogen. Inc.
Automotive parts	Styrene and VOC's	Naturally occurring bacteria	
Polystyrene plant	Styrene	Naturally occurring bacteria	Envirogen, Inc.
Mining, chemical and petrochemical	H_2S, SO_2	Chilobacillus ferroxidans	Dowa Mining Company
Agricultural and animal wastes	Sulphites, H_2S, mercaptans, low mol. weight fatty acids, etc.	*Streptomyces griseus, Streptomyces antibioticus, Thermoactinomyces* spp.	
Livestock production	NH_3		
Animal rendering	Furans, sulphur compounds, H_2S, thiophene, NH_3, etc.	Biofilters	Germany
Resin, pesticides, petrochemical, and textile	Phenol derivatives	*Candida tropicalis, pseudomonas putida, fusarium flocci-ferium, trichosporon cutaneum*	

Thus, physical and chemical treatments for wastes result only in transformation of pollutants from one form to the other. Biotreatments are relatively simple, specific, work at ambient temperatures/pressure, and are less costly. As seen from the above examples biotechnology can offer an effective solution to deodourisation. Biotreatment of waste gases is still considered to be an experimental technology, but its use is on the ascent.

ODOUR POLLUTION IN THE ENVIRONMENT AND THE DETECTION INSTRUMENTATION

Odour or malodour, which refers to unpleasant smells, is nowadays considered an important environmental pollution issue. Odour pollution abatement has involved a number of bodies. A comprehensive description of pollution abatement and the development of the accompanying instrumentation technology are therefore critical links to understand the whole dimension of odour pollution in the environment. In this section, odour pollution in the environment will be reviewed, including its sources and dispersion, the physical and chemical properties of odour, odour emission regulations in selected countries, odour control technologies as well as the state-of-the-art instrumentation and technology that are necessary to monitor odour, e.g. chemical sensors, olfactometry, gas chromatography, and electronic noses.

Odour, which refers to unpleasant smells, is considered as an important environmental pollution issue. Attention to odour as an environmental nuisance has been growing as a result of increasing industrialisation and the awareness of people's need for a clean environment. As a consequence, efforts to abate odour problems are necessary in order to maintain the quality of the environment. In this framework, understanding the odour problem and the origin and dispersion of odours, abatement and detection methods are, therefore, very important aspects of odour pollution in the environment.

One of the challenges when dealing with the odour pollution problem is the technique for the detection of odour emissions. Detection is an important aspect concerning compliance with the environmental regulations, since the detection results will be used as proof of the release of odourous substances to the environment. A successful and excellent detection technique will result in a sequence of accountably data. A reliable instrument, therefore, is necessary.

There is a growing tendency in industry to develop a detection system that enables real-time measurements. In this way, a simple and quick on-line-monitoring system can be established and time-consuming methods avoided. Sampling and conventional analytical procedures are then no longer necessary, since the detection and measurement of the odourous compounds can be carried out quickly and the results presented on demand.

The state-of-the-art method for detecting odour emissions is the classical olfactometry. By this method, odour assessment is based on the sensory panel of a group of selected people (panellists) with 95 per cent probability of average odour sensitive. The method does not exclude that, physiological differences in the smelling abilities of the panel members can lead to subjective results. The olfactometry method is also very costly and requires an exact undertaking in an experienced odour laboratory in order to achieve a reliable result. Moreover, for a continuous monitoring of time-dependent processes, a system based on the human sensory system is not feasible.

A number of researches on the development of odour detection systems are currently being carried out to improve the present systems. The development of new, appropriate systems that are based on devices rather than on the human sensory system are important for increasing the acceptance by stakeholders and avoiding subjectivity in odour measurements. In this section two points will be covered and are devoted to describe the relationship between odour pollution and the detection instrumentation:

1. Survey of the biogenic odour emissions in the environment and their abatement methods.
2. Overview of the current development in odour detection instrumentation

Odour Pollution in the Environment

Sources and dispersion of odours

This description is presented here to point out the relationship between any activity (industrial, agricultural, household, etc.) that can be a source of odours and their odour release. Such a relationship is important and critical in the framework of odour abatement in order to understand any activity that results in odourous gases and the kinds of odour compounds that might be produced. Table 4.5 shows the sources of odour in the environment and the released odour compounds. Table 4.6 lists some major odour compounds and their smell characteristics.

Odour substances emitted from any source will be regarded important in the context of odour pollution if they are dispersed in the surrounding area. This means that odour molecules are distributed from the odour sources into the environment. Without any dispersion process odour production will not result in complaints by the people in the surrounding area.

For that reason, many researchers have studied odour dispersion in the atmosphere, using not only a model but also direct measurements. Successful examples concerning odour emissions, dispersion and dispersion modelling are cited in the following.

Table 4.5. Sources of odour in the environment.

Source	Odourous compounds or group
Chemical and petroleum industries:	
Refineries	Hydrogen sulphide, sulphur dioxide, ammonia, organic acids, hydrocarbons, mercaptans, aldehydes
Inorganic chemicals (fertilisers, phosphates production, soda ash, lime, sulphuric acids, etc.)	Ammonia, aldehydes, hydrogen sulphide, sulphur dioxide
Organic chemicals (paint industry, plastics, rubber, soap, detergents, textiles)	Ammonia, aldehydes, sulphur dioxide, mercaptans, organic acid
Pharmaceutical industry	Aldehydes, aromatic, phenol, ammonia, etc.
Rubber, plastics, glass industries	Nitro compounds (amines, oxides), sulphur oxides, solvents, aldehydes, ketones, phenol, alcohols, etc.
Composting facilities	Ammonia, sulphur containing compounds, terpene, alcohols, aldehydes, ester, ketones, volatile fatty acids (VFA)
Animal feedlots	Ammonia, hydrogen sulphides, alcohol, aldehydes, N_2O
Waste-water treatment plant	Hydrogen sulphides, mercaptan, ammonia, amines, skatoles, indoles, etc.

Table 4.6. Major odour compounds and their senses.

Compound	Formula	Odour sense
Acetaldehyde	CH_3CHO	Pungent
Ammonia	NH_3	Pungent
Butyric acid	$CH_3CH_2CH_2COOH$	Rancid
Diethyl sulphide	$C_2H_5C_2H_5S$	Garlic
Dimethyl amine	CH_3CH_3NH	Fishy
Dimethyl sulphide	CH_3CH_3S	Decayed cabbage
Ethyl mercaptan	C_2H_5SH	Decayed cabbage
Formaldehyde	$HCHO$	Pungent
Hydrogen sulphide	H_2S	Rotten eggs
Methyl mercaptan	CH_3SH	Decayed cabbage
Phenol	C_6H_5OH	Empyreumatic
Propyl mercaptan	C_3H_7SH	Unpleasant
Sulphur dioxide	SO_2	Pungent
Trimethyl amine	$CH_3CH_3CH_3N$	Fishy
Valeric acid	$CH_3CH_2CH_2CH_2COOH$	Body odour

Kuroda and others evaluated the emissions of malodourous compounds (volatile fatty acids, ammonia, and sulphur containing compounds), greenhouse gases (methane [CH_4], and nitrous oxide [N_2O]) from a facility for composting swine feces. They showed a basic emission pattern of malodourous compounds and two greenhouse gases during composting of solid waste. Valsaraj elaborated odour emission modelling and its relationship to meteorology, topography and dispersion; concentration of odour (μg) per cubic meter at any time within the atmosphere; and the odour emission rate at a stack and point sources. Corsi and Olson derived models that are used for estimating volatile organic compound (VOC) emissions from waste-water. They provide a general overview of emissions estimation methods and available computer models. Frechen and Köster proposed a measurement method called 'odour emission capacity (OEC)' to describe a parameter influencing amount and variation of the odour emission mass flow, i.e. amount of odourants present in the liquid. They concluded that the determination of the OEC is a new and very valuable tool when assessing the relevance of different liquids with regard to possible odour emissions. It was also possible to determine the emission capacity of specific compounds of the liquid phase such as hydrogen sulphide or others. McIntyre emphasised that correctly and intelligently applied atmospheric dispersion models are a valuable part of the technical toolkit for tackling odour problems. It was also pointed out that modelling is a good and useful tool for selecting and quantifying the beneficial effects of odour control programs for waste-water treatment facilities. Wallenfang developed a gas dispersion model and verified it experimentally. The numerical model can be used to predict the dispersion pattern of odour molecules in the environment as well as to demonstrate the distribution of odour molecules through a diffused obstacles.

Characteristics of odour molecules

The odours that we identify in the space around us are the result of the interaction between molecules given off by the odourous material and the sensory cells located in our nose. When we sniff a rose, for example, we draw up into our nose volatile molecules that interact with the sensory cells and our interpretation of the nerve impulses generated by this interaction is positive. In the same way, however, an unpleasant odour, e.g. bad egg, is sensed because of the interaction between the odourous molecules of butyl mercaptan present in the nose cavity and the sensory cells.

Odour dimensions

There are four odour dimensions, i.e. detectability, intensity, quality, and hedonic tone:

1. Detectability (or odour threshold) refers to the minimum concentration of odourant stimulus necessary for detection in some specified percentage of the test population. The odour threshold is determined by diluting the odour to the point where 50 per cent of the test population or panel can no longer detect the odour.

2. Intensity is the second dimension of the sensory perception of odourants and refers to the perceived strength or magnitude of the odour sensation. Intensity increases as a function of concentration. The relationship of the perceived intensity and odour concentration is expressed by Stevens as a psychophysical power function as follows:

$$S = k I^n$$

   where,

   S = perceived intensity of odour sensation (empirically determined)
   I = physical intensity (odour concentration)
   k = constant
   n = Stevens exponent

3. Odour quality is the third dimension of odour. It is expressed in descriptors, i.e. words that describe the smell of a substance. This is a qualitative attribute that is expressed in words, such as fruity. A list of smells is already provided in Table 4.6.

4. Hedonic tone is a category judgement of the relative like (pleasantness) or dislike (unpleasantness) of the odour. It can range from 'very pleasant' (high score, positive) to 'unpleasant' (low score, negative).

Understanding odour characteristics

Understanding the odour characteristics is related to the odour pollution control technology. Physical and chemical characteristics of odour molecules should be well understood before a control technique is chosen. Card described an example of a choice between a physical and a chemical separation method for odour control. The method can be physical if the compounds are in different phases or have different particle sizes. If the compounds are dissolved in either gases or liquids, then the separation must be chemically based. The difference in the chemical characteristics of the target compounds to those of the compounds in solution determines the available methods to effect this separation.

The following are examples of the relationship between the odour characteristics and their significance for pollution control:

1. Vapour pressure: Vapour pressure is the gas phase concentration that is in equilibrium with a pure liquid phase at a particular temperature. Knowledge of the volatility of a compound greatly affects the options for odour and VOCs control. As an example, hexane is highly volatile, and adsorption is ineffective since hexane volatilises from the adsorbent. In such cases, thermal oxidation may be the control technology of last resort.

2. Solubility in water: Water solubility is defined as the concentration in the aqueous phase that is in equilibrium with the pure component phase. The ability of a compound to dissolve in water is the critical factor in determining whether the compound is suitable for control by liquid scrubbing. Solubility of any odour compound or odour mixtures in water must also be taken into account, since the sampling technique in the field involves a cooling step where a part of odour compounds will be dissolved in the condensate water and be drawn from the sample.

3. Ionisation: If an odour compound ionises in solution, the performance and economics of liquid scrubbing systems can generally be enhanced. For example, the removal of ammonia and hydrogen sulphide in a gas stream is very dependent on the fact that these gases will ionise in solution. The addition of either acid (for ammonia removal) or caustics (for hydrogen sulphide removal) greatly increases the ability of liquid scrubbers to remove these compounds.

Molecular mass, volatility and functional groups

Typically, odourants have relative molecular masses between 30 and 300 g/mole. Molecules heavier than this have, in general, a vapour pressure at room temperature too low to be active odourants. The volatility of molecules is not, however, solely determined by their molecular weight. The strength of the interactions between the molecules also plays an important role, with non-polar molecules being more volatile than polar ones. A consequence of this is that most odourous molecules tend to have one or at most two polar functional groups. Molecules with more functional groups are in general too involatile to be active odourants.

Observations on two composting facilities in Bonn and Stuttgart, Germany, during field measurements showed that the results are also in accordance. The odour compounds released from a composting facility located near Stuttgart consisted of compounds whose molecular weights are in between 17 g/mole (ammonia) and 152 g/mole (thujone). Another composting facility near Bonn also showed that the molecular masses of odourous compounds are in between 46 g/mole (ethanol) and 136 g/mole (limonene).

Odour as an environmental buisance

A list of unpleasant odour compounds that are seen as environmental nuisances is already presented in Table 4.6. However, agreement on whether an odour is pleasant or unpleasant is sometimes thought of as being very personal. Pleasantness or unpleasantness is a result of emotions in the individuals. The following indicates ideas of pleasantness and unpleasantness and the human response to odours:

1. Human reactions to odours are similar to our reactions to other sense stimuli: involuntary and spontaneous, either liking or disliking, and indifference.
2. Reasons for the above cannot be interpreted, i.e. usually the reasons, if there are any, show no trends or give no explanations.
3. Previous experience with an odour or with similar odours sometimes determines if an odour is liked or disliked.
4. According to bodily needs, food smells are pleasant or unpleasant.
5. Pleasant odours tend to feed those emotions that are affected by 'beautiful' things in the environment.

There is a general agreement on which odours are experienced as unpleasant, e.g. odours that are pungent (ammonia), rotten eggs, stinking (garbage wastes), and rancid odours. Odours that are sweet (flowers), fresh (outdoor odours), and appetising (food), are mostly experienced as pleasant odours. A provisional conclusion can be drawn stating that if an odour is regarded as an environmental nuisance, it means that the odour is an unpleasant one.

Individual sensitivity to the quality and intensity of an odourant can vary significantly, and this variability accounts for the difference in sensory and physical responses experienced by individuals who inhale the same amounts and types of compounds. This distinction between 'odour', which is a sensation, and 'odourant', which is a volatile chemical compound, is important for everyone dealing with the odour issue to recognise. When odourants are emitted into the air, individuals may or may not perceive an odour. When people perceive what they regard as unacceptable amounts or types of odour, odourous emissions can become an 'odour problem'. Simply, an odour problem results from an odour that is unpleasant.

Numerous regulations on control of odour in the environment are being passed in many countries, especially in industrialised countries, where the attention to and demand for clean air is an important aspect of the human environment. This results in odour emission regulations and air quality norms.

In Germany, for example, regulations concerning odour control are very strict due to a high population density and large number of waste treatment plants. Thus, it is almost impossible to find locations for treatment plants without annoying people with odour emissions. Many plants have already been built near residential areas and people complain about odour emissions. A number of statutes, regulations and guidelines concerning odour that in effect regulate air emissions from facilities in Germany, Canada and USA are listed in Table 4.7.

Odour Pollution Reduction Technologies

There are several methods to reduce odour coming from waste gases. However, there is no single treatment technology that can effectively and economically be applied to every industrial or commercial application. The effectiveness of a technology can often be defined by the flow rates and concentrations at which adequate cost-effective treatment can be expected. For all technologies, cost-effectiveness is site specific. Seasonal fluctuations can also be an important parameter for a typical odour controlling method, as reported by Gao who made a technical and economic comparison between biofiltration and wet chemical oxidation (scrubbing) for odour control at waste-water treatment plants. The following parts are overview of the methods currently available.

Table 4.7. Odour-related regulations in selected countries (USA, Germany, and Canada).

Country	Regulations	Remarks
USA	Clean Air Act (CAA)	Regulates stationary sources of volatile organic compounds (VOC)
	Resource Conservation and Recovery Act (RCRA)	Regulates emissions arising from transportation and storage of hazardous waste and disposal
	Toxic Substances Control Act (TSCA)	Limits the distribution, use or disposal of chemicals that can have adverse health and environmental effects
	Comprehensive Environmental Response, Compensation, and Liability Act of 1980 (CERCLA)	Requires states to establish a process for developing local emergency preparedness programs and to receive and disseminate information on hazardous chemicals present at facilities within local communities
	Occupational Safety and Health Act (OSHA)	Provides the basis for regulations protecting workers in the workplace
Germany	VDI 3881	Olfactometry
	GIRL (*Geruchsimmissions-Richtlinie*)	Odour pollutants guidelines
	VDI 3940 [VDI 1991]	Determination of odour in ambient air by field inspections
Canada	The Environmental Protection and Enhancement Act (EAPA) in Alberta Province	Prohibitions against the release of compounds that cause a 'significant adverse effect'
	Waste Management Act in British Columbia Province	Defines an air contaminant as a substance that 'interferes or is capable of interfering with the normal conduct of business'
	The Environment Act in Manitoba Province	Includes odour in its definition of pollutant, where it may 'interfere with or is likely to interfere with the comfort, well-being, livelihood or enjoyment of life by a person'

Biological systems

Biological treatment is effective and economical for low concentrations of contaminants in large quantities of air. On the other hand, chemical treatment requires aggressive additives, causing problems to the environment, whereas physical processes do not eliminate but transfer the pollutants to a new stream to be treated. Biological systems for odour control rely basically on the micro-organism activity that converts odour compounds in the waste air or waste-water to carbon dioxide and water as in a chemical system. Biological systems include biofilters, biological scrubbers (or bioscrubbers), and biological trickling filters (or biotrickling filters). They are often known as bioreactors. Successful biodegradation of odour using biofilters, biotrickling filters and bioscrubbers are listed in Table 4.8. The differences between these bioreactors and the advantages as well as disadvantages are presented in Tables 4.9 and 4.10.

Biofilters are the most widely used and accepted vapour-phase biological treatment systems, and have been systematically applied in various forms throughout many parts of the world for more than 30 years. In biological scrubbers and biological trickling filters, gas contaminants are absorbed in a free liquid phase prior to biodegradation by either suspended or immobilised microbes. In a biotrickling filter, microbes fixed to an inorganic packing material and suspended microbes in the water phase degrade the absorbed contaminants as they pass through the reactor. In bioscrubbers, after initial contaminant absorption, the degradation of the contaminants is performed by a suspended consortium of microbes in a separate vessel.

Table 4.8. Examples of successful odour biodegradation using biofilter, biotrickling filter and bioscrubber.

Abatement method	Biodegraded odour compounds	Process efficiency
Biofilter	BTEX (benzene, toluene, ethylbenzene, o-xylene)	≥90%
	Hydrogen sulphide (H_2S), ammonia (NH_3)	≥95%
	Trichloroethylene (C_2HCl_3)	30–60%
	Ammonia (NH_3)	≥95%
	Acrylonitrile (C_3H_3N)	≥95%
	Toluene (C_7H_8)	84%
		57–99%
Biotrickling filter	Toluene (C_7H_8)	94%
	Styrene (C_8H_8)	97–99%
	Diethyl ether ($C_4H_{10}O$)	72–99%
		95%
Bioscrubber	Hydrogen sulphide (H_2S)	99%
	n-Butanol ($C_4H_{10}O$)	84–100%
Hybrid bioreactor		
Biofilter and bubble column	Benzene (C_6H_6)	65–100%
Biofilter and bioscrubber	Ammonia (NH_3)	83%
	Butanol (C_4H_8O)	80%

Table 4.9. Difference between biofilter, biotrickling filter and bioscrubber in terms of micro-organisms and water phase.

Reactor	Micro-organisms	Water phase
Biofilter	Fixed	Stationary
Biotrickling filter	Fixed	Flowing
Bioscrubber	Suspended	Flowing

Chemical systems and hybrid systems

As regards chemical systems, several technologies are currently available. Some of them function through the addition of chemicals to liquid, thermal oxidation, and chemical scrubbing.

Addition of chemicals to liquids to control odour relies on the reaction of the odourous components with a chemical treatment reagent. The chemical treatment reagent alters the concentration of the odourous components in the aqueous phase and hence lowers the emission of the component. For example, a common odourous component in waste-water is hydrogen sulphide (H_2S). Chemical addition can alter the oxygen balance in the waste-water by: (i) oxidising sulphides, (ii) precipitating dissolved sulphides, and (iii) changing the ability of the sulphate- or organic-sulphides-reducing organisms to generate sulphides. Some examples of oxidants used are chlorine (Cl_2), sodium hypochlorite ($NaOCl$), or potassium permanganate ($KMnO_4$), and hydrogen peroxide (H_2O_2).

In thermal oxidation, a hydrocarbon odour compound is converted to carbon dioxide and water vapour in the presence of oxygen and heat at a temperature of 700° to 1400°C. With catalysts such as platinum, palladium, and rubidium, this process can be achieved at a temperature of 300° to 700°C.

A general equation showing this relationship is:

$$C_nH_{2m} + (n + m/2)\ O_2 \Rightarrow nCO_2 + mH_2O + heat$$

Table 4.10. Relative advantages and disadvantages of air phase bioreactors.

Biofilter	Biotrickling filters	Bioscrubbers
Advantages		
Simple operation	Simple operation	Good process control possible
Low investment costs	Low investment costs	High mass transfer
Low running costs	Low running costs	Suitable for highly contaminated waste air
Degradation of less water-soluble pollutant	Suitable for moderately contaminated waste air	Suitable for process modelling
		High operational stability
Suitable for reduction of odourous pollutants	Ability to control pH	Ability to add nutrients
	Ability to add nutrients	
Disadvantages		
Low waste-air volumetric flow rate	Limited process control	High investment cost
Only low pollutant concentration	Channelling can be a problem	High running cost
Process control impossible	Limit service life of filter bed	Production of excess biomass
Channelling of air flow is normal	Excess biomass not disposable	Disposal of water
Limited service life of filter bed		Possible plugging in adsorption stage
Excess biomass not disposable		

When applying chemical scrubbing, odour compounds are fed in a reaction chamber in which contact between odour compounds and a fog or droplet of chemical occurs. This odour control system removes odour by spraying very fine mist droplets of a controlled diluted chemical solution into an odourous stream that passes through a hollow, cylindrical reaction chamber. Cleaned air leaving the reaction chamber is discharged through the exhaust stack to the atmosphere.

A hybrid system is a combination of different systems. In many industrial applications, this is considered to be more cost-effective than a single standard control. Although hybrid systems can offer improved-cost effectiveness, they require a higher degree of preliminary engineering and understanding of each component of the hybrid system. Therefore, it is important to carefully select the cases in which hybrid control systems are employed. Yeom and Yoo showed a novel hybrid system to remove benzene by using a combination of biofilter and bubble column. It was shown that 65–100 per cent removal efficiency was reached, depending on the airflow rate and benzene concentration.

Odour Pollution Detection Instrumentation

Chemical sensors

In the field of sensor technology, the term 'chemical sensor' addresses a special group of sensors that are different to other sensors, i.e. thermal sensors, magnetic sensors, optical sensors, and mechanical sensors (Fig. 4.9). According to the definition, a chemical sensor is a device that responds to a particular analyte in a selective way through a chemical reaction, and which can be used for the qualitative or quantitative determination of the analyte. It can be seen that such a definition encompasses all sensors based on chemical reactions including biosensors, which make use of highly specific and sensitive biochemicals, and biological reactions for species recognition.

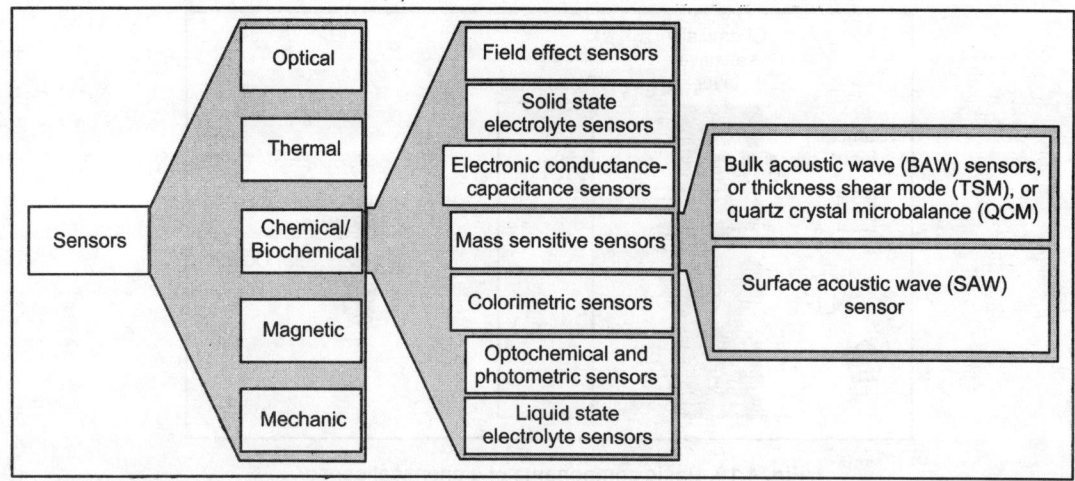

Fig. 4.9. Classification of sensors showing the sensor types, including chemical sensors, mass sensitive sensors and the quartz crystal microbalance (QCM) sensor.

Göpel and Schierbaum proposed another definition. Chemical or biochemical sensors are (miniaturised) devices that convert a chemical state into an electronic signal. A chemical state is determined by the different concentrations, partial pressures, or activities of particles such as atoms, molecules, ions, or biologically relevant compounds to be detected in the gas, liquid, or solid phase. The chemical state of the environment with its different compounds determines the complete analytical information. Cattrall classified the chemical sensors according to the transducer type into the following groups: electrochemical, optical, heat-sensitive, and mass-sensitive. Electrochemical sensors include potentiometric sensors and voltametric/amperometric sensors. Optical sensors, which are often referred to as 'optodes', rely on the association between spectroscopic measurements and the chemical reaction. Heat sensitive sensors are often known as calorimetric sensors in which the heat of a chemical reaction involving the analyte is monitored with a transducer such as a thermistor or a platinum thermometer. Flammable gas sensors make use of this principle.

Mass sensitive sensors make use of the piezoelectric effect and include devices such as the surface acoustic wave (SAW) sensor and are particularly useful as gas sensors. They rely on a change in mass on the surface of an oscillating crystal, which shifts the frequency of oscillation. The extent of the frequency shift is a measure of the amount of material adsorbed on the surface. The bulk acoustic wave sensor (BAW) also belongs to the group of mass sensitive sensors. BAW is also referred to as the quartz crystal microbalance (QCM) or thickness shear mode device (TSM). A more detailed explanation of the QCM is presented in the next sub-chapters. Göpel and Schierbaum classified chemical and biochemical sensors according to the different sensor characteristics used for particle detection. The most commonly used properties are potential (field effect sensors), voltages (solid-state electrolyte sensors), conductivity and capacity (electronic conductance and capacitance sensors), mass (mass sensitive sensors), heat (calorimetric sensors), or optical constant (optochemical and photometric sensors) and voltages (liquid state electrolyte sensors) (Fig. 4.9). The working principles of a chemical sensor are primarily based on the interaction between sample input (e.g. odour molecules) and the chemically sensitive materials on the sensor surface. This interaction results in a change of mass and it is then converted into an electronic signal by a transducer. Figure 4.10 shows the basic components of a chemical sensor.

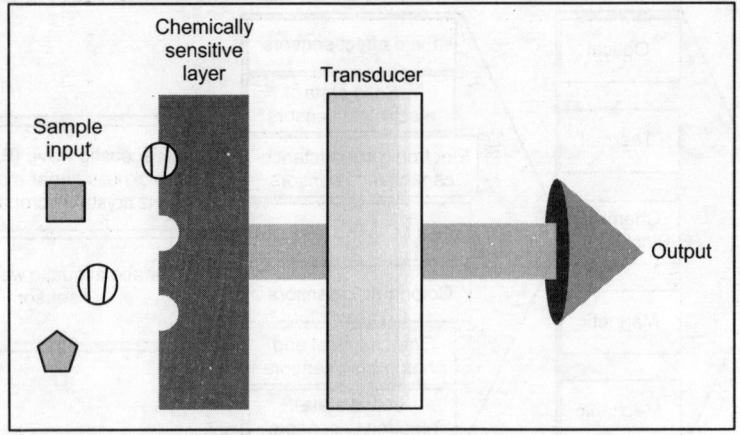

Fig. 4.10. Basic components of a chemical sensor.

The application fields of chemical sensors are very broad. Among these are:

1. Environmental control (air, water, soil).
2. Working area measurements (workplace, household, car, etc.).
3. Emission measurements (car, waste water, etc.).
4. Process control and regulation (biotechnological and chemical plants, fermentation process, etc.).
5. Medical applications (clinical diagnostics, anesthetics, veterinary).
6. Agricultural (analysis in agriculture and gardening, detection of pesticides, etc.).

In the context detection of odour and volatile organic compound (VOC) emissions, a brief list of widespread applications of chemical sensors developed during the past years is summarised in Table 4.11.

Table 4.11. Chemical sensor applications relevant to the odour and volatile organic compound (VOC) emissions detection.

Application fields	Detection objects	Sensors
Environmental control	Propane, propanol	Metal oxides sensor with multivariate analysis
	Solvent vapours (pentane, hexane, heptane, etc.)	QCM with PCA and neural network
Measurements in working areas	Gas mixture analysis	MOSFET sensor with PCA and artificial neural network
	Harmful organic vapours detection	QCM sensors
Emission measurements	Waste-water separation	Polypirrole sensors with multivariate analysis
	Ammonia emission	QCM sensor array
Process control and regulation	Bioreactor off-gas composition monitoring	MOSFET sensor with PCA
	Block milk products classification	Neotronics eNOSE electronic nose

(Contd ...)

Application fields	Detection objects	Sensors
Medical applications	Urine analysis	QCM sensors with PCA
	Human skin odour analysis	QCM sensors with self-organising map (SOM) analysis
	Human breath analysis	Metal oxide sensors with signal pattern evaluation
Agricultural	Vinegar discrimination	AromaScan electronic nose
	Boar taint intensity discrimination	Conducting polymer sensor array with pattern recognition routines

Olfactometry and gas chromatography

Olfactometer is the state-of-the-art odour measurement system. It is used to measure the odour detection threshold (or recognition threshold) and the hedonic tone of an odour substance. The odour detection threshold is the lowest concentration of any odour substance that can be detected by 50 per cent of the test population (known as panellists or assessors), whereas the hedonic tone is a scale based on ratings which measure the degree of pleasure provided by a specific characteristic of an odour substance.

An odour measurement is expressed as an odour unit (OU). In European countries (EU), the unit used is the European Odour Unit (OU_E), a unit that has caused much confusion in the research community because its format differs from those commonly used to describe concentrations, i.e. mass per volume (kg/m^3) or volume per volume (ppm). In 2000, Australia and New Zealand jointly set up a new odour-testing standard essentially identical to the European Standard. By definition, 1 OU_E is the amount of odourants that, when evaporated into 1 m^3 of a neutral gas in standard conditions, elicits a physiological response from a human panel equivalent to that elicited by 123 µg of n-butanol evaporated in 1 m^3 gas in standard conditions. According to the EPA definition, 123 µg of n-butanol is known as one European Reference Odour Mass (EROM). The hedonic tone is a subjective judgement of the relative pleasantness or unpleasantness of any odour. A numbering system can be applied to this scale, ranging from a small number for 'dislike' (or 'unpleasant') and a large number for 'like' (or 'pleasant'). Another quantification system for hedonic tone is the use of a 20-point scale, starting from '−10' for unpleasant and '+10' for pleasant odours. An example of a hedonic tone for any odour substance under assessment can also be defined as follows: 1 = dislike very much; 2 = dislike; 3 = neither like nor dislike; 4 = like; and 5 = like very much.

The problem involved in the use of olfactometry is the subjectivity of the panel's members. An exact replication of a measurement of the same substance is not possible, since the sensitivity of different panels is obviously not the same. Furthermore, for measuring harmful gases, a panel certainly cannot be recommended. Olfactory fatigue is also a common side-effect observed in panel members.

Odour compounds can also be recognised by means of analytical instruments such as gas chromatography. An odour-containing gas sample is fed onto the instrument through the head of the chromatographic column. The sample is then transported through the column by the flow of the inert and gaseous mobile phase of the carrier gas. Later, the detector responds to the compounds but not to the carrier gas. The signal from the detector is expressed as a graph known as a chromatograph. By comparing the respective peaks and the reference graph, the compound present in the sample can be distinguished. Although the measuring system is simple, the costs are high, since the instrumentation is expensive. Gardner and Bartlett added that the use of gas chromatography requires considerable skills. For the above reasons, the technique is not used for routine evaluation.

Electronic noses

An electronic nose (E-nose) is an instrument that is designed to approach or to substitute the function of the biological olfaction system (e.g. human nose). Gardner and Bartlett defined the E-nose as an instrument that comprises an array of electronic, chemical sensors with partial specificity and an appropriate pattern recognition system, capable of recognising simple or complex odours. This definition restricts the term E-nose to those types of sensor array systems that are specifically used to sense odourous molecules in an analogous manner to the human nose. According to another definition by Pearce, the E-nose is a machine that is designed to detect and discriminate among complex odours using a sensor array.

The sensor array consists of broadly tuned (nonspecific) sensors that are treated with a variety of odour-sensitive biological or chemical materials. An odour stimulus generates a characteristic fingerprint (or smell-print) from the sensor array. Patterns or fingerprints from known odours are used to construct a database and train a pattern recognition system so that unknown odours can subsequently be classified and identified. Thus, the E-nose instrument is comprised of hardware components for collecting and transporting odours to the sensor array as well as an electronic circuitry to digitise and store the sensor responses for signal processing. A diagram of the basic components of a typical E-nose is depicted in Fig. 4.11.

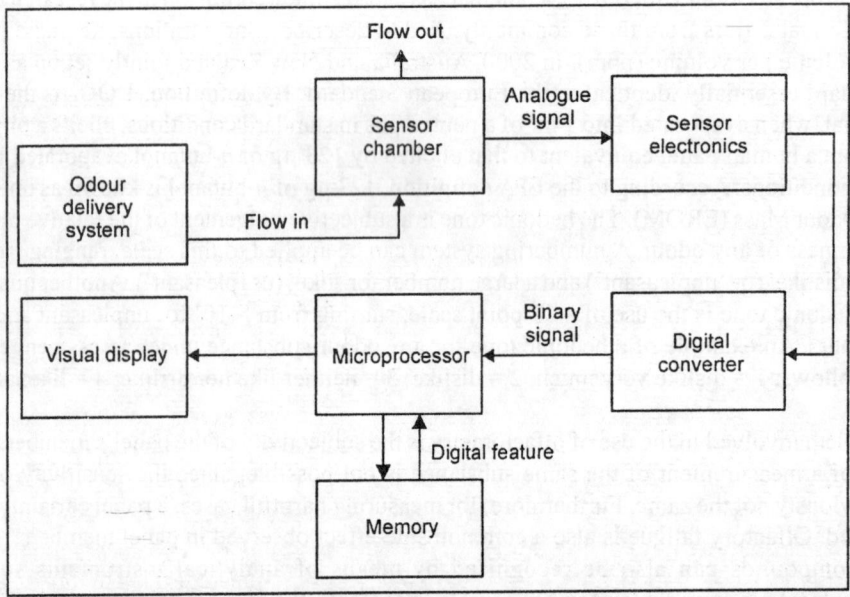

Fig. 4.11. Basic components of an electronic nose (E-nose) instrument system.

Considerable research has been directed towards the development of E-nose instrumentation over the past decade. Numerous research groups now exist in countries such as Australia, Denmark, France, Germany, Japan, Sweden, UK and USA. There is also increasing interest in the research, development and application of E-noses, i.e. of sensors and sensor arrays, with the aim to:

1. Complement techniques of analytical chemistry in order to classify gas mixtures, odours, air quality, or toxicity.

2. Develop cheap and small on-line instruments for fast imaging of specific chemicals, odours, or toxic substances with high spatial and time resolution (including, e.g. instruments required for quality and process control).

3. Develop new materials for odour detection based on molecular recognition principles that are similar to those in the human nose.

Among these, the last point (development of new materials) might be the most difficult problem. This is in line with the fact that the fundamental problem in the application of QCM sensors is to find a suitable coating layer and a method of reproducibility when applying it.

There are a number of records of E-nose applications in daily life, including medicine, agricultural fields, environmental monitoring, etc. In the following, a selection of applications regarding odour detection, monitoring or measurement are listed: identification of odours from reagents (ethanol, ether, acetone, ethyl acetate), liquors (beer, spirit, samshu, wine), and perfumes (phenethyl alcohol, ionone, vanillyl alcohol, ethyl isobutyrate, thymol); measurements of sewage odours; characterisation of olives oil based on their volatile substances; diabetes diagnosis based on the expired breath of diabetics; discrimination of polymer samples used in the automotive industry. A number of electronic nose systems currently available on the market are Alpha MOS, AromaScan, Bloodhound, Lennartz Electronic, Smart Nose, Cyrano Sciences, etc. These utilise a range of sensor technologies either alone or in combination.

Metal oxide sensors (MOS)

Metal oxides sensors are devices that translate the changes in the concentration of gaseous chemical species into electrical signals. They consist basically of a sensitive layer, an insulating layer, two electrodes and a heating heater. A scheme of a MOS is given in Fig. 4.12. The semiconducting layer oxidises the sample compound at a temperature level of 250° to 450°C. When the semiconducting substance absorbs the released electrones, its conductivity changes. In consequence, the change of resistance in the electrical circuit is registered.

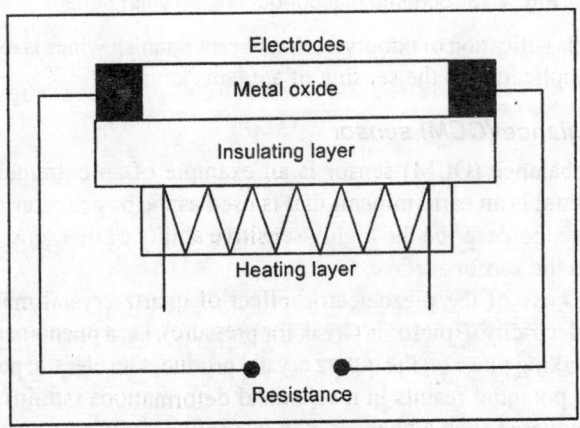

Fig. 4.12. Scheme of a metal oxide sensor.

The sensitivity of the sensor can be adjusted by choosing different operation temperatures and by dotation with noble metals as catalytic dopants. The application of pattern recognition systems is made difficult by the fact that the dependency of the sensor signal on the concentration of the gaseous species is generally not linear (Table 4.12).

Table 4.12. Applications of metal oxide sensors for odour detection.

Detected odour(s)
Five malodours collected in the field: printing houses, paint shop, waste water treatment plant, urban waste composting facilities, rendering plant
Selective detection of CO and NH_3
Organic vapours: benzene, toluene, and methanol
Trimethylamine

Conducting polymer sensors

Conducting polymer sensors (Fig. 4.13) are being widely used for odour sensing in the form of arrays consisting of highly sensitive, scarcely selective, chemoresistive sensors characterised by different sensitivity spectra. The working principle of the sensor is based on the change of the conductivity during the diffusion of gaseous molecules in the polymer layer. Due to the use of pyrrol as a master polymonomer, the sensor is highly sensitive to polar compounds. By an inclusion of different metal ions into the polymer, the sensor can be adjusted for various chemical species.

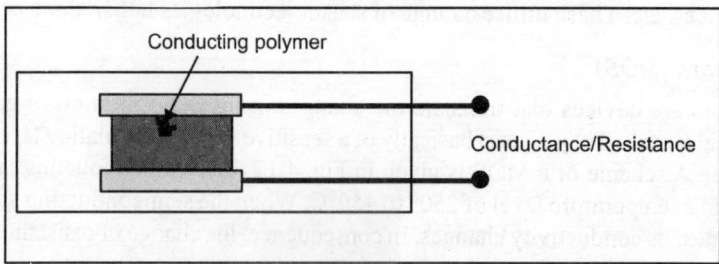

Fig. 4.13. Scheme of a conducting polymer sensor

An application in the classification of odours from different Spanish wines is explained in Guadarrama. Another example of an application is the sensing of aqueous ammonia.

Quartz crystal microbalance (QCM) sensor

The quartz crystal microbalance (QCM) sensor is an example of an extremely sensitive detector of mass changes. Quartz crystal is an earth mineral that is used as the basic material of the sensor, and the term 'microbalance' is used to describe the highly sensitive ability of this sensor to detect a very small ('micro') mass change on the sensor surface.

A QCM sensor makes use of the piezoelectric effect of quartz crystal materials. Piezoelectricity literally means 'pressure electricity' ('piezo' is Greek for pressure), i.e. a phenomenon where a mechanical stress (e.g. compression) taking place on the quartz crystal produces an electric potential, and conversely, an application of electric potential results in mechanical deformations (strain) on the quartz. Jacques and Pierre Curie first discovered such a phenomenon in 1880.

By employing these properties, wave phenomena can be generated. The velocity of the waves and, as a result, their frequencies are influenced by a large number of parameters, including mass effects at the surface of the piezoelectric material.

A QCM sensor is a kind of mass sensitive sensor, a member of the chemical sensors group. The basic material of the QCM sensor consists of quartz crystal, which is equipped with metal electrodes (e.g. gold). A sensitive coating material on the sensor surface is used to enable detection of the measurand (analyte) in the environment. An appropriate electronic circuit is necessary to make conversion of the measured quantity to an electrical signal possible.

The basic working principles of the quartz crystal microbalance sensor are depicted in Fig. 4.14. Analytes that are present in the surrounding space (e.g. a measuring chamber) of a QCM sensor will interact with the sensitive coating material on the sensor surface. In this interaction, analyte molecules are adsorbed into or absorbed onto the sensitive coating material (e.g. polymer). The adsorption or absorption of the analytes by the coating material results in a mass change on the sensor surface. Consequently, the mass change on the sensor surface is converted to the frequency change.

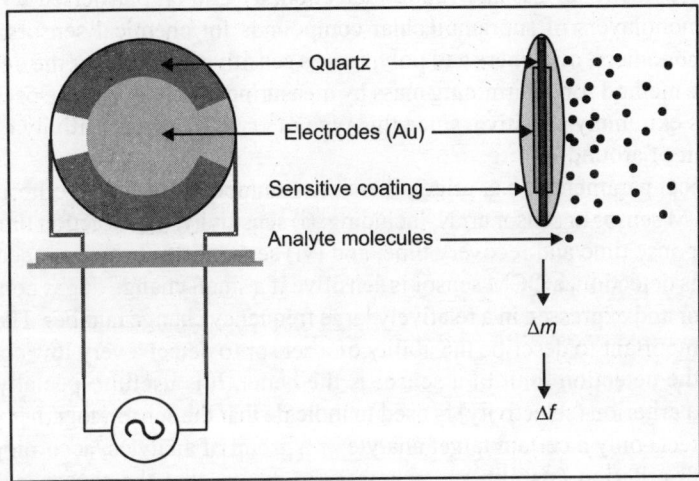

Quartz

Electrodes (Au)

Sensitive coating

Analyte molecules

Δm

Δf

Fig. 4.14. Basic working principles of a quartz crystal microbalance (QCM) sensor.

Using an equation derived by Sauerbrey, a mass change on a QCM sensor surface due to adsorption of any analyte by sensitive coating material can be expressed in a frequency change quantity as follows:

$$\Delta f = -2.3 \times 10^6 \, F^2 \, (\Delta m / A)$$

where,

Δf = the frequency change [Hz].

F = the oscillating frequency of the quartz crystal [MHz] (for a typical AT-Quartz, F = 10 MHz).

Δm = the mass change of the adsorbed analyte, i.e. odour substance [g].

A = the area coated by the film [cm^2].

The interaction between odour molecules and the sensitive coating materials (known as 'guest-host interaction') plays an important role in the detection process. In this interaction, the analyte (i.e. the odour molecules) acts as the guest, whereas the sensitive coating material is the host. There are a number of chemically sensitive material classes, e.g.

1. Polymers (polyethers, polyurethanes, polysiloxanes, polypyroles, nafion, etc.).
2. Molecular crystals (phthalocyanines, porphyrines, etc.).
3. Supramolecular structures (calixarenes, zeolites, cyclodextrines, cyclophanes, etc.).

Because of its importance, special attention has been paid to this guest-host interaction by researchers during the last decades. Studies concerning its energy aspects, for example, have been carried out by

Dickert. They show that the sensor signal of these supramolecular analyte-receptors can be predicted by a method that uses estimated free energies of the guest-host complex formation. Another study demonstrated the application of molecular modelling to provide meaningful structural information on the guest-host interactions of cyclodextrine and chloroform. In this way, computational chemistry helps to achieve a better understanding of what happens during the inclusion process. This saves time- and money-consuming synthesis and makes molecular modelling an excellent tool for the design of sophisticated chemical sensitive layers.

More detailed studies on coating materials have been performed by Buhlmann on clathrates as coating materials for dielectric transducers with regard to organic solvent vapour sensors; by van de Leur and van der Waal on polypyrrolle for gas and vapour detection; by Cao on plasticised PVC coatings; Weib on self-assembled monolayers of supramolecular compounds for chemical sensors; and by Zhou on silicon-containing monomers, oligomers and polymers as sensitive coatings for the detection of organic solvent vapours. The method for determining mass by measuring the change in the oscillation frequency of a quartz crystal is extremely sensitive, since this type of crystal has a sensitivity of about 10^{-9} g/Hz with a detection limit of around 10^{-12} g.

Besides economical parameters (e.g. price), there are a number of technical criteria determining the performance of a QCM sensor or sensor array, including: (i) sensitivity, (ii) detection limit, (iii) selectivity, (iv) stability, (v) response time and recovery time, and (vi) sensor drift. In the perspective of the use of a QCM sensor for gas detection, a QCM sensor is sensitive if a small change of gas concentration can be detected by the sensor and expressed in a relatively large frequency change number. The second criterion (detection limit) is important to describe the ability of a sensor to detect a very low concentration of an analyte. The lower the detection limit of a sensor is the better. It is useful especially for detection of trace gases. The third criterion (selectivity) is used to indicate that the sensor, together with the sensitive coating material, detects only a certain target analyte or a group of analytes, according to the designed objective. The fourth criterion (stability) is necessary to ensure that the sensor is long-term stabile (endure) enough to be implemented in a variety of measurement locations and situations and to show stabile results. The criteria 'response time' is the time required for a sensor to read a certain percentage (e.g. 80 per cent) of a full-scale reading after being exposed to a full-scale concentration of a given gas, whereas 'recovery time' indicates the time required by a sensor to return to normal condition and to be ready for a new measurement after a measurement cycle.

The criterion 'sensor drift' is a phenomenon where an undesired change in output takes place over a period of time that is unrelated to input. Sensor drift can be due to ageing, temperature effects, or sensor 'poisoning', etc. The QCM sensor can be used as a single sensor or as a group of sensors, known as sensor array. A sensor array, however, is not simply a group of a number of discrete sensors that are used together, but rather a set of an integrated sensors that are formed on a common substrate and used as a complete unit. As the field of applications has been developed, attention has moved towards the development of sensors specifically for use in arrays. Furthermore, almost all such arrays have been made up of a single sensor type.

The advantages of the use of sensors in an array form are: (i) technical conditioning, i.e. control of temperature stability, sample mass flow rate, etc. are simpler, (ii) a more compact measuring chamber, i.e. a single measuring chamber is used by all sensors, and (iii) better description of the measurand, i.e. the measurand can be described in a better way by a series of sensors (in form of a pattern) than if it were described by a single sensor. The quartz crystal microbalance sensor has been used in a numerous fields of application including gas mixture analysis, detection of solvent vapours, detection of organic vapours, detection of carbon dioxide (CO_2), discrimination of aromatic optical isomers, discrimination of odourants, detection of mutagenic polycyclic compounds, detection of organic pollutants in water, detection of L-glutamic acid, and discrimination of aromas from various Japanese sake, etc.

LAWS AND REGULATIONS

Generally, most regulatory organisations/municipal authorities resort to nuisance prevention law to abate odour, but it is being followed more in breach than in practice. Perhaps the most effective method of overcoming the problem is to have regulation, which can be directed towards specific emissions from specific industries. For example, regulation of pulp mills odours which limits emission of mercaptans. The development of odour measurement, regulation and control technique has been greatly progressed in most of the countries. In Australia, several states are reviewing the existing regulations on the control of odour from various source based on a forthcoming Australian–New Zealand standard of odour measurement using dynamic olfactomerty.

In the USA, odour has been listed as one of the key area to be dealt with the Agriculture Air Quality Task Force. There are some State regulations in the US to address the odour from some specific sources. In Minnesota, the Feedlot Hydrogen Sulphide Program administered by the Minnesota Pollution Control Agency is in vougue since July 1997. The hydrogen sulphide standard works in the following way: each gas sample represents an average value of the gas over a continuous 30-minute period. A violation occurs if the hydrogen sulphide ambient air quality exceeds 30 and 50 ppb within certain time period. Iowa's 'manure law' was enacted in 1995, to create a system of setback distances for lagoons and buildings, depending on the nature of the surrounding area. Also Iowa Department of Natural Resources has proposed new regulations for swine producers, which require manure injection rather than spreading.

There are more activities in addressing gaseous and odour problems from large-scale swine facilities. Europe has focussed on two primary areas of concern— nitrogen emissions and odour prevention. The Netherlands has 'an extremely strict approach' for regulating nitrogen emissions. By 2013, farmers must reduce their emissions by 70 per cent of 1980 levels, with an assumption that a 10 per cent reduction rate in ammonia emissions will result in a 7 per cent reduction of odour. In Denmark, odour laws were established during 1950 to 1980. These laws required ventilation chimneys and setback distances from houses. By the end of 1980s, it was felt that the general code of good agricultural practice had not reduced odour to acceptable levels. The Ministry of Environment then imposed restrictions on the construction and location of manure storage and swine buildings, as well as on the land application of manure. Germany focuses more on managing nutrients than on paying specific attention to ammonia odour emissions. The Fertiliser Ordinance enacted in July, 1996, requires manure to be worked into non-tilled soils immediately after application. Producers have to send records of both manure storage and cropland application to the government. There are strict controls on lagoons as well – they must be lined, covered and equipped with underground pipes for the detection of leaks.

In India, Schedule II and Schedule VI (General standards for discharge of effluents) under Environmental (Protection) Rules, 1986 prescribes that all efforts shall be made to remove unpleasant odour as far as practicable. Nonetheless, there are only two industries, wherein industry specific standards under Schedule I of these Rules mandate odour removal. These are the fermentation industries and the natural rubber industries. The standards for many other major odour pollution causing industries such as pulp and paper mills, tanneries, meat processing industries, bulk drug and pharmaceutical units, food and fruit processing units, dairies and milk plants, etc. do not specify odour control.

Chapter 5

Indoor Air Pollution

INTRODUCTION

Many of us spend the greater part of our lives indoors. The indoor environment of homes and offices is often more seriously polluted than the outdoor atmosphere. Thus, for many people, the risks to health may be greater due to indoor air pollution than the outdoor air. The major sources of indoor air pollution are those that release gases and particles into air, and inadequate ventilation makes the situation much worse. The major factors that determine the quality of indoor air are:

1. The nature of outdoor air quality around the building.
2. The air exchange rate of the building (ventilation).
3. The materials used in the construction of the building (presence of chemicals).
4. The activities that go on inside the building (cleaning, cooking, heating, etc.).
5. Use of household chemicals.

INDOOR AIR AND YOUR HEALTH

In recent years, there have been an increasing number of complaints about the poor quality of indoor atmosphere from residents, and workers in offices and commercial buildings. The pollutants found in indoor air are responsible for many harmful health effects. The effects may show up immediately after a single exposure, and include irritation of eyes, nose and throat, headache, dizziness and fatigue. Such immediate effects are usually of short duration and treatable.

Sometimes the treatment is simply eliminating the person's exposure to the source of pollution. Symptoms of some diseases such as asthma, hypersensitivity, pneumonitis and humidifier fever may appear soon after exposure to certain indoor air pollutants. Though most of these diseases can be treated, nevertheless, some pose serious risks.

Other health problems may appear either years after a single exposure had occurred or on repeated exposures. These effects, which include some respiratory diseases, heart diseases and cancer, can be severely debilitating or fatal. Further research is needed to better understand the health hazards caused by exposure to average pollutant concentrations found in homes and those by higher concentrations that occur during short periods of time.

Pollutant Sources

There are many sources of indoor air pollution in a home. Most homes have more than one source that contribute to indoor air problems. Many activities go on inside a home which include cleaning, cooking, heating by open or enclosed fires, smoking, etc. Some important indoor air pollutants are generated by burning of oil, gas, kerosene, wood and tobacco products or produced by building materials, furnishings, wet or damp carpets, household chemical products, air conditioners, dehumidifiers and outdoor sources such as radon and pesticides. Inadequate ventilation can increase indoor pollutant level because of insufficient air movement to dilute emissions from indoor sources and carry them out of home.

Carbon monoxide

One of the most acutely toxic indoor air contaminants is carbon monoxide (CO), a colourless, odourless gas that is a by-product of incomplete combustion of fossil fuels. Common sources of carbon monoxide are tobacco smoke, space heaters using fossil fuels, defective central heating furnaces and automobile exhaust. Improvements in indoor levels of CO are systematically improving from increasing numbers of smoke-free restaurants and other legislated non-smoking buildings. By depriving the brain of oxygen, high levels of carbon monoxide can lead to nausea, unconsciousness and death. According to the American Conference of Governmental Industrial Hygienists (ACGIH), the time-weighted average (TWA) limit for carbon monoxide (630–08–0) is 25 ppm.

Carbon dioxide

Carbon dioxide (CO_2) is a surrogate for indoor pollutants emitted by humans and correlates with human metabolic activity. Carbon dioxide at levels that are unusually high indoors may cause occupants to grow drowsy, get headaches, or function at lower activity levels. Humans are the main indoor source of carbon dioxide. Indoor levels are an indicator of the adequacy of outdoor air ventilation relative to indoor occupant density and metabolic activity. To eliminate most Indoor Air Quality complaints, total indoor carbon dioxide should be reduced a difference of less than 600 ppm above outdoor levels. NIOSH considers that indoor air concentrations of carbon dioxide that exceed 1000 ppm are a marker suggesting inadequate ventilation. ASHRAE recommends that carbon dioxide levels not exceed 700 ppm above outdoor ambient levels. The UK standards for schools say that carbon dioxide in all teaching and learning spaces, when measured at seated head height and averaged over the whole day should not exceed 1500 ppm. The whole day refers to normal school hours (i.e. 9.00 am to 3.30 pm) and includes unoccupied periods such as lunch breaks. European standards limit carbon dioxide to 3500 ppm. OSHA limits carbon dioxide concentration in the workplace to 5000 ppm for prolonged periods, and 35,000 ppm for 15 minutes. Exhaust gas leakages can occur from furnace metal exhaust pipes that lead to the chimney when there are leaks in the pipe and the pipe gas flow area diameter has been reduced.

Ozone

Ozone is produced by ultraviolet light from the sun hitting the earth's atmosphere (especially in the ozone layer), lightning, certain electric devices (such as air ionisers), and as a by-product of other types of pollution.

Ozone exists in greater concentrations at altitudes commonly flown by passenger jets. Reactions between ozone and onboard substances, including skin oils and cosmetics, can produce toxic chemicals as by-products. Ozone itself is also irritating to lung tissue and harmful to human health. Larger jets have ozone filters to reduce the cabin concentration to safer and more comfortable levels.

Outdoor air used for ventilation may have sufficient ozone to react with common indoor pollutants as well as skin oils and other common indoor air chemicals or surfaces. Particular concern is warranted when using 'green' cleaning products based on citrus or terpene extracts as these chemicals react very quickly with ozone to form toxic and irritating chemicals as well as fine and ultrafine particles. Ventilation with outdoor air containing elevated ozone concentrations may complicate remediation attempts.

Radioactivity

Most of the radioactivity inside a building is associated with radon, which is emitted from uranium in the soil or rock on which homes are built. Radon is a product of radioactive decay process beginning with uranium — 238 and thorium — 232. Because of their longer half-lives (4.5 and 14 billion years respectively) they are present in trace quantities in many geological materials. Sometimes radon enters into the home through well water. Being a gas, radon escapes from construction material, penetrates through cracks in buildings and is released into the indoor atmosphere where it may be inhaled. The average indoor radon level is 1.3 pCi/l (pico curies per liter).

Moulds and other allergens

These biological chemicals can arise from a host of means, but there are two common classes: (i) moisture induced growth of mould colonies, and (ii) natural substances released into the air such as animal dander and plant pollen. Moisture buildup inside buildings may arise from water penetrating compromised areas of the building envelope or skin, from plumbing leaks, from condensation due to improper ventilation, or from ground moisture penetrating a building part. In areas where cellulosic materials (paper and wood, including drywall) become moist and fail to dry within 48 hours, mould mildew can propagate and release allergenic spores into the air.

In many cases, if materials have failed to dry out several days after the suspected water event, mould growth is suspected within wall cavities even if it is not immediately visible. Through a mould investigation, which may include destructive inspection, one should be able to determine the presence or absence of mould. In a situation where there is visible mould and the indoor air quality may have been compromised, mould remediation may be needed. Mould testing and inspections should be done by an independent investigator to avoid any conflict of interest and to insure accurate results; free mould testing offered by remediation companies is not recommended.

There are some varieties of mould that contain toxic compounds (mycotoxins). However, exposure to hazardous levels of mycotoxin via inhalation is not possible in most cases, as toxins are produced by the fungal body and are not at significant levels in the released spores. The primary hazard of mould growth, as it relates to indoor air quality, comes from the allergenic properties of the spore cell wall. More serious than most allergenic properties is the ability of mould to trigger episodes in persons that already have asthma, a serious respiratory disease.

Mould is always associated with moisture, and its growth can be inhibited by keeping humidity levels below 50 per cent. Moisture problems causing mould growth can be direct such as a water leaks and/or indirect such as condensation due to humidity levels.

Asbestos fibres

The US Federal Government (www.osha.gov) and some States have set standards for acceptable levels of asbestos fibres in indoor air. Many common building materials used before 1975 contain asbestos, such as some floor tiles, ceiling tiles, taping muds, pipe wrap, mastics and other insulation materials. Normally significant releases of asbestos fiber do not occur unless the building materials are disturbed, such as by cutting, sanding, drilling or building remodelling. There are particularly stringent regulations applicable to schools.

Inhalation of asbestos fibres over long exposure times is associated with increased incidence of lung cancer. Asbestos is found in older homes and buildings, but it is most dangerous in schools and industrial settings. It was once widely used in shingles, fireproofing, heating systems and floor and ceiling, tiles in older buildings. When asbestos-containing material is damaged or disintegrates, microscopic fibres are dispersed into the air. The risk of lung cancer from inhaling asbestos fibres is also greater to smokers. The symptoms of the disease do not usually appear until about 20 to 30 years after the first exposure to asbestos. Removal of asbestos-containing materials is not always optimal because the fibres can be spread into the air during the removal process. A management program for intact asbestos-containing materials is often recommended instead.

Volatile organic compounds (VOC)

There are many aliphatic and aromatic compounds contributing to VOC concentrations, with chloroform, acetone, chlorinated compounds and formaldehyde being predominant in many locations. Consumer products used in homes contribute other VOCs to the indoor atmosphere. For example, latex paints contain toluene, ethylbenzene, 2-propanol and butane. Many organic compounds are emitted from construction materials, furnishings and consumer products such as latex paints, cleaning agents, household solvents, detergents, waxes and varnishes.

Formaldehyde in construction materials such as particle boards, plywood and in urea formaldehyde foam insulation, leaks into air when the temperature rises. Combined with other contaminants it can cause headache, respiratory irritations, watery eyes, nausea, skin irritation and heart problems.

Indoor combustion

Combustion of fuels such as oil, gas, kerosene, etc. inside a building contributes to the concentration of VOCs and it is also a source of stable inorganic gases. The common indoor pollutants due to combustion of fuels are particulate matter, oxides of nitrogen, oxides of sulphur, carbon monoxide, hydrocarbons and other odour causing chemicals. The emission quantity of these pollutants depends upon the type of fuel used, fuel/oxidant ratio and other combustion conditions. Proper venting of exhaust gases reduces this problem.

Tobacco smoking

Tobacco smoke contains a complex mixture of over 4000 compounds, more than 40 of which are known to cause cancer, and as many are strong irritants. Tobacco smoking is a source of VOCs including polyaromatic hydrocarbons (PAH), organic bases like nicotine, aldehydes, ketones, organic acids and respirable particulate matter. Smoking inside home is a cause for large amounts of indoor pollutants. Natural or mechanical ventilation techniques do not remove them from the air as quickly as they build up.

Cigarette smoking is another source of formaldehyde. The directly inhaled air drawn through a cigarette may contain formaldehyde concentrations more than 400 times the level of concentration in the indoor atmosphere. Tobacco smoking is responsible for approximately 3000 lung cancer deaths every year in non-smoking adults and causes respiratory infections in hundreds of thousands of children.

Biological contaminants

Biological contaminants include pollens, bacteria, mildew, fungal spores, etc. There are many sources of these pollutants. Pollens originated from plants, and viruses and bacteria are transmitted by people and animals. Biological contaminants cause allergic diseases, pneumonitis, and some types of asthma.

By controlling the relative humidity level (30–50 per cent) in a home, the growth of some of these sources of biological pollutants can be minimised.

Preventive Measures

Preventive measures for indoor air pollution in homes, apartments and offices involve eliminating or controlling the sources of pollution, increasing ventilation and installing air-cleaning devices. Indoor air pollutants their sources and impact on health is given in Table 5.1.

Table 5.1. Air pollutants their sources and impact on health.

	Indoor air pollutants	
Pollutants	*Sources*	*Health effects*
Radon	Construction materials from geological sources	Lung cancer
Formaldehyde	Particle boards, plywood, urea formaldehyde foams, tobacco smoking, furniture, etc.	Headache, respiratory irritations, nausea, skin irritation, watery eyes, and heart problems
Other VOCs (Toluene, 2- propanol, phenols, aldehydes, ketones, esters, etc.)	Paints, solvents, wood preservatives, aerosol sprays, varnishes. cleansers, air fresheners, etc.	ENT irritations, headache; some organic compounds can cause cancer
CO_2, CO, NO_x, SO_x and HC	Combustion of fuels (kerosene heaters, gas stoves, leaking chimneys, tobacco smoking, etc.)	Fatigue, dizziness, confusion, nausea like symptoms that disappear once out of the house
Particulate matter (suspended and respirable) and PAH	Tobacco smoking and combustion of fuels	Burning sensation of eyes and nose, bronchitis asthma, cancer, reduced lung functions
Pesticides	Moth repellants, insecticides and other pesticides, termiticides, disinfectants	Eye, nose, throat irritation, damage to central nervous system, kidneys; risk of cancer
Biological pollutants (pollens, viruses, bacteria, etc.)	Wet or moist walls. ceiling, carpets, poorly maintained humidifiers, air conditioners and house hold pets	ENT irritation, allergy, shortness of breath, humidifier fever and other infectious disease
Lead	Lead based paints, contaminated soil and water	Affects all systems within the body, central nervous system and kidney

Many interventions have been shown to be effective in reducing indoor air pollution. These can be grouped into three categories which are shown in Table 5.2.

Best Practices and Implementation Lessons Learned

Following are the best practices and implementation lessons learned:
1. Effectiveness will depend on three key considerations: (i) the policy and regulatory context, (ii) making sure that all relevant sectors/perspectives are considered in interventions that aim to increase fuel efficiency, reduce health risks and improve local ecology, and (iii) local community involvement in technology design and application, especially with regard to stoves and ventilation.
2. Improved stove design should be needs-based and tailored to the local ecology and socio-economic conditions of the community.

Table 5.2. Methods to reduce air pollution.

Intervention type	Intervention examples	Key monitoring indicators
Technologies which aim at improved cooking/heating devices, improved fuels, or reduced need for heating	Better stove design Better ventilation to reduce IAP Switch to cleaner (but typically more costly) fuels Chemical treatment of some fuels, e.g. coal Reduce size of fuel pieces, e.g. briquettes and pellets instead of large coal lumps Better insulation Solar energy for boiling water	Particulates (PM10) in the air in homes Carbon monoxide (CO in parts per million or ppm) Fluoride levels in food (for households that use coal which contains fluoride) Ideally, have beneficiaries wear a device that measures direct exposure to PM10 over a specified time period (e.g. 24 hrs) before and after the intervention
Technologies aimed at improving the living environment	Partitions, walls or screens in homes to separate cooking and sleeping/living areas Better ventilation or ducts, hoods to carry smoke and particulates outside the house	
Behavioural change to reduce exposure and/or reduce smoke generation	Reduce time spent in kitchen/cooking area Keep lids on pots while cooking proper stove maintenance and cleaning Push fuel (esp plant stalks) deeper into stove so that less smoke 'escapes' into the room Keep children away from the smoke	

3. All interventions (or combinations of interventions) need to be customised to local circumstances.
4. Stove building and maintenance can create income-generating opportunities for the local economy, and enable women especially to make more productive use of their time, earn higher incomes and gain financial independence.
5. Subsidies can be offered as an incentive to adopt improved stoves as long as they are managed effectively and phased out gradually to limit continuous reliance on public funding.
6. Rural credit schemes can be an effective way to promote demand and finance community-based stove improvement schemes.
7. Stove improvement programs need to include a component to inform, educate and communicate the health, environmental, energy and financial consequences of indoor air pollution and how improved stoves together with behavioral change can lead to better health and household finances.
8. Long-term sustainable solutions require full participation from local government, NGOs, the commercial sector and local communities.

EXPOSURE ASSESSMENT

Assessment of exposures provides information on the concentration and distribution of each pollutant. Personal exposure factors include time spent in different indoor environments, activities and behaviour, breathing rates, and so forth. These factors indicate that individuals may experience pollution levels very different form those measured at a nearby fixed monitoring station. The pollutant benzene provides a good example (Fig. 5.1). Personal exposure must also take into account in the major indoor sources; an example is environmental tobacco smoke that produces particulate matter (Fig. 5.2), leading to concentrations that frequently occur in excess of ambient standards.

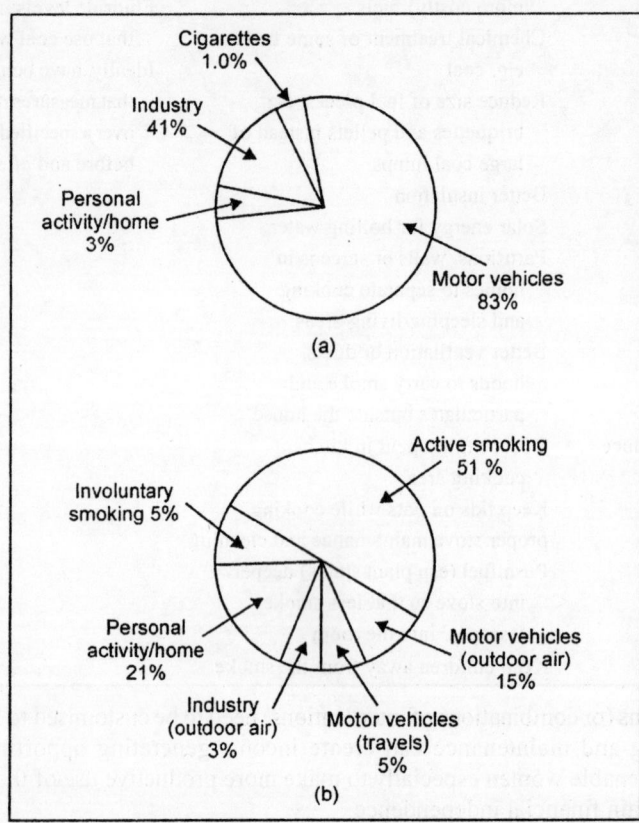

Fig. 5.1. Benzene as a pollutant (a) emission, (b) exposures.

Biological exposures to pollutants are being developed for a variety of toxic materials in the air. Lead (from indoor and outdoors) has well established blood assays. These tests have indicated declines in lead in the body (at least that amount circulating in the blood) that parallel the declines in amount of lead in the atmosphere, the amount of lead in gasoline, and the amount of lead in paint and dust that were mandated by law.

Indoor monitoring has specific requirements related to the type of pollutants and the enclosed spaces. Monitoring protocols are developed with regard to the availability, practicality, and expense of continuous or integrated sampling methods to measure the pollutant over the time periods of interest. These monitors are either fixed or portable. Personal monitors have been developed for some pollutants.

Good surrogate measurements are often utilised to represent more complex exposures (for example, total hydrocarbons for all organics).

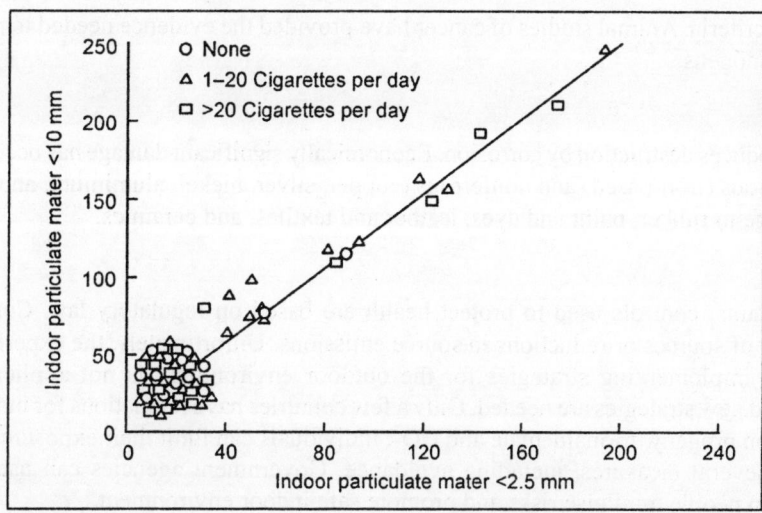

Fig. 5.2. Indoor particulate matter of 2.5 and 10-micrometer diameter related to the amount of environment tobacco smoke.

Effects

Characterisation of the effects of air pollution requires determination of concentration exposure, exposure-dose (for humans and animals), and exposure/dose-response relationships. These studies have involved humans, animals, and materials.

Humans

Effects experienced by humans include acute and chronic symptoms (morbidity), increases in acute respiratory illness, and declines in lung functions, especially in individuals with chronic diseases. There are several indoor pollutants that have similar, and possibly synergistic effects. Particulate matter, NO_2, and formaldehyde (HCHO) produce irritation of the eyes and mucous membrane; various types of particulate matter, HCHO, and allergens produce so called allergic symptoms; NO_2, and HCHO affect cognitive skills; CO, NO, NO_2, nitrates, and methylene chloride (CH_2Cl_2) change carboxyhaemoglobin and methemoglobin levels and affect the heart and brain; and radon progeny especially when associated with tobacco smoke, can promote lung cancer. Lead has been shown to have adverse neurological effects at the concentrations that occurred frequently in the past, especially in infants and children. Benzene has been shown to produce a form of leukemia, and both radon progeny and asbestos have been shown to produce lung cancer in humans. Other types of cancer may occur with exposure to carcinogens found in occupational settings. Aeroallergens are of great importance, as many people have allergic problems, including asthma. It is like that aeroallergens and interactive exposures with other pollutants can produce and aggravate these problems. Thus, a large amount of acute disability is related to exposure to indoor pollutants.

Animals

Animal studies have shown effects similar to those found in humans in regard to most of the same pollutants. These studies used mortality, lung pathology, and changes in pulmonary physiology and immunology as criteria. Animal studies of cancer have provided the evidence needed to protect humans from those carcinogens.

Materials

Air pollution produces destruction by corrosion. Economically significant damage has occurred in various metals, both ferrous (iron-based) and nonferrous (copper, silver, nickel, aluminium, and zinc) and also extensive damage to rubber, paint and dyes, leather and textiles, and ceramics.

Controls

For some pollutants, controls used to protect health are based on regulatory law. Control measures include removal of sources or reductions in source emissions. Unfortunately, the experience gained in developing and implementing strategies for the outdoor environment is not applicable to indoor environment, and new strategies are needed. Only a few countries have regulations for indoor pollutants, specifically radon progeny, formaldehyde and CO_2. Individuals can limit their exposures, especially in residences, by several measures, including avoidance. Government agencies can use public health education to help people minimise risks and promote safe indoor environment.

Source removal is the first and most logical form of control. Reductions in source emissions have been achieved for several products through changes in manufacturing process or construction. Reduction of exposure also includes standard direct methods such as ventilation and cleaning as well good design.

SPECIFIC DISEASES ASSOCIATED WITH INDOOR AIR POLLUTANT EXPOSURE

Respiratory illness, cancer, tuberculosis, perinatal outcomes including low birth weight, and eye diseases are the morbidities associated with indoor air pollution.

Respiratory Illness

The effect of air pollutants in general would depend on the composition of the air that is inhaled which will depend on the type of fuel used and the conditions of combustion, ventilation and duration for which the inhalation occur. The most commonly reported and obvious health effect of indoor air pollutants is the increase in the incidence of respiratory morbidity. Studies by the NIOH on the prevalence of respiratory symptoms in women using traditional fuels (biomass) ($n = 175$) and LPG ($n = 99$), matched for economic status and age, indicated that the relative risk (with 95 per cent C.I.) for cough, and shortness of breath (dyspnoea) was 3.2 (1.6–6.7), and 4.6 (1.2–18.2) respectively.

Childhood acute respiratory infections

Acute lower respiratory infections

Acute respiratory infections (ARIs) are the single most important cause of mortality in children aged less than 5 years, accounting for around 3–5 million deaths annually in this age group. Many studies in developing countries have reported on the association between exposure to indoor air pollution and acute lower respiratory infections. The studies on indoor air pollution from household biomass fuel are reasonably consistent and, as a group, show a significant increase in risk for exposed young children compared with those living in households using cleaner fuels or being otherwise less exposed. Some of the studies carried out in India have reported no association between use of biomass fuels and ARI in children. In a case-control study in children under five years of age in south Kerala, where children with severe pneumonia as ascertained by WHO criteria were compared with those having non-severe ARI attending out patient department, the fuel used for cooking was not a significant risk factor for severe ARI. Non-severe ARI controls may represent the continuum (predecessor) of the cases themselves. Smith in a cross-sectional study in 642 infants dwelling in urban slums of Delhi and using wood and kerosene respectively, did not find a significant difference in the prevalence of acute lower respiratory tract infections and the fuel type.

Upper respiratory tract infections and otitis media

Studies on the relationship between indoor air pollution and acute upper respiratory infections in children both from developed and developing nations have not been able to demonstrate the relationship between the two. However, there is strong evidence that exposure to environmental tobacco smoke causes middle ear disease. A recent meta-analysis reported an odds ratio of 1.48 (1.08–2.04) for recurrent otitis media if either parent smoked, and one of 1.38 (1.23–1.55) for middle ear effusion in the same circumstances. A clinic based case-control study of children in rural New York state reported an adjusted odds ratio for otitis media, involving two or more separate episodes, of 1.73 (1.03–2.89) for exposure to woodburning stoves.

Chronic pulmonary diseases

Chronic obstructive pulmonary disease and chronic cor pulmonale

In developed countries, smoking is responsible for over 80 per cent of cases of chronic bronchitis and for most cases of emphysema and chronic obstructive pulmonary disease. Padmavati and colleagues pointed out to the relationship between exposure to indoor air pollutants and chronic obstructive lung disease leading to chronic cor pulmonale. These studies showed that in India, the incidence of chronic cor pulmonale is similar in men and women despite the fact that 75 per cent of the men and only 10 per cent women are smokers. Further analysis of the cases of chronic cor pulmonale in men and women showed that chronic cor pulmonale was more common in younger women. Chronic cor pulmonale seemed to occur 10–15 years earlier in women. The prevalence of chronic cor pulmonale was lower in the southern states than the northern states of India. This is attributed to higher ambient temperatures during most part of the year allowing for greater ventilation in the houses during cooking. The authors attributed this higher prevalence of chronic cor pulmonale in women to domestic air pollution as a result of the burning of solid biomass fuels leading to chronic bronchitis and emphysema which result in chronic cor pulmonale. Subsequent studies in India confirmed these findings. Numerous studies from other countries, including ones with cross-sectional and case-control designs, have reported on the association between exposure to biomass smoke and chronic bronchitis or chronic obstructive pulmonary disease.

Pneumoconiosis

Pneumoconiosis is a disease of industrial workers occupationally exposed to fine mineral dust particles over a long time. The disease is most frequently seen in miners. Cases of respiratory morbidity who did not respond to routine treatment and whose radiological picture resembled pneumoconiosis have been reported in Ladakh. However, there are no industries or mines in any part of Ladakh and therefore exposure to dust from these sources was ruled out. Two factors considered responsible for the development of this respiratory morbidity were (i) exposure to dust from dust storms. In the spring dust storms occur in many parts of Ladakh. During these storms the affected villages are covered by a thick blanket of fine dust, and the inhabitants are exposed to a considerable amount of dust for several days. The frequency, duration and severity of these dust storms vary considerably from village to village, and (ii) Exposure to soot — due to the severe cold in Ladakh, ventilation in the houses is kept at a minimum. The fire place is used for both cooking and heating purposes. To conserve fuel during non-cooking periods, the wood is not allowed to burn quickly but is kept smouldering to prolong its slow heating effect. The inmates are thus exposed to high concentrations of soot. The clinico-radiological investigations of 449 randomly selected villagers from three villages having mild, moderate and severe dust storms showed prevalence of pneumoconiosis of 2.0, 20.1 and 45.3 per cent respectively. The chest radiographs of the villagers showed radiological characteristics which were indistinguishable from those found in miners and industrial workers suffering from pneumoconiosis. The dust concentrations in the kitchens without chimneys varied from 3.22 to 11.30 mg/m^3 with a mean of 7.50 mg/m^3. The free silica content of these dust samples was below 1 per cent. Dust samples sufficient to allow measurement of the dust concentrations could not be collected during the periods of dust storms. A preliminary analysis of the settled dust samples collected immediately after the storms indicated that about 80 per cent of the dust was respirable and the free silica content ranged between 60 and 70 per cent. Detailed statistical analysis of the data showed that the frequency of dust storms, use of chimney in the houses and age were the most important factors related to the development of pneumoconiosis. Thus, the results of medical and radiological investigations positively established the occurrence of pneumoconiosis in epidemic proportion. Exposure to free silica from dust storms and soot from domestic fuel were suggested as the causes of pneumoconiosis. Low oxygen levels or some other factor associated with high altitude may be an important contributory factor in causation of pneumoconiosis because it has been reported that the miners working at high altitude are more prone to develop pneumoconiosis than their counterparts exposed to the same levels of dust and working in the mines at normal altitude.

Lung Cancer

The link between lung cancer in Chinese women and cooking on an open coal stove has been well established. Smoking is a major risk factor for lung cancer, however, about two-thirds of the lung cancers were reported in nonsmoking women in China, India and Mexico. The presence of previous lung disease, for example tuberculosis which is common in Indian women, is a risk factor for development of lung cancer in non-smokers. The smoke from biomass fuels contain a large number of compounds such as polyaromatic hydrocarbons, formaldehyde, etc. known for their mutagenic and carcinogenic activities, but there is a general lack of epidemiological evidence connecting lung cancer with biomass fuel exposure. The factors associated with rural environment may have a modulating effect on the occurrence of lung cancer and therefore the low incidence of lung cancer in Indian women should not lead to a final conclusion of no link between biomass exposure and lung cancer. It may be concluded that at present there is limited evidence of indoor exposure from coal fires leading to lung cancer and there is no evidence for the biomass fuels. Further investigations are needed to reach definite conclusions.

Pulmonary Tuberculosis

Smith recently reported the association between use of biomass fuels and pulmonary tuberculosis on the basis of analysis of data collected on 2,60,000 Indian adults interviewed during the 2006–2007 National Family Health Survey. Persons living in households burning biomass fuels were reported to have odd ratio of 2.58 (1.98–3.37) compared to the persons using cleaner fuel, with an adjustment for confounding factors such as separate kitchen, indoor overcrowding, age, gender, urban or rural residence and caste. The analysis further indicated that, among persons aged 20 years and above, 51 per cent of the prevalence of active tuberculosis was attributed to smoke from cooking fuel. However, this study has inherent weakness that the cases of tuberculosis were self reported. There is strong possibility of false reporting as no investigation was done to confirm the reliability of the reporting. Gupta and Mathur have reported similar findings from northern India. This study did not control for the confounding factors except for age. There is experimental evidence to show that the exposure to wood smoke may increases susceptibility of the lungs to infections. Exposure to smoke interferes with the mucociliary defences of the lungs and decreases several antibacterial properties of lung macrophages, such as adherence to glass, phagocytic rate and the number of bacteria phagocytosed. Chronic exposure to tobacco smoke also decreases cellular immunity, antibody production and local bronchial immunity, and there is increased susceptibility to infection and cancer. Indeed, tobacco smoke has been associated with tuberculosis. Although the evidence in favour of tuberculosis associated with biomass fuel exposure is extremely weak, there is a theoretical possibility of such an association and considering the public health importance of the problem further experimental and epidemiological studies are necessary.

Cataract

During cooking particularly with biomass fuels, air has to be blown into the fire from time to time especially when the fuel is moist and the fire is smouldering. This causes considerable exposure of the eyes to the emanating smoke. In a hospital-based case-control study in Delhi the use of liquefied petroleum gas was associated with an adjusted odds ratio of 0.62 (0.4–0.98) for cortical, nuclear and mixed, but not posterior sub capsular cataracts in comparison with the use of cow dung and wood. An analysis of over 1,70,000 people in India yielded an adjusted odds ratio for reported partial or complete blindness of 1.32 (1.16–1.50) in respect of persons mainly using biomass fuel compared with other fuels after adjusting for socio-economic, housing and geographical variables; there was a lack of information on smoking, nutritional state, and other factors that might have influenced the prevalence of cataract. It is believed that the toxins from biomass fuel smoke are absorbed systematically and accumulate in the lens resulting in its opacity. The growing evidence that environmental tobacco smoke causes cataracts is supportive.

Adverse Pregnancy Outcome

Low birth weight (LBW) is an important public health problem in developing nations attributed mainly to undernutrition in pregnant women. Low birth weight has serious consequences including increased possibility of death during infancy. Exposure to carbon monoxide from tobacco smoke during pregnancy has been associated with LBW. Levels of carbon monoxide in the houses using biomass fuels are high enough to result in carboxyhaemoglobin levels comparable to those in smokers. In rural Guatemala, babies born to women using wood fuel were 63 g lighter than those born to women using gas and electricity, after adjustment for socio-economic and maternal factors. A study carried out in Ahmedabad reported an excess risk of 50 per cent of stillbirth among women using biomass fuels during pregnancy. An association between exposure to ambient air pollution and adverse pregnancy outcome has been widely reported. Considering the association of LBW with a number of disease conditions later in life, there is a need for further studies.

Intervention

Adequate evidence exists to indicate that indoor air pollution in India is responsible for a high degree of morbidity and mortality warranting immediate steps for intervention. The intervention program should include (i) public awareness, (ii) change in pattern of fuel use, (iii) modification in stove design, (iv) improvement in the ventilation, and (v) multisectoral approach.

Public awareness

The first and the most important step in the prevention of illnesses resulting from biomass fuels is to educate the public, administrators and politicians to ensure their commitment and promoting awareness of the long-term health effects on the part of users. This may lead to people finding ways of minimising exposure through better kitchen management and infant protection.

Change in pattern of fuel use

The choice of fuel is mainly a matter of availability, affordability and habit. The gobar gas plant which uses biomass mainly dung has been successfully demonstrated to produce economically viable quantities of cooking gas and manure. Recently, the Government of Andhra Pradesh has introduced a program called the Deepam Scheme to subsidise the cylinder deposit fee for women from households with incomes below the poverty line to facilitate the switch from biomass to LPG. Such schemes will encourage the rural poor to use cleaner fuels. The use of solar energy for cooking is also recommended.

Modification in stove design

Use of cleaner fuels should be the long-term goal for the intervention. Till this goal is achieved, efforts should be made to modify the stoves to make them fuel efficient and provide them with a mechanism (e.g. chimney) to remove pollutants from the indoor environment. Several designs of such stoves have been produced. NIOH study showed significant decrease in levels of SPM, SO_2, NO_x and formaldehyde with specially designed smokeless stoves in comparison with traditional cooking stoves. However, they have not been accepted widely. Large scale acceptance of improved stoves would require determined efforts. The most important barriers to new stove introduction are not technical but social.

Improvement in ventilation

In many parts of the country poor rural folk are provided with subsidised houses under various government/international agencies aided schemes. Ventilation in the kitchen should be given due priority in the design of the houses. In existing houses, measures such as putting a window above the cooking stove and providing cross ventilation through the door may help in diluting the pollution load.

Multisectoral approach

Effective tackling of indoor air pollution requires collaboration and commitment between agencies responsible for health, energy, environment, housing and rural development.

To sum up indoor air pollution caused by burning traditional fuels such as dung, wood and crop residues causes considerable damage to the health of particularly women and children. There is evidence associating the use of biomass fuel with acute respiratory tract infections in children, chronic obstructive lung diseases, and pneumoconiosis in the residents of Ladakh villages. Lung cancer has been found to be associated with the use of coal in China, however, there is no evidence associating it with the use of biomass fuels. Cataract and adverse pregnancy outcome are the other conditions shown to be associated with the use of biomass fuels. The association of tuberculosis and chronic lung infections with the use of biomass fuels has not been proved.

Finally, there is enough evidence to accept that indoor air pollution in India is responsible for a high degree of morbidity and mortality warranting immediate steps for intervention. The first and the most important step in the prevention of illnesses resulting from the use of biomass fuels is to educate the public, administrators and politicians to ensure their commitment for the improvement of public health. There is utmost requirement to collect better and systematic information about actual exposure levels experienced by households in different districts and climatic zones and develop a model for predicting the exposure levels based on fuel use and other household data therein (exposure atlas) to protect the health of children, women and elderly persons.

Chapter 6

Meteorological Aspects of Air Pollution

INTRODUCTION

The practicing weather forecaster cares very little about the composition of the air aside from its water vapour content. But we are beginning to realise that there is more to meteorology than advising the public whether or not it will rain tomorrow. We live in the atmosphere and it is important that we keep it as clean as possible. Man's dispersal of wastes into the air has become so enormous that in certain areas, particularly urban ones like Los Angeles, the atmosphere is at times incapable of providing adequate local dilution.

Though pollution by cities of their own immediate locality is a major concern, we should keep in mind that most pollutants are not destroyed by disposing of them downwind of a city. Under some circumstances, the pollutants of many cities combine and affect areas far removed from the sources of pollution. Ultimately, if the exponential growth of industry with its waste disposal problems continues, contamination may become global, making further dilution impossible. When pollution becomes a worldwide problem, the allowable releases will depend on the atmosphere's cleansing ability. The consequences of air pollution reach beyond immediate effects such as damage to health and agriculture. They affect the weather as well. There is strong evidence that increased atmospheric contamination reduces visibility and modifies electrical conductivity, precipitation, and the radiation balance. What subtle and far-reaching effects will result from these and other phenomena are now unknown.

Our present aims in studying the meteorology of air pollution are measurement, understanding, and prediction. Unfortunately, none of these reduces the amount of pollutant put into the atmosphere. But through the proper selection of industrial sites, local contamination can be minimised; through meteorological detective work, an offending culprit can be uncovered; and through forecasting, pollutants can, in principle, be withheld pending a return to conditions more conducive to dilution.

The major effort in meteorological research of air pollution is on a microscale—citywide contamination. At present, a meteorological analysis is considered successful if a way is found for the pollutants to be vented into the atmosphere without local difficulties. By the time the effluent reaches the next city, diffusion is expected to reduce concentrations to a satisfactory level. But there are exceptions even today. For example, Dr. C. D. Keeling of the Scripps Institution of Oceanography at La Jolla, Calif., almost 100 miles south of Los Angeles, says that he has often watched the carbon dioxide concentration rise from 310 to 340 ppm as a brown cloud from the north descends on his observation post. Similarly, it has been estimated that 15 per cent of the organic particulates measured at Louisville, Ky, originate outside the area of Greater Louisville.

145

The areawide air pollution problem will be with us in the near future, if it is not here today. Meteorological study of this larger problem has so far succeeded in identifying the weather conditions conducive to large-scale air pollution episodes and confirming the relation between these conditions and high concentrations of pollutants.

DISPERSION OF POLLUTANTS

Dispersion Characteristics of Stack Plumes

Dispersion is the process of spreading out pollution emission over a large area and thus reducing their concentration. Wind speed and environmental lapse rates directly influence the dispersion pattern. Five classifications of plume behaviour, which may occur under some commonly encountered metrological conditions, are shown in Fig. 6.1(a) to 6.1(e) and discussed in the following sections.

Coning

A coning plume, shown in Fig. 6.1(a), occurs under essentially neutral stability, when environmental lapse rate is equal to adiabatic lapse rate, and moderate to strong winds occur. The plume enlarges in the shape of a cone. A major part of pollution may be carried fairly far downwind before reaching ground.

Looping

Under super-adiabatic condition, both upward and downward movement of the plume is possible. Large eddies of a strong wind cause a looping pattern, Fig. 6.1(b). Although the large eddies tend to disperse pollutants over a wide region, high ground level concentrations may occur close to the stack.

Fanning

A fanning plume occurs in the presence of a negative lapse rate when vertical dispersion is restricted, Fig. 6.1(c). The pollutants disperse at the stack height, horizontally in the from of a fanning plume.

Fumigation

As shown in Fig. 6.1(d), when the emission from the stack is under an inversion layer, the movement of the pollutants in the upward direction is restricted. The pollutants move downwards. The resulting fumigation can lead to a high ground level concentration downwind of the stack.

Lofting

When the stack is sufficiently high and the emission is above an inversion layer, Fig. 6.1(e), mixing in the upward direction is uninhibited, but downward motion is restricted. Such lofting plumes do not result in any significant concentration at ground level. However, the pollutants are carried hundreds of kilometres from the source.

Stability Classification

For the purpose of calculation of concentration of pollutants downwind of a source the stability of the atmosphere is classified as:

A = Extremely unstable, B = Moderately unstable, C = Slightly unstable, D = Neutral, E= Slightly stable, F = Stable.

These classifications are arrived at from metrological condition of wind speed, solar insulation and cloudiness, given in Table 6.1.

Table 6.1. Atmospheric stability classification.

Surface[a] wind, m/s	Day solar insulation			Night cloudiness	
	Strong[b]	Moderate[c]	Slight[d]	Cloudy (>4/8)	Clear (<3/8)
<2	A	A-B	B	E	F
2–3	A-B	B	C	E	F
3–5	B	B-C	C	D	E
5–6	C	C-D	D	D	E
>6	C	D	D	D	D

[a]At 10 m above ground, [b]sun higher than 60°C and clear sky, [c]sun 35°–60°C and few broken clouds or clear sky, [d]sun 15°–35°C, cloudy.

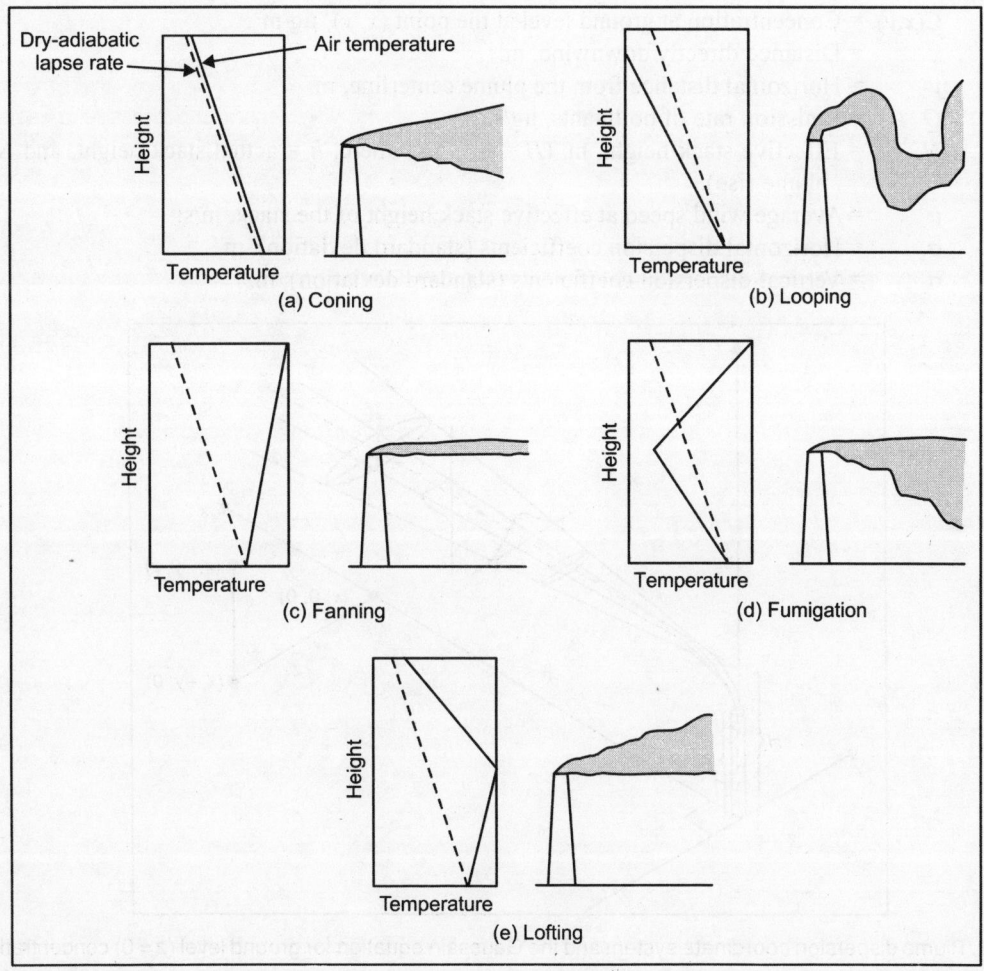

(a) Coning

(b) Looping

(c) Fanning

(d) Fumigation

(e) Lofting

Fig. 6.1. Effect of lapse rates and stack, heights on plume behaviour (adiabatic profile -------, environmental profile ————).

The above described stability classification is known as Pasquill's stability classification which is the most popular one because of its simplicity.

Gaussian Plume Model

Gaussian plume model is used to calculate the concentration of a pollutant downwind of a point source. Figure 6.2 shows the three dimensional coordinates system and the Gaussian plume equation for the case when the value of z is zero, for calculating ground level concentration:

$$C(x,y) = \frac{Q}{\pi u \sigma_y \sigma_z}\left(\exp\frac{-H^2}{2\sigma_z^2}\right)\left(\exp\frac{-y^2}{2\sigma_y^2}\right)$$

where,

$C(x,y)$ = Concentration at ground level at the point (x, y), $\mu g/m^3$.
x = Distance directly downwind, m.
y = Horizontal distance from the plume centerline, m.
Q = Emission rate of pollutants, $\mu g/s$.
H = Effective stack height, m, ($H = h + \Delta h$, where, h = actual stack height, and Δh = plume rise).
u = Average wind speed at effective stack height of the stack, m/s.
σ_y = Horizontal dispersion coefficients (standard deviation), m.
σ_z = Vertical dispersion coefficients (standard deviation), m.

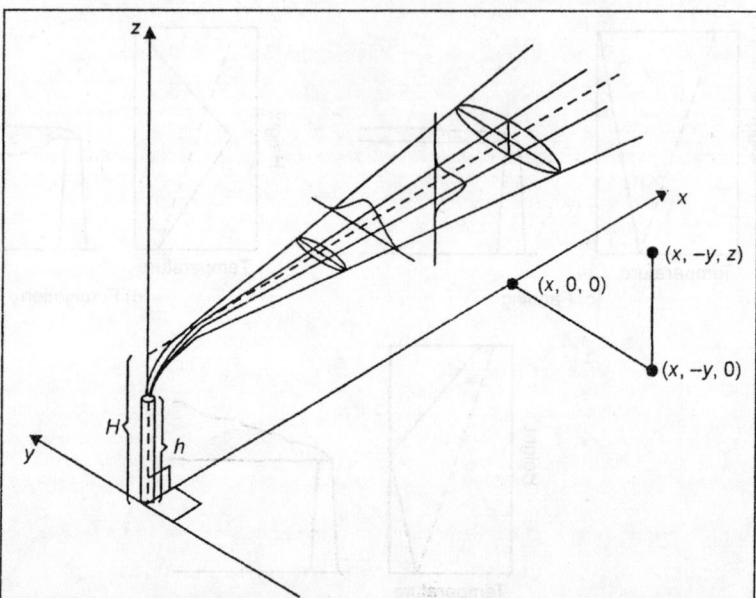

Fig. 6.2. Plume dispersion coordinate system and the Gaussain equation for ground level ($z = 0$) concentration.

The dispersion coefficient depends on the atmospheric stability class and increase with the downwind distance from the source. Figure 6.3 gives values of the dispersion coefficients as a function of distance for various stability classes. The Gaussian plume equation can be used to predict ground level pollutant concentrations under different stability conditions for a given pollution source.

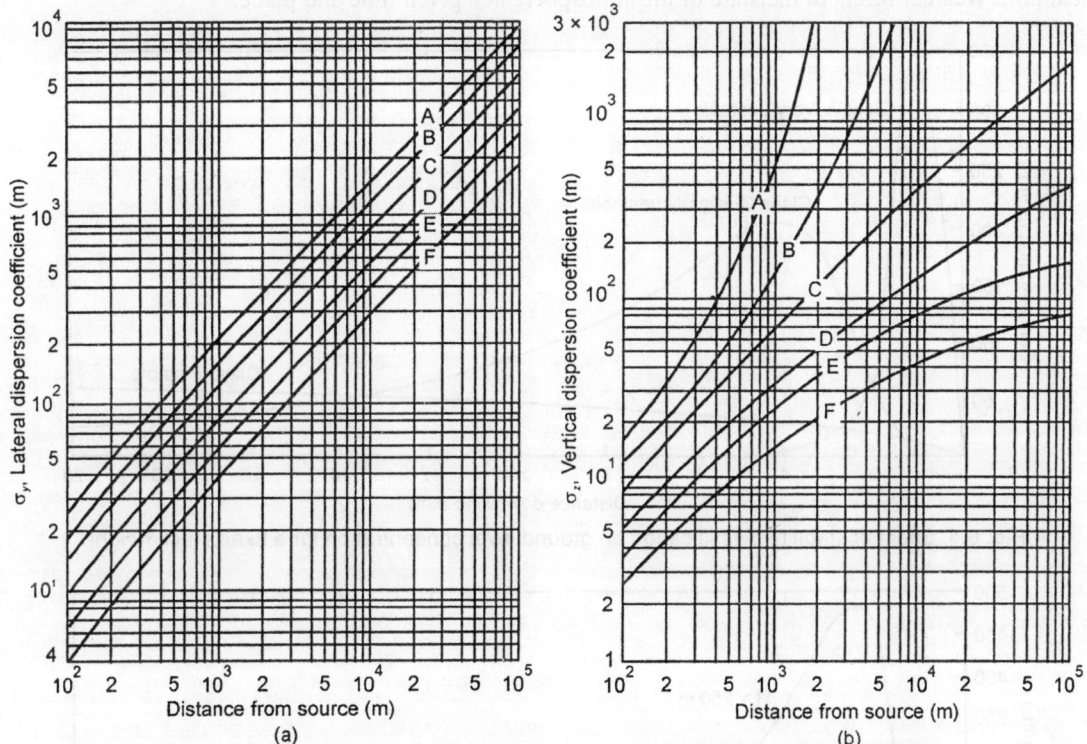

Fig. 6.3. Gaussain dispersion coefficients as a function of downwind distance.

Figure 6.4 shows the downwind ground level concentrations due to emissions from a thermal power plant calculated for different stability classes. Note that the turbulence in an unstable atmosphere results in a high concentration near the stack. Downwind, however, concentrations drop off very quickly. The stable atmosphere, on the other hand, has a much lower peak. However, it continues to be appreciable for a considerable distance downwind.

As shown in Fig. 6.5, the ground level concentration is also quite sensitive to stack height. In order to disperse pollutants over a larger area, stack heights in the range of 200 to 250 m are quite common. The stack heights shown in Fig. 6.5 are effective stack heights, which are higher than the actual stack heights because of the buoyancy of the plume at the emission point as shown in Fig. 6.2.

It is possible to calculate average ground level concentrations over a specified period by integrating values occurring under different stability conditions and wind speed and direction. Such data can be presented in the form of isopleths as shown in Fig. 6.6.

CONTROL EMISSIONS TECHNOLOGIES—TRANSPORT AND DISPERSION OF AIR POLLUTANTS

The transport and dispersion of air pollutants in the ambient air are influenced by many complex factors. Global and regional weather patterns and local topographical conditions affect the way that pollutants are transported and dispersed. For example, the prevailing direction for weather patterns in the United States is from west to east and this is an important factor in the transport of pollutants that contribute to acid rain. Weather refers to the state of the atmosphere at a given time and place.

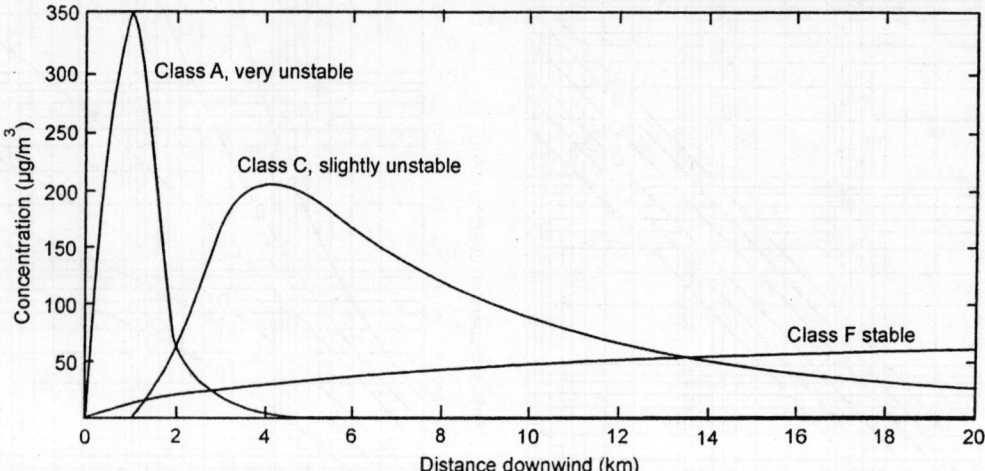

Fig. 6.4. Effect of stability classification on ground level concentration for a fixed stack height.

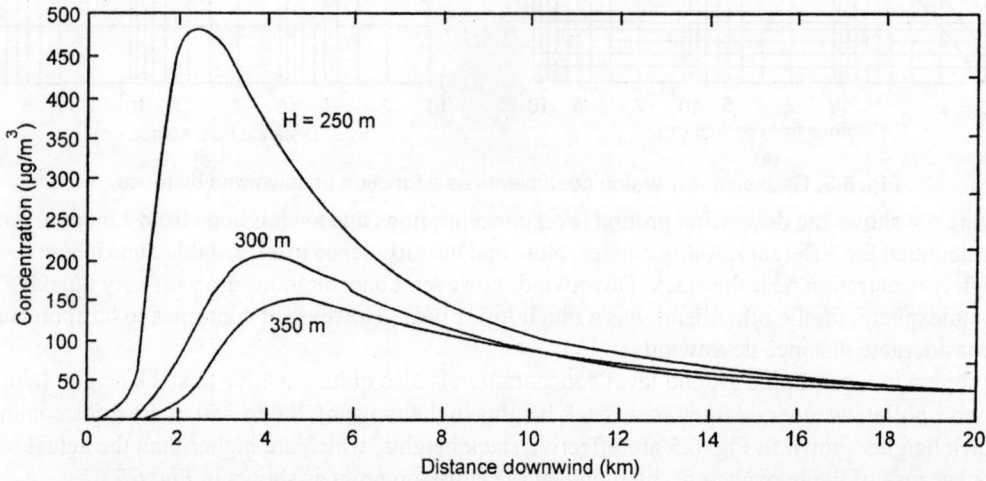

Fig. 6.5. Effect of stack height on ground level concentration for constant stability classification.

On a more local level, the primary factors affecting transport and dispersion of pollutants are wind and stability. Wind is the natural horizontal motion of the atmosphere. It occurs when warm air rises, and cool air comes in to take its place. Wind is caused by differences in pressure in the atmosphere. The pressure is the weight of the atmosphere at a given point. The height and temperature of a column of air determines the atmospheric weight. Because cool air weighs more than warm air, a high pressure mass of air is made up of cool and heavy air. Conversely, a low pressure mass of air is made up of warmer and lighter air. Differences in pressure cause air to move from high pressure areas to low pressure areas, resulting in wind. Wind speed can greatly affect the pollutant concentration in a local area. The higher the wind speed, the lower the pollutant concentration. Wind dilutes pollutants and rapidly disperses them throughout the immediate area.

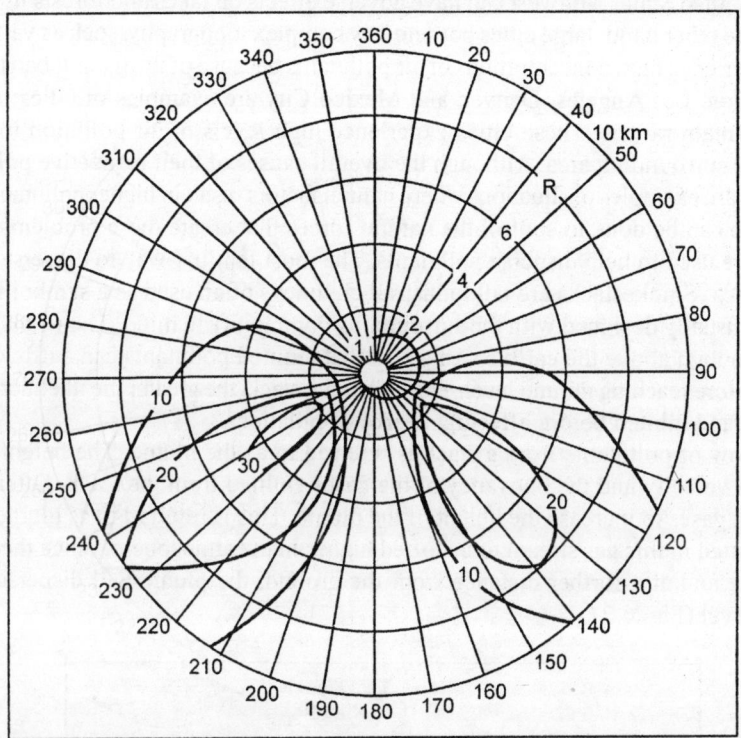

Fig. 6.6. Hypothetical monthly ground level isopleths of pollutant concentration.

Atmospheric stability refers to the vertical motion of the atmosphere. Unstable atmospheric conditions result in a vertical mixing. Typically, the air near the surface of the earth is warmer in the day time because of the absorption of the sun's energy. The warmer and lighter air from the surface then rises and mixes with the cooler and heavier air in the upper atmosphere causing unstable conditions in the atmosphere. This constant turnover also results in dispersal of polluted air. Stable atmospheric conditions usually occur when warm air is above cool air and the mixing depth is significantly restricted. This condition is called a temperature inversion. During a temperature inversion, air pollution released into the atmosphere's lowest layer is trapped there and can be removed only by strong horizontal winds. Because high-pressure systems often combine temperature inversion conditions and low wind speeds, their long residency over an industrial area usually results in episodes of severe smog.

The dispersion of pollutants from a source is also influenced by the amount of turbulence in the atmosphere near the source. Turbulence can be created by both the horizontal and vertical motion of the atmosphere. Other basic meteorological factors that affect concentration of air pollutants in the ambient air are: solar radiation, precipitation, and humidity. Solar radiation contributes to the formation of ozone and acts to create secondary pollutants in the air. Humidity and precipitation can also act on pollutants in the air to create more dangerous secondary pollutants, such as the substances responsible for acid rain. Precipitation can also have a beneficial effect by washing pollutant particles from the air and helping to minimise particulate matter formed by activities such as construction and some industrial processes. Because of the factors responsible for the transport and dispersion of pollutants, air pollution produced in the United States Midwest can have adverse effects on lakes and forests in the East coast of the country. On the other hand, large cities bordered by complex topography, such as valleys or mountain ranges, often experience high concentrations of air pollutants because of the natural barrier that interrupts pollution dispersion. Los Angeles, Denver, and Mexico City are examples of cities located in basins bordered by mountain ranges. These cities experience high levels of air pollution influenced by the topography of the surrounding area. Although the overall causes of their respective pollution problems are complex, they are examples of situations where natural factors result in higher pollutant concentrations.

Although little can be done to control the natural forces that create these problems, there are some techniques that are used to help disperse pollutants. The most familiar way to disperse air pollutants is to use a smokestack. Smokestacks are tall industrial chimneys often used as a symbol for air pollution. Smokestacks are usually designed with the surrounding community in mind. Their function is to release pollutants high enough above the earth's surface so that emitted pollutants can sufficiently disperse in the atmosphere before reaching ground level. The higher the stack, the greater the likelihood that pollutants will be dispersed and diluted before affecting nearby populations.

The visible flow of pollutants from a stack is referred to as the plume. The height of the plume is influenced by the velocity and the buoyancy of the gases emitted from the stack. Often, heat energy is added to the stack gases to increase the height of the plume. This is referred to as plume rise and allows air pollutants emitted in this gas stream to be lofted higher in the atmosphere. Since the plume is higher in the atmosphere and at a further distance from the ground, the plume will disperse more before it reaches ground level (Fig. 6.7).

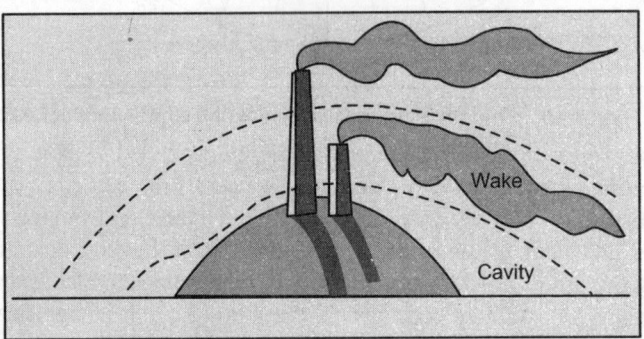

Fig. 6.7. Plume aerodynamic downwash caused by stack height and the immediate surroundings of the stack.

The shape of the plume is affected by the stack height and the immediate surroundings of the stack. As air moves over and around buildings and other structures, turbulent wakes are formed. Depending upon the release height of a plume (stack height) it may be possible for the plume to be pulled down into this wake area. This is referred to as aerodynamic or building downwash of the plume and can lead to elevated pollutant concentrations immediately downwind of the source. On the other hand, stability conditions in the atmosphere also will cause changes in the plume behaviour. Stable conditions will cause the plume to be 'flat' while instable conditions will cause the plume to 'roll' or 'loop'.

Pollutants emitted from smokestacks can be transported over long distances. In general, the concentration of the pollutant decreases as it travels from the point of release and is dispersed by wind and other natural sources. Weather patterns influence the general direction and dispersion of the pollutant. As we mentioned before, pollutants released in the Midwest affect the population and the natural habitat in the East. Weather patterns are also the reason why pollution problems such as acid rain are issues of regional and international concern. Finally, pollutant dispersion and transport can be negatively affected by climate and geographical factors. An example of how climate can affect pollutant dispersion is temperature inversion. An inversion can prevent the rise and dispersal of pollutants from the lower layers of the atmosphere and cause a localised air pollution problem like the episodes that took place in London, England and Donora, Pennsylvania.

ATMOSPHERIC DISPERSION MODELLING

Atmospheric dispersion modelling is the mathematical simulation of how air pollutants disperse in the ambient atmosphere. It is performed with computer programs that solve the mathematical equations and algorithms which simulate the pollutant dispersion. The dispersion models are used to estimate or to predict the downwind concentration of air pollutants or toxins emitted from sources such as industrial plants, vehicular traffic or accidental chemical releases.

Such models are important to governmental agencies tasked with protecting and managing the ambient air quality. The models are typically employed to determine whether existing or proposed new industrial facilities are or will be in compliance with the National Ambient Air Quality Standards (NAAQS) in the United States and other nations. The models also serve to assist in the design of effective control strategies to reduce emissions of harmful air pollutants.

Air dispersion models are also used by public safety responders and emergency management personnel for emergency planning of accidental chemical releases. Models are used to determine the consequences of accidental releases of hazardous or toxic materials. Accidental releases may result fires, spills or explosions that involve hazardous materials, such as chemicals or radionuclides. The results of dispersion modelling, using worst case accidental release source terms and meteorological conditions, can provide an estimate of location impacted areas, ambient concentrations, and be used to determine protective actions appropriate in the event a release occurs. Appropriate protective actions may include evacuation or shelter-in-place for persons in the downwind direction. At industrial facilities, this type of consequence assessment or emergency planning is required under the Clean Air Act (CAA) codified in part 60 of Title 40 of the Code of Federal Regulations.

The dispersion models vary depending on the mathematics used to develop the model, but all require the input of data that may include:

1. Meteorological conditions such as wind speed and direction, the amount of atmospheric turbulence (as characterised by what is called the 'stability class'), the ambient air temperature, the height to the bottom of any inversion aloft that may be present, cloud cover and solar radiation.
2. Source term (the concentration or quantity of toxins in emission or accidental release source terms) and temperature of the material
3. Emissions or release parameters such as source location and height, type of source (i.e. fire, pool or vent stack) and exit velocity, exit temperature and mass flow rate or release rate.
4. Terrain elevations at the source location and at the receptor location(s), such as nearby homes, schools, businesses and hospitals.

5. The location, height and width of any obstructions (such as buildings or other structures) in the path of the emitted gaseous plume, surface roughness or the use of a more generic parameter 'rural' or 'city' terrain.

Many of the modern, advanced dispersion modelling programs include a preprocessor module for the input of meteorological and other data, and many also include a post-processor module for graphing the output data and/or plotting the area impacted by the air pollutants on maps. The plots of areas impacted may also include isopleths showing areas of minimal to high concentrations that define areas of the highest health risk. The isopleths plots are useful in determining protective actions for the public and responders.

The atmospheric dispersion models are also known as atmospheric diffusion models, air dispersion models, air quality models, and air pollution dispersion models.

Atmospheric Layers

Discussion of the layers in the earth's atmosphere is needed to understand where airborne pollutants disperse in the atmosphere. The layer closest to the earth's surface is known as the troposphere. It extends from sea-level to a height of about 18 km and contains about 80 per cent of the mass of the overall atmosphere. The stratosphere is the next layer and extends from 18 km to about 50 km. The third layer is the mesosphere which extends from 50 km to about 80 km. There are other layers above 80 km, but they are insignificant with respect to atmospheric dispersion modelling.

The lowest part of the troposphere is called the atmospheric boundary layer (ABL) or the planetary boundary layer (PBL) and extends from the earth's surface to about 1.5 to 2.0 km in height. The air temperature of the atmospheric boundary layer decreases with increasing altitude until it reaches what is called the inversion layer (where the temperature increases with increasing altitude) that caps the atmospheric boundary layer. The upper part of the troposphere (i.e. above the inversion layer) is called the free troposphere and it extends up to the 18 km height of the troposphere.

The ABL is of the most important with respect to the emission, transport and dispersion of airborne pollutants. The part of the ABL between the earth's surface and the bottom of the inversion layer is known as the mixing layer. Almost all of the airborne pollutants emitted into the ambient atmosphere are transported and dispersed within the mixing layer. Some of the emissions penetrate the inversion layer and enter the free troposphere above the ABL.

In summary, the layers of the earth's atmosphere from the surface of the ground upwards are: the ABL made up of the mixing layer capped by the inversion layer; the free troposphere; the stratosphere; the mesosphere and others. Many atmospheric dispersion models are referred to as boundary layer models because they mainly model air pollutant dispersion within the ABL. To avoid confusion, models referred to as mesoscale models have dispersion modelling capabilities that extend horizontally up to a few hundred kilometres. It does not mean that they model dispersion in the mesosphere.

AIR POLLUTION DISPERSION TERMINOLOGY

Air pollution dispersion terminology includes the words and technical terms that have a special meaning to those who work in the field of air pollution dispersion modelling. Governmental environmental protection agencies (local, state, province and national) of many countries have also adopted and used much of the terminology in their laws and regulations regarding air pollution control.

Some of the words and technical terms in air pollution dispersion terminology quite often have other special meanings when used in fields of activity other than air pollution dispersion modelling.

Air Pollution Emission Plumes

There are three primary types of air pollution emission plumes:
1. Buoyant plumes: Plumes which are lighter than air because they are at a higher temperature and lower density than the ambient air which surrounds them, or because they are at about the same temperature as the ambient air but have a lower molecular weight and hence lower density than the ambient air. For example, the emissions from the flue gas stacks of industrial furnaces are buoyant because they are considerably warmer and less dense than the ambient air. As another example, an emission plume of methane gas at ambient air temperatures is buoyant because methane has a lower molecular weight than the ambient air.
2. Dense gas plumes: Plumes which are heavier than air because they have a higher density than the surrounding ambient air. A plume may have a higher density than air because it has a higher molecular weight than air (for example, a plume of carbon dioxide). A plume may also have a higher density than air if the plume is at a much lower temperature than the air. For example, a plume of evaporated gaseous methane from an accidental release of liquefied natural gas (LNG) may be as cold as $-161°C$.
3. Passive or neutral plumes: Plumes which are neither lighter or heavier than air.

Air Pollutant Emission Sources

The types of air pollutant emission sources are commonly characterised as either point, line, area or volume sources:
1. Point source: A point source is a single, identifiable source of air pollutant emissions (for example, the emissions from a combustion furnace flue gas stack). Point sources are also characterised as being either elevated or at ground-level. A point source has no geometric dimensions.
2. Line sources: A line source is one-dimensional source of air pollutant emissions (for example, the emissions from the vehicular traffic on a roadway).
3. Area source: An area source is a two-dimensional source of diffuse air pollutant emissions (for example, the emissions from a forest fire, a landfill or the evaporated vapours from a large spill of volatile liquid).
4. Volume source: A volume source is a three-dimensional source of diffuse air pollutant emissions. Essentially, it is an area source with a third (height) dimension (for example, the fugitive gaseous emissions from piping flanges, valves and other equipment at various heights within industrial facilities such as oil refineries and petrochemical plants). Another example would be the emissions from an automobile paint shop with multiple roof vents or multiple open windows.

Other air pollutant emission source characterisations are:
1. Sources may be characterised as either stationary or mobile. Flue gas stacks are examples of stationary sources and buses are examples of mobile sources.
2. Sources may be characterised as either urban or rural because urban areas constitute a so-called heat island and the heat rising from an urban area causes the atmosphere above an urban area to be more turbulent than the atmosphere above a rural area.
3. Sources may be characterised by their elevation relative to the ground as either surface or ground-level, near surface or elevated sources.
4. Sources may also be characterised by their time duration:
   (a) Puff or intermittent: short term sources (for example, many accidental emission releases are short term puffs).
   (b) Continuous: a long-term source (for example, most flue gas stack emissions are continuous).

Characterisation of Atmospheric Turbulence

The amount of turbulence in the ambient atmosphere has a major effect on the dispersion of air pollution plumes because turbulence increases the entrainment and mixing of unpolluted air into the plume and thereby acts to reduce the concentration of pollutants in the plume (i.e. enhances the plume dispersion). It is therefore important to categorise the amount of atmospheric turbulence present at any given time.

Pasquill atmospheric stability classes

The oldest and, for a great many years, the most commonly used method of categorising the amount of atmospheric turbulence present was the method developed by Pasquill in 1961. He categorised the atmospheric turbulence into six stability classes named A, B, C, D, E and F with class A being the most unstable or most turbulent class, and class F the most stable or least turbulent class. Table 6.2 lists the six classes and Table 6.3 provides the meteorological conditions that define each class. For air dispersion modelling exercises, the conditions of dual stability classes like A – B, B – C and C – D can be considered as B, C and D respectively.

Table 6.2. The Pasquill stability classes.

Stability class	Definition	Stability class	Definition
A	Very unstable	D	Neutral
B	Unstable	E	Slightly stable
C	Slightly unstable	F	Stable

Table 6.3. Meteorological conditions that define the Pasquill stability classes.

Surface windspeed		Daytime incoming solar radiation			Night-time cloud cover	
m/s	mi/hr	Strong	Moderate	Slight	> 50%	< 50%
< 2	< 5	A	A – B	B	E	F
2 – 3	5 – 7	A – B	B	C	E	F
3 – 5	7 – 11	B	B – C	C	D	E
5 – 6	11 – 13	C	C – D	D	D	D
> 6	> 13	C	D	D	D	D

Note: Class D applies to heavily overcast skies, at any windspeed day or night.

Historical stability class data, known as the Stability Array (STAR) data, for sites within the USA can be purchased from the National Climatic Data Center (NCDC).

Advanced methods of categorising atmospheric turbulence

Many of the more advanced air pollution dispersion models do not categorise atmospheric turbulence by using the simple meteorological parameters commonly used in defining the six Pasquill classes as shown in Table 6.3. The more advanced models use some form of Monin-Obukhov similarity theory.

For example, the US EPA's most advanced model, AERMOD, no longer uses the Pasquill stability classes to categorise atmospheric turbulence. Instead, it uses the surface roughness length and the Monin-Obukhov length. As another example, the United Kingdom's most advanced model, ADMS 3, uses the Monin-Obukhov length, the boundary layer height and the windspeed to categorise the atmospheric turbulence. The detailed explanation of the mathematical formulation for the turbulence categorisation methods used in AERMOD, ADMS 3 and other advanced air pollution dispersion models is very complex and beyond the scope of this chapter.

Miscellaneous Other Terminology

1. Building effects or downwash: When an air pollution plume flows over nearby buildings or other structures, turbulent eddies are formed in the downwind side of the building. Those eddies cause a plume from a stack source located within about five times the height of a nearby building or structure to be forced down to the ground much sooner than it would if a building or structure were not present. The effect can greatly increase the resulting nearby ground-level pollutant concentrations downstream of the building or structure. If the pollutants in the plume are subject to depletion by contact with the ground (particulates, for example), the concentration increase just downstream of the building or structure will decrease the concentrations further downstream.

2. Deposition of the pollution plume components to the underlying surface can be defined as either dry or wet deposition:
   (a) Dry deposition is the removal of gaseous or particulate material from the pollution plume by contact with the ground surface or vegetation (or even water surfaces) through transfer processes such as absorption and gravitational sedimentation. This may be calculated by means of a deposition velocity, which is related to the resistance of the underlying surface to the transfer.
   (b) Wet deposition is the removal of pollution plume components by the action of rain. The wet deposition of radionuclides in a pollution plume by a burst of rain often forms so called hot spots of radioactivity on the underlying surface.

3. Inversion layers: Normally, the air near the earth's surface is warmer than the air above it because the atmosphere is heated from below as solar radiation warms the earth's surface, which in turn then warms the layer of the atmosphere directly above it. Thus, the atmospheric temperature normally decreases with increasing altitude. However, under certain meteorological conditions, atmospheric layers may form in which the temperature increases with increasing altitude. Such layers are called inversion layers. When such a layer forms at the earth's surface, it is called a surface inversion. When an inversion layer forms at some distance above the earth, it is called an inversion aloft (sometimes referred to as a capping inversion). The air within an inversion aloft is very stable with very little vertical motion. Any rising parcel of air within the inversion soon expands, thereby adiabatically cooling to a lower temperature than the surrounding air and the parcel stops rising. Any sinking parcel soon compresses adiabatically to a higher temperature than the surrounding air and the parcel stops sinking. Thus, any air pollution plume that enters an inversion aloft will undergo very little vertical mixing unless it has sufficient momentum to completely pass through the inversion aloft. That is one reason why an inversion aloft is sometimes called a capping inversion.

4. Mixing height: When an inversion aloft is formed, the atmospheric layer between the earth's surface and the bottom of the inversion aloft is known as the mixing layer and the distance between the earth's surface and the bottom of inversion aloft is known as the mixing height. Any air pollution plume dispersing beneath an inversion aloft will be limited in vertical mixing to that which occurs beneath the bottom of the inversion aloft (sometimes called the lid). Even if the pollution plume penetrates the inversion, it will not undergo any further significant vertical mixing. As for a pollution plume passing completely through an inversion layer aloft, that rarely occurs unless the pollution plume's source stack is very tall and the inversion lid is fairly low.

CLEAR AIR TURBULENCE

Clear air turbulence (CAT) is the turbulent movement of air masses in the absence of any visual cues, such as clouds, and is caused when bodies of air moving at widely different speeds meet. The atmospheric region most susceptible to CAT is the high troposphere at altitudes of around 7000–12000 metres (23000–39000 ft) as it meets the tropopause. Here CAT is most frequently encountered in the regions of jet streams. At lower altitudes it may also occur near mountain ranges. Thin cirrus cloud can also indicate high probability of CAT. CAT can be hazardous to the comfort, and even safety, of air travel.

Detection

Clear air turbulence is usually impossible to detect with the naked eye and very difficult to detect with conventional radar, with the result that it is difficult for aircraft pilots to detect and avoid it. However, it can be remotely detected with instruments that can measure turbulence with optical techniques, such as scintillometers or Doppler LIDARs. Although the altitudes near the tropopause are usually cloudless, thin cirrus cloud can form where there are abrupt changes of air velocity, for example associated with jet streams. Lines of cirrus perpendicular to the jet stream indicate possible CAT, especially if the ends of the cirrus are dispersed in which case the direction of dispersal can indicate if the CAT is stronger at the left or at the right of the jet stream.

Factors That Increase Clear Air Turbulence (CAT) Probability

Detecting and predicting CAT is hard for meteorologists because it is at such heights that even when caused by factors that can be measured, intensity and location cannot be determined precisely. However because this turbulence affects long range aircraft that fly near the tropopause, CAT has been intensely studied. Several factors affect that likelihood of CAT. Often more than one factor is present. 64 per cent of the non-light turbulences (not only CAT) are observed less than 150 nautical miles (280 km) away from the core of a jet stream.

Jet stream

A jet stream alone will rarely be the cause of CAT, although there is horizontal wind shear at its edges and within it caused by the different relative air speeds of the stream and the surrounding air. Rossby waves caused by this jet stream shear and the Coriolis force cause it to meander.

Temperature gradient

A temperature gradient is the change of temperature over a distance in some given direction. Where the temperature of a gas changes, so does its density and where the density changes CAT can appear.

Vertical

From the ground upwards through the troposphere temperature decreases with height; from the tropopause upwards through the stratosphere temperature increases with height. Such variations are examples of temperature gradients.

Horizontal

A horizontal temperature gradient may occur, and hence air density variations, where air velocity changes. An example: the speed of the jet stream is not constant along its length; additionally air temperature and hence density will vary between the air within the jet stream and the air outside.

Wind shear

Wind shear is a difference in relative speed between two adjacent air masses. An excessive wind shear produces vortexes, and when the wind shear is of sufficient degree the air will tend to move chaotically. As already explained, temperature and wind velocity decrease with height in the tropospere, and the reverse is true within the stratosphere. These differences cause changes in air density, and hence viscosity. The viscosity of the air thus presents both inertias and accelerations which cannot be determined in advance.

Vertical

Vertical wind shear above the jet stream (i.e. in the stratosphere) is sharper when it is moving upwards, because wind speed increases with height in the stratosphere. This is the reason CAT can be generated above the tropopause, despite the stratosphere otherwise being a region which is vertically stable. On the other hand, vertical wind shear moving downwards within the stratosphere is more moderate (i.e. because downwards wind shear within the stratosphere is effectively moving against the manner in which wind speed changes within the stratosphere) and CAT is never produced in the stratosphere. Similar considerations apply to the troposphere but in reverse.

Horizontal

When strong wind deviates, the change of wind direction implies a change in the wind speed. A stream of wind can change its direction by differences of pressure. CAT appears more frequently when the wind is surrounding a low pressure region, especially with sharp troughs that change the wind direction more than 100°. Extreme CAT has been reported without any other factor than this.

Mountain waves

Mountain waves are formed when four requirements are met (Fig. 6.8). When these factors coincide with jet streams, CAT can occur:
1. A mountain range, not an isolated mountain.
2. Strong perpendicular wind.
3. Wind direction maintained with altitude.
4. Temperature inversion at the top of the mountain range.

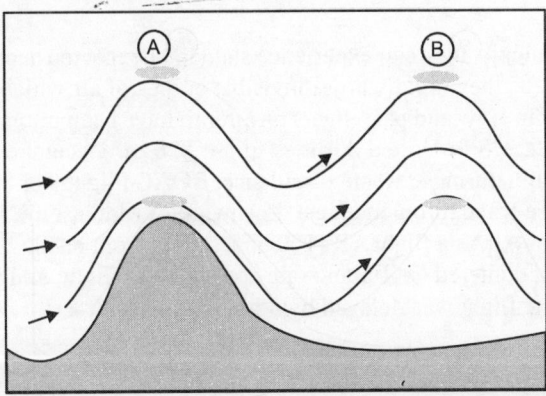

Fig. 6.8. Wind flow over a mountain produces oscillations (A), (B), etc.

Gravity wave wind shear

The tropopause is a layer which separates two very different types of air. Beneath it the air gets colder and the wind gets faster with height. Above it the air warms and wind velocity decreases with height. These changes in temperature and velocity can produce fluctuation in the altitude of the tropopause, called gravity waves.

Effects on Aircraft

In the context of air flight, CAT is sometimes colloquially referred to as 'air pockets'. Standard airplane radars cannot detect CAT, as CAT is not associated with clouds that show unpredictable movement of the air. Airlines and pilots should be aware of factors that cause or indicate CAT in order to reduce the probability of meeting turbulence, as it can injure crew and passengers and even affect flight safety. Aircraft in level flight rely on a constant air density to retain stability. Where air density is significantly different, for instance because of temperature gradient, especially at the tropopause, CAT can occur. Where an aircraft changes its position horizontally from within the jet stream to outside the jet stream, or vice versa, a horizontal temperature gradient may be experienced. Because jet streams meander, such a change of position need not be the result of a change of course by the aircraft. Because the altitude of the tropopause is not constant, an airplane that flies at a constant altitude would traverse it and encounter any associated CAT.

Pilot rules

When a pilot experiences CAT, a number of rules should be applied:
1. The aircraft has to sustain the recommended velocity for turbulence.
2. When following the jet stream to escape from the CAT the aircraft has to change altitude and/or heading.
3. When the CAT arrives from one side of the airplane, the pilot must observe the thermometer to determine if the aircraft is above or below the jet stream and then move away from the tropopause.
4. When the CAT is associated with a sharp trough, the plane has to go through the low pressure region instead of around it.
5. The pilot has to communicate position, altitude and velocity of the turbulence to Air Traffic Control, to warn other aircraft entering the region.

Cases

Because aircraft move so quickly, they can experience sudden unexpected accelerations or bumps from turbulence, including CAT (as they rapidly cross invisible bodies of air which are moving vertically at many different speeds). Cabin crew and passengers on aircraft have been injured (and in a small number of cases, killed, as in the case of a United Airlines Flight 826 on December 28, 1997) when tossed around inside an aircraft cabin during extreme turbulence. BOAC Flight 911 broke-up in flight in 1966 after experiencing severe lee wave turbulence just downwind of Mount Fuji, Japan.

Another case of CAT was AirAsia flight AK5100 on its flight from Kuala Lumpur to Kota Kinabalu (Malaysia). The airplane encountered CAT about one hour into the flight, and part of the crew suffered injuries, such as bruises. The flight was delayed by 5 hours as a result.

Wake turbulence

Wake turbulence is another dangerous type of clear-air turbulence, but in this case the causes are quite different to those set out above, in the case of wake turbulence, the rotating vortex-pair created by the wings of a large aircraft as it travels lingers for a significant amount of time after the passage of the aircraft, sometimes more than a minute.

When this occurs, the lingering turbulence caused by the wake of the wing tips can deflect or even flip a smaller aircraft on the ground or in the air, this phenomenon can also lead to accidents with large aircraft as well. Delta Air Lines Flight 9570 crashed at the Greater Southwest International Airport in 1972 while landing behind a DC-10. This accident lead to new rules for minimum following separation time from 'heavy' aircraft. American Airlines Flight 587 crashed shortly after takeoff from John F. Kennedy International Airport in 2001 due to pilot overreaction to wake turbulence from a Boeing 747 (Fig. 6.9).

Fig. 6.9. This picture from a NASA study on wingtip vortices qualitatively illustrates wake turbulence.

A major component of wake turbulence are the wingtip vortices. Many aircraft are now made with wingtip devices to reduce such turbulence (which also improves both the lift-to-drag ratio and fuel economy).

WIND SPEED

Wind speed is the speed of wind, the movement of air or other gases in an atmosphere. It is a scalar quantity, the magnitude of the vector of motion, wind speed, or wind velocity (when directionality is considered), is a fundamental abiotic factor that affects the growth and metabolism of many plant species.

Wind speed has always meant the movement of air in an outside environment, but the speed of air movement inside is important in many areas, including weather forecasting, aircraft and maritime operations, building and civil engineering. High wind speeds can cause unpleasant side effects, and strong winds often have special names, including gales, hurricanes, and typhoons.

Factors Affecting Wind Speed

Wind speed is affected by a number of factors and situations, operating on varying scales (from micro to macro scales). These include the pressure gradient, Rossby waves and jet streams, and local weather conditions. There are also links to be found between wind speed and wind direction, notably with the pressure gradient and surfaces over which the air is found.

Pressure gradient is a term to describe the difference in air pressure between two points in the atmosphere or on the surface of the earth. It is vital to wind speed, because the greater the difference in pressure, the faster the wind flows (from the high to low pressure) to balance out the variation. The pressure gradient, when combined with the Coriolis effect and friction, also influences wind direction.

Rossby waves are strong winds in the upper troposphere. These operate on a global scale and move from West to East (hence being known as Westerlies). The Rossby waves are themselves a different wind speed from what we experience in the lower troposphere.

Local weather conditions play a key role in influencing wind speed, as the formation of hurricanes, monsoons and cyclones as freak weather conditions can drastically affect the velocity of the wind.

Highest Speed

During the passage of Tropical Cyclone Olivia on 10 April 1996, an automatic weather station on Barrow Island, Australia, registered a maximum wind gust of 408 km/hr (220 kn; 253 mph). The wind gust was evaluated by the WMO Evaluation Panel who found that the anemometer was mechanically sound and the gust was within statistical probability and ratified the measurement in 2010, however, the wind speed was measured inside of the cyclone and not at ground level. During the cyclone several extreme gusts of greater than 300 km/hr (160 kt) were recorded, with a maximum 5 minute mean speed of 176 km/hr (95 kt), the extreme gust factor was in the order of 2.27–2.75 times the mean wind speed. The pattern and scales of the gusts suggests that a mesovortex was embedded in the already strong eyewall of the cyclone.

The second highest surface wind speed ever officially recorded is 372 km/h (231 mph) at the Mount Washington (New Hampshire) Observatory in the US on 12 April 1934, using a heated anemometer. The anemometer, specifically designed for use on Mount Washington, was later tested by the US National Weather Bureau and confirmed to be accurate. The highest surface wind speed ever officially recorded in Asia was recorded in Afghanistan on 14 August 2008: 328 km/hr (204 mph) in Ab-Paran, Ghowr.

Windspeeds within certain atmospheric phenomena (such as tornadoes) may greatly exceed these values but have never been accurately measured. The figure of 509 km/hr (316 mph) during the F5 tornado in Moore, Oklahoma is often quoted as the highest surface wind speed but was measured 30 m (90 feet) above ground.

In 1991, a chase team from the University of Oklahoma chased a tornado in Red Rock, Oklahoma and used a portable Doppler weather radar to measure a wind speed of 460 km/hr (286 mph).

According to Alan F. Arbogast (Discovering Physical Geopgraphy) wind direction and speed are affected by three main factors:

1. Pressure gradient—the difference in barometric pressure between adjacent zones of high and low pressure.
2. Frictional forces—features on the earth's surface which oppose the wind, e.g. mountains, trees, buildings, etc.
3. Coriolis effect—the earth's rotation causes winds to be deflected to the right in the Northern Hemisphere, and in the Southern Hemisphere to the left.

All three of these combined result in the spiral motion of air in both high and low pressure systems. Wind speed is a common factor in the design of structures and buildings around the world. The wind speed is often the governing factor in the 'lateral' design of a structure and is used by professional engineers and designers.

SECTION II

Global Warming

Chapter 7

Impact of Air Pollution on Local and Global Scale

INTRODUCTION

Air pollution is somewhat difficult to define because many air pollutants, at low concentrations, are essential nutrients for the sustainable development of ecosystems. So, air pollution could be defined as: 'A state of the atmosphere, which leads to the exposure of human beings and/or ecosystems to such high levels or loads of specific compounds or mixtures thereof, that damage is caused'. With very few exceptions, all compounds that are considered air pollutants have both natural as well as human-made origins. Air pollution is not a new phenomenon; in Medieval times, the burning of coal was forbidden in London while Parliament was in session. Air pollution problems have dramatically increased in intensity as well as scale due to the increase in emissions since the Industrial Revolution.

LOCAL POLLUTION

All reports on air pollution in the nineteenth and early twentieth centuries indicate that the problems were local, and concentrated in or around industrial centers and big cities. Even the infamous environmental catastrophes in the area of Liege in the 1930s—the first recorded occurrence of death by air pollution or in London in the 1950s, were essentially local phenomena. In the London smog episode, stagnant air accumulated dangerously high sulphur dioxide and sulphuric acid concentrations, killing several thousand inhabitants. Epidemiological research has recently shown several damaging effects of aerosols on the respiratory tract, inducing asthma, bronchitis, and early demise. Figure 7.1 shows effects of air pollution for different scales of space.

REGIONAL POLLUTION

Only in the second half of the 20th century were the effects of air pollution detected on regional (>500 km), continental, and global scales. Around 1960, acid deposition, commonly referred to as acid rain, caused the first observed effects on regional to continental scales. Lakes in Scandinavia, as well as in North America, lost their fish populations as the lakes were acidified by acid deposition to the point that fish eggs would not produce young fish anymore. About 10 years later, acid deposition was found to cause damage to forests and loss of vitality of trees. Smog episodes in US cities, such as Los Angeles, were reported during the same period.

Reactions of volatile organic compounds (VOCs) and nitrogen oxides (NO_x) produced high concentrations of ozone and peroxides that are harmful to human and ecosystem health. Around the same period, the first high oxidant concentrations (the complex mixture of ozone, peroxides and other products of the reactions of organics and nitrogen oxides are called oxidants) were more and more frequently occurring in Europe during stagnant meteorological conditions.

```
Local air pollution, inorganic and organic compounds near
sources, ozone and aerosols

Regional pollution (scale 500 to 1000 km) acid deposition,
eutrophication, regional ozone

Global scale pollution:
  Stratospheric ozone loss, Antartic ozone hole

Climate change
```

Fig. 7.1. Effects of air pollutants for different scales of space.

At the same time, severe eutrophication (damage and changes in ecosystems due to the availability of excessive amounts of nutrients) was encountered in Europe and the US. Deposition of ammonium and nitrates were shown to contribute substantially to high nutrient concentrations in soil and groundwater. Nitrates and ammonium are beneficial, even essential, for development of vegetation, but in too high concentrations they lead to the loss of diversity, especially in oligotrophic (adapted to low nutrient availability) ecosystems.

GLOBAL POLLUTION

The next scale of air pollution is its effect on global dimensions, such as the destruction of stratospheric ozone due to emissions of CFC's (chlorofluorocarbon compounds). This issue was given a lot of attention in the period 1985–1995, as it was revealed that the destruction of stratospheric ozone leads to higher UV-light intensities and a higher incidence of skin cancer. From 1990 onwards, the increase in the concentrations of radiative active substances (compounds which alter the radiative balance of the earth; greenhouse gases; but also aerosols, and water in liquid form, as clouds) and the connected climatic consequences brought about new research in air pollution. The above time sequence of problems related to air pollution could give the impression of sudden increases in air pollution concentrations. However, that is probably not the case, as can be explained in the case of ozone. By carefully characterising old methodologies, Voltz-Thomas and Kley have been able to reconstruct ozone concentrations in the free troposphere (the air not directly influenced by processes taking place at the earth's surface).

The ozone concentrations in Europe slowly increased at a rate of 1 to 2 per cent per year from 10 ppb (parts per billion) to over 50 ppb. It is well-documented that the effects of ozone start at levels of about 40 parts per billion (ppb is a mixing ratio of 1 molecule ozone in a billion molecules of air). Thus, it is not surprising that the effects of ozone were detected in the 1970s, as background continental ozone was already 30 ppb, and additional oxidant formation would increase the ozone concentrations locally or regionally. However, the increase in the continental background concentrations of ozone had already been occurring over a long period of time. In general, the effects of pollution, and of air pollution in particular, are a function of the degree of transgression from the limits over which effects can be expected.

TRANSFORMATION AND DEPOSITION

Pollutants that affect human and ecosystem health in the form in which they are initially emitted are called primary pollutants. Sulphur dioxide is a good example of a primary air pollutant. Ozone, on the other hand, is a good example of a so-called secondary air pollutant; ozone is a product of a large set of atmospheric, chemical reactions involving nitrogen oxides (NO_x) and volatile organic compounds (VOCs). These precursors (compounds which lead to a certain product, in this case, ozone) are emitted by a wide variety of sources. In the US and Europe, they are mainly produced by the transportation sector. These precursors are then transported by air movement and involved in a large and intricate set of atmospheric reactions under the influence of sunlight; ozone is one of the products.

The next stage is the deposition of primary and secondary air pollutants. Deposition can take place in two different ways: wet deposition and dry deposition. In wet deposition, air pollution is first incorporated in clouds or precipitation, and is transported by way of precipitation (acid rain) to the earth's surface. In dry deposition, air pollutants are deposited directly on vegetation and the earth's surface. Effects caused by acid deposition and eutrophication are directly linked to the deposition loads.

The removal of pollutants by chemical transformations and deposition processes is essential. If, for instance, sulphur dioxide emitted by natural sources such as volcanoes were not removed through these processes, a concentration of 1,000 ppb or more would be reached in the atmosphere in only a few months, endangering the survival of ecosystems and the human population.

Once the effects of air pollutants are related to the transgression of critical loads or concentrations, it is possible to derive the necessary emission reductions. Models describing the complete process from emissions via transport, through chemical conversions, to deposition and exposure plus their effects are used for this purpose.

IMPACT OF LOCAL AIR POLLUTION

Environmental Problems Caused by Local Air Pollution

A wide range of compounds can cause local air pollution. The emissions of sulphur dioxide, nitrogen oxides, carbon monoxide, fine particulate matter, organic compounds like benzene, toluene and poly-aromatic hydrocarbons (PAH), and heavy metals in particulate matter (lead, cadmium) can cause local concentrations to reach levels which are harmful to human health. Some compounds like nitrogen oxide and volatile organic compounds cause air pollution problems in stagnant air, as the reactions between these compounds form ozone and other oxidants. Ozone is at present, together with particulate matter, the most serious pollutant in cities in developed countries. It is also becoming very problematic in developing countries. In developed countries, sulphur dioxide concentrations were brought down by replacing coal with cleaner-burning oil or low-sulphur natural gas in the production of energy or by controlling the emissions with sulphur dioxide scrubbers. The black smoke emissions were reduced by using more efficient burners in energy production and through the installation of particulate control systems such as electrostatic precipitators and filtration systems, among other things. The nitrogen oxide emissions from burners and automobiles were also drastically reduced, but the enormous increase in energy production and transportation has compensated for these lower emission factors.

A problem is the differentiation between the contribution of local pollution and pollution of regional origin. In stagnant weather conditions, when regional or inter-regional transport is not taking place, local sources are responsible. Examples of local pollution are the smog in Los Angeles, California and the ozone in Athens, Greece. However, in many cases, a combination of different sources has an impact. For example, five years ago, ozone was attributed to local sources, both in Hong Kong and Guangzhou, China. Recent research has clearly shown that ozone in these locations is often of regional origin. These findings have greatly influenced abatement measures; local measures will reduce the ozone concentrations less than expected if nitrogen oxides and volatile organic compounds are of regional origin. For at least three decades, research has been carried out to find a connection between human health and air pollution. It became clear, around 10 years ago, that such a connection exists. Improvement of inadequate measurement methods for air quality, and especially advances in epidemiological methodology (the study of statistical behaviour of illness), played an important role in determining the link between air quality and public health. Two epidemiological methods are used for this purpose, namely, time series investigations and large Cohort studies:

1. The principle of the time series method is based on comparing the actual mortality with the mortality calculated by way of the known impact of different diseases, accidents, etc. Typically, a small fraction of a few per cent is left that cannot be explained this way. This fraction is correlated with different possible factors, and a good correlation with air pollution is found, while no other known factor can give a satisfactory explanation. More detailed investigations show, at least for Europe and the United States, that no correlation is found with, e.g. sulphur dioxide concentrations, but a statistically significant correlation is found with concentrations of fine particles, PM_{10} or $PM_{2.5}$, and/or ozone.

2. The other epidemiological method used to trace the impact of air pollution is the Cohort Method. In the Cohort Method, a large group of people is characterised and divided into groups according to the parameters under investigation. Groups can be formed, e.g. as smoker/nonsmoker, living near road/living further away, etc. Next, mortality and the frequency of diseases are estimated as a function of these parameters, comparing the different groups. Statistical techniques can 'tease apart' different parameters and determine their influence on the outcome. Cohort studies are needed to quantify the decrease in life-span caused by ozone and aerosols (e.g. a large Dutch Cohort Study estimated this decrease in life-span to be one to three years), but also to discriminate between different factors.

Health Effects of Ozone

The formation of ozone can take several to 24 hours, depending on temperature, amount of received solar radiation, and relative humidity. Episodes of stagnant air are quite often encountered in cities such as Mexico City, Athens, and Barcelona. Under these stagnant conditions, local emissions can lead to high ozone concentrations which can affect human health and damage vegetation. In other cases, as encountered in the Netherlands but also in, for example, the Guangdong Province of China, transport by wind and hence exchange of air generally takes place. Ozone is mainly a regional problem in these areas. High ozone peak values have serious effects on human health, and air quality standards are based on prevention of impact on human health. Damage to crops is also a severe problem but is more dependent on long-term exposure.

A preliminary estimate shows that in the Netherlands, 1900 deaths of 16 million inhabitants are caused by ozone pollution. The impact of ozone on vegetation, trees, and agricultural crops, is well-documented. The damage from ozone to vegetation starts as soon the ozone concentrations exceed 40 ppb. The sum of total of hourly concentrations over 40 ppb is expressed as AOT-40 and damage is expected at values over 3000 to 4000.

Health Effects of Particulate Matter

Epidemiological research has established that particle pollution has a large impact on human health. The large number of very small particles is caused by high-temperature processes. Recently, transportation emissions have increased greatly. Otto (cars with ignition) and diesel engines cause great emissions of very small particles with very little mass and short lifetimes, because they coagulate very quickly. Aerosol particles between 0.2 and 2 micron present a large fraction of the mass; they are present in smaller numbers than the small particles, but they have relatively long lifetimes. Large particles are very low in number, have large mass, and short lifetimes; they 'fall' out of the atmosphere, a process called deposition.

The epidemiological research has not yet revealed, with certainty, which fraction of aerosol, PM_{10} (mass of aerosol particles with a diameter <10 μm), $PM_{2.5}$ (mass of aerosol with a diameter < 2.5 μm) or ultrafine particles or specific compounds in aerosol particles are responsible for the observed health effects. Many indicators point in the direction of ultrafine particles.

The effect of aerosols on human health has been traced in different locations. Increased mortality as a result of this pollution is detected everywhere. Initial studies found that the largest effects were observed in 'clean areas', such as the US. The mortality trends also showed a gradient going from Western to Eastern Europe. In more recent studies, the same sensitivities for aerosol have been found, independent of location. The extra mortality is roughly the same in the US, Western and Eastern Europe, and China with 2–4 per cent more deaths with an increase in concentration of 100 μg × m^{-3}.

Black smoke is a proxy (can be taken as indicative) for transportation emissions. The effect seems to be linear and no effect threshold has been found. Time series epidemiological research indicates that the increased death rate is 2 per cent per 100 μg × m^{-3}, for the Netherlands. This means, at the Dutch ambient yearly PM_{10} level of 32 μg × m^{-3}, that an additional 1700 to 3000 acute deaths per year can be attributed to aerosols, of a population of 16 million. It is much more difficult to assess the chronic effects of exposure to aerosols and the estimates of these chronic effects are quite uncertain, but could be, according to a large Dutch Cohort Study, in the order of 10,000 to 15,000 extra deaths per year for the Netherlands. In comparison, the number of deaths attributed to traffic accidents in the Netherlands is about 1300 per year.

In this large Dutch Cohort Study, groups have been sorted out as a function of living near or further away from roads with heavy traffic. Emissions from traffic clearly generate large gradients in aerosol number concentrations. However, the extra mass contribution to $PM_{2.5}$ or PM_{10}, but also of soot (indicated as EC, elemental carbon), as a function of the distance from roads, is limited.

The results of this Cohort Study indicate that extra mortality (number of deaths per given number of inhabitants) due to cardiopulmonary diseases is twice as high near roads in comparison to locations further away from roads. The implication is that Dutch 'background' aerosol concentrations have a clear impact on human health, but that this risk is twice as high near major roads.

The results of this Dutch Cohort Study indicate that $PM_{2.5}$ or ultrafine particles (aerosol with a diameter <0.1 μm) could have a large impact on human health. Very small particles are retained to a large extent in human lungs while those with the main mass (0.1 to 2 μm) are not very effectively absorbed. It is clear that extra mortality due to aerosols in, for example, Chinese cities could be quite high. Average $PM_{2.5}$ concentrations in large Chinese cities are probably between 50 and 120 μg × m^{-3}.

If the same increased mortality as found in Europe is assumed, and linear extrapolation is made (which are both quite uncertain assumptions but no sufficient data are available to make better estimates), then an extra mortality of 5000 to 10,000 due to acute effects and 20,000 to 50,000 due to chronic effects per year could be expected for a city like Beijing, China. Apart from this extra mortality, a large number of extra cases of non-mortal illnesses, such as asthma and bronchitis as well as heart diseases, can be predicted.

It is obvious that the impact of aerosols on human health will result in a large economic loss. As a result, air quality standards were introduced five years ago in Europe (yearly average PM_{10} of $40\ \mu g \times m^{-3}$) and in the US (annual $PM_{2.5}$ standard of $15\ \mu g \times m^{-3}$) and measures are being taken to meet these standards. The fact that no mechanism has been found to explain the effects of aerosols, and that correct measurements of mass and chemical composition of aerosols pose many problems, are severe hindrances to formulating and implementing effective air quality standards for aerosols. However, these new findings of the impact of aerosols, and especially the important role of transportation emissions, are already leading to changes in the development of road infrastructure, among other things, in Europe.

Recently, overviews have been compiled to analyse the impact of aerosols ($PM_{2.5}$) on human health, in terms of decreased life-span in months, in Europe. The wide-ranging impacts of particulate matter are still greatly uncertain, and as long as the mechanisms of the impact on human health are not known, it will be difficult to reach absolute certainty.

However, preliminary results are quite serious and it is important to determine what these effects could mean in cities of developing nations, where the concentrations of fine particulates are up to 10 times higher than in developed nations.

Visibility and Aerosols

Aerosols have a large impact on visibility, as we see objects by their reflected light. Aerosols of a certain size, between 0.1 and 1.5 μm in diameter, scatter visible light. Not only is the amount of light emitted by the object we are looking diminished due to the presence of aerosols, but the sunlight that is scattered by the particles forms a background haze that also limits visibility. Two types of scattering, 'Mie' and 'Rayleigh' scattering are observed. Rayleigh scattering is scattering that occurs primarily as a result of gas molecules, limiting the horizontal visibility. It is a function of the wavelength of the light and limits the visibility to approximately 130–260 km (80–160 miles) depending on the colour of the light. If the wavelength of the incoming light and the size of a particle are about the same, the light can be scattered in all directions, including backward (back scatter). This scattering by particles is referred to as Mie scatter (Fig. 7.2).

Particles between 0.1 and 1.5 micron, with about the same diameter as the wavelength of incoming solar visible light, scatter the visible light and reduce visibility. In addition, particles can also absorb visible light (such as carbon-based particles). Some light may pass through the particles or aerosols if no absorption takes place. Very small particles have minimum influence on the light. For a given size distribution of particles, a relation can be developed between the visibility and the aerosol mass as shown in Fig. 7.3.

In Fig. 7.3, the uncertainty is given by the gray band. The main source of this uncertainty is variation in size distribution, which can vary greatly in different locations. Thus, one can derive the aerosol concentration from visibility only with rather large uncertainty.

Reduction of visibility is a pollution problem, especially in those locations which are known for their scenic beauty. In mega-cities, such as Mexico City or Beijing, China, visibility can be reduced to a few hundred meters due to the impact of aerosols.

GLOBAL CONCERNS OF AIR POLLUTION

Due to the nature of global weather patterns, pollution released into the air in one country is often transported across boundaries and negatively impacts other countries. Increasing output of pollutants caused by human activities has led to higher global atmospheric concentrations of methane, carbon dioxide, nitrous oxide, HFCs (hydrofluorocarbons) and PFCs (perfluorocarbons). These are all gases that contribute to global warming. The air pollutants which cause acid rain also tend to cross national borders and cause harm to ecosystems located in countries where the pollution did not originate. Additionally, trade between developed and developing countries often encourages increased air pollution in the latter as developing countries such as China pollute the atmosphere as it satiates other countries' needs for inexpensive goods.

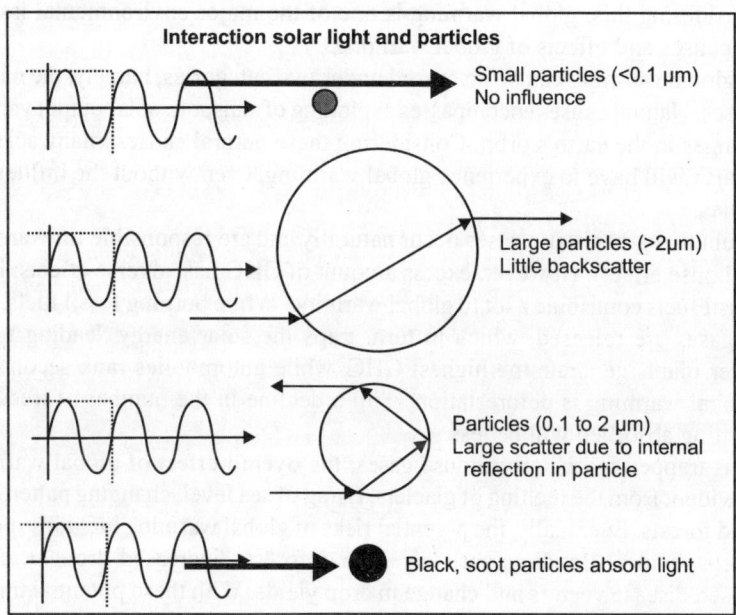

Fig. 7.2. Interaction between aerosols and incoming solar radiation.

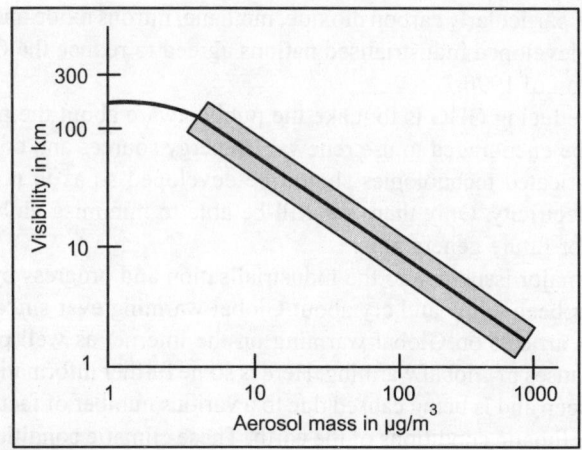

Fig. 7.3. Connection between aerosol mass and visibility, in km.

GLOBAL WARMING

Global warming can be defined as an increase in the average temperature of the earth, including the surface air and oceans. As per studies conducted, it has been stated that the earth's temperature has increased by about 1°F in the last 100 years. This rise in temperature was profound during the industrial revolution, when emission of carbon dioxide and other greenhouse gases in the atmosphere increased. If this statistics continue in the coming future, then the weather pattern of the earth will not be the same as it is today. Considering this, global warming is one of the major environmental issues today. Let's take a look at the causes and effects of global warming.

The causes of global warming can be discussed under two categories, namely, the natural causes and the man-made causes. Natural causes encompasses exploding of sunspots, solar output variations, volcanic eruptions and changes in the earth's orbit. Considering these natural causes, many scientists are of the opinion that the earth will have to experience global warming, even without the influence of industrial and human activities.

There is no doubt that greenhouse gases occur naturally and are responsible for warming the earth to sustain life (greenhouse effect). However, excess amount of GHG has adverse effects; human activities like burning of fossil fuels contribute a lot to global warming. While burning fossil fuels, carbon dioxide, ozone and other gases are released, which in turn, traps the solar energy, leading to an increase in temperature. Power plants generate the highest GHG, while automobiles rank second. Another man-made cause of global warming is deforestation; with a decline in the number of trees, the amount of carbon dioxide getting absorbed will be less.

As more heat is trapped by the greenhouse gases, the overall effect of global warming is climatic change, which is evident from the melting of glaciers, rising of sea level, changing pattern of precipitation and drying of cloud forests. Eventually, the potential risks of global warming includes species extinction, disturbed food webs, coastal flooding, shrinkage of rainforests, increased drought, extreme weather conditions, increased disease vectors and change in crop yields. With these potential threats, there have been many issues regarding the responsibilities that should be taken up against global warming.

A major impact on global warming is the signing of Kyoto Protocol, an environmental treaty, on December 11, 1997 by many nations. The main objective of Kyoto protocol is to reduce the emission of greenhouse gases (GHG), particularly carbon dioxide, methane, nitrous oxide and sulphur hexafluoride. Under this protocol, the developed industrialised nations agreed to reduce the GHG by 5.2 per cent in comparison to the emission of 1990.

An effective way of reducing GHG is to make the public aware about the pros and cons of global warming. Public should be encouraged to use renewable energy sources and conserve energy as much as possible. More sophisticated technologies should be developed so as to make cleaner cars and a better way to generate electricity. Only then, we will be able to minimise global warming and at the same time, save energy for future generations.

Global Warming is a major issue due to the industrialisation and progress by humankind since the past few years. There has been a hue and cry about Global warming ever since the idea was first put forward. There are many articles on Global warming on the internet as well print media which give further details about the causes of Global warming. Here is some further information on Global warming.

Global warming has been and is being caused due to a various number of factors. Global warming is basically a change in the climatic conditions of the earth. These climatic conditions vary due to various reason, external and internal. Changes to climatic conditions and therefore Global warming can be caused to natural or man-made circumstances also. Some of the factors causing Global warming are volcanic emissions and solar activity. According to the solar variation theory, the sun has been gaining strength and is at it's strongest since a sixty years. Therefore, it may now be acting as a cause of global warming. Sunspots are also said to be a cause or catalyst for Global warming. Recent reports suggest that the number of sunspots in an area directly affects the amount of time the nearby earth takes to cool. The sun is the main source of energy to the earth. The earth absorbs about 70 per cent of the earth's solar flux. This solar flux increases the temperature of the earth's atmosphere, land and oceans.

Orbital forcing is also said to be one of the natural causes of Global warming. The reports show the effect of the slow tilting of the earth's axis on the climate of the earth. The greenhouse effect is said to be the most important factor regarding global warming. When infrared radiation from the atmosphere increases the temperature of the earth's surface, it is termed as the greenhouse effect. The greenhouse effect has increased the earth's temperature by about 24 per cent. Carbon dioxide contributes about 12 per cent of the greenhouse effect, while water vapour contributes 36 per cent of the greenhouse effect. Methane causes 5 to 10 per cent of Global warming, while ozone makes around three to 7 per cent of the greenhouse effect possible. Solar variation is said to be another reason of Global warming. The changes in the amount of radiant energy emitted by the sun are known as solar variation. This solar variation has been correlated with the changes in the earth's climate and temperature.

Along with the natural causes of Global warming, scientists have also contributed rapid industrialisation to the increase of Global warming today. Humans had first affected Global warming some eight thousand years ago, with the start of agriculture. Due to the clearing of the forests for agriculture, the amount of carbon dioxide in the atmosphere increased drastically.

Scientists are of the opinion that industrialisation releases various gases like carbon-dioxide and methane which are known to contribute to Global warming. Deforestation is also said to increase global warming. Trees contain a high level of carbon, and therefore their cutting creates an increase of carbon in the atmosphere. Humankind also contributes to the increase of carbon dioxide by the burning of fossil fuels.

The contribution of humankind to global warming due to the burning of fossil fuels has increased by about 80 per cent in the past twenty years. If the greenhouse effect did not exist, the temperature of the earth would be around 27°C less. Some scientists are of the opinion that human life would be impossible on planet earth if the temperature would be so less.

Global warming has effects and consequences on all walks of life. The consequences of global warming can be seen in the atmosphere, weather as well as the health of individuals. As it is obvious by its name, Global warming is a global phenomenon with a number of effects on the global level. Global warming has various effects, ranging from the effects to the atmosphere to the economical, environmental as well as the health life of human beings. There are also a number of effects to the nature and atmosphere. One of the most serious effects of Global warming that humans have to think about are the effects on the health of individuals, nations and therefore civilisations. The rise in temperature due to Global warming is known to be supportive to various viral diseases like the west nile virus and malaria. This will result in economic as well as health effects on human beings. For example, Global warming will increase the incidence of such diseases in poorer countries where these diseases exist. Global warming will also cause countries who have eradicated these diseases to spend more on vaccinations and other ways of eradication like pesticides, etc.

Other than these effects, Global warming has other effects on the health of human beings. Global warming results in a drastic rise in temperature. This rise in temperature will finally result in an increase in the mortality rate of people. A higher temperature causes problems to people with cardiovascular problems. In extreme cases, people are known to have died of heatstroke. People may also have heat exhaustion problems. Respiratory problems are also known to arise out of a high temperature.

High temperature also causes the concentration of ozone in the lower atmosphere. Ozone is a harmful pollutant and causes respiratory problems. Ozone is also known to damage lung tissues and therefore cause more complications for people with asthma. These are some of the health effects of Global warming.

Global warming has many other effects other than the health of individuals. Global warming may also cause a decline in agriculture due to the rise in temperature. The agriculture will also decline due to the role of carbon dioxide in photosynthesis. Carbon dioxide prevents photorespiration and therefore is the cause of the damage of many crops. Global warming also results in increased number and longer droughts. This will result in a increase in the ozone gas at the ground level. The increase of the ozone at the ground level will result in a substantial depletion of crops.

The increase in temperature will also cause various transport infrastructure like roads, bridges, ships to face greater temperature changes. Due to this, the maintainable costs of the transport infrastructure will increase. This may cause broken runways, malformed roads and sunken foundations.

Global warming also results in a rise in the sea levels. Therefore, this will result in an increase of the costs of the coastal defense. This is also dangerous, because the most important trade ports are always at the coast of any area for trade reasons.

Global warming will also have a effect on the daily weather. One of the most important effects is the increase in extreme temperatures. The levels of evaporation will also increase due to Global warming.

Global warming also effects the ecosystems. The effects of Global warming have already been seen in birds. Other effects of global warming, like a decreased snow level, increased temperature and other weather changes will have a effect not only on humans but also entire ecosystems. These ecosystems will change and cause many traditional inhabitants to leave their inhabitants. This may cause extinction of species. Another Global warming consequence is the decline of the ecosystem's productivity. Decline of glaciers, known as glacier retreat is another Global warming consequence. Decline of glaciers directly results in flash floods, landslides and glacial lake overflow.

GLOBAL WARMING CAUSES AND EFFECTS

Global warming is the process wherein the average temperature of the earth's near surface air increases, owing largely to various anthropogenic activities. Though there are some natural causes for this rise in temperature, they stand to be insignificant when compared to the anthropogenic causes. Understanding global warming causes and effects can give us a brief idea of the dreadful phenomena our future generations may have to face. Here are some of the prominent global warming causes and effects.

Causes of Global Warming

The causes of global warming are broadly divided into two categories—natural causes and anthropogenic (man-made) causes.

Natural causes

Natural causes of global warming include the release of methane gas from arctic tundra and wetlands, climate change, volcanoes, etc. Methane, a greenhouse gas which traps the heat within the earth's atmosphere, is let out in large quantities in the arctic tundra and wetlands. In case of volcanoes, when a volcano erupts, tons of ash is let out into the atmosphere. Even though nature contributes to global warming, this contribution is very insignificant when compared to human contribution for this hazard.

Anthropogenic causes

Anthropogenic causes for global warming are those which are caused due to human activities. The most prominent cause being man-made pollution. A large part of this pollution can be attributed to the burning of fossil fuels. This includes burning coal to produce electricity as well as burning gasoline to power internal combustion engine vehicles. When these fossil fuels are burnt, they let out carbon dioxide, which is yet another greenhouse gas which traps heat within the atmosphere of the earth and contributes to global warming. Secondly when the earth is dug to extract these fossil fuels in the process known as mining, the methane inside the earth's crust escapes into the atmosphere and adds to other greenhouse gases such as carbon dioxide. If we start investigating the anthropogenic causes of global warming, we zero in on one of the most important cause of global warming—population.

More population means more requirements, which includes food, electricity and transport. In order to fulfil these requirements, more fossil fuels are consumed, which eventually leads to global warming. Humans breathe out carbon dioxide, and with an increasing population, the amount of carbon dioxide humans breathe out also increases and leads to global warming. Even agriculture contributes to global warming, owing to the extensive use of fertilisers, and the dung produced by cattle which is another prominent source of methane.

Effects of Global Warming

The effects of global warming range from a rise in sea levels to the extinction of certain species of flora and fauna. Basically, global warming means an increase in temperature of the earth's atmosphere. This increase in temperature will trigger a series of events which can cause a lot of destruction on the planet.

Changes in the global sea level

As the temperature will increase, the ice cover on the planet will start melting. The water from these melting glaciers will end up in the oceans, which will lead to a rise in the sea level. Over the last century, sea levels have increased by 4 to 8 inches, and by 2100, it is expected to increase to 35 inches. An additional 2 degree rise in global temperature will lead to the complete melting of the Greenland ice cap, which will cause the sea level to rise by 5 to 6 meters.

Such a rise will cause many of the low lying areas, such as the US Gulf Coast and Bangladesh, as well as islands, such as Lakswadweep, to submerge underwater. If the whole of the Antarctic ice sheet melts, the global sea level is expected to rise by 10.5 meters.

Drastic changes in climate patterns

Global warming will alter the climatic patterns of the planet. As far as precipitation is concerned, it will increase in equatorial, polar and sub-polar reasons, and decrease in subtropics. This change in precipitation pattern will trigger a drought in some regions, while floods in other regions. Warming of the atmosphere will increase the temperature of ocean waters, which will continue being warm for a few centuries. Warm water will lead to frequent natural disasters like hurricanes. Overall, the planet will experience extreme weather conditions, characterised by flood and droughts, heat waves and cold waves, and extreme storms like cyclones and tornadoes.

Widespread extinction of flora and fauna

A rise in global temperature will also hamper the rich biodiversity of various ecosystems. According to the Intergovernmental Panel on (IPCC), an increase in global temperature by 1.5 to 2.5 degrees will make 20 to 30 per cent of species vulnerable to extinction, while a rise of about 3.5 degrees will make 40 to 70 per cent species vulnerable to extinction. Climate change will result in loss of habitat for many animal species like polar bears and tropical frogs. More importantly, any change in the climate patterns will seriously affect the migration patterns of various bird species. Irregular patterns of precipitation will affect animals and humans alike.

Global Warming and Humans

In case of humans, global warming will affect our food and water supplies as well as our health conditions. Changes in precipitation will affect basic necessities such as agriculture, power production etc. Increase in the temperature of ocean waters will hamper fisheries. The sudden change in climate patterns will have a hazardous effect on the human body which won't be able to endure the extreme conditions, a hint of which can be seen in form of frequent heat waves and cold waves. Increase in natural calamities such as storms, will lead to heavy human causalities. Infectious diseases will rise to a great extent as disease transmitting insects will adapt to wet, hot conditions. Many people will die of malnutrition as food production will decrease due to frequent droughts and floods.

These were just a few of the numerous global warming causes and effects. Many people argue that global warming is a slow process, and will take centuries for all these devastating effects to take place. But they forget that the factors which cause global warming are rapidly rising. The rate at which we are contributing to global warming has rose considerably, and is expected to rise at a faster rate in the future. We have already done enough of damage, and hence it's high time we understand the global warming causes, effects and the future repercussions and work out some global warming solutions at the earliest. We may not live to face the dreaded consequences of global warming, but if we don't act fast, it will be our future generations who will have to bear the brunt.

GLOBAL WARMING EFFECTS ON ANIMALS

To get a better understanding of global warming, it is important to understand the greenhouse effect. The greenhouse effect is the rise in the temperatures caused by absorption of the sun's heat and light by the earth's surface (forest, deserts, glaciers, etc.), which then is reflected back and trapped within the earth's atmosphere. The greenhouses gases, help to keep the earth warm, and this is the reason why life on earth has existed, and still thrives.

However, with an increase in the gases like carbon dioxide, ozone, nitrous oxide, methane and water vapour in the atmosphere, as a fallout to growing environmental pollution; industrial, domestic, and loss of vast stretches of grassland and rain forest, earth has gotten nearly 14 per cent hotter than what it used to be 50 years ago, with 2005 being recorded as the hottest year ever. Besides humans and plants, global warming effects on animals is a cause of concern. Let's take a look at the global warming causes, effects and the future vis-a-vis animals.

Animals are essential to maintain the circle of life and the food chain. It is just not the animals alone, insects, reptiles, and the aquatic life are all interdependent on each other, and on the plants and humans as well.

Loss of Habitat and Food

To make room for an ever growing population, many forests, grasslands, and even deserts, have been made habitable, only for humans. Rainforests and grasslands support many life forms; it is home to the tiny insects and also the mighty, grizzly bears. When forests are cleared out to develop more land for domestic constructions, industrial reasons and farming, most of these animals have to adapt themselves to live in shrinking areas, where everything for them becomes less; food, water, hunting and breeding ground. Loss of habitat also makes these animals vulnerable to being hunted down, either in their own little space, or when they come close to human habitats searching for food. With deforestation, many trees and other plants, that provide food to herbivorous animals no longer exist, causing death due to starvation and malnutrition.

This in turn has taken its toll on all other omnivorous and carnivorous animals too, making the entire animal life susceptible to extinction. Many animals, domestic or wild, who venture into the human habitat for food, eat from the garbage, mostly picking up plastic, rusted metal or contaminated food. This too has a devastating effect on their health.

As global warming causes climate change, many great deserts like the Sahara, are no longer able to sustain their animal population. Loss of habitat is most vividly seen in the Arctic, where global warming is melting the glaciers, pushing the polar bears into extinction. The melting glaciers have caused water levels to rise in many oceans, threatening to drown many tropical islands and forests, that teem with animal life. The Gulf war oil spills, along with oil tanker spills, have devastated a large number of aquatic life. The pictures of dead fishes covered in oil on many beaches, is a sad reflection of the future that lies in store for them. Changes in weather patterns and coastlines affect the food patterns of most aquatic creatures.

Hibernation, Breeding and Migration

Studies now indicate a change in the hibernation, breeding, and migration patterns of animals. It is believed that hibernation and egg laying or animal birth, which are important aspects of animal life, are now happening on an average of 5.1 days earlier per decade. This unhealthy pattern affects the newborn, and quite a few are now born with defects, or are stillborn. Early egg laying is one of the reasons why insects like butterflies, and small birds, are disappearing fast in North America, where higher temperatures lead to earlier spring seasons. Many animals and birds, including penguins or flamingos, travel long distances to warmer climates, for breeding purposes. Devastation of the migratory routes and their habitat, has forced many of them to alter or not migrate at all. This forces them to seek alternative migration habitats, where they have to compete for food and shelter with other migratory or resident animals and birds. The same is also happening in case of aquatic mammals, who prefer warmer waters for breeding and hibernation.

Animals that migrate depending on seasonally-linked phenomena, such as the formation of ice, lakes and other water bodies, and the availability of seasonal foods, also suffer, when the environmental conditions around their migratory destination changes. Global warming affects us all. The only difference being, we, humans can have a profound effect on the way global warming issues are tackled. Global warming effects on animals will have serious repercussions on the entire life cycle. Life on the planet thrives, because we thrive on each other and because of each other. Any link broken or weakened will affect all life on earth, and some of the consequences will be irreversible.

ISSUES AGAINST GLOBAL WARMING

The year was somewhere in the future. Man's environment had undergone a dramatic upheaval. Trees were scarce and you could find them in abundance in only carefully maintained sanctuaries. The air was filled with particulate matter. Visibility was greatly diminished. The vast ocean was pumped and the seawater was being processed in desalination plants. Tigers and lions had become extinct. The Himalayas were devoid of snow and ice. Antarctica was a warm continent. Much of Asia and Africa had become uninhabitable because of very hot climate. Cancer and lung diseases were rampant. The weather was characteristically fickle. One didn't know what to expect. A cold winter spell was followed by a scorching summer, which was perhaps followed by a torrential downpour.

The above scenario is not science fiction but a reality which global warming could cause. Global warming is the increase in the temperature of the earth's atmosphere caused by greenhouse gases like water vapour, carbon dioxide, methane and ozone. There has been growing awareness and concern among people about this phenomenon. It has become a global issue although it is hotly debated in international circles.

It has a fair share of supporters and critics. Its proponents warn that if immediate preventive steps are not taken it will have a disastrous effect on future humankind. Its detractors argue that the problem is grossly overestimated and nothing should be done. They also argue that the steps to tackle the problem will hinder economic progress.

Arguments Against Global Warming

Here we take a look at the arguments against global warming:

1. It is argued that global warming is a minor issue because of which major issues like HIV/AIDS, Nuclear proliferation and poverty are not devoted their deserved time and resources.
2. Vested interest of scientists. It is argued that scientists exaggerate the effects of global warming because they receive funds from environmental companies.
3. Unreliability of computer climate models. It is argued that these models are not able to predict tomorrow's weather. So how can they predict long-term climate change?
4. There are other factors involved in global warming. It is argued that human activities are not the only cause of global warming.
5. Newspapers sensationalise global warming in order to sell. It is argued that newspapers distort the picture of global warming when actually that is not the case.
6. Scientists have made wrong predictions before. It is argued that science and scientists are not always right. Perhaps they have made an error in their calculations or drawn incorrect conclusions on available evidence.
7. The science of global warming is not proved. It is argued that we don't have long-term historical records of weather.
8. Water vapour plays a major part in global warming. It is argued that man made emissions like carbon dioxide has only minor effects.
9. The global warming is a natural phenomenon. Man has no role to play in it. Only our environment is responsible.
10. The temperature increase is very small especially when it is spread over a century.
11. The earth was warmer before. That did not have harmful consequences on humans.
12. The increase in temperature will help plants grow in currently cold and uninhabitable areas.
13. The increase in the level of carbon dioxide will stimulate plant growth.
14. Steps to limit global warming will decrease economic growth and hurt the poor.
15. People in fossil fuel industries will lose their jobs.
16. Climate change has been more rapid in the past.
17. Rise in carbon dioxide levels has always come after a temperature change and not before.
18. The upsurge in solar activity in the sun has caused global warming.

One realises that most of the arguments against global warming are factually incorrect or far fetched. People advocating these arguments have a vested interest in activities and industries that contribute to global warming. Unfortunately they are able to sway a section of the population that global warming is not a serious issue. Earnest efforts must be taken to dispel these myths and make the earth a better place to live in.

HOW TO STOP GLOBAL WARMING?

Global warming refers to the increase in average temperature of the earth, particularly at the lower atmosphere due to the abundant increase of greenhouse gases. This is primarily due to the human's intervention and the life style they have adapted in the recent years.

Greenhouse gases in its natural content in the atmosphere are required, for it acts as a greenhouse around the earth. It allows the sun's rays to pass through the atmosphere but doesn't let it to escape and creates a surrounding suitable for life. Life cannot exist without greenhouse gases like carbon dioxide, water vapour, methane, etc. but they should be existing in permissible quantities only. The problem of global warming arises when people started contributing abundant amount of greenhouse gases which traps more heat in the atmosphere thus increasing the temperature.

Ways to Stop Global Warming

Owing to the overall rise in the temperature, the glaciers in the Antarctic region begin to melt which has increased the overall sea level. If this situation continues, many low lying areas will submerge in the near future. Global warming also increases the occurrences of hurricanes. There are many easy solutions to reduce global warming and its impact. First of all, people should understand the problem and take measures accordingly to save the world.

1. People should reduce the usage of electrical appliances which emits green house gases. For e.g. the refrigerator releases chloroflurocarbon (CFC) and the incandescent light lamp emits 300 pounds of carbon dioxide a year. This can be replaced by a compact fluorescent light bulb which saves much energy.

2. Follow RRR-reduce, reuse, recycle: People should not dump waste products in the ground. Plant products, food waste, vegetable dump undergoes anaerobic decomposition i.e. they break down to produce methane, a green house gas instead of oxygen. Hence the product usage and wastage should be reduced or recycled for a healthy atmosphere.

3. Trees absorb a large amount of carbon dioxide: Many trees should be planted since they involve in photosynthesis, food preparation with the help of sunlight. During this process, trees absorb carbon dioxide and exhale oxygen. Also, existing forests should be saved and usage of plant by-products should not be wasted.

4. Usage of green power prevents 300 kg of carbon dioxide to be emitted into the atmosphere. The electricity obtained from the renewable resources like wind and water is called green power. The cost is also low in case of green power.

5. Insulation of the ceiling of a house and power saving is the important factor to reduce global warming. The electric appliances should be switched off instead to hold it in stand by mode. This will save more power since stand by mode consumes 40 per cent of the energy.

6. People should use only energy efficient appliances. Thermostat should be used for air conditioners since it reduces the temperature automatically.

7. Consumption of organic food should be increased because organic soil absorb large amount of carbon dioxide. Buying local food reduces the consumption of fuel. Cows emits large amount

of methane due to their vegetarian diet. Hence meat consumption should be reduced. Also tetra packs should be used instead of tinned food.

8. Periodic maintenance of the vehicles helps in efficient usage of fuel and reduces release of greenhouse gases. Proper inflation of tyres should be done and fuel wastage should be avoided.

9. Teach your neighbourhood and friends about the cause and impacts of global warming and methods to reduce it. Conservation of forests also forms a factor to reduce global warming.

Hence, individuals and government should be concerned about the environment and stop the incoming danger due to global warming.

GLOBAL WARMING MYTHS AND FACTS

Global warming has been taking place ever since the earth was created. However, the growing amount of global warming today has forced people to pay attention to it and press the panic button. Global warming is the increase of the temperature in the earth's oceans and near-air surface. Here are some facts, and more importantly, myths about the Global warming.

The first evidences of Global warming were seen some twenty thousand years ago, just after the last ice age. The earth's temperature has risen from some ten to eight degrees Celsius. Also, the sea-level has risen by five to twenty meters. Human actions have been instrumental in causing Global warming. Massive industrialisation have injected huge amounts of carbon-dioxide and other gases causing Global warming. The temperature on the planet has risen by 0.6 of a degree Celsius every one hundred years. Greenhouse gases are another major reason of Global warming. The year 2005 was the warmest year since the 1800s.

Carbon dioxide and methane are the biggest criminals in the causing of Global warming. Methane drastically increases the amount of Global warming in the atmosphere. It can increase the atmosphere by about 23 times the normal strength. The emission of methane into the atmosphere has large effects for a short span of time, say ten years.

Carbon has lesser effects over a large span of time. Therefore, the global warming potential of methane is about ten times more as compared to carbon dioxide. Other gases, which are commonly given out by Industrial processes are nitrous oxide and ozone. Another human factor causing Global warming is deforestation. Today, deforestation occurs at the rate of fifty three thousand square miles per year. Trees consist of 50 per cent carbon, and therefore deforestation is a major contributing factor to Global warming. The other causes of Global warming are pollution caused by cars and trucks.

Increase in Global warming can result in other dangerous situations for humankind. The increasing Global warming can bring about more extreme weather situations, a reduction of stream flows during summer, a rising sea level and glacier retreats. Global warming is also known to create decrease in agricultural yield and in some cases, species extinction.

A recent study has shown that China has the most emissions of greenhouse gases. The study says that while China's emission of greenhouse gases were only 43 per cent of that compared to the United States of America in 2001, they have shot up to 93 per cent today. China has also increased its consumption of fossil fuels by 9.3 per cent. China has now decreed that the progress and promotion of government officials will not be only judged by their economic progress, but also their environmental progress. The above facts about Global warming therefore are a warning about the ecology of the world. However, there is a debate whether the entire concept of Global warming is a fact or fiction. While scientists generally say that Global warming is a fact, some judge Global warming itself to be a myth. Global warming is a debatable issue, and there is a dearth of facts about Global warming. Some scientists even say that the entire concept of Global warming is a myth. Scientists say that other scientists have prepared the Global warming myth for various political reasons. The scientists are of the opinion that though Global warming may be a reality, there is really very little that the common man can do about it. They say that the earth has had atmospheric changes ever since it was born.

They also say that the atmospheric changes had nothing to do with humans and their inventions because this was too long ago before our cell phones or our computers. Scientists also say that the emissions of carbon dioxide into the atmosphere do not cause any global warming, because of the simple reason that warming causes carbon retention in the atmosphere. Some eminent fiction authors have created novels about global warming myths. One such book making the rounds says that Global warming cannot be either stopped or increased by humankind.

GREENHOUSE GASES AND GLOBAL WARMING

There are numerous environmental issues which threaten the very existence of life on earth, and global warming is perhaps the most severe of them all. Many people assume that the greenhouse effect and global warming are one and the same thing, which is absolutely incorrect. A closer look at the list of greenhouse gases and global warming causes and effects reveals that the concentration of greenhouses gases in the atmosphere is one of the numerous causes of global warming. That being said, the relationship between greenhouse effect and global warming can be best defined as cause and effect relationship.

Difference Between Greenhouse Gases and Global Warming

The term 'greenhouse gases' refers to various gases in the earth's atmosphere, which are typically characterised by their ability to absorb infrared radiations coming from the Sun. The entire process wherein the sun's infrared radiations are trapped within the atmosphere by these greenhouse gases is referred to as the 'greenhouse effect'. Greenhouse gases list includes gases such as carbon dioxide, carbon monoxide, methane, chlorofluorocarbons, etc. some of which stay in the atmosphere for several years and contribute to the greenhouse effect on the planet. The atmospheric concentration of these gases is one of the main causes of the greenhouse effect. Global warming, on the other hand, refers to an incessant rise in global average temperature triggered by various natural and anthropogenic causes— greenhouse gases being one of them.

Relationship Between Greenhouse Gases and Global Warming

Even though we say that the atmospheric concentration of greenhouse gases has a key role to play when it comes to global warming, these gases are not the only causes of this hazardous phenomenon. Other than the atmospheric concentration of these gases, global warming causes also include numerous other natural occurrences and anthropogenic activities. For instance, solar radiations (a natural cause) and deforestation (an anthropogenic cause) are not at all related to greenhouse gases, but they do play a crucial role in causing the global temperature to rise. On the contrary, if it were not for these greenhouse gases, the earth would have been freezing cold and devoid of any of the present life-forms which inhabit it. The fact that these gases play a crucial role in maintaining the necessary balance in global temperature makes their presence on the planet very important.

If greenhouse gases are so important, why are they blamed for global warming? Actually, the problem arises when the amount of these gases in the earth's atmosphere exceeds the amount required to maintain temperature balance. This increase in greenhouse gases atmospheric concentration results in trapping of more infrared radiations within the earth's atmosphere, and contributes to rise in global average temperature. When it comes to natural causes of global warming that are closely related to greenhouse effect—methane gas release is perhaps the most prominent one. Similarly, anthropogenic causes of global warming which are associated with greenhouse effect include—use of vehicles, stationary sources such as industries, activities such as mining and agricultural, etc.

While naturally occurring greenhouse gases have been playing the important role of regulating the temperature on earth since several centuries, those gases that are released as a result of human activities have changed the overall picture. These greenhouse gases include carbon dioxide (with a lifetime of 200 years), nitrous oxide (120 years), various CFC's (with their lifetime ranging between 5–1000 years) and gases such as perfluoropentane and perfluorohexane (with lifetime exceeding 1000 years). These numbers can be deceiving though, and the relationship between methane and global warming which reveals that the damage caused by methane in its lifetime of 12 years far exceeds the same caused by carbon dioxide in its lifetime of 200 years, highlights this fact very well.

That was a significant bit of information about greenhouse gases and global warming, which stressed on the fact that greenhouse gases do fuel global warming and climate change. This relationship between global warming and the greenhouse effect has to be taken into consideration when we talk about various ways to stop global warming. Taking into consideration the role played by these gases in triggering what is referred to as human induced global warming, doing away with them can help us ease the various global warming problems on the planet by a certain extent.

GREENHOUSE EFFECT

The earth is heated by sunlight and some of the heat that is absorbed by the earth is radiated back into space. However, some of the gases in the lower atmosphere, acting like glass in a greenhouse, allow solar radiations (in the range 300 to 2500 nm, i.e. near UV, visible and near infrared region, while filtering the dangerous UV radiations, i.e. <300 nm) but do not allow the earth to reradiate the heat into space. In other words, these gases in the atmosphere are transparent to the sunlight coming in, but they strongly absorb infrared radiation, which the earth sends back as heat. A part of the heat so trapped in these atmospheric gases is re-emitted to the earth's surface. The net result is the heating of the earth's surface by this phenomenon, called the 'greenhouse effect'. The gases that are responsible for this greenhouse effect are CO_2, water vapour, CH_4 and man-made chlorofluorocarbons (CFCs). Water vapour strongly absorbs infrared radiations in the range 4000 to 8000 nm and CO_2 in the range 12,000 to 16,300 nm. The radiations in the range 8000 to 12,000 nm escape unabsorbed and this is known as the region of atmospheric window. Carbon dioxide is released by volcanoes, oceans, decaying plants as well as human activities, such as deforestation and combustion of fossil fuels. Automobile exhausts account for 30 per cent of CO_2 emissions in developed countries.

Methane is released from coal mines, decomposition of organic matter in swamps, rice paddy cultivation, guts of termites in forest debris and stomachs of ruminants. Chlorofluorocarbons (CFCs) are used as coolants in refrigerators, propellants in aerosol sprays, plastic foam materials like 'thermocoles' or 'styrofoam' and in automobile air-conditioners. In fact, the 'greenhouse gases' (particularly CO_2 and water vapour) are responsible for keeping our planet warm and thus sustaining life on the earth. If the greenhouse gases were very less or totally absent then the average temperature on the earth would have been at sub-zero levels.

But, however, if the concentration of greenhouse gases increases, they may trap too much of heat, which may threaten the very existence of life on earth. For instance, the CO_2 present in the atmosphere of the planet *Venus*, is about 60,000 times more than that on earth. Hence, the average temperature of *Venus* is about 425°C, making the existence of life impossible there.

Oceans and bio-mass are the major sinks for atmospheric CO_2. Oceans convert CO_2 into soluble bicarbonates. The photosynthetic activity in the green plants increases with the increase in CO_2 level in the atmosphere. Forests are the places where lot of photosynthetic activity occurs. They also act as vast reservoirs of fixed but readily oxidisable carbon in the form of vegetation, wood and humus. Hence, forests maintain a balance in the atmospheric CO_2 level, and deforestation upsets this balance and increases the atmospheric CO_2 level.

It is estimated that the atmospheric CO_2 content has increased by 25 per cent during the last two centuries. This is mostly attributed to the industrial revolution and is one of the reasons for the slight increase in the global temperature (about 0.5°C). Since the concentrations of greenhouse gases have been continuously increasing because of deforestation, industrialisation, increased burning of fossil fuels, mining, exhausts from increasing number of automobiles and other anthropogenic activities, there is an increasing concern about the possible 'global warming'. Some scientists fear that if proper precautions are not taken, the concentration of greenhouse gases in the atmosphere may double within the next 50–100 years. If this happens, the average global temperature may increase by 4°–5°C. This will increase the evaporation of surface waters, which may influence climatic changes depending upon the pattern of cloud formation. For instance, low-level dense clouds may exert cooling effect whereas high-level thin cloud formation may exert heating effect due to increased greenhouse effect. The projections from computer modelling regarding the climatic changes that could be triggered off due to 'global warming' reveal alarming scenarios. Even a 1.5°C rise in surface temperature can adversely affect food production in the world.

Thus, the wheat growing zones in the northern latitude may be shifted from the USSR and Canada to the polar regions, i.e. from fertile soils to poor soils near the North Pole. The biological productivity of the ocean would also decrease due to warming of the earth's surface layer, which in turn, may reduce the transport of nutrients from deeper layers to the surface by vertical circulation. Computer modelling also indicates the following effects due to 'global warming': melting of the polar ice caps; dry areas becoming drier; humid areas like the Amazon suffering more intense tropical storms; drastic drop in food production, particularly in lands within 35 degrees north and south of the Equator; increased breeding of pests and diseases due to more humid conditions; shorter, wetter and warmer winters and longer, hotter and drier summers, particularly in mid-continental areas. Global warming may also trigger increased thermal expansion of oceans and melting of glaciers, which may result in an increase in the sea-level by 20 cm to 1.5 metres by the latter part of the 21st century. Thus, cities like Mumbai, Miami, London, Venice, Bangkok and Leningrad may become extremely vulnerable. Defences against the rising sea-levels and expanding oceans are very difficult and expensive, which many nations cannot afford. Further, a global temperature rise, is likely to cause more floods, hurricanes, and tornadoes.

There are differences of opinion among experts regarding the dynamics and effects of 'global warming' due to the complexity of natural phenomena that might be operating simultaneously. More accurate future climatic projections will be possible with better super-computer models, based on greater understanding of the complex natural climatic forces involved. But until that time, the possible devastating effect due to 'global warming' by the 'greenhouse effect' cannot be underestimated. Some of the steps suggested to minimise the 'greenhouse effect' include reduction in the use of fossil fuels, encouraging the use of alternative sources of energy (e.g. solar, geothermal, wind, bio-gas, etc.), conservation of forests, extensive afforestation, encouraging community forestry, reduction in the use of automobiles, research in the development of more efficient automobile engines, ban on CFCs and nuclear explosions, development of environmentally compatible technologies with the help of intensive inter-disciplinary research, effective check on the growth of population and imparting of non-formal and formal environmental education.

IMPACT OF GLOBAL WARMING

The incessant rise in average near-surface temperature of the planet, as a result of various natural occurrences and anthropogenic activities, is known as global warming. That definition of global warming may be technically correct, but whether it highlights the seriousness of the issue is a question that is difficult to answer.

There is a lot more to know about this environmental issue which is acting as a catalyst for several other natural disasters on the planet. In our attempt to find out what are the impacts of global warming on the environment, we came across quite a few global warming facts which threaten our very existence on earth.

Over the last century, the average global temperature has increased by 0.32°F, and if the climatic models prepared by Intergovernmental Panel on Climate Change (IPCC) are to be believed, by the end of this century the earth will be warmer by 2° to 11.5°F. A mere look at evidence of global warming or evaluation of global warming statistics is more than enough to substantiate these claims. The fact that this problem has reached such proportion can be attributed to our inability to differentiate between global warming myths and facts to a certain extent.

Global warming causes can be broadly categorised into two groups—the natural causes, which include volcanic eruption, solar radiations, methane release, etc. and the anthropogenic causes (man-made causes) include burning of fossil fuels, deforestation, mining, etc. While natural causes of global warming have always been there, and play a crucial role in maintaining the balance in global temperature, the significant rise in various anthropogenic causes of global warming, and rise in concentration of greenhouse gases in the atmosphere, is turning out to be of grave concern. Trees play a crucial role in the environment by absorbing carbon dioxide in the atmosphere and therefore global warming and deforestation are also closely associated.

Going through the following section will help you get a better understanding of various other causes of global warming:

1. Global warming and carbon dioxide.
2. Methane and global warming.
3. Fossil fuels and global warming.
4. Greenhouse effect and global warming.

So what will happen if global warming continues unabated? As the temperature continues to soar, it will trigger numerous changes on the planet—quite of few of which will affect us directly, while others will affect us indirectly. In order to make it simple, we have divided the environmental impacts of global warming into three categories—on the basis of who will be at the suffering end.

Global Warming Impact on Plants

Global warming will trigger numerous changes in climatic conditions, and this climate change will in turn trigger the extinction of various plant species. These plants have taken years together to adapt to the conditions in which they thrive today, any changes in these climatic conditions now will result in adverse effects on these species. Climate change, i.e. change in pattern of seasons to be precise, will also alter the growing season of these plants, which will in turn affect their reproduction cycle. It will be difficult for these species to undergo the entire adaptation process all over, and in this process many of these species will become extinct. Irregularities in precipitation, irrespective of whether it is drought or flood, will affect various plant species including agricultural crops. Going through endangered plants list will give you a better idea of the seriousness of the issue.

Global Warming Impact on Animals

If you thought the impact of global warming on plants was severe, the same will be even more harsh on numerous animal species. In fact, the global warming effects on animals have already begun, and a look at the list of extinct animals is full of evidence of the same. The Monteverde golden toad endemic to the rainforests of Cost Rica was one of the first causalities of global warming. It became extinct as a result of habitat loss, when extremely hot summers dried up the lake which this species inhabited. The list of endangered animals also includes a number of animal species which have lost their natural habitat as a result of global warming. The relationship between polar bears and global warming best explains the effects and consequences of global warming on animals. Melting of polar ice as a result of global warming has resulted in loss of habitat for this species, and this has forced the species to move further north. The same trend has been witnessed in almost all the biomes of the world, including the ocean biome wherein coral reefs are fighting for their basic existence in warm ocean water.

Global Warming Impact on Humans

Without plants and animals, it will be impossible for us to survive on this planet. We are dependent on these plant and animal species for everything—including the basics such as oxygen and food. Only plants have the ability to prepare their own food by resorting to the process of photosynthesis—the same energy is eventually transferred to us when we eat plants or eat animals which feed on these plants. We are just a part of the food chain, if any link of this chain is broken everybody, including us, will have to bear the brunt of the same. That was about our dependence on plants and animals, but our woes are not just restricted to global warming effects on plants and animals. When this hazardous phenomenon will reach its peak, we will have to face a number of problems. Global warming and melting glaciers are closely associated, and so are their impacts on humans. Glaciers act as freshwater reservoirs on the planet, and if they melt we will be left devoid of water to drinking.

Similarly, glacial melting at high altitude will also trigger flash floods in the surrounding. All this water will be drained into the oceans, and that in turn will result in sea level rise. As sea level continues to rise all the low lying area and tiny islands will get submerged. Maldives islands sinking is one of the best evidence of this global warming problems. When water from melting glaciers enters the sea, the warm and cold water currents will result in conditions ideal for formation of hurricanes. As a result of this, the frequency of hurricanes will increase. Warm climate will also come as a blessing for various disease spreading insects, and thus some diseases which are restricted to tropical areas as of today will spread to various other regions. Extreme weather—drought, floods and lengthy or short growing seasons, will destroy the agricultural sector, and that will leave us devoid of food.

WORST EFFECTS OF GLOBAL WARMING

There was a time, when scientists said that its high time when we develop a serious understanding for the global warming effects, and make efforts to lessen the severity that they are causing to our earth. But we humans have crossed over that high time and have still not made ourselves aware of the disaster that we are going to face in near future, because of the ever-increasing global warming effects. Do we make a conscious effort to use only the required amount of water? Have we maintained the habit of switching off electrical appliances when they are not in use? Do we try to make minimal use of our vehicles so as to create less pollution? Have we ever sacrificed our wish for burning fancy crackers to celebrate different occasions? Or, have we thought even for a minute before felling one more tree to the ground?

Global warming, is one of the hottest topics on the platter of scientific studies. And by now, everyone must have come to know about this undesired effect that is taking place in nature. Carbon dioxide, methane, water vapour, nitrous oxide and ozone are the main greenhouse gases. Now what do these gases do is, they allow the sun's radiation to enter the earth's atmosphere. However, they create a blockage to stop the heat to get radiated back to the atmosphere. This process is known as the greenhouse effect, and is actually necessary for the sustainability of life on earth. Now what has been happening is that the level of greenhouse gases has been increasing to alarming levels. And this is causing the atmosphere to trap more heat than required. This process is increasing the average temperature of the earth. And this very increase, which has been in progress since the mid-20th century, is termed as 'global warming'.

Top Ten Worst Effects of Global Warming

Generally speaking, there could be no count of the number of global warming effects on earth we are encountering or have to encounter in future. However, we have tried putting down the below description, which would let you know about the worst effects which have come into existence.

1. One of the most obvious effects of global warming is associated with the polar ice caps. Due to the increased heat, the ice caps have already started melting. Over time, this will increase the water levels in seas and oceans. Experts have estimated that if all the glaciers are caused to melt, it would result in the water level to rise to about 230 feet. Also, it must be known that ice caps are supposed to be a source of fresh water. Once they start melting in large amounts, they would begin mixing with the seas and oceans thus, making them less saltier. And when this happens, there would be a huge impact on the ocean currents, eventually jeopardising the regulation of temperature. This is one of the severe global warming effects on weather that is predicted by scientists.

2. The second effect that increasing levels of global warming would have is related to the survival of the wild animal species. The increasing level of sea water would cause the lowland coastal areas to incur flood. The most obvious repercussion of this would be the destruction of plant species, which ultimately would impact the diet of the herbivores, and eventually the carnivores. One example which can represent the global warming effects on animals could be the extinction of polar bears. These animals use the sea ice for hunting. And if there is no ice, there would be no hunting, and eventually, no polar bears.

3. With the temperature on the rise, more and more places would face incidences of drought and diseases.

4. Another one of the severe: The effects and consequences of global warming is again related to the melting glaciers and ice caps. The meltdown would cause the oceans and seas to get totally exposed to the sun's heat. This would result in more absorption of heat thus, warmer waters. The end result of this may be the creation of stronger and frequent hurricanes.

5. Increasing temperature, elevated levels of carbon dioxide, malfunctioning precipitation and many other factors would jointly contribute to drastic climatic change. This would have its effects on agriculture and food production around the world.
6. People who love to eat salmon, would also feel the global warming effects when this very fish species is wiped out completely from the waters of the earth. Scientific studies have shown the disappearance of the pacific salmon from 40 per cent of their habitats.
7. Those who fancy growing wildflower species, may have to part with their love with the wild plants. It has been predicted that in the Western United States, there could occur a wipe out a fifth of the wildflower species, replacing them with grasses.
8. There could be no Christmas trees in near future. There is this bug known as the pine bark beetle which munches on these kinds of trees. But their population is kept under control by winter temperatures. So if the temperature goes on rising, these insects would thrive thus, killing more and more of such trees.
9. There would be an increase in heat-related heart and respiratory medical conditions, if the temperature keeps on rising. This would result in more deaths over time.
10. As there could be an instability in food and water supply, one of the worst repercussions to this could be threats for war and conflict. It may be an obvious fact that countries which suffer water shortages and crop loss, could trigger instances which might lead to next world war.

ACTIONS TO STOP GLOBAL WARMING

Looking at the ill effects of global warming, such as rising sea levels, unpredictable climate changes, increased frequency of floods and droughts, stronger and deadlier hurricanes, lesser availability of fresh water, outbreak of diseases such as malaria, extinction of many animal and plant species and irreversible changes and damages to ecosystem, many of us are compelled to ask, 'how to take action to reduce global warming?'. Well, although politically, many steps are being taken to tackle this environmental issue, still as individuals, we too can do our bit to save our planet from these impending dangers. Given below is a list of such effective actions to stop global warming.

Ten Ways to Help Stop Global Warming

Be a vegetarian

Eating fruits and vegetables is not only healthy for your body, but it helps in reducing greenhouse gas emissions to a large extent too. Livestock are mainly fed on corn, which requires ammonium nitrate fertiliser to grow. To create this fertiliser, large quantities of petroleum is used, thus leading to more greenhouse gas emissions and global warming.

Recycle

Recycling is very important so make sure that you recycle plastic bottles, glass bottles, organic wastes in your garden, etc. Also, while purchasing, either of glass or paper, make sure that you buy recycled stuff, as it saves lots of energy and prevents deforestation.

Use CFLs (compact fluorescent bulbs)

One of the ways to stop global warming is to replace the bulbs in your house with CFLs. It will save you a lot of money and at the same time, save a hell lot of carbon dioxide too. If you are really serious about this cause, then one of the most effective actions to stop global warming that you can take is to gift CFLs to your neighbours, friends, colleagues, etc.

Go local

Do you know how much energy can be saved if you buy locally produced products? For the simple reason that transportation involved with importing goods can be greatly cut down. So, reduce your carbon footprint by going local.

Buy products with less-packaging

Goods which come with minimal packaging are a win-win for everybody. The consumer's garbage is reduced. The producer's manufacturing costs is reduced. And most importantly, carbon dioxide usage is reduced. So, always buy products which come with less packaging.

Get a programmable thermostat

Get a programmable thermostat installed in your house as they automatically lower and raise the air conditioning and heat. Or alternately, you yourself can lower the thermostat by two degrees during winters and raise it by two degrees during summers.

Insulate your home

If you insulate and weatherise your home, you will be saving lots of carbon dioxide as well as lots of money, which is otherwise spent on home heating and cooling. So, insulate the walls and ceilings of your house and if you already have one in place which is old, replace it!

Grow trees

Do you know that a single tree can absorb as much as one ton of carbon dioxide? So, one of the simplest actions to stop global warming that you can take is to plant trees in your garden.

Drive less

To reduce carbon dioxide emissions, drive less. Go in for car pooling with colleagues for travelling to work. Take your cycle or walk down, if you have to go somewhere near and travel by public transport, whenever you can.

Carry reusable shopping bags

Another of the stop global warming tips is, instead of taking a disposable shopping bag from every shop you make purchases from, carry a reusable shopping bag. It will reduce wastage and the resultant carbon dioxide, methane emissions as well as environmental pollution.

Covering the utensils while cooking, not keeping the electrical appliances on stand-by, taking shower and not a bath, flying less frequently—as you can see, there is no dearth of actions to stop global warming, which an individual can take, provided he is serious about making a difference to the society, the world and the planet. As Mahatma Gandhi once said, 'Be the change you want to see in the world'.

STOP GLOBAL WARMING TIPS

Thanks to the global warming statistics compiled and put forth by eminent scientists from all over the world and organisations such as the Intergovernmental Panel on Climate Change (IPCC), today everybody knows what will happen if global warming continues unabated. While all of us understand the seriousness of the issue, not many of us are willing to devote hours together for this cause. Frankly speaking, hours of devotion is not needed at all, as just following a few tips and measures can help ensure that all the hazards of global warming and climate change are kept at bay. If you are wondering how a few minutes from your busy schedule can help stop global warming, tips and measures given below will come as an eye opener for you. But before we move on to these tips and measures, let's go through a brief write-up about statistical evidence of global warming in order to get a better understanding of the issue.

Stop Global Warming: Facts

The term 'global warming' refers to the incessant rise in average near surface temperature of planet earth, to an extent wherein it triggers numerous changes in the climate and physical properties of the planet. At present, the average global temperature is rising at the rate of 0.36°F per decade. Global warming effects on earth exist in plenty, and each of these effects are related to each other. For instance, rising temperature would result in glacial melting, and melting glaciers would add to the ocean water and result in submerging of low lying areas and tiny islands. This is just one of the numerous examples of ecological imbalance which highlights the fact that global warming is occurring. Other evidences include extreme weather conditions, increase in frequency of hurricanes, extinction of animal species, etc. This might come as a surprise for many, but we are responsible for this natural hazard to a significant extent, and the anthropogenic causes of global warming make it more than evident.

Stop Global Warming: Tips to Follow to Curb Global Warming

While there exist numerous environmental issues which threaten the basic existence of life on the planet, the issue of global warming is the most threatening among them all. That being said, we need to understand that we have already done a significant bit of irreversible damage to the earth. Even though we can't bring back all those glaciers which have melted and animals which have become extinct, we can at least try and get into a damage control mode in order to make sure that we don't create more problems for the future generations.

If you are wondering how can we stop global warming, you can follow some simple tips, or actually some simple alterations in your daily activities, given below.

1. Do not use incandescent light bulbs, they emit 300 lbs of carbon dioxide in the atmosphere every year. Instead you can use environment friendly compact fluorescent light bulbs.
2. Do not waste electricity by leaving electronic appliances on (or on a stand-by mode). Even better, unplug appliances when they are not in use.
3. One of the simplest ways to stop global warming is to plant trees. They don't just release oxygen in the atmosphere but also help in absorbing carbon dioxide present in the atmosphere, and thus prove to be very useful for the environment.
4. If and whenever possible, resort to walking or cycling instead of using your car. This has numerous benefits—it will reduce pollution, save fuel and make sure that you stay healthy.
5. Drive your vehicle efficiently, ideally within the speed limits recommended by the manufacturer, and put the engine off whenever you take a halt—even on traffic signals or at the gas station.
6. Resort to recycle and reuse formula, especially when it comes to items of daily use—such as plastic and paper. This will not just save the tons of raw material used to produce them, but also save energy which goes into their production.

7. Change your dietary habits and switch over to organic foods. While going vegetarian is the best bet, even doing away with a single non-vegetarian meal a week can be of great help.
8. Instead of resorting to energy sources which require burning of fossil fuels, opt for alternative energy sources which are environment friendly in nature.
9. If possible, try and give up on unnecessary luxuries of life—such as use of air conditioners, a car for every member in the family, etc. if not, make sure that you use them efficiently.
10. Most important of all, understand the seriousness of the issue and resolve that you will follow these tips to stop global warming in the future.

That covered all you needed to know about how to stop global warming. Tips given in this section will surely help you prevent this impending disaster from befalling our race, but only when they are followed religiously. The number of extinct animals in the last 100 years, melting of the Himalayan glaciers and Maldives islands sinking—all come as the warning signs of this impending danger that we have been ignoring for quite some time now, and if we continue to do the same it may turn out to be too late for us to save the planet. The need of the hour is to come up with global warming solutions, and implement them at the earliest.

STOP GLOBAL WARMING: TIPS TO PREVENT GLOBAL WARMING

It may come as a surprise for many, but you can help in saving this planet from the menace of global warming—which is regarded to be one of the most fierce environmental issue that our planet is facing today. Over the last century, the average global temperature has soared by $1.5°F$, and the implications of this have already started surfacing in various parts of the world in the form of evidence of global warming. We have been turning a blind eye towards this issue, and any more delay and the damage caused to the planet would be irreversible.

While the administration is putting in efforts to find different ways to prevent global warming, we can work at the grass root level and contribute to the same. If we are to save our planet, we need to follow the stop global warming tips given below religiously.

Extreme weather, melting glaciers, extinction of species and many more. The lengthy list of global warming effects on earth shows how exactly all the life-forms on the planet, including humans, would bear the brunt of global warming. Nobody can deny the fact that we have been responsible in making the situation worse, and therefore the onus is on us to put in efforts and repair the damage done. So how can we stop global warming? Some simple, yet effective, things we can do to help global warming effects abate are given below:

1. Switch to compact fluorescent light bulbs from regular incandescent light bulbs. This will help you reduce your energy consumption by 60 per cent.
2. Avoid deforestation and plant more trees. These trees will help in reducing the carbon dioxide concentration in the earth's atmosphere by absorbing it.
3. Switch to public transport or environment friendly means of transport such as cycling or walking instead of using vehicles and contributing to air pollution.
4. If you don't have any option but to use a vehicle, make sure that you drive efficiently. Simple measures like switching the engine off on red light or driving at normal speeds can help you save a significant amount of fuel and avoid air pollution.
5. Install programmable thermostat in your house. Lowering the thermostat settings by $2°$ in winter and increasing it by $2°$ in summer will make sure that you don't waste energy in heating or cooling the interiors.

6. Never keep the electrical appliances in your home or office on stand-by mode because these appliances use electricity even when in stand by mode.
7. Unplug appliances when they are not in use, as plugged appliances tend to extract power from the grid even when they are switched off.
8. Replace your old electrical appliances with new energy efficient appliances, which don't just help you save energy but also go easy on your pocket.
9. By resorting to reuse and recycle method you can save a significant amount of energy and raw material which is used in production of paper, plastic bottles, etc.
10. You can switch over to environment friendly alternative energy sources, such as solar energy and wind energy, to meet your daily energy requirements.

Those were some stop global warming tips which will guide you to do your bit to save the planet. While some of the anthropogenic causes of global warming, like pollution or deforestation, can be curbed, it's nearly impossible to curb every single cause of the same. Similarly, natural causes, such as volcanic eruptions and methane release, will continue to contribute to this phenomenon. That being said, the onus is on us to cut down on our direct or indirect contribution to global warming, and this is something which can be done by resorting to some simple prevent global warming tips given above. If we inculcate these simple, yet effective ways to stop global warming in our daily life, we will be able to reduce the impact of this natural disaster on planet earth, if not curb it totally.

GLOBAL WARMING STATISTICS

Whether global warming is really happening or it's just a hype created by some people for their selfish gains is one of the most intricate questions in the world today. On one hand, we have the critics of global warming continuously rambling that the computer models used to study global warming and climate change are too crude to be taken seriously. On the other, we have the scientists who claim that this phenomenon is really occurring and we need to find a solution for the same as soon as possible. The end result is confusion all around, especially for the layman who is not at all aware of various geographical occurrences and their implications on the planet. One of the easiest ways to understand that 'global warming is occurring' is to go through the compilation of global warming statistics. Owing to our ignorance about global warming facts and statistics, we have been contributing to this hazardous phenomena indirectly since a long time. It's high time we take a note of these statistics on global warming and put in efforts to reduce the intensity of global warming effects on earth.

Global Warming and Rising Temperature

Global warming statistics compiled by NASA's Goddard Institute for Space Studies (GISS) has revealed that the average global temperatures have soared by 0.8°F over the last century alone. Even more chilling is the fact that temperatures are increasing at an alarming rate of 0.36°F per decade. The first decade of this century has been the warmest decade ever, with 2005 getting the distinction of being the hottest year in the history of the planet. A report compiled by the Intergovernmental Panel on Climate Change (IPCC) revealed that as many as 11 of the last 12 years feature in the list of 12 hottest years that the planet has ever experienced.

Global Warming and Carbon Dioxide

When it comes to interpretation of global warming statistics, charts and graphs tend to play a crucial role. Irrespective of which chart or graph you resort to, you will find carbon dioxide concentration topping the causes of global warming. An enormous amount of carbon dioxide is released in the atmosphere as a result of numerous anthropogenic activities, including use of vehicles and industrial waste. The statistics which highlight the global warming and carbon dioxide relationship reveal that coal alone constitutes for about 90 per cent of the carbon concentration in the earth's atmosphere. On a serious note, carbon dioxide is just one of the numerous greenhouse gases which have the tendency to cause global warming.

Global Warming and Methane Gas

Other than carbon dioxide, the relationship between methane and global warming has also become quite prominent over the last few decades. Methane concentration in the earth's atmosphere has increased from 700 parts per billion (PPB) in 1750s to 1745 parts per billion (PPB) in 1990s. With a life span of 7 years, methane plays a significant role in inducing global warming. If we stop all the activities which release methane in the atmosphere this very moment, it will take us 7 years to get rid of all the methane in the earth's atmosphere.

Global Warming and Melting Glaciers

Melting glaciers is one of the most prominent evidence of global warming. The Glacier National Park in Montana, USA, which boasted of 150 glaciers in 1910 is left with a mere 27 glaciers today. The same condition is experienced in various other parts of the world, including the polar regions. Various regions along the Himalayan mountain range have been subjected to frequent flash floods as a result of increase in glacier melting here. A recent study related to global warming and melting glaciers by the IPCC revealed that the Himalayan glaciers will be gone by 2035 if global warming continues at the rate at which it is happening.

Global Warming and Extreme Weather

Nothing highlights the occurrence of global warming as the prevalence of extremities in the weather does. The most obvious abnormalities in weather can be seen in context of drought and precipitation patterns. Estimates suggest that 30–60 per cent of the United States is subjected to drought at some or the other point of the time every year. While the world has experienced a rise in the amount of precipitation by 2 per cent per decade, the same for the United States has been approximately 6 per cent per decade. Similarly, the frequency of heat waves in the United States which had gone down in 1960s and 1970s has suddenly undergone a rapid increase.

Global Warming and Hurricanes

The rise in frequency of hurricanes is yet another obvious evidence of global warming. This rise suggests that increasing near-surface temperature of the planet is providing ideal conditions for hurricanes to flourish. Global warming statistics published in the environmental journal—nature, revealed that the frequency as well as the intensity of hurricane storms has increased by a significant extent over the last 30 years. It is estimated that every 1.8°F rise in the surface temperature of the ocean will result in an increase in the intensity of hurricanes by 5 per cent.

Global Warming and Environmental Changes

Everybody is bearing the brunt of climate change, and plant and animal species on the planet are no exception. Over the last 150 years, the planet has lost around 40 per cent of its forest cover, and experienced a rise of 30 per cent in terms of desertification. A look at the list of extinct animals in the last 100 years reveals that more than 40 animal species have become extinct over the last century alone. With global warming continued unabated, more of such causalities in Kingdom Animalia cannot be ruled out in near future. More on global warming and deforestation.

For some people this compilation of global warming statistics would come as a real eye opener about what will happen if global warming continues. While many people may dismiss these statistics as mere presumptions or publicity stunts, the fact is that global warming is happening and it's becoming more and more obvious with every passing day. That leaves us with only one option—save the planet. There are numerous ways to stop global warming, and resorting to these very global warming solutions can make any difference for us as well as the various other life-forms on the planet. It's high time we understand that if we don't curb global warming right now, doomsday for us is just round the corner.

WHAT WILL HAPPEN IF GLOBAL WARMING CONTINUES?

Global warming is happening, and there are numerous evidences to prove it. We are left with two options—either turn a blind eye towards the evidence of global warming or acknowledge them and put in efforts to lessen the impact of this disaster. Among the various evidences, the most crucial is the fact that the global average temperature of the planet has increased by 1.8°C over the last century. While those who have turned a blind eye towards this phenomenon are least bothered, those who know the seriousness of this environmental issue are left wondering what will happen if global warming continues unabated? In order to answer this question we need to take into consideration the various global warming effects on earth and its life forms.

What is Going to Happen if Global Warming Continues Unabated?

It is predicted that the soaring surface temperature of the earth will melt the water stored in the form of huge glaciers, in the Polar Regions as well as those at the high altitudes. These melting glaciers will affect the humans as well as animal species on the planet. A significant amount of freshwater is stored in these glaciers and if they melt, all the freshwater will be drained into the oceans, thus leaving us devoid of water to drink. Animals will be affected as large scale melting of glaciers will result in loss of habitat for several species. These signs of habitat loss for various species has also left environmentalists pondering about what will happen if global warming continues.

In fact, the Polar bears and Arctic foxes are already facing the threat of habitat loss due to excessive melting of ice in their natural habitat. Extinction of plants and animals will also occur as these species will not be able to sustain the climate change triggered by global warming. In fact, a look at the number of extinct animals in the last 100 years just adds to the grave concerns about the ability of various species to adapt to the rapid change in climate. Yet another example of animals extinct due to rising temperature is that of Monte Verde toad which was endemic to the tropical rainforests of Costa Rica.

The fact that water stored in these glaciers will be drained into the oceans will add to the volume of water in the ocean basins and result in abnormal rise in sea level. Melting ice may be the major contributor to the rise in sea level, but factors such as thermal expansion (i.e. expansion of water by heating) of the upper layer of oceans is also doing its bit to add to this global warming problem. Rising sea level will submerge the low lying coastal areas, as increasing water in the various oceans of the world will encroach upon the land and flood these areas. A large number of countries will be affected by this phenomenon, including low lying countries such as Bangladesh and tiny Islands such as Maldives which will meet a watery grave. In fact, Maldives islands sinking is the best example of the looming threat of global warming on the low lying areas. Those who are not yet concerned about what will happen to the earth if global warming continues also need to take into consideration the fact that a significant percentage of the world population lives in the big cities, such as New York and Mumbai, of these coastal areas. If ocean water encroaches upon land, these cities won't be spared either. In other words this will be a habitat loss for humans, as well. More importantly, the economy of these countries, as well as the whole world, will face a major setback if these cities go underwater. Yet another bad news for humans would be the spread of tropical diseases due to global warming. As this rise in temperature of the planet continues, the difference between the climate in polar areas and tropical areas will be diminished. As whole of planet will experience a tropical climate the diseases, such as Malaria, which are predominant to the tropical region will spread out to the subtropical and polar areas. Warm climate will also help the disease spreading insects, such as mosquitoes, to flourish in these regions. Not to forget, the planet will also be subjected to various extreme weather conditions as a result of global warming. The climate pattern will undergo a drastic change, and untimely rains and droughts will become more frequent. Similarly, the frequency of hurricanes will increase as the rise in temperature will heat the ocean water, thus providing a nourishing environment for these natural disasters to occur. If you doubt this fact then the rise in the number of hurricanes over the last decade is yet another striking evidence of global warming. At the end of the day, the species which will be the most affected due to global warming will be human beings. We may be the most intelligent species on the planet, but we are also the most dependent on it.

These are just some assumptions about what will happen if global warming continues at the ongoing rate. At ground level, the effects and consequences of global warming can be less severe than this (as we will adapt to the climate change) or much worse than this. The fact that we are dependent on the nature to such an extent makes us most vulnerable to the various threats of global warming. That being said, the onus is on us to put in some efforts to save the planet earth. There are a number of ways to prevent global warming; we just need to implement them in our day to day life. If these global warming solutions don't help in tackling the problems, they will at least help in making sure that the impact of these hazards of global warming on the planet are minimal.

PHOTOCHEMICAL SMOG

Photochemical smog is initiated by the photochemical dissociation of NO_2 and the consequent secondary reactions involving unsaturated hydrocarbons, other organic compounds and free radicals, leading to the formation of organic peroxides and ozone. This phenomenon takes place during sunny days with low winds and low level inversion. Photochemical smog and the consequent formation of aerosols reduce visibility, cause irritation to eyes and damage plants and rubber goods.

The oxidation of SO_2 can also take place by interaction with the free radical $HO\cdot$ present in photochemical smog

$$SO_2 + HO\cdot \longrightarrow HOSO_2\cdot$$
$$HOSO_2\cdot + O_2 \longrightarrow HOSO_2 O_2\cdot$$
$$HOSO_2 O_2\cdot \longrightarrow HOSO_2 O\cdot + NO_2\cdot$$
$$\text{(sulphate)}$$

Chemical oxidation of SO_2 may also take place in water droplets, present in aerosols. This reaction is accelerated in the presence of NH_3 and catalysts, e.g. oxides of Mn, Fe, Cu, Ni. Solid particles, such as soot, bring about catalytic oxidation of SO_2 by providing a heterogeneous phase for contact. Soot is formed during combustion of solid and liquid fuels in domestic and industrial operations, and automobile emissions. Sulphur dioxide is a pollutant responsible for smog formation, acid rains and corrosion of metals and alloys.

Oxidation of Organic Compounds

Organic compounds such as hydrocarbons, aldehydes and ketones absorb solar radiation and undergo various photochemical and chemical reactions involving free radicals. Some of these reactions are catalysed by particulate matter such as soot and metal oxides. Some of the resultant intermediates and final products contribute to photochemical smog formation.

FORMATION AND DEPLETION OF OZONE IN THE STRATOSPHERE

Ozone is an important chemical species present in the stratosphere. At an altitude of about 30 km, its concentration is about 10 ppm. The ozone layer present in the stratosphere acts as a protective shield for life on earth. It strongly absorbs ultraviolet radiations from the sun in the region 220–330 nm and thereby protects life on earth from severe radiation damage, such as DNA mutation and skin cancer. Thus only a small fraction of UV radiation reaches the lower atmosphere and the earth's surface. Ozone is formed in the stratosphere by photochemical reaction:

$$O_2 + h\nu \ (242 \ nm) \longrightarrow O + O$$
$$O + O_2 + M \ (third \ body, such \ as \ N_2 \ or \ O_2) \longrightarrow O_3 + M$$

The third body absorbs the excess energy liberated by the above reaction and thereby the ozone molecule is stabilised. Thus, ozone is constantly formed in the stratosphere. However, it is also destroyed by chlorine, released due to volcanic activity and also by reaction with: (i) nitric oxide, (ii) atomic oxygen, and (iii) reactive hydroxyl radical, which are also present in the atmosphere. In the atmosphere, nitrogen oxide (NO) comes from chemical and photochemical reactions, supersonic jets, nuclear explosions, etc. Cl_2 comes from CFC's and volcanoes; and OH comes from biomass burning and from natural water systems by the following reactions:

1. $O_3 + NO \longrightarrow NO_2 + O_2$
2. $O_3 + O \longrightarrow O_2 + O_2$
3. $O_3 + HO\cdot \longrightarrow HO_2 + HOO\cdot$
4. $HOO\cdot + O \longrightarrow HO\cdot + O_2$

Ozone, in the stratosphere, is also destroyed by man-made chlorofluorocarbons (CFCs), which are used as coolants in refrigerators, air-conditioners, propellants in aerosol sprays and in plastic foams, such as 'thermocole' or 'styrofoam'. The CFC molecules, escaping into the atmosphere, decompose to release chlorine in the ozone layer (by photo-dissociation) and each atom of chlorine, thus liberated is capable of attacking several ozone molecules.

$$Cl + O_3 \longrightarrow ClO + O_2$$

This reaction is followed by:

$$ClO + O \longrightarrow Cl + O_2$$

which regenerates Cl atoms, so that a long chain process is involved, which conserves Cl atoms. The environmental hazards of CFCs were recognised as early as 1970. In fact, temporary thinning in the stratospheric ozone layer, leading to the formation of 'ozone hole' was actually detected over the Antarctica during September to November 1985. Reported increase in cases of skin cancer in South Australia are also attributed to UV radiations reaching the earth, due to depletion of the ozone layer.

The detection of the 'ozone hole' over Antarctica in 1985 attracted the attention of global scientific community. The US immediately banned the use of CFCs in spray cans. Further, in the year 1987, 24 nations signed the Montreal Protocol, which aimed at 35 per cent reduction in the global production of the CFCs by the year 1999. Simultaneously, efforts to produce chlorine-free substitutes have also started. In fact, synthesis of a product called HFC-134a has already been reported as an effective substitute for CFC. The use of hydrofluorocarbons (HFCs), hydrochlorofluorocarbons (HCFCs), and methyl cyclohexane (MCH) as substitutes for CFCs is envisaged for several applications.

Almost all the sulphur present in liquid and gaseous fuels and about 80 per cent of sulphur present in solid fuels appears as SO_2 in the flue gases. Depending on the sulphur content of the fuel burnt and the conditions of combustion (e.g. percentage of excess air used), the concentration of SO_x in flue gases varies from 0.05 to 0.4 per cent. However, in metallurgical operations such as smelting of sulphide ores, the SO_2 concentration in stack gases may be 5 to 10 per cent. SO_2 is oxidised to SO_3 in atmospheric air by photolytic and catalytic processes involving ozone, NO_x and hydrocarbons, giving rise to the formation of photochemical smog. Oxidation of SO_2 can take place in presence of catalysts such as NO_x, metal oxides, soot and dust. Under normal humid conditions, SO_3 reacts with water vapour to produce droplets of H_2SO_4 aerosol which gives rise to the so-called 'acid rain'. The sulphuric acid and sulphate aerosols present in urban air are smaller than 2 μ and hence can easily reach the pulmonary region of lungs, causing serious respiratory problems, particularly in older people.

$$SO_2 + O_3 \longrightarrow SO_3 + O_2$$
$$SO_3 + H_2O > H_2SO_4 > (H_2SO_4)_n$$

$$SO_2 + 1/2\ O_2 + H_2O \xrightarrow[\text{metal oxide, soot, etc.}]{\text{Catalyst such as}} H_2SO_4 \xrightarrow{\text{Aerosol}} (H_2SO_4)_n$$

Control of SO_x emissions from the anthropogenic activities is contemplated along the following lines:

1. Removing SO_x from flue gases before letting them out into the atmosphere: Chemical scrubbers such as (i) Lime stone or (ii) Citric acid are suggested to absorb SO_2 from the flue gases.
   (a) $2CaCO_3 + 2SO_2 + O_2 \longrightarrow 2CaSO_4 + CO_2$
   (b) $SO_2 + H_2O \longrightarrow HSO_3^- + H^+$
   $HSO_3^- + H_2\ cit^- \longrightarrow (HSO_3 \cdot H_2\ cit)^{-2}$

2. Removing sulphur from the fuels used for combustion: Pyritic sulphur in coal can be removed by grinding and washing in coal washeries. However, organically bound sulphur cannot be easily removed from coals. Research is in progress to synthesise special type of micro-organisms using biotechnology, which are capable of converting organically bound sulphur into soluble form.

3. Utilising low-sulphur fuels.

4. Generation of power by alternative energy sources and discouraging fossil-fuel based thermal power-plants.

ACID RAIN

Rain has always been valued by mankind, because good crops and abundant water supplies are possible only due to timely and plentiful rainfall. Summer rains refresh people. Spring rains recharge the aquifers and cleanse the groundwater. Autumn rains and winter snow help clean the air. Rain, in general, brings with it a sense of hope, vitality and a promise for the future.

Over the last few decades, simple rainfall has taken on a threatening complexity in some parts of the world. In these locales, rain must pass through an atmosphere polluted with oxides of sulphur (SO_x) and of nitrogen (NO_x). The falling rain and snow often react with these oxide pollutants to produce often a mixture of sulphuric acid, nitric acid and water. This is known as acid precipitation or acid rain. Rain tends to be naturally acidic with a pH of 5.6 to 5.7 due to the reaction of atmospheric CO_2 with water to produce carbonic acid. This small amount of acidity is sufficient to dissolve minerals the earth's crust and make them available to plant and animal life, but it is not acidic enough to inflict any major damage. Other atmospheric substances from volcanic eruptions, forest fires and other similar natural phenomena also contribute to the acidity in rain. Thus, even with the enormous amounts of acids created by nature annually, normal rainfall is able to assimilate them to the point where they cause little, if any, known damage. But, it is the contributions of SO_x, NO_x, etc. from anthropogenic activities that disturb this acid balance and convert natural and mildly acidic rain into precipitation with far-reaching environmental consequences. Acid rain represents one of the major consequences of air pollution, because of large SO_x and NO_x emissions from big industrial areas. The longer the SO_x and NO_x remain in the atmosphere, the greater are the chances of their oxidation to H_2SO_4 and HNO_3 due to photochemical and catalytic chemical reactions. Acid rains may cause extensive damage to materials and terrestrial ecosystems, such as water, fish, vegetation, stone, steel, paint, soil and mankind.

The only practical approach to counter the problem of acid rain is to reduce SO_x and NO_x emissions. The following three general options are considered for this purpose:

1. Energy conservation resulting in reduced fuel consumption and hence slower emissions of SO_x and NO_x. Conservation via more efficient fuel use and through improved thermal insulation is also being studied.
2. Desulphurisation and denitrification of fuels of stack gases and increased use of fuels naturally low in sulphur content or use of technologies that reduce SO_x and NO_x emissions. Desulphurisation and use of low NO_x-producing technologies are the only viable control options today and will perhaps continue to be so for some more time.
3. Substitutions for fossil fuels by other alternative energy forms may offer future solutions to this problem.

Reduction of SO_x emissions can be accomplished by: (i) removing the sulphur content before the fuel is burnt with the help of techniques such as coal cleaning, coal gasification and desulphurisation of liquid fuels, (ii) removing the sulphur content during combustion, as in fluidised-bed combustion, and (iii) removal of sulphur emissions after combustion, as in stack or flue gas desulphurisation systems or scrubbers. The future of SO_x control from traditional fuel sources lies in the perfection of these techniques.

Reduction of NO_x emissions from stationary combustion sources can be achieved by modification of furnace and burner design, and/or modification of operating conditions. The combustion modification techniques available now include using two-stage combustion, precisely controlling air, injecting water during combustion, recirculating flue gases, and/or by altering design of firing chambers. Reductions in NO_x emissions from mobile combustion sources may be achieved by lowering the combustion temperatures in the engine and catalytic removal of NO_x from exhaust gases using devices such as a three-way system that simultaneously reduces carbon monoxide, hydrocarbons and NO_x.

METHANE AND GLOBAL WARMING

Even though we have the tendency of relating global warming directly to the amount of carbon dioxide released in the atmosphere, the fact is that other greenhouse gases are equally responsible for this hazardous phenomenon. Among the other greenhouse gases which are responsible for global warming, one of the most prominent gas is methane. Global warming effects on earth are becoming more and more obvious, and scientists are trying their best to come up with proper explanations for the same. A recent study at NASA revealed that the effects of methane on global warming are thrice than what was estimated before.

Methane Release and Global Warming

Enormous amount of methane gas is trapped beneath the earth's crust. This gas is released in the atmosphere as a result of anthropogenic activities, such as mining and oil drilling. Other than such anthropogenic activities, methane trapped beneath the earth's crust is released through volcanoes and geological faults as well. Methane is also produced anaerobically by the process of methanogenesis. Other then these sources of methane, one of the most prominent source is cattle, or ruminating animals to be precise. Methane is produced within the body of these ruminating animals in a process referred to as enteric fermentation. Methane, when released in the atmosphere, tends to trap the heat coming from the sun, and sends it back to the surface of the earth, thus contributing to climate change. Read more on other causes of global warming.

Methane Global Warming Potential

When it comes to global warming, methane is 21 times more potent than carbon dioxide. Add to it the fact that the amount of methane released in the atmosphere has increased significantly over the last few years. In 1750, methane concentration in the earth's atmosphere was 700 parts per billion (ppb), which soared up to 1745 parts per billion (ppb) in 1998, and continues to soar. Methane has a life-span of approximately 7 years. Some may argue that this is negligent as compared to carbon dioxide with a life-span of a hundred years. However, we also need to take into consideration the fact that methane is 21 times stronger than carbon dioxide, and hence has the tendency to cause more harm than the later.

Curb Methane Release: Curtail Global Warming

Scientists believe that the very fact that methane is responsible for global warming to such a great extent can work in our favour. Carbon dioxide has a life-span of a 100 years, while methane has a life-span of just 7 years. This means that if the release of methane is stopped at this very moment, it will take 7 years to clean up its concentration in the atmosphere. In context of carbon dioxide, it would take 100 years to clean up from the moment its release is stopped. Completely stopping the release of methane may not be practical, but no one will deny the fact that it can be lessened gradually. At the same time, the fact that methane is 21 times more potent also means that cleaning up the atmosphere of this gas will help in reducing the impact of global warming on earth by a significant extent.

Scientists are trying their best to come up with some concrete global warming solutions in order to curb the damage caused by this phenomenon. Understanding methane global warming potential is one thing, and acting upon it is another. Global warming is undoubtedly a major problem which we have been taking lightly all this while. This delay has just fuelled the menace as well as other disasters associated with global warming.

Evidence of global warming, such as melting glaciers, rising sea levels and frequent hurricanes, are the most prominent signs of the approaching disaster. The fact of the matter is that we are losing time, and if we still continue turning a blind eye towards the hazards of global warming, we might lose a lot, if not everything.

EVIDENCE OF GLOBAL WARMING

Among the various environmental issues we are facing today, global warming is by far the most grievous one. That, however, does not mean that other issues we face are not serious. In fact, all these environmental issues are related to each other, and thus pose a big threat to the planet on the whole. The bad news, though, is that we are not able to understand the seriousness of the issue of global warming. Standing at loggerheads, global warming skeptics and scientists, who believe that global warming is happening, are busy in the tussle to get an upper hand on the other. While the scientists continue to provide the evidence of global warming one after another, the skeptics are refuting them at an equally fast pace. In the meanwhile, we are only losing time, and inching towards a disaster with every passing day.

What is the Evidence of Global Warming?

Over the last decade, scientists have provided substantial scientific evidence of global warming, but the skeptics continue their refuting spree, citing that the evidence provided is based on computer models which are too crude to trust. As laymen, we might not understand anything about these computer models and climatic models, but the evidence of global warming we can see with our own eyes is too obvious to ignore. Melting glaciers, submerging low lands, warm decades the list of global warming evidence is quite long. The only problem, though, is that we have decided to turn a blind eye towards them.

Soaring Temperature

It won't take an Einstein to realise that its getting hotter by the day. If the data compiled by the scientists at NASA is to be believed, the first decade of the 21st century was the warmest decade ever. In fact, this decade had nine out of the ten hottest years the planet has ever witnessed. A report compiled by NASA's Goddard Institute for Space Studies revealed that the global temperatures have increased by 1.4 degrees Fahrenheit (i.e. 0.8 degree Celsius) over the last century. More recent studies have also revealed that the rate at which global warming is occurring is also increasing.

The last two decades of the 20th century were the hottest indeed, but these records have already been broken by the first decade of the 21st century, and the rise is expected to continue, and even worsen.

Rising Levels of Carbon Dioxide

A major evidence of global warming caused by humans, the rising levels of carbon dioxide in the atmosphere speaks in volumes about the, till now debated, anthropogenic causes of global warming. The international energy statistics reveal that human carbon emissions have also gone up dramatically. Greenhouse gases, such as carbon dioxide and methane, get trapped in the atmosphere, and increase the temperature of the planet. A large part of these gases is the atmosphere are released as a result of human activities. Carbon dioxide is released in the atmosphere by human activities such as use of vehicles and production of electricity, while methane is released in the atmosphere as a result of activities such as mining and cattle rearing.

Melting Glaciers

Yet another prominent evidence of global warming, glacial melting is also occurring at a rapid pace. These glaciers have been retreating since a long time, however, the rate at which we are losing them over the last few decades is indeed a matter of grave concern. The problem of melting glaciers is not just restricted to tropical regions, but is also quite prominent in the polar areas. The soaring temperature in Arctic region has been melting the ice in this region, thus resulting in loss of habitat for various animal species inhabiting this region. Glacier melting in tropical areas, such as Himalayas and South America, pose a threat of flash floods to the areas in vicinity. Melting glaciers are a matter of grave concern, also because they are important sources of fresh water the world over.

Submerging of Low Lying Areas

Melting glaciers is contributing to the rise in sea level, which is continuously increasing and threatening the low lying areas across the world. Rising water is slowly and steadily encroaching upon the low lying coastal areas and small islands, and submerging them. Maldives islands sinking is by far the best example of ocean water encroachment. This is a serious threat to a number of large cities, which lie on the coast, including Mumbai in India, London in England and San Francisco in the United States. Other then these cities, countries like Bangladesh are also bound to bear the brunt of rising sea level.

Rising Number of Storms

Yet another scientific evidence of global warming is the rising number of storms in the tropical areas. Large scale melting of glaciers and constant heating of ocean water, due to soaring temperature, provides as ideal environment for storm formation. A recent study by a journal dealing in environmental issues revealed that the intensity of storms has increased over the last three decades. The problem is that they have not just become stronger, but have become long lasting as well. Researchers state that the intensity of hurricanes increases by 5 per cent for every 1°C rise in temperature. If this trend continues, the storms are just going to be more intense, and cause more destruction.

IS GLOBAL WARMING A HOAX?

Global warming is one of the most serious problems the world is facing today. It is, in fact, threatening the basic survival of several life forms on the planet. The extreme weather we are experiencing today is in itself a major evidence that global warming is happening, but not all the people out there seem to be convinced. A significant part of the population is still not able to figure out whether is global warming a hoax or is global warming really happening?

In spite of the evidence that goes to support global warming, there are many who are still of the opinion that global warming is a hoax. However, global warming myths and facts, if studied, shows that global warming is indeed happening. The following are the major arguments that help decide if global warming is fact or fiction.

Arguments Against Global Warming

The arguments put forth by global warming skeptics revolve around two aspects—unreliability of the data and previous climate changes. They argue that most of the data which suggests that the planet is threatened by global warming is based on mere assumptions. The only reliable source of data is the satellites based in space, and these satellites which monitor the climate of the entire planet don't give any evidence of climate change in the lower troposphere. On the other hand, the data collected by sources on the earth is subjected to alteration due to several reasons, including the presence of industries nearby and human error. Arguments against global warming also stress on the fact that the atmosphere-ocean dynamics are quite unpredictable, and hence it is difficult to predict the detailed evolution of weather in the future. The computerised climatic models that are used to predict the climate trend in future are too crude to be trusted. Global warming skeptics also believe that the expenses that are being incurred on various global warming solutions, in the absence of any concrete evidence of global warming, would be futile and result in heavy loss for the developing as well as the developed countries. And lastly, if at all the global temperatures rise by modest amounts; it would be beneficial for us in several ways. The growing season will be prolonged, we will have more usable land and warmer winters will mean safer climate.

So is global warming a hoax? In spite of the arguments above, there is a lot of evidence to that supports the existence of global warming as well, that somehow skeptics have decided to turn a blind eye to or ignore. Average global temperatures have increased by 1°F over the last 100 years. This is evident by the study of air bubbles trapped in polar ice. According to NASA, the last decade was the warmest decade ever experienced. Glaciers at high altitudes and ice sheets in polar region are melting rapidly. Melting glaciers and polar ice is causing sea water to rise, which in turn is threatening the low lying areas with a watery grave. Maldives islands sinking is perhaps one of the best examples of the danger posed by the encroaching sea. Sources of drinking water are also hampered as a result of melting of the glaciers at high altitude.

Loss of habitat is driving animals to extinction. Polar bears and the Arctic fox have been forced to migrate to the North Pole in search of a suitable habitat. These are just a few of the several global warming effects on earth. If someone asks any questions like is global warming a hoax, or is global warming real, even after going through all the above evidence, then it is absolutely impossible to change his mindset, which is hell bent on proving that global warming is a myth. A rise in our body temperature is an indicator of something being wrong with our body. So a rise in the temperature of the planet would naturally be an indicator of something being wrong with the planet. The earlier we diagnose it, the better it will be for us.

IS GLOBAL WARMING REAL?

Retreating glaciers, rising sea levels, extinction of species, extreme weather conditions evidences of global warming exist in plenty. However, not all the people out there are convinced that these trends are any indicators of global warming. Even, eminent scientists and researchers all over the world are divided over this issue. All the evidence provided by researchers, who feel that the planet is heading for a disastrous climate change, is dismissed by global warming skeptics as mere exaggerations. Amidst all this confusion, the layman is left wondering whether the phenomenon of global warming is really happening, or is it just another large-scale controversy triggered for some personal gains of a few people sitting at the crux.

When researchers pitching for global warming talk about rising temperature, they refer to the average global temperature, and not that of a particular region or a country. On an average, the global temperature has increased by 1°F over the last century. In some regions, the temperature rise is less than the global average, while in some places it is more. For instance, the temperature in Arctic region has increased twice the global average in the same time-frame.

Global Warming is Real: Proofs to Substantiate the Claims

This is one question that haunts many people who wonder is global warming a real threat, or just a hoax. There are quite a few measures by which temperature trends can be studied. The simplest measure is to refer to the climatic data compiled over the past few years. The fact that 8 years (2001–2008) of 21st century, feature among the 20 warmest years, since 1901, in itself, is a clear indicator that the earth is getting warmer. These studies, about the rising temperature trends, are not just restricted to written records. Some scientific methods, like measuring the temperature of air bubbles caught in polar ice sheets and studying the rings of trees living for thousands of years, also help in studying climate change. As far as future predictions are concerned, scientists are left with no other option but to depend on the probability and computerised models, something which skeptics of global warming disapprove of.

Skeptic's Take: Is Global Warming Real

In order to prove that global warming is real, scientists also, bring to notice the minor alterations the planet has witnessed over the last decade or two, which are expected to worsen with time. This includes extreme temperatures (scorching summers and chilled winters), unnatural precipitation pattern causing floods and droughts, melting glaciers and polar ice caps, rising sea levels threatening to submerge low-lying areas etc. However, all these examples seem to be of least importance to global warming skeptics. According to them, the data used to predict the warming trend is insufficient, and hence, untrustworthy. They feel that the scientists, who are raising the alarm, are interpreting the data incorrectly and coming up with false conclusions. In their arguments against global warming, skeptics state that the changes, observed over the last few years, are part of natural climate shift, and all the life-forms, subsisting on the planet, will adapt to these changes with time.

The catastrophe of global warming seems to be stuck between two groups, neither of which is able to figure out is global warming real or not. Although ignorance about global warming myths facts has made this concept even more difficult to resolve, it would be foolish to throw caution to the winds and continue fuelling the disaster. There is no point in coming up with foolish arguments like 'our generation will not face these problems'. It is time to open our eyes and do a reality check. This will make us realise what a disaster we are heading for. We need to come up with some concrete global warming solutions and implement them at the earliest.

One thing people ought to understand is that these global warming effects on earth are not going to crop up overnight, they will start surfacing gradually, over the years. More importantly, the effects will be more severe during the gradual rise, as various life forms will find it difficult to adjust to the changing environment.

Some of them have already started to surface, but we have been deliberately turning a blind eye towards them. If we continue to do so, by the time we open our eyes, we may realise that the time to ask 'Is global warming real?', would be long gone.

GLOBAL WARMING: FACT OR FICTION

Although, environmentalists the world over are afraid that the hazards of global warming will sweep away life on planet earth sometime in near future, global warming skeptics still refuse to budge. The whole concept of global warming revolves around the fact that the earth is getting warmer, and a significant part of this rise in temperature can be attributed to anthropogenic causes, i.e. man-made causes. Time and again global warming skeptics have come up with arguments against global warming. According to them, the predictions made by various environmentalists and scientists are based on climatic models, which are actually computer generated programs, difficult to rely upon. Before we move on to see whether global warming is fact or fiction, let's have a brief look at the anthropogenic as well as the natural causes of global warming.

Man-Made vs Natural Causes of Global Warming

Whenever the subject 'global warming is fact or fiction' comes to the table, the blame game between human induced causes and natural causes of global warming are at their peak. Although the scientists from all over the world agree that global warming is happening, there is no consensus on the fact that the majority of the causes of global warming today are anthropogenic. Natural causes of global warming do exist, but they have always been there. In fact, they are necessary to maintain the ideal balance in the climate of the earth. They trap the sunlight and help in keeping the planet warm enough for humans to thrive. Without these natural factors causing global warming, earth would have been much more colder, perhaps not even suitable for human inhabitation. The major problem the planet is facing today is human induced global warming. Over the last few decades, i.e. since the industrial revolution to be precise, the use of fossil fuels for power generation has been on a tremendous rise. Burning of these fossil fuels produces greenhouse gases, like carbon dioxide, which remain in the atmosphere for thousands of years and trap sunlight to cause global warming. Other anthropogenic activities which lead to global warming include deforestation, mining, use of vehicles, etc.

Not everyone is ready to believe that global warming is actually happening. In fact, the world is divided into two groups—people who understand the hazards of global warming and people who feel that the whole issue of global warming is a hoax. Both are armed with substantial amount of claims and arguments to support their stand and oppose the stand of others. The biggest global warming fiction is perhaps that human induced global warming has a negligible share in increasing global temperatures and climate change.

The truth is that the amount of carbon dioxide generated by human actions is too much to term it 'insignificant' or 'negligible' as the global warming skeptics do. Greenhouse gases, such as carbon dioxide and methane, have formed a layer in the atmosphere. As the sunlight enters the earth's atmosphere, it is trapped by these gases and due to this trapping of sunlight, global temperatures are rising. The expected temperature rise is predicted to be somewhere between 1.4° to 5.8°C by the end of this century. This is a significant increase considering that a fall of 5°C triggered an ice age on this planet some centuries ago. A rise by 5°C is bound to trigger some calamities on the planet by causing a drastic climate change that will wipe off several plant and animal species from the planet.

From rising sea level to increasing frequency of hurricanes—the evidence of global warming exist in plenty. Surprisingly though, we are trying our best to turn a blind eye towards the fact that earth is getting warmer, and this warming is threatening our basic existence. The evidence mentioned below are perhaps the best possible examples to claim that global warming is happening and it's going to affect us sooner or later.

Climate Change Due to Global Warming

The increasing number of hurricanes is perhaps the biggest evidence of climate change induced by global warming. More importantly, a look at the data of the last 10 years shows that summers have been warmer and winters colder than usual. In fact, 11 of the last 13 years feature in top 25 warmest years on the planet. These variations in global temperatures are also causing droughts, floods, heat waves, cold waves and changes in the precipitation patterns around the world. The rise in global temperatures also mean that certain disease carrying insects, which thrive in warm climates, will spread to newer regions on the warm planet, thus propagating the spread of the diseases they carry.

Extinctions Triggered by Global Warming

The list of extinct animals in the last 100 years include mammals, birds, reptiles, amphibians and insects as well. A large number of this extinct animals were wiped off the planet due to sudden change in temperature levels and loss of habitat due to climate change. One such example is the Monteverde toad, endemic to the tropical rainforests of Costa Rica. It became extinct in 1989 as a result of unusually warm climate, wherein sudden evaporation of water bodies killed all the tadpoles before they matured. Another example of an animal fighting for its basic survival against the odds of global warming is the polar bear. Melting ice caps is resulting loss of habitat for these bears, forcing them to migrate further north towards the pole.

The hazards of global warming haven't yet started to affect us directly, but the indirect effects of global warming have already started to surface. Basically, all the things on the planet are related to each other, and therefore the global warming effects on earth are also interrelated. We humans are most dependent on nature, which we have been exploiting continuously over the years, so any alterations in nature are bound to impact us either way. Crop failure induced by climate change will leave us without food. Extinction of tigers will increase the number of herbivores, which will feed on the vegetation cover, and loss of vegetation cover in turn will impact the rains as well as the ground water source. Global warming induced sea level rise and melting glaciers are bound to disrupt the fresh water supply, thus leaving us with no water to drink. Rise in sea levels will also submerge coastal areas and tiny islands. More importantly, most of the prominent cities in the world are located in the coastal areas. Submerging of coastal areas of Bangladesh, Maldives islands sinking, flash floods in the Himalayas, etc. are just some examples of how global warming is affecting the various regions of the world. Such evidence is more than enough to prove that global warming is fact, and not fiction.

These global warming facts can help you in realising how serious the issue is. As the global warming skeptics rightly point out, it is difficult to rely on climatic models to predict what the future has in store for us. Even though we are armed with some of the most sophisticated machinery, it is difficult for us to predict tomorrows weather, so how can we predict what would be happening 10 or 100 years down the lane. That, however, doesn't mean that we don't take any precautionary measures to save planet earth. We can't just pray that the global warming myth never comes true, nor can we take some stance like we will act only when its starts affecting us.

While the bigger nations are trying to figure out whether global warming is fact or fiction, small island nations have already started to bear the brunt of global warming. If some global warming solutions are not implemented soon, next would be the turn of bigger nations. If greenhouse gas emissions continue to occur at the current rate, then doomsday will be just round the corner, by the end of this century. Global warming has been occurring since the earth was formed, but lately the rate at which this is taking place has been alarming enough to raise some concerns. Over the last decade environmentalists have been coming up with facts about global warming, while skeptics have been refuting them as mere myths. These global warming myths and facts will continue to trigger heated debates for some time to come, we can just hope that these debates or arguments don't continue even when the calamity itself knocks on our door.

FOSSIL FUELS AND GLOBAL WARMING

Coal, oil and natural gas are the three different forms of fossil fuels that are widely used. They are formed by the process of anaerobic decomposition of organic matter under the surface of the earth for millions of years. Large scale use of fossil fuels started since industrial revolution. Today, they are the most cheap sources of energy available for the use of both personal as well as commercial purposes. Petroleum is used to fuel our vehicles while coal and natural gas are used to produce electricity for our homes and offices. Statistics show that almost three-fourth of the demands of the energy in the world is fulfilled by fossil fuels. Let us find out the relationship between fossil fuels and global warming.

Burning Fossil Fuels Contribution to Global Warming

The main function of the carbon dioxide present in the atmosphere is to trap the heat obtained from sunlight and do not let it go beyond the atmosphere. It is because of the carbon dioxide in the atmosphere that our planet is warmer than any other planet of the solar system. When there is a rise in the percentage of carbon dioxide in the air, the amount of heat captured by the carbon dioxide also increases. This in turn contributes towards overall rise in the surface temperature of the earth which is also known as global warming. On burning of fossil fuels, various types of gases like carbon dioxide, carbon monoxide, methane, nitrous oxide, etc. are released. A significant proportion of the carbon dioxide emitted into the atmosphere is by burning of the fossil fuels. Evidence obtained from various research studies suggest that since the middle of the nineteenth century, there have been at least 25 per cent increase in the carbon dioxide content in the atmospheric air. This is all because of extensive use of fossil fuels across the globe. As a result, in the last 150 years, the temperature of the earth has already increased more than 1 degree Fahrenheit. Moreover, it is going to increase further in the next hundred years. Thus our planet will be much more hotter in the next century.

This will have a severe effect on the climate of the earth. The weather conditions of various places of the earth will change drastically. Droughts and floods will occur more frequently in many inland areas that have extreme weather condition which will badly affect the agriculture. All the glaciers of the earth will be melting at a much faster pace. As a result, the areas nearby the water bodies like the coastal regions and the banks of the river will get submerged under water. Many deltas, islands, thickly populated cities are likely to go under water. Thus you can see that the issues of fossil fuels and global warming and climate change are all interwoven with each other.

Fossil Fuels and the Environment

Combustion of fossil fuels not only gives out carbon dioxide into the air, it also releases many other harmful acidic substances like sulphuric acid and carbonic acid and cause air pollution. When in air, these gases undergoes some chemical changes and return to the surface of the earth in the form of acid rain. This has a huge impact on the entire environment. It affects soil and plant life and causes water pollution. Large areas of land surfaces are dug up for the purpose of extraction of fossil fuels from their deposits under the earth. After the removal, this land becomes unusable and thus causes a permanent damage to the land and causes frequent earthquakes. Fossil fuels are transported from one place to another by tankers and ships. Any leakage in these tankers causes oil spills. Such type of accidents have occurred a number of times in the past. This not only leads to water pollution but also poses a serious threat to marine lives.

Hope you have understood the relation between fossil fuels and global warming. We all have a role to play in controlling global warming. If we take some small measures from our side, then we can save our planet from disaster. For this, we have to reduce our huge demand for energy. We have to decrease energy consumption in our homes, use our vehicles only for travelling short distances. We also have to stop cutting trees and plant more number of trees because they can absorb carbon dioxide from the atmosphere and thus check its level from rising.

GLOBAL WARMING EFFECTS ON EARTH

The menace of global warming has multiplied over the years and unfortunately, it continues to multiply, as the tussle between the developed and developing nations to implement emission cuts continues.

Global warming is a very serious issue, as the effect of global warming on one particular component of earth triggers a series of ill-effects on other related components. For instance, melting polar ice will raise the water level, which will in turn submerge the low lying areas around the world.

Who is Affected by Global Warming?

The question will be very difficult to answer. Instead, a question like who won't be affected by global warming would be much easier to answer, because the answer will be short—No One! Humans, animals, plants, climate, land... you name it and it will be affected by global warming. In fact, some species of plants and animals are already on the verge of extinction. Studies indicate that around 15 to 37 per cent of plant and animal species will be wiped off the planet by the year 2050. Changes in climatic patterns have already started to show, sea level is rising... to make it short, we are just heading for destruction.

Global warming effects on the environment

The average temperature of the planet for each year, over the last decade, has been featuring in top 25 high temperatures of all time. In fact, 1998 and 2005 were the hottest years in the history of the earth. The planet is getting warmer by the day, and just because the change is happening gradually we can't ignore the fact that we are vulnerable to the threat of global warming. The effects of global warming on earth are legion. In fact, the appropriate phrase would be—'the entire planet is threatened by the hazards of global warming'. Global warming effects on earth are many, and to understand the overall effects of global warming on earth, we have to understand the effects of global warming on each component of the planet.

A large number of animal species will disappear from the planet, owing to the loss of habitat triggered by global warming. There is no doubt that many animals will bear the brunt of climate change sparked by global warming. In fact, it's feared that global warming is bound to trigger a mass extinction very soon and one-third of the animal species will become extinct by 2050.

Global Warming effects on polar bears: Polar bears depend on ice formed on the sea when hunting. If the ice melts, the range of polar bear will decrease to a great extent, and this loss of habitat will in turn lead to a decline in the polar bear population.

Global warming effects on penguins: Melting sea ice will also result in decline in the growth of algae, which in turn will result in decline of tiny organisms, such as krill shrimp, which constitute a very important part of the penguin diet. And thus, scarcity of food and loss of habitat will eventually drive the penguins towards extinction.

Global warming effects on arctic fox: Warm temperature has been driving the Arctic fox further north in search of cooler habitat, but the rate at which we are losing colder regions, the Arctic fox is bound to lose the battle for survival sometime soon.

These were just a few of the animals which are threatened by global warming. The long list includes caribou, butterflies, hibernating animals, migratory birds and various fish species as well.

Global warming effects on plants

Even plants will not be spared from the brunt of global warming effects on earth. Owing to the drastic changes in temperature levels, various plant species have been experiencing difficulties in adapting to the areas wherein once they flourished. The growing season of some plant species has also been altered, which in turn has disturbed the reproduction cycle of the species, thus giving a drastic blow to the plant population. Even the changes in precipitation patterns can lead to hazardous effects on various plants species. Global warming effects on agriculture is the best possible example one can give to explain the effects of global warming on plants. Frequent rains will lead to flooding, whereas less rains will result in drought, both of which will only lead to the destruction of agricultural fields. Other natural disasters, such as hurricanes, which are also caused due to global warming, can have a disastrous effects on plant life. Furthermore extinction of animals due to global warming will also lead to negative impacts on the plant life, either directly or indirectly. For instance, extinction of tiger will result in increase in number of herbivores and excessive feeding by herbivores will result in depletion of the forest cover. Extinction of birds will affect the pollination process and hamper reproduction in plants. Owing to all these factors, various plant species are also expected to become extinct by the end of this century.

Global warming effects on weather

Increasing temperatures will lead to adverse effects on weather as well. Even minor alterations in global temperatures will trigger a series of weather extremities, and alter the climatic patterns of the planet. The number of natural calamities have increased over the period of time. Last three decades have witnessed a rise in number of category 4 and category 5 storms. The Intergovernmental Panel on Climate Change (IPCC) acknowledges the fact that the frequency of intense rains has increased over the last 50 years. On one hand, heating of the ocean due to global warming gives rise to ferocious hurricanes, while more than the normal temperature on land gives rise to intense heat waves. Higher temperature leads to faster evaporation of water and leads to drought in one part, and brings in heavy rainfalls and causes flooding in other part of the world. Although we can't conclude the serious effects of global warming on weather by taking into consideration a single drought year or a single devastating hurricane, the trend of these natural occurrences speak in volumes for themselves.

Global warming effects on glaciers

One of the more severe effects of global warming on earth is the melting of perennial and permanent ice covers on the planet. There are several thousands of glaciers spread all over the world which form an important source of fresh water. Monitoring of these glaciers, by projects such as Global Land Ice Measurement from Space (GLIMS), has revealed that these glaciers are disappearing at an alarming rate. This is viewed as one of the most prominent factor for rising sea levels. The glaciers at the Patagonian ice fields of Argentina have receded 1.5 kilometres overs the last two decades. The number of glaciers in Glacier National Park in Montana had dropped from an estimated figure of 150 to 50 within a span of 150 years, and is expected to drop further, eventually leading to disappearance of all the glaciers by 2030. Melting glaciers can trigger severe natural calamities, such as flash floods, in the surrounding regions. More importantly, the melted water flows into the oceans thus causing the sea level to rise, which eventually leads to submerging of low lying areas such as Bangladesh and Maldives.

Global warming effects on sea levels

One of the most grievous among the various global warming effects on earth is the rise in sea levels, which are threatening to encroach up on land. If the sea levels rise it will result in a watery grave to several low lying areas, tiny islands and reclaimed portions of land. So how exactly is global warming affecting sea levels? Basics of geographical studies suggests that water expands when heated. In case global warming, rising global temperatures are causing the water bodies to heat, expand and thus encroach on land. Another prominent reason for sea level rise due to global warming is melting ice from the glaciers and polar ice sheets. These ice stores are far massive than we can imagine. In fact, melting of

West Antarctic Ice Sheet alone can possibly cause the sea to rise by a whopping 10 meters. According to the Intergovernmental Panel on Climate Change (IPCC) the sea levels have seen a rise of 6.7 inches in the last century, and if the alarming rate of global warming continues, the sea water levels may rise up to 22 inches by 2100. This will mean that island like Maldives and Tuvalu and low lying areas like Bangladesh will go underwater, and important cities like Mumbai, Shanghai and Florida will become vulnerable to the water grave like the legendary city of Atlantis. In fact, Maldives islands sinking is the best possible example of destruction due to rise in sea levels.

Global warming effects on coral reefs

The effects of global warming on the coral reefs are devastating, in fact, the phenomenon of global warming is on the verge of making coral reefs the first major ecosystem to be wiped off the planet very soon. When the ocean water gets warm, the algae in the ocean tends to produce toxic oxygen compounds called superoxides which is damaging for the corals. As a defense mechanism, the corals eject their algal lodgers, which leaves the reefs starved for nutrients and their colour turns white. This process is referred to as bleaching. Global warming is threatening the coral reefs to a great extent, and the fact is that if coral reefs are wiped off the planet, it will affect one-third of planets marine biodiversity, as well as other ecosystems related to the coral reefs directly or indirectly.

Global warming effects on humans

When the whole environment will experience the effects of global warming, naturally humans won't be an exception. In fact, we will be the worst affected beings on the planet because directly or indirectly we are dependent on all the above mentioned components of the environment. Animals and plants are related to each other, extinction of either will put tremendous pressure on other, eventually leading to its extinction. Humans, in turn, are dependent on both for many purposes, so extinction of animals or plants will also affect humans to a great extent. Irregular weather will have a severe impact on several human activities. Warmer summers will mean more allergies and even more disease spreading insects. Unnatural precipitation will lead to destruction of crops and hamper agriculture. Rising temperatures will lead to warming of ocean bodies, which will in turn increase the frequency of hurricanes. Destruction of coral reefs will lead to loss of marine life including fish which is an important constituent of human diet. Coastal areas around the world are highly populated, so any rise in sea level will lead to a heavy impact on the people residing in coastal areas. Glaciologist's estimate that if melting of glaciers at the present rate continues, around 20 per cent of Bangladesh will get submerged in the sea by 2020. Countries like Maldives, with the highest point of 2.4 meters above mean sea level, will get submerged if the sea water levels rise by 3 meters. Loss of glaciers will hamper the water supply for millions. Rise in sea level will alter the coastlines thus affecting tourism sector. The saline water from oceans will flow into river beds thus making the river water unusable.

Global warming effects on economy

The repercussions of various effects of global warming on earth will also be felt on the economy of various countries. The most affected would be the countries with agriculture-led economy. Global warming will trigger a series of changes in weather conditions which will take a toll on agriculture and allied activities. Owing to unnatural precipitation pattern, crop failure will become a very common phenomenon. Economies dependent on tourism, such as Maldives, will also bear the brunt of global warming. As the water levels rise all the coastal area will get submerged leaving the world devoid of all the beautiful beaches. Considering that the world has become a global village, the domino effect of global warming will also be seen on other countries and more importantly on the world economy.

These were some of the damaging global warming effects on earth. Not everything about global warming would be bad though. Global warming will mean a longer growing season and hence an increase in production. In the United States, it will melt the polar caps along the Northwest Passage way which will lower the shipping costs. The problem is that the negative effects of global warming on earth far exceed the positive effects and thus the whole world is concerned about the future and trying to find our some convincing global warming solutions.

Global warming has man-made as well as natural causes. The natural causes of global warming which includes water vapour and volcanoes are beyond our reach, but we can make sure that man-made causes of global warming are reduced. If we don't initiate the necessary steps soon we will have to face the wrath of global warming sometime in near future.

GLOBAL WARMING AND CARBON DIOXIDE (CO_2)

Global warming results from the effect of what is called the 'greenhouse effect'. Carbon dioxide (CO_2), methane, water vapour, nitrous oxide and ozone are the main greenhouse gases. The main purpose of a greenhouse is to allow sun's radiation (heat) to enter the enclosure easily, however, does not allow the heat to be radiated back to the atmosphere. Similar to a greenhouse, our atmosphere allows the sun's radiation to heat the earth and slows down the radiated heat to go back into the space. This greenhouse effect is essential to sustain life on earth.

Now, due to the increased level of greenhouse gases, the atmosphere's ability to trap and hold the heat increases more than required. As a result, the average temperature of the earth increases and this natural phenomenon is known as 'global warming'. It is also defined as a sustained increase in the atmosphere that causes abnormal climate change. In the following description about global warming and carbon dioxide, a brief explanation regarding effects of global warming and CO_2 emission in the atmosphere, has been provided.

Global Warming and Carbon Dioxide Emissions

Power plants

Power plants are the major contributors to the increased level of carbon dioxide emissions in the atmosphere. The plants work for generating electricity by burning fossil fuels in a massive scale, which produces large amounts of CO_2. Coal is known to be responsible for about 93 per cent of the emissions in the power plants. Natural gas produces 80 per cent less carbon per unit of energy than coal, and hence so much of impact on the environment results in the ever-increasing pace of global warming.

Vehicles

Gasoline-burning engines produce about 20 per cent of carbon dioxide emitted in the atmosphere. In the United States, 33 per cent of emissions is from vehicles. Sports bikes and vehicles result in more emission than general vehicles designed for normal roads.

CO_2 due to deforestation

The ever-increasing deforestation is also one of the main culprits of global warming. As we are all aware of the fact that, trees take in huge amount of CO_2 from the atmosphere and releases oxygen. So, if there are no trees, there will be no absorption of this global warming gas and thus, the situation worsens further. This fact is important when it comes to the reason as to why one of the global warming causes is carbon dioxide? As the world is progressing its way towards development, forests are vanishing away from the face of the earth. More urbanisation results in more deforestation for land and timber requirements, and all these factors boil down to one 'dead end', called global warming.

Some Major Negative Effects of Global Warming

Increase in sea level

As the earth warms further, polar ice caps would melt and increase the water levels in seas and oceans. It is estimated that melting down of all glaciers will cause the water level to rise to about 230 feet. So you can imagine what it would be like? Further, ice caps are fresh water and their melting and mixing with the seas and oceans will greatly imbalance the ecosystem. This is because, the ice caps will desalinate the water and make it less salty. This process will disturb the ocean currents and hence their regulation of temperature. Ice caps are known to reflect much of sunlight back into space, further cooling the earth. Melting of these reflectors will leave oceans and seas to be totally exposed to sun's radiation. These water bodies cannot reflect light but absorb it as they are dark in colour. This will further contribute to the warming of the earth.

Health and disease

Places of the earth that will drastically warm up due to global warming, will face more incidents of diseases and drought. Such areas will have increased infestation by disease-carrying insects and heat waves.

It is a known fact that people today are inflicted by many new diseases unheard of in the last century. Increase in CO_2 levels in the cities is the cause of pollution, and the effects of pollution are well documented. Global warming is also responsible for the introduction of some new diseases. Bacteria are known to be more effective and multiply much faster in warmer temperatures compared to cold temperatures. The increase in temperature has led to increase in the microbes that cause diseases.

Hurricanes

Due to the melting of ice caps, oceans will absorb more heat resulting in warmer waters. This will favour stronger and more frequent incidents of hurricanes. Increasing emissions of CO_2 into the atmosphere is making matters worse than ever when it comes to dealing with 'global warming and carbon dioxide'. We can also form an idea about the 'most probable' aftermaths of the always-increasing temperature of out planet earth, and the idea is definitely not a pleasing one. All other factors, along with those responsible for increased amount of carbon dioxide causing global warming, are perhaps impossible to do away with. These factors are arising only due to human needs; development being an apt example, and so they can be regarded as 'inevitable'. However, we can always help, if not in eliminating such factors totally, but reducing their emergence as much as possible. We should always remember that whatever 'good' we do for our mother earth, we are doing it for ourselves and the future generation to come.

Global Warming and The Greenhouse Effect

The greenhouse effect is a natural warming process of the earth. The theory is that when the sun's energy reaches the earth, some of it is reflected back to space while the rest is absorbed. The absorbed energy warms the earth's surface which then emits heat energy back toward space in the form of long wave radiation. This outgoing radiation is partially trapped by greenhouse gases such as carbon dioxide, methane and water vapour which then radiate the energy in all directions, warming the earth's surface and atmosphere. Troubling facts about the greenhouse effect include that the increasing amounts of greenhouse gases intensify the greenhouse effect. Higher concentrations of CO_2 and other greenhouse gases trap more infrared energy in the atmosphere than what occurs naturally. The additional heat further warms the atmosphere and earth's surface. Climate models suggest this natural warming is being enhanced by human activities that increase concentrations of greenhouse gases in the atmosphere and thus, in turn, intensify the greenhouse effect. The increase in atmospheric concentration of greenhouse gases, accentuating the natural greenhouse effect globally, results in global warming.

Global warming affects many different facets of life on earth. There will be winners and losers, even within a single region. But globally, the losses are expected to far outweigh the benefits. The regions that will get most severely affected are often the regions that emit the least green house gases. This is one of the challenges that policy makers face in finding fair international responses to the problem.

Water resources

Many of the major rivers in Europe and Asia emanate from the glaciers in the mountains. For example, the whole of the grain producing northern belt of India is fed by rivers which originate from glaciers in the great Himalayas; electricity is also produced wherever the rivers are dammed. Over the years it has been seen that because of global warming the glaciers are receding at an alarming rate, impacting the flow of the rivers, causing a reduction in production of foodgrain and electricity.

Agriculture

It is said that should there be a few degree heating up of the earth it would have a potentially negative effect on the production of corn in North America, where much of the world's foodgrain comes from. This would result in higher prices of foodgrain causing starvation in third world countries. On the other hand, it may so happen that colder regions further north would be able to grow crops that have never been cultivable before. However, it is a known fact that the availability of cultivable land decreases as you go north.

Ecosystems

We are aware of the fact that the ecosystem of the world is fragile and delicately balanced. Any change brought about by global warming could result in disastrous consequences. Some of these the world has experienced in the last few decades by way of floods, droughts, frequent hurricanes, tornadoes, and severe cold waves. There have also been an increased number of forest fires in the recent past. There is, however, a very divergent view on global warming facts. Many skeptics are of the view that global warming is a good and natural phenomenon. There are various benefits of global warming, and as per them it will increase humidity in tropical deserts, melting of snow bound areas will result in vast tracks of land getting available for agriculture, the increase in carbon dioxide in the atmosphere will trigger plant growth.

Another argument of global warming is that the earth has been gradually warming up through history, and nature will continue to adapt to these changes. Despite the views of some skeptics mentioned above, there is a recorded negative impact of global warming. Due to the direct link between global warming and the greenhouse effect, the only way to turn the situation around is for everyone to do their bit to reduce the emission of greenhouse gases. Reduction of use of fossil fuels, use of alternate sources of energy, restricting the emission of pollutants by industries, reducing forest degradation, eliminating the use of CFCs and planting more trees will help us overcome this seemingly insurmountable problem.

NATURAL CAUSES OF GLOBAL WARMING

Global warming is a major environmental issue today. All over the world there is a projected continuation of deterioration, with respect to surface temperature and subsequent, drastic climatic variations. The increase of global surface temperature is alarming and not totally the result of the greenhouse gases that are anthropogenic in nature. The temperature increase has been screaming attention since the middle of the last century. Natural phenomena like solar variation and the 'Feedback' effect are also responsible for the occurrence. The gigantic steps taken in the field of industrialisation and greenhouse gas emissions have added to the carbon dioxide (CO_2) in the atmosphere.

Solar Variation

Climate changes in orbital forcing and solar luminosity are also responsible for variation in surface temperatures. Variations in solar activity is being held responsible for most of the global warming, since the 1950s. According to recent research, solar output variations are possibly amplified by galactic cosmic rays. The magnetic activity in and around the sun plays a major role in deflecting cosmic rays. These in turn are responsible for cloud condensation and climatic changes. Increased solar activity could be the result of excessive warming of the stratosphere. Solar cycles lead to an increase in brightness and heat generation. The light and heat get trapped in the stratosphere. The relation between solar radiation and global warming however, is yet to be proven and is one of the most speculated topics deliberated upon in recent times. The link between changes in cosmic rays and temperature is being studied alongside human induced reasons for the phenomenon.

Feedback Effect

It has been observed that warming only results in further warming and the resultant 'positive feedback' culminates in global warming. The opposite reaction is referred to as 'negative feedback'. Water vapour is the outcome of positive feedback. An increase in saturation of vapour pressure results in a subsequent increase in the quantity of atmospheric water vapour. This makes the atmosphere warmer and a stagnancy in the relative humidity. Feedback results in the emission of infrared radiation from clouds back to the earth's surface.

This facilitates further warming, depending on the type of cloud and the altitude. Longwave radiation that is emitted or released from the upper atmosphere is much less than the lower layers. The radiation that is emitted from the lower atmosphere thus gets reabsorbed by the surface. Subsequently, when global temperatures increase, the ice caps near the poles melt and add to the sea water level. Land and open water absorb solar radiation extensively and this has led to the arctic shrinkage.

Implications of Global Warming

Scientists work on fluid dynamics and transfer of radioactivity to study climate sensitivity. Human-caused emissions add to positive feedback and this results in long and short-term changes in the natural climate cycles. The causes of climate change in recent times is largely due to natural and human induced phenomena. It is now clearly observed that specific global weather events are related to global warming. Increase in global temperature has surfaced in the form of melting glaciers and a rise in sea level, variations in precipitation patterns and flooding and numerous other extreme weather events. Neglect on our part to address reduced streamflows and disease vectors, related to global warming, will have a profound effect on our natural environment. It is essential to study the natural causes for this phenomenon and since little or nothing can be done about them, reduce the human-generated causes for environmental pollution that add to the problem. Geo-engineering is a very essential part of environmental science, for quick remediation of the effects of global warming. Ocean acidification, extinction concerns and disrupted food webs need to be sorted and treated amidst this awareness.

GLOBAL WARMING AND DEFORESTATION

Trees play a vital role in the equilibrium of the ecosystem. Deforestation is a process of cutting trees to make space for pastures or for industries and households of the ever-increasing human population. Excessive cutting of trees for urban use and other purposes is detrimental to the environmental balance. It is needless to say that deforestation has several adverse effects on the environment.

One of the major disadvantages of deforestation is that it disrupts the water cycle. Trees are responsible for drawing up water from the soil and releasing moisture into the atmosphere. Deforestation causes a disturbance in the water cycle and makes the environment drier. Climate change is a severe outcome of excessive cutting down of trees. Forests lock up atmospheric carbon during the process of photosynthesis. Trees contain a major portion of carbon from the atmosphere.

Clearing of the forest cover has a contrary effect on the environment. It results in an increase in the amount of carbon and other greenhouse gases in the environment. Burning of forests results in the emission of a large amount of carbon dioxide into the air. Carbon dioxide and other greenhouse gases like the oxides of nitrogen and methane are known to trap atmospheric heat, thus increasing the average temperature of the earth's surface. This increase in the temperature near the earth's surface and oceans is termed as global warming. The rise in the average temperature of our planet is bound to cause the sea level to increase. Global warming has already begun causing the melting of glaciers and of the ice at the poles, thus adding to the rise in the sea level. This phenomenon is a serious threat to the life on earth and it is we, who need to take the right measures to prevent it from happening.

We should not forget that trees add to the biodiversity in nature. Animal life thrives on vegetation. By cutting down trees, we deprive animals of their sources of food and cause the destruction of animal life. It can lead to the extinction of a variety of animal species. Global warming that is largely caused by deforestation further endangers plant and animal life, thereby disturbing the balance in nature.

It is believed that the use of fossil fuels and the burning of oil and gas cause global warming. It is true that pollution caused by the burning of oil and gas and the release of pollutants causes global warming. But research has revealed that deforestation is one of its major causes. It is the main reason behind the rise in the level of greenhouse gases in the atmosphere, leading to the greenhouse effect.

Extreme weather conditions, changing agricultural yields and increase in the disease vectors are some of the other effects of global warming. Deforestation, being the primary reason behind global warming, we need to show greater concern towards the felling of trees. We need to take quick measures on preventing deforestation so that we can hope for an environment conducive to live in.

POLAR BEARS AND GLOBAL WARMING

Global warming is the phenomenon wherein the temperature of the earth is increasing because of various factors and the result is diverse climactic changes all across the world and that is affecting the life of flora and fauna of all parts of the world. The pollution is increasing everyday and the ways to cope with it are limited. As a result of which the effects and counter effects have started showing drastically. It is actually considered that the rise in temperature would lead to the acidification of ocean waters and also ozone layer depletion. The major concern is the melting of glaciers, which implies that there is a fair chance of the world drowning one day and before that the extinction of animals that can only exist in the snow capped mountains and glaciers. Here we would discuss one such species, which is meeting a sad fate due to Global warming. Polar bears, which are native of the Arctic, are known as the world's largest land carnivore.

Why are Polar Bears Affected?

Being native to the Arctic, these animals are used to live on the ice and the sea and the melting of ice is taking away their habitat in a way. Not only are they losing their habitat but also are also being exposed to starvation since these carnivorous animals mostly feed on seals, walruses and whales which are also disappearing with the change in climate. Moreover the food that they get has high degree of pollutant, which leads to birth defects and reduction in the immunity of these bears.

How are These Polar Bears Affected?

As the ice floes are melting, most of the Polar bears are drowning and most of the time these bears drown because the ice caps retreat before their seasonal time to which the animals are not habituated. The drowning is not due to the fact that these Polar bears cannot swim, they can actually swim for miles together but they are used to swimming between ice sheets and the absence of ice sheets make them susceptible. The death rate of Polar bears has increased manifold with rate of ice retreat being on a constant increase. The distance between survival and extinction is ever increasing for these Polar bears and that has also started showing statistically as the number of Polar bears are decreasing.

Most of the time it happens so that before these bears can actually hoard enough food for themselves and come back to a safe place the ice melts and catches them at a vulnerable position. Polar bears are also used to hunt seals, the number of which is also declining. The next point is that due to acidification of oceans the marine animals are consuming contaminated food, which in turn is passed onto the Polar Bears who feed on them, thus weakening their immune systems and also leading to birth defects.

Is there Any Way of Saving These Polar Bears?

The answer is 'No' of we think from the point of view of finding a solution for Global warming but there is definitely a gleam of hope that can be seen since these animals are trying to adapt to the changing climatic conditions. The younger bears seem to be able to sustain on the beaches and are also found near the carcasses of whales butchered by humans, which is very unlike Polar bears. They would hunt for themselves and their preference has always been seals. So this can be seen as a way of adaptation for these creatures and we hope that they do adapt well to the change and sustain it.

GLOBAL WARMING AND PLANETS

According to the scientific community, the climate of the earth is changing rapidly. Over the last century, the earth has become hotter by 1°F, and it is projected that it will continue to do so, even more quickly, in the next few decades. The reason being, the atmosphere being polluted by greenhouse gases like carbon dioxide, most of it from industrial activity, the burning of fossil fuels, and deforestation. In fact, according to climate researchers, most of this global warming that has occurred over the past 50 years is due to the consequences of human activity.

What are the Consequences of Global Warming on Human Life and the Environment?

It goes without saying that the phenomenon of global warming is highly complex, and therefore difficult to predict what exactly the full-scale consequences on human life and the environment will be. However, with each passing year scientists are learning more and more about how global warming is impacting the earth, and many of them now agree that if there is a continuation of the current trends, then some of the effects of global warming will be.

Changes in climatic conditions

There will be an increase in the average temperatures all over the world, leading to more frequent heat waves. This will lead to drought-like conditions, increasing the chances of forest fires that are larger in scale, more intense, and harder to put out. Rising temperatures can also affect the climatic system of a region by increasing the energy of storm conditions, resulting in more severe rainfalls. Likewise, when the waters in the oceans heat up, it increases the energy of tropical storms, resulting in hurricanes that are more powerful and destructive.

Reduction of agricultural productivity

It is said that if the global warming continues at its current rate, the resulting increase in the temperature will lead to a drastic reduction in agricultural productivity, especially in northern continental regions. It is projected that the soil in these areas will be drier in the summer, partly because of the snow melting earlier in the spring, and more cloudless and hotter summers, which will in turn result in the moisture in the soil evaporating extensively, causing severe droughts. This will be further exacerbated by water levels in rivers and lakes falling because of less rainfall.

Inundation of coastal areas and low-lying regions due to rising sea levels

There is evidence that the sea levels are already rising, and this phenomenon is likely to increase because of glaciers melting in the mountains and the ice caps in the Arctic and Antarctic regions melting. This will have an impact on coastal areas that are low-lying like the estuaries of Chesapeake Bay and along the Gulf of Mexico. According to a report that the United Nations published, if global warming continues, the coastal areas of the Netherlands and Bangladesh will be flooded by 2100. Plus, it is also said that Maldives will disappear completely, which can happen by the sea level rising just by two feet.

The disruption of ecosystems

It is expected the global warming will cause extensive disruptions in ecosystems, with the consequential loss of a number of species, particularly those that are unable to adapt to the changes. According to some assessments, it is predicted that by the year 2050 about a million species will be extinct, if the current rates of global warming continues. Some of the ecosystems that are most vulnerable are alpine meadows found in the Rocky Mountains, the delicate ecosystem of the Himalayan ranges, mangrove forests in coastal areas, coral reefs, and so on.

Effects on health

Global warming will lead to heat waves occurring more intensively and more frequently, resulting heat-related health effects and deaths. These effects may also exacerbate problems caused by poor air quality in various regions, which are already affecting people around the world. It is also projected that the virulence and wider range of tropical diseases will also increase due to global warming.

Dealing with Global Warming

Dealing with and mitigating the adverse effects of global warming is not going to be an easy task. Ultimately, we will have to reduce the emissions of greenhouse gases like carbon dioxide dramatically if we want to protect the planet, the economy, as well as our health. In order to accomplish this, we will have to find the will to make fundamental changes to the way our global economy is powered, switching from using fossil fuels to renewable and more efficient sources of energy.

WAYS TO PREVENT GLOBAL WARMING

The one single concern that is threatening the very existence of lives in this planet is global warming. The drastic changes in the climatic conditions pose serious threat to our future generation. Owing to the results of global warming glaciers is retreating, sea levels are increasing, Polar bears and other rare cold climate species are slowly becoming extinct. Navigators maintain that the Icebergs that once dominated the Atlantic and Pacific oceans have now vanished. Scientists and environmentalists conducted various studies and concluded that the past 30 years has been the warmest period in the global history. The main causes attributed towards global warming are human activities. The burning of fossil fuel is one of the major factors that contribute to global warming.

Reasons for Global Warming

Global warming can be minimised to a great extent, if we eliminate the causes which are mostly human made. The responsibility of preventing global warming rests both on individual as well as the state. In individual level we can change our practices such as minimise the usage of fossil based fuel, reduce the electricity consumption by using energy efficient appliances. Vehicular pollution can be minimised by using the public transport system. The nucleus goal is to beget the global warming under control by restricting the carbon dioxide release and other heat ensnaring greenhouse gases into the environment. On an average nearly 10,000 pounds of carbon dioxide is released per year in significant countries like Canada and US. This can be immediately curtailed by becoming energy efficient. Reducing the usage of oil, coal and gasoline are one of the effective ways of preventing global warming.

Home Appliances Contribution to Global Warming

Regrettably, it is noticed that an average home contributes to global warming more than a car. The increase of contribution is at home because the energy utilised in our homes is received from power plants that burn fossil fuel to provide power for our electric products. We can save energy by substituting compact fluorescent bulbs in the place of incandescent light bulbs. This will save your money as well as your energy bill will be minimal. The CFL bulbs last longer and the energy consumed is also less. LED bulbs are efficient and energy savers. Home appliances also contribute a lot in elevating the energy bill. The higher is the energy efficient appliance, the lesser is the cost of running the appliance. So purchasing energy efficient appliances will help in reducing the utility bill and also in protecting the environment. Similarly, purchase major appliances such as dishwasher, air conditioner or refrigerator with maximum energy efficiency. This reduces the carbon dioxide pollution. As per the US energy report, heating as well as cooling systems emit maximum carbon dioxide in the atmosphere. The energy used for our homes for heating goes in vain though the prevention is inexpensive and simple. This energy that goes vain can be saved by reducing the need for air conditioners. This can help improve the environment from pollution. The largest source is transportation that adds to greenhouse gases. Vehicles are responsible and poor maintenance of vehicle contributes to pollution and global warming.

Maintaining Vehicle Efficiency

This can be protected by increasing the overall fuel efficiency of your vehicle and paying attention to your driving style and maintenance. Buying fuel efficient hybrid cars that allows using gas electric engines and thereby cutting global warming pollution to a great extent. Driving less and making use of public transports, walking or riding a bike will save the environment from pollution. Besides this, consolidating trips and encouraging car-pooling is one of the effective ways of preventing global warming. Recycling maximum products, eating local foods and vegetarian meals, painting home in light colour, purchasing energy certificates as well as carbon offsets are few of the ways of preventing our wonderful planet, the earth from the disastrous global warming. Though, there are many ways to prevent initiating and following it with determination will yield the desired results. This is a unison effort and so all the hands have to join together with force to push the effects of global warming back beyond sight.

So much has been said and written about global warming. With news reports that continuously inform readers about the harmful effects of global warming, it is sure most reactions would be, 'something needs to be done about global warming'. But in reality there are only a handful of people who actually go out of their way and incorporate ways to stop global warming in their daily lives. Apart from this, they educate people to use various ways to stop global warming as well. Scientists believe that various human actions are leading to global warming which is why we see strange changes in the weather and the extinction of most species as well. Well, many of us are not aware how we can contribute in different ways to prevent global warming. Have you really wondered how you can do your bit to stop global warming? If you haven't, it is high time you do so and if you are looking for some global warming solutions, spare a few minutes from your busy schedule and read on to know more about ways to stop global warming. We can make a difference in various ways to prevent global warming by implementing some of these steps in your everyday life.

1. Many of us cannot really give up luxuries such as your personal car and not everybody would resort to the use of public transport. But the least you can do is look for a car that is highly fuel-efficient. Every single gallon of gasoline that is burnt, will result in at least 20 pounds of carbon dioxide being emitted into the atmosphere. If you are the only person in the car, you can use the concept of car pool; join hands with your colleagues and come together to work. Whenever possible, use a cycle or walk to any nearby destinations.

2. When it comes to the purchase of appliances, always go in for a highly efficient model. Look for the symbol of the Energy Star; this is awarded by the Environmental Protection Agency. Although, such appliances may take a toll on your wallet, it hardly matters if you consider how useful it can be to the environment.

3. Avoid leaving appliances on a stand by mode; for e.g. use the power on and off buttons for your television set rather than leaving it on for the entire day.

4. If you do not require that extra freezer, unplug it right away. This would help to prevent carbon dioxide emissions into the atmosphere. Keeping the fridge next to the boiler would make it consume more energy. Give a thought to where you would need to place the refrigerator.

5. Use the washing machine and the dishwasher only when they are fully loaded. Also use the low temperature settings to conserve energy. Did you know that reducing the use of the dryer would save energy and also reduce the emissions of carbon dioxide? Try to air-dry your clothes whenever possible.

6. Use an energy saving light bulb to reduce the global warming. This would save energy to a great extent and even reduce your electricity bills as well.

7. Try to buy organic food. Look for locally grown food and avoid buying frozen foods. (Frozen food uses ten times more energy.)

8. Plant a tree; it could be in your backyard or you can gather with a group of like-minded people and plant one in your neighbourhood. We all know how trees are essential to reduce the effects of global warming. Look after your locality and encourage friends and family as well.

Spread the word: Share these global warming solutions and create an awareness to save our planet earth. Remember, it is our responsibility and every little effort counts. Each one of us can help to make a difference.

GLOBAL WARMING AND MELTING GLACIERS

The average global temperature has risen more than expected in the past few decades. Many prefer to use the milder term 'climate change' instead of the harsher 'global warming' to describe this change in average global temperature. The main cause of global warming is thought to be the 'greenhouse effect' that is mainly caused by us humans. With an increase in temperature glaciers worldwide are melting faster than the time taken for new ice layers to form, sea water is getting hotter and expanding causing sea levels to rise, rivers overflow due to melting glaciers causing floods, forest fires are on the rise, and innumerable undesired effects are taking place due to global warming.

The 'greenhouse effect' takes place when certain gases in the atmosphere of the earth trap heat. The term 'greenhouse' is used because light is allowed to reach the earth, but most of the heat generated is not allowed to escape, just as in a greenhouse. The more the greenhouse gases in the atmosphere, the more heat will be trapped within the earth's atmosphere, causing average earth temperatures to rise.

The greenhouse effect was first described by Joseph Fourier way back in 1824. The earth's temperature has increased by half a degree celsius over the past century due to an increase in greenhouse gases. This slight increase may seem negligible, but the earth's ecosystem is very fragile, and even such small changes can prove disastrous.

Greenhouse gases are a natural part of the atmosphere and the main sources of these greenhouse gases are carbon dioxide, methane, nitrous oxide, and fluorocarbons. Increased greenhouse gases in the past century can be attributed to human activity such as burning of fossil fuels such as coal, oil, and natural gas, reduced forest cover due to deforestation, increase in atmospheric methane gas due to mass rearing of cattle (in the process of digestion cattle and sheep produce and release methane into the atmosphere).

Glaciers

Glaciers are formed by snow that gets compressed and forms a thick ice mass over time. This ice begins melting when the temperature rises, and is again replaced by a fresh layer of snow. This process goes on and the glacier keeps getting bigger over time. The problem with any glacier begins when the ice melts at a faster rate than the snow that replaces it. The glacier will keep receding over time and will finally vanish.

Melting Glaciers

The melting of glaciers is a natural process. Many communities worldwide depend on the fresh water from these melting glaciers for their domestic use. Some countries depend on the melting water from glaciers for their production of electricity. Agriculture in many nations depends primarily on melting glacier water that flows in their rivers. All this melting water is constantly replaced by fresh snow that compresses into ice over time and will subsequently melt into water. This cycle goes on and on maintaining a perfect balance in the generation of fresh water and size of the glacier.

Glaciers Worldwide are Melting Fast

The last century has been a problem for glaciers across the globe. They are melting, but at an alarming rate. Fresh snow that replaces the melting ice is not able to maintain the size of almost any glacier worldwide. One of the main causes for this is thought to be 'global warming'. As the average global temperature keeps on increasing, ice from glaciers keep melting faster. The effects of ice glaciers melting more than required can cause catastrophes of unimaginable proportions. If global warming is causing ice glaciers to melt faster, the reduced ice cover over earth in turn is causing temperatures to rise further. Ice glaciers deflect almost 80 per cent of the heat from the sun and absorb about 20 per cent of the heat. When an ice glacier vanishes and exposes the earth below, 80 per cent of the heat from the sun is absorbed by the earth, and only about 20 per cent of this heat is deflected back. This increases the temperature of the earth, which increases the temperature of sea water. Sea water expands with an increase in water temperature and causes sea levels to rise. Melting water from glaciers will finally empty into the sea, causing a further increase in sea levels. All low lying areas near the sea will go under water and humans living here will be displaced. At the rate at which sea levels are rising, it is estimated that many South American and Asian countries will be the first to suffer from this effect. There are many more effects that rapidly melting glaciers cause. While some areas will witness unprecedented floods, other areas will witness severe draught. Whether witnessing floods or draught, agriculture will be severely hit, causing scarcity of foodgrains. Nations depending on hydroelectricity will have to switch over to other sources to generate their electricity, in effect further polluting the atmosphere. Forest fires will happen more frequently (they already are in Australia and the US) causing great stress to humans living in the vicinity. The bad effects of rapidly melting ice glaciers are limitless.

Help Control Global Warming

We are all responsible for the position we are in today as far as 'global warming' is concerned. We must all try to reduce any further damage so that further generations can lead better lives.

Try to reduce burning fossil fuels wherever possible (automobile exhaust is a major cause). Cycle instead of using the car for short distances; it will not only help the environment, but will make you fitter too. Make use of energy saving electrical and electronic equipment. Switch off lights and air-conditioners when not required (even if it is for a very short duration). Do not use any product that contains chlorofluorocarbons (CFCs). CFCs in any products are banned in the US, but could be an ingredient in aerosols and cleaners manufactured in other countries. CFCs deplete the atmospheric ozone layer and allow harmful ultraviolet rays to penetrate and reach the earth. Deforestation must be curbed as much as possible. Plant as many trees as and when possible. However small our contribution today, it will help make the earth a better place in the future.

NASA: GLOBAL WARMING CONTINUES UNABATED

According to NASA's Goddard Institute for Space Studies (GISS) , the decade January 2000 to December 2009 was the warmest on record. 2005 was the warmest year, while 2009 came in second in a tie with four more years (2002, 2003, 2006 and 2007) in the same decade. However, in 2009 the Southern Hemisphere was the warmest, since 1880, when temperatures were recorded for the first time.

GISS director James Hansen said, 'There's always interest in the annual numbers and a given year's ranking, but the ranking often misses the point. There's substantial year-to-year variability of global temperature caused by the tropical El-Niño-La Niña cycle. When we average temperature over five or ten years to minimise the variability, we find that global warming is continuing unabated.'

Though 2008 was the coolest year in the decade due to the La Niña event over the Pacific Ocean, warm temperatures made a comeback in 2009. 'Of course, the contiguous 48 states cover only 1.5 per cent of the world area, so the US temperature does not affect the global temperature much', Hansen said. Though the winter may have been colder in the United States, China and Europe, the Southern Hemisphere and the Arctic remained notably warm. Climatologists have recently emphasised on the importance of understanding the difference between weather, i.e. day-to-day local events, and climate, i.e. long-term global trends. According to NASA, the average global temperatures have been seeing a rise by 0.2°C or 0.36°F in the past three decades. In entirety, an increase by about 0.8°C or 1.5°F was seen in the average global temperatures since 1880. The analysis was based on data sourced from meteorological stations around the world, satellite observations and Antarctica research station, NASA said.

IT'S NOT JUST GAS THAT TRIGGERS WARMING

Recently a scientific paper released by the Union environment ministry has challenged the popular notion that global warming is caused mostly because of gases such as carbon dioxide, chemical compound, CO_2, a colourless, odourless, tasteless gas that is about one and one-half times as dense as air under ordinary conditions of temperature and pressure, nitrous oxide, methane and water vapour.

Cosmic Rays Too Make Temperature Rise

The paper 'contribution of changing galactic cosmic ray flux to global warming' by U.R. Rao, former chairman of the Indian Space Research Organisation, has argued that cosmic rays, which cannot be controlled, can have a much larger impact on climate change than previously thought.

New experiments provide persuasive evidence to show that the changes in the intensity of galactic cosmic rays can significantly affect global temperature. The Inter-Governmental Panel for Climate Change (IPCC), the UN climate body, needs to take a relook into their future prediction of global warming by factoring in the long-term changes in cosmic ray intensity, the paper says.

With the increase in solar activity, the primary cosmic ray intensity has decreased by 9 per cent during the past 150 years leading to lesser cloud cover and reduced albedo reflectivity of the surface of a planet, moon, asteroid, or other celestial body that does not shine by its own light. Albedo is measured as the fraction of incident light that the surface reflects back in all directions. Radiation being reflected back into space. This, in turn, causes an increase in the earth's surface temperature. The IPCC working group report has predicted an increase in the earth's surface temperature and sea level rise to be between 1.8°C (under best scenario) and 4°C (under worst case scenario) by the end of the 21st century. The effect of cosmic ray intensity over long periods, however, could add or subtract to the global warming. This means predicting future global warming and sea level rise is not simple, since it also significantly depends on the unpredictability of cosmic ray intensity.

'We conclude that the contribution to climate change due to the change in galactic cosmic ray intensity is quite significant and needs to be factored into the prediction of global warming and its effect on sea level raise and weather prediction', the paper states. The paper has been accepted for publication by the *Current Science*, a journal published in collaboration with the Indian Academy of Science.

Chapter 8

Hurricanes

INTRODUCTION

In meteorology, a tropical cyclone (or tropical storm, typhoon or hurricane, depending on strength and location) is a type of low-pressure system which generally forms in the tropics. While some, particularly those that make landfall in populated areas, are regarded as highly destructive, tropical cyclones are an important part of the atmospheric circulation system, which moves heat from the equatorial region toward the higher latitudes.

Structurally, a tropical cyclone is a large, rotating area of clouds, wind, and thunderstorm activity. The primary energy source of a tropical cyclone is the release of heat of condensation from water vapour condensing at high altitudes. Because of this, a tropical cyclone can be thought of as a giant vertical heat engine.

The ingredients for a tropical cyclone include a pre-existing weather disturbance, warm tropical oceans, moisture, and relatively light winds aloft. If the right conditions persist long enough, they can combine to produce the violent winds, incredible waves, torrential rains, and floods associated with this phenomenon (Fig. 8.1).

This use of condensation as a driving force is the primary difference setting tropical cyclones apart from other meteorological phenomena, such as mid-latitude cyclones, which draw energy mostly from pre-existing temperature gradients in the atmosphere. To drive its heat engine, a tropical cyclone must stay over warm water, which provides the atmospheric moisture needed. The evaporation of this moisture is driven by the high winds and reduced atmospheric pressure present in the storm, resulting in a sustaining cycle.

CLASSIFICATION AND TERMINOLOGY

Tropical cyclones are classified into three main groups: tropical depressions, tropical storms, and a third group whose name depends on the region.

A tropical depression is an organised system of clouds and thunderstorms with a defined surface circulation and maximum sustained winds of less than 17 metres per second (33 knots, 38 mph, or 62 km/hr). It has no eye, and does not typically have the spiral shape of more powerful storms.

A tropical storm is an organised system of strong thunderstorms with a defined surface circulation and maximum sustained winds between 17 and 33 meters per second (34 to 63 knots, 39 to 73 mph, or 62 to 117 km/hr). At this point, the distinctive cyclonic shape starts to develop, though an eye is usually not present (Fig. 8.2).

Fig. 8.1. Hurricane Ivan viewed from the International Space Station.

Fig. 8.2. Eye of Typhoon Odessa, Pacific Ocean.

The term used to describe tropical cyclones with maximum sustained winds exceeding 33 meters per second (63 knots, 73 mph, or 117 km/hr) varies depending on region of origin, as follows:

1. Hurricane in the North Atlantic Ocean, North Pacific Ocean east of the dateline, and the South Pacific Ocean east of 160°E.
2. Typhoon in the Northwest Pacific Ocean west of the dateline.
3. Severe tropical cyclone in the Southwest Pacific Ocean west of 160°E or Southeast Indian Ocean east of 90°E.

4. Severe cyclonic storm in the North Indian Ocean.

5. Tropical cyclone in the Southwest Indian Ocean.

This is the intensity at which tropical cyclones tend to develop an eye, which is an area of relative calm surrounded by the strongest winds of the storm, in the eyewall. The strongest of these storms have had maximum sustained windspeeds recorded at 85 meters per second (165 knot, 190 mph, 305 km/hr). In other places in the world, hurricanes have been called Bagyo in the Philippines, Chubasco in Mexico, and Taino in Haiti. Hurricanes are categorised on a 1-to-5 scale according to the strength of their winds, using the Saffir-Simpson Hurricane Scale. A Category 1 storm has the lowest wind speeds, while a Category 5 hurricane has the strongest. These are relative terms, because lower category storms can sometimes inflict greater damage than higher category storms, depending on where they strike and the particular hazards they bring. In fact, tropical storms can also produce significant damage and loss of life, mainly due to flooding. The US National Hurricane Center classifies hurricanes of Category 3 or above as Major Hurricanes. The Joint Typhoon Warning Center classifies typhoons with wind speeds of at least 150 mi/hr (67 m/s or 241 km/hr; a strong Category 4 storm) as Super Typhoons.

The definition of sustained winds recommended by the World Meteorological Organisation (WMO) is that of a ten-minute average, and that definition is adopted by most countries. However, a few countries use different definitions: the United States, for example, defines sustained winds based on a one minute average wind measured at about 10 meters (33 ft) above the surface. An extratropical cyclone is a storm that was once tropical in nature. However, once it passed over land or cool waters, its energy source changed from released heat from condensing water to the difference in temperature between air masses. From space, these storms resemble a comma. Extratropical cyclones still can be dangerous because their continuing low pressure causes powerful winds. In the United Kingdom and Europe, some severe northeast Atlantic cyclonic depressions are referred to as 'hurricanes', even though they rarely originate in the tropics. These European windstorms can generate hurricane-force windspeeds but are not given individual names. In British shipping forecasts, winds of force 12 on the Beaufort scale are described as 'hurricane force'. There is also a polar counterpart to the tropical cyclone, called an arctic cyclone.

LOCATION

Nearly all tropical cyclones form within 30° of the equator and 87 per cent form within 20° of it. Since the Coriolis effect initiates and maintains tropical cyclone rotation, such cyclones almost never form or move within about 10° of the equator. However, it is possible for tropical cyclones to form within this boundary if another source of initial rotation is provided. These conditions are extremely rare and such storms are believed to form at a rate of less than one a century. Most tropical cyclones form in a worldwide band of thunderstorm activity known as the Intertropical convergence zone (ITCZ). Worldwide, an average of 80 tropical cyclones form each year.

Major Basins

There are seven main basins of tropical cyclone formation:

1. Western North Pacific Ocean: Tropical storm activity in this region frequently affects China, Japan, the Philippines, and Taiwan. This is by far the most active basin, accounting for one-third of all tropical cyclone activity in the world. National meteorology organisations, as well as the Joint Typhoon Warning Center (JTWC) are responsible for issuing forecasts and warnings in this basin.

2. Eastern North Pacific Ocean: This is the second most active basin in the world, and is also the most dense (a large number of storms for a small area of ocean). Storms which form in this

basin can affect western Mexico, Hawaii and on extremely rare occasions, California. The Central Pacific Hurricane Center is responsible for forecasting the western part of this area, and the National Hurricane Center for the eastern part.

3. South Western Pacific Ocean: Tropical activity in this region largely affects Australia and Oceania, and is forecast by Australia and New Guinea.

4. Northern Indian Ocean: This basin is actually divided into two areas, the Bay of Bengal and the Arabian Sea, with the Bay of Bengal dominating (5 to 6 times more activity). Hurricanes which form in this basin have historically cost the most lives — most notably, the 1970 Bhola cyclone killed 2,00,000. Nations affected by this basin include India, Bangladesh, Sri Lanka, Thailand, Burma, and Pakistan, and all of these countries issue region forecasts and warnings. Rarely, a tropical cyclone formed in this basin will affect the Arabian Peninsula.

5. Southeastern Indian Ocean: Tropical activity in this region affects Australia and Indonesia, and is forecast by those nations.

6. Southwestern Indian Ocean: This basin is the least understood, due to a lack of historical data. Cyclones forming here impact Madagascar, Mozambique, Mauritius, and Kenya, and these nations issue forecasts and warnings for the basin.

7. North Atlantic Basin: The most well studied of all tropical basins, the North Atlantic includes the Atlantic Ocean, the Caribbean Sea, and the Gulf of Mexico. Tropical cyclone formation here varies widely year to year, ranging from over twenty to just one. The average is ten. The United States, Mexico, Central America, the Caribbean Islands and Canada are affected by storms in this basin. Forecasts for all storms are issued by the National Hurricane Center based in Miami, Florida; the Canadian Hurricane Centre, based in Halifax, Nova Scotia, also issues forecasts and warnings for storms expected to affect Canadian territory and waters. Hurricanes that strike Mexico, Central America, and Caribbean island nations, often do intense damage: they are deadlier when over warmer water, and the United States is better able to evacuate people from threatened areas than many other nations. Many of the more intense Atlantic storms are Cape Verde-type hurricanes, forming just west of Africa near the Cape Verde islands.

Unusual Formation Areas

The following areas spawn tropical cyclones only very rarely.

1. Southern Atlantic Ocean: A combination of cooler waters, the lack of an Inter-tropical Convergence Zone, and wind shear makes it very difficult for the Southern Atlantic to support tropical activity. However, three tropical cyclones have been observed here — a weak tropical storm in 1991 off the coast of Africa, Cyclone Catarina (sometimes also referred to as Aldonça), which made landfall in Brazil in 2004, and a smaller storm in January of 2004, east of Salvador, Brazil as a Category 1 hurricane. The January storm is thought to have reached tropical storm intensity based on scatterometre winds.

2. Central North Pacific: Shear in this area of the Pacific Ocean severely limits tropical development. However, this region is commonly frequented by tropical cyclones that form in the much more favorable Eastern North Pacific Basin.

4. Mediterranean Sea: Storms which appear similar to tropical cyclones in structure sometimes occur in the Mediterranean basin. Such cyclones formed in September 1947, September 1969, January 1982, September 1983, and January 1995. There is debate on whether these storms were tropical in nature.

Timing

Worldwide, tropical cyclone activity peaks in late summer when water temperatures are warmest. However, each particular basin has its own seasonal patterns.

In the north Atlantic, a distinct hurricane season occurs from June 1 to November 30, sharply peaking in early September. The northeast Pacific has a broader period of activity, but in a similar timeframe to the Atlantic. The northwest Pacific sees tropical cyclones year-round, with a minimum in February and a peak in early September. In the north Indian basin, storms are most common from April to December, with peaks in May and November. In the southern hemisphere, tropical cyclone activity begins in late October, and ends in May. Southern hemisphere activity peaks in mid-February to early March.

Structure

A strong tropical cyclone consists of the following components:

1. Surface low: All tropical cyclones rotate around an area of low atmospheric pressure near the earth's surface. The pressures recorded at the centers of tropical cyclones are among the lowest that occur on earth's surface at sea level.
2. Warm core: tropical cyclones are characterised and driven by the release of large amounts of latent heat of condensation as moist air is carried upwards and its water vapour condenses. This heat is distributed vertically, around the center of the storm. Thus, at any given altitude (except close to the surface where water temperature dictates air temperature) the environment inside the cyclone is warmer than its outer surroundings.
3. Central Dense Overcast (CDO): The Central Dense Overcast is a dense shield of rain bands and thunderstorm activity surrounding the central low. Tropical cyclones with symmetrical CDO tend to be strong and well developed.
4. Eye: A strong tropical cyclone will harbor an area of sinking air at the center of circulation. Weather in the eye is normally calm and free of clouds (however, the sea may be extremely violent). Eyes are home to the coldest temperatures of the storm at the surface, and the warmest temperatures at the upper levels. The eye is normally circular in shape, and may range in size from 8 km to 200 km (5 miles to 125 miles) in diameter. In weaker cyclones, the CDO covers the circulation center, resulting in no visible eye.
5. Eyewall: The eyewall is a circular band of intense convection and winds immediately surrounding the eye. It has the most severe conditions in a tropical cyclone. Intense cyclones show eye-wall replacement cycles, in which outer eye walls form to replace inner ones. The mechanisms which make this occur are still not fully understood.
6. Outflow: The upper levels of a tropical cyclone feature winds headed away from the center of the storm with an anticyclonic rotation. Winds at the surface are strongly cyclonic, weaken with height, and eventually reverse themselves. Tropical cyclones owe this unique characteristic to the warm core at the center of the storm.

FORMATION AND DEVELOPMENT

The formation of tropical cyclones is still the topic of extensive research, and is still not fully understood. Five factors are necessary to make tropical cyclone formation possible:

1. Sea surface temperatures above 26.5° Celsius to at least a depth of 50 meters. Warm waters are the energy source for tropical cyclones. When these storms move over land or cooler areas of water they weaken rapidly.

2. Upper level conditions must be conducive to thunderstorm formation. Temperatures in the atmosphere must decrease quickly with height, and the mid-troposphere must be relatively moist.
3. A pre-existing weather disturbance. This is most frequently provided by tropical waves— nonrotating areas of thunderstorms which move through the world's tropical oceans.
4. A distance of approximately 10 degrees or more from the equator (2004's Hurricane Ivan, the strongest storm to be so far south, started its formation at 9.7° north). The Coriolis Effect initiates and helps maintain the rotation of a tropical cyclone. The absence of this effect at and near the equator prohibits development.
5. Lack of vertical wind shear (change in wind velocity over height). High levels of wind shear can break apart the vertical structure of a tropical cyclone, prohibiting development.

Tropical cyclones can occasionally form despite not meeting these conditions. A combination of a pre-existing disturbance, upper level divergence, and a monsoon related cold spell led to the creation of Typhoon Vamei at only 1.5° north of the equator in 2001. It is estimated that the factors leading to the formation of this typhoon occur only once every 400 years.

When a tropical cyclone reaches higher latitudes or passes over land, it may merge with weather fronts or develop into a frontal cyclone, also called extratropical cyclone. In the Atlantic ocean, such tropical-derived cyclones of higher latitudes can be violent and may occasionally remain at hurricane-force wind speeds when they reach Europe as a European windstorm.

Dissipation

A tropical cyclone can cease to have tropical characteristics in several ways:

1. It moves over land, thus depriving it of the warm water it needs to power itself, and quickly loses strength. Most strong storms become disorganised areas of low pressure within a day or two of landfall. There is, however, a chance they could regenerate if they manage to get back over open warm water. If a storm is over mountains for even a short-time, it can rapidly lose strength. This is, however, the cause of many storm fatalities, as the dying storm unleashes torrential rainfall, and in mountainous areas, this can lead to deadly mudslides.
2. It remains in the same area of ocean for too long, consuming all the heat available and dissipating.
3. It experiences wind shear, causing the convection to lose direction and the heat engine breaks down.
4. It can be weak enough to be consumed by another area of low pressure, disrupting it and joining to become a large area of non-cyclonic thunderstorms.
5. It enters colder waters. This does not necessarily mean the death of the storm, but the storm will lose its tropical characteristics. These storms are extratropical cyclones.

Even after a tropical cyclone is said to be extratropical or dissipated, it can still have gale-force winds and drop several inches of rainfall.

Attempts to dissipate cyclones

In the 1960's and 70's, the United States government attempted to weaken hurricanes in its Project Stormfury by seeding with silver iodide. It was thought that the seeding would cause changes in the structure of the hurricane, essentially disrupting the eyewall to collapse and thus reduce the winds. However, it was later determined that these eyewall replacement cycles happen naturally, and so the success of the program was impossible to gauge.

Observations

Intense tropical cyclones pose a particular observation challenge. As they are a dangerous oceanic phenomenon, weather stations are rarely available on the site of the storm itself, unless it is passing over an island or a coastal area, or an unfortunate ship is caught in the storm. Even in these cases, real-time measurement taking is generally only possible in the periphery of the cyclone, where conditions are less catastrophic. It is however possible to take *in situ* measurements, in real-time, by sending specially equipped reconnaissance flights into the cyclone. These are flown by four-engine turboprop aircraft, which take direct and remote-sensing measurements and launch dropsondes inside the cyclone. The cyclone can also be imaged remotely by radar, and by weather satellites using visible light and infrared.

Effects

A mature tropical cyclone can release heat at a rate upwards of 6×10^{14} watts. This is two hundred times the total rate of human electrical production, and is equivalent to detonating a 10 megaton nuclear bomb every 20 minutes. Tropical cyclones on the open sea cause large waves, heavy rain, and high winds, disrupting international shipping and sometimes sinking ships. However, the most devastating effects of a tropical cyclone occur when they cross coastlines, making landfall. A tropical cyclone moving over land can do direct damage in four ways.

1. High winds: Hurricane strength winds can damage or destroy vehicles, buildings, bridges, etc. High winds also turn loose debris into flying projectiles, making the outdoor environment even more dangerous.
2. Storm surge: Tropical cyclones cause an increase in sea level which can flood coastal communities. This is the worst effect, as cyclones claim 80 per cent of their victims when they first strike shore.
3. Heavy rain: The thunderstorm activity in a tropical cyclone causes intense rainfall. Rivers and streams flood, roads become impassable, and landslides can occur.
4. Tornado activity: The broad rotation of a hurricane often spawns tornadoes. While these tornadoes are normally not as strong as their non-tropical counterparts, they can still cause tremendous damage.

Often, the secondary effects of a tropical cyclone are equally damaging. They include:

1. Disease: The wet environment in the aftermath of a tropical cyclone, combined with the destruction of sanitation facilities and a warm tropical climate can induce epidemics of disease which claim lives long after the storm passes.
2. Power outages: Tropical cyclones often knock out power to tens of thousands of people, prohibiting vital communication and hampering rescue efforts.
3. Transportation difficulties: Tropical cyclones often destroy key bridges, overpasses, and roads, complicating efforts to transport food, clean water, and medicine to the areas that need it.

Beneficial effects of tropical cyclones

The human toll of cyclones cannot have a price put on it. However, cyclones may bring much-needed precipitation to otherwise dry regions. An appreciable percentage of Japan's rainfall is due to typhoons. Hurricane Camille averted drought conditions and ended water deficits along much of its path. Additionally, the destruction caused by Camille on the Gulf coast spurred redevelopment, multiplying many times the land values that existed before the storm. However, disaster officials point out that this is not necessarily a good thing; it just encourages more people to live in what is clearly a danger area for deadly storms.

Additionally, hurricanes actually help to maintain global heat balance by moving warm, moist tropical air northward to the mid-latitudes and polar regions.

Notable Cyclones

Tropical cyclones that cause massive destruction are fortunately rare, but when they happen, they can cause damage in the thousands of lives and the billions of dollars. The deadliest tropical cyclone on record is a 100 mph (160 km/hr Category 2) storm that hit the densely populated Ganges Delta region of East Pakistan (now Bangladesh) on November 13, 1970. It killed anywhere from 2,00,000 to 5,00,000 people. The Indian Ocean basin has historically been the deadliest, with three storms since 1900 killing over 1,00,000 people each in Bangladesh.

In the Atlantic basin, three storms have killed more than 10,000 people. Hurricane Mitch during the 1998 Atlantic hurricane season caused severe flooding and mudslides in Honduras, killing at least 10,000 people and changing the landscape enough that entirely new maps of the nation were needed. The Galveston Hurricane of 1900, which made landfall at Galveston, Texas as an estimated category 4 storm, killed 6000 to 12,000 people and remains the deadliest natural disaster in the history of the United States. The deadliest Atlantic storm on record was the Great Hurricane of 1780, which killed between 20 and 30 thousand people in the Antilles.

The most costly storm was 1992's Hurricane Andrew, which caused an estimated $25 billion in damage in Florida and the U.S. Gulf Coast, and remains the most destructive natural disaster in United States history. The second most costly storm was 2004's Hurricane Charley, that stuck the Southwest Coast of Florida on August 13, 2004 as a Category 4 hurricane with winds of 150 mph. Charley caused over 14 billion dollars damage.

The most intense storm on record was Typhoon Tip in the northwestern Pacific Ocean in 1979, which had a minimum pressure of only 870 mb and maximum sustained windspeeds of 190 mph (305 km/hr). Fortunately, it weakened before striking Japan. Tip does not, however, hold alone the record for fastest sustained winds in a cyclone; Typhoon Keith in the Pacific, and Hurricane Camille and Hurricane Allen in the North Atlantic currently share this record as well, although recorded windspeeds that fast are suspect, since most monitoring equipment is likely to be destroyed by such conditions.

Camille was the only storm to actually strike land while at that intensity, making it, with 190 mph (305 km/hr) sustained winds and 210 mph (335 km/hr) gusts, the strongest tropical cyclone of record to ever hit land. For comparison, these speeds are encounted at the center of a strong tornado, but Camille was much larger than any tornado.

Typhoon Nancy in 1961 had recorded windspeeds of 213 mph (343 km/hr), but recent research indicates that windspeeds from the 1940s to the 1960s were gauged too high, and this is no longer considered the fastest storm on record. Similarly, a gust caused by Typhoon Paka over Guam was recorded at 236 mph (380 km/hr); however, this reading had to be discarded, since the anemometer was damaged by the storm. Had it been confirmed, this would be the strongest wind ever recorded at the earth's surface. (The current record is held by a non-hurricane wind registering 231 mph (372 km/hr) at Mount Washington in New Hampshire.) Tip was also the largest cyclone on record, with a circulation of 1350 miles (2170 km) wide. The average tropical cyclone is only 300 miles (480 km) wide.

Hurricane Iniki in 1992 was the most powerful storm to strike Hawaii in recorded history, hitting Kauai as a Category 4 hurricane, killing six and causing $3 billion in damage.

On December 25, 1974, Tropical Cyclone Tracy hit Darwin, Australia. It was the most devastating natural disaster to have ever hit an Australian city, destroying around 70 per cent of the homes in

Darwin. Fifty people died in Darwin, and 16 at sea. Authorities had managed to evacuate most of Darwin. Although Cyclone Tracy was quite small—the smallest cyclone on record, in fact, little larger than 30 miles (48 km) wide—it was very severe, with winds of up to 135 mph (217 km/hr). The damage was estimated to be close to $A400 million, which (at 2004 exchange rates) is approximately equal to $ US 280 million.

On March 26, 2004, Cyclone Catarina became the first recorded South Atlantic hurricane. Previous South Atlantic cyclones in 1991 and 2004 reached only tropical storm strength. Hurricanes may have formed there prior to 1960 but were not observed until weather satellites began monitoring the earth's oceans in that year. A tropical cyclone need not be particularly strong to cause memorable damage; Tropical Storm Allison in 2001 had its name retired for killing 41 people and causing over $5 billion damage in Texas, even though it never became a hurricane. Hurricane Jeanne in 2004 was only a tropical storm when it made a glancing blow on Haiti, but flooding and mudslides killed over 2000 people.

Naming of Tropical Cyclones

Tropical cyclones with winds exceeding 17 metres per second are given names, to assist in recording insurance claims, to assist in warning people of the coming storm, and to further indicate that these are important storms that should not be ignored. These names are taken from lists which vary from region to region and are drafted a few years ahead of time. The lists are decided upon, depending on the regions, either by committees of the World Meteorological Organisation (called primarily to discuss many other issues), or by national weather services involved in the forecasting of the storms. Each year, the names of particularly destructive storms are 'retired' and new names are chosen to take their place.

TROPICAL CYCLONE

A tropical cyclone is a storm system characterised by a large low-pressure center and numerous thunderstorms that produce strong winds and heavy rain. Tropical cyclones strengthen when water evaporated from the ocean is released as the saturated air rises, resulting in condensation of water vapour contained in the moist air. They are fueled by a different heat mechanism than other cyclonic windstorms such as nor'easters, European windstorms, and polar lows. The characteristic that separates tropical cyclones from other cyclonic systems is that any height in the atmosphere, the center of a tropical cyclone will be warmer than its surrounds; a phenomenon called 'warm core' storm systems.

The term 'tropical' refers to both the geographic origin of these systems, which form almost exclusively in tropical regions of the globe, and their formation in maritime tropical air masses. The term 'cyclone' refers to such storms' cyclonic nature, with counterclockwise rotation in the Northern Hemisphere and clockwise rotation in the Southern Hemisphere. The opposite direction of spin is a result of the Coriolis force. Depending on its location and strength, a tropical cyclone is referred to by names such as hurricane, typhoon, tropical storm, cyclonic storm, tropical depression, and simply cyclone.

While tropical cyclones can produce extremely powerful winds and torrential rain, they are also able to produce high waves and damaging storm surge as well as spawning tornadoes. They develop over large bodies of warm water, and lose their strength if they move over land due to increased surface friction and loss of the warm ocean as an energy source. This is why coastal regions can receive significant damage from a tropical cyclone, while inland regions are relatively safe from receiving strong winds. Heavy rains, however, can produce significant flooding inland, and storm surges can produce extensive coastal flooding up to 40 kilometres (25 mi) from the coastline. Although their effects on human populations can be devastating, tropical cyclones can also relieve drought conditions. They also carry heat and energy away from the tropics and transport it toward temperate latitudes, which makes them an important part of the global atmospheric circulation mechanism. As a result, tropical cyclones help to maintain equilibrium in the earth's troposphere, and to maintain a relatively stable and warm temperature worldwide. Many tropical cyclones develop when the atmospheric conditions around a weak disturbance in the atmosphere are favourable. The background environment is modulated by climatological cycles and patterns such as the Madden-Julian oscillation, El Niño-Southern Oscillation, and the Atlantic multidecadal oscillation. Others form when other types of cyclones acquire tropical characteristics. Tropical systems are then moved by steering winds in the troposphere; if the conditions remain favourable, the tropical disturbance intensifies, and can even develop an eye. On the other end of the spectrum, if the conditions around the system deteriorate or the tropical cyclone makes landfall, the system weakens and eventually dissipates. It is not possible to artificially induce the dissipation of these systems with current technology.

Physical Structure

All tropical cyclones are areas of low atmospheric pressure in the earth's atmosphere. The pressures recorded at the centers of tropical cyclones are among the lowest that occur on earth's surface at sea level. Tropical cyclones are characterised and driven by the release of large amounts of latent heat of condensation, which occurs when moist air is carried upwards and its water vapour condenses. This heat is distributed vertically around the center of the storm. Thus, at any given altitude (except close to the surface, where water temperature dictates air temperature) the environment inside the cyclone is warmer than its outer surroundings.

Eye and center

A strong tropical cyclone will harbor an area of sinking air at the center of circulation. If this area is strong enough, it can develop into a large 'eye'. Weather in the eye is normally calm and free of clouds, although the sea may be extremely violent. The eye is normally circular in shape, and may range in size from 3 km (1.9 mi) to 370 km (230 mi) in diameter. Intense, mature tropical cyclones can sometimes exhibit an outward curving of the eyewall's top, making it resemble a football stadium; this phenomenon is thus sometimes referred to as the stadium effect.

There are other features that either surround the eye, or cover it. The central dense overcast is the concentrated area of strong thunderstorm activity near the center of a tropical cyclone; in weaker tropical cyclones, the CDO may cover the center completely. The eyewall is a circle of strong thunderstorms that surrounds the eye; here is where the greatest wind speeds are found, where clouds reach the highest, and precipitation is the heaviest. The heaviest wind damage occurs where a tropical cyclone's eyewall passes over land. Eyewall replacement cycles occur naturally in intense tropical cyclones. When cyclones reach peak intensity they usually have an eyewall and radius of maximum winds that contract to a very small size, around 10 km (6.2 mi) to 25 km (16 mi). Outer rainbands can organise into an outer ring of thunderstorms that slowly moves inward and robs the inner eyewall of its needed moisture and angular momentum. When the inner eyewall weakens, the tropical cyclone weakens (in other words, the maximum sustained winds weaken and the central pressure rises). The outer eyewall replaces the inner one completely at the end of the cycle. The storm can be of the same intensity as it was previously or even stronger after the eyewall replacement cycle finishes. The storm may strengthen again as it builds a new outer ring for the next eyewall replacement.

Size

One measure of the size of a tropical cyclone is determined by measuring the distance from its center of circulation to its outermost closed isobar, also known as its ROCI. If the radius is less than two degrees of latitude or 222 km (138 mi), then the cyclone is 'very small' or a 'midget'. A radius between 3 and 6 latitude degrees or 333 km (207 mi) to 670 km (420 mi) are considered 'average-sized'. 'Very large' tropical cyclones have a radius of greater than 8° or 888 km (552 mi). Use of this measure has objectively determined that tropical cyclones in the northwest Pacific Ocean are the largest on earth on average, with Atlantic tropical cyclones roughly half their size. Other methods of determining a tropical cyclone's size include measuring the radius of gale force winds and measuring the radius at which its relative vorticity field decreases to 1×10^{-5} s^{-1} from its center (Table 8.1).

Table 8.1. Size descriptions of tropical cyclones.

ROCI	Type
Less than 2° latitude	Very small/midget
2° to 3° of latitude	Small
3° to 6 ° of latitude	Medium/Average
6° to 8° of latitude	Large anti-dwarf
Over 8° of latitude	Very large

Mechanics

A tropical cyclone's primary energy source is the release of the heat of condensation from water vapour condensing, with solar heating being the initial source for evaporation. Therefore, a tropical cyclone can be visualised as a giant vertical heat engine supported by mechanics driven by physical forces such as the rotation and gravity of the earth. In another way, tropical cyclones could be viewed as a special type of mesoscale convective complex, which continues to develop over a vast source of relative warmth and moisture. While an initial warm core system, such as an organised thunderstorm complex, is necessary for the formation of a tropical cyclone, a large flux of energy is needed to lower atmospheric pressure more than a few millibars (0.10 inch of mercury). The inflow of warmth and moisture from the underlying ocean surface is critical for tropical cyclone strengthening. A significant amount of the inflow in the cyclone is in the lowest 1 km (3300 ft) of the atmosphere.

Condensation leads to higher wind speeds, as a tiny fraction of the released energy is converted into mechanical energy; the faster winds and lower pressure associated with them in turn cause increased surface evaporation and thus even more condensation. Much of the released energy drives updrafts that increase the height of the storm clouds, speeding up condensation. This positive feedback loop, called the Wind-induced surface heat exchange, continues for as long as conditions are favourable for tropical cyclone development. Factors such as a continued lack of equilibrium in air mass distribution would also give supporting energy to the cyclone. The rotation of the earth causes the system to spin, an effect known as the Coriolis effect, giving it a cyclonic characteristic and affecting the trajectory of the storm.

What primarily distinguishes tropical cyclones from other meteorological phenomena is deep convection as a driving force. Because convection is strongest in a tropical climate, it defines the initial domain of the tropical cyclone. By contrast, mid-latitude cyclones draw their energy mostly from pre-existing horizontal temperature gradients in the atmosphere. To continue to drive its heat engine, a tropical cyclone must remain over warm water, which provides the needed atmospheric moisture to keep the positive feedback loop running. When a tropical cyclone passes over land, it is cut off from its heat source and its strength diminishes rapidly.

The passage of a tropical cyclone over the ocean causes the upper layers of the ocean to cool substantially, which can influence subsequent cyclone development. This cooling is primarily caused by wind-driven mixing of cold water from deeper in the ocean and the warm surface waters. This effect results in a negative feedback process which can inhibit further development or lead to weakening. Additional cooling may come in the form of cold water from falling raindrops (this is because the atmosphere is cooler at higher altitudes). Cloud cover may also play a role in cooling the ocean, by shielding the ocean surface from direct sunlight before and slightly after the storm passage. All these effects can combine to produce a dramatic drop in sea surface temperature over a large area in just a few days.

Scientists estimate that a tropical cyclone releases heat energy at the rate of 50 to 200 exajoules (10^{18} J) per day, equivalent to about 1 PW (10^{15} watt). This rate of energy release is equivalent to 70 times the world energy consumption of humans and 200 times the worldwide electrical generating capacity, or to exploding a 10 megaton nuclear bomb every 20 minutes.

In the lower troposphere, the most obvious motion of clouds is toward the center. However tropical cyclones also develop an upper-level (high-altitude) outward flow of clouds. These originate from air that has released its moisture and is expelled at high altitude through the 'chimney' of the storm engine. This outflow produces high, cirrus clouds that spiral away from the center. The clouds thin as they move outwards from the center of the system and are evaporated. They may be thin enough for the sun to be visible through them. These high cirrus clouds may be the first signs of an approaching tropical cyclone. As air parcels are lifted within the eye of the storm the vorticity is reduced, causing the outflow from a tropical cyclone to have anti-cyclonic motion.

Major Basins and Related Warning Centers

There are six Regional Specialised Meteorological Centers (RSMCs) worldwide. These organisations are designated by the World Meteorological Organisation and are responsible for tracking and issuing bulletins, warnings, and advisories about tropical cyclones in their designated areas of responsibility. Additionally, there are six Tropical Cyclone Warning Centers (TCWCs) that provide information to smaller regions. The RSMCs and TCWCs are not the only organisations that provide information about tropical cyclones to the public. The Joint Typhoon Warning Center (JTWC) issues advisories in all basins except the Northern Atlantic for the purposes of the United States Government. The Philippine Atmospheric, Geophysical and Astronomical Services Administration (PAGASA) issues advisories and names for tropical cyclones that approach the Philippines in the Northwestern Pacific to protect the life and property of its citizens. The Canadian Hurricane Center (CHC) issues advisories on hurricanes and their remnants for Canadian citizens when they affect Canada.

On 26, March 2004, Cyclone Catarina became the first recorded South Atlantic cyclone and subsequently struck southern Brazil with winds equivalent to Category 2 on the Saffir-Simpson Hurricane Scale. As the cyclone formed outside the authority of another warning center, Brazilian meteorologists initially treated the system as an extratropical cyclone, although subsequently classified it as tropical.

Formation

Worldwide, tropical cyclone activity peaks in late summer, when the difference between temperatures aloft and sea surface temperatures is the greatest. However, each particular basin has its own seasonal patterns. On a worldwide scale, May is the least active month, while September is the most active whilst November is the only month with all the tropical cyclone basins active.

Times

In the Northern Atlantic Ocean, a distinct cyclone season occurs from June 1 to November 30, sharply peaking from late August through September. The statistical peak of the Atlantic hurricane season is 10 September. The Northeast Pacific Ocean has a broader period of activity, but in a similar time frame to the Atlantic. The Northwest Pacific sees tropical cyclones year-round, with a minimum in February and March and a peak in early September. In the North Indian basin, storms are most common from April to December, with peaks in May and November. In the Southern Hemisphere, the tropical cyclone year begins on July 1 and runs all year round and encompasses the tropical cyclone seasons which run from November 1 until the end of April with peaks in mid-February to early March (Table 8.2).

Table 8.2. Season lengths and seasonal averages.

Basin	Season start	Season end	Tropical storms (>34 knots)	Tropical cyclones (>63 knots)	Category 3+ TCs (>95 knots)
Northwest Pacific	April	January	26.7	16.9	8.5
South Indian	November	April	20.6	10.3	4.3
Northeast Pacific	May	November	16.3	9.0	4.1
North Atlantic	June	November	10.6	5.9	2.0
Australia Southwest Pacific	November	April	9	4.8	1.9
North Indian	April	December	5.4	2.2	0.4

Factors

The formation of tropical cyclones is the topic of extensive ongoing research and is still not fully understood. While six factors appear to be generally necessary, tropical cyclones may occasionally form without meeting all of the following conditions. In most situations, water temperatures of at least 26.5°C (79.7°F) are needed down to a depth of at least 50 m (160 ft); waters of this temperature cause the overlying atmosphere to be unstable enough to sustain convection and thunderstorms. Another factor is rapid cooling with height, which allows the release of the heat of condensation that powers a tropical cyclone. High humidity is needed, especially in the lower-to-mid troposphere; when there is a great deal of moisture in the atmosphere, conditions are more favourable for disturbances to develop. Low amounts of wind shear are needed, as high shear is disruptive to the storm's circulation. Tropical cyclones generally need to form more than 555 km (345 mi) or 5° of latitude away from the equator, allowing the Coriolis effect to deflect winds blowing towards the low pressure center and creating a circulation. Lastly, a formative tropical cyclone needs a pre-existing system of disturbed weather, although without a circulation no cyclonic development will take place. Low-latitude and low-level westerly wind bursts associated with the Madden-Julian oscillation can create favourable conditions for tropical cyclogenesis by initiating tropical disturbances.

Locations

Most tropical cyclones form in a worldwide band of thunderstorm activity called by several names: the Intertropical Front (ITF), the Intertropical Convergence Zone (ITCZ), or the monsoon trough. Another important source of atmospheric instability is found in tropical waves, which cause about 85 per cent of intense tropical cyclones in the Atlantic ocean, and become most of the tropical cyclones in the Eastern Pacific basin.

Tropical cyclones move westward when equatorward of the subtropical ridge, intensifying as they move. Most of these systems form between 10 and 30 degrees away of the equator, and 87 per cent form no farther away than 20° of latitude, north or south. Because the Coriolis effect initiates and maintains tropical cyclone rotation, tropical cyclones rarely form or move within about 5° of the equator, where the Coriolis effect is weakest. However, it is possible for tropical cyclones to form within this boundary as Tropical Storm Vamei did in 2001 and Cyclone Agni in 2004.

Movement and Track

Steering winds

Although tropical cyclones are large systems generating enormous energy, their movements over the earth's surface are controlled by large-scale winds — the streams in the earth's atmosphere. The path of motion is referred to as a tropical cyclone's track and has been compared by Dr. Neil Frank, former director of the National Hurricane Center, to 'leaves carried along by a stream'.

Tropical systems, while generally located equatorward of the 20th parallel, are steered primarily westward by the east-to-west winds on the equatorward side of the subtropical ridge — a persistent high pressure area over the world's oceans. In the tropical North Atlantic and Northeast Pacific oceans, trade winds — another name for the westward-moving wind currents — steer tropical waves westward from the African coast and towards the Caribbean Sea, North America, and ultimately into the central Pacific ocean before the waves dampen out. These waves are the precursors to many tropical cyclones within this region. In the Indian Ocean and Western Pacific (both north and south of the equator), tropical cyclogenesis is strongly influenced by the seasonal movement of the Intertropical Convergence Zone and the monsoon trough, rather than by easterly waves. Tropical cyclones can also be steered by other systems, such as other low pressure systems, high pressure systems, warm fronts, and cold fronts.

Coriolis effect

The earth's rotation imparts an acceleration known as the Coriolis effect, Coriolis acceleration, or colloquially, Coriolis force. This acceleration causes cyclonic systems to turn towards the poles in the absence of strong steering currents. The poleward portion of a tropical cyclone contains easterly winds, and the Coriolis effect pulls them slightly more poleward. The westerly winds on the equatorward portion of the cyclone pull slightly towards the equator, but, because the Coriolis effect weakens toward the equator, the net drag on the cyclone is poleward. Thus, tropical cyclones in the Northern Hemisphere usually turn north (before being blown east), and tropical cyclones in the Southern Hemisphere usually urn south (before being blown east) when no other effects counteract the Coriolis effect.

The Coriolis effect also initiates cyclonic rotation, but it is not the driving force that brings this rotation to high speeds – that force is the heat of condensation.

Interaction with the mid-latitude westerlies

When a tropical cyclone crosses the subtropical ridge axis, its general track around the high-pressure area is deflected significantly by winds moving towards the general low-pressure area to its north. When the cyclone track becomes strongly poleward with an easterly component, the cyclone has begun recurvature. A typhoon moving through the Pacific Ocean towards Asia, for example, will recurve offshore of Japan to the north, and then to the northeast, if the typhoon encounters southwesterly winds (blowing northeastward) around a low-pressure system passing over China or Siberia. Many tropical cyclones are eventually forced toward the northeast by extratropical cyclones in this manner, which move from west to east to the north of the subtropical ridge. An example of a tropical cyclone in recurvature was Typhoon Ioke in 2006, which took a similar trajectory.

Landfall

Officially, landfall is when a storm's center (the center of its circulation, not its edge) crosses the coastline. Storm conditions may be experienced on the coast and inland hours before landfall; in fact, a tropical cyclone can launch its strongest winds over land, yet not make landfall; if this occurs, then it is said that the storm made a direct hit on the coast. As a result of the narrowness of this definition, the landfall area experiences half of a land-bound storm by the time the actual landfall occurs. For emergency preparedness, actions should be timed from when a certain wind speed or intensity of rainfall will reach land, not from when landfall will occur.

Multiple storm interaction

When two cyclones approach one another, their centers will begin orbiting cyclonically about a point between the two systems. The two vortices will be attracted to each other, and eventually spiral into the center point and merge. When the two vortices are of unequal size, the larger vortex will tend to dominate the interaction, and the smaller vortex will orbit around it. This phenomenon is called the Fujiwhara effect, after Sakuhei Fujiwhara.

Dissipation

Factors

A tropical cyclone can cease to have tropical characteristics in several different ways. One such way is if it moves over land, thus depriving it of the warm water it needs to power itself, quickly losing strength. Most strong storms lose their strength very rapidly after landfall and become disorganised areas of low pressure within a day or two, or evolve into extratropical cyclones. There is a chance a tropical cyclone could regenerate if it managed to get back over open warm water, such as with Hurricane Ivan. If it remains over mountains for even a short time, weakening will accelerate. Many storm fatalities occur in mountainous terrain, as the dying storm unleashes torrential rainfall, leading to deadly floods and mudslides, similar to those that happened with Hurricane Mitch in 1998. Additionally, dissipation can occur if a storm remains in the same area of ocean for too long, mixing the upper 60 metres (200 ft) of water, dropping sea surface temperatures more than 5°C (9°F). Without warm surface water, the storm cannot survive.

A tropical cyclone can dissipate when it moves over waters significantly below 26.5°C (79.7°F). This will cause the storm to lose its tropical characteristics (i.e. thunderstorms near the center and warm core) and become a remnant low pressure area, which can persist for several days. This is the main dissipation mechanism in the Northeast Pacific ocean. Weakening or dissipation can occur if it experiences vertical wind shear, causing the convection and heat engine to move away from the center; this normally ceases development of a tropical cyclone. Additionally, its interaction with the main belt of the Westerlies, by means of merging with a nearby frontal zone, can cause tropical cyclones to evolve into extratropical cyclones. This transition can take 1–3 days. Even after a tropical cyclone is said to be extratropical or dissipated, it can still have tropical storm force (or occasionally hurricane/typhoon force) winds and drop several inches of rainfall. In the Pacific ocean and Atlantic ocean, such tropical-derived cyclones of higher latitudes can be violent and may occasionally remain at hurricane or typhoon-force wind speeds when they reach the west coast of North America. These phenomena can also affect Europe, where they are known as European windstorms; Hurricane Iris's extratropical remnants are an example of such a windstorm from 1995. Additionally, a cyclone can merge with another area of low pressure, becoming a larger area of low pressure. This can strengthen the resultant system, although it may no longer be a tropical cyclone. Studies in the 2007 have given rise to the hypothesis that large amounts of dust reduce the strength of tropical cyclones.

Artificial dissipation

In the 1960s and 1970s, the United States government attempted to weaken hurricanes through Project Stormfury by seeding selected storms with silver iodide. It was thought that the seeding would cause supercooled water in the outer rainbands to freeze, causing the inner eyewall to collapse and thus reducing the winds. The winds of Hurricane Debbie — a hurricane seeded in Project Stormfury — dropped as much as 31 per cent, but Debbie regained its strength after each of two seeding forays. In an earlier episode in 1947, disaster struck when a hurricane east of Jacksonville, Florida promptly changed its course after being seeded, and smashed into Savannah, Georgia. Because there was so much uncertainty about the behaviour of these storms, the federal government would not approve seeding operations unless the hurricane had a less than 10 per cent chance of making landfall within 48 hours, greatly reducing the number of possible test storms. The project was dropped after it was discovered that eyewall replacement cycles occur naturally in strong hurricanes, casting doubt on the result of the earlier attempts. Today, it is known that silver iodide seeding is not likely to have an effect because the amount of supercooled water in the rainbands of a tropical cyclone is too low.

Other approaches have been suggested over time, including cooling the water under a tropical cyclone by towing icebergs into the tropical oceans. Other ideas range from covering the ocean in a substance that inhibits evaporation, dropping large quantities of ice into the eye at very early stages of development (so that the latent heat is absorbed by the ice, instead of being converted to kinetic energy that would feed the positive feedback loop), or blasting the cyclone apart with nuclear weapons. Project Cirrus even involved throwing dry ice on a cyclone. These approaches all suffer from one flaw above many others: tropical cyclones are simply too large and short-lived for any of the weakening techniques to be practical.

Effects

Tropical cyclones out at sea cause large waves, heavy rain, and high winds, disrupting international shipping and, at times, causing shipwrecks. Tropical cyclones stir up water, leaving a cool wake behind them, which causes the region to be less favourable for subsequent tropical cyclones. On land, strong winds can damage or destroy vehicles, buildings, bridges, and other outside objects, turning loose debris into deadly flying projectiles. The storm surge, or the increase in sea level due to the cyclone, is typically the worst effect from landfalling tropical cyclones, historically resulting in 90 per cent of tropical cyclone deaths. The broad rotation of a landfalling tropical cyclone, and vertical wind shear at its periphery, spawns tornadoes. Tornadoes can also be spawned as a result of eyewall mesovortices, which persist until landfall.

Over the past two centuries, tropical cyclones have been responsible for the deaths of about 1.9 million people worldwide. Large areas of standing water caused by flooding lead to infection, as well as contributing to mosquito-borne illnesses. Crowded evacuees in shelters increase the risk of disease propagation. Tropical cyclones significantly interrupt infrastructure, leading to power outages, bridge destruction, and the hampering of reconstruction efforts.

Although cyclones take an enormous toll in lives and personal property, there may be important factors in the precipitation regimes of places they impact, as they may bring much-needed precipitation to otherwise dry regions. Tropical cyclones also help maintain the global heat balance by moving warm, moist tropical air to the middle latitudes and polar regions. The storm surge and winds of hurricanes may be destructive to human-made structures, but they also stir up the waters of coastal estuaries, which are typically important fish breeding locales. Tropical cyclone destruction spurs redevelopment, greatly increasing local property values.

Observation and Forecasting

Observation

Intense tropical cyclones pose a particular observation challenge, as they are a dangerous oceanic phenomenon, and weather stations, being relatively sparse, are rarely available on the site of the storm itself. Surface observations are generally available only if the storm is passing over an island or a coastal area, or if there is a nearby ship. Usually, real-time measurements are taken in the periphery of the cyclone, where conditions are less catastrophic and its true strength cannot be evaluated. For this reason, there are teams of meteorologists that move into the path of tropical cyclones to help evaluate their strength at the point of landfall.

Tropical cyclones far from land are tracked by weather satellites capturing visible and infrared images from space, usually at half-hour to quarter-hour intervals. As a storm approaches land, it can be observed by land-based Doppler radar. Radar plays a crucial role around landfall by showing a storm's location and intensity every several minutes.

In situ measurements, in real-time, can be taken by sending specially equipped reconnaissance flights into the cyclone. In the Atlantic basin, these flights are regularly flown by United States government hurricane hunters. The aircraft used are WC-130 Hercules and WP-3D Orions, both four-engine turboprop cargo aircraft. These aircraft fly directly into the cyclone and take direct and remote-sensing measurements. The aircraft also launch GPS dropsondes inside the cyclone. These sondes measure temperature, humidity, pressure, and especially winds between flight level and the ocean's surface. A new era in hurricane observation began when a remotely piloted Aerosonde, a small drone aircraft, was flown through Tropical Storm Ophelia as it passed Virginia's Eastern Shore during the 2005 hurricane season. A similar mission was also completed successfully in the western Pacific ocean. This demonstrated a new way to probe the storms at low altitudes that human pilots seldom dare.

Forecasting

Because of the forces that affect tropical cyclone tracks, accurate track predictions depend on determining the position and strength of high- and low-pressure areas, and predicting how those areas will change during the life of a tropical system. The deep layer mean flow, or average wind through the depth of the troposphere, is considered the best tool in determining track direction and speed. If storms are significantly sheared, use of wind speed measurements at a lower altitude, such as at the 700 hPa pressure surface (3000 metres/9800 feet above sea level) will produce better predictions. Tropical forecasters also consider smoothing out short-term wobbles of the storm as it allows them to determine a more accurate long-term trajectory. High-speed computers and sophisticated simulation software allow forecasters to produce computer models that predict tropical cyclone tracks based on the future position and strength of high- and low-pressure systems. Combining forecast models with increased understanding of the forces that act on tropical cyclones, as well as with a wealth of data from earth-orbiting satellites and other sensors, scientists have increased the accuracy of track forecasts over recent decades. However, scientists are not as skillful at predicting the intensity of tropical cyclones. The lack of improvement in intensity forecasting is attributed to the complexity of tropical systems and an incomplete understanding of factors that affect their development.

Classifications, Terminology, and Naming

Intensity classifications

Tropical cyclones are classified into three main groups, based on intensity: tropical depressions, tropical storms, and a third group of more intense storms, whose name depends on the region. For example, if a tropical storm in the Northwestern Pacific reaches hurricane-strength winds on the Beaufort scale, it is referred to as a *typhoon*; if a tropical storm passes the same benchmark in the Northeast Pacific Basin, or in the Atlantic, it is called a *hurricane*. Neither 'hurricane' nor 'typhoon' is used in either the Southern Hemisphere or the Indian Ocean. In these basins, storms of tropical nature are referred to simply as 'cyclones'.

Additionally, as indicated in the table below, each basin uses a separate system of terminology, making comparisons between different basins difficult. In the Pacific Ocean, hurricanes from the Central North Pacific sometimes cross the International Date Line into the Northwest Pacific, becoming typhoons; on rare occasions, the reverse will occur. It should also be noted that typhoons with sustained winds greater than 67 metres per second (130 kn) or 150 miles per hour (240 km/hr) are called *Super Typhoons* by the Joint Typhoon Warning Center.

Tropical depression

A tropical depression is an organised system of clouds and thunderstorms with a defined, closed surface circulation and maximum sustained winds of less than 17 metres per second (33 kn) or 38 miles per hour (61 km/hr). It has no eye and does not typically have the organisation or the spiral shape of more powerful storms. However, it is already a low-pressure system, hence the name 'depression'. The practice of the Philippines is to name tropical depressions from their own naming convention when the depressions are within the Philippines' area of responsibility.

Tropical storm

A tropical storm is an organised system of strong thunderstorms with a defined surface circulation and maximum sustained winds between 17 metres per second (33 kn) [39 miles per hour (63 km/hr)] and 32 metres per second (62 kn) [73 miles per hour (117 km/hr)]. At this point, the distinctive cyclonic shape starts to develop, although an eye is not usually present. Government weather services, other than the Philippines, first assign names to systems that reach this intensity (thus the term named storm).

Hurricane or typhoon

A hurricane or typhoon (sometimes simply referred to as a tropical cyclone, as opposed to a depression or storm) is a system with sustained winds of at least 33 metres per second (64 kn) or 74 miles per hour (119 km/hr). A cyclone of this intensity tends to develop an eye, an area of relative calm (and lowest atmospheric pressure) at the center of circulation. The eye is often visible in satellite images as a small, circular, cloud-free spot. Surrounding the eye is the eyewall, an area about 16 km (9.9 mi) to 80 km (50 mi) wide in which the strongest thunderstorms and winds circulate around the storm's center. Maximum sustained winds in the strongest tropical cyclones have been estimated at about 85 metres per second (165 kn) or 195 miles per hour (314 km/hr).

Origin of storm terms

The word *typhoon*, which is used today in the Northwest Pacific, may be derived from Urdu, Persian and Arabic *tufan*, which in turn originates from Greek *Typhon*, a monster from Greek mythology associated with storms. The related Portuguese word *tufão*, used in Portuguese for typhoons, is also derived from *Typhon*. The word is also similar to Chinese 'taifeng' ('toifung' in Cantonese) (literally great winds), and also to the Japanese 'taifu', which may explain why 'typhoon' came to be used for East Asian cyclones.

The word *hurricane*, used in the North Atlantic and Northeast Pacific, is probably derived from the name of a Mayan storm god, Huracan, via the Spanish, *huracán*. Huracan is also the source of the word *Orcan*, another word for a European windstorm. Another possible source is Hyrrokkin, a Jotun or giantess in Norse mythology, called upon by the Aesir to launch the ship bearing the body of the god Balder, which was too heavy for even the gods to move.

Naming

Storms reaching tropical storm strength were initially given names to eliminate confusion when there are multiple systems in any individual basin at the same time, which assists in warning people of the coming storm. In most cases, a tropical cyclone retains its name throughout its life; however, under special circumstances, tropical cyclones may be renamed while active. These names are taken from lists that vary from region to region and are usually drafted a few years ahead of time. The lists are decided on, depending on the regions, either by committees of the World Meteorological Organisation (called primarily to discuss many other issues), or by national weather offices involved in the forecasting of the storms. Each year, the names of particularly destructive storms (if there are any) are 'retired' and new names are chosen to take their place. Different countries have different local conventions; for example, in Japan, storms are referred to by number (each year), such as (Typhoon 9).

Notable Tropical Cyclones

Tropical cyclones that cause extreme destruction are rare, although when they occur, they can cause great amounts of damage or thousands of fatalities. The 1970 Bhola cyclone is the deadliest tropical cyclone on record, killing more than 3,00,000 people and potentially as many as 1 million after striking the densely populated Ganges Delta region of Bangladesh on 13 November, 1970. Its powerful storm surge was responsible for the high death toll. The North Indian cyclone basin has historically been the deadliest basin. Elsewhere, Typhoon Nina killed nearly 1,00,000 in China in 1975 due to a 100 year flood that caused 62 dams including the Banqiao Dam to fail. The Great Hurricane of 1780 is the deadliest Atlantic hurricane on record, killing about 22,000 people in the Lesser Antilles. A tropical cyclone does need not be particularly strong to cause memorable damage, primarily if the deaths are from rainfall or mudslides. Tropical Storm Thelma in November 1991 killed thousands in the Philippines, while in 1982, the unnamed tropical depression that eventually became Hurricane Paul killed around 1000 people in Central America.

Hurricane Katrina is estimated as the costliest tropical cyclone worldwide, causing $81.2 billion in property damage with overall damage estimates exceeding $100 billion. Katrina killed at least 1836 people after striking Louisiana and Mississippi as a major hurricane in August 2005. Hurricane Andrew is the second most destructive tropical cyclone in US history, with damages totaling $40.7 billion, and with damage costs at $31.5 billion, Hurricane Ike is the third most destructive tropical cyclone in US history. The Galveston Hurricane of 1900 is the deadliest natural disaster in the United States, killing an estimated 6000 to 12,000 people in Galveston, Texas. Hurricane Mitch caused more than 10,000 fatalities in Latin America. Hurricane Iniki in 1992 was the most powerful storm to strike Hawaii in recorded history, hitting Kauai as a Category 4 hurricane, killing six people, and causing US $3 billion in damage. Other destructive Eastern Pacific hurricanes include Pauline and Kenna, both causing severe damage after striking Mexico as major hurricanes. In March 2004, Cyclone Gafilo struck northeastern Madagascar as a powerful cyclone, killing 74, affecting more than 2,00,000, and becoming the worst cyclone to affect the nation for more than 20 years.

The most intense storm on record was Typhoon Tip in the northwestern Pacific Ocean in 1979, which reached a minimum pressure of 870 mbar (25.69 inHg) and maximum sustained wind speeds of 165 knots (85 m/s) or 190 miles per hour (310 km/hr). Tip, however, does not solely hold the record for fastest sustained winds in a cyclone. Typhoon Keith in the Pacific and Hurricanes Camille and Allen in the North Atlantic currently share this record with Tip. Camille was the only storm to actually strike land while at that intensity, making it, with 165 knots (85 m/s) or 190 miles per hour (310 km/hr) sustained winds and 183 knots (94 m/s) or 210 miles per hour (340 km/hr) gusts, the strongest tropical cyclone on record at landfall. Typhoon Nancy in 1961 had recorded wind speeds of 185 knots (95 m/s) or 215 miles per hour (346 km/hr), but recent research indicates that wind speeds from the 1940s to the 1960s were gauged too high, and this is no longer considered the storm with the highest wind speeds on record. Similarly, a surface-level gust caused by Typhoon Paka on Guam was recorded at 205 knots (105 m/s) or 235 miles per hour (378 km/hr). Had it been confirmed, it would be the strongest non-tornadic wind ever recorded on the earth's surface, but the reading had to be discarded since the anemometer was damaged by the storm. In addition to being the most intense tropical cyclone on record, Tip was the largest cyclone on record, with tropical storm-force winds 2,170 kms (1350 mi) in diameter. The smallest storm on record, Tropical Storm Marco, formed during October 2008, and made landfall in Veracruz. Marco generated tropical storm-force winds only 37 kms (23 mi) in diameter. Hurricane John is the longest-lasting tropical cyclone on record, lasting 31 days in 1994. Before the advent of satellite imagery in 1961, however, many tropical cyclones were underestimated in their durations. John is also the longest-tracked tropical cyclone in the Northern Hemisphere on record, which had a path of 7165 miles (13,280 km). Reliable data for Southern Hemisphere cyclones is unavailable.

Changes due to El Niño-Southern Oscillation

Most tropical cyclones form on the side of the subtropical ridge closer to the equator, then move poleward past the ridge axis before recurving into the main belt of the Westerlies. When the subtropical ridge position shifts due to El Niño, so will the preferred tropical cyclone tracks. Areas west of Japan and Korea tend to experience much fewer September–November tropical cyclone impacts during El Niño and neutral years. During El Niño years, the break in the subtropical ridge tends to lie near 130°E which would favour the Japanese archipelago. During El Niño years, Guam's chance of a tropical cyclone impact is one-third of the long-term average. The tropical Atlantic ocean experiences depressed activity due to increased vertical wind shear across the region during El Niño years. During La Niña years, the formation of tropical cyclones, along with the subtropical ridge position, shifts westward across the western Pacific ocean, which increases the landfall threat to China.

Long-term Activity Trends

While the number of storms in the Atlantic has increased since 1995, there is no obvious global trend; the annual number of tropical cyclones worldwide remains about 87 ± 10. However, the ability of climatologists to make long-term data analysis in certain basins is limited by the lack of reliable historical data in some basins, primarily in the Southern Hemisphere. In spite of that, there is some evidence that the intensity of hurricanes is increasing. Kerry Emanuel stated, 'Records of hurricane activity worldwide show an upswing of both the maximum wind speed in and the duration of hurricanes. The energy released by the average hurricane (again considering all hurricanes worldwide) seems to have increased by around 70 per cent in the past 30 years or so, corresponding to about a 15 per cent increase in the maximum wind speed and a 60 per cent increase in storm lifetime.'

Atlantic storms are becoming more destructive financially, since five of the ten most expensive storms in United States history have occurred since 1990. According to the World Meteorological Organisation, 'recent increase in societal impact from tropical cyclones has largely been caused by rising concentrations of population and infrastructure in coastal regions'. Pielke normalised mainland US hurricane damage from 1900 to 2005 values and found no remaining trend of increasing absolute damage. The 1970s and 1980s were notable because of the extremely low amounts of damage compared to other decades. The decade 1996–2005 was the second most damaging among the past 11 decades, with only the decade 1926–1935 surpassing its costs. The most damaging single storm is the 1926 Miami hurricane, with $157 billion of normalised damage.

Often in part because of the threat of hurricanes, many coastal regions had sparse population between major ports until the advent of automobile tourism, therefore, the most severe portions of hurricanes striking the coast may have gone unmeasured in some instances. The combined effects of ship destruction and remote landfall severely limit the number of intense hurricanes in the official record before the era of hurricane reconnaissance aircraft and satellite meteorology. Although the record shows a distinct increase in the number and strength of intense hurricanes, therefore, experts regard the early data as suspect. The number and strength of Atlantic hurricanes may undergo a 50–70 year cycle, also known as the Atlantic Multidecadal Oscillation. Nyberg reconstructed Atlantic major hurricane activity back to the early 18th century and found five periods averaging 3–5 major hurricanes per year and lasting 40–60 years, and six other averaging 1.5–2.5 major hurricanes per year and lasting 10–20 years. These periods are associated with the Atlantic multidecadal oscillation. Throughout, a decadal oscillation related to solar irradiance was responsible for enhancing/dampening the number of major hurricanes by 1–2 per year.

Although more common since 1995, few above-normal hurricane seasons occurred during 1970–94. Destructive hurricanes struck frequently from 1926–60, including many major New England hurricanes. Twenty-one Atlantic tropical storms formed in 1933, a record only recently exceeded in 2005, which saw 28 storms. Tropical hurricanes occurred infrequently during the seasons of 1900–1925; however, many intense storms formed during 1870–99. During the 1887 season, 19 tropical storms formed, of which a record 4 occurred after 1st November and 11 strengthened into hurricanes. Few hurricanes occurred in the 1840s to 1860s; however, many struck in the early 19th century, including a 1821 storm that made a direct hit on New York City. Some historical weather experts say these storms may have been as high as Category 4 in strength.

These active hurricane seasons predated satellite coverage of the Atlantic basin. Before the satellite era began in 1960, tropical storms or hurricanes went undetected unless a reconnaissance aircraft encountered one, a ship reported a voyage through the storm, or a storm hit land in a populated area. The official record, therefore, could miss storms in which no ship experienced gale-force winds, recognised it as a tropical storm (as opposed to a high-latitude extra-tropical cyclone, a tropical wave, or a brief squall), returned to port, and reported the experience.

Proxy records based on paleotempestological research have revealed that major hurricane activity along the Gulf of Mexico coast varies on timescales of centuries to millennia. Few major hurricanes struck the Gulf coast during 3000–1400 BC and again during the most recent millennium. These quiescent intervals were separated by a hyperactive period during 1400 BC and 1000 AD, when the Gulf coast was struck frequently by catastrophic hurricanes and their landfall probabilities increased by 3–5 times. This millennial-scale variability has been attributed to long-term shifts in the position of the Azores High, which may also be linked to changes in the strength of the North Atlantic Oscillation.

According to the Azores High hypothesis, an anti-phase pattern is expected to exist between the Gulf of Mexico coast and the Atlantic coast. During the quiescent periods, a more northeasterly position of the Azores High would result in more hurricanes being steered towards the Atlantic coast. During the hyperactive period, more hurricanes were steered towards the Gulf coast as the Azores High was shifted to a more southwesterly position near the Caribbean. Such a displacement of the Azores High is consistent with paleoclimatic evidence that shows an abrupt onset of a drier climate in Haiti around 3200 ^{14}C years BP, and a change towards more humid conditions in the Great Plains during the late-Holocene as more moisture was pumped up the Mississippi Valley through the Gulf coast. Preliminary data from the northern Atlantic coast seem to support the Azores High hypothesis. A 3000 year proxy record from a coastal lake in Cape Cod suggests that hurricane activity increased significantly during the past 500–1000 years, just as the Gulf coast was amid a quiescent period of the last millennium.

Related Cyclone Types

In addition to tropical cyclones, there are two other classes of cyclones within the spectrum of cyclone types. These kinds of cyclones, known as extratropical cyclones and subtropical cyclones, can be stages a tropical cyclone passes through during its formation or dissipation. An extratropical cyclone is a storm that derives energy from horizontal temperature differences, which are typical in higher latitudes. A tropical cyclone can become extratropical as it moves toward higher latitudes if its energy source changes from heat released by condensation to differences in temperature between air masses; additionally, although not as frequently, an extratropical cyclone can transform into a subtropical storm, and from there into a tropical cyclone. From space, extratropical storms have a characteristic 'comma-shaped' cloud pattern. Extratropical cyclones can also be dangerous when their low-pressure centers cause powerful winds and high seas. A subtropical cyclone is a weather system that has some characteristics of a tropical cyclone and some characteristics of an extratropical cyclone. They can form in a wide band of latitudes, from the equator to 50°. Although subtropical storms rarely have hurricane-force winds, they may become tropical in nature as their cores warm. From an operational standpoint, a tropical cyclone is usually not considered to become subtropical during its extratropical transition.

Tropical Cyclones in Popular Culture

In popular culture, tropical cyclones have made appearances in different types of media, including films, books, television, music, and electronic games. The media can have tropical cyclones that are entirely fictional, or can be based on real events. For example, George Rippey Stewart's Storm, a best-seller published in 1941, is thought to have influenced meteorologists into giving female names to Pacific tropical cyclones. Another example is the hurricane in The Perfect Storm, which describes the sinking of the Andrea Gail by the 1991 Perfect Storm. Also, hypothetical hurricanes have been featured in parts of the plots of series such as The Simpsons, Invasion, Family Guy, Seinfeld, Dawson's Creek, and CSI Miami. The 2004 film The Day After Tomorrow includes several mentions of actual tropical cyclones as well as featuring fantastical 'hurricane-like' non-tropical Arctic storms.

Chapter 9

Volcanoes

INTRODUCTION

A volcano is an opening or rupture, in a planet's surface or crust, which allows hot magma, volcanic ash and gases to escape from below the surface. Volcanoes are generally found where tectonic plates are diverging or converging. A mid-oceanic ridge, for example the Mid-Atlantic Ridge, has examples of volcanoes caused by divergent tectonic plates pulling apart; the Pacific Ring of Fire has examples of volcanoes caused by convergent tectonic plates coming together. By contrast, volcanoes are usually not created where two tectonic plates slide past one another. Volcanoes can also form where there is stretching and thinning of the earth's crust (called 'non-hotspot intraplate volcanism'), such as in the East African Rift, the Wells Gray-Clearwater volcanic field and the Rio Grande Rift in North America.

Volcanoes can be caused by mantle plumes. These so-called hotspots, for example at Hawaii, can occur far from plate boundaries. Hotspot volcanoes are also found elsewhere in the solar system, especially on rocky planets and moons (Fig. 9.1).

Fig. 9.1. Pinatubo ash plume reaching a height of 19 km, 3 days before the climactic eruption of 15 June 1991.

PLATE TECTONICS AND HOTSPOTS

Divergent Plate Boundaries

Divergent plate boundaries are locations where plates are moving away from one another. This occurs above rising convection currents. The rising current pushes up on the bottom of the lithosphere, lifting it and flowing laterally beneath it. This lateral flow causes the plate material above to be dragged along in the direction of flow. At the crest of the uplift, the overlying plate is stretched thin, breaks and pulls apart (Fig. 9.2).

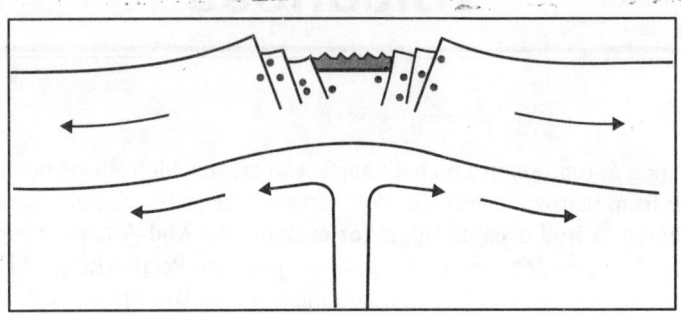

Fig. 9.2. Divergent plate boundaries.

At the mid-oceanic ridges, two tectonic plates diverge from one another. New oceanic crust is being formed by hot molten rock slowly cooling and solidifying. The crust is very thin at mid-oceanic ridges due to the pull of the tectonic plates. The release of pressure due to the thinning of the crust leads to adiabatic expansion, and the partial melting of the mantle causing volcanism and creating new oceanic crust. Most divergent plate boundaries are at the bottom of the oceans, therefore most volcanic activity is submarine, forming new seafloor. Black smokers or deep sea vents are an example of this kind of volcanic activity. Where the mid-oceanic ridge is above sea-level, volcanic islands are formed, for example, Iceland.

Convergent Plate Boundaries

Subduction zones are places where two plates, usually an oceanic plate and a continental plate, collide. In this case, the oceanic plate subducts, or submerges under the continental plate forming a deep ocean trench just offshore. Water released from the subducting plate lowers the melting temperature of the overlying mantle wedge, creating magma. This magma tends to be very viscous due to its high silica content, so often does not reach the surface and cools at depth. When it does reach the surface, a volcano is formed. Typical examples for this kind of volcano are Mount Etna and the volcanoes in the Pacific Ring of Fire (Fig. 9.3).

Hotspots

Hotspots are not usually located on the ridges of tectonic plates, but above mantle plumes, where the convection of the earth's mantle creates a column of hot material that rises until it reaches the crust, which tends to be thinner than in other areas of the earth. The temperature of the plume causes the crust to melt and form pipes, which can vent magma. Because the tectonic plates move whereas the mantle plume remains in the same place, each volcano becomes dormant after a while and a new volcano is then formed as the plate shifts over the hotspot. The Hawaiian Islands are thought to be formed in such a manner, as well as the Snake River Plain, with the Yellowstone Caldera being the part of the North American plate currently above the hot spot (Fig. 9.4).

Fig. 9.3. Mount Rinjani eruption in 1994, in Lombok, Indonesia.

Fig. 9.4. Lava enters the Pacific at the big island of Hawaii.

Volcanic Features

The most common perception of a volcano is of a conical mountain, spewing lava and poisonous gases from a crater at its summit. This describes just one of many types of volcano, and the features of volcanoes are much more complicated. The structure and behaviour of volcanoes depends on a number of factors. Some volcanoes have rugged peaks formed by lava domes rather than a summit crater, whereas others present landscape features such as massive plateaus. Vents that issue volcanic material (lava, which is what magma is called once it has escaped to the surface, and ash) and gases (mainly steam and magmatic gases) can be located anywhere on the landform (Fig. 9.5).

Fig. 9.5. Conical Mount Fuji in Japan, at sunrise from Lake Kawaguchi.

Other types of volcano include cryovolcanoes (or ice volcanoes), particularly on some moons of Jupiter, Saturn and Neptune; and mud volcanoes, which are formations often not associated with known magmatic activity.

Active mud volcanoes tend to involve temperatures much lower than those of igneous volcanoes, except when a mud volcano is actually a vent of an igneous volcano.

Fissure vents

Volcanic fissure vents are flat, linear cracks through which lava emerges.

Shield volcanoes

Shield volcanoes, so named for their broad, shield-like profiles, are formed by the eruption of low-viscosity lava that can flow a great distance from a vent, but not generally explode catastrophically. Since low-viscosity magma is typically low in silica, shield volcanoes are more common in oceanic than continental settings. The Hawaiian volcanic chain is a series of shield cones, and they are common in Iceland, as well.

Lava domes

Lava domes are built by slow eruptions of highly viscous lavas. They are sometimes formed within the crater of a previous volcanic eruption (as in Mount Saint Helens), but can also form independently, as in the case of Lassen Peak. Like stratovolcanoes, they can produce violent, explosive eruptions, but their lavas generally do not flow far from the originating vent.

Cryptodomes

Cryptodomes are formed when viscous lava forces its way up and causes a bulge. The 1980 eruption of Mount St. Helens was an example. Lava was under great pressure and forced a bulge in the mountain, which was unstable and slid down the north side.

Volcanic cones (cinder cones)

Volcanic cones or cinder cones are the result from eruptions that erupt mostly small pieces of scoria and pyroclastics (both resemble cinders, hence the name of this volcano type) that build up around the vent. These can be relatively short-lived eruptions that produce a cone-shaped hill perhaps 30 to 400 meters high. Most cinder cones erupt only once. Cinder cones may form as flank vents on larger volcanoes, or occur on their own. Parícutin in Mexico and Sunset Crater in Arizona are examples of cinder cones. In New Mexico, Caja del Rio is a volcanic field of over 60 cinder cones.

Stratovolcanoes (composite volcanoes)

Stratovolcanoes or composite volcanoes are tall conical mountains composed of lava flows and other ejecta in alternate layers, the strata that give rise to the name. Stratovolcanoes are also known as composite volcanoes, created from several structures during different kinds of eruptions. Strato/composite volcanoes are made of cinders, ash and lava. Cinders and ash pile on top of each other, lava flows on top of the ash, where it cools and hardens, and then the process begins again. Classic examples include Mt. Fuji in Japan, Mayon Volcano in the Philippines, and Mount Vesuvius and Stromboli in Italy.

In recorded history, explosive eruptions by stratovolcanoes have posed the greatest hazard to civilisations, as ash is produced by an explosive eruption. No supervolcano erupted in recorded history. Shield volcanoes have not an enormous pressure build up from the lava flow. Fissure vents and monogenetic volcanic fields (volcanic cones) have not powerful explosive eruptions, as they are many times under extension. Stratovolcanoes (30°–35°) are steeper than shield volcanoes (generally 5°–10°), their lose tephra are material for dangerous lahars.

Supervolcanoes

A supervolcano is a large volcano that usually has a large caldera and can potentially produce devastation on an enormous, sometimes continental, scale. Such eruptions would be able to cause severe cooling of global temperatures for many years afterwards because of the huge volumes of sulphur and ash erupted. They are the most dangerous type of volcano. Examples include Yellowstone Caldera in Yellowstone National Park and Valles Caldera in New Mexico (both western United States), Lake Taupo in New Zealand, Lake Toba in Sumatra, Indonesia and Ngorogoro Crater in Tanzania, Krakatoa near Java and Sumatra, Indonesia. Supervolcanoes are hard to identify centuries later, given the enormous areas they cover. Large igneous provinces are also considered supervolcanoes because of the vast amount of basalt lava erupted, but are non-explosive.

Submarine volcanoes

Submarine volcanoes are common features on the ocean floor. Some are active and, in shallow water, disclose their presence by blasting steam and rocky debris high above the surface of the sea. Many others lie at such great depths that the tremendous weight of the water above them prevents the explosive release of steam and gases, although they can be detected by hydrophones and discolouration of water because of volcanic gases. Pumice rafts may also appear. Even large submarine eruptions may not disturb the ocean surface. Because of the rapid cooling effect of water as compared to air, and increased buoyancy, submarine volcanoes often form rather steep pillars over their volcanic vents as compared to above-surface volcanoes. They may become so large that they break the ocean surface as new islands. Pillow lava is a common eruptive product of submarine volcanoes. Hydrothermal vents are common near these volcanoes, and some support peculiar ecosystems based on dissolved minerals.

Subglacial volcanoes

Subglacial volcanoes develop underneath icecaps. They are made up of flat lava which flows at the top of extensive pillow lavas and palagonite. When the icecap melts, the lavas on the top collapse, leaving a flat-topped mountain.

These volcanoes are also called table mountains, tuyas or (uncommonly) mobergs. Very good examples of this type of volcano can be seen in Iceland, however, there are also tuyas in British Columbia. The origin of the term comes from Tuya Butte, which is one of the several tuyas in the area of the Tuya River and Tuya Range in northern British Columbia.

Tuya Butte was the first such landform analysed and so its name has entered the geological literature for this kind of volcanic formation. The Tuya Mountains Provincial Park was recently established to protect this unusual landscape, which lies north of Tuya Lake and south of the Jennings River near the boundary with the Yukon Territory.

Mud volcanoes

Mud volcanoes or mud domes are formations created by geo-excreted liquids and gases, although there are several different processes which may cause such activity. The largest structures are 10 kilometres in diameter and reach 700 meters high.

Erupted Material

Lava composition

Another way of classifying volcanoes is by the composition of material erupted (lava), since this affects the shape of the volcano. Lava can be broadly classified into four different compositions:

1. If the erupted magma contains a high percentage (>63 per cent) of silica, the lava is called felsic.
   (a) Felsic lavas (dacites or rhyolites) tend to be highly viscous (not very fluid) and are erupted as domes or short, stubby flows. Viscous lavas tend to form stratovolcanoes or lava domes. Lassen Peak in California is an example of a volcano formed from felsic lava and is actually a large lava dome.
   (b) Because siliceous magmas are so viscous, they tend to trap volatiles (gases) that are present, which cause the magma to erupt catastrophically, eventually forming stratovolcanoes. Pyroclastic flows (ignimbrites) are highly hazardous products of such volcanoes, since they are composed of molten volcanic ash too heavy to go up into the atmosphere, so they hug the volcano's slopes and travel far from their vents during large eruptions. Temperatures as high as 1200°C are known to occur in pyroclastic flows, which will incinerate everything flammable in their path and thick layers of hot pyroclastic flow deposits can be laid down, often up to many meters thick. Alaska's Valley of Ten Thousand Smokes, formed by the eruption of Novarupta near Katmai in 1912, is an example of a thick pyroclastic flow or ignimbrite deposit. Volcanic ash that is light enough to be erupted high into the earth's atmosphere may travel many kilometres before it falls back to ground as a tuff (Fig. 9.6).
2. If the erupted magma contains 52–63 per cent silica, the lava is of intermediate composition.
   (a) These 'andesitic' volcanoes generally only occur above subduction zones (e.g. Mount Merapi in Indonesia).

(b) Andesitic lava is typically formed at convergent boundary margins of tectonic plates, by several processes such as:

(i) Hydration melting of peridotite and fractional crystallisation.

(ii) Melting of subducted slab containing sediments.

(iii) Magma mixing between felsic rhyolitic and mafic basaltic magmas in an intermediate reservoir prior to emplacement or lava flow.

Fig. 9.6. The Stromboli volcano off the coast of Sicily has erupted continuously for thousands of years, giving rise to the term strombolian eruption.

3. If the erupted magma contains <52 and >45 per cent silica, the lava is called mafic (because it contains higher percentages of magnesium (Mg) and iron (Fe)) or basaltic. These lavas are usually much less viscous than rhyolitic lavas, depending on their eruption temperature; they also tend to be hotter than felsic lavas. Mafic lavas occur in a wide range of settings:

(a) At mid-ocean ridges, where two oceanic plates are pulling apart, basaltic lava erupts as pillows to fill the gap.

(b) Shield volcanoes (e.g. the Hawaiian Islands, including Mauna Loa and Kilauea), on both oceanic and continental crust.

(c) As continental flood basalts.

4. Some erupted magmas contain <= 45 per cent silica and produce ultramafic lava. Ultramafic flows, also known as komatiites, are very rare; indeed, very few have been erupted at the earth's surface since the Proterozoic, when the planet's heat flow was higher. They are (or were) the hottest lavas, and probably more fluid than common mafic lavas.

Lava texture

Two types of lava are named according to the surface texture: 'A 'a and pahoehoe, both Hawaiian words. 'A 'a is characterised by a rough, clinkery surface and is the typical texture of viscous lava flows. However, even basaltic or mafic flows can be erupted as 'a' a flows, particularly if the eruption rate is high and the slope is steep.

Pahoehoe is characterised by its smooth and often ropey or wrinkly surface and is generally formed from more fluid lava flows. Usually, only mafic flows will erupt as pahoehoe, since they often erupt at higher temperatures or have the proper chemical make-up to allow them to flow with greater fluidity.

Volcanic Activity

Popular classification of volcanoes

Active

A popular way of classifying magmatic volcanoes is by their frequency of eruption, with those that erupt regularly called active, those that have erupted in historical times but are now quiet called dormant, and those that have not erupted in historical times called extinct. However, these popular classifications—extinct in particular—are practically meaningless to scientists. They use classifications which refer to a particular volcano's formative and eruptive processes and resulting shapes (Fig. 9.7).

Fig. 9.7. Active volcano Mount St. Helens shortly after the eruption of 18 May 1980.

There is no real consensus among volcanologists on how to define an 'active' volcano. The lifespan of a volcano can vary from months to several million years, making such a distinction sometimes meaningless when compared to the lifespans of humans or even civilisations. For example, many of earth's volcanoes have erupted dozens of times in the past few thousand years but are not currently showing signs of eruption. Given the long lifespan of such volcanoes, they are very active. By human lifespans, however, they are not (Fig. 9.8). Scientists usually consider a volcano to be erupting or likely to erupt if it is currently erupting, or showing signs of unrest such as unusual earthquake activity or significant new gas emissions. Most scientists consider a volcano active if it has erupted in holocene times. Historic times is another timeframe for active. But it is important to note that the span of recorded history differs from region to region. In China and the Mediterranean, recorded history reaches back more than 3000 years but in the Pacific Northwest of the United States and Canada, it reaches back less than 300 years, and in Hawaii and New Zealand, only around 200 years. The Smithsonian Global Volcanism Program's definition of active is having erupted within the last 10,000 years (the 'holocene' period). Presently there are about 500 active volcanoes in the world—the majority following along the Pacific 'Ring of Fire'—and around 50 of these erupt each year. The United States is home to 50 active volcanoes. There are more than 1500 potentially active volcanoes. An estimated 500 million people live near active volcanoes.

Fig. 9.8. Damavand, the highest volcano in Asia, is a potentially active volcano with fumaroles and solfatara near its summit.

Extinct

Extinct volcanoes are those that scientists consider unlikely to erupt again, because the volcano no longer has a lava supply. Examples of extinct volcanoes are many volcanoes on the Hawaiian–Emperor seamount chain in the Pacific Ocean (extinct because the Hawaii hotspot is centered near the Big Island), Hohentwiel, Shiprock and the Zuidwal volcano in the Netherlands, Edinburgh Castle Scotland is located atop an Extinct Volcano also. Otherwise, whether a volcano is truly extinct is often difficult to determine. Since 'supervolcano' calderas can have eruptive lifespans sometimes measured in millions of years, a caldera that has not produced an eruption in tens of thousands of years is likely to be considered dormant instead of extinct.

Dormant

It is difficult to distinguish an extinct volcano from a dormant one. Volcanoes are often considered to be extinct if there are no written records of its activity. Nevertheless volcanoes may remain dormant for a long period of time, Yellowstone has a repose/recharge period of around 700 ka and Toba of around 380 ka. Vesuvius was described by Roman writers as having been covered with gardens and vineyards before its famous eruption of AD 79, which destroyed the towns of Herculaneum and Pompeii. Before the catastrophic eruption of 1991, Pinatubo was an inconspicuous volcano, unknown to most people in the surrounding areas. More recently, the long-dormant Soufrière Hills volcano on the island of Montserrat was thought to be extinct before activity resumed in 1995. Another recent example is Fourpeaked Mountain in Alaska, which, prior to its eruption in September 2006, had not erupted since before 8000 BC and was long thought to be extinct.

NOTABLE VOLCANOES

The 16 current decade volcanoes are:
1. Avachinsky-Koryaksky, Kamchatka, Russia.

2. Nevado de Colima, Jalisco and Colima, Mexico.
3. Mount Etna, Sicily, Italy.
4. Galeras, Nariño, Colombia.
5. Mauna Loa, Hawaii, USA.
6. Mount Merapi, Central Java, Indonesia.
7. Mount Nyiragongo, Democratic Republic of the Congo.
8. Mount Rainier, Washington, USA.
9. Sakurajima, Kagoshima Prefecture, Japan.
10. Santa Maria/Santiaguito, Guatemala.
11. Santorini, Cyclades, Greece.
12. Taal Volcano, Luzon, Philippines.
13. Teide, Canary Islands, Spain.
14. Ulawun, New Britain, Papua New Guinea.
15. Mount Unzen, Nagasaki Prefecture, Japan.
16. Vesuvius, Naples, Italy.

EFFECTS OF VOLCANOES

There are many different types of volcanic eruptions and associated activity: phreatic eruptions (steam-generated eruptions), explosive eruption of high-silica lava (e.g. rhyolite), effusive eruption of low-silica lava (e.g., basalt), pyroclastic flows, lahars (debris flow) and carbon dioxide emission. All of these activities can pose a hazard to humans. Earthquakes, hot springs, fumaroles, mud pots and geysers often accompany volcanic activity (Fig. 9.9).

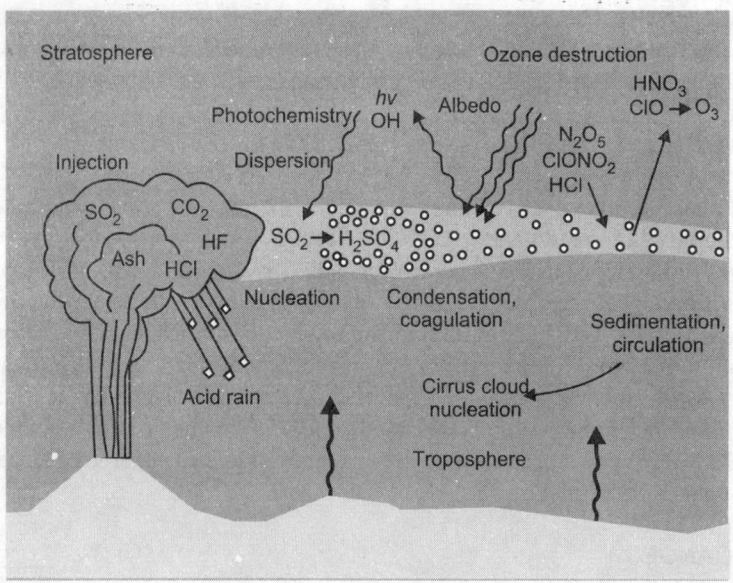

Fig. 9.9. Volcanic 'injection'.

The concentrations of different volcanic gases can vary considerably from one volcano to the next. Water vapour is typically the most abundant volcanic gas, followed by carbon dioxide and sulphur dioxide. Figure 9.10 shows solar radiation reduction from volcanic eruptions.

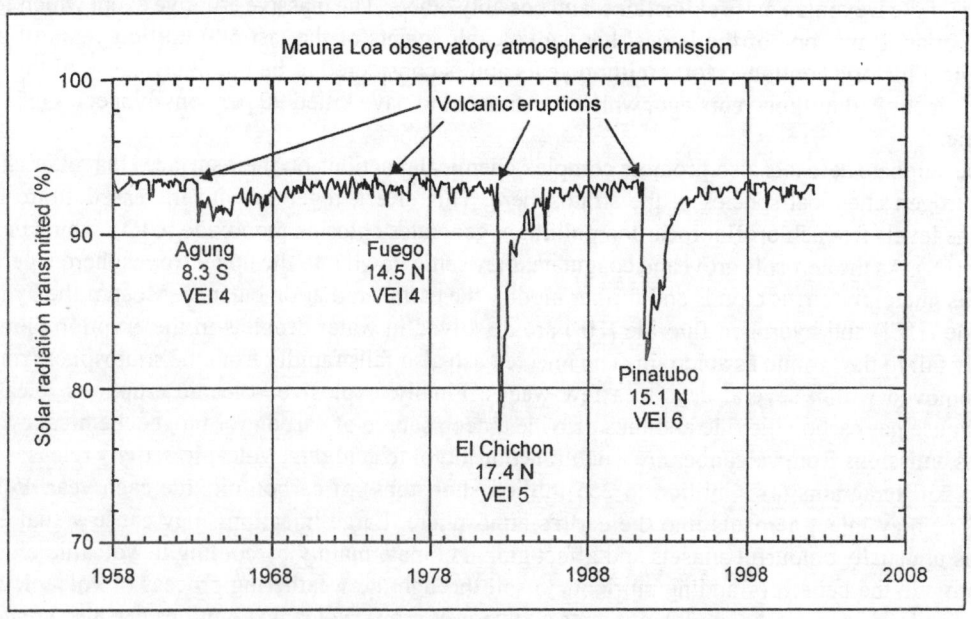

9.10. Solar radiation reduction from volcanic eruptions.

Other principal volcanic gases include hydrogen sulphide, hydrogen chloride, and hydrogen fluoride. A large number of minor and trace gases are also found in volcanic emissions, for example hydrogen, carbon monoxide, halocarbons, organic compounds, and volatile metal chlorides. Large, explosive volcanic eruptions inject water vapour (H_2O), carbon dioxide (CO_2), sulphur dioxide (SO_2), hydrogen chloride (HCl), hydrogen fluoride (HF) and ash (pulverised rock and pumice) into the stratosphere to heights of 16–32 kilometres (10–20 mi) above the earth's surface. The most significant impacts from these injections come from the conversion of sulphur dioxide to sulphuric acid (H_2SO_4), which condenses rapidly in the stratosphere to form fine sulphate aerosols. The aerosols increase the earth's albedo—its reflection of radiation from the Sun back into space—and thus cool the earth's lower atmosphere or troposphere; however, they also absorb heat radiated up from the earth, thereby warming the stratosphere. Several eruptions during the past century have caused a decline in the average temperature at the earth's surface of up to half a degree (Fahrenheit scale) for periods of one to three years—sulphur dioxide from the eruption of Huaynaputina probably caused the Russian famine of 1601–1603.

One proposed volcanic winter happened c. 70,000 years ago following the supereruption of Lake Toba on Sumatra island in Indonesia. According to the Toba catastrophe theory to which some anthropologists and archeologists subscribe, it had global consequences, killing most humans then alive and creating a population bottleneck that affected the genetic inheritance of all humans today. The 1815 eruption of Mount Tambora created global climate anomalies that became known as the 'Year Without a Summer' because of the effect on North American and European weather. Agricultural crops failed and livestock died in much of the Northern Hemisphere, resulting in one of the worst famines of the

19th century. The freezing winter of 1740–41, which led to widespread famine in northern Europe, may also owe its origins to a volcanic eruption.

It has been suggested that volcanic activity caused or contributed to the End-Ordovician, Permian-Triassic, Late Devonian mass extinctions, and possibly others. The massive eruptive event which formed the Siberian Traps, one of the largest known volcanic events of the last 500 million years of earth's geological history, continued for a million years and is considered to be the likely cause of the 'Great Dying' about 250 million years ago, which is estimated to have killed 90 per cent of species existing at the time.

The sulphate aerosols also promote complex chemical reactions on their surfaces that alter chlorine and nitrogen chemical species in the stratosphere. This effect, together with increased stratospheric chlorine levels from chlorofluorocarbon pollution, generates chlorine monoxide (ClO), which destroys ozone (O_3). As the aerosols grow and coagulate, they settle down into the upper troposphere where they serve as nuclei for cirrus clouds and further modify the earth's radiation balance. Most of the hydrogen chloride (HCl) and hydrogen fluoride (HF) are dissolved in water droplets in the eruption cloud and quickly fall to the ground as acid rain. The injected ash also falls rapidly from the stratosphere; most of it is removed within several days to a few weeks. Finally, explosive volcanic eruptions release the greenhouse gas carbon dioxide and thus provide a deep source of carbon for biogeochemical cycles.

Gas emissions from volcanoes are a natural contributor to acid rain. Volcanic activity releases about 130 to 230 teragrams (145 million to 255 million short tons) of carbon dioxide each year. Volcanic eruptions may inject aerosols into the earth's atmosphere. Large injections may cause visual effects such as unusually colourful sunsets and affect global climate mainly by cooling it. Volcanic eruptions also provide the benefit of adding nutrients to soil through the weathering process of volcanic rocks. These fertile soils assist the growth of plants and various crops. Volcanic eruptions can also create new islands, as the magma cools and solidifies upon contact with the water.

Ash thrown into the air by eruptions can present a hazard to aircraft, especially jet aircraft where the particles can be melted by the high operating temperature. Dangerous encounters in 1982 after the eruption of Galunggung in Indonesia, and 1989 after the eruption of Mount Redoubt in Alaska raised awareness of this phenomenon. Nine Volcanic Ash Advisory Centers were established by the International Civil Aviation Organisation to monitor ash clouds and advise pilots accordingly.

VOLCANOES ON OTHER PLANETARY BODIES

The earth's Moon has no large volcanoes and no current volcanic activity, although recent evidence suggests it may still possess a partially molten core. However, the Moon does have many volcanic features such as maria (the darker patches seen on the moon), rilles and domes.

The planet Venus has a surface that is 90 per cent basalt, indicating that volcanism played a major role in shaping its surface. The planet may have had a major global resurfacing event about 500 million years ago, from what scientists can tell from the density of impact craters on the surface. Lava flows are widespread and forms of volcanism not present on earth occur as well. Changes in the planet's atmosphere and observations of lightning have been attributed to ongoing volcanic eruptions, although there is no confirmation of whether or not Venus is still volcanically active. However, radar sounding by the Magellan probe revealed evidence for comparatively recent volcanic activity at Venus's highest volcano Maat Mons, in the form of ash flows near the summit and on the northern flank.

There are several extinct volcanoes on Mars, four of which are vast shield volcanoes far bigger than any on earth. They include Arsia Mons, Ascraeus Mons, Hecates Tholus, Olympus Mons, and Pavonis

Mons. These volcanoes have been extinct for many millions of years, but the European Mars Express spacecraft has found evidence that volcanic activity may have occurred on Mars in the recent past as well. Jupiter's moon Io is the most volcanically active object in the solar system because of tidal interaction with Jupiter. It is covered with volcanoes that erupt sulphur, sulphur dioxide and silicate rock, and as a result, Io is constantly being resurfaced. Its lavas are the hottest known anywhere in the solar system, with temperatures exceeding 1800 K (1500°C). In February 2001, the largest recorded volcanic eruptions in the solar system occurred on Io. Europa, the smallest of Jupiter's Galilean moons, also appears to have an active volcanic system, except that its volcanic activity is entirely in the form of water, which freezes into ice on the frigid surface. This process is known as cryovolcanism, and is apparently most common on the moons of the outer planets of the solar system (Fig. 9.11).

Fig. 9.11. The Tvashtar volcano erupts a plume 330 km (205 mi) above the surface of Jupiter's moon Io.

In 1989 the Voyager 2 spacecraft observed cryovolcanoes (ice volcanoes) on Triton, a moon of Neptune, and in 2005 the Cassini-Huygens probe photographed fountains of frozen particles erupting from Enceladus, a moon of Saturn. The ejecta may be composed of water, liquid nitrogen, dust, or methane compounds. Cassini-Huygens also found evidence of a methane-spewing cryovolcano on the Saturnian moon Titan, which is believed to be a significant source of the methane found in its atmosphere. It is theorised that cryovolcanism may also be present on the Kuiper Belt Object Quaoar.

A 2010 study of the exoplanet COROT-7b, which was detected by transit in 2009, studied that tidal heating from the host star very close to the planet and neighbouring planets could generate intense volcanic activity similar to Io.

Traditional Beliefs about Volcanoes

Many ancient accounts ascribe volcanic eruptions to supernatural causes, such as the actions of gods or demigods. To the ancient Greeks, volcanoes' capricious power could only be explained as acts of the gods, while 16th/17th-century German astronomer Johannes Kepler believed they were ducts for the earth's tears. One early idea counter to this was proposed by Jesuit Athanasius Kircher (1602–1680), who witnessed eruptions of Mount Etna and Stromboli, then visited the crater of Vesuvius and published his view of an earth with a central fire connected to numerous others caused by the burning of sulphur, bitumen and coal.

Various explanations were proposed for volcano behaviour before the modern understanding of the earth's mantle structure as a semisolid material was developed. For decades after awareness that compression and radioactive materials may be heat sources, their contributions were specifically discounted. Volcanic action was often attributed to chemical reactions and a thin layer of molten rock near the surface.

Panoramas

A panorama is any wide-angle view or representation of a physical space, whether in painting, drawing, photography, film/video, or a three-dimensional model. Figures 9.12 to 9.17 shows different types of panoramas.

Fig. 9.12. Mount Bromo, East Java, Indonesia.

Fig. 9.13. Crater of Mount Tangkuban Perahu, West Java, Indonesia.

Fig. 9.14. Irazú Volcano, Costa Rica.

Fig. 9.15. Taal Volcano, Philippines.

Fig. 9.16. Crater of Sierra Negra volcano, Isabela island, Galapagos, Ecuador.

Fig. 9.17. Vulcano island with the north coast of Sicily in the background.

PREDICTION OF VOLCANIC ACTIVITY

Prediction of volcanic eruption (volcanic eruption forecasting) is an interdisciplinary scientific and engineering approach to natural catastrophic event forecasting. Volcanic activity prediction has not been perfected, but significant progress has been made in recent decades. Significant amounts are spent monitoring and prediction of volcanic activity by the Italian government through the Istituto Nazionale di Geofisica e Vulcanologia INGV, by the United States Geological Survey (USGS), and by the Geological Survey of Japan. These are the largest institutions that invest significant resources monitoring and researching volcanos (as well as other geological phenomena). Many countries operate volcano observatories at a lesser level of funding, all of which are members of the World Organisation of Volcano Observatories (WOVO).

General Principles

Various methods including the following sections are used to help predict eruptions. In using these methods, five major principles form the basis of eruption forecasting is as follows:

1. The principle of inflection points in trends states that with unknown rates of change, a point in time is reached at which the volcanic system becomes unstable and likely will erupt.
2. The principle of coinciding change states that one monitored parameter alone may not yield significant symptoms to diagnose an imminent eruption, but unrelated trends of several monitored parameters may start co-evolving as the system approaches a state of instability.
3. The principle of known behaviour treats a volcano as if it were a medical patient, assuming that responses to changes in the underground may be highly individual to a volcano's particular internal structure and can become better known by understanding its past eruptive characteristics.
4. The principle of unexpected behaviour treats volcanoes, the public, and decision-makers alike as inherently inconsistent systems—leading to unexpected eruptions (e.g. fast magma ascent from unexpected depth), and mitigation failures.

5. The principle of symptom-based short-term forecast as with all the other principles is similar to an epidemiological diagnosis, whereby forecasts are based on symptoms and patient history.

Volcanic eruptions can to date not be predicted by stochastic methods, but only by catching early symptoms before an imminent eruption. Therefore, continuous monitoring even of dormant volcanoes, though costly, is the only way to enable eruptive behaviour forecasts. The following sections describe individual groups of methods typically deployed in monitoring volcanoes and the symptomatic evolution of their activity

Methods

The most widely used method is studying the geographical area of the volcano. Taking seismic readings, measuring poison gases, and using satellites

Seismicity

General principles of volcano seismology

Seismic activity (earthquakes and tremors) always occurs as volcanoes awaken and prepare to erupt and are a very important link to eruptions. Some volcanoes normally have continuing low-level seismic activity, but an increase may signal a greater likelihood of an eruption. The types of earthquakes that occur and where they start and end are also key signs. Volcanic seismicity has three major forms: short-period earthquake, long-period earthquake, and harmonic tremor.

1. Short-period earthquakes are like normal fault-generated earthquakes. They are caused by the fracturing of brittle rock as magma forces its way upward. These short-period earthquakes signify the growth of a magma body near the surface and are known as 'A' waves. These type of seismic events are often also referred to as Volcano-Tectonic (or VT) events or earthquakes.

2. Long-period earthquakes are believed to indicate increased gas pressure in a volcano's plumbing system. They are similar to the clanging sometimes heard in a house's plumbing system, which is known as 'water hammer'. These oscillations are the equivalent of acoustic vibrations in a chamber, in the context of magma chambers within the volcanic dome and are known as 'B' waves. These are also known as resonance waves and long period resonance events.

3. Harmonic tremors are often the result of magma pushing against the overlying rock below the surface. They can sometimes be strong enough to be felt as humming or buzzing by people and animals, hence the name.

Patterns of seismicity are complex and often difficult to interpret; however, increasing seismic activity is a good indicator of increasing eruption risk, especially if long-period events become dominant and episodes of harmonic tremor appear.

Using a similar method, researchers can detect volcanic eruptions by monitoring infra-sound—sub-audible sound below 20 Hz. The IMS Global Infrasound Network, originally set up to verify compliance with nuclear test ban treaties, has 60 stations around the world that work to detect and locate erupting volcanoes.

Seismic case studies

A relation between long-period events and imminent volcanic eruptions was first observed in the seismic records of the 1985 eruption of Nevado del Ruiz in Colombia. The occurrence of long-period events were then used to predict the 1989 eruption of Mount Redoubt in Alaska and the 1993 eruption of Galeras in Colombia. In December 2000, scientists at the National Center for Prevention of Disasters in

Mexico City predicted an eruption within two days at Popocatépetl, on the outskirts of Mexico City. Their prediction used research that had been done by Bernard Chouet, a Swiss volcanologist who was working at the United States Geological Survey and who first observed a relation between long-period events and an imminent eruption. The government evacuated tens of thousands of people; 48 hours later, the volcano erupted as predicted. It was Popocatépetl's largest eruption for a thousand years, yet no one was hurt.

Iceberg tremors

It has recently been published that the striking similarities between iceberg tremors, which occur when they run aground, and volcanic tremors may help experts develop a better method for predicting volcanic eruptions. Although icebergs have much simpler structures than volcanoes, they are physically easier to work with. The similarities between volcanic and iceberg tremors include long durations and amplitudes, as well as common shifts in frequencies.

Gas emissions

As magma nears the surface and its pressure decreases, gases escape. This process is much like what happens when you open a bottle of soda and carbon dioxide escapes. Sulphur dioxide is one of the main components of volcanic gases, and increasing amounts of it herald the arrival of increasing amounts of magma near the surface. For example, on May 13, 1991, an increasing amount of sulphur dioxide was released from Mount Pinatubo in the Philippines. On May 28, just two weeks later, sulphur dioxide emissions had increased to 5000 tons, ten times the earlier amount. Mount Pinatubo later erupted on June 12, 1991. On several occasions, such as before the Mount Pinatubo eruption and the 1993 Galeras, Colombia eruption, sulphur dioxide emissions have dropped to low levels prior to eruptions. Most scientists believe that this drop in gas levels is caused by the sealing of gas passages by hardened magma. Such an event leads to increased pressure in the volcano's plumbing system and an increased chance of an explosive eruption.

Ground deformation

Swelling of the volcano signals that magma has accumulated near the surface. Scientists monitoring an active volcano will often measure the tilt of the slope and track changes in the rate of swelling. An increased rate of swelling, especially if accompanied by an increase in sulphur dioxide emissions and harmonic tremors is a high probability sign of an impending event. The deformation of Mount St. Helens prior to the May 18, 1980 eruption was a classic example of deformation, as the north side of the volcano was bulging upwards as magma was building up underneath. Most cases of ground deformation are usually detectable only by sophisticated equipment used by scientists, but they can still predict future eruptions this way. The Hawaiian Volcanoes show significant ground deformation; there is inflation of the ground prior to an eruption and then an obvious deflation post-eruption. This is due to the shallow magma chamber of the Hawaiian Volcanoes; movement of the magma is easily noticed on the ground above.

Thermal monitoring

Both magma movement, changes in gas release and hydrothermal activity can lead to thermal emissivity changes at the volcano's surface. These can be measured using several techniques:

1. Forward looking infrared radiometry (FLIR) from hand-held devices installed on-site, at a distance or airborne.

2. Infrared band satellite imagery.
3. *In situ* thermometry (hot springs, fumaroles).
4. Heat flux maps.
5. Geothermal well enthalpy changes.

Hydrology

There are four main methods that can be used to predict a volcanic eruption through the use of hydrology:

1. Borehole and well hydrologic and hydraulic measurements are increasingly used to monitor changes in a volcanoes subsurface gas pressure and thermal regime. Increased gas pressure will make water levels rise and suddenly drop right before an eruption, and thermal focusing (increased local heat flow) can reduce or dry out acquifers.
2. Detection of lahars and other debris flows close to their sources. USGS scientists have developed an inexpensive, durable, portable and easily installed system to detect and continuously monitor the arrival and passage of debris flows and floods in river valleys that drain active volcanoes.
3. Pre-eruption sediment may be picked up by a river channel surrounding the volcano that shows that the actual eruption may be imminent. Most sediment is transported from volcanically disturbed watersheds during periods of heavy rainfall. This can be an indication of morphological changes and increased hydrothermal activity in absence of instrumental monitoring techniques.
4. Volcanic deposit that may be placed on a river bank can easily be eroded which will dramatically widen or deepen the river channel. Therefore, monitoring of the river channels width and depth can be used to assess the likelihood of a future volcanic eruption.

Remote Sensing

Remote sensing is the detection by a satellite's sensors of electromagnetic energy that is absorbed, reflected, radiated or scattered from the surface of a volcano or from its erupted material in an eruption cloud.

1. Cloud sensing: Scientists can monitor the unusually cold eruption clouds from volcanoes using data from two different thermal wavelengths to enhance the visibility of eruption clouds and discriminate them from meteorological clouds
2. Gas sensing: Sulphur dioxide can also be measured by remote sensing at some of the same wavelengths as ozone. TOMS (Total Ozone Mapping Spectrometer) can measure the amount of sulphur dioxide gas released by volcanoes in eruptions
3. Thermal sensing: The presence of new significant thermal signatures or 'hot spots' may indicate new heating of the ground before an eruption, represent an eruption in progress or the presence of a very recent volcanic deposit, including lava flows or pyroclastic flows.
4. Deformation sensing: Satellite-borne spatial radar data can be used to detect long-term geometric changes in the volcanic edifice, such as uplift and depression. In this method, called InSAR (Interferometric Synthetic Aperture Radar), DEMs generated from radar imagery are subtracted from each other to yield a differential image, displaying rates of topographic change.
5. Forest monitoring: In recent period it has been demonstrated the location of eruptive fractures could be predicted, months to years before the eruptions, by the monitoring of forest growth. This tool based on the monitoring of the trees growth has been validated at both *Mt. Niyragongo* and *Mt. Etna* during the 2002–2003 volcano eruptive events.

Mass movements and mass failures

Monitoring mass movements and failures uses techniques lending from seismology (geophones), deformation, and meteorology. Landslides, rock falls, pyroclastic flows, and mud flows (lahars) are example of mass failures of volcanic material before, during, and after eruptions.

The most famous volcanic landslide was probably the failure of a bulge that built up from intruding magma before the Mt. St. Helens eruption in 1980, this landslide 'uncorked' the shallow magmatic intrusion causing catastrophic failure and an unexpected lateral eruption blast. Rock falls often occur during periods of increased deformation and can be a sign of increased activity in absence of instrumental monitoring. Mud flows (lahars) are remobilised hydrated ash deposits from pyroclastic flows and ash fall deposits, moving downslope even at very shallow angles at high speed. Because of their high density they are capable of moving large objects such as loaded logging trucks, houses, bridges, and boulders. Their deposits usually form a second ring of debris fans around volcanic edifices, the inner fan being primary ash deposits. Downstream of the deposition of their finest load, lahars can still pose a sheet flood hazard from the residual water. Lahar deposits can take many months to dry out, until they can be walked on. The hazards derived from lahar activity can last several years after a large explosive eruption. A team of US scientists developed a method of predicting lahars. Their method was developed by analysing rocks on *Mt. Rainier* in Washington. The warning system depends on noting the differences between fresh rocks and older ones. Fresh rocks are poor conductors of electricity and become hydrothermically altered by water and heat. Therefore, if they know the age of the rocks, and therefore the strength of them, they can predict the pathways of a lahar. A system of acoustic flow monitors (AFM) has also been emplaced on Mount Rainier to analyse ground tremors that could result in a lahar, providing an earlier warning.

LOCAL CASE STUDIES

Nyiragongo

The eruption of Mt. Nyiragongo on January 17, 2002 was predicted a week earlier by a local expert who had been watching the volcanoes for years. He informed the local authorities and a UN survey team was dispatched to the area; however, it was declared safe. Unfortunately, when the volcano erupted, 40 per cent of the city of Goma was destroyed along with many people's livelihoods. The expert claimed that he had noticed small changes in the local relief and had monitored the eruption of a much smaller volcano two years earlier. Since he knew that these two volcanoes were connected by a small fissure, he knew that Mt. Nyiragongo would erupt soon.

Mt. Etna

British geologists have developed a method of predicting future eruptions of Mt. Etna. They have discovered that there is a time lag of 25 years between events that happen below the surface and events that happen on the surface, i.e. a volcanic eruption. The careful monitoring of deep crust events can help predict accurately what will happen in the years to come. So far they have predicted that between 2007 and 2015, volcanic activity will be half of what it was in 1972.

Sakurajima, Japan

Sakurajima is possibly one of the most monitored areas on earth. The Sakurajima Volcano lies near Kagoshima City, which has a population of 5,00,000 people. Both the Japanese Meteorological Agency

(JMA) and Kyoto University's Sakurajima Volcanological Observatory (SVO) monitors the volcano's activity. Since 1995, Sakurajima has only erupted from its summit with no release of lava. Monitoring techniques at Sakurajima:

1. Likely activity is signalled by swelling of the land around the volcano as magma below begins to build up. At Sakurajima, this is marked by a rise in the sea-bed in Kagoshima Bay—tide levels rise as a result.

2. As magma begins to flow, melting and splitting base rock can be detected as volcanic earthquakes. At Sakurajima, they occur two to five kilometres beneath the surface. An underground observation tunnel is used to detect volcanic earthquakes more reliably.

3. Groundwater levels begin to change, the temperature of hot springs may rise and the chemical composition and amount of gases released may alter. Temperature sensors are placed in bore holes which are used to detect ground water temp. Remotes sensing is used on Sakurajima since the gases are highly toxic—the ratio of HCl gas to SO_2 gas increases significantly shortly before an eruption.

4. As an eruption approaches, tiltmetre systems measure minute movements of the mountain. Data is relayed in real-time to monitoring systems at SVO.

5. Seismometers detect earthquakes which occur immediately beneath the crater, signalling the onset of the eruption. They occur 1 to 1.5 seconds before the explosion.

6. With the passing of an explosion, the tiltmeter system records the settling of the volcano.

MARITIME IMPACTS OF VOLCANIC ERUPTIONS

Less commonly publicised than the effects on aviation—and with less potential for catastrophe—maritime Impacts of volcanic eruptions are also dangerous. When a volcano erupts, large amounts of noxious gases, steam, rock, and ash are released into the atmosphere; fine ash can be transported thousands of miles from the volcano, while high concentrations of coarse particles fall out of the air near the volcano. The high concentrations of hazardous toxic gases are localised in the immediate vicinity of the volcano (Fig. 9.18).

Fig. 9.18. A pumice raft from an undersea eruption in Tonga.

Until more recently public focus has mainly been on effects on aviation effects—ash, which can be undetectable, can cause an aircraft's engine to cut out with catastrophic potential. However, the July 2008 eruption of Okmok Volcano in Alaska triggered attention to the maritime effects. Employees at the National Weather Service Ocean Prediction Center's Ocean Applications Branch examined this event and partnered with the Alaska Volcano Observatory to compile information on the topic.

Ash can affect marine transportation in many ways:

1. Volcanic ash can clog air intake filters in a matter of minutes, crippling airflow to vital machinery. Ash particles are very abrasive and, if they get into an engine's moving parts, can cause severe damage very quickly.
2. Water is the main component in volcanic eruptions; it is what makes them so explosive. Through chemical reactions, toxic gases that are released in eruptions can bond or adsorb to ashfall particles. As the particles land on skin, metal, or other exposed shipboard equipment, they can begin to corrode.
3. Certain types of volcanic ash do not dissolve easily in water. Instead, they clump on the surface of the ocean in pumice rafts. These rafts can clog salt water intake strainers very quickly, which can result in overheating of shipboard machinery dependent on sea water service cooling.
4. Heavy amounts of volcanic ash reduce visibility to less than ½ mi, which is a hazard to navigation. This, combined with the above three other main impacts make sailing in the vicinity of volcanic ash very dangerous for mariners.

VOLCANIC EXPLOSIVITY INDEX

The Volcanic Explosivity Index (VEI) was devised by Chris Newhall of the US Geological Survey and Stephen Self at the University of Hawaii in 1982 to provide a relative measure of the explosiveness of volcanic eruptions. Volume of products, eruption cloud height, and qualitative observations (using terms ranging from 'gentle' to 'mega-colossal') are used to determine the explosivity value. The scale is open-ended with the largest volcanoes in history given magnitude 8.

A value of 0 is given for non-explosive eruptions (less than 10^4 cubic metres of tephra ejected) with 8 representing a mega-colossal explosive eruption that can eject 10^{12} cubic metres of tephra and have a cloud column height of over 25 km (16 mi). The scale is logarithmic, with each interval on the scale representing a tenfold increase in observed eruption criteria (exception: between VEI 0 and VEI 1). Figure 9.19 shows VEI and ejecta volume correlation.

Note that ash, volcanic bombs, and ignimbrite are all treated alike. Density and vesicularity (gas bubbling) of the volcanic products in question is not taken into account. In contrast, the DRE (Dense-Rock Equivalent) is sometimes calculated to give the actual amount of magma erupted. Another weakness of the VEI is that it does not take into account the power output of an eruption, which makes it extremely difficult to determine with prehistoric or unobserved eruptions.

Classification

Scientists indicate how powerful volcanic eruptions are using the VEI. It records how much volcanic material is thrown out, how high the eruption goes, and how long it lasts. The scale goes from 0 to 8. An increase of 1 indicates a 10 times more powerful eruption (Table 9.1).

Note: There is a discontinuity in the definition of the VEI between indices 1 and 2. The lower border of the volume of ejecta jumps by a factor of 100 from 10,000 to 1,000,000 m^3 while the factor is 10 between all higher indices.

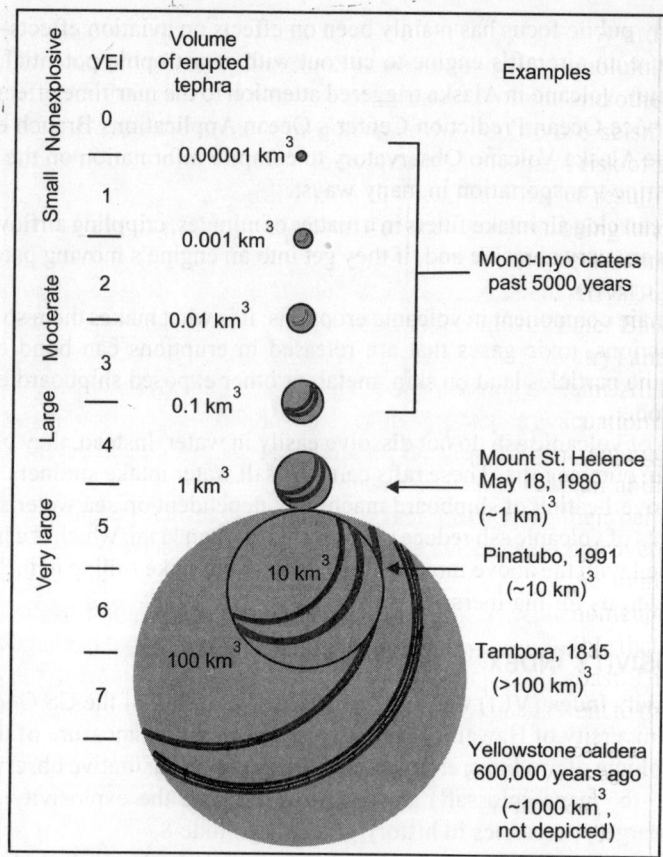

Fig. 9.19. VEI and ejecta volume correlation.

Table 9.1. Volcanic explosivity index.

VEI	Ejecta volume	Classification	Description	Plume	Frequency	Example	Occurrences in last 10,000 years*
0	<10,000 m³	Hawaiian	Nonexplosive	<100 m	Constant	Mauna Loa	Many
1	>10,000 m³	Hawaiian/ Strombolian	Gentle	100–1000 m	Daily	Stromboli	Many
2	> 1,000,000 m³	Strombolian/ Vulcanian	Explosive	1–5 km	Weekly	Galeras	3477*
3	> 10,000,000 m³	Vulcanian/ Peléan	Severe	3–15 km	Yearly	Cordón Caulle	868
4	> 0.1 km³	Peléan/Plinian	Cataclysmic	10–25 km	≥10 yrs	Eyjafjallajökull	421
5	> 1 km³	Plinian	Paroxysmal	>25 km	≥50 yrs	Mount St. Helens	166

(Contd ...)

VEI	Ejecta volume	Classification	Description	Plume	Frequency	Example	Occurrences in last 10,000 years*
6	> 10 km³	Plinian/ Ultra-Plinian	Colossal	>25 km	≥100 yrs	Krakatoa	51
7	> 100 km³	Plinian/ Ultra-Plinian	Super-colossal	>25 km	≥1000 yrs	Tambora	5 (+2 suspected)
8	> 1000 km³	Ultra-Plinian	Mega-colossal	>25 km	≥10,000 yrs	Taupo (26,500 BP)	0

*Count of VEI 2 and VEI 3 eruptions in the last 10,000 years are based on 1994 figures maintained by the Global Volcanism Program of the Smithsonian Institution. Count of eruptions greater than VEI 3 in the last 10,000 years are based on its 2010 figures. There are also 58 plinian eruptions, and 13 caldera-forming eruptions, of large, but unknown magnitudes.

A total of 47 eruptions of VEI–8 magnitude or above, ranging in age from Ordovician to Pleistocene, are identified, of which 42 eruptions are known from the past 36 million years. The most recent one is Lake Taupo's Oruanui eruption, occurring 26,500 years ago, which means that there have not been any Holocene (within the last 10,000 years) eruptions with a VEI of 8.

Examples of Eruptions by VEI

Table 9.2 gives examples of eruptions by VEI.

Table 9.2. Examples of eruptions by VEI.

VEI	Volcano (eruption)	Year
0	Hoodoo mountain	7050 BC?
0	Mauna Loa	1984
0	Lake Nyos	1986
0	Piton de la Fournaise	2004
1	Wells Gray-Clearwater volcanic field	1500?
1	Kilauea	1983 - present
1	Nyiragongo	2002
2	Mount Hood	1865-1866
2	Kilauea	1924
2	Tristan da Cunha	1961
2	Mount Usu	2000–2001
2	Whakaari/White Island	2001
3	Mount Garibaldi	9300 BP
3	Nazko Cone	7,200 BP
3	Mount Edziza	950 AD ± 1000 years
3	Mount Vesuvius	1913–1944
3	Surtsey	1963–1967
3	Eldfell	1973

(Contd ...)

VEI	Volcano (eruption)	Year
3	Nevado del Ruiz	1985
3	Mount Etna	2002–2003
4	Mount Pelée	1902
4	Parícutin	1943–1952
4	Hekla	1947
4	Galunggung	1982
4	Mount Spurr	1992
4	Mount Okmok	2008
4	Eyjafjallajökull	2010
4	Mount Merapi	2010
5	Hekla (Hekla 3 eruption)	1021 +130/−100 BC
5	Mount Meager	≈400 BC (2350 BP)
5	Mount Vesuvius (Pompeian eruption)	79
5	Mount Edgecumbe/Putauaki	c. 300
5	Mount Tarumae	1739
5	Mount Mayon	1814
5	Mount Tarawera	1886
5	Katla	1918
5	Mount Agung	1963
5	Mount St. Helens	1980
5	El Chichón	1982
5	Mount Hudson	1991
5	Chaiten	2008
6	Morne Diablotins	30,000 BP
6	Laacher See	12,900 BP?
6	Nevado de Toluca	10,500 BP
6	Mount Okmok	8300 BP
6	Mount Etna	8000 BP?
6	Mount Veniaminof	1750 BC
6	Mount Vesuvius (Avellino eruption)	1660 BC ± 43 years
6	Grímsvötn	8230 BC ± 50 years
6	Mount Aniakchak	≈1645 BC
6	Mount Okmok	c. 400 BC
6	Ambrym	c. AD 100
6	Ilopango	450 ± 30 years
6	Mount Churchill (White River Ash)	≈750 (1200 BP)
6	Katla (Eldgjá)	934
6	Baekdu Mountain (Tianchi eruption)	969 ± 20 years

(Contd ...)

VEI	Volcano (eruption)	Year
6	Kuwae	1452 or 1453
6	Bárðarbunga	1477
6	Huaynaputina	1600
6	Laki	1783
6	Krakatoa	1883
6	Santa María	1902
6	Novarupta	1912
6	Mount Pinatubo	1991
7	Sesia Valley caldera	280 Ma
7	Bennett Lake Volcanic Complex	50 Ma
7	Valles (Lower Bandelier eruption)	1.47 Ma
7	Yellowstone (Mesa Falls eruption)	1.3 Ma
7	Valles (Upper Bandelier eruption)	1.15 Ma
7	Long Valley Caldera (Bishop eruption)	759,000 BP
7	Maninjau	280,000 BP
7	Atitlán (Los Chocoyos eruption)	84,000 BP
7	Kurile (Golygin eruption)	41,500 BP
7	Campi Flegrei	37,000 BP
7	Aira Caldera	22,000 BP
7	Kurile (Ilinsky eruption)	≈6400 BC
7	Crater Lake (Mount Mazama eruption)	≈5700 BC
7	Kikai (Akahoya eruption)	≈5300 BC
7	Thera (Minoan eruption)	1620s BC
7	Taupo (Hatepe eruption)	186
7	Mount Tambora (1815 eruption)	1815
8	Scafells	Ordovician
8	Glen Coe	420 Ma
8	La Garita Caldera	27 Ma
8	Yellowstone (Huckleberry Ridge eruption)	2.2 Ma
8	Galán	2.2 Ma
8	Yellowstone (Lava Creek eruption)	640,000 BP
8	Whakamaru (Whakamaru Ignimbrite/Mount Curl Tephra)	254,000 BP
8	Toba	69,000–77,000 BP
8	Taupo (Oruanui eruption)	26,500 BP

LIST OF LARGEST VOLCANIC ERUPTIONS

In a volcanic eruption, lava, tephra (volcanic bombs, lapilli, and ash), and various gases are expelled from a volcanic vent or fissure. While many eruptions only pose dangers to the immediately surrounding area, earth's largest eruptions can have a major regional or even global impact, with some affecting the climate and contributing to mass extinctions. Volcanic eruptions can generally be characterised as either explosive eruptions, sudden ejections of rock and ash, or effusive eruptions, relatively gentle outpourings of lava. A separate list is given below for each type. All of the eruptions listed below have produced at least 1000 km³ (240 cu mi) of lava and tephra; for explosive eruptions, this corresponds to a Volcanic Explosivity Index (or VEI) of 8. They are at least a thousand times larger than the 1980 eruption of Mount St. Helens which produced only 1 km³ (0.2 cu mi) of material, and at least six times larger than the 1815 eruption of Mount Tambora, the largest eruption in recent history, which produced 160 km³ (38 cu mi) of volcanic deposits. There have probably been many such eruptions during earth's history beyond those shown in these lists. However erosion and plate tectonics have taken their toll, and many eruptions have not left enough evidence for geologists to establish their size. Even for the eruptions listed here, estimates of the volume erupted can be subject to considerable uncertainty.

Explosive Eruptions

In explosive eruptions, the eruption of magma is driven by the rapid release of pressure, often involving the explosion of gas previously dissolved within the material. The most famous and destructive historical eruptions are mainly of this type. An eruptive phase can consist of a single eruption, or a sequence of several eruptions spread over several days, weeks or months. Explosive eruptions usually involve thick, highly viscous felsic magma, high in volatiles like water vapour and carbon dioxide. Pyroclastic materials are the primary product, typically in the form of tuff. Eruptions the size of that at Lake Toba 74 thousand years ago (2800 km³ or more) occur worldwide every 50,000 to 1,00,000 years (Table 9.3).

Table 9.3. Explosive eruptions.

Volcano—Eruption	Age (Ma)	Location	Volume (km³)	Notes
Guarapuava—Tamarana—Sarusas	132	Paraná and Etendeka traps	8600	–
Santa Maria—Fria	~132	Paraná and Etendeka traps	7800	–
Guarapuava—Ventura	~132	Paraná and Etendeka traps	7600	–
Sam Ignimbrite and	29.5	Yemen	6800	Volume includes 5550 km³ of distal tuffs. This estimate is uncertain to a factor of 2 or 3.
Goboboseb–Messum volcanic centre—Springbok quartz latite unit	132	Paraná and Etendeka traps, Brazil and Namibia	6340	–
Caxias do Sul—Grootberg	~132	Paraná and Etendeka traps	5650	–
La Garita Caldera—Fish Canyon tuff	27.8	San Juan volcanic field, Colorado	5,000	Commonly regarded as the largest tuff ever measured on earth, or largest confidently-measured tuff on earth. It is part of at least 20 large caldera-forming eruptions in the San Juan volcanic field and surrounding area that formed around 26 to 35 Ma.

(Contd ...)

Volcano—Eruption	Age (Ma)	Location	Volume (km³)	Notes
Jacui—Goboboseb II	~132	Paraná and Etendeka traps	4350	–
Ourinhos—Khoraseb	~132	Paraná and Etendeka traps	3900	–
Jabal Kura'a Ignimbrite	29.6	Yemen	3800	Volume estimate is uncertain to a factor of 2 or 3.
Windows Butte tuff	31.4	William's Ridge, central Nevada	3500	Part of the Mid-Tertiary ignimbrite flare-up
Anita Garibaldi—Beacon	~132	Paraná and Etendeka traps	3450	–
Indian Peak Caldera Complex—Wah Wah Springs tuff	29.5	Eastern Nevada/ Western Utah	3200	Indian Peak Caldera Complex total volume over 10,000 cubic km, Wah Wah Springs tuff being the largest
Oxaya ignimbrites	19	Chile	3000	Really a regional correlation of many ignimbrites originally thought to be distinct
Lund Tuff	29	Great Basin, USA	3000	Similar in composition to the Fish Canyon Tuff
Lake Toba—Youngest Toba Tuff	0.073	Sunda Arc, Indonesia	2800	Largest eruption on earth in at least the last 25 million years, responsible for the Toba catastrophe theory, a population bottleneck of the human species
Pacana Caldera— Atana ignimbrite	4	Chile	2800	Forms a resurgent caldera.
Iftar Alkalb—Tephra	29.5	Afro-Arabian	2700	–
Yellowstone caldera— Huckleberry Ridge Tuff	2.059	Yellowstone hotspot	2450	Largest Yellowstone eruption on record
Whakamaru	0.254	Taupo Volcanic Zone, New Zealand	2000	Largest in the Southern Hemisphere in the Late Quaternary
Palmas BRA-21— Wereldsend	29.5	Paraná and Etendeka traps	1900	–
Kilgore tuff	4.3	Near Kilgore, Idaho	1800	Last of the eruptions from the Heise volcanic field
Sana'a Ignimbrite— Tephra 2W63	29.5	Afro-Arabian	1600	–
Millbrig eruptions— Bentonites	454	England, exposed in Northern Europe and Eastern US	1509	One of the oldest large eruptions preserved
Blacktail tuff	6.5	Blacktail, Idaho	1500	First of several eruptions from the Heise volcanic field
Emory Caldera— Kneeling Nun tuff	33	Southwestern New Mexico	1310	–

(Contd ...)

Volcano—Eruption	Age (Ma)	Location	Volume (km³)	Notes
Timber Mountain tuff	11.6	Southwestern Nevada	1200	Also includes a 900 cubic km tuff as a second member in the tuff
Paintbrush tuff (Topopah Spring Member)	12.8	Southwestern Nevada	1200	Related to a 1000 cubic km tuff (Tiva Canyon Member) as another member in the Paintbrush tuff
Bachelor—Carpenter Ridge tuff	28	San Juan volcanic field	1200	Part of at least 20 large caldera-forming eruptions, including the world's largest, the Fish Canyon tuff in the San Juan volcanic field and surrounding area that formed around 26 to 35 Ma
Bursum—Apache Springs Tuff	28.5	Southern New Mexico	1200	Related to a 1050 cubic km tuff, the Bloodgood Canyon tuff
Taupo Volcano—Oruanui eruption	0.027	Taupo volcanic zone, New Zealand	1170	Most recent VEI 8 eruption
Huaylillas Ignimbrite	15	Bolivia	1100	Predates half of the uplift of the central Andes
Bursum—Bloodgood Canyon tuff	28.5	Southern New Mexico	1050	Related to a 1200 cubic km tuff, the Apache Springs tuff
Yellowstone Caldera—Lava Creek Tuff	0.639	Yellowstone hotspot	1000	Last large eruption in the Yellowstone National Park area
Cerro Galán	2.2	Catamarca Province, Argentina	1000	Elliptical caldera is ~35 km wide
Paintbrush tuff (Tiva Canyon Member)	12.7	Southwestern Nevada	1000	Related to a 1200 cubic km tuff (Topopah Spring Member) as another member in the Paintbrush tuff
San Juan—Sapinero Mesa Tuff	28	San Juan volcanic field	1000	Part of at least 20 large caldera-forming eruptions, including the world's largest, the Fish Canyon tuff in the San Juan volcanic field and surrounding area that formed around 26 to 35 Ma
Uncompahgre—Dillon and Sapinero Mesa Tuffs	28.1	San Juan volcanic field	1000	Part of at least 20 large caldera-forming eruptions, including the world's largest, the Fish Canyon tuff in the San Juan volcanic field and surrounding area that formed around 26 to 35 Ma

(Contd ...)

Volcano—Eruption	Age (Ma)	Location	Volume (km³)	Notes
Platoro—Chiquito Peak tuff	28.2	San Juan volcanic field	1000	Part of at least 20 large caldera-forming eruptions, including the world's largest, the Fish Canyon tuff in the San Juan volcanic field and surrounding area that formed around 26 to 35 Ma
Mount Princeton— Wall Mountain tuff	35.3	Thirtynine Mile, volcanicarea Colorado	1000	Helped cause the exceptional preservation at Florissant Fossil Beds National Monument

Effusive Eruptions

Effusive eruptions involve a relatively gentle, steady outpouring of lava rather than large explosions. They can continue for years or decades, producing extensive fluid mafic lava flows. For example, Kilauea on Hawaii has continued erupting from 1983 to the present, producing 2.7 km³ (1 cu mi) of lava covering more than 100 km² (40 sq mi). The largest effusive eruption in history occurred in Iceland during the 1783–1784 eruption of Laki, which produced about 15 km³ (4 cu mi) of lava and killed one fifth of Iceland's population. The ensuing disruptions to the climate may also have killed millions elsewhere (Table 9.4).

Table 9.4. Effusive eruptions.

Eruption	Age (Ma)	Location	Volume (km³)	Notes
Mahabaleshwar–Rajahmundry Traps (Upper)	64.8	Deccan traps, India	9300	–
Wapshilla Ridge flows	~15.5	Columbia River Basalt Group, United States	5000–10,000	Member comprises 8–10 flows with a total volume of ~50,000 km³
McCoy Canyon flow	15.6	Columbia River Basalt Group, United States	4300	–
Umtanum flows	~15.6	Columbia River Basalt Group, United States	2750	Two flows with a total volume of 5500 km³
Sand Hollow flow	15.3	Columbia River Basalt Group, United States	2660	–
Pruitt Draw flow	16.5	Columbia River Basalt Group, United States	2350	–
Museum flow	15.6	Columbia River Basalt Group, United States	2350	–
Moonaree Dacite	1591	Gawler Range Volcanics, Australia	2050	One of the oldest large eruptions preserved
Rosalia flow	14.5	Columbia River Basalt Group, United States	1900	–

(Contd...)

Eruption	Age (Ma)	Location	Volume (km³)	Notes
Joseph Creek flow	16.5	Columbia River Basalt Group, United States	1850	–
Ginkgo Basalt	15.3	Columbia River Basalt Group, United States	1600	–
California Creek— Airway Heights flow	15.6	Columbia River Basalt Group, United States	1500	–
Stember Creek flow	15.6	Columbia River Basalt Group, United States	1200	–

Large Igneous Provinces

Highly active periods of volcanism in what are called large igneous provinces have produced huge oceanic plateaus and flood basalts in the past. These can comprise hundreds of large eruptions, producing millions of cubic kilometres of lava in total. No large flood basalt type eruptions have occurred in human history, the most recent having occurred over 10 million years ago. They are often associated with break-up of supercontinents such as Pangea in the geologic record, and may have contributed to a number of mass extinctions. Most large igneous provinces have either not been studied thoroughly enough to establish the size of their component eruptions, or are not preserved well enough to make this possible. Many of the eruptions listed above thus come from just two large igneous provinces: the Paraná and Etendeka traps and the Columbia River Basalt Group. The latter is the most recent large igneous province, and also one of the smallest. A list of large igneous provinces follows to provide some indication of how many large eruptions may be missing from the lists given in Table 9.5.

Table 9.5. Large igneous provinces.

Igneous province	Age (Ma)	Location	Volume (km³)	Notes
Ontong Java– Manihiki–Hikurangi Plateau	121	Southwest Pacific Ocean	59–77	Largest igneous body on earth, later split into three widely separated oceanic plateaus, with a fourth component perhaps now accreted onto South America. Possibly linked to the Louisville hotspot.
Kerguelen Plateau– Broken Ridge	112	South Indian Ocean, Kerguelen Islands	17	Linked to the Kerguelen hotspot. Volume includes Broken Ridge and the Southern and Central Kerguelen Plateau (produced 120–95 Ma), but not the Northern Kerguelen Plateau (produced after 40 Ma).
North Atlantic Igneous Province	55.5	North Atlantic Ocean	6.6	Linked to the Iceland hotspot.
Mid-Tertiary ignimbrite flare-up	32.5	Southwest United States: mainly in Colorado, Nevada, Utah, and New Mexico	5.5	Mostly andesite to rhyolite explosive (0.5 km³) to effusive (5 km³) eruptions, 25–40 Ma. Includes many volcanic centers, including the San Juan volcanic field.

(Contd ...)

Igneous province	Age (Ma)	Location	Volume (km³)	Notes
Caribbean large igneous province	88	Caribbean-Colombian oceanic plateau	4	Linked to the Galápagos hotspot.
Siberian Traps	249.4	Siberia, Russia	1–4	Possibly the largest outpouring of lava on land ever recorded, thought to have caused Permian-Triassic extinction, largest mass extinction event ever.
Karoo-Ferrar	183	Southern Africa, Antarctica	2.5	Formed as Gondwana broke up
Paraná and Etendeka traps	133	Brazil/Angola and Namibia	2.3	Linked to the Tristan hotspot
Central Atlantic Magmatic Province	200	Laurasia continents	2	Formed as Pangea broke up
Deccan Traps	65.5	Deccan Plateau, India	1.5	May have helped kill the dinosaurs.
Emeishan Traps	256.5	Southwestern China	1	Along with Siberian Traps, may have contributed to the Permian–Triassic extinction event.
Coppermine River Group	1267	Mackenzie Large Igneous Province/ Canadian Shield	0.65	Consists of at least 150 individual flows.
Afro-Arabian flood volcanism	28.5	Ethiopia/Yemen/Afar, Arabian-Nubian Shield	0.35	Associated with silicic, explosive tuffs
Columbia River Basalt Group	16	Pacific Northwest, United States	0.18	Well exposed by Missoula Floods in the Channeled Scablands.

JAPAN: VOLCANO ERUPTION PROMPTS EVACUATION WARNING

Officials from Tokyo urged more than 1000 residents to seek safer ground on 31st January, 2011 and expanded a no-access zone around a volcano that has exploded back to life in southern Japan. The 4662-foot (1421-meter) Shinmoedake volcano erupted on 28th January, 2011 for the first time in 52 years. The volcano is located in a remote part of the Kirishima range on the southern Japan island of Kyushu. No injuries have been reported. On 31st January, 2011 five days after it burst back to life, the volcano was still spewing a spectacular plume into the air, sending a blanket of ash out over a wide area and prompting several hundred residents to seek shelter in evacuation centers. Officials in the town of Takaharu urged about 1100 residents to go to evacuation centers because of the danger of debris, ash and landslides. The warning was not mandatory, however, and some residents were returning to their homes instead. The Meteorological Agency, meanwhile, broadened a no-access danger zone to two miles (three kilometers) from the peak and was planning to send in helicopters to monitor activity near the crater. Small rocks ejected from the eruptions have broken windows in buildings and cars near Shinmoedake. The eruption has also disrupted train service, closed schools and forced some domestic flight cancelations. Most transportation had been restored by Monday.

Experts said a dome of lava was growing larger inside the volcano's crater, but it was not certain whether the dome would grow enough to spill over the rim and create large flows down the volcano's sides. Avalanches of superheated gas, ash and rock have already been observed. The Japanese islands are volcanic in origin and dozens of active volcanos continue to erupt with some regularity across the country. In 1991, 43 people died in the eruption of Mount Unzen, also on Kyushu island. Figure 9.20 shows the ash and smoke and Fig. 9.21 shows volcanic lightning or dirty thunderstorm in Shinmoedake peak—Japan in January, 2011.

Fig. 9.20. Ash and smoke: An aerial view of Japan's Shinmoedake volcano on 28 January, 2011.

Fig. 9.21. Volcanic lightning or a dirty thunderstorm is seen above Shinmoedake peak, one of Mount Kirishima's many calderas, in this photo taken from Kirishima city and released by Minami-Nippon Shimbun January 28, 2011.

Chapter 10

Tornado

INTRODUCTION

A tornado is a violent, dangerous, rotating column of air that is in contact with both the surface of the earth and a cumulonimbus cloud or in rare cases, the base of a cumulus cloud. Tornadoes come in many shapes and sizes, but are typically in the form of a visible condensation funnel, whose narrow end touches the earth and is often encircled by a cloud of debris and dust. Most tornadoes have wind speeds less than 110 miles per hour (177 km/hr), are approximately 250 feet (80 m) across, and travel a few miles (several kilometers) before dissipating. The most extreme can attain wind speeds of more than 300 mph (480 km/hr), stretch more than two miles (3 km) across, and stay on the ground for dozens of miles (more than 100 km).

Various types of tornadoes include the landspout, multiple vortex tornado, and waterspout. Waterspouts are characterised by a spiraling funnel-shaped wind current, connecting to a large cumulus or cumulonimbus cloud. They are generally classified as non-supercellular tornadoes that develop over bodies of water. These spiraling columns of air frequently develop in tropical areas close to the equator, and are less common at high latitudes. Other tornado-like phenomena that exist in nature include the gustnado, dust devil, fire whirls, and steam devil (Fig. 10.1).

Fig. 10.1. A tornado near Anadarko, Oklahoma. The funnel itself is the thin tube reaching from the cloud to the ground. The lower part of this Tornado is surrounded by a translucent dust cloud, kicked up by the Tornado's strong winds at the surface. Note that the actual wind of the Tornado has a much wider radius than the funnel.

Tornadoes have been observed on every continent except Antarctica. However, the vast majority of Tornadoes in the world occur in the Tornado Alley region of the United States, although they can occur nearly anywhere in North America. They also occasionally occur in south-central and eastern Asia, the Philippines, northern and east-central South America, Southern Africa, northwestern and southeast Europe, western and southeastern Australia, and New Zealand. Tornadoes can be detected before or as they occur through the use of Pulse-Doppler radar by recognising patterns in velocity and reflectivity data, such as hook echoes, as well as by the efforts of storm spotters.

There are several different scales for rating the strength of Tornadoes. The Fujita scale rates Tornadoes by damage caused, and has been replaced in some countries by the updated Enhanced Fujita Scale. An F0 or EF0 Tornado, the weakest category, damages trees, but not substantial structures. An F5 or EF5 tornado, the strongest category, rips buildings off their foundations and can deform large skyscrapers. The similar TORRO scale ranges from a T0 for extremely weak Tornadoes to T11 for the most powerful known Tornadoes. Doppler radar data, photogrammetry, and ground swirl patterns (cycloidal marks) may also be analysed to determine intensity and assign a rating.

ETYMOLOGY

The word tornado is an altered form of the Spanish word tronada, which means 'thunderstorm'. This in turn was taken from the Latin tonare, meaning 'to thunder'. It most likely reached its present form through a combination of the Spanish Tronada and tornar ('to turn'); however, this may be a folk etymology. A Tornado is also commonly referred to as a 'twister', and is also sometimes referred to by the old-fashioned colloquial term cyclone. The term 'cyclone' is used as a synonym for 'Tornado' in the often-aired 1939 film, *The Wizard of Oz*. The term 'twister' is also used in that film, along with being the title of the 1996 Tornado-related film *Twister*.

DEFINITIONS

A Tornado is 'a violently rotating column of air, in contact with the ground, either pendant from a cumuliform cloud or underneath a cumuliform cloud, and often (but not always) visible as a funnel cloud'. For a vortex to be classified as a Tornado, it must be in contact with both the ground and the cloud base. Scientists have not yet created a complete definition of the word; for example, there is disagreement as to whether separate touchdowns of the same funnel constitute separate Tornadoes. Tornado refers to the vortex of wind, not the condensation cloud.

Funnel Bloud

A Tornado is not necessarily visible; however, the intense low pressure causing the high wind speeds (as described by Bernoulli's principle) and rapid rotation (due to cyclostrophic balance) usually causes water vapour in the air to become visible as a funnel cloud or condensation funnel.

There is some disagreement over the definition of funnel cloud and condensation funnel. According to the glossary of meteorology, a funnel cloud is any rotating cloud pendant from a cumulus or cumulonimbus, and thus most Tornadoes are included under this definition. Among many meteorologists, the funnel cloud term is strictly defined as a rotating cloud which is not associated with strong winds at the surface, and condensation funnel is a broad term for any rotating cloud below a cumuliform cloud (Fig. 10.2).

Fig. 10.2. This Tornado has no funnel cloud; however, the rotating dust cloud indicates that strong winds are occurring at the surface, and thus it is a true Tornado.

Tornadoes often begin as funnel clouds with no associated strong winds at the surface, although not all evolve into a Tornado. However, many Tornadoes are preceded by a funnel cloud. Most Tornadoes produce strong winds at the surface while the visible funnel is still above the ground, so it is difficult to discern the difference between a funnel cloud and a Tornado from a distance.

Outbreaks and Families

Occasionally, a single storm will produce more than one Tornado, either simultaneously or in succession. Multiple tornadoes produced by the same storm cell are referred to as a 'tornado family'. Several Tornadoes are sometimes spawned from the same large-scale storm system. If there is no break in activity, this is considered a Tornado outbreak, although there are various definitions. A period of several successive days with Tornado outbreaks in the same general area (spawned by multiple weather systems) is a Tornado outbreak sequence, occasionally called an extended Tornado outbreak.

CHARACTERISTICS

Size and Shape

Most Tornadoes take on the appearance of a narrow funnel, a few hundred yards (meters) across, with a small cloud of debris near the ground. Tornadoes may be obscured completely by rain or dust. These Tornadoes are especially dangerous, as even experienced meteorologists might not see them. Tornadoes can appear in many shapes and sizes. Small, relatively weak landspouts may be visible only as a small swirl of dust on the ground. Although the condensation funnel may not extend all the way to the ground, if associated surface winds are greater than 40 mph (64 km/hr), the circulation is considered a tornado. A tornado with a nearly cylindrical profile and relative low height is sometimes referred to as a 'stovepipe' Tornado. Large single-vortex Tornadoes can look like large wedges stuck into the ground, and so are known as 'wedge Tornadoes' or 'wedges'. The 'stovepipe' classification is also used for this type of tornado, if it otherwise fits that profile. A wedge can be so wide that it appears to be a block of dark clouds, wider than the distance from the cloud base to the ground. Even experienced storm observers may not be able to tell the difference between a low-hanging cloud and a wedge Tornado from a distance. Many, but not all major Tornadoes are wedges.

Tornadoes in the dissipating stage can resemble narrow tubes or ropes, and often curl or twist into complex shapes. These Tornadoes are said to be 'roping out', or becoming a 'rope Tornado'. When they rope out, the length of their funnel increases, which forces the winds within the funnel to weaken due to conservation of angular momentum. Multiple-vortex Tornadoes can appear as a family of swirls circling a common center, or may be completely obscured by condensation, dust, and debris, appearing to be a single funnel.

In the United States, Tornadoes are around 500 feet (150 m) across on average and stay on the ground for 5 miles (8 km). Yet, there is a wide range of Tornado sizes. Weak Tornadoes, or strong yet dissipating Tornadoes, can be exceedingly narrow, sometimes only a few feet or couple meters across. One Tornado was reported to have a damage path only 7 feet (2 m) long. On the other end of the spectrum, wedge Tornadoes can have a damage path a mile (1.6 km) wide or more. A Tornado that affected Hallam, Nebraska on May 22, 2004, was up to 2.5 miles (4 km) wide at the ground.

In terms of path length, the Tri-State Tornado, which affected parts of Missouri, Illinois, and Indiana on March 18, 1925, was on the ground continuously for 219 miles (352 km). Many Tornadoes which appear to have path lengths of 100 miles (160 km) or longer are composed of a family of Tornadoes which have formed in quick succession; however, there is no substantial evidence that this occurred in the case of the Tri-State Tornado. Modern reanalysis of the path suggests that the Tornado may have begun 15 miles (24 km) further west than previously thought, lengthening its track.

Appearance

Tornadoes can have a wide range of colours, depending on the environment in which they form. Those which form in a dry environment can be nearly invisible, marked only by swirling debris at the base of the funnel. Condensation funnels which pick up little or no debris can be grey to white. While traveling over a body of water as a waterspout, they can turn very white or even blue. Funnels which move slowly, ingesting a lot of debris and dirt, are usually darker, taking on the colour of debris. Tornadoes in the Great Plains can turn red because of the reddish tint of the soil, and Tornadoes in mountainous areas can travel over snow-covered ground, turning white.

Lighting conditions are a major factor in the appearance of a Tornado. A Tornado which is 'back-lit' (viewed with the sun behind it) appears very dark. The same Tornado, viewed with the sun at the observer's back, may appear gray or brilliant white. Tornadoes which occur near the time of sunset can be many different colours, appearing in hues of yellow, orange, and pink.

Dust kicked up by the winds of the parent thunderstorm, heavy rain and hail, and the darkness of night are all factors which can reduce the visibility of Tornadoes. Tornadoes occurring in these conditions are especially dangerous, since only weather radar observations, or possibly the sound of an approaching Tornado, serve as any warning to those in the storm's path. Most significant Tornadoes form under the storm's updraft base, which is rain-free, making them visible. Also, most Tornadoes occur in the late afternoon, when the bright sun can penetrate even the thickest clouds. Night-time Tornadoes are often illuminated by frequent lightning.

There is mounting evidence, including Doppler On Wheels mobile radar images and eyewitness accounts, that most Tornadoes have a clear, calm center with extremely low pressure, akin to the eye of tropical cyclones. This area would be clear (possibly full of dust), have relatively light winds, and be very dark, since the light would be blocked by swirling debris on the outside of the tornado. Lightning is said to be the source of illumination for those who claim to have seen the interior of a Tornado.

Rotation

Tornadoes normally rotate cyclonically in direction (counterclockwise in the northern hemisphere, clockwise in the southern). While large-scale storms always rotate cyclonically due to the Coriolis effect, thunderstorms and tornadoes are so small that the direct influence of the Coriolis effect is unimportant, as indicated by their large Rossby numbers. Supercells and Tornadoes rotate cyclonically in numerical simulations even when the Coriolis effect is neglected. Low-level mesocyclones and Tornadoes owe their rotation to complex processes within the supercell and ambient environment.

Approximately 1 per cent of Tornadoes rotate in an anticyclonic direction in the northern hemisphere. Typically, systems as weak as landspouts and gustnadoes can rotate anticyclonically, and usually only those which form on the anticyclonic shear side of the descending rear flank downdraft in a cyclonic supercell. On rare occasions, anticyclonic Tornadoes form in association with the mesoanticyclone of an anticyclonic supercell, in the same manner as the typical cyclonic Tornado, or as a companion Tornado either as a satellite tornado or associated with anticyclonic eddies within a supercell.

Sound and Seismology

Tornadoes emit widely on the acoustics spectrum and the sounds are caused by multiple mechanisms. Various sounds of tornadoes have been reported throughout time, mostly related to familiar sounds for the witness and generally some variation of a whooshing roar. Popularly reported sounds include a freight train, rushing rapids or waterfall, a nearby jet engine, or combinations of these. Many tornadoes are not audible from much distance; the nature and propagation distance of the audible sound depends on atmospheric conditions and topography.

The winds of the Tornado vortex and of constituent turbulent eddies, as well as airflow interaction with the surface and debris, contribute to the sounds. Funnel clouds also produce sounds. Funnel clouds and small tornadoes are reported as whistling, whining, humming, or the buzzing of innumerable bees or electricity, or more or less harmonic, whereas many Tornadoes are reported as a continuous, deep rumbling, or an irregular sound of 'noise'. Since many tornadoes are audible only when very near, sound is not reliable warning of a Tornado. And, any strong, damaging wind, even a severe hail volley or continuous thunder in a thunderstorm may produce a roaring sound.

Tornadoes also produce identifiable inaudible infrasonic signatures. Unlike audible signatures, Tornadic signatures have been isolated; due to the long distance propagation of low-frequency sound, efforts are ongoing to develop Tornado prediction and detection devices with additional value in understanding Tornado morphology, dynamics, and creation. Tornadoes also produce a detectable seismic signature, and research continues on isolating it and understanding the process.

Electromagnetic, Lightning and Other Effects

Tornadoes emit on the electromagnetic spectrum, with Sferics and E-field effects detected. There are observed correlations between Tornadoes and patterns of lightning. Tornadic storms do not contain more lightning than other storms and some tornadic cells never produce lightning. More often than not, overall cloud-to-ground (CG) lightning activity decreases as a tornado reaches the surface and returns to the baseline level when the tornado lifts. In many cases, intense tornadoes and thunderstorms exhibit an increased and anomalous dominance of positive polarity CG discharges. Electromagnetics and lightning have little or nothing to do directly with what drives tornadoes (tornadoes are basically a thermodynamic phenomenon), although there are likely connections with the storm and environment affecting both phenomena.

Luminosity has been reported in the past and is probably due to misidentification of external light sources such as lightning, city lights, and power flashes from broken lines, as internal sources are now uncommonly reported and are not known to ever have been recorded. In addition to winds, tornadoes also exhibit changes in atmospheric variables such as temperature, moisture, and pressure. For example, on June 24, 2003 near Manchester, South Dakota, a probe measured a 100 mbar (hPa) (2.95 inHg) pressure decrease. The pressure dropped gradually as the vortex approached then dropped extremely rapidly to 850 mbar (hPa) (25.10 inHg) in the core of the violent tornado before rising rapidly as the vortex moved away, resulting in a V-shape pressure trace. Temperature tends to decrease and moisture content to increase in the immediate vicinity of a tornado.

LIFE CYCLE

Supercell Relationship

Tornadoes often develop from a class of thunderstorms known as supercells. Supercells contain mesocyclones, an area of organised rotation a few miles up in the atmosphere, usually 1–6 miles (2–10 km) across. Most intense tornadoes (EF3 to EF5 on the Enhanced Fujita Scale) develop from supercells. In addition to tornadoes, very heavy rain, frequent lightning, strong wind gusts, and hail are common in such storms.

Most tornadoes from supercells follow a recognisable life cycle. That begins when increasing rainfall drags with it an area of quickly descending air known as the rear flank downdraft (RFD). This downdraft accelerates as it approaches the ground, and drags the supercell's rotating mesocyclone towards the ground with it.

Formation

As the mesocyclone approaches the ground, a visible condensation funnel appears to descend from the base of the storm, often from a rotating wall cloud. As the funnel descends, the RFD also reaches the ground, creating a gust front that can cause damage a good distance from the tornado. Usually, the funnel cloud becomes a tornado within minutes of the RFD reaching the ground.

Maturity

Initially, the tornado has a good source of warm, moist inflow to power it, so it grows until it reaches the 'mature stage'. This can last anywhere from a few minutes to more than an hour, and during that time a tornado often causes the most damage, and in rare cases can be more than one mile (1.6 km) across. Meanwhile, the RFD, now an area of cool surface winds, begins to wrap around the tornado, cutting off the inflow of warm air which feeds the tornado.

Demise

As the RFD completely wraps around and chokes off the tornado's air supply, the vortex begins to weaken, and become thin and rope-like. This is the 'dissipating stage'; often lasting no more than a few minutes, after which the tornado fizzles. During this stage the shape of the tornado becomes highly influenced by the winds of the parent storm, and can be blown into fantastic patterns. Even though the tornado is dissipating, it is still capable of causing damage. The storm is contracting into a rope-like tube and, like the ice skater who pulls her arms into spin faster, winds can increase at this point (Fig. 10.3).

Fig. 10.3. A sequence of images showing the birth of a tornado. First, the rotating cloud base lowers. This lowering becomes a funnel, which continues descending while winds build near the surface, kicking up dust and other debris. Finally, the visible funnel extends to the ground, and the tornado begins causing major damage. This Tornado, near Dimmitt, Texas, was one of the best-observed violent tornadoes in history.

As the Tornado enters the dissipating stage, its associated mesocyclone often weakens as well, as the rear flank downdraft cuts off the inflow powering it. In particular, intense supercells tornadoes can develop cyclically. As the first mesocyclone and associated Tornado dissipate, the storm's inflow may be concentrated into a new area closer to the center of the storm. If a new mesocyclone develops, the cycle may start again, producing one or more new tornadoes. Occasionally, the old (occluded) mesocyclone and the new mesocyclone produce a tornado at the same time.

Although this is a widely accepted theory for how most tornadoes form, live, and die, it does not explain the formation of smaller Tornadoes, such as landspouts, long-lived tornadoes, or tornadoes with multiple vortices. These each have different mechanisms which influence their development—however, most Tornadoes follow a pattern similar to this one.

TYPES OF TORNADO

Multiple Vortex

A multiple-vortex tornado is a type of tornado in which two or more columns of spinning air rotate around a common center. Multivortex structure can occur in almost any circulation, but is very often observed in intense tornadoes. These vortices often create small areas of heavier damage along the main tornado path. This is a distinct phenomenon from a satellite tornado, which is a weaker tornado which forms very near a large, strong tornado contained within the same mesocyclone. The satellite tornado may appear to 'orbit' the larger tornado (hence the name), giving the appearance of one, large multi-vortex tornado. However, a satellite tornado is a distinct circulation, and is much smaller than the main funnel.

Waterspout

A waterspout is defined by the National Weather Service as a tornado over water. However, researchers typically distinguish 'fair weather' waterspouts from tornadic waterspouts. Fair weather waterspouts are less severe but far more common, and are similar to dust devils and landspouts. They form at the bases of cumulus congestus clouds over tropical and subtropical waters. They have relatively weak winds, smooth laminar walls, and typically travel very slowly. They occur most commonly in the Florida Keys and in the northern Adriatic Sea. In contrast, tornadic waterspouts are stronger tornadoes over water. They form over water similarly to mesocyclonic tornadoes, or are stronger tornadoes which cross over water. Since they form from severe thunderstorms and can be far more intense, faster, and longer-lived than fair weather waterspouts, they are more dangerous (Fig. 10.4).

Fig. 10.4. A waterspout near the Florida Keys.

Landspout

A landspout, or dust-tube tornado, is a tornado not associated with a mesocyclone. The name stems from their characterisation as a 'fair weather waterspout on land'. Waterspouts and landspouts share many defining characteristics, including relative weakness, short lifespan, and a small, smooth condensation funnel which often does not reach the surface. Landspouts also create a distinctively laminar cloud of dust when they make contact with the ground, due to their differing mechanics from true mesoform tornadoes. Though usually weaker than classic tornadoes, they can produce strong winds which could cause serious damage.

Similar Circulations

Gustnado

A gustnado, or gust front tornado, is a small, vertical swirl associated with a gust front or downburst. Because they are not connected with a cloud base, there is some debate as to whether or not gustnadoes are tornadoes. They are formed when fast moving cold, dry outflow air from a thunderstorm is blown through a mass of stationary, warm, moist air near the outflow boundary, resulting in a 'rolling' effect (often exemplified through a roll cloud). If low level wind shear is strong enough, the rotation can be turned vertically or diagonally and make contact with the ground. The result is a gustnado. They usually cause small areas of heavier rotational wind damage among areas of straight-line wind damage (Fig. 10.5).

Fig. 10.5. A dust devil in Nevada.

Dust devil

A dust devil resembles a tornado in that it is a vertical swirling column of air. However, they form under clear skies and are no stronger than the weakest tornadoes. They form when a strong convective updraft is formed near the ground on a hot day. If there is enough low level wind shear, the column of hot, rising air can develop a small cyclonic motion that can be seen near the ground. They are not considered tornadoes because they form during fair weather and are not associated with any clouds. However, they can, on occasion, result in major damage in arid areas.

Fire whirls and steam devils

Small-scale, tornado-like circulations can occur near any intense surface heat source. Those that occur near intense wildfires are called fire whirls. They are not considered tornadoes, except in the rare case where they connect to a pyrocumulus or other cumuliform cloud above. Fire whirls usually are not as strong as tornadoes associated with thunderstorms. They can, however, produce significant damage. A steam devil is a rotating updraft that involves steam or smoke. Steam devils are very rare. They most often form from smoke issuing from a power plant smokestack. Hot springs and deserts may also be suitable locations for a steam devil to form. The phenomenon can occur over water, when cold arctic air passes over relatively warm water.

INTENSITY AND DAMAGE

The Fujita scale and the Enhanced Fujita Scale rate tornadoes by damage caused. The Enhanced Fujita (EF) Scale was an upgrade to the older Fujita scale, by expert elicitation, using engineered wind estimates and better damage descriptions. The EF Scale was designed so that a tornado rated on the Fujita scale would receive the same numerical rating, and was implemented starting in the United States in 2007. An EF0 tornado will probably damage trees but not substantial structures, whereas an EF5 tornado can rip buildings off their foundations leaving them bare and even deform large skyscrapers. The similar TORRO scale ranges from a T0 for extremely weak tornadoes to T11 for the most powerful known tornadoes. Doppler radar data, photogrammetry, and ground swirl patterns (cycloidal marks) may also be analysed to determine intensity and award a rating. Tornadoes vary in intensity regardless of shape, size, and location, though strong tornadoes are typically larger than weak tornadoes. The association with track length and duration also varies, although longer track tornadoes tend to be stronger. In the case of violent tornadoes, only a small portion of the path is of violent intensity, most of the higher intensity from subvortices.

In the United States, 80 per cent of tornadoes are EF0 and EF1 (T0 through T3) tornadoes. The rate of occurrence drops off quickly with increasing strength — less than 1 per cent are violent tornadoes (EF4, T8 or stronger). Outside Tornado Alley, and North America in general, violent tornadoes are extremely rare. This is apparently mostly due to the lesser number of tornadoes overall, as research shows that tornado intensity distributions are fairly similar worldwide. A few significant tornadoes occur annually in Europe, Asia, southern Africa, and southeastern South America, respectively.

CLIMATOLOGY

The United States has the most tornadoes of any country, nearly four times more than estimated in all of Europe, excluding waterspouts. This is mostly due to the unique geography of the continent. North America is a large continent that extends from the tropics north into arctic areas, and has no major east-west mountain range to block air flow between these two areas. In the middle latitudes, where most tornadoes of the world occur, the Rocky Mountains block moisture and buckle the atmospheric flow, forcing drier air at mid-levels of the troposphere due to downsloped winds, and causing the formation of a low pressure area downwind to the east of the mountains. Increased westerly flow off the Rockies force the formation of a dry line when the flow aloft is strong, while the Gulf of Mexico fuels abundant low-level moisture in the southerly flow to its east. This unique topography allows for frequent collisions of warm and cold air, the conditions that breed strong, long-lived storms throughout the year. A large portion of these tornadoes form in an area of the central United States known as Tornado Alley. This area extends into Canada, particularly Ontario and the Prairie Provinces, although southeast Quebec, interior British Columbia, and western New Brunswick are also tornado-prone. Tornadoes also occur across northeastern Mexico.

The United States averages about 1200 tornadoes per year. The Netherlands has the highest average number of recorded tornadoes per area of any country [more than 20, or 0.0013 per sq mi (0.00048 per km^2), annually], followed by the UK [around 33, or 0.00035 per sq mi (0.00013 per km²), per year], but most are small and cause minor damage. In absolute number of events, ignoring area, the UK experiences more tornadoes than any other European country, excluding waterspouts.

Tornadoes kill an average of 179 people per year in Bangladesh, the most in the world. This is due to high population density, poor quality of construction and lack of tornado safety knowledge, as well as other factors. Other areas of the world that have frequent tornadoes include South Africa, parts of Argentina, Paraguay, and southern Brazil, as well as portions of Europe, Australia and New Zealand, and far eastern Asia.

Tornadoes are most common in spring and least common in winter. Spring and fall experience peaks of activity as those are the seasons when stronger winds, wind shear, and atmospheric instability are present. Tornadoes are focused in the right front quadrant of landfalling tropical cyclones, which tend to occur in the late summer and autumn. Tornadoes can also be spawned as a result of eyewall mesovortices, which persist until landfall. Favourable conditions can occur any time of the year.

Tornado occurrence is highly dependent on the time of day, because of solar heating. Worldwide, most tornadoes occur in the late afternoon, between 3 pm and 7 pm local time, with a peak near 5 pm. Destructive tornadoes can occur at any time of day. The Gainesville Tornado of 1936, one of the deadliest tornadoes in history, occurred at 8:30 am local time.

Associations with Climate and Climate Change

Associations to various climate and environmental trends exist. For example, an increase in the sea surface temperature of a source region increases atmospheric moisture content. Increased moisture can fuel an increase in severe weather and tornado activity, particularly in the cool season.

Some evidence does suggest that the Southern Oscillation is weakly correlated with changes in tornado activity, which vary by season and region, as well as whether the ENSO phase is that of El Niño or La Niña.

Climatic shifts may affect tornadoes via teleconnections in shifting the jet stream and the larger weather patterns. The climate-tornado link is confounded by the forces affecting larger patterns and by the local, nuanced nature of tornadoes. Although it is reasonable that global warming may affect trends in tornado activity, any such effect is not yet identifiable due to the complexity, local nature of the storms, and database quality issues. Any effect would vary by region.

DETECTION

Rigorous attempts to warn of tornadoes began in the United States in the mid-20th century. Before the 1950s, the only method of detecting a tornado was by someone seeing it on the ground. Often, news of a tornado would reach a local weather office after the storm. However, with the advent of weather radar, areas near a local office could get advance warning of severe weather. The first public tornado warnings were issued in 1950 and the first tornado watches and convective outlooks in 1952. In 1953 it was confirmed that hook echoes are associated with tornadoes. By recognising these radar signatures, meteorologists could detect thunderstorms probably producing tornadoes from dozens of miles away.

Radar

Today, most developed countries have a network of weather radars, which remains the main method of detecting signatures probably associated with tornadoes. In the United States and a few other countries, Doppler weather radar stations are used. These devices measure the velocity and radial direction (towards or away from the radar) of the winds in a storm, and so can spot evidence of rotation in storms from more than a hundred miles (160 km) away. When storms are distant from a radar, only areas high within the storm are observed and the important areas below are not sampled. Data resolution also decreases with distance from the radar. Some meteorological situations leading to tornadogenesis are not readily detectable by radar and on occasion tornado development may occur more quickly than radar can complete a scan and send the batch of data. Also, most populated areas on earth are now visible from the Geostationary Operational Environmental Satellites (GOES), which aid in the nowcasting of tornadic storms.

Storm Spotting

In the mid-1970s, the US National Weather Service (NWS) increased its efforts to train storm spotters to spot key features of storms which indicate severe hail, damaging winds, and tornadoes, as well as damage itself and flash flooding. The program was called Skywarn, and the spotters were local sheriff's deputies, state troopers, firefighters, ambulance drivers, amateur radio operators, civil defense (now emergency management) spotters, storm chasers, and ordinary citizens. When severe weather is anticipated, local weather service offices request that these spotters look out for severe weather, and report any tornadoes immediately, so that the office can warn of the hazard.

Usually spotters are trained by the NWS on behalf of their respective organisations, and report to them. The organisations activate public warning systems such as sirens and the emergency alert system, and forward the report to the NWS. There are more than 2,30,000 trained Skywarn weather spotters across the United States. In Canada, a similar network of volunteer weather watchers, called Canwarn, helps spot severe weather, with more than 1000 volunteers. In Europe, several nations are organising spotter networks under the auspices of Skywarn Europe and the Tornado and Storm Research Organisation (TORRO) has maintained a network of spotters in the United Kingdom since 1974. Storm spotters are needed because radar systems such as NEXRAD do not detect a tornado; merely signatures which hint at the presence of tornadoes. Radar may give a warning before there is any visual evidence of a tornado or imminent tornado, but ground truth from an observer can either verify the threat or determine that a tornado is not imminent. The spotter's ability to see what radar cannot is especially important as distance from the radar site increases, because the radar beam becomes progressively higher in altitude further away from the radar, chiefly due to curvature of earth, and the beam also spreads out.

Visual Evidence

Storm spotters are trained to discern whether a storm seen from a distance is a supercell. They typically look to its rear, the main region of updraft and inflow. Under the updraft is a rain-free base, and the next step of tornadogenesis is the formation of a rotating wall cloud. The vast majority of intense tornadoes occur with a wall cloud on the backside of a supercell.

Evidence of a supercell comes from the storm's shape and structure, and cloud tower features such as a hard and vigorous updraft tower, a persistent, large overshooting top, a hard anvil (especially when backsheared against strong upper level winds), and a corkscrew look or striations. Under the storm and closer to where most tornadoes are found, evidence of a supercell and likelihood of a tornado includes inflow bands (particularly when curved) such as a 'beaver tail', and other clues such as strength of inflow, warmth and moistness of inflow air, how outflow- or inflow-dominant a storm appears, and how far is the front flank precipitation core from the wall cloud. Tornadogenesis is most likely at the interface of the updraft and rear flank downdraft, and requires a balance between the outflow and inflow.

Only wall clouds that rotate spawn tornadoes, and usually precede the tornado by five to thirty minutes. Rotating wall clouds are the visual manifestation of a mesocyclone. Barring a low-level boundary, tornadogenesis is highly unlikely unless a rear flank downdraft occurs, which is usually visibly evidenced by evaporation of cloud adjacent to a corner of a wall cloud. A tornado often occurs as this happens or shortly after; first, a funnel cloud dips and in nearly all cases by the time it reaches halfway down, a surface swirl has already developed, signifying a tornado is on the ground before condensation connects the surface circulation to the storm. Tornadoes may also occur without wall clouds, under flanking lines, and on the leading edge. Spotters watch all areas of a storm, and the cloud base and surface.

EXTREMES

The most extreme tornado in recorded history was the Tri-State Tornado, which roared through parts of Missouri, Illinois, and Indiana on March 18, 1925. It was likely an F5, though tornadoes were not ranked on any scale in that era. It holds records for longest path length (219 miles, 352 km), longest duration (about 3.5 hrs), and fastest forward speed for a significant tornado (73 mph, 117 km/hr) anywhere on earth. In addition, it is the deadliest single tornado in United States history (695 dead). The tornado was also the second costliest tornado in history at the time, but in the years since has been surpassed by several others if population changes over time are not considered. When costs are normalised for wealth and inflation, it ranks third today. The deadliest tornado in world history was the Daultipur-Salturia Tornado in Bangladesh on April 26, 1989, which killed approximately 1300 people. Bangladesh has had at least 19 tornadoes in its history kill more than 100 people, almost half of the total in the rest of the world.

The most extensive tornado outbreak on record was the Super Outbreak, which affected a large area of the central United States and extreme southern Ontario in Canada on April 3 and 4, 1974. Not only did this outbreak feature 148 tornadoes in 18 hours, but many were violent; six were of F5 intensity, and twenty-four peaked at F4 strength. This outbreak had sixteen tornadoes on the ground at the same time during its peak. More than 300 people, possibly as many as 330, were killed by tornadoes during this outbreak.

While it is nearly impossible to directly measure the most violent tornado wind speeds (conventional anemometers would be destroyed by the intense winds), some tornadoes have been scanned by mobile Doppler radar units, which can provide a good estimate of the tornado's winds.

The highest wind speed ever measured in a tornado, which is also the highest wind speed ever recorded on the planet, is 301 ± 20 mph (484 ± 32 km/hr) in the F5 Bridge Creek-Moore, Oklahoma tornado which killed 36 people. Though the reading was taken about 100 feet (30 m) above the ground, this is a testament to the power of the strongest tornadoes.

Storms that produce tornadoes can feature intense updrafts, sometimes exceeding 150 mph (240 km/hr). Debris from a tornado can be lofted into the parent storm and carried a very long distance. A tornado which affected Great Bend, Kansas in November 1915, was an extreme case, where a 'rain of debris' occurred 80 miles (130 km) from the town, a sack of flour was found 110 miles (177 km) away, and a cancelled check from the Great Bend bank was found in a field outside of Palmyra, Nebraska, 305 miles (491 km) to the northeast. Waterspouts and tornadoes have been advanced as an explanation for instances of raining fish and other animals.

SAFETY

Though tornadoes can strike in an instant, there are precautions and preventative measures that people can take to increase the chances of surviving a tornado. Authorities such as the storm prediction center advise having a pre-determined plan should a tornado warning be issued. When a warning is issued, going to a basement or an interior first-floor room of a sturdy building greatly increases chances of survival. In tornado-prone areas, many buildings have storm cellars on the property. These underground refuges have saved thousands of lives.

Some countries have meteorological agencies which distribute tornado forecasts and increase levels of alert of a possible tornado (such as tornado watches and warnings in the United States and Canada). Weather radios provide an alarm when a severe weather advisory is issued for the local area, though these are mainly available only in the United States. Unless the tornado is far away and highly visible, meteorologists advise that drivers park their vehicles far to the side of the road (so as not to block emergency traffic), and find a sturdy shelter. If no sturdy shelter is nearby, getting low in a ditch is the next best option. Highway overpasses are one of the worst places to take shelter during tornadoes, as they are believed to create a venturi effect, increasing the danger from the tornado by increasing the wind speed and funneling debris underneath the overpass.

MYTHS AND MISCONCEPTIONS

Folklore often identifies a green sky with tornadoes, and though the phenomenon may be associated with severe weather, there is no evidence linking it specifically with tornadoes. It is often thought that opening windows will lessen the damage caused by the tornado. While there is a large drop in atmospheric pressure inside a strong tornado, it is unlikely that the pressure drop would be enough to cause the house to explode. Some research indicates that opening windows may actually increase the severity of the tornado's damage. A violent tornado can destroy a house whether its windows are open or closed (Fig. 10.6).

Fig. 10.6. Salt Lake City Tornado, August 11, 1999. This tornado disproved several misconceptions, including the idea that tornadoes cannot occur in areas like Utah or in cities.

Another commonly held belief is that highway overpasses provide adequate shelter from tornadoes. On the contrary, a highway overpass is a dangerous place during a tornado. In the 1999 Oklahoma tornado outbreak of May 3, 1999, three highway overpasses were directly struck by tornadoes, and at all three locations there was a fatality, along with many life-threatening injuries. The small area under the overpasses is believed to cause a Venturi effect. By comparison, during the same tornado outbreak, more than 2000 homes were completely destroyed, with another 7000 damaged, and yet only a few dozen people died in their homes.

An old belief is that the southwest corner of a basement provides the most protection during a tornado. The safest place is the side or corner of an underground room opposite the tornado's direction of approach (usually the northeast corner) or the central-most room on the lowest floor. Taking shelter in a basement, under a staircase, or under a sturdy piece of furniture such as a workbench further increases chances of survival. Finally, there are areas which people believe to be protected from tornadoes, whether by being in a city, near a major river, hill, or mountain, or even protected by supernatural forces. Tornadoes have been known to cross major rivers, climb mountains, affect valleys, and have damaged several city centers. As a general rule, no area is 'safe' from tornadoes, though some areas are more susceptible than others.

ONGOING RESEARCH

Meteorology is a relatively young science and the study of tornadoes is newer still. Although researched for about 140 years and intensively for around 60 years, there are still aspects of tornadoes which remain a mystery. Scientists have a fairly good understanding of the development of thunderstorms and mesocyclones, and the meteorological conditions conducive to their formation. However, the step from supercell (or other respective formative processes) to tornadogenesis and predicting tornadic vs. non-tornadic mesocyclones is not yet well known and is the focus of much research (Fig. 10.7).

Fig. 10.7. A Doppler on wheels unit observing a tornado near Attica, Kansas.

Also under study are the low-level mesocyclone and the stretching of low-level vorticity which tightens into a tornado, namely, what are the processes and what is the relationship of the environment and the convective storm. Intense tornadoes have been observed forming simultaneously with a mesocyclone aloft (rather than succeeding mesocyclogenesis) and some intense tornadoes have occurred without a mid-level mesocyclone. In particular, the role of downdrafts, particularly the rear-flank downdraft, and the role of baroclinic boundaries, are intense areas of study.

Reliably predicting tornado intensity and longevity remains a problem, as do details affecting characteristics of a tornado during its life cycle and tornadolysis. Other rich areas of research are tornadoes associated with mesovortices within linear thunderstorm structures and within tropical cyclones.

Scientists still do not know the exact mechanisms by which most tornadoes form, and occasional tornadoes still strike without a tornado warning being issued. Analysis of observations including both stationary and mobile (surface and aerial) *in situ* and remote sensing (passive and active) instruments generates new ideas and refines existing notions. Numerical modelling also provides new insights as observations and new discoveries are integrated into our physical understanding and then tested in computer simulations which validate new notions as well as produce entirely new theoretical findings, many of which are otherwise unattainable. Importantly, development of new observation technologies and installation of finer spatial and temporal resolution observation networks have aided increased understanding and better predictions.

Research programs, including field projects such as VORTEX (Verification of the Origins of Rotation in Tornadoes Experiment), deployment of TOTO (the TOtable Tornado Observatory), Doppler on wheels (DOW), and dozens of other programs, hope to solve many questions that still plague meteorologists. Universities, government agencies such as the National Severe Storms Laboratory, private-sector meteorologists, and the National Center for Atmospheric Research are some of the organisations very active in research; with various sources of funding, both private and public, a chief entity being the National Science Foundation.

Chapter 11

Tsunami

INTRODUCTION

A tsunami or tidal wave is a series of water waves (called a tsunami wave train) caused by the displacement of a large volume of a body of water, usually an ocean, but can occur in large lakes. Tsunamis are a frequent occurrence in Japan; approximately 195 events have been recorded. Due to the immense volumes of water and energy involved, tsunamis can devastate coastal regions.

Earthquakes, volcanic eruptions and other underwater explosions (including detonations of underwater nuclear devices), landslides and other mass movements, meteorite ocean impacts or similar impact events, and other disturbances above or below water all have the potential to generate a tsunami.

The Greek historian Thucydides was the first to relate tsunami to submarine earthquakes, but understanding of tsunami's nature remained slim until the 20th century and is the subject of ongoing research. Many early geological, geographical, and oceanographic texts refer to tsunamis as 'seismic sea waves'.

Some meteorological conditions, such as deep depressions that cause tropical cyclones, can generate a storm surge, called a meteotsunami, which can raise tides several metres above normal levels. The displacement comes from low atmospheric pressure within the centre of the depression. As these storm surges reach shore, they may resemble (though are not) tsunamis, inundating vast areas of land. Such a storm surge inundated Burma in May 2008.

ETYMOLOGY

Tsunami are sometimes referred to as tidal waves. In recent years, this term has fallen out of favour, especially in the scientific community, because tsunami actually have nothing to do with tides. The once-popular term derives from their most common appearance, which is that of an extraordinarily high tidal bore. Tsunami and tides both produce waves of water that move inland, but in the case of tsunami the inland movement of water is much greater and lasts for a longer period, giving the impression of an incredibly high tide. Although the meanings of 'tidal' include 'resembling' or 'having the form or character of' the tides, and the term tsunami is no more accurate because tsunami are not limited to harbours, use of the term tidal wave is discouraged by geologists and oceanographers.

GENERATION MECHANISMS

The principal generation mechanism (or cause) of a tsunami is the displacement of a substantial volume of water or perturbation of the sea. This displacement of water is usually attributed to either earthquakes, landslides, volcanic eruptions, or more rarely by meteorites and nuclear tests. The waves formed in this way are then sustained by gravity.

It is important to note that tides do not play any part in the generation of tsunamis, hence referring to tsunamis as 'tidal waves' is inaccurate.

Seismicity Generated Tsunamis

Tsunamis can be generated when the sea floor abruptly deforms and vertically displaces the overlying water. Tectonic earthquakes are a particular kind of earthquake that are associated with the earth's crustal deformation; when these earthquakes occur beneath the sea, the water above the deformed area is displaced from its equilibrium position. More specifically, a tsunami can be generated when thrust faults associated with convergent or destructive plate boundaries move abruptly, resulting in water displacement, due to the vertical component of movement involved. Movement on normal faults will also cause displacement of the seabed, but the size of the largest of such events is normally too small to give rise to a significant tsunami.

Tsunamis have a small amplitude (wave height) offshore, and a very long wavelength (often hundreds of kilometers long), which is why they generally pass unnoticed at sea, forming only a slight swell usually about 300 millimetres (12 in) above the normal sea surface. They grow in height when they reach shallower water, in a wave shoaling process described below. A tsunami can occur in any tidal state and even at low tide can still inundate coastal areas.

On April 1, 1946, a magnitude-7.8 (Richter Scale) earthquake occurred near the Aleutian Islands, Alaska. It generated a tsunami which inundated Hilo on the island of Hawai'i with a 14 metres (46 ft) high surge. The area where the earthquake occurred is where the Pacific Ocean floor is subducting (or being pushed downwards) under Alaska.

Examples of tsunami at locations away from convergent boundaries include Storegga about 8000 years ago, Grand Banks 1929, Papua New Guinea 1998 (Tappin, 2001). The Grand Banks and Papua New Guinea tsunamis came from earthquakes which destabilised sediments, causing them to flow into the ocean and generate a tsunami. They dissipated before traveling transoceanic distances.

The cause of the Storegga sediment failure is unknown. Possibilities include an overloading of the sediments, an earthquake or a release of gas hydrates (methane, etc.).

The 1960 Valdivia earthquake (M_w 9.5) (19:11 hrs UTC), 1964 Alaska earthquake (M_w 9.2), and 2004 Indian Ocean earthquake (M_w 9.2) (00:58:53 UTC) are recent examples of powerful megathrust earthquakes that generated tsunamis (known as teletsunamis) that can cross entire oceans. Smaller (M_w 4.2) earthquakes in Japan can trigger tsunamis (called local and regional tsunamis) that can only devastate nearby coasts, but can do so in only a few minutes.

In the 1950s, it was discovered that larger tsunamis than had previously been believed possible could be caused by giant landslides. These phenomena rapidly displace large water volumes, as energy from falling debris or expansion transfers to the water at a rate faster than the water can absorb. Their existence was confirmed in 1958, when a giant landslide in Lituya Bay, Alaska, caused the highest wave ever recorded, which had a height of 524 metres (over 1700 feet). The wave didn't travel far, as it struck land almost immediately.

Two people fishing in the bay were killed, but another boat amazingly managed to ride the wave. Scientists named these waves megatsunami. Scientists discovered that extremely large landslides from volcanic island collapses can generate megatsunami, that can travel trans-oceanic distances. Some of Figures, of tsunamis are shown in Figs 11.1 and 11.2.

Fig. 11.1. Destruction made by tsunami.

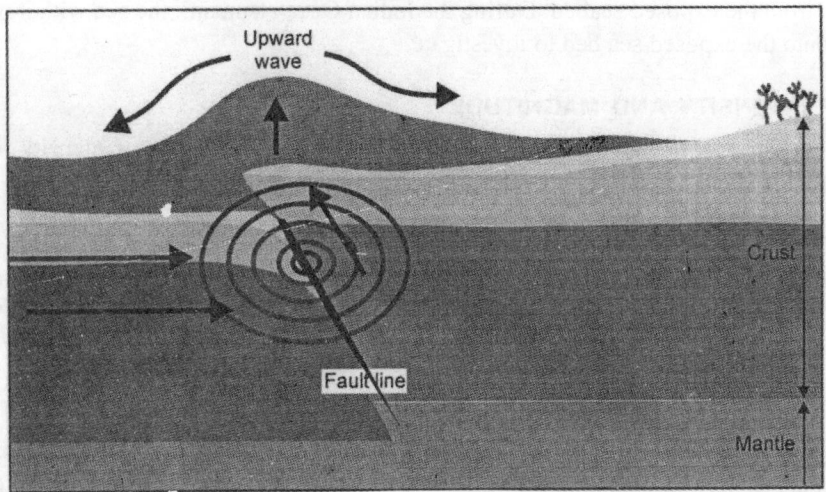

Fig. 11.2. Advance alarm system by tsunami.

CHARACTERISTICS OF TSUNAMI

While everyday wind waves have a wavelength (from crest to crest) of about 100 metres (330 ft) and a height of roughly 2 metres (6.6 ft), a tsunami in the deep ocean has a wavelength of about 200 km (120 mi). Such a wave travels at well over 800 km per hour (500 mph), but due to the enormous wavelength the wave oscillation at any given point takes 20 or 30 minutes to complete a cycle and has an amplitude of only about 1 metre (3.3 ft). This makes tsunamis difficult to detect over deep water. Ships rarely notice their passage. As the tsunami approaches the coast and the waters become shallow, wave shoaling compresses the wave and its velocity slows below 80 kilometres per hour (50 mph). Its wavelength diminishes to less than 20 km (12 mi) and its amplitude grows enormously, producing a distinctly visible wave. Since the wave still has such a long wavelength, the tsunami may take minutes to reach full height. Except for the very largest tsunamis, the approaching wave does not break (like a surf break), but rather appears like a fast moving tidal bore. Open bays and coastlines adjacent to very deep water may shape the tsunami further into a step-like wave with a steep-breaking front. When the tsunami's wave peak reaches the shore, the resulting temporary rise in sea level is termed 'run up'. Run up is measured in metres above a reference sea level. A large tsunami may feature multiple waves arriving over a period of hours, with significant time between the wave crests. The first wave to reach the shore may not have the highest run up. About 80 per cent of tsunamis occur in the Pacific Ocean, but are possible wherever there are large bodies of water, including lakes. They are caused by earthquakes, landslides, volcanic explosions, and bolides.

DRAWBACK OF TSUNAMI

If the first part of a tsunami to reach land is a trough—called a drawback—rather than a wave crest, the water along the shoreline recedes dramatically, exposing normally submerged areas. A drawback occurs because the water propagates outwards with the trough of the wave at its front. Drawback begins before the wave arrives at an interval equal to half of the wave's period. Drawback can exceed hundreds of metres, and people unaware of the danger sometimes remain near the shore to satisfy their curiosity or to collect fish from the exposed seabed. During the Indian Ocean tsunami, the sea withdrew and many people went onto the exposed sea bed to investigate.

SCALES OF INTENSITY AND MAGNITUDE

As with earthquakes, several attempts have been made to set up scales of tsunami intensity or magnitude to allow comparison between different events.

Intensity Scales

The first scales used routinely to measure the intensity of tsunami were the Sieberg-Ambraseys scale, used in the Mediterranean Sea and the *Imamura-Iida intensity scale*, used in the Pacific Ocean. The latter scale was modified by Soloviev, who calculated the Tsunami intensity I according to the formula:

$$I = \frac{1}{2} + \log_2 H_{av}$$

where, H_{av} is the average wave height along the nearest coast. This scale, known as the *Soloviev-Imamura tsunami intensity scale*, is used in the global tsunami catalogues compiled by the NGDC/NOAA and the Novosibirsk Tsunami Laboratory as the main parameter for the size of the tsunami.

Magnitude Scales

The first scale that genuinely calculated a magnitude for a tsunami, rather than an intensity at a particular location was the ML scale proposed by Murty and Loomis based on the potential energy. Difficulties in calculating the potential energy of the tsunami mean that this scale is rarely used. Abe introduced the *tsunami magnitude scale M_t*, calculated from,

$$M_t = a \log h + b \log R = D$$

where, h is the maximum tsunami-wave amplitude (in m) measured by a tide gauge at a distance R from the epicenter, a, b and D are constants used to make the M_t scale match as closely as possible with the moment magnitude scale.

WARNINGS AND PREDICTIONS

Drawbacks can serve as a brief warning. People who observe drawback (many survivors report an accompanying sucking sound), can survive only if they immediately run for high ground or seek the upper floors of nearby buildings. In 2004, ten-year old Tilly Smith of Surrey, England, was on Maikhao beach in Phuket, Thailand with her parents and sister, and having learned about tsunamis recently in school, told her family that a tsunami might be imminent. Her parents warned others minutes before the wave arrived, saving dozens of lives. She credited her geography teacher, Andrew Kearney (Fig. 11.3).

Fig. 11.3. One of the deep water buoys used in the DART tsunami warning system.

In the 2004 Indian Ocean tsunami drawback was not reported on the African coast or any other eastern coasts it reached. This was because the wave moved downwards on the eastern side of the fault line and upwards on the western side. The western pulse hit coastal Africa and other western areas. A tsunami cannot be precisely predicted, even if the magnitude and location of an earthquake is known. Geologists, oceanographers, and seismologists analyse each earthquake and based on many factors may or may not issue a tsunami warning. However, there are some warning signs of an impending tsunami, and automated systems can provide warnings immediately after an earthquake in time to save lives. One of the most successful systems uses bottom pressure sensors that are attached to buoys. The sensors constantly monitor the pressure of the overlying water column. This is deduced through the calculation:

$$P = \rho g h$$

where,

P = the overlying pressure in newtons per metre square.

ρ = the density of the seawater = 1.1×10^3 kg/m^3.

g = the acceleration due to gravity = 9.8 m/s^2.

h = the height of the water column in metres.

Hence for a water column of 5000 m depth the overlying pressure is equal to:

$$P = \rho g h \left(1.1 \times 10^3 \frac{kg}{m^3} \right) \left(9.8 \frac{m}{s^2} \right) \left(5.0 \times 10^3 m \right) = 5.4 \times 10^7 \frac{N}{m^2} = 54 MPa$$

or about 5500 tons-force per square metre.

Regions with a high tsunami risk typically use tsunami warning systems to warn the population before the wave reaches land. On the west coast of the United States, which is prone to Pacific Ocean tsunami, warning signs indicate evacuation routes. In Japan, the community is well-educated about earthquakes and tsunamis, and along the Japanese shorelines the tsunami warning signs are reminders of the natural hazards together with a network of warning sirens, typically at the top of the cliff of surroundings hills. The Pacific Tsunami Warning System is based in Honolulu, Hawai'i. It monitors Pacific Ocean seismic activity. A sufficiently large earthquake magnitude and other information triggers a tsunami warning. While the subduction zones around the Pacific are seismically active, not all earthquakes generate tsunami. Computers assist in analysing the tsunami risk of every earthquake that occurs in the Pacific Ocean and the adjoining land masses. As a direct result of the Indian Ocean tsunami, a re-appraisal of the tsunami threat for all coastal areas is being undertaken by national governments and the United Nations Disaster Mitigation Committee. A tsunami warning system is being installed in the Indian Ocean. Computer models can predict tsunami arrival, usually within minutes of the arrival time. Bottom pressure sensors relay information in real time. Based on these pressure readings and other seismic information and the seafloor's shape (bathymetry) and coastal topography, the models estimate the amplitude and surge height of the approaching tsunami. All Pacific Rim countries collaborate in the Tsunami warning system and most regularly practice evacuation and other procedures. In Japan, such preparation is mandatory for government, local authorities, emergency services and the population. Some zoologists hypothesise that some animal species have an ability to sense subsonic Rayleigh waves from an earthquake or a tsunami. If correct, monitoring their behaviour could provide advance warning of earthquakes, tsunami, etc. However, the evidence is controversial and is not widely accepted. There are unsubstantiated claims about the Lisbon quake that some animals escaped to higher ground, while many other animals in the same areas drowned. The phenomenon was also noted by media sources in Sri Lanka in the 2004 Indian Ocean earthquake. It is possible that certain animals (e.g. elephants) may have heard the sounds of the tsunami as it approached the coast. The elephants' reaction was to move away from the approaching noise. By contrast, some humans went to the shore to investigate and many drowned as a result (Fig. 11.4).

Fig. 11.4. A seawall at Tsu, Japan.

It is not possible to prevent a tsunami. However, in some tsunami-prone countries some earthquake engineering measures have been taken to reduce the damage caused on shore. Japan built many tsunami walls of up to 4.5 metres (15 ft) to protect populated coastal areas. Other localities have built floodgates and channels to redirect the water from incoming tsunami. However, their effectiveness has been questioned, as tsunami often overtop the barriers.

For instance, the Okushiri, Hokkaido tsunami which struck Okushiri Island of Hokkaido within two to five minutes of the earthquake on July 12, 1993 created waves as much as 30 metres (100 ft) tall—as high as a 10-storey building. The port town of Aonae was completely surrounded by a tsunami wall, but the waves washed right over the wall and destroyed all the wood-framed structures in the area. The wall may have succeeded in slowing down and moderating the height of the tsunami, but it did not prevent major destruction and loss of life. Natural factors such as shoreline tree cover can mitigate tsunami effects. Some locations in the path of the 2004 Indian Ocean tsunami escaped almost unscathed because trees such as coconut palms and mangroves absorbed the tsunami's energy. In one striking example, the village of Naluvedapathy in India's Tamil Nadu region suffered only minimal damage and few deaths because the wave broke against a forest of 80,244 trees planted along the shoreline in 2002 in a bid to enter the Guinness Book of Records. Environmentalists have suggested tree planting along tsunami-prone seacoasts. Trees require years to grow to a useful size, but such plantations could offer a much cheaper and longer-lasting means of tsunami mitigation than artificial barriers.

Mitigation

Natural barriers

A report published by the United Nations Environment Programme (UNEP) suggests that the tsunami of 26th December 2004 caused less damage in the areas where natural barriers were present, such as mangroves, coral reefs or coastal vegetation. A Japanese study of this tsunami in Sri Lanka used satellite imagery modelling to establish the parameters of coastal resistance as a function of different types of trees.

TSUNAMIS IN LAKES

A tsunami is defined as a series of water waves caused by the displacement of a large volume of a body of water, such as an ocean. This is misleading as destructive water waves are not restricted to the ocean, in the case of this article the body of water being investigated will be a lake rather than an ocean. Tsunamis in lakes are becoming increasingly important to investigate as a hazard, due to the increasing popularity for recreational uses, and increasing populations that inhabit the shores of lakes. Tsunamis generated in lakes and reservoirs are of high concern because it is associated with a near field source region which means a decrease in warning times to minutes or hours.

CAUSES

Inland tsunami hazards can be generated by many different types of earth movement, these are earthquakes in or around lake systems, landslides, debris flow, rock avalanches and volcanogenic processes such as gas or mass flow characteristics, these are discussed in more detail below.

Earthquakes

Tsunamis in lakes can be generated by fault displacement beneath or around lake systems. Faulting shifts the ground in a vertical motion through reverse, normal or oblique strike slip faulting processes, this displaces the water above causing a tsunami (Fig. 11.5). The reason strike-slip faulting does not cause tsunamis is because there is no vertical displacement within the fault movement, only lateral movement resulting in no displacement of the water. In an enclosed basin such as a lake, tsunamis are referred to as the initial wave produced by coseismic displacement from an earthquake, and the seiche as the harmonic resonance within the lake.

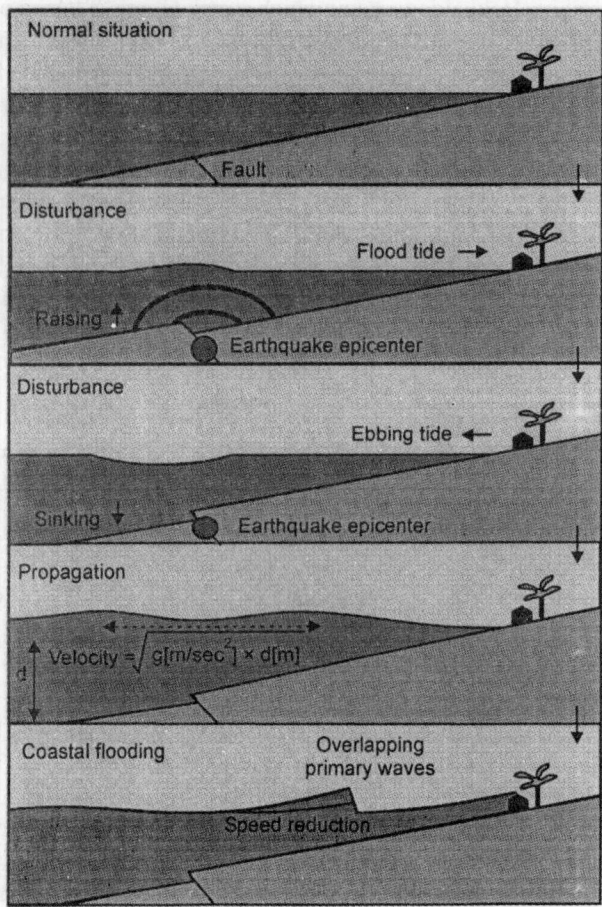

Fig. 11.5. Diagram showing how earthquakes can generate a tsunami.

In order for a tsunami to be generated certain criteria is required:

1. Needs to occur just below the lake bottom.
2. Earthquake is of high or moderate magnitude typically over magnitude four.
3. Displaces a large enough volume of water to generate a tsunami.

These tsunamis are of high damage potential due to being within a lake, making them of a near field source. This means a vast decrease in warning times, resulting in organised emergency evacuations after the generation of the tsunami being virtually impossible, and due to low lying shores even small waves lead to substantial flooding. Planning and education of residents needs to be done beforehand, so that when an earthquake is felt they know to head to higher ground and what routes to take to get there.

Lake Tahoe

Lake Tahoe is an example of a lake that is in danger of having a tsunami due to faulting processes. Lake Tahoe in Nevada USA lies within an intermountain basin bounded by faults, with most of these faults submerged at the lake bottom or hidden in glaicofluvial deposits. Lake Tahoe has had many prehistoric eruptions and in studies done of the lake bottom sediments, a 10m high scarp has displaced the lake bottom sediments, this indicates that the water was displaced by the same magnitude, as well as generating a tsunami. A tsunami and seiche in Lake Tahoe can be treated as shallow-water long waves as the maximum water depth is much smaller than the wavelength. This demonstrates the interesting impact that lakes have on the tsunami wave characteristics, as it is very different to ocean tsunami wave characteristics due to the ocean being deeper, and lakes being relatively shallow in comparison. With ocean tsunami waves amplitudes only increase when the tsunami gets close to shore, in lake tsunami waves are generated and stay in a shallow environment.

This would have a huge and devastating impact on the 34,000 permanent residences that live along the lake side, not to mention the impact on the booming tourism industry in the area. Damage wise tsunami run ups would leave areas of the lake completely inundated due to permanent ground subsidence attributed to the earthquake, with the highest run ups and amplitudes being attributed to the seiches rather than the actual tsunami. The reason seiches cause so much damage is due to resonance within the bays reflecting the waves were they combine to make larger standing waves, for more information see seiches.

Sub-aerial Mass Flows

Sub aerial mass flows happen when a large amount of sediment becomes unstable, this can happen for example from the shaking from an earthquake, or saturation of the sediment initiating a sliding layer. This volume of sediment then flows into the lake giving a sudden large displacement of water. Tsunamis generated by sub aerial mass flows are defined in terms of the first initial wave being the tsunami wave and any tsunamis in terms of sub aerial mass flows are characterised into three zones. A splash zone or wave generation zone, this is the region were landslides and water motion are coupled and it extends as far as the landslide travels. Near field area, were the concern is based on the characteristics of the tsunami wave such as amplitude and wave length which are crucial for predictive purposes. Far field area, the process is influenced mainly by dispersion characteristics and is not often used when investigating tsunamis in lakes, as most lake tsunamis are related only to near field processes.

New Zealand example

In the event of the Alpine fault in New Zealand rupturing in the South Island, it is predicted that there would be shaking of approximately magnitude eight in the lake side towns of Queenstown (Lake Wakatipu) and Wanaka (Lake Wanaka). These could possibly cause sub-aerial mass flows that could generate tsunamis within the lakes, this would have a devastating impact on the 15,453 residents (2006 Census) who occupy these lake towns, not only in the potential losses of life and property, but the damage to the booming tourism industry would take years to rebuild.

The Otago Regional Council, responsible for the area, has recognised that in such an event, tsunamis could occur in both lakes but have yet to learn any more about it. This is interesting as if an event was to happen in peak seasons such as summer around Christmas, or in the snow season when the population of these areas nearly double, it would pose a huge management issue. The reason for this huge management issue is the volume of people in these areas would be unknown and supplies and aid would be insufficient. In summer a lot of temporary structures such as tents are put up on the lake shore, these tents provide no protection from a tsunami wave. This is why it is so important to investigate what would happen in such an event so that people can be educated in order to reduce vulnerability in these areas.

Volcanogenic Processes

In this section the focus is on tsunamis generated in lakes by volcanogenic processes in terms of gas build up causing violent lake over turns, with other processes such as pyroclastic flows not accounted for, as it requires more complex modelling. Lake overturns can be incredibly dangerous and occur when gas trapped at the bottom of the lake is heated by rising magma causing an explosion and lake overturn, an example of this is Lake Kivu.

Lake Kivu

Lake Kivu as seen in Fig. 11.6 of Africa's great lakes, it lies on the border between the Democratic Republic of the Congo and Rwanda, and is part of the Great Rift Valley. Being part of the Great Rift Valley means it is affected by volcanic activity beneath the lake, this has led to a build up of methane and carbon dioxide at the bottom of the lake, this build up can lead to violent lake overturns. Lake over turns are due to volcanic interaction with the water at the bottom of the lake that has high gas concentrations, this leads to heating of the lake and this rapid rise in temperature would spark a methane explosion displacing a large amount of water, followed nearly simultaneously by a release of carbon dioxide. This carbon dioxide would suffocate large numbers of people, with a possible tsunami generated from water displaced by the gas explosion effecting all of the 2 million people who occupy the shores of Lake Kivu. This is incredibly important as the warning times for an event such as a lake overturn is incredibly short in the order of minutes and the event itself may not even be noticed. Education of locals and preparation is crucial in this case and a lot of research in this area has been done in order to try understand what is happening within the lake, in order to try reduce the effects when this phenomenon does happen.

Fig. 11.6. Satellite image of Lake Kivu in Africa.

A lake turn-over in Lake Kivu occurs from one of two scenarios. Either (i) up to another hundred years of gas accumulation leads to gas saturation in the lake, resulting in a spontaneous outburst of gas originating at the depth at which gas saturation has exceeded 100 per cent, and (ii) a volcanic or even seismic event triggers a turn-over. In either case a strong vertical lift of a large body of water results in a plume of gas bubbles and water rising up to and through the water surface. As the bubbling water column draws in fresh gas-laden water, the bubbling water column widens and becomes more energetic as a virtual 'chain reaction' occurs which would look like a watery volcano. Very large volumes of water are displaced, vertically at first, then horizontally away from the centre at surface and horizontally inwards to the bottom of the bubbling water column, feeding in fresh gas-laden water. The speed of the rising column of water increases until it has the potential to rise 25 m or more in the centre above lake level. The water column has the potential to widen to well in excess of a kilometre, in a violent disturbance of the whole lake. The watery volcano may take as much as a day to fully develop while it releases upwards of 400 billion cubic metres of gas (~12 tcf). Some of these parameters are uncertain, particularly the time taken to release the gas and the height to which the water column can rise. As a secondary effect, particularly if the water column behaves irregularly with a series of surges, the lake surface will both rise by up to several metres and create a series of tsunamis or waves radiating away from the epicentre of the eruption. Surface waters may simultaneously race away from the epicentre at speeds as high as 20–40 m/s, slowing as distances from the centre increase. The size of the waves created is unpredictable. Wave heights will be highest if the water column surges periodically, resulting in wave heights is great as 10–20 m.

This is caused by the ever-shifting pathway that the vertical column takes to the surface. No reliable model exists to predict this overall turnover behaviour. For tsunami precautions it will be necessary for people to move to high ground, at least 20m above lake level. A worse situation may pertain in the Ruzizi River where a surge in lake level would cause flash-flooding of the steeply sloping river valley dropping 700 m to Lake Tanganyika, where it is possible that a wall of water from 20–50 m high may race down the gorge. Water is not the only problem for residents of the Kivu basin; the more than 400 billion cubic metres of gas released creates a denser-than-air cloud which may blanket the whole valley to a depth of 300 m or more. The presence of this opaque gas cloud, which would suffocate any living creatures with its mixture of carbon dioxide and methane laced with hydrogen sulphide, would cause the majority of casualties. Residents would be advised to climb to at least 400m above the lake level to ensure their safety. Strangely the risk of a gas explosion is not great as the gas cloud is only about 20 per cent methane in carbon dioxide, a mixture that is difficult to ignite.

HAZARD MITIGATION

Hazard mitigation for tsunamis in lakes is immensely important in the preservation of life, infrastructure and property. In order for hazard management of tsunamis in lakes to function at full capacity there are four aspects that need to be balanced and interacted with each other, these are:

1. Readiness (preparedness for a tsunami in the lake).
   (a) Evacuation plans.
   (b) Evacuation plans making sure equipment and supplies are on standby in case of a tsunami.
   (c) Education of locals on what hazard is posed to them and what they need to do in the event of a tsunami in the lake such as seen in Fig. 11.7.
2. Response to the tsunami event in the lake.
   (a) Rescue operations.
   (b) Getting aid into the area such as food and medical equipment.
   (c) Providing temporary housing for people who have been displaced.
3. Recovery from the tsunami.
   (a) Re-establishing damaged road networks and infrastructure.

    (b) Re-building and/or relocation for damaged buildings.

    (c) Clean up of debris and flooded areas of land.

4. Reduction (plans to reduce the effects of the next tsunami).

    (a) Putting in place land use zoning to provide a buffer for tsunami run ups, meaning that buildings cannot be built right on the lake shore.

When all these aspects are taken into consideration and continually managed and maintained, the vulnerability of an area to a tsunami within the lake decreases. This is not because the hazard its self has decreased but the awareness of the people who would be affected makes them more prepared to deal with the situation when it does occur. This reduces recovery and response times for an area, decreasing the amount of disruption and in turn the effect the disaster has on the community.

Fig. 11.7. Photo of a Tsunami Hazard Zone warning sign in Bamfield.

FUTURE RESEARCH

Investigation into the phenomena of tsunamis in lakes for this section was restricted by certain limitations. Internationally there has been a fair amount of research into certain lakes but not all lakes that can be affected by the phenomenon have been covered.

This is especially true for New Zealand with the possible occurrence of tsunamis in our major lakes recognised as a hazard, but with no further research completed.

The reports found from international examples were extremely useful as individual case studies, but due to different definitions for lake tsunamis depending on generation source and different equations for predictions.

This made it hard to correlate the different events to see if some general rules can be obtained in terms of wave dynamics and propagation. It is increasingly important that further knowledge is gained on tsunamis in lakes, not only for scientific purposes but also to relay this information to the people who would be affected if a lake tsunami was to occur. The hazard of tsunamis and associated seiches will not stop increasing, more and more people are making themselves vulnerable to this hazard, and due to lack of knowledge and growing populations the devastation when one does occur will be incomprehendable. Increasing knowledge of tsunamis in lakes and what to do in an event of a tsunami will not only makes us more prepared, it will save lives.

TSUNAMI WARNING SYSTEM

A Tsunami Warning System (TWS) is a system to detect tsunamis and issue warnings to prevent loss of life and property. It consists of two equally important components: a network of sensors to detect tsunamis and a communications infrastructure to issue timely alarms to permit evacuation of coastal areas (Fig. 11.8).

Fig. 11.8. Evacuation route sign in a low-lying coastal area on the West Coast of the United States.

There are two distinct types of tsunami warning systems: international and regional. Both depend on the fact that, while tsunamis travel at between 500 and 1000 km/hr (around 0.14 and 0.28 km/s) in open water, earthquakes can be detected almost at once as seismic waves travel with a typical speed of 4 km/s (around 14,400 km/hr). This gives time for a possible tsunami forecast to be made and warnings to be issued to threatened areas, if warranted. Unfortunately, until a reliable model is able to predict which earthquakes will produce significant tsunamis, this approach will produce many more false alarms than verified warnings. In the correct operational paradigm, the seismic alerts are used to send out the watches and warnings. Then, data from observed sea level height (either shore-based tide gauges or DART buoys) are used to verify the existence of a tsunami.

Other systems have been proposed to augment the warning paradigm. For example, it has been suggested that the duration and frequency content of t-wave energy (which is earthquake energy trapped in the ocean SOFAR channel) is indicative of an earthquake's tsunami potential. The first rudimentary system to alert communities of an impending tsunami was attempted in Hawaii in the 1920s. More advanced systems were developed in the wake of the April 1, 1946 (caused by the 1946 Aleutian Islands earthquake) and May 23, 1960 (caused by the 1960 Valdivia earthquake) tsunamis which caused massive devastation in Hilo, Hawaii.

ANIMAL INFRASOUND

The elephants in the 2004 Indian Ocean Tsunami fled for the hills. They sensed the tsunami due to their ability to hear the soundwaves outside of the range of human hearing range of 20 Hz and 20,000 Hz. In addition to hearing within the human range, elephants can also hear in the range of Infrasound, beneath 20 Hz down to 0.001 Hz.

INTERNATIONAL WARNING SYSTEMS (IWS)

Pacific Ocean

Tsunami warnings for most of the Pacific Ocean are issued by the Pacific Tsunami Warning Center (PTWC), operated by the United States's NOAA in Ewa Beach, Hawaii. NOAA's West Coast and

Alaska Tsunami Warning Center (WCATWC) in Palmer, Alaska issues warnings for the west coast of North America, including Alaska, Canada, and the western coterminous United States. PTWC was established in 1949, following the 1946 Aleutian Island earthquake and a tsunami that resulted in 165 casualties on Hawaii and in Alaska; WCATWC was founded in 1967. International coordination is achieved through the International Coordination Group for the Tsunami Warning System in the Pacific, established by the Intergovernmental Oceanographic Commission of UNESCO.

Indian Ocean (ICG/IOTWS)

After the 2004 Indian Ocean Tsunami which killed almost 230,000 people, a United Nations conference was held in January 2005 in Kobe, Japan, and decided that as an initial step towards an International Early Warning Programme, the UN should establish an Indian Ocean Tsunami Warning System. This then resulted in a system of warnings in Indonesia. This will also save the lives and the livelihood of the people.

North Eastern Atlantic, the Mediterranean and Connected Seas (ICG/NEAMTWS)

The First United Session of the Inter-governmental Coordination Group for the Tsunami Early Warning and Mitigation System in the North Eastern Atlantic, the Mediterranean and connected Seas (ICG/NEAMTWS), established by the Intergovernmental Oceanographic Commission of UNESCO Assembly during its 23rd Session in June 2005, through Resolution XXIII.14, took place in Rome on 21st and 22nd November, 2005.

The Meeting, hosted by the Government of Italy (Italian Ministry of Foreign Affairs and Ministry for Environment and Protection of the Territory), was attended by more than 150 participants from 24 countries, 13 organisations and numerous observers.

Caribbean

A Caribbean wide tsunami warning system has been planned to be set up by 2012 by member nations representatives who met in Panama City in March 2008. Panama's last major tsunami killed 4500 people in 1882. Barbados has said it will review or test its Tsunami protocol in February 2010 as a regional pilot.

REGIONAL WARNING SYSTEMS

Regional (or local) warning system centres use seismic data about nearby earthquakes to determine if there is a possible local threat of a tsunami. Such systems are capable of issuing warnings to the general public (via public address systems and sirens) in less than 15 minutes. Although the epicenter and moment magnitude of an underwater quake and the probable tsunami arrival times can be quickly calculated, it is almost always impossible to know whether underwater ground shifts have occurred which will result in tsunami waves. As a result, false alarms can occur with these systems, but due to the highly localised nature of these extremely quick warnings, disruption is small.

CONVEYING THE WARNING

Detection and prediction of tsunamis is only half the work of the system. Of equal importance is the ability to warn the populations of the areas that will be affected. All tsunami warning systems feature multiple lines of communications (such as SMS, e-mail, fax, radio, texting and telex, often using hardened dedicated systems) enabling emergency messages to be sent to the emergency services and armed forces, as well to population alerting systems (e.g. sirens).

SHORTCOMINGS

No system can protect against a very sudden tsunami, where the coast in question is too close to the epicenter. A devastating tsunami occurred off the coast of Hokkaido in Japan as a result of an earthquake on July 12, 1993. As a result, 202 people on the small island of Okushiri, Hokkaido lost their lives, and hundreds more were missing or injured. This tsunami struck just three to five minutes after the quake, and most victims were caught while fleeing for higher ground and secure places after surviving the earthquake.

While there remains the potential for sudden devastation from a tsunami, warning systems can be effective. For example if there were a very large subduction zone earthquake (moment magnitude 9.0) off the west coast of the United States, people in Japan, for example, would have more than 12 hrs (and likely warnings from warning systems in Hawaii and elsewhere) before any tsunami arrived, giving them some time to evacuate areas likely to be affected.

TSUNAMI-PROOF BUILDING

A tsunami-proof building is a purposefully designed building which will through its design integrity withstand and survive the forces of a tsunami wave or extreme storm surge. Hydrodynamically shaped to offer protection from high waves. The laminar flow around it will protect the walls. The structure rests on a hollow masonry block that can hold 75 m^3 (4.2^{3}m) of water. Enough to supply a family for 3 months. Designed with battered walls, cantilever steps and a wooden superstructure with the walls jutting out. Bamboo ply panels covers the sides. Eight solar panels powers the lights, water pumps and most of the kitchen equipment. Concomitant with its mechanical strength, it provides its occupants with independent potable water storage for a minimum of two months and has an independent electrical power source provided by generator or solar panel array. The first example known has been constructed at Poovar Island in southern Kerala, India.

IMPACT OF TSUNAMI IN INDIA

Almost all the countries situated around the Bay of Bengal were affected by the tsunami waves in the morning hours of 26 December 2004 (between 0900–1030 hrs IST). The killer waves were triggered by an earthquake measuring 8.9 on the Richter scale that had an epicenter near the west coast of Sumatra in Indonesia. The first recorded tsunami in India dates back to 31st December 1881. An earthquake of magnitude 7.5 on the Richter scale, with its epicenter believed to have been under the sea off the coast of Car Nicobar Island, caused the tsunami. The last recorded tsunami in India occurred on 26th June 1941, caused by an earthquake with magnitude exceeding 8.5. This caused extensive damage to the Andaman Islands. There are no other well-documented records of Tsunami in India.

It was all quiet on the waterfront on the Sunday morning after Christmas in 2004 at Kanyakumari, the famous Marina Beach in Chennai and elsewhere on the Kerala coast and Andaman Nicober Islands. There was the excitement of a holyday with an offbeat mood with swarms of people on the sea front: children playing cricket and man and women on their morning work at the Marina. Elsewhere, fishermen were putting out to sea for the day's catch. Then all on a sudden, a curious thing happened. The holidaymakers at Kanyakumari were awestruck when the sea receded from the shores. In the present tsunami, India was the third country severely battered after Indonesia and Srilanka. In India the State severely affected by tsunami are Tamilnadu, Pondicherry, Andhra Pradesh, Kerala and Andaman and Nicobar Island. The following Table.11.1 shows the average scenario of tsunami devastation in the respective areas. The data relating to the Andaman and Nicober are yet to be assessed, for which it does appear in the Table 11.1.

Table 11.1. Tsunami damage in India.

Factor	Tsunami damage in India				
	Andhra Pradesh	Kerala	Tamil Nadu	Pondcherry	Total
Population affected	2,11,000	2,470,000	6,91,000	43,000	3,415,000
Area affected (Ha)	790	Unknown	2487	790	4067
Length of coast affected (Km)	985	250	1000	25	2260
Extent of penetration (Km)	0.5–2.0	1–2	1–1.5	0.30–3.0	
Reported height of tsunami (m)	5	3–5	7–10	10	
Villages affected	301	187	362	26	876
Dwelling units	1557	11832	91037	6403	1,10,829
Cattle lost	195	Unknown	5476	3445	9116

Tamil Nadu

The state of Tamil Nadu has been the worst affected on the mainland, with a death toll of 7793. Nagapattinam district has had 5,525 casualties, with entire villages having been destroyed. Kanyakumari district has had 808 deaths, Cuddalore district 599, the state capital Chennai 206 and Kancheepuram district 124. The death tolls in other districts were Pudukkottai, Ramanathapuram, Tirunelveli, Thoothukudi, Tiruvallur, Thanjavur, Tiruvarur and Viluppuram.

Those killed in Kanyakumari include pilgrims taking a holy dip in the sea. Of about 700 people trapped at the Vivekananda Rock Memorial off Kanyakumari, 650 were rescued. In Chennai, people playing on the Marina beach and those who taking a Sunday morning stroll were washed away, in addition to the fisherfolk who lived along the shore and those out at sea. The death toll at Velankanni in Nagapattinam district is currently 1500. Most of these people were visiting the Basilica of the Virgin Mary for Christmas, while others were residents of the town. The nuclear power station at Kalpakkam was shut down after sea water rushed into a pump station.

Pondicherry

An estimated 30,000 people are homeless in the Union territory of Pondicherry. The current official toll is 560. The affected districts are Pondicherry (107 dead), Karaikal (453 dead). Kariakal is the most devastrated area from the Pondicherry Union territory. Where massive destruction and loss of casualities accure. This mishalp occured because of uncover stone block. Mostly fisherfolk are affected due to location and distance between sea and their basti (village). Fishing peoples are just preparing for venturing into sea and within fraction of seconds everything wash away and their boats are damaged they lost every thing in terms of life and property. More than 453 people are died so far and still some are missing.

Kerala

The current official toll is 168. The affected districts are Kollam (131 dead), Alappuzha (32 dead), Ernakulam (5 dead). The tsunami that hit the Kerala coast on December 26, 2004, were three to five metres high, according to the National Institute of Disaster Management, (NIDM) which functions under the ministry of home affairs. The Tidal upsurge had affected 250 killometers of the Kerala coastline and entered between one or two kilometers inland pounded 187 villages affecting 24.70 lakh persons in the state. As many as 6,280 dwelling units were destroyed. As many as 84,773 persons were evacuated from the coastal areas and accomodated in 142 Relief Camps opened in Kollam, Alappuzha and Ernakulam

Districts. According to NIDM,131 lives were lost in Kollam, 32 in Alappuzha and five in Ernakulam, taking the official death toll to 168. High wave swept the coast along a 40-km stretch, from Sakthikulangare in the south to Thrikunnapuzha in the north. This stretch has two narrow strips of land sandwiched between the sea and back water.

Andhra Pradesh

The current official toll is 105. The affected districts are Krishna (35 dead), Prakasam (35 dead), Nellore (20 dead), Guntur (4 dead), West Godavari (8 dead) and East Godavari (3 dead).

Andaman and Nicobar

The Andaman and Nicobar Islands comprise 572 islands (all land masses in both low and high tides) out of which 38 are inhabited, both by people from the mainland and indigenous tribes. The islands lie just north of the earthquake epicentre, and the tsunami reached a height of 15 m in the southern Nicobar Islands. The official death toll is 812, and about 7000 are still missing. The unofficial death toll (including those missing and presumed dead) is estimated to be about 7000. The Great Nicobar and Car Nicobar islands were the worst hit among all the islands because of their proximity to the quake and relative flatness. Aftershocks continue to rock the area. One-fifth of the population of the Nicobar Islands is said to be dead, injured or missing. Chowra Island has lost two-thirds of its population of 1500. Entire islands have been washed away, and the island of Trinket has been split in two. Communications was restored after 15 days with the Nancowry group of islands, some of which have been completely submerged, with the total number of the population still out of contact exceeding 7000. Among the casualties in Car Nicobar, 100 Indian Air Force personnel and their family members were washed away when the wave hit their air base, which was reported to have been severely damaged. The St. Thomas Cathedral (also known as the John Richardson church after John Richardson, a missionary and member of Parliament) was washed away.

The church, established in 1930 was one of the oldest and prominent churches in the region. A cricket stadium named after John Richardson and a statue dedicated to him were also washed away. The majority of the population of Andaman Islands is made up of people from the mainland, mostly from West Bengal and Tamil Nadu. The natives of Andaman and Nicobar Islands are endangered tribal groups, such as such as the Jarawa, the Sentinelese, the Shompen, the Onge and the Andamanese. They are regarded as anthropologically significant as they are some of the world's most primitive tribes and considered the world's only link to ancient civilisation. Most of these tribes have maintained their aboriginal lifestyle for centuries, and government policy has been to not interfere with them unless absolutely essential. It is reported that most of the native islanders survived the tsunami because they live on higher ground or far from the coast. The Onge, Jarawa, Sentinelese and Andamanese have been reached by survey teams and are confirmed to be safe although the number of dead is unknown. The Sentinelese live on a reserved island and are hostile to outsiders which is making it difficult for Indian officials to visit the island. They have shot arrows at helicopters sent to check on them. In the Nicobar Islands, the Nicobarese, a Mongoloid tribe, have lost about 656 lives with 3000 still missing. Surveys are being conducted on the Shompen located on Great Nicobar Island.

India's only active volcano, Barren 1, located at Barren Island 135 km (80 miles) northeast of the capital Port Blair, erupted because of increased seismic activity on 30th December, 2004. People have been evacuated since then and there have been no reports of any casualties.

Chapter 12

El Nino and La Niña

INTRODUCTION

El Niño is also known as the southern oscillation. The El Niño (ENSO) phenomenon can be described as a type of abnormal warming that occurs on the surface ocean waters in the part of the eastern tropical pacific that is known as the southern oscillation. The southern oscillation operates in a type of seesaw pattern that occurs when the surface air pressure between the western and the eastern tropical pacific is reversed. When this happens, the surface pressure is reversed as well as the ocean warming. When these two reversals are reversed in a mostly simultaneous manner, then weather forecasters and meteorologists tend to call this phenomenon as the southern oscillation or the El Niño phenomenon.

This nickname was given to the weather phenomenon by South American fisherman who began to notice the El Niño phenomenon in the previous decade. But what does the El Niño phenomenon name mean? The El Niño name specifically translates as the child, but in this context the El Niño name translates as 'the Christ child'. It was named thusly by the South American fishermen because the weather phenomenon occurs at roughly the same time as the celebration of the Christ Child, that is, during the Christmas season. However, although the El Niño generally begins during these months (winter months in the United States and North America) it can last anywhere from a few weeks to several months.

What causes the condition described as El Niño to start in the first place: The phenomenon known as El Niño is still a bit of a mystery to the scientists. In general, most scientists cannot pinpoint the exact reason for why the El Niño develops. However, most scientists believe that the El Niño has contributed to some of the most devastating weather that has ever occurred in the United States. For example, most researchers believe that the El Niño phenomenon is at least partially responsible for the disaster and devastation wrought by several large weather events such as Mississippi floods of 1993 and the California floods of 1995. The El Niño phenomenon is also thought to contribute to the drought conditions that have plagued many parts of the world, including areas in Australia, South America, and Africa.

The phenomenon of El Niño is also held at least partly responsible for the movement of storms that have spared such areas as Florida in North America. It is also thought that the El Niño phenomenon has helped contribute to the lack of serious hurricanes and storms in areas of the North Atlantic. However, just because El Niño has not caused serious storm damage in these areas, it does not mean that we can predict the future behavior of El Niño. Scientists are striving to figure out how the El Niño phenomenon works. The NASA Earth scientists are taking part in a worldwide struggle to understand how the El Niño events work. However, although scientists are now striving to understand the different patterns of

El Niño, we still do not understand exactly how it works. For now, it seems, the El Niño phenomenon will remain a mystery.

ENSO causes extreme weather such as floods, droughts and other weather disturbances in many regions of the world. Developing countries dependent upon agriculture and fishing, particularly those bordering the Pacific Ocean, are the most affected. In popular usage, the El-Niño-Southern Oscillation is often called just 'El Niño'. El Niño is Spanish for 'the boy' and refers to the Christ child, because periodic warming in the Pacific near South America is usually noticed around Christmas. The expression of ENSO is potentially subject to dramatic changes as a result of global warming, and is a target for research in this regard (Fig. 12.1).

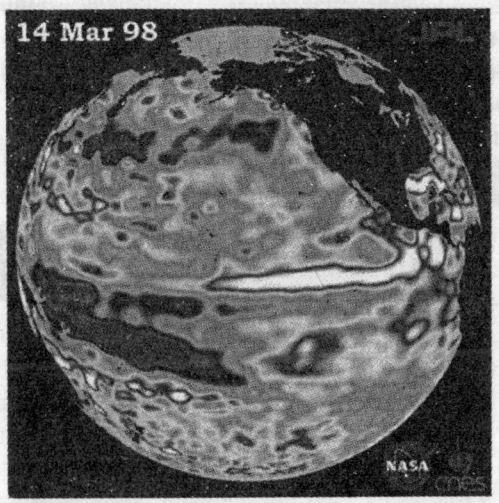

Fig. 12.1. The 1997 El Niño observed by TOPEX/Poseidon. The white areas off the tropical coasts of South and North America indicate the pool of warm water.

El Niño is defined by prolonged differences in Pacific Ocean surface temperatures when compared with the average value. The accepted definition is a warming or cooling of at least 0.5°C (0.9°F) averaged over the east-central tropical Pacific Ocean. Typically, this anomaly happens at irregular intervals of 2–7 years and lasts nine months to two years. The average period length is 5 years. When this warming or cooling occurs for only seven to nine months, it is classified as El Niño/La Niña 'conditions'; when it occurs for more than that period, it is classified as El Niño/La Niña 'episodes'.

The first signs of an El Niño are:

1. Rise in surface pressure over the Indian Ocean, Indonesia, and Australia.
2. Fall in air pressure over Tahiti and the rest of the central and eastern Pacific Ocean.
3. Trade winds in the south Pacific weaken or head east.
4. Warm air rises near Peru, causing rain in the northern Peruvian deserts.
5. Warm water spreads from the west Pacific and the Indian Ocean to the east Pacific. It takes the rain with it, causing extensive drought in the western Pacific and rainfall in the normally dry eastern Pacific.

El Niño's warm rush of nutrient-poor tropical water, heated by its eastward passage in the Equatorial Current, replaces the cold, nutrient-rich surface water of the Humboldt Current. When El Niño conditions last for many months, extensive ocean warming and the reduction in Easterly Trade winds limits upwelling of cold nutrient-rich deep water and its economic impact to local fishing for an international market can be serious.

EARLY STAGES AND CHARACTERISTICS OF EL NIÑO

Although its causes are still being investigated, El Niño events begin when trade winds, part of the Walker circulation, falter for many months. A series of Kelvin waves—relatively warm subsurface waves of water a few centimetres high and hundreds of kilometres wide—cross the Pacific along the equator and create a pool of warm water near South America, where ocean temperatures are normally cold due to upwelling. The weakening of the winds can also create twin cyclones, another sign of a future El Niño. The Pacific Ocean is a heat reservoir that drives global wind patterns, and the resulting change in its temperature alters weather on a global scale. Rainfall shifts from the western Pacific toward the Americas, while Indonesia and India become drier.

Jacob Bjerknes in 1969 helped toward an understanding of ENSO, by suggesting that an anomalously warm spot in the eastern Pacific can weaken the east-west temperature difference, disrupting trade winds that push warm water to the west. The result is increasingly warm water toward the east. Several mechanisms have been proposed through which warmth builds up in equatorial Pacific surface waters, and is then dispersed to lower depths by an El Niño event. The resulting cooler area then has to 'recharge' warmth for several years before another event can take place.

While not a direct cause of El Niño, the Madden-Julian Oscillation, or MJO, propagates rainfall anomalies eastward around the global tropics in a cycle of 30–60 days, and may influence the speed of development and intensity of El Niño and La Niña in several ways. For example, westerly flows between MJO-induced areas of low pressure may cause cyclonic circulations north and south of the equator. When the circulations intensify, the westerly winds within the equatorial Pacific can further increase and shift eastward, playing a role in El Niño development. Madden-Julian activity can also produce eastward-propagating oceanic Kelvin waves, which may in turn be influenced by a developing El Niño, leading to a positive feedback loop.

SOUTHERN OSCILLATION

The Southern Oscillation is the atmospheric component of El Niño. This component is an oscillation in surface air pressure between the tropical eastern and the western Pacific Ocean waters. The strength of the Southern Oscillation is measured by the *Southern Oscillation Index* (SOI). The SOI is computed from fluctuations in the surface air pressure difference between Tahiti and Darwin, Australia. El Niño episodes are associated with negative values of the SOI, meaning that the pressure at Tahiti is relatively low compared to Darwin (Fig. 12.2).

Low atmospheric pressure tends to occur over warm water and high pressure occurs over cold water, in part because deep convection over the warm water acts to transport air. El Niño episodes are defined as sustained warming of the central and eastern tropical Pacific Ocean. This results in a decrease in the strength of the Pacific trade winds, and a reduction in rainfall over eastern and northern Australia.

Walker Circulation

During non-El Niño conditions, the Walker circulation is seen at the surface as easterly trade winds which move water and air warmed by the sun towards the west. This also creates ocean upwelling off the coasts of Peru and Ecuador and brings nutrient-rich cold water to the surface, increasing fishing stocks. The western side of the equatorial Pacific is characterised by warm, wet low pressure weather as the collected moisture is dumped in the form of typhoons and thunderstorms. The ocean is some 60 centimetres (24 in) higher in the western Pacific as the result of this motion (Fig. 12.3).

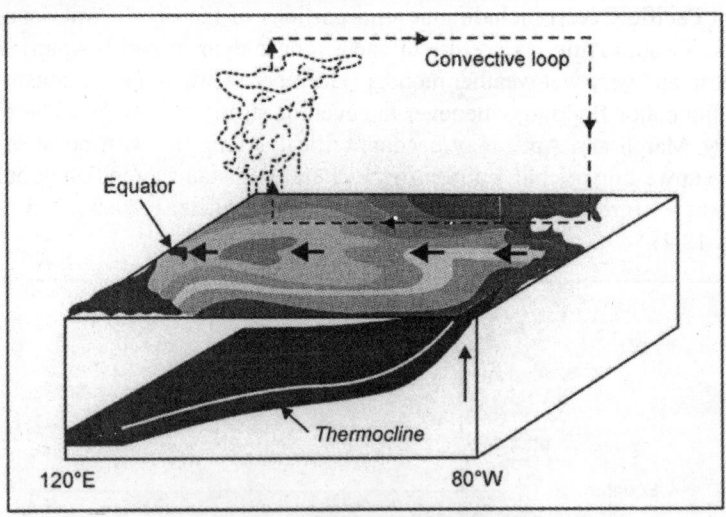

Fig. 12.2. Normal Pacific pattern. Equatorial winds gather warm water pool toward west. Cold water upwells along South American coast.

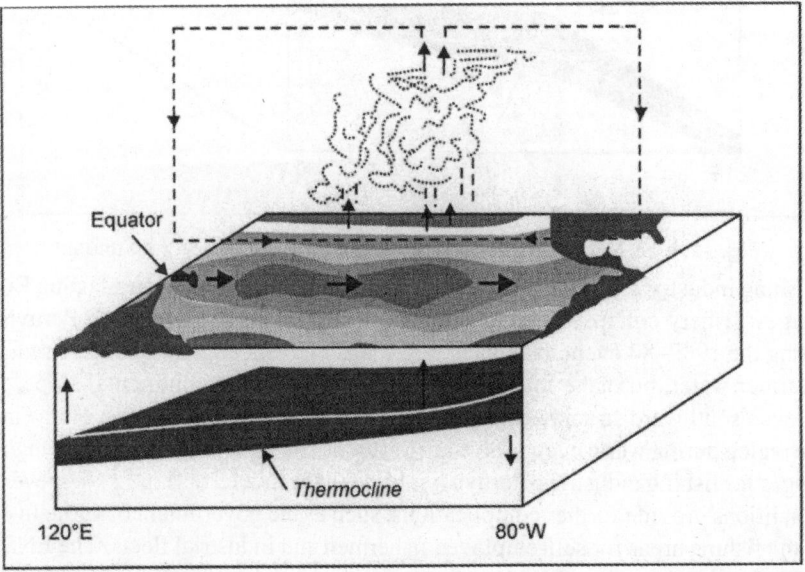

Fig. 12.3. El Niño Conditions. Warm water pool approaches South American coast. Absence of cold upwelling increases warming.

EFFECTS OF ENSO'S WARM PHASE (EL NIÑO)

South America

Because El Niño's warm pool feeds thunderstorms above, it creates increased rainfall across the east-central and eastern Pacific Ocean including several portions of the South American west coast. The effects of El Niño in South America are direct and stronger than in North America. An El Niño is associated with warm and very wet weather months December–April along the coasts of northern Peru and Ecuador, causing major flooding whenever the event is strong or extreme. The effects during the months of February, March and April may become critical. Along the west coast of South America, El Niño reduces the upwelling of cold, nutrient-rich water that sustains large fish populations, which in turn sustain abundant sea birds, whose droppings support the fertiliser industry. This leads to fish kills offshore Peru (Fig. 12.4).

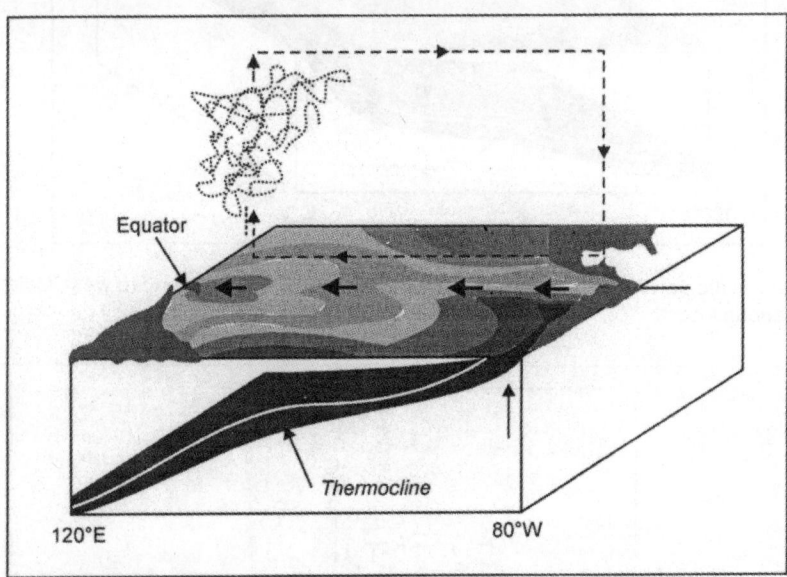

Fig. 12.4. La Niña Conditions. Warm water is further west than usual.

The local fishing industry along the affected coastline can suffer during long-lasting El Niño events. The world's largest fishery collapsed due to overfishing during the 1972 El Niño Peruvian anchoveta reduction. During the 1982–83 event, jack mackerel and anchoveta populations were reduced, scallops increased in warmer water, but hake followed cooler water down the continental slope, while shrimp and sardines moved southward so some catches decreased while others increased. Horse mackerel have increased in the region during warm events. Shifting locations and types of fish due to changing conditions provide challenges for fishing industries. Peruvian sardines have moved during El Niño events to Chilean areas. Other conditions provide further complications, such as the government of Chile in 1991 creating restrictions on the fishing areas for self-employed fishermen and industrial fleets. The ENSO variability may contribute to the great success of small fast-growing species along the Peruvian coast, as periods of low population removes predators in the area. Similar effects benefit migratory birds that travel each spring from predator-rich tropical areas to distant winter-stressed nesting areas.

Southern Brazil and northern Argentina also experience wetter than normal conditions but mainly during the spring and early summer. Central Chile receives a mild winter with large rainfall, and the Peruvian-Bolivian Altiplano is sometimes exposed to unusual winter snowfall events. Drier and hotter weather occurs in parts of the Amazon River Basin, Colombia and Central America.

North America

Winters, during the El Niño effect, are warmer and drier than average in the Northwest, Northmidwest, and Northmideast United States, and therefore those regions experience reduced snowfalls. Meanwhile, significantly wetter winters are present in northwest Mexico and the southwest United States including central and southern California, while both cooler and wetter than average winters in northeast Mexico and the southeast United States (including the Tidewater region of Virginia) occur during the El Niño phase of the oscillation.

In Canada, both warmer and drier winters over much of the country occur, although less variation from normal is seen in the Maritime Provinces. The following summer is warmer and sometimes drier creating a more active than average forest fire season over Central/Eastern Canada. Some believed that the ice-storm in January 1998, which devastated parts of Southern Ontario and Southern Quebec, was caused or accentuated by El Niño's warming effects. El Niño warmed Vancouver for the 2010 Winter Olympics, such that the area experienced a subtropical-like winter during the games.

Summers, during the El Niño effect, are wetter than average in the Northwest, Northmidwest, Northmideast, and mountain regions of the United States.

El Niño is credited with suppressing hurricanes and made the 2009 hurricane season the least active in twelve years. El Niño is also associated with increased wave-caused coastal erosion along the United States Pacific Coast. There is some evidence that El Niño activity is correlated with incidence of red tides off the Pacific coast of California.

Tropical Cyclones

Most tropical cyclones form on the side of the subtropical ridge closer to the equator, then move poleward past the ridge axis before recurving into the main belt of the Westerlies. When the subtropical ridge position shifts due to El Niño, so will the preferred tropical cyclone tracks. Areas west of Japan and Korea tend to experience much fewer September–November tropical cyclone impacts during El Niño and neutral years. During El Niño years, the break in the subtropical ridge tends to lie near 130°E, which would favour the Japanese archipelago. During El Niño years, Guam's chance of a tropical cyclone impact is one-third of the long-term average. The tropical Atlantic ocean experiences depressed activity due to increased vertical wind shear across the region during El Niño years.

Elsewhere

In Africa, East Africa, including Kenya, Tanzania and the White Nile basin experiences, in the long rains from March to May, wetter than normal conditions. There are also drier than normal conditions from December to February in south-central Africa, mainly in Zambia, Zimbabwe, Mozambique and Botswana. Direct effects of El Niño resulting in drier conditions occur in parts of Southeast Asia and Northern Australia, increasing bush fires and worsening haze and decreasing air quality dramatically. Drier than normal conditions are also generally observed in Queensland, inland Victoria, inland New South Wales and eastern Tasmania from June to August. West of the Antarctic Peninsula, the Ross, Bellingshausen, and Amundsen Sea sectors have more sea ice during El Niño. The latter two and the

Weddell Sea also become warmer and have higher atmospheric pressure. El Niño's effects on Europe are not entirely clear, but certainly it is not nearly as affected as at least large parts of other continents. There is some evidence that an El Niño may cause a wetter, cloudier winter in Northern Europe and a milder, drier winter in the Mediterranean Sea region. The El Niño winter of 2006/2007 was unusually mild in Europe, and the Alps recorded very little snow coverage that season.

Most recently, Singapore experienced the driest February in 2010 since records begins in 1869. With only 6.3 millimetres of rain fell in the month and temperatures hitting as high as 35°C on 26 February, 1968 and 2005 had the next driest Februaries when 8.4 mm of rain fell.

LA NIÑA

La Niña is a coupled ocean-atmosphere phenomenon that is the counterpart of El Niño as part of the broader El Niño-Southern Oscillation climate pattern. During a period of La Niña, the sea surface temperature across the equatorial Eastern Central Pacific Ocean will be lower than normal by 3°–5°C. In the United States, an episode of La Niña is defined as a period of at least 5 months of La Niña conditions. The name La Niña originates from Spanish, meaning 'the girl', analogous to El Niño meaning 'the boy'. La Niña, sometimes informally called 'anti-El Niño', is the opposite of El Niño, where the latter corresponds instead to a higher sea surface temperature by a deviation of at least 0.5 °C, and its effects are often the reverse of those of El Niño. El Niño is famous due to its potentially catastrophic impact on the weather along both the Chilean, Peruvian and Australian coasts, among others. La Niña is often preceded by a strong El Niño.

Effects of La Niña

La Niña causes mostly the opposite effects of El Niño, for example, El Niño would cause a wet period in the Midwestern U.S., while La Niña would typically cause a dry period in this area. At the other side of the Pacific La Niña can cause heavy rains. For India, an El Niño is often a cause for concern because of its adverse impact on the south-west monsoon; this happened in 2009. A La Niña, on the other hand, is often beneficial for the monsoon, especially in the latter half. The La Niña that appeared in the Pacific in 2010 probably helped last year's south-west monsoon end on the favourable note. But then, it also contributed to the deluge in Australia, which resulted in one of that country's worst natural disasters with large parts of the north-east under water. It wreaked similar havoc in south-eastern Brazil and played a part in the heavy rains and consequent flooding that have affected Sri Lanka.

Recent Occurrences

There was a strong La Niña episode during 1988–1989. La Niña also formed in 1995, and in 1999–2000. A minor La Niña occurred 2000–2001. The last La Niña was a moderate one, which developed in mid 2007 and lasted until early 2009. NOAA confirmed that a moderate La Niña developed in their November El Niño/Southern Oscillation Diagnostic Discussion, and that it would likely continue into 2008. According to NOAA, Expected La Niña impacts during November—January include a continuation of above-average precipitation over Indonesia and below-average precipitation over the central equatorial Pacific. For the contiguous United States, potential impacts include above average precipitation in the Northern Rockies, Northern California, and in southern and eastern regions of the Pacific Northwest. Below-average precipitation is expected across the southern tier, particularly in the southwestern and southeastern states.

However, an El Niño returned in May/June 2009 and lasted until April 2010. The effects of El Niño in 2009 were already being seen in the fall of 2009 as the remnants of Tropical Storm Ida strengthened into a powerful coastal storm. A new La Niña episode developed quite quickly in the eastern and central tropical Pacific in mid-2010, and lasted at least until early 2011. This La Niña, combined with record high ocean temperatures in the north-eastern Indian Ocean, has been responsible for the 2010–2011 Queensland floods.

CURRENT SITUATION AND OUTLOOK

At present, warmer than normal sea-surface temperatures are observed in the central equatorial Pacific, and most computer models are currently forecasting some warming to prevail in the central and eastern equatorial Pacific for the remainder of the year. Development of an El Niño event in the second half of the year would not be unprecedented, but would be unusual. While the chances of an El Niño have increased, expert opinion currently favours a range of possible outcomes for the basin-wide state of the tropical Pacific from near-neutral to El Niño for the remainder of the year, giving El Niño about as much chance as not of developing. La Niña is not considered likely.

The previous statement indicated that slightly warmer than normal conditions prevailed in the central/western equatorial Pacific. Since then, the area of anomalous warmth has migrated and expanded slowly toward the east, with equatorial sea-surface temperatures being over 1°C warmer than normal from around the dateline to about 140°W over the last month. If this were to persist, it would satisfy one condition for an El Niño. However, other conditions are not yet indicative of a basin-wide El Niño, particularly the presence of below normal seasurface temperatures in the far eastern equatorial Pacific, from about 120°W to the South American coast.

The increase in the likelihood of an El Niño developing during the remainder of the year also increases the likelihood of development of the characteristic climate patterns that accompany such events. Indeed, sea-surface temperature conditions in the central and western equatorial Pacific are already of a structure similar to that of an El Niño. Sea-surface temperatures in the western equatorial Pacific are near to or slightly below normal. If this condition were to continue in combination with the above normal conditions in the central equatorial Pacific, climate patterns symptomatic of El Niño conditions could arise in the central and western tropical Pacific region and surrounding continents.

The atypical circumstances at this time make it important to consider carefully the conditions prevailing in other tropical ocean basins, as well. Regional climate fluctuations can also be driven by sea-surface temperature patterns in the tropical Atlantic and tropical Indian Oceans. Monitoring of conditions at and beneath the ocean surface in these regions is in early development, and as yet there is incomplete understanding of the mechanisms of systematic sea-surface temperature changes in these ocean basins. Nonetheless, correlations between observed anomalies in the Atlantic or Indian Oceans and local and regional seasonal climate fluctuations are important factors in making detailed interpretations of possible regional consequences of the current state of the climate system. When considering response strategies, it is important to consult National Meteorological and Hydrological Services for local and regional information.

In summary:

1. An unusual situation currently prevails: sea-surface temperatures in the central equatorial Pacific are warmer than normal and at levels typically associated with El Niño, but a basin-wide pattern is not yet established. Eastern equatorial Pacific temperatures are actually below normal, while basin-wide atmospheric patterns are not characteristic of El Niño.

2. Forecast models and expert interpretation indicate that surface temperatures in the eastern equatorial Pacific are expected to rise over the next few months, but uncertainty in the magnitude of the rise is such that outcomes for the basin-wide state of the tropical Pacific range from near-neutral to El Niño for the remainder of the year.

3. Whether or not a basin-wide El Niño develops, the unusual conditions developing in the tropical Pacific do provide important information on the range of possible climate patterns to expect for surrounding regions during the coming months.

The situation in the tropical Pacific will continue to be carefully monitored. More detailed interpretations of regional climate fluctuations will be generated routinely by the climate forecasting community over the coming months and will be made available through National Meteorological and Hydrological Services.

Climate Patterns in the Pacific

Research conducted over the past few decades has shed considerable light on the important role played by interactions between the atmosphere and ocean in the tropical belt of the Pacific Ocean in altering global weather and climate patterns. During El Niño events, for example, sea temperature at the surface in the central and eastern tropical Pacific Ocean becomes substantially higher than normal. During La Niña events, the sea surface temperatures in these regions become lower than normal. These temperature changes are strongly linked to major climate fluctuations around the globe and, once initiated, such events can last for 12 months or more. The strong El Niño event of 1997–1998 was followed by a prolonged La Niña phase that extended from mid-1998 to early 2001. The El Niño phase of 2002–2003 was not as strong as that in 1997–1998. El Niño events change the likelihood of particular climate patterns around the globe, but the outcomes of each such event are never exactly the same. Furthermore, while there is generally a relationship between the global impacts of an El Niño event and its intensity, there is always potential for an event to generate serious impacts in some regions irrespective of its intensity.

Forecasting and Monitoring the El Niño/La Niña Phenomenon

The forecasting of Pacific Ocean developments is undertaken in a number of ways. Complex computer models project the evolution of the tropical Pacific Ocean from its currently observed state. Statistical forecast models can also capture some of the precursors of such developments. Expert analysis of the current situation adds further value, especially in interpreting the implications of the evolving situation below the ocean surface. All forecast methods try to incorporate the effects of ocean-atmosphere interactions within the climate system. The meteorological and oceanographic data that allow El Niño and La Niña episodes to be monitored and forecast are drawn from national and international observing systems. The exchange and processing of the data are carried out under programs coordinated by the World Meteorological Organisation.

Chapter 13

Kyoto Protocol

INTRODUCTION

From December 1, through 11, 1997, more than 160 nations met in Kyoto, Japan, to negotiate binding limitations on greenhouse gases for the developed nations, pursuant to the objectives of the Framework Convention on Climate Change of 1992. The outcome of the meeting was the Kyoto Protocol, in which the developed nations agreed to limit their greenhouse gas emissions, relative to the levels emitted in 1990. The United States agreed to reduce emissions from 1990 levels by 7 per cent during the period 2008 to 2012.

Depending on who you talk to, the Kyoto Protocol is either: (i) an expensive, bureaucratic solution to fix a problem that may not even exist, and (ii) the last, best chance to save the world from the 'time bomb' of global warming. Those are the extremes in what has become a polarising debate that has engaged governments, consumers, environmental groups and industry all over the world for more than 20 years. The problem the Kyoto Protocol is trying to address is climate change, and more specifically, the speed at which the earth is warming up. Whether Kyoto can accomplish this is very much a matter of debate.

For the record, when the Kyoto Protocol went into effect February 16, 2005, 141 countries had ratified it, including every major industrialised country—except the United States, Australia and Monaco. The US is responsible for about a quarter of the emissions that have been blamed for global warming.

WORLD'S FASTEST GROWING POLLUTERS

Two of the world's fastest growing polluters—India and China—have signed on. But because they are considered developing countries, with other serious problems to overcome, they have been given a pass on the first Kyoto round and do not have to begin making emissions cuts until after 2012.

1. Is the climate changing?
2. What are the very long-term climate predictions?
3. What is causing the world to warm up?
4. Isn't there a lot of debate over the whole issue of climate change?
5. What does the Kyoto Protocol require?
6. Does the American decision to pull out of the Kyoto Protocol doom the deal?
7. How are emission targets met?
8. Is Canada still planning to meet its Kyoto commitments?
9. What happens if a country fails to reach its Kyoto emissions target?

Is the Climate Changing?

The United Nations certainly thinks so. And so do most (but not all) scientists who study climate. In February 2007, the United Nations Intergovernmental Panel on Climate Change (IPCC) released a report that said global warming was 'very likely'—meaning an at least 90 per cent certainty—caused by human activity. The report has some telling predictions. The document forecasts that the average temperature will rise 1.8° to 4°C by the year 2100 and sea levels will creep up by 17.8 centimetres to 58.4 centimetres by the end of the century. If polar sheets continue to melt, another rise of 9.9 centimetres to 19.8 centimetres is possible.

Past reports from the organisation have examined the changes in the previous century. In a 2001 report, the IPCC said the average global surface temperature had risen by about 0.6 degrees since 1900, with much of that rise coming in the 1990s—likely the warmest decade in 1000 years.

The IPCC also found that snow cover since the late 1960s has decreased by about 10 per cent and lakes and rivers in the Northern Hemisphere are frozen over about two weeks less each year than they were in the late 1960s. Mountain glaciers in non-polar regions have also been in 'noticeable retreat' in the 20th century, and the average global sea level has risen between 0.1 and 0.2 metres since 1900. Simply put, the world is getting warmer and the temperature is rising faster than ever.

What are the Very Long-term Climate Predictions?

The IPCC predicts more floods, intense storms, heat waves and droughts. Its study forecasts a rise of 1.4 to 5.8 degrees Celsius in the global mean surface temperature over the next 100 years, with developing countries most vulnerable. Other studies are even more apocalyptic. A report commissioned by the World Wildlife Fund predicts 'dangerous' warming of the earth's surface in as little as 20 years, with the Arctic warming so much that its polar ice could melt in the summer by the year 2100, pushing polar bears close to extinction. The Arctic Climate Impact Assessment predicts that caribou, musk ox and reindeer would find their habitats severely reduced. Northern aboriginal peoples around the world would find their way of life changed forever, the study said.

What is Causing the World to Warm Up?

Most scientists blame industrialisation. Since the 19th century, the richer countries of the Northern Hemisphere have been pumping out ever-increasing volumes of heat-trapping greenhouse gases like carbon dioxide. Industrial societies burn fossil fuels in their power plants, homes, factories and cars. They clear forests (trees absorb carbon dioxide) and they build big cities.

Greenhouse gases allow solar radiation to pass through the earth's atmosphere. But after the earth absorbs part of that radiation, it reflects the rest back. That's where the problem lies. Particles of greenhouse gas absorb the radiation, heating up, and warming the atmosphere. The increasing levels of greenhouse gases are causing too much energy to be trapped—the so-called greenhouse effect.

Is not There a Lot of Debate Over the Whole Issue of Climate Change?

While scientists tend to agree that the earth is warming, not all agree that rising greenhouse gas emissions are the culprits. A vocal minority say the earth's climate warms and cools in long cycles that have nothing to do with greenhouse gases. Some dispute the data concerning rising sea levels and rising temperatures. Others dispute the projections, which are based on computer models. But again, those views are those of a minority. Most climatologists agree that global warming is causing unprecedented climate change and that things will get worse unless something is done.

What does the Kyoto Protocol Require?

The Kyoto Protocol was adopted in late 1997 to address the problem of global warming by reducing the world's greenhouse gas emissions. It is considered a first step and is not expected to solve the world's climate change problems by the time its first commitment period ends in 2012.

Kyoto sets out an agenda for reducing greenhouse gas emissions by 5.2 per cent from 1990 levels (although 'economies in transition', like Russia, can pick different base years). Some reports say the lower target is to be met by 2010. But that's shorthand for the actual target date, which is to achieve those emission cuts over a five-year average (2008 to 2012).

All countries are not treated equally by Kyoto. Canada, for instance, has committed to chopping its greenhouse gas emissions by six per cent. The US target was a seven per cent reduction. But in 2001, one of the first acts of newly-elected President George W. Bush was to formally withdraw the US from Kyoto. Bush said the US would not ratify the treaty because it would damage the US economy and major developing nations like China and India were not covered by its provisions.

Kyoto also allows some industrialised countries to make no cuts, or even to emit more greenhouse gases than they did in 1990. Russia's and New Zealand's emission levels are capped at their 1990 levels. Iceland can emit up to 10 per cent more greenhouse gases, Australia eight per cent more. (Like the US, Australia has announced it won't ratify Kyoto). Developing nations are not subject to any emissions reduction caps under Kyoto.

Much of the criticism around the Kyoto Protocol is over political realities and the limitations of the treaty. Critics say a five per cent cut will accomplish little, especially with the United States not on board. Some Canadian critics say our economy will pay a heavy price for meeting our Kyoto commitments because we'll have to compete with an American economy that faces no such restrictions. Many doubt that Canada's target cuts can be reached in Kyoto's first phase that ends in 2012.

Others say the money to implement Kyoto would be much better spent on improving land usage and infrastructure in poor countries.

Does the American Decision to Pull Out of the Kyoto Protocol Doom the Deal?

The American decision was not enough to kill Kyoto. One of President Bush's first acts was to announce that he would not send Kyoto to the Senate for ratification—mainly because the deal had little chance of being passed. He also argued Kyoto would be bad for the US economy and would be ineffective, because major developing nations like India and China were not covered by its provisions.

But that didn't stop world ratification of the protocol. Russia came onboard on Sept. 30, 2004. That gave the deal enough support to come into effect on Feb. 16, 2005. Still, no country on the planet is responsible for producing as much greenhouse gas as the United States. Without significant action from the Americans, Kyoto's targets would be difficult to reach.

How are Emission Targets Met?

Emission targets can be met in several ways. The most obvious way is to actually reduce greenhouse gas emissions—more fuel-efficient cars, fewer coal-fired power plants. But Kyoto also allows for three other mechanisms. Countries can buy emissions credits from countries that don't need them to stay below their emissions quotas. A country can also earn emissions credits through something called joint implementation, which allows a country to benefit by carrying out something like a reforestation project in another industrialised country or 'economy in transition'. There's also what's called a clean development mechanism that encourages investment in developing countries by promoting the transfer of environmentally-friendly technologies. Each developed country must develop its own strategy to meet its Kyoto commitments. Industrial countries that ratify Kyoto are legally bound to see that their emissions do not exceed their 2008/2012 targets.

Is Canada Still Planning to Meet Its Kyoto Commitments?

In a word—no. The election of a Conservative government in 2006 brought about a reversal in Canada's climate change policy. The specific emissions reduction targets of the Kyoto Protocol—at least as far as Canada was concerned—would be abandoned.

In April 2005, then Prime Minister Paul Martin and his Liberal government unveiled what they called Moving Forward on Climate Change: A Plan for Honouring Our Kyoto Commitment. Under their revised plan, the Liberals pledged to spend $10 billion over seven years to help Canada cut its average greenhouse gas emissions by 270 megatons a year from 2008 to 2012.

However, when Prime Minister Stephen Harper and the Conservative government tabled the federal budget in May 2006, there wasn't a single mention of the Kyoto Protocol. Finance Minister Jim Flaherty repeated his pledge to develop a $2 billion, five-year 'made-in-Canada' climate change plan, but there were no details. The budget also set aside $370 million over two years for a new tax credit that would benefit commuters who buy monthly transit passes. In September 2006, Environment Minister Rona Ambrose said Canada had no chance of meeting its targets under the Kyoto Protocol. She accused the Liberals of wasting $1 billion on emission-reduction efforts without keeping the country on track to meet its promises under the international agreement. 'Kyoto did not fail this country', Ambrose said. 'The Liberal Party of Canada failed Kyoto'. Ambrose said the government would instead act on greenhouse gases and other pollution with new targets in a proposed clean air act, announced in October 2006. The Clean Air Act targets would be 'intensity-based', meaning that environmental emissions would be relative to the economic output of various industries. That means even though individual emission limits for each barrel of oil or piece of coal could be lowered, if production increases, the overall amount of greenhouse gas emissions and air pollutants could grow.

Critics of intensity-based targets say the approach allows heavily polluting industries, such as Alberta's oilsands, to continue to grow and pollute while remaining under government-imposed limitations.

The bill does not set short-term targets to cut greenhouse gas emissions, and its emissions regulations on large polluters don't take effect until 2010.

What Happens if a Country Fails to Reach Its Kyoto Emissions Target?

The Kyoto protocol contains measures to assess performance and progress. It also contains some penalties. Countries that fail to meet their emissions targets by the end of the first commitment period (2012) must make up the difference plus a penalty of 30 per cent in the second commitment period. Their ability to sell credits under emissions trading will also be suspended.

The protocol was developed under the United Nations Framework Convention on Climate Change (UNFCCC). Participating countries that have ratified the Kyoto Protocol have committed to cut emissions of not only carbon dioxide, but of also other greenhouse gases, being:

1. Methane (CH_4).
2. Nitrous oxide (N_2O).
3. Hydrofluorocarbons (HFCs).
4. Perfluorocarbons (PFCs).
5. Sulphur hexafluoride (SF_6).

If participant countries continue with emissions above the targets, then they are required to engage in emissions trading, i.e. buying 'credits' from other participant countries who are able to exceed their reduction targets in order to offset.

The goals of Kyoto were to see participants collectively reducing emissions of greenhouse gases by 5.2 per cent below the emission levels of 1990 by 2012. While the 5.2 per cent figure is a collective one, individual countries were assigned higher or lower targets and some countries were permitted increases. For example, the USA was expected to reduce emissions by 7 per cent. This chart gives you an idea why different countries were apportioned different targets (Fig. 13.1).

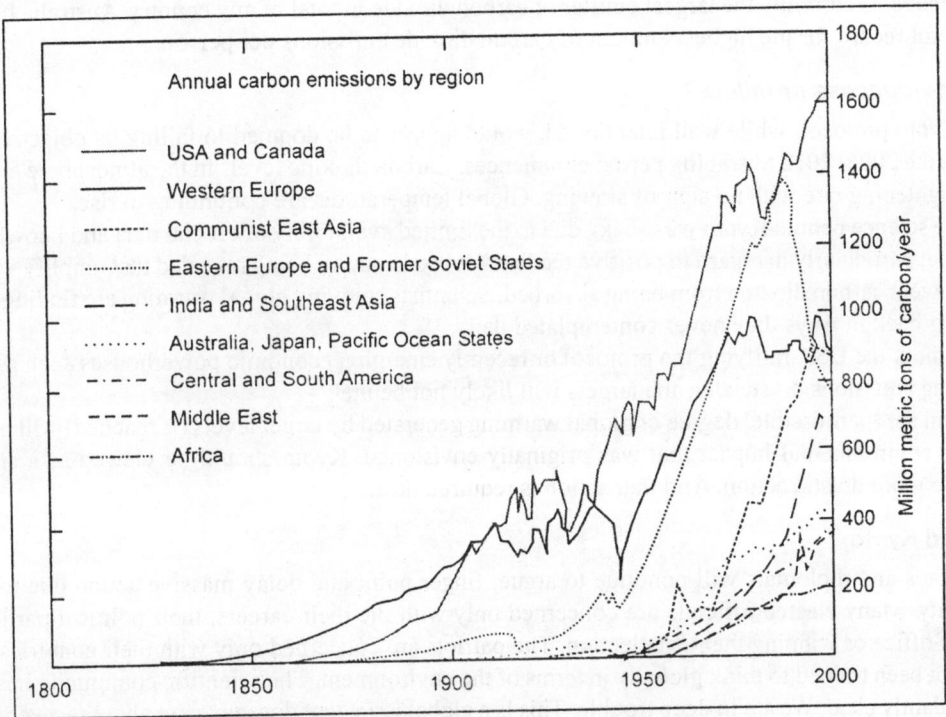

Fig. 13.1. Graph by Robert A. Rohde.

India and China, which have ratified the Kyoto Protocol, are not obligated to reduce greenhouse gas production at the moment as they are developing countries, i.e. they weren't seen as the main culprits for emissions during the period of industrialisation thought to be the cause for the global warming of today. This is a little odd given that China is about to overtake the USA in emissions, but take into account the major differences in population and that much of the production in these countries is fuelled by demand from the West and influence from the West on their own culture. As a result of this loophole, the West has effectively outsourced much of its carbon emissions to China and India.

This phenomenon, whether intended or coincidental is a major hole in the Kyoto Protocol.

Signing versus Ratification

While almost every country in the world has signed the Kyoto Protocol, the signature alone is symbolic; a token gesture of support. Ratification carries legal obligations and effectively becomes a contractual arrangement. 169 countries have ratified the agreement. Of the signatories, only 2 refused to ratify Kyoto up until December of 2007—Australia and the USA.

Australia negotiated hard when the Kyoto Protocol was being developed; in fact it was to be allowed an 8 per cent increase in emissions. Even so, Australia refused to ratify the agreement until a change in government in late 2007. The excuse—it will be bad for Australia's economy, the same reasoning the USA uses. But regardless of even that, in order to have a health economy, you need a reasonably healthy environment to support it.

What makes the USA and Australia's (previous) position even more untenable is that the USA, as seen above, is currently the largest emitter of carbon dioxide in total of any country. Australia holds the shameful record for the highest amount of carbon dioxide emissions per person.

Kyoto—success or failure?

The Kyoto protocol, while well intentioned, would appear to be doomed to failing its objectives even before the 2008–2012 averaging period commences. Carbon dioxide levels in the atmosphere are rising at a frightening rate with no sign of slowing. Global temperatures are continuing to rise.

The science behind Kyoto was shaky due to the limited availability of crucial data and knowledge at the time; particularly in regard to positive feedback loops in nature being revealed that amplify warming and prevent carbon dioxide from being absorbed. Scientists studying global warming are finding Nature fighting back in ways they never contemplated daily.

Without the USA ratifying the protocol or recently emerging economic powerhouses such as China reducing emissions drastically; the targets will likely not be met.

Even the 'permissible' degree of global warming generated by target levels (if reached) will have far greater environmental impact that was originally envisioned. Kyoto should be viewed as a stepping stone to more drastic action. And that action is required now.

Beyond Kyoto

Politicians and diplomats will continue to argue, finger point and delay massive action due to a silo mentality. Many elected officials are concerned only with the their careers, their political parties, the term of office or winning the next election. The patriots are concerned only with their countries. They have not been trained to think globally in terms of the environment. The scientific community has made it abundantly clear. We are in deep trouble. This is a global issue that does not care about race, colour or creed, nor political affiliation, although ironically the people who produce the least emissions will be the ones to suffer the most. That's always been the way of humanity.

It's down to us as individuals to not only do what we can to reduce our own carbon emissions, but to raise the awareness of others until collectively our shouts are such a mighty voice that no politician can ignore it. Better they hear the shouts of protest now than the screams of agony from wars over natural resources or the wailing of starvation in the future and it may well be their own voices amongst the anguish; that's how little time we have left.

OBJECTIVES

The objective is the 'stabilisation and reconstruction of greenhouse gas concentrations in the atmosphere at a level that would prevent dangerous anthropogenic interference with the climate system'. The objective of the Kyoto climate change conference was to establish a legally binding international agreement, whereby all the participating nations commit themselves to tackling the issue of global warming and greenhouse gas emissions (Fig. 13.2). The target agreed upon was an average reduction of 5.2 per cent from 1990 levels by the year 2012. According to the treaty, in 2012, Annex I (Table 13.1) countries must have fulfilled their obligations of reduction of greenhouse gases emissions established for the first commitment period (2008–2012). The Protocol expires at the end of 2012.

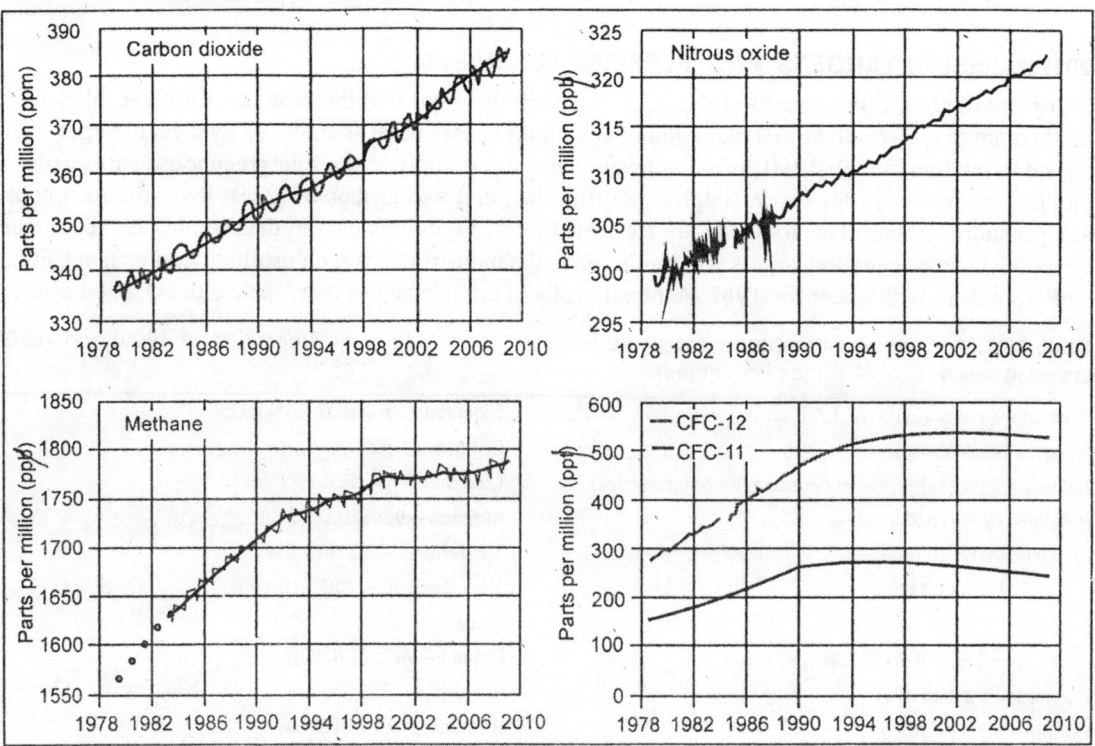

Fig. 13.2. Kyoto is intended to cut global emissions of greenhouse gases.

Principal Concepts of Kyoto Protocol

The five principal concepts of the Kyoto Protocol are:

1. Commitments to the Annex-countries: The heart of the Protocol lies in establishing commitments for the reduction of greenhouse gases that are legally binding for Annex I countries. Dividing the countries in different groups is one of the key concepts in making commitments possible, where only the Annex I countries in 1997, where seen as having the economic capacity to commit themselves and their industry. Making only the few nations in the Annex 1 group committed to the protocols limitations.

2. Implementation: In order to meet the objectives of the Protocol, Annex I countries are required to prepare policies and measures for the reduction of greenhouse gases in their respective countries. In addition, they are required to increase the absorption of these gases and utilise all mechanisms available, such as joint implementation, the clean development mechanism and emissions trading, in order to be rewarded with credits that would allow more greenhouse gas emissions at home.

3. Minimising impacts on developing countries by establishing an adaptation fund for climate change.
4. Accounting, Reporting and Review in order to ensure the integrity of the Protocol.
5. Compliance: Establishing a Compliance Committee to enforce compliance with the commitments under the Protocol.

2012 EMISSION TARGETS AND 'FLEXIBLE MECHANISMS'

Thirty-nine of the forty Annex I (Table 13.1) countries have ratified the protocol. Of these thirty-four have committed themselves to a reduction of greenhouse gases (GHG) produced by them to targets that are set in relation to their 1990 emission levels. The targets apply to the four greenhouse gases carbon dioxide, methane, nitrous oxide, sulphur hexafluoride, and two groups of gases, hydrofluorocarbons and perfluorocarbons. The six GHG are translated into CO_2 equivalents in determining reductions in emissions. These reduction targets are in addition to the industrial gases, chlorofluorocarbons, or CFCs, which are dealt with under the 1987 Montreal Protocol on Substances that Deplete the Ozone Layer.

Table 13.1. Annex I countries under the Kyoto Protocol, their 2012 commitments (% of 1990) and 1990 emission levels (% of all Annex I countries).

Australia – 108% (2.1% of 1990 emissions)	Liechtenstein – (0.0015%) 92%
Austria – 92% (0.4%)	Lithuania – 92%
Belarus – 95% (subject to acceptance by other parties)	Luxembourg – 92% (0.1%)
Belgium – 92% (0.8%)	Monaco – 92% (0.0015%)
Bulgaria – 92% (0.6%)	Netherlands – 92% (1.2%)
Canada – 94% (3.33%)	New Zealand – 100% (0.19%)
Croatia – 95%	Norway – 101% (0.26%)
Czech Republic – 92% (1.24%)	Poland – 94% (3.02%)
Denmark – 92% (0.4%)	Portugal – 92% (0.3%)
Estonia – 92% (0.28%)	Romania – 92% (1.24%)
Finland – 92% (0.4%)	Russian Federation – 100% (17.4%)
France – 92% (2.7%)	Slovakia – 92% (0.42%)
Germany – 92% (7.4%)	Slovenia – 92%
Greece – 92% (0.6%)	Spain – 92% (1.9%)
Hungary – 94% (0.52%)	Sweden – 92% (0.4%)
Iceland – 110% (0.02%)	Switzerland – 92% (0.32%)
Ireland – 92% (0.2%)	Turkey
Italy – 92% (3.1%)	Ukraine – 100%
Japan – 94% (8.55%)	United Kingdom – 92% (4.3%)
Latvia – 92% (0.17%)	United States of America – 93% (36.1%) (nonparty)

Under the Protocol, only the Annex I countries have committed themselves to national or joint reduction targets, [formally called 'quantified emission limitation and reduction objectives' (QELRO)—Article 4.1] that range from a joint reduction of 8 per cent for the European Union and others, to 7 per cent for the United States (nonbinding as the US is not a signatory), 6 per cent for Japan and 0 per cent for Russia. The treaty permits emission increases of 8 per cent for Australia and 10 per cent for Iceland.

Emission limits do not include emissions by international aviation and shipping. Annex I countries can achieve their targets by allocating reduced annual allowances to major operators within their borders, or by allowing these operators to exceed their allocations by offsetting any excess through a mechanism that is agreed by all the parties to the UNFCCC, such as by buying emission allowances from other operators which have excess emissions credits.

38 of the 39 Annex I countries have agreed to cap their emissions in this way, two others are required to do so under their conditions of accession into the EU, and one more (Belarus) is seeking to become an Annex I country.

Flexible Mechanisms

The Protocol defines three 'flexibility mechanisms' that can be used by Annex I countries in meeting their emission reduction commitments. The flexibility mechanisms are International Emissions Trading (IET), the Clean Development Mechanism (CDM), and Joint Implementation (JI). IET allows Annex I countries to 'trade' their emissions (Assigned Amount Units, AAUs, or 'allowances' for short). For IET, the economic basis for providing this flexibility is that the marginal cost of emission abatement differs among countries. Trade could potentially allow the Annex I countries to meet their emission reduction commitments at a reduced cost. This is because trade allows emissions to be abated first in countries where the costs of abatement are lowest, thus increasing the efficiency of the Kyoto agreement.

The CDM and JI are called 'project-based mechanisms', in that they generate emission reductions from projects. The difference between IET and the project-based mechanisms is that IET is based on the setting of a quantitative restriction of emissions, while the CDM and JI are based on the idea of 'production' of emission reductions. The CDM is designed to encourage production of emission reductions in non-Annex I countries, while JI encourages production of emission reductions in Annex I countries.

The production of emission reductions generated by the CDM and JI can be used by Annex B countries in meeting their emission reduction commitments. The emission reductions produced by the CDM and JI are both measured against a hypothetical baseline of emissions that would have occurred in the absence of a particular emission reduction project. The emission reductions produced by the CDM are called Certified Emission Reductions (CERs); reductions produced by JI are called Emission Reduction Units (ERUs). The reductions are called 'credits' because they are emission reductions credited against a hypothetical baseline of emissions.

International emissions trading

The most advanced emissions trading system (ETS) is the one developed by the EU. Ellerman and Buchner suggested that during its first two years in operation, the EU ETS turned an expected increase in emissions of 1–2 per cent per year into a small absolute decline. Grubb suggested that a reasonable estimate for the emissions cut achieved during its first two years of operation was 50–100 $MtCO_2$ per year, or 2.5–5 per cent.

Clean development mechanism

Between 2001, which was the first year Clean Development Mechanism (CDM) projects could be registered, and 2012, the end of the Kyoto commitment period, the CDM is expected to produce some 1.5 billion tons of carbon dioxide equivalent (CO_2e) in emission reductions. Most of these reductions are through renewable energy, energy efficiency, and fuel switching. By 2012, the largest potential for production of CERs are estimated in China (52 per cent of total CERs) and India (16 per cent). CERs produced in Latin America and the Caribbean make up 15 per cent of the potential total, with Brazil as the largest producer in the region (7 per cent).

Joint implementation

The formal crediting period for joint implementation (JI) was aligned with the first commitment period of the Kyoto Protocol, and did not start until January 2008. In November 2008, only 22 JI projects had been officially approved and registered. The total projected emission savings from JI by 2012 are about one-tenth that of the CDM. Russia accounts for about two-thirds of these savings, with the remainder divided up roughly equally between the Ukraine and the EU's New Member States. Emission savings include cuts in methane, HFC, and N_2O emissions.

DETAILS OF THE AGREEMENT

According to a press release from the United Nations Environment Program:
> 'After 10 days of tough negotiations, ministers and other high-level officials from 160 countries reached agreement this morning on a legally binding Protocol under which industrialised countries will reduce their collective emissions of greenhouse gases by 5.2 per cent. The agreement aims to lower overall emissions from a group of six greenhouse gases by 2008–12, calculated as an average over these five years. Cuts in the three most important gases—carbon dioxide (CO_2), methane (CH_4), and nitrous oxide (N_2O)—will be measured against a base year of 1990. Cuts in three long-lived industrial gases—hydrofluorocarbons (HFCs), perfluorocarbons (PFCs), and sulphur hexafluoride (SF_6)—can be measured against either a 1990 or 1995 baseline.'

National limitations range from 8 per cent reductions for the European Union and others, to 7 per cent for the US, 6 per cent for Japan, 0 per cent for Russia, and permitted increases of 8 per cent for Australia and 10 per cent for Iceland.

The agreement supplements the United Nations Framework Convention on Climate Change (UNFCCC) adopted at the Earth Summit in Rio de Janeiro in 1992, which did not set any limitations or enforcement mechanisms. All parties to UNFCCC can sign or ratify the Kyoto Protocol, while non-parties to UNFCCC cannot. The Kyoto Protocol was adopted at the third session of the Conference of Parties to the UNFCCC (COP 3) in 1997 in Kyoto, Japan. Most provisions of the Kyoto Protocol apply to developed countries, listed in Annex I to UNFCCC.

National emission targets exclude international aviation and shipping. Kyoto Parties can use land use, land use change, and forestry (LULUCF) in meeting their targets. LULUCF activities are also called 'sink' activities. Changes in sinks and land use can have an effect on the climate (IPCC, 2007). Particular criteria apply to the definition of forestry under the Kyoto Protocol.

Forest management, cropland management, grazing land management, and revegetation are all eligible LULUCF activities under the protocol. Annex I parties use of forestry management in meeting their targets is capped.

Common but Differentiated Responsibility

UNFCCC adopts a principle of 'common but differentiated responsibilities'. The parties agreed that:
1. The largest share of historical and current global emissions of greenhouse gases originated in developed countries.
2. Per capita emissions in developing countries are still relatively low.

3. The share of global emissions originating in developing countries will grow to meet social and development needs.

Emissions

Per-capita emissions are a country's total emissions divided by its population. Per-capita emissions in the industrialised countries are typically as much as ten times the average in developing countries. This is one reason industrialised countries accepted responsibility for leading climate change efforts in the Kyoto negotiations. In Kyoto, the countries that took on quantified commitments for the first period (2008–12) corresponded roughly to those with per-capita emissions in 1990 of two tons of carbon or higher. In 2005, the top-20 emitters comprised 80 per cent of total GHG emissions. Countries with a Kyoto target made up 20 per cent of total GHG emissions.

Another way of measuring GHG emissions is to measure the total emissions that have accumulated in the atmosphere over time. Over a long time period, cumulative emissions provide an indication of a country's total contribution to GHG concentrations in the atmosphere. Over the 1900–2005 period, the US was the world's largest cumulative emitter of energy-related CO_2 emissions, and accounted for 30 per cent of total cumulative emissions. The second largest emitter was the EU, at 23 per cent; the third largest was China, at 8 per cent; fourth was Japan, at 4 per cent; fifth was India, at 2 per cent. The rest of the world accounted for 33 per cent of global, cumulative, energy-related CO_2 emissions.

Top-ten emitters

What follows is a ranking of the world's top ten emitters of GHGs for 2005. The first figure is the country's or region's emissions as a percentage of the global total. The second figure is the country's/ region's per-capita emissions, in units of tons of GHG per-capita:

1. China* – 17%, 5.8.
2. United States*** – 16%, 24.1.
3. European Union-27*** – 11%, 10.6.
4. Indonesia** - 6%, 12.9.
5. India – 5%, 2.1.
6. Russia*** – 5%, 14.9.
7. Brazil – 4%, 10.0.
8. Japan*** – 3%, 10.6.
9. Canada*** – 2%, 23.2.
10. Mexico – 2%, 6.4.

Notes

1. These values are for the GHG emissions from fossil fuel use and cement production. Calculations are for carbon dioxide (CO_2), methane (CH_4), nitrous oxide (N_2O) and gases containing fluorine (the F-gases HFCs, PFCs and SF_6).
2. These estimates are subject to large uncertainties regarding CO_2 emissions from deforestation; and the per country emissions of other GHGs (e.g. methane). There are also other large uncertainties which mean that small differences between countries are not significant. CO_2 emissions from the decay of remaining biomass after biomass burning/deforestation are not included.
3. *Excluding underground fires.

4. **Including an estimate of 2000 million tons CO_2 from peat fires and decomposition of peat soils after draining. However, the uncertainty range is very large.
5. ***Industrialised countries: official country data reported to UNFCCC.

Financial commitments

The protocol also reaffirms the principle that developed countries have to pay billions of dollars, and supply technology to other countries for climate-related studies and projects. The principle was originally agreed in UNFCCC.

Revisions

The protocol left several issues open to be decided later by the sixth Conference of Parties (COP). COP6 attempted to resolve these issues at its meeting in the Hague in late 2000, but was unable to reach an agreement due to disputes between the European Union on the one hand (which favoured a tougher agreement) and the United States, Canada, Japan and Australia on the other (which wanted the agreement to be less demanding and more flexible). In 2001, a continuation of the previous meeting (COP6bis) was held in Bonn where the required decisions were adopted. After some concessions, the supporters of the protocol (led by the European Union) managed to get Japan and Russia in as well by allowing more use of carbon dioxide sinks. COP7 was held from 29 October 2001 through 9 November 2001 in Marrakech to establish the final details of the protocol.

The first Meeting of the Parties to the Kyoto Protocol (MOP1) was held in Montreal from 28 November to 9 December 2005, along with the 11th conference of the Parties to the UNFCCC (COP11). The 3, December 2007, Australia ratified the protocol during the first day of the COP13 in Bali.

Of the signatories, 36 developed C.G. countries (plus the EU as a party in the European Union) agreed to a 10 per cent emissions increase for Iceland; but, since the EU's member states each have individual obligations, much larger increases (up to 27 per cent) are allowed for some of the less developed EU countries. Reduction limitations expire in 2013.

Enforcement

If the enforcement branch determines that an annex I country is not in compliance with its emissions limitation, then that country is required to make up the difference plus an additional 30 per cent. In addition, that country will be suspended from making transfers under an emissions trading program.

Negotiations

Article 4.2 of the UNFCCC commits industrialised countries to '[take] the lead' in reducing emissions. The initial aim was for industrialised countries to stabilise their emissions at 1990 levels by the year 2000. The failure of key industrialised countries to move in this direction was a principal reason why Kyoto moved to binding commitments.

At the first UNFCCC Conference of the Parties in Berlin, the G77 (a coalition of 77 developing nations within the UN) was able to push for a mandate where it was recognized that:

1. Developed nations had contributed most to the then-current concentrations of GHGs in the atmosphere.
2. Developing country emissions per-capita were still relatively low.
3. And that the share of global emissions from developing countries would grow to meet their development needs.

This mandate was recognised in the Kyoto Protocol in that developing countries were not subject to emission reduction commitments in the first Kyoto commitment period. However, the large potential for growth in developing country emissions made negotiations on this issue tense. In the final agreement, the Clean Development Mechanism was designed to limit emissions in developing countries, but in such a way that developing countries do not bear the costs for limiting emissions. The general assumption was that developing countries would face quantitative commitments in later commitment periods, and at the same time, developed countries would meet their first round commitments.

Base year

The choice of the 1990 main base year remains in Kyoto, as it does in the original Framework Convention. The desire to move to historical emissions was rejected on the basis that good data was not available prior to 1990. The 1990 base year also favoured several powerful interests including the UK, Germany and Russia. This is because the UK and Germany had high CO_2 emissions in 1990.

In the UK following 1990, emissions had declined because of a switch from coal to gas (dash for gas), which has lower emissions than coal. This was due to the UK's privatisation of coal mining and its switch to natural gas supported by North sea reserves. Germany benefited from the 1990 base year because of its reunification between West and East Germany. East Germany's emissions fell dramatically following the collapse of East German industry after the fall of the Berlin Wall. Germany could therefore take credit for the resultant decline in emissions.

Japan promoted the idea of flexible baselines, and favoured a base year of 1995 for HFCs. Their HFC emissions had grown in the early 1990s as a substitute for CFCs banned in the Montreal Protocol. Some of the former Soviet satellites wanted a base year to reflect their highest emissions prior to their industrial collapse. EIT countries are privileged by being able to choose their base-year nearly freely. However the oldest base-year accepted is 1986.

Emissions cuts

The G77 wanted strong uniform emission cuts across the developed world of 15 per cent. Countries, such as the US, made suggestions to reduce their responsibility to reduce emissions. These suggestions included:

1. The inclusion of carbon sinks (e.g. by including forests, that absorb CO_2 from the atmosphere).
2. And having net current emissions as the basis for responsibility, i.e. ignoring historical emissions.

The US originally proposed for the second round of negotiations on Kyoto commitments to follow the negotiations of the first. In the end, negotiations on the second period were set to open no later than 2005. Countries over-achieving in their first period commitments can 'bank' their unused allowances for use in the subsequent period.

The EU initially argued for only three GHGs to be included—CO_2, CH_4, and N_2O—with other gases such as HFCs regulated separately. The EU also wanted to have a 'bubble' commitment, whereby it could make a collective commitment that allowed some EU members to increase their emissions, while others cut theirs. The most vulnerable nations—the Association of Small Island States (AOSIS)—pushed for deep uniform cuts by developed nations, with the goal of having emissions reduced to the greatest possible extent.

The final days of negotiation of the Protocol saw a clash between the EU and the US and Japan. The EU aimed for flat-rate reductions in the range of 10–15 per cent below 1990 levels, while the US and Japan supported reductions of 0–5 per cent. Countries that had supported differentiation had different ideas as to how it should be calculated, and many different indicators were proposed: relating to GDP, energy intensity (energy use per unit of economic output), etc. According to Grubb, the only common theme of these indicators was that each proposal suited the interests of the country making the proposal.

The final commitments negotiated in the protocol are the result of last minute political compromises. These include an 8 per cent cut from the 1990 base year for the EU, 7 per cent for the US, 6 per cent for Canada and Japan, no cut for Russia, and an 8 per cent increase for Australia. This sums to an overall cut of 5.2 per cent below 1990 levels. Since Australia and the US did not ratify the treaty (although Australia has since done), the cut is reduced from 5.2 per cent to about 2 per cent.

Considering the growth of some economies and the collapse of others since 1990, the range of implicit targets is much greater. The US faced a cut of about 30 per cent below 'business-as-usual' (BAU) emissions (i.e. predicted emissions should there be no attempt to limit emissions), while Russia and other economies in transition faced targets that allowed substantial increases in their emissions above BAU. On the other hand, Grubb pointed out that the US, having per-capita emissions twice that of most other OECD countries, was vulnerable to the suggestion that it had huge potential for making reductions. From this viewpoint, the US was obliged to cut emissions back more than other countries.

Flexibility mechanisms

Negotiations over the flexibility mechanisms included in the Protocol proved controversial. Japan and some EU member states wanted to ensure that any emissions trading would be competitive and transparent. Their intention was to prevent the US from using its political leverage to gain preferential access to the likely surplus in Russian emission allowances. The EU was also anxious to prevent the US from avoiding domestic action to reduce its emissions. Developing countries were concerned that the US would use flexibility to its own advantage, over the interests of weaker countries.

Compliance

The protocol defines a mechanism of 'compliance' as a 'monitoring compliance with the commitments and penalties for noncompliance'. According to Grubb, the explicit consequences of noncompliance of the treaty are weak compared to domestic law. Yet, the compliance section of the treaty was highly contested in the Marrakesh Accords. According to Grubb, Japan made some unsuccessful efforts to 'water-down' the compliance package.

GOVERNMENT ACTION AND EMISSIONS

Annex I

In total, Annex I Parties to the UNFCCC (including the US) managed a cut of 3.3 per cent in GHG emissions between 1990 and 2004. In 2007, projections indicated rising emissions of 4.2 per cent between 1990 and 2010. This projection assumed that no further mitigation action would be taken. The reduction in the 1990s was driven significantly by economic restructuring in the economies-in-transition (EITs). Emission reductions in the EITs had little to do with climate change policy. Some reductions in Annex I emissions have occurred due to policy measures, such as promoting energy efficiency.

Progress towards targets

Progress toward the emission reduction commitments set in the Kyoto Protocol has been mixed. World Bank reported that there were significant differences in performance across individual countries:

1. For the Annex I non-Economies-in-Transition (non-EIT) Kyoto Protocol (KP) Parties, emissions in 2005 were 5 per cent higher than 1990 levels. Their Kyoto target for 2008–2012 is for a 6 per cent reduction in emissions. The Annex I non-EITs KP Parties are Australia, Austria, Belgium, Canada, Denmark, Finland, France, Germany, Greece, Iceland, Ireland, Italy, Japan, Liechtenstein, Luxembourg, Monaco, Netherlands, New Zealand, Norway, Portugal, Spain, Sweden, Switzerland, and the United Kingdom.
2. The Annex I Economies in Transition (EIT) KP Parties emissions in 2005 were 35 per cent below 1990 levels. Their Kyoto target is for a 2 per cent reduction. The Annex I EIT KP Parties are Belarus, Bulgaria, Croatia, Czech Republic, Estonia, Hungary, Latvia, Lithuania, Poland, Romania, Russian Federation, Slovakia, Slovenia, and Ukraine.
3. In 2005, the Annex I non-KP Parties emissions were 18 per cent above their 1990 levels. The Annex I non-KP Parties are Turkey and the United States.
4. In total, the Annex I KP Parties emissions for 2005 were 14 per cent below their 1990 levels. Their Kyoto target is for a 4 per cent reduction.

KP parties

According to the Netherlands Environmental Assessment Agency, the industrialised countries with a Kyoto target will, as a group, probably meet their emission limitation requirements. Collectively, this was for a 4 per cent reduction relative to 1990 levels. A linear extrapolation of the 2000–2005 emissions trend led to a projected emission reduction in 2010 of almost 11 per cent. Including the potential contribution of CDM projects, which may account for emissions reductions of approximately 500 megatons CO_2-eq per year, the reduction might be as large as 15 per cent.

The expected reduction of 11 per cent was attributed to the limited increase in emissions in OECD countries, but was particularly due to the large reduction of about 40 per cent until 1999 in the EITs. The reduction in emissions for the smaller EITs aids the EU-27 in meeting their collective target. The EU expects that it will meet its collective target of an 8 per cent reduction for the EU-15. This reduction includes:

1. CDM and JI projects, which are planned to contribute 2.5 per cent towards the target.
2. Carbon storage in forests and soils (carbon sinks), which contribute another 0.9 per cent.

Japan expects to meet its Kyoto target, which includes a 1.6 per cent reduction from CDM projects and a 3.9 per cent reduction from carbon storage, contributing to a total reduction of 5.5 per cent. In other OECD countries, emissions have increased. In Canada, Australia, New Zealand and Switzerland, emissions have increased by 25 per cent compared to the base year, while in Norway, the increase was 9 per cent. In the view of PBL, these countries will only be able to meet their targets by purchasing sufficient CDM credits or by buying emissions (hot air) from EIT countries.

Non-KP parties

Emissions in the US have increased 16 per cent since 1990. According to PBL, the US will not meet its original Kyoto target of a 6 per cent reduction in emissions.

Non-Annex I

UNFCCC compiled and synthesised information reported to it by non-Annex I Parties. Most non-Annex I Parties belonged in the low-income group, with very few classified as middle-income. They are not obligated by the limits of emissions in the Kyoto Protocol. Fast growing economy countries like China, South Africa, India and Brazil are still in this non-obligated group. Most Parties included information on policies relating to sustainable development. Sustainable development priorities mentioned by non-Annex I Parties included poverty alleviation and access to basic education and health care. Many non-Annex I Parties are making efforts to amend and update their environmental legislation to include global concerns such as climate change.

A few parties, e.g. South Africa and Iran, stated their concern over how efforts to reduce emissions could affect their economies. The economies of these countries are highly dependent on income generated from the production, processing, and export of fossil fuels.

As the non-Annex 1 countries aren't obligated to any commitment on emissions some critics argue that their signatures on the protocol have been free and unsignificant.

Emissions

GHG emissions, excluding land use change and forestry (LUCF), reported by 122 non-Annex I Parties for the year 1994 or the closest year reported, totalled 11.7 billion tons (billion = 1,000,000,000) of CO_2-eq. CO_2 was the largest proportion of emissions (63 per cent), followed by methane (26 per cent) and nitrous oxide (N_2O) (11 per cent).

The energy sector was the largest source of emissions for 70 Parties, whereas for 45 Parties the agriculture sector was the largest. Per capita emissions (in tons of CO_2-eq, excluding LUCF) averaged 2.8 tons for the 122 non-Annex I Parties.

1. The Africa region's aggregate emissions were 1.6 billion tons, with per capita emissions of 2.4 tons.
2. The Asia and Pacific region's aggregate emissions were 7.9 billion tons, with per capita emissions of 2.6 tons.
3. The Latin America and Caribbean region's aggregate emissions were 2 billion tons, with per capita emissions of 4.6 tons.
4. The 'other' region includes Albania, Armenia, Azerbaijan, Georgia, Malta, Republic of Moldova, and the former Yugoslav Republic of Macedonia. Their aggregate emissions were 0.1 billion tons, with per capita emissions of 5.1 tons.

Parties reported a high level of uncertainty in LUCF emissions, but in aggregate, there appeared to only be a small difference of 1.7 per cent with and without LUCF. With LUCF, emissions were 11.9 billion tons, without LUCF, total aggregate emissions were 11.7 billion tons.

Trends

In several large developing countries and fast growing economies (China, India, Thailand, Indonesia, Egypt, and Iran) GHG emissions have increased rapidly. For example, emissions in China have risen strongly over the 1990–2005 period, often by more than 10 per cent year. Emissions per-capita in non-Annex I countries are still, for the most part, much lower than in industrialised countries. Non-Annex I countries do not have quantitative emission reduction commitments, but they are committed to mitigation actions. China, for example, has had a national policy program to reduce emissions growth, which included the closure of old, less efficient coal-fired power plants.

VIEWS ON THE PROTOCOL

Smith and others assessed the literature on climate change policy. They found that no authoritative assessments of the UNFCCC or its Protocol asserted that these agreements had, or will, succeed in solving the climate problem. In these assessments, it was assumed that the UNFCCC or its Protocol would not be changed. The Framework Convention and its Protocol include provisions for future policy actions to be taken.

World Bank commented on how the Kyoto Protocol had only had a slight effect on curbing global emissions growth. The treaty was negotiated in 1997, but by 2005, energy-related emissions had grown 24 per cent. World Bank also stated that the treaty had provided only limited financial support to developing countries to assist them in reducing their emissions and adapting to climate change.

Some of the criticism of the protocol has been based on the idea of climate justice. This has particularly centred on the balance between the low emissions and high vulnerability of the developing world to climate change, compared to high emissions in the developed world.

Some environmentalists have supported the Kyoto Protocol because it is 'the only game in town', and possibly because they expect that future emission reduction commitments may demand more stringent emission reductions. In 2001, sixteen national science academies stated that ratification of the protocol represented a 'small but essential first step towards stabilising atmospheric concentrations of greenhouse gases'. Some environmentalists and scientists have criticised the existing commitments for being too weak. Many economists think that the commitments are stronger than is justified. The lack of quantitative emission commitments for developing countries led the US and Australia (under Prime Minister John Howard) to decide not to ratify the treaty. Australia, under former Prime Minister Kevin Rudd, has since ratified the treaty. Despite ratification, Australia has thus far not implemented legislation to bring itself into compliance.

An editor has expressed a concern that this paragraph lends undue weight to certain ideas, incidents, controversies or matters relative to the article subject as a whole. Please help to create a more balanced presentation. Discuss and resolve this issue before removing this message.

In May 2010 the Hartwell Paper was published by the London School of Economics with funding from the Japan Iron and Steel Federation, Tokyo, Japan and Japan Automobile Manufacturers Association, Inc., Tokyo, Japan. The authors argued that after what they regard as the failure of the 2009 Copenhagen Climate Summit, the Kyoto Protocol crashed and they claimed that it 'has failed to produce any discernable real world reductions in emissions of greenhouse gases in fifteen years'. They argued that this failure opened an opportunity to set climate policy free from Kyoto and the paper advocates a controversial and piecemeal approach to decarbonisation of the global economy.

SUCCESSOR

In the nonbinding 'Washington Declaration' agreed on 16 February 2007, Heads of governments from Canada, France, Germany, Italy, Japan, Russia, United Kingdom, the United States, Brazil, China, India, Mexico and South Africa agreed in principle on the outline of a successor to the Kyoto Protocol. They envisage a global cap-and-trade system that would apply to both industrialised nations and developing countries, and hoped that this would be in place by 2009.

On 7 June 2007, leaders at the 33rd G8 summit agreed that the G8 nations would 'aim to at least halve global CO_2 emissions by 2050'. The details enabling this to be achieved would be negotiated by environment ministers within the United Nations Framework Convention on Climate Change in a process that would also include the major emerging economies.

A round of climate change talks under the auspices of the United Nations Framework Convention on Climate Change (UNFCCC) concluded in 31st August, 2007 with agreement on key elements for an effective international response to climate change. A key feature of the talks was a United Nations report that showed how efficient energy use could yield significant cuts in emissions at low cost. The talks were meant to set the stage for a major international meeting to be held in Nusa Dua, Bali, which started on 3rd December, 2007.

The Conference was held in December 2008 in Poznan, Poland. One of the main topics on this meeting was the discussion of a possible implementation of avoided deforestation also known as Reducing emissions from deforestation and forest degradation (REDD) into the future Kyoto Protocol.

After the lack of progress leading to a binding commitment or an extension of the Kyoto commitment period in climate talks at COP 15 in Copenhagen, Denmark in 2009, there are several further rounds of negotiation COP 16 in Cancun, Mexico in 2010, South Africa in 2011 (COP 17), and in either Qatar or South Korea in 2012 (COP 18). Because any treaty change will require the ratification of the text by various countries' legislatures before the end of the commitment period Dec. 31, 2012, it is likely that agreements in South Africa or South Korea/Qatar will be too late to prevent a gap between the commitment periods.

RUSSIA BREAKS THE ICE ON KYOTO

Russia has sold quotas on noxious and greenhouse emissions, for the first time since the coming into force of the Kyoto Protocol. Japan's Mitsubishi and Nippon Oil have purchased quotas on 290 thousand tons of greenhouses gases from the Russian 'Gazprom Neft'.

The Kyoto Protocol became the first globally signed agreement to protect the environment against greenhouse gases. The activity of man is one of the causes of global warming, and global warming is largely to blame for droughts, floods and hurricanes. Under the Kyoto Protocol, countries can reduce greenhouse emissions through so-called 'green' investments, when foreign investors finance projects to boost energy efficiency, or they can sell quotas. In case of 'green' investments, the reduction is ascribed to the investing party. This was the case with the sale of Russian quotas to Japan. Japanese companies financed the development of three Russian oilfields thereby radically reducing petroleum gas emissions into the atmosphere. President Medvedev sees a reduction in greenhouse emissions as one of the priority issues on the country's economic agenda. A Voice of Russia correspondent asked the opinion of Boris Profiriyev, an employee with the Institute of Economic Analysis.

'Russia has broken the ice by selling quotas to Japan. Besides, Russia has stockpiled on quotas and desperately needs revamping its own economy, the energy sector in particular. Projects of this kind could yield Russia 2–2,5 billion dollars a year. Russia's Economic Development Ministry and Sberbank, which monitor this process, have approved about one hundred emissions-related projects.'

Unlike Russia, other signatories to the Kyoto Protocol are in no hurry to act on their commitments. Opposition from other countries disrupted the signing of new climate agreements on Copenhagen and Cancun summits. But Nature, as we see from the calamities of late, does not tolerate disregard and neglect on the part of humanity.

SECTION III

Air Pollution from Chemical, Metallurgical and Miscellaneous Industries

Chapter 14

Metallurgical and Mining Industries

INTRODUCTION

The metallurgical industries can be broadly divided into primary, secondary, and miscellaneous metal production operations. 'Primary metals' refers to the production of metals from ore. 'Secondary metals' refers to the manufacturing of alloys by utilising metals from scrap and salvage, as well as ingots. 'Miscellaneous metal' production encompasses industries with operations that produce or use metals for final products. Metallurgical industries include the following:

Primary aluminium	Secondary aluminium
Metallurgical coke	Secondary brass and bronze melting processes
Copper smelting	Iron foundries
Ferroalloy industry	Secondary lead smelting
Steel industry	Steel foundries
Primary lead smelting	Secondary zinc
Zinc smelting	

IRON AND STEEL INDUSTRY

The iron and steel industry is one of the oldest and largest manufacturing operations in India. There are three basic types of steel manufacturers: integrated mills, mini-mills, and speciality-steel mills. Integrated steel mills use iron ore and coal as the raw materials for manufacturing various steel products. Mini-mills produce steel from scrap, typically using electric arc furnaces. Speciality-steel mills are similar in operation to mini-mills; they manufacture stainless, tool and high-alloy steels.

Few people outside the steel industry are aware of the enormous volumes of water and air required to produce iron and steel from raw materials (Fig. 14.1) is a graphic representation of the materials required in a primary iron producing facility, the Blast furnace, in producing one ton of finished iron. About 3.5 tons of air must be supplied to the blast furnace to reduce the ore to metal, and this amount of air produces about 5 tons of blast furnace gas. So, it is apparent that the tonnage of air and gas far exceeds the solid charge to the furnace and the products tapped from the furnace.

Steelmaking Process

There are two process routes for making steel in the world today: the electric arc furnace and the basic oxygen converter. The latter requires a charge of molten iron, which is produced in Blast furnaces. The raw materials for producing molten iron are iron ore, coking coal, and fluxes (materials that help the

337

chemical process)—mainly limestone. Blended coal is first heated in coke ovens to produce coke. This process is known as carbonisation. The gas produced during carbonisation is extracted and used for fuel elsewhere in the steelworks. Other by-products (such as tar and benzole) are also extracted for further refining and sale. Once carbonised, the coke is pushed out of the ovens and allowed to cool.

Fine-sized ore is first mixed with coke and fluxes and heated in a sinter plant. This is a continuous moving belt on which the coke is ignited. The high temperatures generated fuse the ore particles and fluxes together to form a porous clinker called sinter. The use of sinter in the Blast furnace helps make the ironmaking process more efficient. Iron ore lumps and pellets, coke, sinter; and possibly extra flux are carried to the top of the Blast furnace on a conveyor or in skips and then tipped, or charged, into the furnace. Hot air (ca. 900°C) is blasted into the bottom of the furnace through nozzles called tuyeres. The oxygen in the air combusts with the coke to form carbon monoxide gas, and this generates a great deal of heat. Frequently oil or coal is injected with the air, which enables less (relatively expensive) coke to be used. The carbon monoxide flows up through the Blast furnace and removes oxygen from the iron ores on their way down, thereby leaving iron. The heat in the furnace melts the iron, and the resulting liquid iron (or hot metal as it is called in the industry) is tapped at regular intervals by opening a hole in the bottom of the furnace and allowing it to flow out.

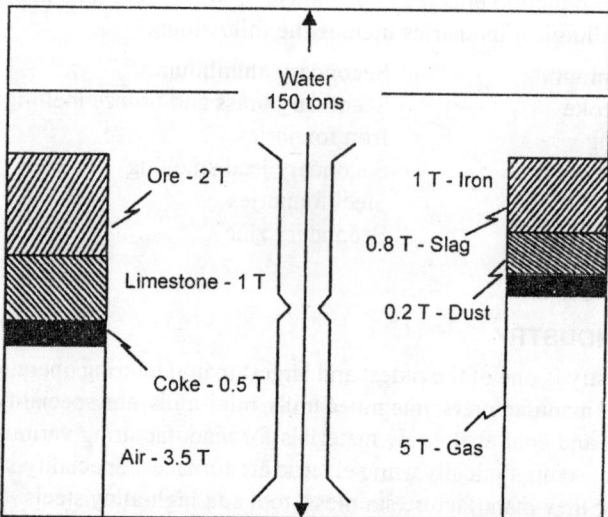

Fig. 14.1. In order to produce 1 ton of finished iron in a normal iron producing facility, the Blast furnace requires 150 tons of water. A troublesome by-product is 0.2 ton of dust that contributes to air pollution.

The fluxes combine with the impurities in the coke and ore to form a molten slag, which floats on the iron and is also removed (tapped) at regular intervals. The hot metal flows into torpedo ladles. These are specially constructed railway containers which transport iron, still in liquid form, to the steel furnace. An example of furnace tapping and hot molten steel being poured into a ladle. The process described above goes on continuously for ten years or more (this is known as a campaign). If the furnace were allowed to cool, damage could be caused to its lining of refractory bricks as a result of their contracting as they cooled.

Eventually the refractory brick linings are worn away, and at that stage the process is stopped and the furnace relined with new bricks, ready to begin its next campaign.

The iron produced by the Blast furnace has a carbon content of 4 to 4.5 per cent as well as a number of other impurities. This makes it relatively brittle. Steelmaking refines iron, amongst other things by reducing its carbon content, to make it a stronger and more manipulable product.

Air requirements and contaminants

As indicated earlier, the largest use of air is in the conversion of iron ore to pig iron, with additional large volumes being required by all the fuel-burning operations as combustion air. The blast air for the blast furnace operation is usually compressed by turbo-blowers, although older mills continue to use reciprocating compressors. The varieties of contaminants encountered in a steel mill are shown in Table 14.1.

Basic oxygen converter

The basic oxygen steelmaking (BOS) process is the major modern process for making bulk steels. Apart from special quality steels (such as stainless steel), all flat products, and long products over a certain size, are rolled from steel made by the BOS process.

Table 14.1. Typical air contaminants.

Contaminants	Sources
Particulates	Stacks on boiler plants, steel-making furnaces, sinter plant, coke ovens, foundries, slag plant, ore preparation, skip hoist area.
Gases	
Sulphur compounds	Combustion of sulphur-bearing fuel.
Carbon monoxide	Blast furnace gas.
Cyanides	Blast furnace and coke oven gas.
Fluorides	Steelmaking furnaces.
Benzene compounds	By-products plant.

The BOS vessel is first tilted to allow materials to be tipped into it (charged). Scrap steel is first charged into the vessel, followed by hot metal (liquid iron) from the Blast furnace. A water-cooled lance is lowered into the vessel through which very pure oxygen is blown at high pressure. The oxygen, through a process known as oxidation, combines with the carbon, and with other unwanted elements, separating them from the metal, leaving steel. Lime-based fluxes (materials that help the chemical process) are charged, and they combine with the 'impurities' to form slag. The main gas formed as a by-product of the oxidation process is carbon monoxide, and this is sometimes collected for use as a fuel elsewhere in the works.

A careful balance between the amounts of hot metal and scrap charged into the converter is maintained as a means of controlling the temperature and to ensure that steel of the required specification is produced. After a sample has been taken to check that the chemical content of the steel is correct, the vessel is again tilted to allow the molten steel to flow out. This is known as tapping. The steel is tapped into a ladle, in which secondary steelmaking frequently takes place. During tapping small quantities of other metals and fluxes are often added to control the state of oxidation and to meet customer requirements for particular grades of steel.

Finally the vessel is turned upside down and the slag tipped out into a container. Steelmaking slag is sometimes recycled to make road building materials. The modern BOS vessel makes up to 350 tons of steel at a time, and the whole process takes about 40 minutes.

Electric arc furnace

The electric arc furnace (EAF) (together with the basic oxygen vessel) is one of the two modern ways of making steel. EAFs are used to produce special quality steels (steels alloyed with other metals) and some ordinary (non-alloy) quality steels—the lighter long products such as those used for reinforcing concrete. Unlike the basic oxygen route, the EAF does not use hot metal. It is charged with 'cold' material. This is normally steel scrap (recycled goods made from steel which have reached the end of their useful life). Other forms of raw material are, however, available which have been produced from iron ore. These include direct reduced iron (DRI) and iron carbide, as well as pig iron, which is iron from a blast furnace which has been cast and allowed to go cold, instead of being charged straight into a basic oxygen vessel. Steel scrap (or other ferrous material) is first tipped into the EAF from an overhead crane. A lid is then swung into position over the furnace. This lid contains electrodes which are lowered into the furnace. An electric current is passed through the electrodes to form an arc. The heat generated by this arc melts the scrap. The electricity needed for this process is enough to power a town with a population of 1,00,000. During the melting process, other metals (ferro-alloys) are added to the steel to give it the required chemical composition. As with the basic oxygen process, oxygen is blown in to the furnace to purify the steel and lime and fluorspar are added to combine with the impurities and form slag. After samples have been taken to check the chemical composition of the steel, the furnace is tilted to allow the slag, which is floating on the surface of the molten steel, to be poured off. The furnace is then tilted in the other direction and the molten steel poured (tapped) into a ladle, where it either undergoes secondary steelmaking or is transported to the caster. The modern electric arc furnace typically makes 150 tons in each melt, which takes around 90 minutes.

Special quality steels

A vast range of special quality steels is made in electric arc furnaces by adding other metals to form steel alloys. The most commonly known of these is stainless steel, which has chromium and nickel added to form a corrosion-resistant steel. There are very many others however, the very hard steels used to make machine tools, the steels specially formulated to make them suitable for engineering, steels developed to survive for decades the hostile environment of nuclear reactors, light but strong steels used in aerospace, extra tough steels for armour plating—to name but a few.

Secondary steelmaking

Increasingly today, steels after they have been tapped (poured) from the furnace undergo a further stage of processing called secondary steelmaking before the steel is cast. This applies to both the basic oxygen process route and to the electric arc furnace route.

The molten steel is tapped from the furnace into a ladle. A lid is placed over the ladle to conserve heat. A range of different processes is then available, such as stirring with argon, adding alloys, vacuum degassing or powder injection. The objective in all cases is to fine tune the chemical composition of the steel and/or to improve homogenisation of temperature (making sure that the steel is the same temperature throughout) and remove impurities. Ladle arc heating is a process used to ensure that the molten steel is at exactly the correct temperature for casting.

Continuous casting

Not that many years ago, molten steel used to be poured (teemed) into a large mould where it would be allowed to cool and solidify to form an ingot. The ingot was then put into an oven called a soaking pit, where it would be gently heated to the correct and uniform temperature. This red hot ingot would then be rolled in primary mills, in the first stage of its transformation into a usable steel product, into one of three forms of semifinished product: a slab (a long, thick, flat piece of steel, with a rectangular cross-section) a bloom (a long piece of steel with a square cross-section) or a billet (like a bloom, but with smaller cross-section). In modern plants, this process has largely been superseded by the continuous casting process (concaster), although the ingot route is retained for certain applications where it is the most suitable way of producing the steel required. Note that there are still many steelmaking plants in parts of Eastern Europe that rely heavily on the old ingot route. In a continuous casting machine, molten steel is poured into a reservoir at the top of the machine. It passes at a controlled rate into a water cooled mould where the outer shell of the steel becomes solidified. The steel is drawn down into a series of rolls and water sprays, which ensure that it is both rolled into shape and fully solidified at the same time. At the end of the machine, it is straightened and cut to the required length. Fully formed slabs, blooms, and billets emerge from the end of this continuous process. Thus the concaster combines in a single process what previously took two separate processes. This is both highly energy-efficient and produces a better quality product. The slabs, blooms, or billets are then transported to the hot rolling mill for rolling into steel products which can be used by the manufacturing industry.

Hot rolling

Semifinished products called blooms, billets and slabs are transported from the steelmaking plant to the rolling mills. In many plants steelmaking and rolling are both carried out on the same site. However, there are also many stand-alone rolling mills (some are independently owned while others are typically a part of a larger group but located away from the steelmaking works). Steel products can be classified into two basic types according to their shape: flat products and long products. Slabs are used to roll flat products, while blooms and billets are mostly used to roll long products. Billets are smaller than blooms, and therefore are used for the smaller type of long product. In some cases slabs are used to roll large long products (such as beams). Semifinished products are first heated in a reheat furnace until they are red hot (around 1200°C). On all types of mill the semifinished products go first to a roughing stand. A stand is a collection of steel rolls (or drums) on which pressure can be applied to squeeze the hot steel passing through them, and arranged so as to form the steel into the required shape. The roughing stand is the first part of the rolling mill. The large semifinished product is often passed backwards and forwards through it several times. Each pass gradually changes the shape and dimension of the steel closer to that of the required finished product.

Plate mills

Slabs are used to make plate. Typically, after leaving the plate mill's roughing stand, they are passed through a finishing stand. This is a reversing mill: like on the roughing stand, the steel is passed backwards and forwards through the mill. It is also turned 90° and rolled sideways at one stage during the process. Plate is a large, flat piece of steel perhaps 10 mm or 20 mm thick (although it can be up to 50 mm thick) and up to 5 meters wide. It is used for example to make the hulls and decks of ships or to make large tanks and industrial boilers. It can also be rolled up and welded to form a large steel tube, used for oil and gas pipelines.

Strip mills

Slabs are also used to make steel strip, normally called hot rolled coil. After leaving the roughing stand, the slab passes continuously through a series of finishing stands which progressively squeeze the steel to make it thinner. As the steel becomes thinner, it also of course becomes longer, and starts moving faster. And because the single piece of steel will be a whole range of different thicknesses along its length as each section of it passes through a different stand, different parts of the same piece of steel are travelling at different speeds. This requires very close control of the speeds at which each individual stand rolls. In modern plants the entire process is computer controlled. By the time it reaches the end of the mill, the steel is travelling at about 40 miles per hour. Finally the long strip of steel is coiled and allowed to cool. Hot rolled strip is a flat product which has been coiled to make storage and handling easier. It is a lot thinner than plate, typically a few millimetres thick, although it can be as thin as 1 mm. Its width can vary from 150 mm to nearly 2 meters. It frequently goes through further stages of processing such as cold rolling and is also used to make tubes (smaller tubes than those made from plate).

Long product mills

Blooms and billets are used to make long products. After leaving the roughing stand, the piece of steel passes through a succession of stands which do not just reduce the size of the steel, but also change its shape. In a universal mill, all faces of the piece of steel are rolled at the same time. In other mills, only two sides of the steel are rolled at anyone time, the piece of steel being turned over to allow the other two sides to be rolled. Long products are so-called because they come off the mill as long bars of steel. They are however produced in a vast range of different shapes and sizes. They can have cross-sections shaped like an H or I (called joists, beams, and columns), a U (channels) or a T. These types of steel section are used for construction.

Bars can have cross-sections the shape of squares, rectangles, circles, hexagons, angles. These bars can also be used for construction, but many types of bars are also used for engineering purposes. Rod is coiled up after use and is used for drawing into wire or for fabricating into products used to reinforce concrete buildings, as are some types of bars. Other types of long products include railway rails and piling. Some long product mills make unique shapes of steel to a customer's individual specification. These are known as special sections.

Cooling

In all rolling processes, cooling the steel is a critical factor. The speed at which the rolled product is cooled will affect the mechanical properties of the steel. Cooling speed is controlled normally by spraying water on the steel as it passes through and/or leaves the mill, although occasionally the rolled steel is air-cooled using large fans.

Further processing

Hot rolled products can undergo many forms of further processing before they are finally used to make an end-product (such as a steel-framed building or a consumer product). Such processing includes:

1. Cold rolling and drawing.
2. Fabricating: Steel sections are cut, welded and otherwise prepared to form the steel frame of a building. Rods and bars are similarly cut and shaped to form the steel reinforcement for concrete buildings.
3. Coating.
4. Cutting and slitting: Service centers cut steel into many complex shapes.
5. Profiling: Sheet steel may be pressed into the correct shape for crash barriers or the cladding of buildings (known as profiling).

Cold rolling and drawing

After hot rolling, many steel products undergo a further processing in the cold state. This stage of processing does not alter the shape of the steel product, but it does reduce it in thickness and significantly improve its performance characteristics. Hot rolled coil is commonly cold rolled (also known as cold reduced). The strip is first de-coiled (uncoiled) and then passes through a series of rolling mill stands which apply pressure to the strip and progressively reduce its thickness—down to as low as 0.15 mm. The strip is then recoiled.

Cold rolling processes are also used to improve the surface quality of the steel. Cold rolling also has the effect of hardening steel, so cold reduced strip is subsequently annealed: a process of very carefully controlled heating and cooling to soften it. Another form of cold processing is cold drawing. Steel rod is dragged at pressure (drawn) through a series of dies which progressively reduce the rod's circumference to produce wire. The drawing process substantially increases the steel's tensile strength—steel wires can be spun into huge ropes strong enough to support the world's largest suspension bridges.

Coating

Most steels when exposed to air will gradually rust. (This does not apply to all steels: stainless steel for example was invented in Sheffield—specifically to resist rust.) Steel has therefore always been covered frequently painted for example, by its users in order to protect it. Nowadays the steelmaker can improve steel's corrosion resistance by coating it in the factory prior to delivery to the end-user. A wide range of different coatings is available, including:

1. Zinc coating, normally called galvanising. The zinc can be applied either electrolytically (which gives a thinner coating) or by dipping the steel in a bath of molten zinc. Much of the sheet used to produce car bodies is zinc coated. This has enabled thinner steels to be used for car bodies, thus saving weight and improving fuel consumption. (Without this coating, the thinner steels would have rusted, shortening the car's life.) Wire is also frequently galvanised to extend the product's life.
2. Organic coatings (plastic and paint) can be applied to extend the steel's life, while at the same time giving it an attractive appearance. The walls of many industrial and commercial buildings are made from pre-painted steel sheets. Frequently, a combination of galvanising and organic coatings is used.
3. Tinplate is thin steel sheet with a minute coating of tin applied. This is used for producing food and drinks cans, where rust must be prevented at all costs.
4. Other metals used are chromium, lead, and aluminium. Electrochromium coated sheet is used for the tops of steel drinks cans. Aluminium coated sheet provides a combination of corrosion and heat resistance ideal for car exhaust pipes.

The iron and steel industry also ranks as one of the top five releasers for NO_2, PM_{10}, and SO_2. Carbon monoxide releases occur during ironmaking (in the burning of coke, CO produced reduces iron oxide ore), and during steelmaking (in either the basic oxygen furnace or the electric arc furnace). Nitrogen dioxide is generated during steelmaking.

Particulate matter may be emitted from the cokemaking (particularly in quenching operations), ironmaking, basic oxygen furnace (as oxides of iron that are emitted as submicron dust) or from the electric arc furnace (as metal dust containing iron particulate, zinc, and other materials associated with the scrap). Sulphur dioxide can be released in ironmaking or sintering.

Air Pollution

The important air pollutants in any integrated steel plant will be smoke, dust, fume SO_2 and H_2S. The sources from which these are generated are shown in Tables 14.2 and 14.3.

Table 14.2. Sources of air pollution in an integrated steel plant.

Operation units	Types of pollutants
Coking operations such as charging, discharging, carbonising, and by-products	Smoke, dust and hydrogen sulphide
Coke preparation, crushing, screening, ore drying, sintering, blast furnace charging, gas cleaning, slag disposal and pig casting	Dust and SO_2
Steel making furnaces, rolling mills, ancillaries and mill furnaces	Dust, fume and SO_2
During ancillary operations, involved in the working of iron and steel foundries, boilers and locos.	Dust, SO_2 and smoke
Thermal power plant	Fly ash and SO_2

Table 14.3. Dust emission in different areas of steel plants.

Area	Approximate amount of dust emission
Sinter plant area	15 to 20 kg/T of sinter, SO_2 in combustion gases—0.2 to 2.0 kg/T of sinter
Blast furnace	10 gms/m^3 of Blast furnace gas after coarse separation.
Steel melting shop	
LD, LDAC and Kaldo converters	2 per cent of the charge weight
Electric arc furnace and open hearth furnaces	5–20 kg/T of steel during oxygen blowing period

The location and climatic condition around the industry have a fundamental influence on the transport and diffusion of air pollutants. Among these, meteorological factors, wind direction and speed, duration, frequency and intensity of rainfall, humidity and temperature are of significance.

Preliminary studies in the Durgapur (West Bengal) industrial area indicate a dust fall of 51.6 T/sq. km/month (total solids) and 32.1 tons/sq. km/month (volatile solids). But it is interesting to point out that there are other industries in Durgapur that also contribute towards the pollution loads.

Air pollution control

In order to combat the menace of air pollution and control harmful fumes and other gaseous discharges produced by the processing and production units in an integrated steel plant, several types of equipment are being employed. Air and gas cleaning equipment such as dust collectors, cyclones, settlers and scrubbers, smoke stacks with leak proof vents and shutters, quenching baffles and towers, fans, precipitators, suction and vibrating screening devices, filters, diffusers, desulphurisers, absorbers, spray washers, etc. are being extensively used throughout. Table 14.4 lists important gas cleaning equipment. If proper maintenance and repair of this equipment is not undertaken, the cleaning efficiency will be low thus permitting more pollutants into the atmosphere. In the context of the Indian steel Industry, emphasis should be given to the technical know how in operating this vital equipment which will go a long way in controlling air pollution.

Table 14.4. Types of gas cleaning equipment in steel industry.

Area	Types of gas cleaning equipment used
Sinter plant area	Electrostatic precipitators, bag houses, cyclone and bag filter collectors.
Blast furnace area	Dust catchers, cyclones, electrostatic precipitators, scrubbers.
Converter shops	Cyclone dust collectors, electrostatic precipitators, bag houses and wet collectors.
Electric arc furnace shops	Wet type collectors.
Open hearth furnace shops	Cyclone dust collectors, electrostatic precipitators, etc.
Lime and dolomite calcining plant	Cyclone and bag type collectors.

As the incorporation of air pollution equipment at the design stage would be much more economical and efficient, planners should give due consideration to this aspect.

Corrosion of air pollution control equipment

Corrosion resistance of materials of construction has an important bearing on the design of air pollution control equipment. This is particularly important in view of the fact that corrosion is accelerated by high gas inlet temperature, high velocities, etc. commonly encountered. New methods like use of activated carbon in air pollution control techniques may result in a better environment.

Sinter plants

The dust generated in the sintering plant can be returned to the process. Because of this, most plants are atleast equipped with cyclones. Dry type cleaners are best suited for the cleaning because the sulphur content of gas streams can lead to corrosion problem in wet system. Cyclones, electrostatic precipitators, venture scrubbers, and bag houses are used in various combinations at the various points of emission.

Coke ovens

Gaseous and particulate matter released in the by-product coking operation except that which escapes from ovens to the atmosphere, are conveyed in ducts to a coal chemical processing plant for recovery of chemicals. Emissions occurring from handling operations present dust contaminant problems which are difficult to control. Emissions during charging can be minimised by minimising the openings through which smoke can escape and by creating a slight vacuum inside the oven during charging so that air flows into the openings instead of smoke having come out. The quenching of coke produces a rising cloud of steam in the chimney which lifts coke dust into the atmosphere. Most of this dust appears to fall out in the vicinity of the quench tower. Baffles installed in a quench tower can reduce the dust emission into the atmosphere by about 75 per cent.

Blast furnace

Blast furnace gases contain about 0.2 ton of dust per ton of pig-iron produced. Blast furnace gas is cleaned in three stages:

1. Preliminary cleaning: Settling chamber or dry type cyclones.
2. Primary cleaning: Wet scrubbers.
3. Secondary cleaning: Electrostatic precipitators or high energy scrubbers.

Open hearth furnace

The small size of the particles emitted from open hearth furnace requires high efficiency collection equipment such as venturi scrubbers and electrostatic precipitators. Because of the cost involved and the growing obsolescence of open hearth furnaces, industry has been reluctant to invest money in the required control equipment.

Basic oxygen furnace

The basic oxygen furnace creates more emissions than the open hearth furnace and the particles are smaller. All basic oxygen furnaces are generally equipped with high efficiency electrostatic precipitators or venturi scrubbers. In some countries fabric filters have also been in use.

Electric arc furnace

Electric arc furnaces are becoming more popular for many metal melting operations. Particulate emissions from electric arc furnaces are difficult to collect because of their small size and because of a strong tendency to adhere to fabric surfaces, a high angle of repose, and high resistivity. However, over 95 per cent collection can be achieved with appropriate hooding and high efficiency collection equipment. The characteristically small particle size of electric arc furnace fume precludes the use of dry centrifugal collectors, settling chambers, etc. High efficiency scrubbing systems, electrostatic precipitators and baghouses are used to control fumes in electric arc furnaces.

The majority of the world investment in air pollution control has been directed mostly towards the collection of suspended particles. The reason for this could be that collection techniques for these are simpler and also that such collecting equipment has been well developed and proven. When compared to particulate air pollutants, the gaseous air pollutants are more dangerous. The gaseous air pollutants are sulphur oxides, nitrogen oxides, hydrogen sulphide, carbon monoxide and carbon-dioxide, etc. In the steel industry coke-oven gas and blast furnace gas are cleaned for their further use in the plant. In recent years more attention has been paid to control the emissions of these gaseous pollutants particularly SO_x and NO_x into the atmosphere by adopting desulphurising and denitration technology.

Objectionable properties of air pollutants and general effects on humans

  1. Toxicological and sensory effects on humans:
     (a) Acute toxicity.
     (b) Chronic toxicity.
     (c) Irritation of eyes, lungs, skin, etc.
     (d) Odour.
  2. Effects on plant and animal life:
     (a) Toxic to agricultural crops, timber, grass, etc.
     (b) Toxic to livestock and wild life.

Mini Steel Mills

Mini steel mills normally use the electric arc furnace (EAF) to produce steel from returned steel, scrap, and direct reduced iron. EAF is a batch process with a cycle time of about two to three hours. Since the process uses scrap metal instead of molten iron, cokemaking and ironmaking operations are eliminated. EAFs can economically serve small, local markets. Further processing of steel can include continuous casting, hot rolling and forming, cold rolling, wire drawing, coating, and pickling. As already noted, the continuous casting process by-passes several steps of the conventional ingot teeming process by casting steel directly into semifinished shapes. The casting, rolling, and steel finishing processes are also used in iron and steel manufacturing. Hot steel is transformed in size and shape through a series of hot rolling and forming steps to manufacture semifinished and finished steel products. The hot rolling process consists of slabheating (as well as billet and bloom), rolling, and forming operations. Several types of hot forming mills (primary, section, flat, pipe and tube, wire, rebar, and profile) manufacture a variety of steel products. For the manufacture of a very thin strip or a strip with a high-quality finish, cold rolling must follow the hot rolling operations. Lubricants emulsified in water are usually used to achieve high surface quality and to prevent overheating of the product.

Wire drawing includes heat treatment of rods, cleaning, and sometimes coating. Water, oil or lead baths are used for cooling and to impart desired features. To prepare the steel for cold rolling or drawing, acid pickling is performed to chemically remove oxides and scale from the surface of the steel through use of inorganic acid water solutions. Mixed acids (nitric and hydrofluoric) are used for stainless steel pickling; sulphuric or hydrochloric acid is used for other steels.

Other methods for removing scale include salt pickling, electrolytic pickling, and blasting; blasting is environmentally desirable, where feasible. EAFs produce metal dusts, slag, and gaseous emissions.

The primary hazardous components of EAF dust are zinc, lead, and cadmium; nickel and chromium are present when stainless steels are manufactured. Generally, an EAF produces 10 kilograms of dust per metric ton (kg/T) of steel, with a range of 5 to 30 kg/T, depending on factors such as furnace characteristics and scrap quality. Major pollutants present in the air emissions include particulates (1000 milligrams per normal cubic meter, mg/Nm^3), nitrogen oxides from cutting, scarfing, and pickling operations, and acid fumes (3000 mg/Nm^3) from pickling operations. Both nitrogen oxides and acid fumes vary with steel quality.

Table 14.5 provides a list of pollution prevention practices for reducing air emissions in mini steel mills. Standard treatment technologies for air emissions are as follows. Dust emission control technologies include cyclones, baghouses, and ESPs. Scrubbers are used to control acid mists.

Table 14.5. Pollution prevention (P2) practices for reducing air emissions in mini steel mills.

Locate EAFs in enclosed buildings.
Improve feed quality by using selected scrap to reduce the release of pollutants to the environment.
Use dry dust collection methods such as fabric filters.
Replace ingot teeming with continuous casting.
Use continuous casting for semifinished and finished products wherever feasible. In some cases, continuous charging may be feasible and effective for controlling dust emissions.
Use bottom tapping of EAFs to prevent dust emissions.
Use acid-free methods (mechanical methods such as blasting) for descaling, where feasible.
In the pickling process, use countercurrent flow of rinse water; use indirect methods for heating and pickling baths.
Use closed-loop systems for pickling; regenerate and recover acids from spent pickling liquor using resin bed, retorting, or other regeneration methods such as vacuum crystallisation of sulphuric acid baths.
Reduce nitrogen oxide emissions by use of natural gas as fuel, use low-NO_x burners, and use hydrogen peroxide and urea in stainless steel pickling baths.
Recover zinc from EAF dust containing more than 15 per cent total zinc; recycle EAF dust to the extent feasible.

Fugitive emissions from charging and tapping of EAFs should be controlled by locating the EAF in an enclosed building or using hoods and by evacuating the dust to dust arrestment equipment to achieve an emissions level of less than 0.25 kg/T.

SUMMARY OF POLLUTION PREVENTION AND CONTROL

Steelmaking is a sophisticated and complex process, with many secondary production operations, each having unique air pollution problems to varying degrees. There are a large number of outputs that are produced as a result of the manufacturing of coke, iron, and steel, the forming of metals into basic shapes, and the cleaning and scaling of metal surfaces. For example:

1. Sintering operations can emit significant 'dust' levels of about 20 kilograms per metric ton (kg/T) of steel.

2. Pelletising operations can emit dust levels of about 15 kg/T of steel.

3. Air emissions from pig iron manufacturing in a blast furnace include PM, ranging from less than 10 kg/T of steel manufactured to 40 kg/T; sulphur oxides SO_x mostly from sintering or pelletising operations (1.5 kg/T of steel); nitrogen oxides NO_x mainly from sintering and heating (1.2 kg/T of steel); hydrocarbons; carbon monoxide; in some cases dioxins (mostly from sintering operations); and hydrogen fluoride.

4. Air emissions from steel manufacturing using the BOF may include PM (ranging from less than 15 kg/T to 30 kg/T of steel).

5. For closed systems, emissions come from the desulphurisation step between the Blast furnace and the BOF; the particulate matter emissions are about 10 kg/T of steel.

Coke oven: Pollution prevention in cokemaking has focused on two areas—reducing coke oven emissions and developing coke-less ironmaking techniques. Although these processes have not yet been widely demonstrated on a commercial scale, they may provide important benefits, especially for the integrated segment of the industry, by potentially lowering air emissions and waste-water discharges. Several technologies are available or are under development to reduce the emissions from coke ovens. Typically, these technologies reduce the quantity of coke needed by changing the method by which coke is added to the blast furnace or by substituting a portion of the coke with other fuels. The reduction in the amount of coke produced proportionally reduces the coking emissions. Some of the most prevalent or promising coke reduction technologies are listed in the side-bar discussion that follows. Coke-less, technologies substitute coal for coke in the blast furnace, hence eliminating the need for cokemaking. Such technologies have enormous potential to reduce pollution generated during the steelmaking process. The drawbacks with these technologies are:

1. The capital investment required for retrofits is very significant.

2. Some countries whose economies are dependent upon the steel industry need to undergo significant industry rationalisation and restructuring in order to justify investments into these technologies. For example, Russia and Ukraine, which have significant steel production and export capabilities heavily depend on a labour intensive and 'dirty' coking industry. The elimination of the coking industry in these countries would likely result in significant social implications, such as mass unemployment in certain regions of these countries. Additionally, there are implications of domino effects on other industries, such as coal mining.

Reducing coke over emissions with other technologies: The use of pulverised coal injection technology substitutes pulverised coal for a portion of the coke in the Blast furnace. Use of pulverised coal injection can replace about 25 to 40 per cent of coke in the Blast furnace, substantially reducing emissions associated with cokemaking operations. This reduction ultimately depends on the fuel injection rate applied to the Blast furnaces which will, in turn, be dictated by the ageing of existing coking facilities, fuel costs, oxygen availability, capital requirements for fuel injection, and available hot blast temperature.

Another novel approach is the use of a nonrecovery coke battery. As opposed to the by-product recovery coke plant, the nonrecovery coke battery is designed to allow combustion of the gases from the coking process, thus consuming the by-products that are typically recovered. The process results in lower air emissions and substantial reductions in coking process waste-water discharges. A third option is the Davy Still Auto-process. In this pre combustion cleaning process for coke ovens, coke oven battery process water is utilised to strip ammonia and hydrogen sulphide from coke oven emissions. Still another option is the use of alternative fuels. Steel producers can inject other fuels, such as natural gas, oil, and tar/pitch, instead of coke into the Blast furnace, but these fuels can only replace coke in limited amounts.

Table 14.6 provides some examples of pollution prevention practices aimed at reducing air emissions and capturing energy credits. Remember that this is a very energy intensive manufacturing process, so any efforts aimed at reducing energy requirements are of particular interest to plant managers.

Table 14.6. Examples of pollution prevention (P2) practices in the iron and steel industry.

Process	Recommended pollution prevention practice
Pig iron manufacturing	Improve Blast furnace efficiency by using coal and other fuels (such as oil or gas) for heating instead of coke, thereby minimising air emissions.
	Recover the thermal energy in the gas from the Blast furnace before using it as a fuel.
	Increase fuel efficiency and reduce emissions by improving Blast furnace charge distribution.
	Recover energy from sinter coolers and exhaust gases.
	Use dry SO_x removal systems such as carbon absorption for sinter plants or lime spraying in flue gases.
	Recycle iron-rich materials such as iron ore fines, pollution control dust and scale in a sinter plant.
	Use low-NO_x burners to reduce NO_x emissions from burning fuel in ancillary operations.
	Improve productivity by screening the charge and using better taphole practices.
	Reduce dust emissions at furnaces by covering iron runners when tapping the Blast furnace and by using nitrogen blankets during tapping.
	Use pneumatic transport, enclosed conveyor belts, or self-closing conveyor belts, as well as wind barriers and other dust suppression measures, to reduce the formation of fugitive dust.
Steel manufacturing	Use dry dust collection and removal systems to avoid the generation of waste-water. Recycle collected dust.
	Use BOF gas as fuel.
	Use enclosures for BOF.
	Use a continuous process for casting steel to reduce energy consumption.

ALUMINIUM

The production of aluminium begins with the mining and beneficiation of bauxite. At the mine (usually of the surface type), bauxite ore is removed to a crusher. The crushed ore is then screened and stockpiled, ready for delivery to an alumina plant. In some cases, ore is upgraded by beneficiation (washing, size classification, and separation of liquids and solids) to remove unwanted materials such as clay and silica. At the alumina plant, the bauxite ore is further crushed to the correct particle size for efficient extraction of the alumina through digestion by hot sodium hydroxide liquor. After removal of 'red mud' (the insoluble part of the bauxite) and fine solids from the process liquor, aluminium trihydrate crystals are precipitated and calcined in rotary kilns or fluidised bed calciners to produce alumina (Al_2O_3). Some alumina processes include a liquor purification step. Primary aluminium is produced by the electrolytic reduction of the alumina. The alumina is dissolved in a molten bath of fluoride compounds (the electrolyte), and an electric current is passed through the bath, causing the alumina to dissociate to form liquid aluminium and oxygen. The oxygen reacts with carbon in the electrode to produce carbon dioxide and carbon monoxide. Molten aluminium collects in the bottom of the individual cells or pots and is removed under vacuum into tapping crucibles. There are two prominent technologies for aluminium smelting: prebake and Soderberg. The following discussion focuses on the prebake technology, with its associated reduced air emissions and energy efficiencies. Raw materials for secondary aluminium production are scrap, chips, and dross. Pretreatment of scrap by shredding, sieving, magnetic separation, drying, and so on is designed to remove undesirable substances that affect both aluminium quality and air emissions. The prevailing process for secondary aluminium production is smelting in rotary kilns under a salt cover. Salt slag can be processed and reutilised. Other processes (smelting in induction furnaces and hearth furnaces) need no or substantially less salt and are associated with lower energy demand, but they are only suitable for high-grade scrap. Depending on the desired application, additional refining may be necessary. For damaging (removal of magnesium from the melt), hazardous substances such as chlorine and hexachloroethane are often used, which may produce dioxins and dibenzofurans. Other, less hazardous methods, such as adding chlorine salts, are available. Because it is difficult to remove alloying elements such as copper and zinc from an aluminium melt, separate collection and separate reutilisation of different grades of aluminium scrap are necessary. Note that secondary aluminium production uses substantially less energy than primary production.

Air Pollution and Its Control—Reduction Plant

Fluorides

The main feature of environmental problems in the aluminium smelting industry concerns air pollution caused by fluorides emitted from the aluminium reduction cell. The excessive intake of fluoride can cause fluorosis (skeletal disorders). However, no such cases have been reported of workers in aluminium smelters or of inhabitants around the smelter. Farm animals (cattle, sheep, goats, etc.) can also be affected by fluorosis by ingestion of fluoride through contaminated fodder.

Substantial progress has been made over the last 30 years through use of latest technologies and environmental control measures, in reducing the amount of fluorides emitted from aluminium smelting plants. With sound environmental management involving proper control facilities and technology, the aluminium smelter can be operated with minimum environmental problems and any previously unfavourable image of the industry no longer holds true. However, the latest accepted methods include the provision of dry scrubbing system for the smelter gases which not only controls pollution but also allows recycling of fluorides which ensures economy in consumption of cryolite and aluminium fluoride.

Coaltar pitch volatiles

Tar fumes are generated in the carbon (anode) plant and as the soderberg anodes are consumed in the electrolytic cells. Such processes include pitch smelting, mixing, cooling of the soderberg paste, forming and baking of the anodes, etc. Tar fumes are also given off from the cathode lining mix and carbon ramming mix during the relining of cells. These tar fumes are harmful if inhaled continuously.

Dusts

The exhaust gas from the electrolytic cell contains carbon, alumina and fluoride dust particles. The carbon dust is also given off from the grinding process of coke and pitch, and from anode baking. The alumina dust is generated as fugitive dust from the alumina feeding operation in certain types of cells. When the hot metal is treated by gaseous chloride in the cast-house furnace, aluminium chloride dust is generated which creates a white smoke. Inhalation of these dusts would prove hazardous to the workers' health and need to be controlled.

Sulphur dioxide

Although the amount may be very small, emissions of sulphur dioxide are expected from the electrolytic cell as petroleum coke (anodes) containing sulphur is consumed. The control of this sulphur dioxide needs to be considered only when the smelter is located in an area where in concentration of sulphur dioxide in the ambient air is already very high because of other emission sources. Sulphur dioxide is also emitted from the anode baking furnace and cast-house furnace where sulphur containing fuels are burnt. It goes without saying that sulphur dioxide is considered to be an important air pollutant where the public safety is concerned. However, this is not a serious issue in the case of aluminium smelters.

Nitrogen oxides

Nitrogen oxides are emitted from anode baking and cast-house furnaces. This problem is also not of a serious nature in the aluminium smelters. However, in the case of sulphur dioxide, some control might be needed in accordance with the local conditions of the plant site.

Pollution Prevention and Control

At the bauxite production facilities, dust is emitted to the atmosphere from dryers and materials-handling equipment, through vehicular movement, and from blasting. Although the dust is not hazardous, it can be a nuisance if containment systems are not in place, especially on the dryers and handling equipment. Other air emissions could include NO_x, SO_x, and other products of combustion from the bauxite dryers. Ore washing and beneficiation yield process waste waters containing suspended solids. Runoff from precipitation may also contain suspended solids. At the alumina plant, air emissions can include bauxite dust from handling and processing, limestone dust from limestone handling, burnt lime dust from conveyors and bins, alumina dust from materials handling, red mud dust and sodium salts from red mud stacks (impoundments), caustic aerosols from cooling towers, and products of combustion such as sulphur dioxide and nitrogen oxides from boilers, calciners, various mobile equipment, and kilns. The calciners may also emit alumina dust and the kilns, and burnt lime dust.

In the aluminium smelter, air emissions include alumina dust from handling facilities; coke dust from coke handling; gaseous and particulate fluorides; sulphur and carbon dioxides and various dusts from the electrolytic reduction cells; gaseous and particulate fluorides; sulphur dioxide; tar vapour and carbon particulates from the baking furnace; coke dust, tars, and polynuclear aromatic hydrocarbons (PAHs) from the green carbon and anode-forming plant; carbon dust from the rodding room; and fluxing emissions and carbon oxides from smelting, anode production, casting, and finishing operations. The electrolytic reduction cells (pot line) are the major source of the air emissions, with the gaseous and particulate fluorides being of prime concern. The anode effect associated with electrolysis also results in emissions of carbon tetrafluoride (CF_4) and carbon hexafluoride (C_2F_6), which are greenhouse gases of concern because of their potential for global warming. Emissions numbers that have been reported for uncontrolled gases from smelters are 20 to 80 kg/T for particulates, 6 to 12 kg/T for hydrogen fluoride, and 6 to 10 kg/T for fluoride particulates. Corresponding concentrations are 200 to 800 mg/m^3; 60 to 120 mg/m^3; and 60 to 100 mg/m^3. An aluminium smelter produces 40 to 60 kg of mixed solid wastes per ton of product, with spent cathodes (spent pot and cell linings) being the major fraction. The linings consist of 50 per cent refractory material and 50 per cent carbon. Over the useful life of the linings, the carbon becomes impregnated with aluminium and silicon oxides (averaging 16 per cent of the carbon lining), fluorides (34 per cent of the lining), and cyanide compounds (about 400 parts per million). Contaminant levels in the refractories portion of linings that have failed are generally low. Other by-products for disposal include skim, dross, fluxing slags, and road sweepings.

Atmospheric emissions from secondary aluminium melting include hydrogen chloride and fluorine compounds. Damaging may lead to emissions of chlorine, hexachloroethane, chlorinated benzenes, and dioxins and furans. Chlorinated compounds may also result from the melting of aluminium scrap that is coated with plastic. Salt slag processing emits hydrogen and methane. Solid wastes from the production of secondary aluminium include particulates, pot lining refractory material, and salt slag. Particulate emissions containing heavy metals are also associated with secondary aluminium production.

Pollution prevention is always preferred to the use of end-of-pipe pollution control facilities. Therefore every attempt should be made to incorporate cleaner production processes and facilities to limit, at source, the quantity of pollutants generated. Using the prebake technology rather than the Soderberg technology for aluminium smelting is a significant pollution prevention measure. In the smelter, computer controls and point feeding of aluminium oxide to the centerline of the cell help reduce emissions, including emissions of organic fluorides such as CF_4, which can be held at less than 0.1 kg/T aluminium. Energy consumption is typically 14 megawatt hours per ton (MWh/T) of aluminium, with prebake technology. Soderberg technology uses 17.5 MWh/T. Gas collection efficiencies for the prebake process is better than for the Soderberg process: 98 per cent vs. 90 per cent. Dry scrubber systems using aluminium oxide as the adsorbent for the cell gas permits the recycling of fluorides. The use of low-sulphur tars for baking anodes helps control SO_2 emissions. Spent pot linings are removed after they fail, typically because of cracking or heaving of the lining. The age of the pot linings can vary from 3 to 10 years. By improving the life of die lining through better construction and operating techniques, discharge of pollutants can be reduced. Note that part of the pot lining carbon can be recycled when the pots are relined. Emissions of organic compounds from secondary aluminium production can be reduced by thoroughly removing coatings, paint, oils, greases, and the like from raw feed materials before they enter the melt process. At bauxite facilities, the major sources of dust emissions are the dryers, and emissions are controlled with electrostatic precipitators (ESPs) or baghouses. Removal efficiencies of 99 per cent are achievable. Dust from conveyors and material transfer points is controlled by hoods and enclosures. Dust from truck movement can be minimised by treating road surfaces and by ensuring that vehicles do not drop material as they travel. Dusting from stockpiled material can be minimised by the use of water sprays or by enclosure in a building. At the alumina plant, pollution control for the various production and service areas is implemented as follows:

1. Bauxite and limestone handling and storage: Dust emissions are controlled by baghouses.
2. Lime kilns: Dust emissions are controlled by baghouse systems. Kiln fuels can be selected to reduce SO_x emissions; however, this is not normally a problem, since most of the sulphur dioxide that is formed is absorbed in the kiln.
3. Calciners: Alumina dust losses are controlled by ESPs; SO_2 and NO_x emissions are reduced to acceptable levels by contact with the alumina.

4. Red mud disposal: The mud impoundment area must be lined with impervious clay prior to use to prevent leakage. Water spraying of the mud stack may be required to prevent fine dust from being blown off the stack. Longer-term treatment of the mud may include reclamation of the mud, neutralisation, covering with topsoil, and planting with vegetation.

In the smelter, primary emissions from the reduction cells are controlled by collection and treatment using dry sorbent injection; fabric filters or ESPs are used for controlling particulate matter. Primary emissions comprise 97.5 per cent of total cell emissions; the balance consists of secondary emissions that escape into the potroom and leave the building through roof ventilators. Wet scrubbing of the primary emissions can also be used, but large volumes of toxic waste liquors will need to be treated or disposed of. Secondary emissions result from the periodic replacement of anodes and other operations; the fumes escape when the cell hood panels have been temporarily removed.

While wet scrubbing 'can be used to control the release of secondary fumes, the high-volume, low-concentration gases offer low scrubbing efficiencies, have, high capital and operating costs, and produce large volumes of liquid effluents for treatment. Wet scrubbing is seldom used for secondary fume control in the prebake process. When anodes are baked on site, the dry scrubbing system using aluminium oxide as the adsorbent is used. It has the advantage of being free of waste products, and all enriched alumina and absorbed material are recycled directly to the reduction cells. Dry scrubbing may be combined with incineration for controlling emissions of tar and volatile organic compounds (VOCs) and to recover energy. Wet scrubbing can also be used but is not recommended, since a liquid effluent, high in fluorides and hydrocarbons, will require treatment and disposal. Dry scrubber systems applied to the pot fumes and to the anode baking furnace result in the capture of 97 per cent of all fluorides from the process. The aluminium smelter solid wastes, in the form of spent pot lining, are disposed of in engineered landfills that feature clay or synthetic lining of disposal pits, provision of soil layers for covering and sealing, and control and treatment of any leachate. Treatment processes are available to reduce hazards associated with spent pot lining prior to disposal of the lining in a landfill. Other solid wastes such as bath skimmings are sold for recycling, while spalled refractories and other chemically stable materials are disposed of in landfill sites. Modern smelters using good industrial practices are able to achieve the following in terms of pollutant loads (all values are expressed on an annualised basis): hydrogen fluoride, 0.2 to 0.4 kg/T; total fluoride, 0.3 to 0.6 kg/T; particulates, 1 kg/T; sulphur dioxide, 1 kg/T; and nitrogen oxides, 0.5 kg/T. CF_4 emissions should be less than 0.1 kg/T. For secondary aluminium production, the principal treatment technology downstream of the melting furnace is dry sorbent injection using lime, followed by fabric filters. Waste gases from salt slag processing should be filtered as well. Waste gases from aluminium scrap pretreatment that contain organic compounds of concern may be treated by post-combustion practices. Air emissions should be monitored regularly for particulate matter and fluorides. Hydrocarbon emissions should be monitored annually on the anode plant and baking furnaces. Liquid effluents should be monitored weekly for pH, total suspended solids, fluoride, and aluminium and at least monthly for other parameters. Monitoring data should be analysed and reviewed at regular intervals and compared with the operating standards so that any necessary corrective actions can be taken.

Emission regulations for Indian aluminium smelters

1. *Air Pollution:* Central Pollution Control Board has recommended the following emission standards for aluminium smelter:
   (a) Fluoride – 1 kg/T of Al produced
   (b) Particulate matter – 150 mg/Nm^3

The main pollutants of aluminium smelter are shown in Table 14.7.

Table 14.7. Pollutants of aluminium smelter.

Type of pollution	Pollutants produced
Gaseous emissions	Carbon monoxide, carbon dioxide, hydrogen fluoride, sulphur dioxide, carbon disulphide silicon tetrafluoride, carbon tetrafluoride, hexafluoroethane, water vapour, NO_x, etc.
Solid emissions	Aluminia, cryolite, aluminium fluoride, calcium fluoride, carbon, iron oxide.
Liquid effluents	Fluorine compounds, hydrocarbons (from soderberg plants), entrained water, cyanide.
Smelter wastes	Spent potlinings, anode butts from prebaked anodes, dust from gas cleaning, sludges from cleaning scrubbing water, material from pot skimmings, spills.
Paste preparation emissions and wastes	Coke dust, coal dust fines, hydrocarbon fumes.
Anode baking emissions and wastes	Hydrocarbons, fluorides, sulphur dioxide.
Cast house emission and wastes.	Fluxing fumes (primarily aluminium chloride), trace fluorine, sulphur dioxide.
Emissions and wastes from ancilliary operations	Dust from bulk materials handling, demolition of old pots and cleaning of prebaked anode butts to recover carbon.

POT emissions and collection systems

Many types of pots, having different sizes, design, current rating, etc. are used in aluminium smelters. Three broad categories of these pots are:
1. Multi-anode prebacked anode pots (PB Pots).
2. Horizontal stud soderberg pots (HSS Pots).
3. Vertical stud soderberg pots (VSS Pots).

Primary low volume gas from the smelter is collected with the help of a hood provided on individual pots. The secondary gas stream which escapes the reach of pot hood system, gets diluted by the cell house air and is discharged to the atmosphere through the cell house roof and this diluted secondary stream may not require any treatment before it is discharged to atmosphere.

Pollution control (fume treatment) systems

The air pollution control (fume treatment) system used in aluminium smelters can be divided into two categories:
1. Those which control/collect the harmful constituents from the primary gas stream collected from hoods/skirts.
2. The other type which treat the secondary gas stream (ventilation air).

In the early periods of development of fume treatment methods, attention was first paid to treat ventilation gases, i.e. secondary emission control equipment was developed first to treat the gases of existing smelters. These are normally low energy wet scrubbers capable of treating large volume of gases which contain low concentration of pollutants. Typical scrubbers include spray streams, cross flow packed beds, floating beds, multi venture grids and perforated type plate units. The efficiency of the system is roughly proportional to the energy utilised to contact liquid with gas stream. Removal efficiency of secondary systems, where primary system is not installed, is roughly 40–60 per cent of particulate matter and 70–85 per cent of gaseous fluorides. Roof scrubbers are found in most of the old aluminium smelters where there is no scope for installation of primary pollution control system. Though simple to install, these are least amenable to maintenance. Corrosion problems necessitate use of expensive fibre glass materials or handling of corrosive effluents. The secondary emission control systems do not satisfy statutory regulations. The important pollution control system in vogue for the treatment of pot emissions/gases collected through hoods and skirts include multicyclone collectors, electrostatic precipitators (dry and wet), scrubbing systems (wet and dry).

Wet and dry scrubbing systems

In wet scrubbing systems gases are passed first through cyclones and electrostatic precipitators to remove alumina, carbon dust, bath particles and condensed hydrocarbon particulates. Coarse particles are separated and some finer particles are left in the gas stream. The gases are then sent to scrubbers where there is counter-current flow of gases and scrubbing medium. The scrubbing liquor (generally sodium carbonate solution) is sprayed through a number of nozzles to create a cloud of fine droplets. Conditions for intimate contact between liquid and gases are created. By the time gases reach the outlet of scrubbers, most of the fluorine gas compounds and solid particulates are removed from the gases. The gases free from polluting constituents are released through tall stacks. The liquor in which fluorine gas compounds are dissolved, is treated to produce either cryolite or aluminium fluoride. The fluorine collection efficiency is around 98 per cent. Dry scrubbing system is based on the reaction between certain types of alumina with hydrogen fluoride gas compounds. Potroom gases collected from each individual pot are fed to venturi reactors or fluidised bed reactors to cause an intimate mixing/contact between alumina and gases. During the contact period hydrogen fluoride is chemisorbed on the surface of alumina particles. The solid–gas mixture is then conveyed to dust collectors (bag filters). Here all the solid particulates consisting of alumina, fluoride particulates, bath material, carbon debris and condensed hydro-carbons are separated and collected. Fine pores of bag filters facilitate the recovery of even very fine particulates. The gases free from hydrogen fluoride and dust are discharged into the atmosphere through tall stacks. The materials collected in bag filters are fed to pots. Major advantages of the dry scrubbing system over the wet scrubbing system are:

1. Excellent collection efficiency of total fluorides (over 98 per cent) at a reasonable power consumption.
2. Almost total recovery and direct recycling of the fluorides and alumina emitted from the pots.
3. Hot dry exhaust gases get easily dispersed into the atmosphere.
4. No corrosion problems are experienced even though equipment is made of steel.
5. There are no effluent disposal problems.
6. Operation and maintenance costs are low.

LEAD AND ZINC

Lead and zinc can be produced pyrometallurgically or hydrometallurgically, depending on the type of ore used as a charge. In the pyrometallurgical process, ore concentrate containing lead, zinc, or both is fed, in some cases after sintering, into a primary smelter. Lead concentrations can be 50 to 70 per cent, and the sulphur content of sulphidic ores is in the range of 15 to 20 per cent. Zinc concentration is in the range of 40 to 60 per cent, with sulphur content in sulphidic ores in the range of 26 to 34 per cent. Ores with a mixture of lead and zinc concentrate usually have lower respective metal concentrations. During sintering, a blast of hot air or oxygen is used to oxidise the sulphur present in the feed to sulphur dioxide. Blast furnaces are used in conventional processes for reduction and refining of lead compounds to produce lead. In the most common hydrometallurgical process for zinc manufacturing, the ore is leached with sulphuric acid to extract the lead/zinc. These processes can operate at atmospheric pressure or as pressure leach circuits. Lead/zinc is recovered from solution by electrowinning, a process similar to electrolytic refining. The process most commonly used for low-grade deposits is heap leaching. Imperial smelting is also used for zinc ores.

Primary Lead Processing

The conventional pyrometallurgical primary lead production process consists of four steps: sintering, smelting, drossing, and refining. A feedstock made up mainly of lead concentrate is fed into a sintering machine. Other raw materials may be added, including iron, silica, limestone flux, coke, soda-ash, pyrite, zinc, caustic, and particulates gathered from pollution control devices. The sintering feed, along with coke, is fed into a blast furnace for reducing, where, the carbon also acts as a fuel and smelts the lead-containing materials. The molten lead flows to the bottom of the furnace, where four layers form: 'speiss' (the lightest material, basically arsenic and antimony), 'matte' (copper sulphide and other metal sulphides), Blast furnace slag (primarily silicates), and lead bullion (98 per cent by weight). All layers are then drained off. The speiss and matte are sold to copper smelters for recovery of copper and precious metals. The Blast furnace slag, which contains zinc, iron, silica, and lime, is stored in piles and is partially recycled. Sulphur oxide emissions are generated in blast furnaces from small quantities of residual lead sulphide and lead sulphates in the sinter feed. Rough lead bullion from the Blast furnace usually requires preliminary treatment in kettles before undergoing refining operations. During drossing, the bullion is agitated in a drossing kettle and cooled to just above its freezing point, 370° to 425°C (700° to 800°F). A dross composed of lead oxide, along with copper, antimony, and other elements, floats to the top and solidifies above the molten lead. The dross is removed and is fed into a dross furnace for recovery of the nonlead mineral values. The lead bullion is refined using pyrometallurgical methods to remove any remaining nonlead materials (e.g. gold, silver, bismuth, zinc, and metal oxides such as oxides of antimony, arsenic, tin, and copper). The lead is refined in a cast-iron kettle in five stages. First, antimony, tin, and arsenic are removed. Next, gold and silver are removed by adding zinc. The lead is then refined by vacuum removal of zinc. Refining continues with the addition of calcium and magnesium, which combine with bismuth to form an insoluble compound that is skimmed from the kettle. In the final step, caustic soda, nitrates, or both may be added to remove any remaining traces of metal impurities. The refined lead will have a purity of 99.90 to 99.99 per cent. It may be mixed with other metals to form alloys, or it may be directly cast into shapes.

Secondary Lead Processing

The secondary production of lead begins with the recovery of old scrap from worn-out, damaged, or obsolete products and with new scrap. The chief source of old scrap is lead-acid batteries; other sources include cable coverings, pipe, sheet, and other lead-bearing metals. Solder, a tin-based alloy, may be recovered from the processing of circuit boards for use as lead charge.

Prior to smelting, batteries are usually broken up and sorted into their constituent products. Fractions of cleaned plastic (such as polypropylene) case are recycled into battery cases or other products. The dilute sulphuric acid is either neutralised for disposal or recycled to the local acid market. One of the three main smelting processes is then used to reduce the lead fractions and produce lead bullion.

Most domestic battery scrap is processed in Blast furnaces, rotary furnaces, or reverberatory furnaces. A reverberatory furnace is more suitable for processing fine particles and may be operated in conjunction with a Blast furnace. Blast furnaces produce hard lead from charges containing siliceous slag from previous runs (about 4.5 per cent of the charge), scrap iron (about 4.5 per cent), limestone (about 3 per cent), and coke (about 5.5 per cent).

The remaining 82.5 per cent of the charge is made up of oxides, pot furnace refining drosses, and reverberatory slag. The proportions of rerun slags, limestone, and coke vary but can run as high as 8 per cent for slags, 10 per cent for limestone, and 8 per cent for coke. The processing capacity of the blast furnace ranges from 20 to 80 metric tons per day (tpd).

Newer secondary recovery plants use lead paste desulphurisation to reduce sulphur dioxide emissions and generation of waste sludge during smelting. Battery paste containing lead sulphate and lead oxide is desulphurised with soda ash, yielding market-grade sodium sulphate as a by-product. The desulphurised paste is processed in a reverberatory furnace, and the lead carbonate product may then be treated in a short rotary furnace. The battery grids and posts are processed separately in a rotary smelter.

Pollution Prevention and Control

The principal air pollutants emitted from the processes are particulate matter and sulphur dioxide. Fugitive emissions occur at furnace openings and from launders, casting moulds, and ladles carrying molten materials, which release sulphur dioxide and volatile substances into the working environment. Additional fugitive particulate emissions occur from materials handling and transport of ores and concentrates. Some vapours are produced in hydrometallurgy and in various refining processes.

The principal constituents of the particulate matter are lead/zinc and iron oxides, but oxides of metals such as arsenic, antimony, cadmium, copper, and mercury are also present, along with metallic sulphates. Dust from raw materials handling contains metals, mainly in sulphidic form, although chlorides, fluorides, and metals in other chemical forms may be present. Off-gases contain fine dust particles and volatile impurities such as arsenic, fluorine, and mercury.

Air emissions for processes with few controls may be of the order of 30 kilograms lead or zinc per metric ton (kg/T) of lead or zinc produced. The presence of metals in vapour form is dependent on temperature. Leaching processes will generate acid vapours, while refining processes result in products of incomplete combustion (PICs). Emissions of arsine, chlorine, and hydrogen chloride vapours and acid mists are associated with electrorefining.

The most effective pollution prevention option is to choose a process that entails lower energy usage and lower emissions. Modern flash-smelting processes save energy, compared with the conventional sintering and blast furnace processes. Process gas streams containing over 5 per cent sulphur dioxide are usually used to manufacture sulphuric acid. The smelting furnace will generate gas streams with SO_2 concentrations ranging from 0.5 to 10 per cent, depending on the method used. It is important, therefore, to select a process that uses oxygen-enriched air or pure oxygen. The aim is to save energy and raise the SO_2 content of the process gas stream by reducing the total volume of the stream, thus permitting efficient fixation of sulphur dioxide. Processes should be operated to maximise the concentration of the sulphur dioxide. An added benefit is the reduction (or elimination) of nitrogen oxides NO_x. Regarding standard treatment technologies, ESPs and baghouses are used for product recovery and for the control of particulate emissions. Dust that is captured but not recycled will need to be disposed of in a secure landfill or in another acceptable manner (note: this is an additional cost to control and an added incentive for pollution prevention).

Arsenic trioxide or pentoxide is in vapour form because of the high gas temperatures and must be condensed by gas cooling so that it can be removed in fabric filters. Collection and treatment of vent gases by alkali scrubbing may be required when sulphur dioxide is not being recovered in an acid plant. These also represent high cost control options with post-disposal of waste problems. Table 14.8 provides some common examples of pollution prevention practices, along with energy efficiency programs. All of these fit into the category of relatively low-cost investments.

Table 14.8. Examples of pollution prevention and energy saving practices.

Use doghouse enclosures where appropriate; use hoods to collect fugitive emissions.

Mix strong acidic gases with weak ones to facilitate production of sulphuric acid from sulphur oxides, thereby avoiding the release of weak acidic gases.

Maximise the recovery of sulphur by operating the furnaces to increase the SO_x content of the flue gas and by providing efficient sulphur conversion. Use a double-contact, double-absorption process..

Desulphurise paste with caustic soda or soda ash to reduce SO_2 emissions.

Use energy-efficient measures such as waste heat recovery from process gases to reduce fuel usage and associated emissions.

Recycle condensates, rainwater, and excess process water for washing, for dust control, for gas scrubbing, and for other process applications where water quality is not of particular concern.

Give preference to natural gas over heavy fuel oil for use as fuel and to coke with lower sulphur content.

Use low-NO_x burners.

Use suspension or fluidised bed roasters, where appropriate, to achieve high SO_2 concentrations when roasting zinc sulphides.

Give preference to fabric filters over wet scrubbers or wet electrostatic precipitators (ESPs) for dust control.

Good housekeeping practices are key to minimising losses and preventing fugitive emissions. Losses and emissions are minimised by enclosed buildings, covered conveyors and transfer points, and dust collection equipment. Yards should be paved and runoff water routed to settling ponds.

Air Pollution Control Systems

Two types of gaseous emissions in the sintering process may be identified: (i) process gas from the desulphurising operation containing amongst other things, sulphur dioxide and (ii) hygiene ventilation air from the work places and containing dust and fume. Process gas contains carbon monoxide and carbon dioxide and the hygiene ventilation air contains dust and fume of lead, zinc and cadmium metal/ oxides. Elaborate gas cleaning and hygiene ventilation systems have been proposed for meeting the laid down standards and maintaining safe work atmosphere.

Process gases

Sinter plant rich gases will be cleaned by combination of hot electro-static precipitators, venturi scrubbers and wet electro-static precipitators. The lean gases from tail end are recirculated in sinter plant to enrich SO_2 concentration upto 6 per cent. A DCDA sulphuric acid plant will be installed for converting SO_2. The plant will meet the environmental standards for gaseous emissions of SO_2 as per IS-635-1972 to a limit of 4 kg of SO_2 per ton of sulphuric acid produced.

Furnace gases, after condensation of zinc vapours, will be scrubbed clean in a tower and passed through a Theisen disintegrator to remove all particulate matter to yield clean low calorific value off-gas (18–23 per cent CO) to be utilised as a fuel in preheating of coke and air and in zinc refluxing. The surplus gas will be used to operate a boiler to generate steam for providing power. A flare stack provided would burn any surplus LCV gas after utilisation so that no CO is allowed to join the atmosphere.

Ventilation gases

Baghouses will be used for cleaning dry gases and venturi scrubbers for wet gases with large particulates. There will be a main stack (about 60 m tall) for acid Plant. The height and design of various stacks will be such that ground level concentration of the gaseous pollutants remain within the allowable limits prescribed by the Air (Control & Prevention of Pollution Act, 1981). Lead content in the ambient work place will not exceed 150 mcg/Nm^3. The individual stack dust emissions will not exceed 50 mg/Nm^3 with a lead content of not more than 11.5 mg/Nm^3.

Mercury removal

Hindustan Zinc Limited (HZL) has foreseen the installation of mercury removal plant to keep mercury below 1 ppm in the acid and liquid effluents so as to avoid mercury passage to the biocycle through acid being used for fertiliser manufacture.

Lead refining

In the case of lead refinery, the company had an option between pyre and electro-routes. Although HZL is currently using pyre-methods for refining lead in existing units at Tundoo and Vizag, electro-refining is opted for International Specialty Products (ISP) whereby the dust/fume emissions will be minimised.

Monitoring of in plant atmosphere

Monitoring of in plant atmosphere will be carried out continuously using static sampling equipment at different locations. A 300 metre wide green belt will be developed around the smelter complex to facilitate monitoring pollution effect. Surface and ground water is monitored right from now to provide a base line data.

NICKEL

Primary nickel is produced from two very different ores, lateritic and sulphidic. Lateritic ores are normally found in tropical climates where weathering, with time, extracts and deposits the ore in layers at varying depths below the surface. Lateritic ores are excavated using large earth-moving equipment and are screened to remove boulders. Sulphidic ores, often found in conjunction with copper-bearing ores, are mined from underground. Following is a description of the processing steps used for the two types of ores.

Lateritic Ore Processing

Lateritic ores have a high percentage of free and combined moisture, which must be removed. Drying removes free moisture; chemically bound water is removed by a reduction furnace, which also reduces the nickel oxide. Lateritic ores have no significant fuel value, and an electric furnace is needed to obtain the high temperatures required to accommodate the high magnesia content of the ore. Some laterite smelters add sulphur to the furnace to produce a matte for processing. Most laterite nickel processors run the furnaces so as to reduce the iron content sufficiently to produce ferronickel products. Hydrometallurgical processes based on ammonia or sulphuric acid leach are also used. Ammonia leach is usually applied to the ore after the reduction roast step.

Sulphidic Ore Processing

Flash smelting is the most common process, but electric smelting is used for more complex raw materials when increased flexibility is needed. Both processes use dried concentrates. Electric smelting requires a roasting step before smelting to reduce sulphur content and volatiles. Older nickel-smelting processes, such as blast or reverberatory furnaces, are no longer acceptable because of low energy efficiencies and environmental concerns.

In flash smelting, dry sulphide ore containing less than 1 per cent moisture is fed to the furnace along with preheated air, oxygen-enriched air (30 to 40 per cent oxygen), or pure oxygen. Iron and sulphur are oxidised. The heat that results from exothermic reactions is adequate to smelt concentrate, producing a liquid matte (up to 45 per cent nickel) and a fluid slag. Furnace matte still contains .iron and sulphur, and these are oxidised in the converting step to sulphur dioxide and iron oxide by injecting air or oxygen into the molten bath. Oxides form a slag, which is skimmed off. Slags are processed in an electric furnace prior to discard to recover nickel. Process gases are cooled, and particulates are then removed by gas-cleaning devices.

Nickel Refining

Various processes are used to refine nickel matte. Fluid bed roasting and chlorine-hydrogen reduction produce high-grade nickel oxides (more than 95 per cent nickel). Vapour processes such as the carbonyl process can be used to produce high-purity nickel pellets. In this process, copper and precious metals remain as a pyrophoric residue that requires separate treatment. Use of electrical cells equipped with inert cathodes is the most common technology for nickel refining. Electrowinning, in which nickel is removed from solution in cells equipped with inert anodes, is the more common refining process. Sulphuric acid solutions or, less commonly, chloride electrolytes are used.

Pollution Prevention and Control

Sulphur dioxide is a major air pollutant emitted in the roasting, smelting, and converting of sulphide ores. (Nickel sulphide concentrates contain 6 to 20 per cent nickel and up to 30 per cent sulphur.) SO_2 releases can be as high as 4 MT of sulphur dioxide per metric ton of nickel produced, before controls. Reverberatory furnaces and electric furnaces produce SO_2 concentrations of 0.5 to 2.0 per cent, while flash furnaces produce SO_2 concentrations of over 10 per cent—a distinct advantage for the conversion of the sulphur dioxide to sulphuric acid. Particulate emission loads for various process steps include 2.0 to 5.0 kilograms per metric ton (kg/T) for the multiple hearth roaster; 0.5 to 2.0 kg/T for the fluid bed roaster; 0.2 to 1.0 kg/T for the electric furnace; 1.0 to 2.0 kg/T for the Pierce-Smith converter; and 0.4 kg/T for the dryer upstream of the flash furnace.

Ammonia and hydrogen sulphide are pollutants associated with the ammonia leach process; hydrogen sulphide emissions are associated with acid leaching processes. Highly toxic nickel carbonyl is a contaminant of concern in the carbonyl refining process. Various process off-gases contain fine dust particles and volatilised impurities. Fugitive emissions occur at furnace openings, launders, casting moulds, and ladles that carry molten product. The transport and handling of ores and concentrates produce windborne dust.

Pyrometallurgical processes for processing sulphidic ores are generally dry, and effluents are of minor importance, although wet ESPs are often used for gas treatment, and the resulting waste-water could have high metal concentrations. Process bleed streams may contain antimony, arsenic, or mercury. Large quantities of water are used for slag granulation, but most of this water should be recycled.

Pollution prevention is always preferred to the use of end-of-pipe pollution control facilities. Therefore, every attempt should be made to incorporate cleaner production processes and facilities to limit, at source, the quantity of pollutants generated.

The choice of flash smelting over older technologies is the most significant means of reducing pollution at source. Sulphur dioxide emissions can be controlled by:

1. Recovery as sulphuric acid.
2. Recovery as liquid sulphur dioxide (absorption of clean dry off-gas in water or chemical absorbtion by ammonium bisulphite or dimethyl aniline).
3. Recovery as elemental sulphur, using reductants, such as hydrocarbons, carbon or hydrogen sulphide.

Toxic nickel carbonyl gas is normally not emitted from the refining process because it is broken down in decomposer towers. However, very strict precautions throughout the refining process are required to prevent the escape of the nickel carbonyl into the workplace. Continuous monitoring for the gas, with automatic isolation of any area of the plant where the gas is detected, is required. Impervious clothing is used to protect workers against contact of liquid nickel carbonyl with skin. Preventive measures for reducing emissions of particulate matter include encapsulation of furnaces and conveyors to avoid fugitive emissions. Covered storage of raw materials should be considered. Wet scrubbing should be avoided, and cooling waters should be recirculated. Stormwaters should be collected and used in the process. Process water used to transport granulated slag should be recycled. To the extent possible, all process effluents should be returned to the process. The discharge of particulate matter emitted during drying, screening, roasting, smelting, and converting is controlled by using cyclones followed by wet scrubbers, ESPs, or bag filters. Fabric filters may require reduction of gas temperatures by, for example, dilution with low temperature gases from hoods used for fugitive dust control. Preference should be given to the use of fabric filters over wet scrubbers.

COPPER SMELTING

Copper can be produced either pyrometallurgically or hydrometallurgically. The hydrometallurgical route is used only for a very limited amount of the world's copper production and is normally only considered in connection with *in situ* leaching of copper ores. From an environmental point of view, this is a questionable production route. Several different processes can be used for copper production. The traditional process is based on roasting, smelting in reverbatory furnaces (or electric furnaces for more complex ores), producing matte (copper-iron sulphide), and converting for production of blister copper, which is further refined to cathode copper. This route for production of cathode copper requires large amounts of energy per ton of copper: 30 to 40 million British thermal units (Btu) per ton cathode copper. It also produces furnace gases with low sulphur dioxide concentrations from which the production of sulphuric acid or other products is less efficient. The sulphur dioxide concentration in the exhaust gas from a reverbatory furnace is about 0.5 to 1.5 per cent; that from an electric furnace is about 2 to 4 per cent. So-called flash smelting techniques have therefore been developed that utilise the energy released during oxidation of the sulphur in the ore. The flash techniques reduce the energy demand to about 20 million Btu/T of produced cathode copper.

The SO_2 concentration in the off-gases from flash furnaces is also higher, over 30 per cent, and is less expensive to convert to sulphuric acid. The INCO process results in 80 per cent sulphur dioxide in the off gas. Flash processes have been in use since the early 1950s.

In addition to the above processes, there are a number of newer processes such as Noranda, Mitsubishi, and Contop, which replace roasting, smelting, and converting, or processes such as ISA-SMELT and KIVCET, which replace roasting and smelting. For converting, the Pierce-Smith and Hoboken converters' are the most common processes.

The matte from the furnace is charged to converters, where the molten material is oxidised in the presence of air to remove the iron and sulphur impurities (as converter slag) and to form blister copper. Blister copper is further refined as either fire-refined copper or anode copper (99.5 per cent pure copper), which is used in subsequent electrolytic refining. In fire refining, molten blister copper is placed in a fire-refining furnace, a flux may be added, and air is blown through the molten mixture to remove residual sulphur. Air blowing results in residual oxygen, which is removed by the addition of natural gas, propane, ammonia, or wood. The fire-refined copper is then cast into anodes for further refining by electrolytic processes or is cast into shapes for sale.

In the most common hydrometallurgical process, the ore is leached with ammonia or sulphuric acid to extract the copper. These processes can operate at atmospheric pressure or as pressure leach circuits. Copper is recovered from solution by electrowinning, a process similar to electrolytic refining. The process is most commonly used for leaching low-grade deposits *in situ* or as heaps.

Recovery of copper metal and alloys from copper-bearing scrap metal and smelting residues requires preparation of the scrap (e.g. removal of insulation) prior to feeding into the primary process. Electric arc furnaces using scrap as feed are also common.

Sulphur Dioxide Emissions from Copper Smelters and Their Control

One of the serious problems facing the copper smelting industry is control of gaseous SO_2 emissions. Sulphur is a major component in the concentrate fed to the smelting processes varying in the general range of 25–35 per cent. The mass ratio of S to Cu varies between 0.8–1.6 in copper concentrates. Sulphur also occurs in all fossil fuels used in copper smelting. During various unit operations, sulphur is oxidised in SO_2, although small amount of SO_3 may be formed under certain conditions of temperature and oxygen partial pressure. SO_2 may also get oxidised in the atmosphere to form sulphates and particulate forms of sulphur compounds. These are major contributors to air pollution.

Effects on environment

SO_2 emissions, through the formation of acids and other compounds in the atmosphere can cause damage to human health, vegetation and property. High levels of sulphate concentrations and long-term exposure to sulphates are said to aggravate asthma, lung and heart diseases. Sensitive vegetation can be severely damaged even by low levels of SO_2 in the atmosphere. Studies made in USSR during the past decade showed that pine trees growing in an atmosphere having SO_2 concentrations of 500 mg/m^3 had a growth loss of 48 per cent in comparison with pine trees growing in an atmosphere free of SO_2. Many materials such as metal surfaces, marble, brick, stone work, plastics, rubber, paper become discoloured and brittle when exposed to SO_2. Thus buildings, bridges, steel girders, automobiles and highways are all affected by excessive SO_2 emissions.

Air pollution control regulations

Many countries where copper smelters are located have formulated strict air pollution control standards for SO_2 both for ambient air quality and for allowable emissions and there is increased awareness regarding the harmful effects and dangers of such pollution. In USA, primary and secondary standards have been established. Primary standards (80 mg/m^3 annual arithmetic mean) which protect the public health, define how clean the ambient air must be so that it will not be harmful to human health. Secondary standards (60 mg/m^3 annual arithmatic mean), which protect the public welfare, define how clean the air must be in order to protect against the known or anticipated effects of air pollution on property, materials, climate, economic values and personal comfort.

Emission control technologies

It is first necessary to remove the dust or particulate content as completely as possible before recovery of SO_2 from the smelter gases. In older smelters, primary collection of dust is effected in settling chambers or balloon flues where the low gas velocity and hence long residence time, allows gravity settling of large particles. In modern smelters, high velocity flues and cyclones have replaced the balloon flues. Final gas cleaning in the smelter is carried out in electrostatic precipitators, where over 99 per cent collection efficiency is achieved. SO_2 emission control technology in the copper smelting industry has been developed essentially to treat two types of gas streams: (i) concentrated gas streams which arise from fluid bed roasters, primary smelting furnaces and converters where the gas strength is more than 4.5 per cent SO_2, and (ii) dilute gas streams (containing generally less than 2 per cent SO_2) which arise from multi-hearth roasters, reverberatory furnaces, fugitive emissions and tail gases.

Concentrated gas streams

Concentrated SO_2 gas may be captured and fixed as elemental sulphur or alternatively used in the production of liquid SO_2 or sulphuric acid.

Weak gas streams

These originate from the following sources:
1. Process gas streams (SO_2 content 0.5 to 2 per cent) such as from multiple hearth roasters, reverberatory furnaces, fire refining furnaces etc.
2. Tail gas emissions are from sulphur fixation plants treating, in general, concentrated gas streams.
3. Fugitive emissions (1 to 2 per cent of the total S) arising from transfer of hot calcine from multiple hearth roasters to smelters and leakage through furnace refractories.

Technology for the control of weak SO_2 gas streams has been largely developed for thermal power plants, and has not yet become popular in metallurgical industry for the following reasons :
1. The flue gas desulphurisation (FGD) systems are not yet commercially proven.
2. A stable product generally cannot be produced economically.
3. Systems are expensive to install.
4. SO_2 control conditions are not that stringent at some of the places.

The major source of weak uncontrolled SO_2 emissions from smelters is the off gas generated by reverberatory furnace. For any given plant, this can be from 9 to 34 per cent of the total sulphur input depending on whether the furnace is charged with calcine or concentrate, respectively. Construction, operation and maintenance of reverberatory furnace are important factors in concentrating SO_2 content of the gas. Minimising dilution effects by sealing all openings in the furnace, oxygen enrichment of combustion air and more uniform charging practice can increase the gas strength to about 2.5 per cent SO_2. Using oxygen enrichment to 60 per cent O_2 in air, the El Teniente Smelter in Chile has increased reverberatory furnace gas strength to between 5.8–7.3 per cent SO_2.

In the context of control of weak gas emissions, two approaches have been adopted :
1. Increasing the concentration of SO_2 by using a regenerative system for subsequent processing to sulphuric acid, elemental sulphur or liquid SO_2. Absorbing media used are MgO, citric acid

plus sodium citerate, low temperature water or ammonium bisulphite. The Cominico process which uses ammonium bisulphite as absorbent can achieve high efficiencies of SO_2 removal over a wide range of SO_2 concentrations well within that encountered by copper reverberatory furnaces.

2. Non-regenerative absorption system based on neutralising SO_2 by scrubbing to produce a stable waste product. Important scrubbing systems developed are the lime scrubbing, Palabora scrubbing and double alkali processes. In lime scrubbing system, the inlet gas containing 2.5 to 3.2 per cent SO_2 is washed and cooled in sea water fed gas coolers thereby forming sulphurous acid which then reacts with lime to form calcium sulphite slurry. This is then oxidised to yield calcium sulphate (gypsum) which can be utilised for cement manufacture. In Palabora process, the hydro-separator over-flow from the concentrator, which contains alkaline calcium and magnesium carbonates, is used as a scrubber. Vallerite containing around 22.9 per cent Cu is recovered. Several technologies go under the designation of the double alkali process. One is the 'concentrated double alkali process' which is installed at various small industrial steam plants. The scrubbing medium 'sodium sulphite' is converted to sodium bisulphite by reaction with SO_2. It is then reacted with slaked lime to produce calcium sulphite which is centrifuged to a fine cake with 60–70 per cent solids. 'Sodium-lime double alkali' process uses lime to regenerate the scrubber and also produces a waste calcium sulphite cake.

Economic considerations

The choice of by-product produced i.e. elemental sulphur, liquid SO_2 or H_2SO_4 will largely be determined by the market available for the particular by-product and the economics of production. It is desirable that the sale of by-product should offset at least a part of the pollution control cost. Transportation cost is an important factor in determining the economics, competitiveness and marketability of by-products. It is likely that neutralisation of SO_2 in exhaust gas streams (from utilities) by lime scrubbing to produce a discardable material may be the most suitable in view of low concentrations of SO_2 in these gases.

Sulphur produced from metallurgical gases often contains undesirable impurities such as As, Se and Hg which have to be removed. At present, sulphur produced from metallurgical gases is not significant and the cost of production of sulphur is also high.

Demand for liquid SO_2 is very limited. The single largest use of liquid SO_2 is in the production of sodium hydrosulphite which is used as a bleaching agent in textile and paper industry. It is also expensive to transport liquid SO_2 because it requires special pressure tank cars which must return empty to the supply sources. Thus the scope for marketing of liquid SO_2 is limited.

Sulphuric acid is the most common by-product recovered from metallurgical gases and its production is the accepted pollution control approach for concentrated gas streams in copper smelters. The smelters, in general, have the option of manufacturing sulphuric acid and then either making phosphoric acid, phosphate fertilisers, potassium sulphate, or marketing sulphuric acid itself.

Thus, old unit processes such as multiple hearth roasting and reverberatory smelting are only suitable for recovering 70 per cent of the sulphur in the feed of the smelter. Future emission standards are likely to demand that at least 90 per cent of the sulphur is recovered from the off take gases. The use of oxygen enriched air in newer processes results in lesser energy consumption and generation of high strength SO_2 gas streams. As the matte and slag flow through the enclosed launders in the continuous processes, fugitive emissions of SO_2 are reduced. The use of double contact and absorption process in acid plant and installation of equipment for scrubbing of acid plant tail gas and fugitive emissions, make it possible to obtain better than 90 per cent sulphur recovery. In near future, production of sulphuric acid is likely to remain the accepted central approach for sulphur recovery in copper industry. Key factors involved in the modernisation of old and inefficient unit processes and in installation of air pollution control equipment, would be capital cost and availability of markets for by-products of sulphur fixation processes.

Pollution Prevention and Control

The principal air pollutants emitted from the processes are sulphur dioxide and particulate matter. The amount of sulphur dioxide released depends on the characteristics of the ore-complex ores which may contain lead, zinc, nickel, and other metals, and on whether facilities are in place for capturing and converting the sulphur dioxide. SO_2 emissions may range from less than 4 kilograms per metric ton (kg/T) of copper to 2000 kg/T of copper. Particulate emissions can range from 0.1 kg/T of copper to as high as 20 kg/T of copper. Fugitive emissions occur at furnace openings and from launders, casting moulds, and ladles carrying molten materials. Additional fugitive particulate emissions occur from materials handling and transport of ores and concentrates. Some vapours, such as arsine, are produced in hydrometallurgy and various refining processes. Dioxins can be formed from plastic and other organic material when scrap is melted. The principal constituents of the particulate matter are copper and iron oxides. Other copper and iron compounds, as well as sulphides, sulphates, oxides, chlorides, and fluorides of arsenic, antimony, cadmium, lead, mercury, and zinc, may also be present. Mercury can also be present in metallic form. At higher temperatures, mercury and arsenic could be present in vapour form. Leaching processes will generate acid vapours, while fire-refining processes result in copper and SO_2 emissions. Emissions of arsine, hydrogen vapours, and acid mists are associated with electrorefining. Waste-water from primary copper production contains dissolved and suspended solids that may include concentrations of copper, lead, cadmium, zinc, arsenic, and mercury and residues from mould release agents (lime or aluminium oxides).

Fluoride may also be present, and the effluent may have a low pH. Normally there is no liquid effluent from the smelter other the cooling water; waste waters do originate in scrubbers (if used), wet electrostatic precipitators, cooling of copper cathodes, and so on. In the electrolytic refining process, by-products such as gold and silver are collected as slimes that are subsequently recovered.

Process gas streams containing sulphur dioxide are processed to produce sulphuric acid, liquid sulphur dioxide, or sulphur. The smelting furnace will generate process gas streams with SO_2 concentrations ranging from 0.5 to 80 per cent, depending on the process used. It is important, therefore, that a process, be selected that uses oxygen-enriched air (or pure oxygen) to raise the SO_2 content of the process gas stream and reduce the total volume of the stream, thus permitting efficient fixation of sulphur dioxide. Processes should be operated to maximise the concentration of the sulphur dioxide. An added benefit is the reduction of NO_x. Some pollution prevention practices for this industry include the following:

1. Closed-loop electrolysis plants will contribute to prevention of pollution.
2. Furnaces should be enclosed to reduce fugitive emissions, and dust from dust control equipment should be returned to the process.
3. Energy efficiency measures (such as waste heat recovery from process gases) should be applied to reduce fuel usage and associated emissions.
4. Recycling should be practiced for dust control and gas scrubbing, as well as for cooling water, condensates, rainwater, and excess process water used for washing.

5. Good housekeeping practices are key to minimising losses and preventing fugitive emissions. Such losses and emissions are minimised by enclosed buildings, covered or enclosed conveyors and transfer points, and dust collection equipment. Yards should be paved and runoff water routed to settling ponds. Regular sweeping of yards and indoor storage or coverage of concentrates and other raw materials also reduces materials losses and emissions.

Pollution control technologies acceptable for this industry are as follows. Fabric filters are used to control particulate emissions. Dust that is captured but not recycled will need to be disposed of i.. a secure landfill or other acceptable manner. Vapours of arsenic and mercury present at high gas temperatures are condensed by gas cooling and removed. Additional scrubbing may be required. Effluent treatment by precipitation, filtration, and so on, of process bleed streams, filter backwash waters, boiler blowdown, and other streams may be required to reduce suspended and dissolved solids and heavy metals. Residues that result from treatment are sent for metals recovery or to sedimentation basins. Stormwaters should be treated for suspended solids and heavy metals reduction. Slag should be landfilled or granulated and sold.

Modern plants using good industrial practices should set as targets total dust releases of 0.5 to 1.0 kg/T of copper and SO_2 discharges of 25 kg/T of copper. A double-contact, double-absorption plant should emit no more than 0.2 kg of sulphur dioxide per ton of sulphuric acid produced (based on a conversion efficiency of 99.7 per cent).

SECONDARY MAGNESIUM SMELTING

Secondary magnesium smelters process scrap which contains magnesium to produce magnesium alloys. Sources of scrap for magnesium smelting include automobile crankcase and transmission housings, beverage cans, scrap from product manufacture, and sludges from various magnesium-melting operations. This form of recovery is becoming an important factor in magnesium production.

Magnesium scrap is sorted and charged into a steel crucible maintained at approximately 675°C (1247°F). As the charge begins to burn, flux must be added to control oxidation. Fluxes usually contain chloride salts of potassium, magnesium, barium, and magnesium oxide and calcium fluoride. Fluxes are floated on top of the melt to prevent contact with air. The method of heating the crucible causes the bottom layer of scrap to melt first while the top remains solid. This semi-molten state allows cold castings to be added without danger of 'shooting', a violent reaction that occurs when cold metals are added to hot liquid metals. As soon as the surface of the feed becomes liquid, a crusting flux must be added to inhibit surface burning.

The composition of the melt is carefully monitored. Steel, salts, and oxides coagulate at the bottom of the furnace. Additional metals are added as needed to reach specifications. Once the molten metal reaches the desired levels of key components, it is poured, pumped, or ladled into ingots.

Emissions and Controls

Emissions from magnesium smelting include particulate magnesium oxides (MgO) and from the melting and fluxing processes, and nitrogen oxides from the fixation of atmospheric nitrogen by the furnace temperatures. Carbon monoxide and nonmethane hydrocarbons have also been detected. The type of flux used on the molten material, the amount of contamination of the scrap (especially oil and other hydrocarbons), and the type and extent of control equipment affect the amount of emissions produced.

FOUNDRIES

A foundry is the first among industrial plants producing air pollutants. Air contamination arises from both around and inside the plant. Therefore, it should be controlled to maintain hygienic working conditions. Obviously, if all the pollution could be limited only to inside the foundry, or its generation prevented, pollution of the surrounding atmosphere will not occur. Likewise, if every place inside the foundry where an air contaminant is generated is so controlled that dirt cannot escape to the inside atmosphere, pollution will not occur. Process exhaust systems are designed to effect this control. Regardless of the efficiency of a plant layout, dull surroundings can dull a worker's attitude. Conversely, pleasant surroundings can reduce stress, monotony and stimulate productivity. Creating an appealing work space, therefore, is rewarding to both the worker and the manager. A clean foundry can be maintained by the provision and operation of the following amenities:

1. Process exhaust ventilation.
2. Adequate make-up air.
3. Good housekeeping.

All these factors and environmental control activities can be achieved by upgrading the existing equipment or installing new pollution control facilities. Specialised experience is required for planning and implementing a pollution abatement program that will meet the current and future needs.

Classification of Foundry Areas

Air pollution problems are concentrated in six foundry areas—melting, mould, dumping, shakeout, grinding and tumbling, oil-sand core ovens and shell core areas. To orient the master plan and the construction program, foundry modernisation was developed in certain centres, namely melting, moulding, core-making, casting, finishing and heat-treating sections. Though not mentioned, the atmosphere in the pattern shop is charged with fine particles of sawdust. The following are some of the sections of the foundry facing the problem of pollution and its control:

Melting

The problems of pollution and its control in the melting section vary with the melting furnaces used.

Cupola

If one item of foundry equipment were to be singled out as the largest cause of air pollution, it undoubtedly would be the cupola. This is borne out by the fact that air pollution control officers, throughout the country, have chosen the cupola as their primary target. This is not surprising in view of the fact that the cupola is usually the tallest object to be seen in the plant and seems to be emitting the most smoke. The emission from the cupola consists of a variety of contaminants, the most bothersome from a control standpoint being the metallic oxides, which may be submicronic in size, unburned hydrocarbons, and carbon monoxide. The traditional cupola control contrivance is the familiar 'wet cap' set at the top of the open top cupola. The gases are directed through a water curtain by thermal impetus. Only the larger particles are removed. The resultant discharge is quite dense, even though it may be of a lighter colour.

Because the industry did not seem to be inclined to do a better job of cleaning the cupola effluent, air pollution control authorities have been inclined to write the regulations stringent enough to outlaw the 'wet cap' and demand a more efficient control system.

Various types of systems have been developed for the grey iron cupola by pollution control equipment manufacturers, who have gone to considerable research and obtained valuable practical experience in the process. Mechanical separators, scrubbers, and fabric arresters have been applied, the latter two attaining control to meet the most stringent codes. The mechanical separators have satisfied the less severe regulations allowed for cupolas that are used in the 'jobbing foundry' and are in use only a few hours each day. One typical code limits the particulate matter in the exhaust to 0.4 lb/1000 lb of gas. At this rate, such a cupola will not contribute as much pollution to the air as a 'production' cupola that is operated on a continuous basis and emits the maximum allowed by the same code of 0.1 lb/1000 lb of gas.

To satisfy the more stringent code, however, requires a much more elaborate system. Where a fabric arrester is used, the most successful systems have employed an afterburner to complete the combustion of the carbon monoxide and other combustibles, a cooling section where water sprays are used to lower the temperature of the gases to around 500°F, fibre glass cloth tubes in the baghouse, and an elaborate control system capable of maintaining the baghouse temperature. Equipment safeguards, such as a baghouse bypass which would open to protect the fabric if the temperature exceeded the fabric temperature limit, are normally included. This might be expected to happen at 'burn-down' time. The air pollution control is lost during these periods, but the equipment is protected. Where the fibre-glass bags are used, it is important to limit the use of fluorspar in the cupola charge to periods when the baghouse is not in use lest the fluorine ruin the bags. Fortunately, the need for using fluorspar can usually be limited to 'burn-down' periods. Because this fabric cannot endure stress without danger of rupturing, the dirt removal mechanism must be such as to shake the bags as gently as possible. Even so, if the high efficiencies of air pollution control are to be met on a continuing basis, the operator must be continually vigilant to detect and repair leaks in the fabric. One of the dangers in the use of fabric arresters on a cupola discharge system is the possibility, particularly in colder climates, of condensing out water in the baghouse. The entering airstream has generally been cooled by water sprays and consequently has a high relative humidity and dew point. The controls must hold the gas temperature at a point above the condensation level and below the maximum allowed by the fabric. Where temperature reduction can be done without using water, the danger of plugging up the baghouse is greatly reduced.

Because of the small particulate size in the cupola discharge, it is necessary to apply high energy where the wet scrubber is used. The venturi scrubber is uniquely suited to this application because it can be readily designed for the required pressure drop necessary to attain the cleaning efficiency desired and can be automatically adjusted to maintain, as the gas volume varies. Other designs, such as the flooded disk and units that apply the energy to the water rather than the gas stream, can be used. Normally the gases are cooled to reduce the volume that must be handled. Again, this is done by water sprays, but with the scrubber downstream, the water carryover causes no problem.

It is automatic that the initial cost of the pollution control system will vary with its volumetric capacity. The volume of the gases handled depends on the temperature of the gases, the position on the cupola from which the gases are withdrawn, and whether afterburning is employed. Where the take-off position is above the charging door, there seldom is any dependable restriction to airflow into the door. Consequently, a volume of air approximately equal to the blast air volume will be added to the cupola blast air and to the products of combustion. Under these conditions afterburning, to complete carbon monoxide combustion in order to minimise the possibility of explosions, should be provided. Water vapour is added to the gas by the cooling sprays. The resultant volume, corrected for temperature, must then be handled through the emission control equipment, fan, and stack. As large volumes require large pollution control equipment, and result in high initial and operating costs, the trend, at least for new cupolas, is to locate the take-off position for wet scrubbing control systems below the charging door.

Where the take-off position is below the charging door, the unburned charge built up above the melting level acts as a restriction to the air entering into the door. The volume of gases is thus reduced to the sum of the blast air, the products of combustion, and the volume from the charge door, which can be maintained at about 10 per cent of the blast air. If adequate controls are employed, afterburning, which increases the air volume, can be omitted and the resultant volume, corrected for temperature, can be passed through the scrubber, fan, and stack. In locations where the carbon monoxide must be controlled, afterburning as a final stage is employed.

There are many facets in the cupola pollution control system that must be considered which vary with each individual cupola. The composition of the charge, the condition during burn-down time, the height of the stack, the available space, the climate, and many other variables tend to make each emission control problem unique.

Metal melting with the cupola can be controlled but equipment costs are higher. Considering the pollution problem, with no collection, larger particles deposit nearby while small particles can travel miles together, resulting in an odour problem and visibility interference. The control consists of a wet cap collector, removing larger particles from the effluent gas stream benefiting the immediate plant property and area. But dirt, visibility interference and odour remain problems. Therefore, to meet current emission standards, venturi scrubbers or fabric collectors, are often used in conjunction with mechanical separators. Wet scrubbers can also be employed. Because of the large dust load, electrostatic precipitators are not usually economical. Fabric collectors also may not be employed since not only should the discharge be kept at a sufficiently cool temperature but in colder climates, the high-humidity dust also allows water condensation and clogging of the filter fabric.

Electric furnace

This is less dirty than the cupola but still some air pollutants may be formed. The smoke is emitted from the furnace openings, particularly when power is applied and must be vented. The particulates from the electric furnaces tend to be larger than those from the cupolas and thus they are easier to collect. However, they also possess a characteristic colour and appearance, and finer particles readily scatter light, making them very visible. Therefore, extremely high collection efficiency is required to obtain a clear discharge. The exhaust gases are usually taken to fabric or wet collectors or to electrostatic precipitators for particulate removal. In climates where makeup air need not be heated, arc furnace buildings have been exhausted from the roof level. However, abnormally large exhaust air volumes must be moved to attain satisfactory in-plant conditions. And, as the air should then be cleaned before it is discharged to the atmosphere, the initial and operating costs of control equipment are high. In most cases the exhausted air volume is kept to a minimum for economic reasons. Local hooding will attain effective control of the furnace emissions with much lower exhaust volumes than general exhaust.

The arc furnace emits smoke, particularly when the power is applied, from the openings in the furnace. Local hooding should control the smoke as it comes from the three electrode openings, the pouring spout, and slagging openings.

Various types of hoods are commercially available to control the smoke emissions. They usually employ a combination of the canopy and side draft principles, and are designed to exhaust enough air to keep the exhaust temperature in the 250°F range. The hood may be made in two parts with mating breakaway flanges, the one part mounted on the furnace roof ring and the other on the furnace shell. This type is used on furnaces where the roof is moved aside for charging. An alternate to using a hood on the arc furnace is to draw the fumes directly from the furnace interior. A duct take-off, made of material capable of withstanding the high fume temperatures, is inserted through the furnace roof. This, in effect, makes the furnace become its own hood, as an in-draft through all the openings is induced. It is important to maintain a stable negative static pressure in the furnace in the range of 0.02 to 0.05 in. (water guage) to keep the inflow adequate but as low as possible. Controls that sense this pressure and automatically operate a damper in the exhaust duct to correct the ever-changing condition are required. These controls must be extremely fast-acting and capable of withstanding the high temperatures involved.

It is best to have the exhaust duct connected to the furnace hood at all times. However, as the furnace must tip for pouring and slagging, revolving duct connections and possibly even telescoping duct sections may be required. A simpler method is to use mating breakaway flanges between the furnace hooding and the exhaust duct. As this disconnects the exhaust system from the hood, it results in loss of control during the pouring operation. This is not serious on furnaces where the heating is cut off before the metal is poured.

The exhaust is most often cleaned with fabric arrester equipment, provided that it is of the continuously operating type and that the temperature of the air is below the maximum temperature that the fabric can withstand. Fabrics such as Dacron, which can withstand 275°F, or silicone coated glass fibre cloth if higher temperatures are expected, should be used.

Safeguards to protect the fabric from overly hot gases are normally employed. They may be a simple damper arrangement to bypass the baghouse until the danger is over, or a dampered opening in the duct to bleed in atmospheric air for cooling when the need arises. The furnace operator prefers the former because it does not reduce the volume of air removed from the hood; the air pollution control officer prefers the latter. Where the dust collector can be located a good distance from the furnace, the long connecting duct will act as a heat exchanger. This diminishes the danger of overheated gases and reduces the volume that must be handled by the equipment.

Induction furnace

Electric induction furnaces have been used for years, mostly for melting in the non-ferrous industries. Recently they have been installed in iron foundries as well, finding use as holders, melters, and duplexers. Holding furnaces normally require no exhaust ventilation because the method of melting is clean and the metal itself has had most of the impurities removed during the already performed melting process.

Melters are subject to considerable fume emission, particularly when they are charged. The amount depends on the condition of the charge material. Either canopy or side draft hooding to control the emission during following the charging operation can be used. The type of hooding is determined by local conditions and the required exhaust volume by the type of hooding.

Inoculation of the metal in the duplexing furnace may require some attention. In the nodular iron inoculation process, for instance, where, among other ingredients, magnesium is placed in the inoculation ladle and the metal is poured thereon, there is a definite possibility of excessive fume emission, especially if the operation is not performed properly. Inasmuch as the ladle is usually on a monorail or bridge crane, the side draft type hood, with a relatively high exhaust volume, is required.

As with the arc furnaces, the fume emission may be cleaned by fabric arresters, and the same design criteria would, in general, apply.

Other than cupolas and electric furnaces, there are the direct flame and indirect flame fuel fired furnaces that emit smoke. Hooding of some type designed for the particular condition of the furnace installation is necessary. Usually an adaptation of the canopy type hood can be used. As in all hooding designs, the equipment should be enclosed as much as possible, consistent with the operations involved.

In many instances, the furnace hoods have been served by a gravity stack, discharging the smoke to the atmosphere. In those instances, however, where cleaning the effluent is deemed necessary, exhaust fans capable of overcoming the resistance of the dust collector must be provided.

Melting in this furnace is clean and normally no exhaust ventilation is required. During charging, fume emission may occur and the furnaces must be hooded. All the furnaces dealt so far are mainly applicable to ferrous alloys and in the melting of non-ferrous alloys, certain specific problems may arise. For example, alloys of Al and Mg normally require fluxing agents to aid in the melting. These fluxing agents create the majority of the pollution, which consists primarily of either chlorides, fluorides, sulphur dioxide or oxides of the alkali-metals. Also the oxides of the materials comprising the melt may be present. Gaseous fluorides are the most difficult of the pollutants to handle.

Some wet scrubbers are effective in removing the fluorides but corrosion is a problem. The effluent water must be chemically treated in many instances and an absorbant reagent is simultaneously introduced into the air stream to absorb the fluorides. Dry collectors or tall stacks may be used to reduce pollution.

Pouring and mould cooling department

The molten metal is generally transported to the pouring area in a transfer ladle, where it is poured into the pouring ladle, if not directly into the moulds. The poured moulds are then allowed to cool for a period of time, and then the solidified castings are removed from the mould, rough cleaned of mould material, and allowed to cool until the cast metal is cold enough to handle. During this time the metal emits a light smoke, and except on production type mould lines where the work is done at definite stations, the smoke is best removed by exhausting from above the pouring monorails. Foundries differ widely in the method by which these operations are carried out, and the hooding must be designed for the particular condition.

On production type mould reels, the pouring station is provided with a hood and the fumes are discharged from the area. The mould reel, depending on the type of conveyor used, is also covered or provided with a suitable exhaust hood for the length of travel when the poured mould is on it. These hoods are usually discharged directly to the atmosphere without cleaning, inasmuch as so little pollutant is present that dust collection equipment that could remove the fine particulate involved could not be economically justified. After the casting is removed from the mould, very little, if any, pollutant is forthcoming. Many times, a cooling conveyor carrying hot castings is provided to expose them to outside conditions, possibly above the roof until they are ready for the cleaning room.

From the first place the poured mould is broken, and at every subsequent point where work is done to remove the solidified casting from the mould, exhaust ventilation should be provided. The route that the hot, heat-fractured sand follows on its return trip to the mixer (where it is again conditioned for moulding) must be ventilated at each point where the sand is disturbed. Some of the points, such as the shake-out, flows, and mixer, need special hooding to accommodate workmen required on the process. Other points, such as transfer points where the sand falls from one conveyor to another, can be enclosed as much as possible.

Throughout the design, care should be exercised to locate the duct take-off point of the hood so that the conveyed material is not thrown into the take-off. The hood take-off should also be tapered to keep the point where the exhaust air reaches the duct velocity away from the disturbed material. Too much of the fine sand, so important to the moulder, may otherwise be removed.

On the other hand, the exhaust system may be required to serve a drum screen cooler, an aerator, or some other equipment that naturally will tend to introduce excessive sand in the airstream. A trap should be placed in the exhaust duct as close as possible to the hood takeoff to remove all but the fine dust. A means of removing the collected material or, better, returning it to the sand return system is required. This should be considered when the trap is located.

Either wet scrubbing or dry mechanical collection should be used for this type of system. Where good collection efficiency is desired, the latter type may prove to be inadequate. Fabric arrester equipment may tend to clog because of moisture condensing from the airstream.

Moulding

The sand handling system generates dust as well as the parting compound mist. Usually, systems recirculate used sand for which conveyors, elevators, screens, mixers, etc. are required. From the place where poured mould is broken and at every subsequent point where work is done to remove the solidified casting from the mould, exhaust ventilation should be provided. ·

Casting cleaning department

When the castings have cooled enough to be handled, the sprues and cores must be removed. This is often done by simply knocking them off (or out of the casting) by workmen using handtools. A vibrating table is sometimes provided to aid the operation for higher production. In any event, the station where this is done must be exhausted to protect the workmen and remove the dust. The refuse generated at this operation is often removed by conveyor, either to a sprue mill where the metal is prepared for remelting or to a point where the waste material is deposited. The sprue mill and the conveyors carrying the dusty waste material require dust exhaust. The exhaust air cleaning device for the service may be either a medium-efficiency wet scrubber or a fabric arrester.

Following the knockout, the casting usually requires further cleaning. Depending on the type and the size of casting involved, this may be done by grinding, abrasive blasting, tumble mills, chipping, sawing, cutting, powder washing, etc. All these methods tend to create a dusty condition and require exhaust ventilation.

Grinding

Grinding wheels, whether they are mounted on a stand, supported from a swinging frame, or portable, need hooding that is designed with care, because the rotating wheel, in a sense, acts like a fan and tends to throw dust in all directions. Most of the dust is thrown, however, in the first quadrant from the point where the grinding wheel contacts the castings. It is good, in this case, to locate the duct takeoff point so that the dust is thrown, as much as possible, directly into the take-off. A trap is usually included to drop out the heavy particles and reduce the abrasive erosion on the exhaust ductwork. Swing frame grinders should be arranged so that the dust is thrown into an exhausted booth, or where practical, a local hood may be mounted on the frame and exhausted using a flexible duct. Hand grinders are possibly the most difficult to hood. Where possible, grinding bench hoods with down- and back-draft exhaust will afford environmental dust control.

Abrasive blasting

Relatively large exhaust air volumes are usually associated with abrasive blasting equipment. The equipment will vary in size, method of operation, type of abrasive used, how the abrasive is impelled, how the castings are handled, whether continuous or batch, and in other ways to suit the particular work to be done. The manufactures of this equipment normally recommend the air volume that should be exhausted from each of the connections provided. Inasmuch as no moisture is involved in the process, the exhaust airstream will not have excessively high relative humidity, and the fabric arrester type of dust collector is usually used. Mechanical precleaners, usually simple traps, are recommended to reclaim the usable abrasive and at the same time reduce the dust load on the dust collectors.

Where steel shot is used for such abrasive cleaning, certain types of wet scrubbers should not be used to clean the exhaust air. The type that depends on the impingement principle, where the airsteam must suddenly change direction in order to wet the particulate, will tend to build up ferrous oxide deposits at the points where the change of direction occurs. Cleaning the scrubber then becomes a periodic maintenance chore. The dust collection equipment that is deemed suitable for the cleaning room exhaust systems mentioned above will be equally suitable for systems serving tumble mills, chipping, and other cleaning room operations.

Core-making

Core-making operations generate solvent vapours, resin odours and dust from graphite as well as silica. In the core room, a new vertical tower core oven was proposed originally to replace the existing batch oven, improve core handling and eliminate smoke. Further, it is suggested to use chemically bonded sand in place of oils and for core-making. To evaluate this replacement possibility, an experimental no-bake core unit was set-up. On the basis of success of this test installation, the use of chemically bonded sand in the core room was adopted and core ovens were eliminated.

Finishing section

Considering the various problems associated with the finishing section, a master plan is scheduled by relocation and addition of casting finishing equipment, such as swing grinders, stand grinders, inspection benches and tumbling mills. Swing grinding stations are housed in individual booths with exhaust ducts leading to a bag-type dust collector. Another dust collector is installed to receive emissions from the stand grinders, the tumbling mill and several portable grinding benches. The tumbling mill is fitted with an armoured rubber liner to reduce noise.

Heat treatment

Oil is the only likely contaminant to arise in the heat treatment section. The temperature of the quench oil or water quenching tank must be carefully controlled and the existing furnaces may be fitted with recirculating burners.

Pollution Control in Ferrous Foundries

The production-oriented economy has been expanding continuously the world over and this has triggered off the necessity for a continuously increasing per capita consumption pattern of raw materials and energy. This, in turn, has led to the situation of environmental crisis and a low 'quality of life'. About 3.5 billion tonnes of coal and 2 billion tons of petroleum are being burnt every year on this planet and about 4 billion tons of metalliferrous and non-metallic minerals are being processed annually. Moreover, 300–350 million tons of aerosols and 400 million tonnes of SO_2 along with a large amount of other harmful gases like CO, NO, etc. are polluting the atmosphere. Thus, the quality of air has been eroded mainly due to the role of technology with its present momentum. Air pollution with all its procedures of monitoring, assessment, prevention and control, has already made a definite headway in the scientific, technical, academic and legal activities in India since the 1970s. The Air (Prevention and Control of Pollution) Act, 1981, has been promulgated so as to restore the ecological balance and quality of life.

Essentially, air pollution is through the presence of contaminants like dust, fumes gas, mist, odour, smoke, etc. in quantities and of characteristics and duration that it proves injurious to life (human, plant or animal) or to property, or else it interferes with the comforts of life and safety of property. The harmful substances are mainly evolved in the melting furnaces, during preparation of the moulding sands, during teeming of the molten metal into sand moulds, during extraction of the metals from the moulds and during cleaning operations. Table 14.9 gives the extent of air pollutants generated through various sources and Table 14.10 gives the classification of air-pollutants.

Table 14.9. Air pollutants generated through various sources (Mt/year).

Source	Sulphur oxides	Particulate matter	Carbon monoxide	Hydro- carbons	Nitrogen oxides	Total
Transportation	1.1	0.8	111.5	19.8	11.2	144.4
Fuel combustion in stationary sources	24.4	7.2	1.8	0.9	10.0	44.3
Industrial processes	7.5	14.4	12.0	5.5	0.2	39.6
Solid waste disposal	0.2	1.4	7.9	2.0	0.4	11.9
Miscellaneous	0.2	11.4	18.2	9.2	2.0	41.0
Total	33.4	35.2	151.4	37.4	23.8	281.2

Table 14.10. Classification of air pollutants.

Major classes	Subclasses	Typical members of subclasses
	Hydrocarbons	Methane, butane, octane, benzene, acetylene, ethylene, butadiene
Organic	Aldehydes and ketones	Formaldehyde, acetone
Gases	Other organics	Chlorinated hydrocarbons, benzopyrene, alcohols, organic acid
	Oxides of nitrogen	Nitrogen dioxide, nitric oxide, nitrous oxide
Inorganic gases	Oxides of sulphur	Sulphur dioxide, sulphur trioxide
	Oxides of carbon	Carbon monoxide, carbon dioxide
	Other inorganics	Hydrogen sulphide, hydrogen fluoride, ammonia, chlorine
Particulates	Solid particulates	Dust, smoke
	Liquid particulates	Mist, spray

The following three approaches could be used to control pollution:
1. Define acceptable levels of pollutants in outdoor air and control pollution to such levels accordingly.
2. Develop laws that pollutant emissions from an equipment do not exceed the specified levels.
3. Install the best control available devices to contain the pollutants within the prescribed limits.

Ferrous foundries

Ferrous foundries, big or small, are well-distributed in large numbers throughout India. All of these foundries use various fuels, sand, clay, organic and inorganic binders resins, SiO_2-flour, graphite and asbestos powder, cushioning agents, sea-coal, metal borings and chippings with lubricant and other contaminations, fluxes, grain refiners, alloying elements, inoculants, etc.

As a result, lot of dust, fumes and gases will be generated within the foundry premises. The major constituent of the particulates generated in a foundry is silica of very fine size, which is very injurious to health and may cause 'Silicosis'. Other harmful constituents present in foundry-gases and fumes include CO, NO, SO_2 and cyanides. A brief description of their effects is given below:

Carbon monoxide

Major source of CO is the combustion of fuels in baking ovens and heat-treatment furnaces, oxidation refining during steel-making, teeming of steel in ingot moulds, sand-casting, cupola operation, etc. It passes through the lungs into the blood stream and combines with haemoglobin in the blood and causes the blood to carry less oxygen to the heart and brain. When present in high concentrations, it may cause death by paralysing normal brain functions. At lower levels, it may impair brain functions and may also cause heart disease and other complications.

Sulphur dioxide

Its major source is combustion of fuels, binders used in sand moulding, etc. SO_2 combines with moisture and oxygen, and can form acids which may have corrosive action on different materials in the neighbourhood. In lower concentrations in the form of particulates, it can do greater harm by injuring lung tissues. At higher concentrations, it causes irritation in the upper respiratory duct and results in other complications.

Nitrogen oxide

Oxide of nitrogen is generated through fuel combustion involving adequate air use. It causes lung irritation. In the presence of sunlight, it reacts with hydrocarbons to form photochemical oxidants and smog, which causes respiratory irritation and even cancer.

Cyanides

This is a poisonous gas and must not be allowed to form inside the foundry premises. Ferrous foundries mainly deal with sand, inorganic and organic binders, special additives, pattern materials, metal scraps, alloying elements, moltern metals, slag, inoculants, etc. The metal casting operation, in general, is represented in Fig. 14.2. Amount of dust evolved in various principal working sections of a typical ferrous foundry is given in Tables 14.11, 14.12, 14.13, 14.14 and 14.15. In order to improve the working conditions, the sands and the red-hot castings are also continuously removed by underground gratings and convey systems. Adequate ventilation, however, has to be applied. The operation of knocking castings out of the mould-boxes is highly detrimental to health due to the evolution of harmful gases and dust containing 68 per cent of particles below 2 μm and the rest in the range 2–10 μm.

Table 14.11. Typical analyses of the dust evolved in various sections of ferrous foundry.

Foundry section	Dust generated								
	Fraction (%)							Amount density	
Dia. of particles	<5	5–10	10–20	20–40	40–60	>60	kg/hr	g/cm
Moulders section								
Ordinary edge runner mill (air flow 3000 m³/hr)	–	12.0	1.9	10.0	1.4	74.7	22.5	–
Centrifugal edge runner mill (air flow 15000 m³/hr)	4.7	6.0	20.0	23.3	16.0	30.0	6.0	3.14
Fettling and cleaning section								
(Gratings area where castings are removed from the mould–								
Air flow 12000 m³/hr)	–	4.7	6.0	20.0	23.3	30.0	30	2.3
Wire brushing area	6.0	8.0	22.0	25.0	26.0	13.0	–	–

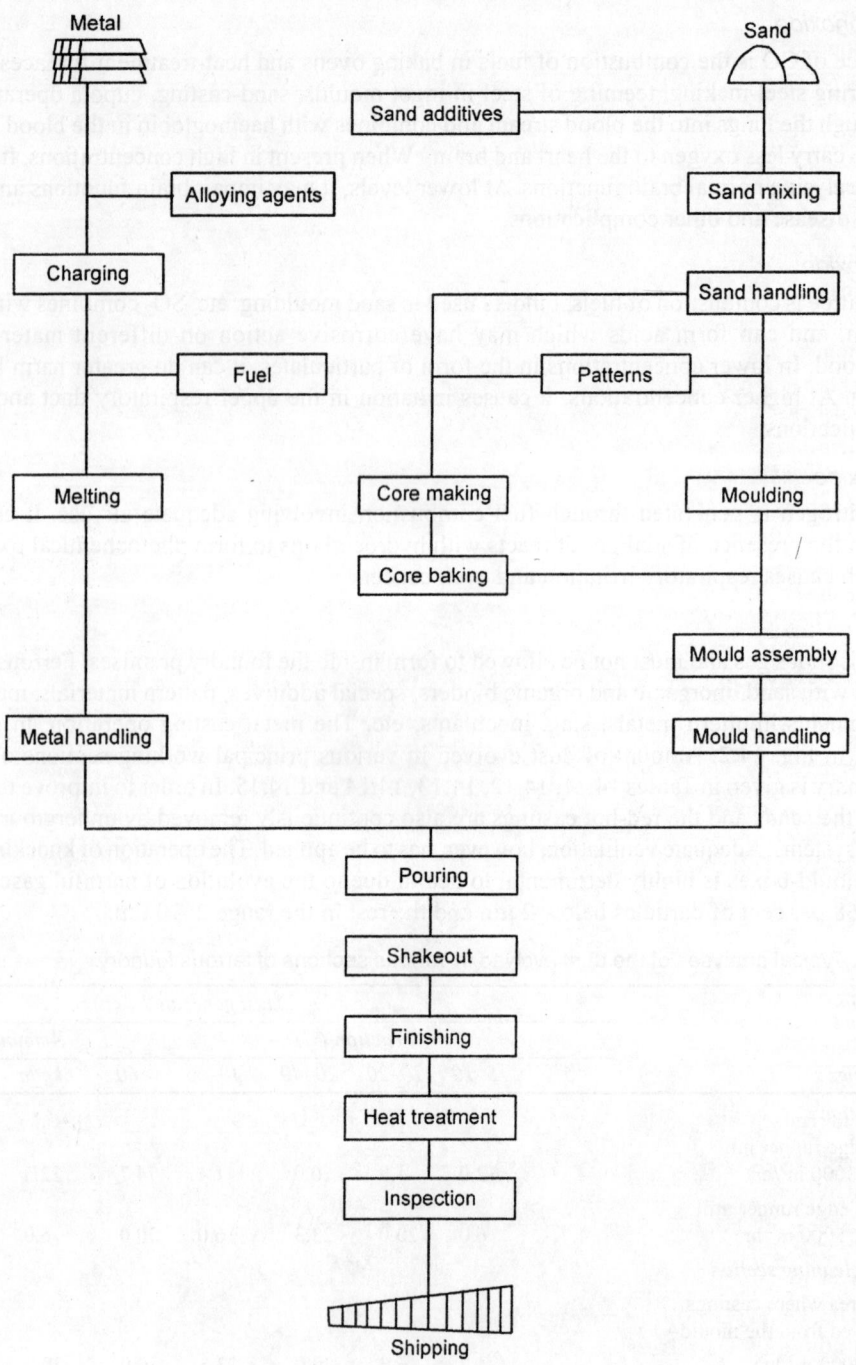

Fig. 14.2. Flowsheet of typical metal casting operations.

Table 14.12. Content of dust in air at principal working places where castings are removed from moulds in large steel foundries.

Operation and working place	Average concentration of dust, mg/m³	Amount of air drawn off, m³/min	Amount of metallic dust, kg per ton
Removal of sand crust from dry castings	80	30–40	0.3–0.25
Remval of sand crust from wet castings	20	30–40	0.07
Cleaning castings from burnt soil	80	30–40	0.30
Cleaning castings with air blast	450	–	1.40–0.75
Automatic machine for removing sand crust from castings	20	30–40	0.06
Cleaning with blast of metal shot in basement under chamber	200	50–60	0.35–0.4
Operation for knocking out castings onto grating without hood	400	110–120	2.4
Operation for knocking out castings onto open grating	100	110–120	0.8
Operation for knocking out castings onto partly closed grating	50.0	110–120	0.35–0.4

Table 14.13. Amount of dust evolved when finishing steel castings.

Equipment used	Amount of air drawn off, m³/hr	Amount of dust in air, g/m³	Amount of dust evolved per ton of cast steel
Cleaning drums	1800 D[1]	4–7[2]	4–7
Emery wheels	16,065	0.025–0.440	0.4–7
Apparatus for cleaning with blast of metal shot	16,750	0.005–0.06	0.08

[1]D–diameter of drum, m.
[2]The first figure refers to castings of a mass less than 20 kg; the second to those of a mass greater than 20 kg.

Table 14.14. Amount of dust in air in the vicinity of hydraulic sandblasting of castings.

Place where sample was taken	Concentration of dust, mg/m³		
	Minimum	Maximum	Average
Working place of hydraulic jet operator	2.2	2.7	2.4
After aperture for hands	20.6	48.6	34.6
Air pipe for blaster	114.6	281.5	198.0
Blaster in front of open aperture	32.0	44.3	38.1
Inside the chamber	390.7	606.8	498.7

The type of exhaust-hood employed has a great effect on the efficiency of the dust-removing equipment. The most convenient hood is one that covers the grating from all sides and is provided with apertures for admitting the mould boxes and removing the castings. Air, which is drawn off, will be in the order of 5000–8000 m³/hr. For gratings up to 2.5 m width, air drawn-off will be 10,000 m³/hr. Hydraulic cleaning of ferrous castings is also widely practiced.

The amount of dust generated in ferrous foundries using hydraulic cleaning is 2.5 to 2.7 times less than those not using hydraulic cleaning.

Table 14.15. Amount of dust evolved during cleaning of cast steel with blast of metal shot.

Working place	mg/m³			Concentration of dust kg per ton of cast steel		
	Minimum	Max.	Average	Minimum	Max.	Average
Metal-shot blast chamber isolated from shop (when feeding shot into apparatus)	53	76	66	0.04	0.06	0.05
Without feeding of shot into apparatus	10	23	14	0.008	0.02	0.016
Near chamber for blasting with metal shot	15	18	17	0.001	0.015	0.015
In shop during operation of chamber	5	8	6	0.0004	0.0005	0.0005
Near drum for blasting with metal shot	2	10	5	–	–	–

Harmful gases evolved in ferrous foundries

In steel foundries, usually direct arc and other electrical furnaces are used whereas cast iron foundries use cupolas. In addition, for the heat-treatment of ferrous castings, oil-fired/gas-fired furnaces are usually used. CO and other harmful constituents evolved in the melting and heat-treatment sections are given in Tables 14.16 and 14.17. The particle size distribution and the chemical composition of cupola dust are given in Tables 14.18 and 14.19. In steel foundries, CO is mainly produced during the melting operation inside the arc furnaces due to C-oxidation reaction at the time of steel-making.

In addition, a huge quality of CO is mainly evolved when pouring molten steel into the sand-moulds. The amount of CO evolved is directly related to the mass of the castings. When molten cast iron is tapped off from a cupola, 120 g of CO and 20 g of carbon dust are exhausted per ton of cast iron.

Table 14.16. Concentration of carbon monoxide in air of steel foundries.

Place where sample was taken	Number of samples taken	Concentration of CO, mg/m³		
		Maximum	Minimum	Average
Core department				
In cabin of monorail crane	2	40	30	35
Near drying furnace	2	15	15	15
Moulding department				
Near semi-automatic machine for removal of castings from mould	14	15	10	12.5
Near stand for heating ladles	8	60	20	50
Near pouring conveyer	12	60	15	30

(Contd ...)

Place where sample was taken	Number of samples taken	Concentration of CO, mg/m³		
		Maximum	Minimum	Average
Working place of steel worker				
When loading charge	10	30	0	20
When melting	8	15	15	15
In cabin of overhead crane	4	60	45	54

Table 14.17. Content of carbon monoxide and dust in waste gases of cupola furnaces.

Capacity of furnace, ton/hr	Amount of gases exhausted into atmosphere, m³/hr	CO		Dust[1]	
		m³/hr	kg per ton of metal	g/m³	kg per ton of metal
5	4500–5400	680–810	170–200	$\frac{20}{13}$	$\frac{18-22}{12-14}$
10	9000–10,000	1350–1500	170–190	$\frac{20}{13}$	$\frac{18-20}{12-13}$
15	12,000–14,000	1800–2100	150–175	$\frac{20}{13}$	$\frac{16-18}{11-12}$
20	17,000–20,000	2500–3000	155–190	$\frac{20}{13}$	$\frac{17-20}{10-13}$

[1] The numerator denotes the amount of dust in the gases before the spark extinguisher; the denominator, after the spark extinguisher.

Table 14.18. Particle-size distribution in dust evolved in cupola furnaces.

Blast	Content of fraction, per cent (by pass), for diameters of particles, µm						
	0–5	5–10	10–25	25–50	50–75	75–150	150
Hot	16.6	13.3	16.0	13.2	12.5	18.4	10.0
Cold	—	2.4	6.2	21.8	26.4	29.9	13.3

Table 14.19. Chemical composition of dust evolved in cupola furnaces.

Blast	Content of components, per cent (by mass)										
	Fe_2O_3	MnO_2	Al_2O_3	SiO_2	CaO	MgO	P_2O_5	SO_2	Na_2O	K_2O	LOC
Hot	24.45	1.90	2.78	30.35	4.28	0.74	0.25	0.7	0.28	0.69	32.82
Cold	22.18	1.52	3.04	28.0	3.66	1.0	0.54	0.74	1.01	1.89	28.98

The content of SO_2 in the furnace gases depends on the amount of sulphur present in the charge and coke. The relative amount of sulphur in the metal, slag and gaseous phases at various slag basicities may be found from the material balance of the respective heats assuming that 12–13 per cent of the input sulphur is going out along with the furnace gases in the form of SO_2.

Moreover, CO and SO_2 may evolve in the drying plants and heat-treatment plants. Volume of gases increases with the increased capacity of the furnaces. Tables 14.20, 14.21 and 14.22 indicate the amounts of various gases evolved during drying and melting operations.

Table 14.20. Amount of carbon monoxide and sulphur dioxide evolved in drying plants.

When and where evolved	Carbon monoxide			Sulphur dioxide		
	mg/l	g/kg	%	mg/l	g/kg	%
During surface drying of moulds[1] where gases are emitted into shop through gaps between mould and top plate	25–3.0	–	–	None	–	–
When solid fuel is burnt	–	250–30	–	–	15–20	–
When liquid fuel is burnt	–	30–50	–	–	40–60	–
In case of complete combustion of gases	–	Not more than 0.75	–	–	None	–
Together with escaping products of combustion of coal	–	–	–	3.0	–	None

[1]Pollution of the atmoshphere by carbon monoxide can be completely eliminated if hot air is used instead of flue gases for surface drying of moulds.

Table 14.21. Amount of gases escaping from electric-arc furnaces when gas cleaning apparatus is functioning and gases are withdrawn by suction through apertures in roofs of furnaces.

Capacity of furnace, tons	Mass of charge, tons	Capacity of tranformer, kVA	Amount of gases exhausted from furnace, m^3/hr		Amount of gases exhausted per ton of steel, m^3
			Measured	Calculated	
5	7.0	3000	700	680	From 350
10	11.5	5000	1100	1030	to 450
20	25.0	7000	22000	2300	
40	45.0	15,000	3900	4030	
100	110.0	25,000	7800	8000	

Table 14.22. Content of harmful gaseous products in process gases exhausted from electric furnaces.

Harmful substances	Average, concentration mg/m^3	Amount of products exhausted, g per ton of steel
Oxides of carbon	13.5×10^3	1350.0
Oxides of nitrogen	550.0	270.0
Oxides of sulphur	5.0	1.60
Cyanides	60.0	28.40
Fluorides	1.2	0.56

Ambient air quality standards are maintained by monitoring for which samplers are located at strategic points to collect air samples after regular intervals of time. Compliance with emission standards is determined by stack testing. In order to provide better working conditions, it is necessary to take all possible measures for determining the contaminants, improving the manufacturing process, perfecting the equipment used and introducing effective apparatus to remove or minimise the pollutants, etc. and introducing adequate ventilation in different sections of the foundries. The castings should be extracted from the moulds and cleaned in separate shops provided with efficient exhaust fans. The conveyor belts should be equipped with specially designed hoods so as to extract maximum of dust and gases evolved.

Non-dispersive infrared analysis (NDIR) set-up should preferably be introduced for detecting CO present in the ambient air. This equipment is costly but has the following advantages:

1. Can be operated by inexperienced personnel.
2. Needs no chemicals.
3. Is reasonably insensitive to changes in the flow rate and ambient temperature.

A number of CO-sinks may be tried as a cheaper process for minimising CO in the ambient air. It is claimed that soil bacteria converts CO to CO_2/CH_4 and thus soil may be used as a cheaper absorbent for CO from the ambient air. Moreover, another possible sink is the reaction of CO with hydroxyl-radicals which may account for 50 per cent of the ambient CO present in foundries.

For the purpose of monitoring NO_x, Jacob's Hochheiser's method, is adopted as a standard by the US EPA. This is a wet chemical method for the quantitative estimation of NO_2 and other nitrogen oxides. Even with catalytic converters, removal of NO_x is not simple and at its high temperature of operation, S will be converted to more harmful SO_3. Hence, the EPA is proposing converters for the time being.

Analysis of particulates consisting of metals, hydrocarbons, etc. can be done by wet chemical methods. Atomic Absorption Spectro-photometer (AAS) may be a very useful instrument in this regard for the estimation of most metals in the ppm level or at lesser levels. Initial cost for such instruments is high but it is worth paying, considering the harmful effects of some biologically non-essential metals like CO, V, Ag, B, As, Sb, etc. When the TLVs are exceeded, they create different extent of toxicity resulting in various diseases. Various control techniques are available for minimising the ambient pollution in ferrous foundries, which include:

1. Electrostatic precipitator.
2. Wet scrubber.
3. Cyclone collector.
4. Bag collector.
5. Settling chamber.

Although unacceptable as a legitimate control device, a tall chimney can perform an air pollution control function by discharging gases at a point high enough to reduce the ground-level concentrations of toxic substances. Suspended particulate matters (SPM) can also be dispersed by tall chimneys. The main factors influencing the concentrations of air pollutants at the ground-level are wind velocity, discharge rate and effective chimney height. Frequently, air pollution can be reduced significantly by simply using good housekeeping practices. A regular cleaning of the premises can do much towards reducing wind-entrained dust. This is particularly true in the sand moulding and core-making shops. Energy substitution or changing of fuels is another way of combating air pollution. For example, substituting natural gas for oil in the pre-heating or baking sections. Moreover, the use of *in situ* processes will help in getting less pollutant atmospheres inside the foundries.

Chapter 15

Chemical and Allied Industries

INTRODUCTION

In any country, economic development and environment protection should go hand in hand. There cannot be economic development (i.e. the improvement of the standard of living of the common man) without increasing the production of goods and services. Every such activity of production has its own associated problem in the balance of ecology and environment in general. Chemical industries, in particular, are being looked at with awe and suspicion in this respect. It should be the responsibility of the chemical manufacturers to use only such technologies which contribute the least to the upsetting or downgradation of the ecology and environment. Efforts are to be made in improving the technologies to better the performance.

The environmental impact of an existing process can be reduced by using clean technologies, which will, in turn, mean higher costs. Manufacturers wanting to determine the environmental implication of their activities need to look beyond the factory fences and consider the impact of obtaining the raw materials and subsequent use of their products.

Air pollution—general control methods: The air pollution control measures applied in chemical manufacturing are 'for the most part' the everyday application of chemical engineering unit operations using mechanical collectors, bag filters, electrostatic precipitators, wet collectors, catalytic oxidation or reduction units, and direct flame incinerators or adsorbers, either alone or in combination.

The selection of the control device for the particular process is a function of the degree of control necessary, the nature of the effluent, the capital and operating costs of the control equipment, the effect of the control system on the process itself, reliability, and the ultimate disposal of the collected waste material.

In addition, process or equipment modifications may be specifically incorporated to eliminate or minimise the emission of air pollutants.

This is the preferred procedure and is particularly applicable to new plants or processes. Thus, the control device alone may not fully indicate the attention given to air pollution control.

PETROLEUM REFINERY

The term petroleum is used to describe the mixture of hydrocarbons in oil, including the gases above the liquid in oil wells and the gases and solids which are dissolved in the liquid. Petroleum was formed in remote periods of geological time from the remains of living organisms. It is, therefore, a fossil fuel.

Control of Air Emissions in Petroleum Refinery

Air pollution control measures

Control of refinery air emissions can be accomplished by process change, equipment change, procedure change, installation of control equipment, improved housekeeping, increased monitoring, and better equipment maintenance. Specific control devices are discussed below for the refining sector. Some combination of these often proves to be the most effective solution. Several of these controls also result in some form of saving.

Vapour control systems

Vapour control systems involve the collection of process or fugitive vapour emissions and their recycling or recovery for hydrocarbon content or fuel value or their destruction. In some instances, where economically feasible, the emission stream itself can be recycled to refinery fuel gas production or the fuel value of the emission stream can be recovered via combustion. Hydrocarbons in vapours can be destroyed via flares or incinerators. The specific configuration of vapour control systems depends on the specific configuration and economics of the refinery.

Flares

Flares are commonly used for the disposal of waste gases during process upsets (e.g. start-up, shutdown) and emergencies. They are basically safety devices that also are used to destroy organic constituents in waste emission streams. Flares can be used for controlling almost any nonhalogenated VOC emission stream. They are designed and operated to handle large fluctuations in flow rate and VOC content. There are several different types of flares, but the prevalent refinery flare, illustrated in Fig. 15.1, is an elevated steam-assisted flare. Flaring generally is considered a control option when the heating value of the emission stream cannot be recovered because of uncertain or intermittent flow, as in process upsets or emergencies. If the waste gas to be flared does not have sufficient heating value to sustain combustion, auxiliary fuel may be required. According to studies conducted by the EPA, 98 per cent destruction efficiency can be achieved by steam-assisted flares when controlling emission streams with heat contents greater than 300 Btu/scf.

Incinerators

There are two types of incinerators—thermal and catalytic. Thermal incinerators are used to control a wide variety of continuous VOC emission streams. Thermal incineration is preferable to flaring when halogenated or sulphur-bearing compounds are present owing to the corrosive properties of these compounds. Destruction efficiencies up to 98–99 per cent are achievable with thermal incineration. Although they accommodate minor fluctuations in flow, thermal incinerators are not well suited to streams with highly variable flow because of reduced residence time and poor mixing during increased flow conditions. Thermal incineration is typically applied to emission streams that are dilute mixtures of VOCs and air. In such cases, due to safety considerations, the concentration of the VOCs is limited in some instances to 25 per cent of the lower explosive limit (LEL). Thus if the VOC concentration is high, dilution may be required. When emission streams controlled by thermal incineration are dilute (i.e. low heat, content), supplementary fuel is required to maintain the desired combustion temperature. Fuel requirements may be reduced by recovering the energy contained in the hot flue gases from the incinerator.

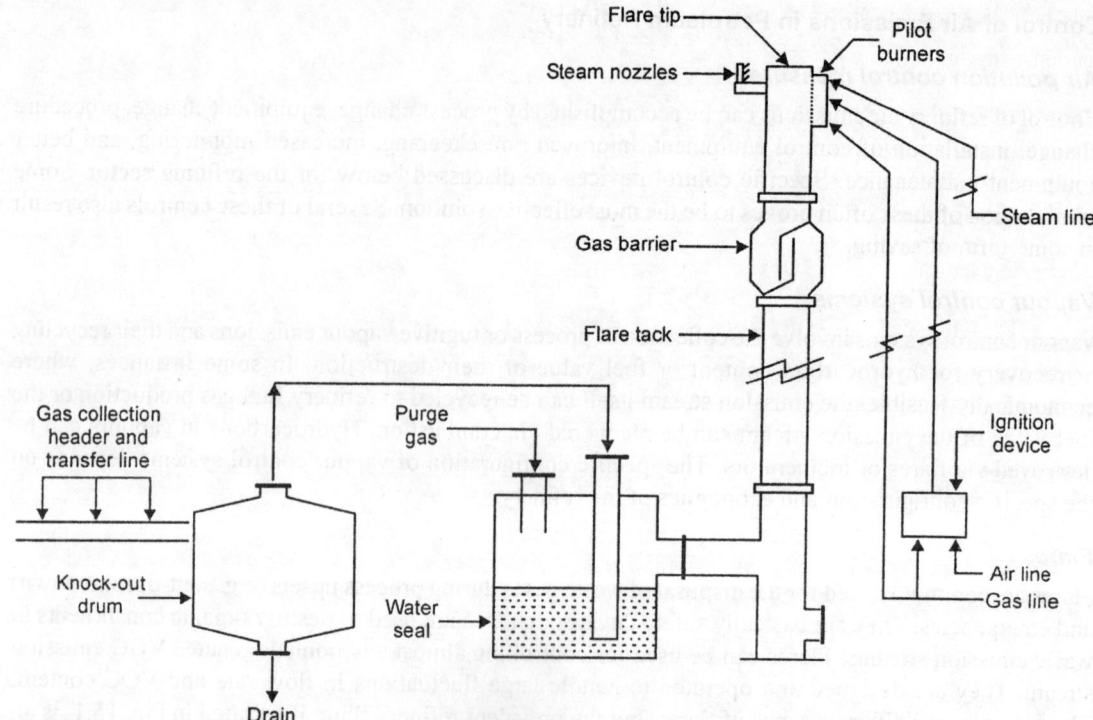

Fig. 15.1. Steam-assisted elevated flare system.

Catalytic incinerators are similar to thermal incinerators in design and operation, but they employ a catalyst to increase the reaction rate. Since the catalyst allows the reaction to take place at lower temperatures, significant fuel savings are possible. Destruction efficiencies of up to 95 per cent are generally achieved with catalytic incineration. Catalytic incineration is not as broadly applicable as thermal incineration because of catalyst sensitivity to pollutants and process conditions.

Refinery fuel gas

Existing boilers or process heaters can be used to control emission streams by recovering the fuel value while destroying the VOCs. This is accomplished by diverting the streams to the refinery fuel gas system or directly into the firebox. When used as emission control devices, boilers or process heaters can provide destruction efficiencies of greater than 98 per cent at a small capital cost. Operating costs are reduced as the recovery of the emission stream heat content reduces fuel requirements. Typically, emission streams are controlled in boilers or process heaters and used as supplemental fuel only if they have sufficient heating value (greater than 150 Btu/scf). In some instances, emission streams with high heat content may be the main fuel to the process heater or boiler. Note that emission streams with low heat content can also be burned in boilers or process heaters when the flow rate of the emission stream is small compared with that of the fuel-air mixture.

There are some limitations on the application of boilers or process heaters as emission control devices. Because these combustion devices are essential to the operation of the refinery, only those emission streams that will not reduce their performance or reliability can be controlled. Streams not suitable for control include those with varying flow rate and/or heating value, high-volume/low-heating-value streams, and streams with corrosive compounds.

Hydrocarbon reuse

The ultimate vapour-recovery process recycles the emission stream to the light ends recovery unit (gas processing), where it is recovered as a salable product, usable feedstock, or gasoline blending component. This requires compressing and separating, by distillation, the stream into its respective components.

Covering a liquid space

Covering a liquid surface or the vapour space above a liquid surface suppresses VOC emissions to the atmosphere. Examples of control by covering are floating roof tanks (external and internal) and waste-water systems. The control of waste-water treatment systems requires covering sources where emission generation is greatest, namely, process drains, junction boxes, and oil — water separators. By suppressing emissions through the separator, more hydrocarbon can be recovered as a liquid and recycled back to the refinery. Covering a drain involves either a physical cover at ground level or a liquid seal in the drain pipe. Emission reductions of 40–50 per cent are achievable by drain seals. Junction boxes require venting to prevent siphoning and vapour locks.

Vents on junction boxes should be at least 4 inches in diameter and 3 ft. in length. To minimise VOC emissions, the vent pipe must also have a water seal. Control efficiencies can be assumed to be equal to those of drain seals.

Any VOC emissions from oil–water separators are controlled by installing a fixed or floating roof. A vapour space under a fixed roof may constitute an explosion or fire hazard. In order to eliminate this problem, the vapour space can be blanketed with either plant gas or an inert gas such as nitrogen. Floating roofs eliminate much of the vapour space, thus greatly reducing the potential for volatilisation from the oil layer. Emission reduction estimates based on qualitative information range from 90 to 98 per cent. Data developed by Litchfield indicate that 85 per cent is representative of the emission reduction achievable by a simple fixed or floating roof. The most obvious factor affecting performance is the degree of maintenance.

In addition to covering components in the waste-water system, the system can be completely closed and vented to a combustion device, such as a flare. Emission reductions depend on the efficiency of the control device; for instance, flares would result in 98 per cent reduction.

Inspection and maintenance

Improved maintenance — including scheduled inspection and monitoring, improved housekeeping, and improved employee training — is a very practical method for reducing hydrocarbon emissions and alleviating odour problems. Moreover, it is often the only control method for some sources, such as valves, relief valves, seals, cooling towers (heat exchanger leaks), and sampling operations. For nearly all sources, especially process drains, waste-water separators, treating units, blind changing, and loading operations, employee awareness of the problem will reduce emissions.

In-line sampler

In-line sampling allows the collection of a representative process sample without having to flush any hydrocarbon to the process drain. Typically, small piping/tubing is run parallel to the process piping. A three-way valve is used for sample collection, thus eliminating any 'dead legs'.

Rupture disks

Rupture disks as a control device are used to protect relief valves from the corrosive process environment. They are typically thin metal disks located on the pressure side of the relief valve. They are designed to burst at the relief valve setting. Owing to their 'one-time' use, rupture disks are applicable for relief valves that are expected to be vented in emergencies only. They also can be used in place of relief valves in certain applications.

New valve technology

New valve technology has been developed that can significantly reduce pollutant emissions from the valve stem packing. It includes new types of packing material for valve stems and new seal technologies. New valve designs have been developed for some applications to prevent leaks by eliminating packing material leak sources around valve stems or by encasing potential leak sources in a 'bellows' structure to trap emissions.

The new 'leakless' valve designs do have limitations of applicability. Some designs cannot withstand the high temperatures and pressures of many refinery process streams. Other designs are limited to valve sizes of 6–8 inches in diameter and, therefore, are applicable only to smaller line sizes in refineries. For refineries, the most widely used control technique to minimise leakage from valves is an effective schedule of inspection and preventative maintenance.

Mechanical seals

Mechanical seals are used to control product emissions from centrifugal pump and compressor glands. A simple mechanical seal consists of two rings with wearing surfaces at right angles to the shaft. One ring is stationary, while the other is attached to the shaft and rotates with it. A spring and the action of the fluid pressure keep the two faces in contact. Lubrication of the wearing faces is effected by a thin film of the material being pumped. The wearing faces are precisely finished to ensure perfectly flat surfaces.

A pressure seal can further reduce emissions. A liquid that is less volatile or dangerous than the product being pumped is introduced between a dual set of seals. Since this liquid is maintained at a higher pressure than the product, some of it passes by the seal and into the product. The pressure differential prevents the product from leaking outward and the sealing liquid provides lubrication. As some of the sealing liquid passes the outer seal (hence the need for low volatility), a means should be provided for its disposal.

Quick change blind/manifold design

Several devices have been developed to reduce spillage, such as Hamer and Greenwood blinds. These 'line' blinds do not require a complete break of the flange connection as 'slip' blinds do, but rely on a gear mechanism to release the plate. Combinations of these devices in conjunction with gate valves allow changing of the line blind while the line is under pressure from either direction.

In addition, during design stages, the manifold can be designed to minimise the need for blinding, as well as the quantity of liquid that will be spilled. With highly volatile products, water or nitrogen can be used to displace the product, precluding any product spillage during the blind change.

New gaskets, bolts and welding

Replacement of leaking gaskets on existing pipe and valve flanges and minimising the number of flanges with the use of welded pipe are methods of reducing VOC emissions from flanges.

Steam stripping

Steam stripping is an effective control for removing VOCs from process streams. Two examples within a refinery include FCC catalyst stripping and sour-water treating. Spent FCC catalyst is steam stripped to remove absorbed hydrocarbons at the reactor exit. Process waste-water is stripped with steam to remove H_2S, ammonia, and light gases. The process is generally conducted in the sour-water stripper. The sour water and steam are fed into the column. The stripped water and condensed steam are routed to the waste-water system and the undesirable constituents are incinerated.

Carbon monoxide boiler

Depending on the refinery, the catalytic cracker may or may not have a CO boiler. Carbon monoxide boilers are used to recover the energy contained in the catalyst regeneration off-gases. The recovered energy is used as process heat for various refinery processes. The fuel used in the CO boiler consists of the process gas from the catalyst regenerator and an auxiliary fuel source. The process gas may contain up to 5–10 per cent CO. Combusting the gas in the boiler produces heat and reduces emissions of CO and VOCs to negligible levels.

Cyclone

Cyclone separators are used for catalyst dust collection in the upper section of both FCC unit reactors and regenerators. The cyclones are employed as a single unit or in multiple two- or three-stage series units. In general, high-efficiency cyclones have dust collection efficiencies of over 90 per cent for particle sizes of more than 15 µm. The efficiency drops off rapidly for particles less than 10 µm.

Electrostatic precipitators

Electrostatic precipitator (ESP) particle removal is accomplished by charging the particles, collecting the particles, and transporting the collected particles into a hopper. An ESP is very sensitive to the aerosol density and the electrical resistivity of the material, but is less sensitive to particle size. The electrical resistivity of the particles influences the drift velocity or the attraction between the particles and the collecting plate. A high resistivity will cause a low drift velocity, which will decrease the overall collection efficiency. Used to control PM emissions on FCC units, ESPs achieve efficiencies as high as 80–85 per cent.

Scrubbers

Scrubbers are used in refineries to separate and purify gaseous streams containing high concentrations of VOCs, SO_2, and PM. Examples of scrubbing within a refinery include applications on the asphalt blowing airstream and on the FCC regenerator off-gas. A common system on FCC regenerators is a caustic scrubber, where particulate removal and sulphur oxides absorption take place in a venturi scrubber. Particulate is removed by inertial impaction of the scrubbing liquid with the entrained particles. The sulphur oxides absorption reactions that take place are as follows:

$$2NaOH + SO_2 \longrightarrow Na_2SO_3 + H_2O$$

Sulphur dioxide and PM removal efficiencies of 93–98 per cent have been recorded.

Combustion controls

All the criteria pollutants and several potential hazardous air pollutants (HAPs) are emitted from combustion sources within refineries. Sulphur oxides can be controlled by fuel selection, fuel desulphurisation, or flue gas treatment. Carbon monoxide, hydrocarbons, and PM can be minimised by more efficient combustion. Nitrous oxides can be controlled by combustion modification or by flue gas treatment. Controls specific to each refinery combustion source are described below.

Internal-combustion engine

The VOC, CO, and NO_x emissions can be controlled by a three-way catalyst (similar to a catalytic converter in a motor vehicle) on a rich-burn engine (fuel-rich combustion). The NO_x emissions can be controlled by selective catalytic reduction (SCR) on a lean-burn engine. Catalyst systems generally are designed for 80 per cent reduction efficiencies. Precombustion chambers, which control the air-to-fuel ratio in stages, are another NO_x-reduction technique for gas-fired engines. Injection timing retard also reduces NO_x production. The magnitude of the reduction may vary considerably among engine types. In general, reductions of 20–34 per cent, with corresponding 1–4 per cent fuel consumption penalty and slight increases of VOCs and CO, are obtained.

Combustion turbines

Wet injection using either water or steam is the most prevalent NO_x-reduction technique for turbines. Depending on turbine type and size, NO, emission levels of 25–50 ppm at 15 per cent oxygen are obtainable. Selective catalytic reduction is also applicable to turbines for NO_x reductions of generally 80 per cent. Coupling SCR with wet injection can result in NO_x emission levels as low as 5 ppm.

Heaters and boilers

In addition to three-way catalysts and SCR, selective noncatalytic reduction processes can be applied to process heater and boiler flue gas for NO_x control. The process using ammonia can achieve 40–60 per cent reduction, and that using urea can achieve 20–80 per cent reduction, depending on temperature. Low-NO_x, burners, which stage the mixing of air and fuel to reduce flame temperature, result in NO_x reductions of 30–40 per cent over conventional burners.

Control of Air Emissions in Petrochemicals

Phthalic anhydride

The major end use for PAN is in the manufacture of plasticisers. Commercially, PAN is made by the catalytic air oxidation of either o-xylene or naphthalene (Fig. 15.2). o-Xylene is obtained from petroleum sources and typically is about 98 per cent pure, with the major impurity being p-xylene. The naphthalene is obtained from coal tar sources, and the purity typically exceeds 95 per cent. The catalyst is a vanadium pentoxide-titanium dioxide mixture containing small quantities of promoters. This active material is surface coated on a support that is inert at the operating temperature levels. The reaction is strongly exothermic.

Air emissions characterisation

Air emission sources from the process include the following:

1. Incinerator or scrubber effluent.
   (a) Oxygenated hydrocarbons.
   (b) Particulates (Note: most are organic solids that eventually vapourise on exposure to air at ambient conditions): (i) primarily due to PAN, which desublimates when cooled below its freezing point of 268°F, (ii) minor quantities of ash, primarily from corrosion of carbon

steel heat-transfer equipment, (iii) in the case of scrubbers, liquid particles from entrainment of the scrubber solution, and (iv) residue particulates if residue is incinerated.

(c) Sulphur dioxide (SO_2) is optionally injected into the reactor as a catalyst promoter: it passes essentially unchanged to the incinerator or scrubber. Also, SO_2/SO_3 is produced by combustion of sulphur containing fuels used in process heaters and thermal incinerators.

(d) Carbon monoxide (CO) is produced in the process reaction, as well as in thermal incinerators and process heaters.

(e) Nitrogen oxides (NO_x are not produced in the process, but are produced in the thermal incineration and in process furnaces from the combustion of the fuel used to provide the heat).

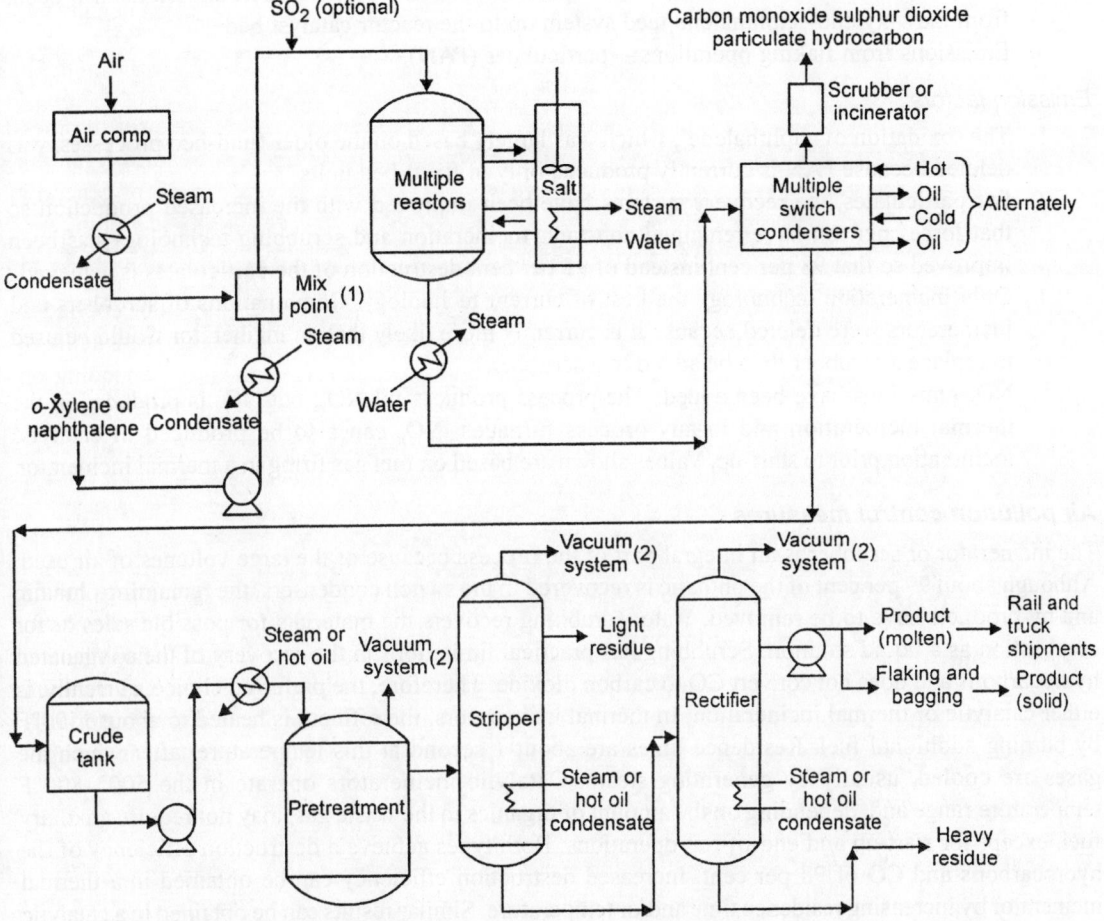

Fig. 15.2. Phthalic anhydride process.

2. Vacuum system vents.

(a) Steam jets with barometric condensers—similar to scrubber effluent, but with very little CO and maleic anhydride present.

(b) Vents discharged to a furnace — part of furnace emissions.

(c) Recycle to switch condensers part of switch condenser effluent.

(d) If the pretreatment is operated above atmospheric pressure, a vacuum system is not required to vent it to, for example, the switch condenser effluent circuit.

3. Storage tanks.

(a) Crude and product storage and transfers — particulates (PAN).

(b) Feed tankage — o-xylene or naphthalene hydrocarbon emissions.

(c) Heat transfer oils — weathering of light ends if degradation occurs.

4. Fugitive emissions valves, flanges, and so on. Very small quantities are involved because most of the process operates close to atmospheric pressure or under a vacuum. Values are based on heavy liquid because of the low vapour pressure of PAN. Hydrocarbon emissions also occur from the o-xylene–naphthalene feed system up to the reactor catalyst bed.

5. Emissions from flaking operations — particulates (PAN).

Emission factors

1. The 'oxidation of naphthalene', which was largely based on the older fluid-bed processes, was deleted because PAN is currently produced only in fixed-bed units.

3. For particulates, the recovery systems have been improved with the increased production so that losses per ton have remained constant. Incineration and scrubbing technology has been improved so that 98 per cent instead of 95 per cent destruction of the particulates is achieved.

4. Only incineration technology the best of current technology. Combinations of scrubbers and incinerators were deleted because it is currently more likely that an incinerator would be used to replace a scrubber than be added to a scrubber.

5. NO_x emissions have been added. The process produces nil NO_x, but NO_x is produced in the thermal incineration and in any process furnaces. NO_x can also be produced in catalytic incineration prior to start-up. Values shown are based on fuel gas firing in a thermal incinerator.

Air pollution control measures

The incinerator or scrubber is an integral part of the process because of the large volumes of air used. Although about 99 per cent of the phthalic is recovered in the switch condensers, the remaining phthalic and by-products have to be removed. Water scrubbing recovers the materials for possible sales or for combustion as a liquid solution. Scrubbing has practical limitations in the recovery of the oxygenated hydrocarbons and does not convert CO to carbon dioxide. Therefore, the preferred choice, currently, is either catalytic or thermal incineration. In thermal incinerators, the effluent is heated to about 1500°F by burning additional fuel. Residence times are about 1 second at this temperature, after which the gases are cooled, usually by generating steam. Catalytic incinerators operate in the 500°–800°F temperature range and, depending on the amount of organics in the waste gas, may not require auxiliary fuel, except for start-up and end-of-run operations. Both types achieve a destruction efficiency of the hydrocarbons and CO of 98 per cent. Increased destruction efficiency can be obtained in a thermal incinerator by increasing residence time and/or temperature. Similar results can be obtained in a catalytic system by increasing either inlet temperature or the amount of catalyst (i.e. residence time in the catalyst bed). Higher temperature in a thermal incinerator increases NO_x formation and requires more steam generation to remain efficient. Catalytic incinerators are limited on catalyst bed outlet temperatures because of potential catalyst damage at elevated temperatures.

Vacuum systems

Because PAN solidifies at 268°F and reacts with water to form solid phthalic acid, vacuum pumps with water seals normally are not used to provide a vacuum. Instead, vacuum jets are used. The effluent from the jets can be treated in one of three ways: scrubbed with water and the resultant solution treated in a biox system, burned as fuel, or recycled back to the process. In the last case, the noncondensibles in the stream are eventually treated in the switch condenser effluent incinerator or scrubber. Alternatively, condensers can be installed ahead of the jets to reduce the organic levels to the point where further treatment of the jet effluent is not required.

Tank vents and transportation

Vent control consists of two types: desublimators or a vacuum system using jets (as described above and including any vents at elevated pressure that can by-pass the jets and go directly to the downstream treatment steps), the simplest type of desublimator is a large sheet-metal box that provides sufficient surface area to cool the vented gases. On cooling, the PAN desublimes from the vapour and collects in the box. Suitable baffles are provided to prevent the vapour from exiting the box until cooled and for periodic removal and recycling of the solid PAN crystals. The vent gas composition corresponds to the vapour pressure of PAN at the exit temperature. Because PAN has a low vapour pressure (0.0035 mm Hg at 100°F), the concentration is low. The total amount of PAN vented will depend on tank filling rates and the rate of the gas blanketing on the tank.

Desublimators mounted on the top of the tank and equipped with heating and cooling facilities also are used. These can be operated to lower outlet temperatures depending on the cooling-medium temperature and do not require manual transfer of the recovered solid Phthalic.

PESTICIDES

Pesticide is a composite term that includes all chemicals that are used to kill or control pests. In agriculture, this includes herbicides (weeds), insecticides (insects), fungicides (fungi), nematicides (nematodes), and rodenticides (vertebrate or rodent poisons). The environment is being polluted by industrial wastes, animal and plant wastes, automobile exhausts and agricultural chemicals, which include pesticides and fertilisers. The major source of environmental contamination by pesticides is the deposits resulting from the application of these chemicals to control agricultural pests and pest causing public health problems. What concerns the society the most is the biological effects of these pesticide derivatives, which at various stages of environmental alteration can come in contact with many biological systems.

Classification of Pesticides

By definition, a pesticide is a pest-killing agent. The term usually refers to one or more materials developed and used to destroy a broad range of specific pests. In legal terminology, pesticides may be defined as 'any substance used for controlling, preventing, destroying, repelling, or mitigating any pest'. Hence, even chemicals that do not actually kill pests may, for practical and legal reasons, be considered pesticides. Included under the term are compounds used as repellents, attractants, antifeedants, etc. A pesticide product consists of an active material in a certain concentrated formulation, such as the above emulsifiable concentrate, seed dispersing in water, designed to enable its safe and effective application. A great many pesticides are available today for the control of unwanted organisms..The insecticides, fungicides, herbicides and microbial agents are classified by chemical nature, which stresses the relationship in chemical structure; this shows the principle involved for these well-known compounds.

Pesticides/Fertilisers causing pollution

Pesticides, by their very nature, are toxic compounds and, as such, besides controlling pests and diseases, they also have potentialities of affecting life and the environment adversely. Pesticides are classified as insecticides, herbicides, rodenticides, fungicides and molluscides.

Fate of pesticides in the environment

Pesticides are generally applied to the soil, plant, water bodies and human settlements by man-mounted equipment, by tractors or by aircrafts, either as liquids, dusts or granules. If these pesticides hit only the target species, there would be least pollution. Often, as little as 25–50 per cent of the pesticide formulations land in the crop area when applied by aircraft and the remaining drift in the atmosphere to contaminate very remote areas of the abiotic environment. The migration of pesticides and their fate, and the routes for the loss of pesticides in the environment are given in Figs 15.3 and 15.4.

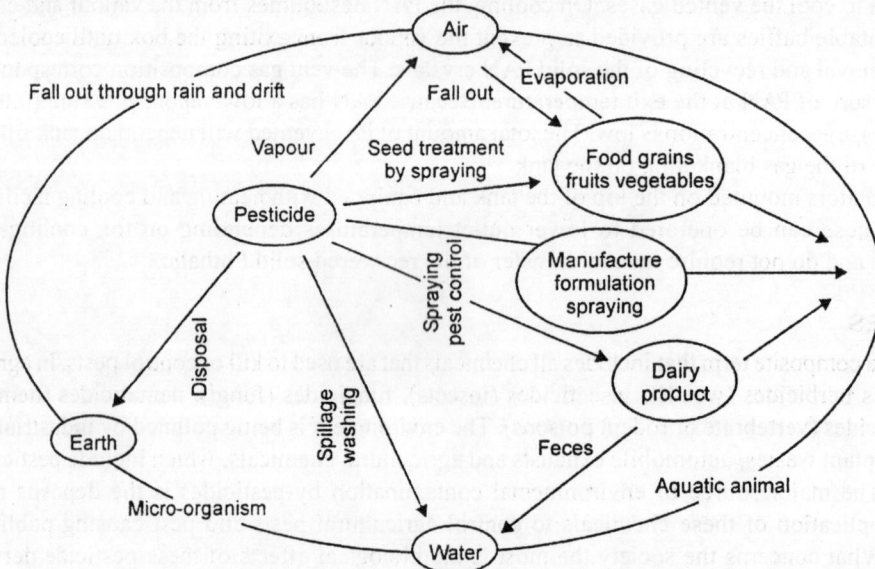

Fig. 15.3. Pesticide cycle in the environment.

    Although most of the problems arising from pesticide drift are confined to within a few kilometres of their application sites, large scale effects are invariably produced either directly or indirectly.

    Furthermore, the persistence of pesticides in the environment is dependent upon many factors such as the nature of the pesticide, temperature, light, humidity, air movement, activity of micro-organism and factors which influence the breakdown and mechanical dispersion of these chemicals.

Effects on air

Several physio-chemical factors such as sunlight, etc. influence the persistence of pesticides in air. This is a highly complex phenomenon and varies from one geographical region to another. The pollution of the atmosphere may occur either by volatilisation of chemicals where filling, loading, mixing and spraying operations are done, through drift and wind erosion of soil particles laden with absorbed residues. The hazard is greater when finer particle size is used, as in case of ultra low volume application and in aerial spray application. Traces of organo-chlorine have been reported in air samples taken from both urban and rural areas of developed countries of the world.

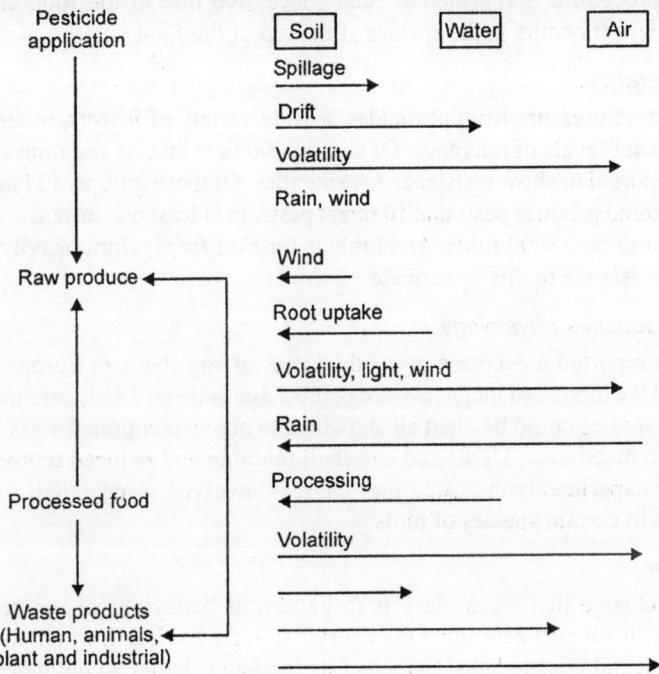

Fig. 15.4. Routes for the loss of pesticides to the environment.

Effects on biotic systems

Disturbance in equilibrium

Pesticides disturb the equilibrium existing between insect pests and their parasites or predators. Perhaps 90 per cent of all the animals live as predators (parasites and herbivores) and feed on living matter; only a few live as true saprophytes. These predators and prey populations generally evolve a balanced supply-demand economy. This balanced economy is in dynamic equilibrium, which, however, if altered by pesticides, will disturb the equilibrium between predators and prey populations.

Increased disease susceptibility

Pesticides are also known to increase the disease susceptibility in hosts. For example, 2, 4-D increased the size of tobacco mosaic virus (TMV) lesions on hypersensitive tobacco.

Bioaccumulation

It is an interesting phenomenon that has significant impact on the dynamic equilibrium that exists between predators and prey. Both animals and plants have the ability to concentrate many types of pesticides in their body tissues. The chlorinated hydrocarbon insecticides have the greatest tendency for their biological concentration. The other pesticides having tendency for bioaccumulation are diazinon, mirex, diquate, paraquate and 2,4-D.

Fish (Golden shiners) are reported to take DDT at 0.265 ppt in water and concentrate it 1,00,000 times in their bodies. In general, however, the capacity for bioaccumulation is usually not as great as this, but when the procedure is repeated at each successive link in the food chain, extremely high concentrations of toxicant occurs in the species at the top of the food chain.

Development of tolerances

Due to intense selective pressure from pesticides, a wide variety of insects, mites, animals and plants have evolved significant levels of tolerance. Of about 2000 peat insects and mite species, a total of 225 species have been reported to show resistance to pesticides. Of these groups, 121 are crop pests, 97 man and animal pests, 6 stored produce pests and 10 forest pests. In at least one instance, the level of increased resistance is reported to be 25000 folds. Amphibians (cricket frogs) from heavily DDT treated cotton field have evolved resistance to this insecticide.

Disturbance in reproductive physiology

Several studies have reported a decrease in the thickness of egg shells in Europe and North America. This partly explained the increased incidence of egg breakage and egg disappearance observed in several species. Many more studies could be cited all showing the above mentioned association between DDT (in particular its main metabolite, DDE) and egg shell thinning and reduced reproductive success. The results from control experiments indicated that DDE is involved in egg shell thinning and reduced reproductive success in certain species of birds.

Effects on behaviour

Another significant change that takes place is in pattern of behaviour of non-targeted species. The patterns of behaviour in animals determine their survival in the habitat in which they live. Each animal responds to environmental cues to help it to select its food and shelter, avoid natural enemies, select a mate and so forth. Hence, any change in an animal's behavioural pattern may significantly influence the survival of the population.

For example, the herbicide 2,4-D made beetles to become sluggish in their behaviour. When this happens they are less successful as predators than unaffected beetles and some of their prey escape capture.

Effects on wildlife, fish and other aquatic organisms

Wildlife biologists believe that DDT and other pesticide residues are potentially toxic to fish and other aquatic organisms. Known effects of pesticides on fish mainly refer to cases of mortality which may occur after accidental spillage or after the use of fish toxic pesticides in places from which the compounds could enter waterways, fish ponds, lakes or important habitats for fish such as rice paddies.

Two pesticides, in particular, have become well known in this respect, viz. Endrin and Endosulfan. However, there are more pesticides, which may kill fish or birds when they are applied or disposed of in the neighbourhood of aquatic habitats. Examples of these are DDT, dieldrin, toxaphase, methyl mercury and niclosimine. Indiscriminate spraying of broad spectrum pesticides such as phosphamidon against spruce budworms and application of DDT, dieldrin and endosulfan may affect a wide variety of animal groups as was shown in Africa where mammals, birds, reptiles, amphibians, fish and invertebrate died in often large numbers during the tsetse control operation.

Effects on plants

Together with target pests, pesticides also alter the chemical make up of plants and, in turn, affect the animals' dependent upon them for food. The changes that occur are specific for both the plants and the pesticides involved. For instance, heptachlor in soil has caused significant changes in the elements N, P, K, Ca, Mg, Mn, Fe, Cu, Al and Zn. Similarly, the protein content of wheat and the KNO_3 content of sugar beet was increased after 2,4-D. This nitrate level was highly toxic to cattle.

Another chemical changes is the increased sugar content seen in plants like ragwort, a weed toxic to cattle, exposed to sublethal doses of 2,4-D. The high sugar content makes it attractive to cattle and sheep, with disastrous results because the toxicity of the plant remains high.

Reduced seed production has been reported after the use of pesticides. For example, when the herbicide 2,4-D was applied, a 10–15 per cent reduction in seed production has been reported.

Effects on foods

Pesticides residues have been detected in food commodities. Low level of residues in foods may result from growing of crops, especially root crops, in soil containing residues of persistent pesticides.

A number of pesticides have an affinity for lipid materials and get accumulated in animal systems, thus contaminating animal products with residues.

Levels of DDT and BHC have been reported at more than the tolerance limits prescribed by WHO and FAO. Apart from these, residues of endrin, heptachlor and endosulphan have been detected in major food crops.

Pesticide residues as high as 35 ppm DDT and 60 ppm BHC in spinach; 36 ppm DDT in Brinjal; 50 ppm DDT in cauliflower, cabbage and radish and nearly 26 ppm in oil have been detected in the Delhi market. Many farmers even spray pesticides and vegetables just before harvesting them, in order to give them a fresh appearance; such causes are more alarming.

The pesticide residue levels detected in meat and poultry, etc. are equally high and often higher. According to a study, the average daily intact (ADI) of DDT and BHC for nonvegetarian needs is respectively 24 and 10 per cent higher. Even this average DDT intake exceeds the maximum ADI of 5 mg/kg of body per day. Thus, there is probably nothing that an Indian eats, which is not contaminated by pesticides.

Possible Health Hazards of Pesticides Residues

Since our food articles contain excessive amount of DDT and BHC, it is not surprising that the Indian population has an unusually high amount of these insecticides in the body. Other sources such as air, water and direct contact with contaminated surface etc. may also contribute to some extent to the DDT or BHC burden in human body.

BHC has been shown to cause higher incidence of liver tumours than DDT and the life span of individual exposed to these insecticides is reduced. The most likely effect of low level residues of insecticides is the impact on enzymes. DDT is known to alter the efficacy of some drugs in the body, by interfering with enzymatic activity of the drug detoxification mechanism. High infant mortality was observed in Guatemala where high residues of DDT were found in mother's milk.

Effects of Fertilisers in the Environment

Although fertiliser loss is economically important to the farmers, the more important concern is its effect on the environment.

There are two major problems that have been associated with high nutrient levels.

1. Algal blooms have been attributed to soluble nitrates and phosphates in water. The causes of such growth are incompletely understood, but the levels of nutrients which it is claimed limits

algal growth are below 0.3 ppm N (nitrogen) and 0.01 ppm P (phosphorus), respectively. In most low land water in the UK, nutrient concentrations have been above these limits for decades, but blooms have been rare and short lived. It is thought that factors such as water, temperature, CO_2 concentrations and the presence of organic matters are important and, in general, it seems unlikely that marginal increase in nutrient levels would initiate algal growth.

2. Increased nitrate levels can cause a health hazard in drinking water for very young babies (up to age of about three months). The nitrate in the water can be reduced in the baby's stomach to nitrate, which combines with the haemoglobin to form methaemoglobin, preventing the transport of oxygen around the body. This is one of the causes of 'blue babies' and about 10 cases of methaemoglobinaemia have been confirmed since the condition was recognised 20 years ago. These have all been associated with ground waters, usually with water from shallow wells, which may have been contaminated with sewage.

Large scale use of nitrate fertilisers could also lead to depletion of ozone. It has been estimated that a 10-fold increase in the use of nitrate fertilisers by the year 2010 could generate enough additional nitrate (N_2O) to reduce the ozone cover significantly. Reduction ranging from 1–20 per cent has been suggested for this source.

Pesticide Transformations

Pesticides undergo transformations both by biotic and abiotic environmental agents. Of the biotic agents (animals, birds, fish, insects, plants, microbes) microbes are probably most important specially as it is known that 99.9 per cent of pesticides sprayed in the environment end up in soil teaming with microbes under our tropical conditions. Amongst the abiotic agents (air, water, soil, sunlight) in our tropical environment, sunlight is probably the most important transforming agent specially on plant, soil and water surfaces.

Microbial transformations have special ecotoxic significance and will be discussed first. Other biotic agents (animals, insects and plants) transform pesticides by producing safer water soluble polar conjugates often in an attempt to remove it from the body system. Microbes on the other hand use xenobiotics and other organic chemicals thrown in the environment as food and as energy source. They are also more adaptable to different energy sources through mutation and inductions. Pesticides which are initially toxic to them become their food due to this adaptation which also helps them to evolve a variety of chemical pathways not found in other biotic agents.

Thus besides oxidation and hydrolysis, used by animals and insects, microbes cause anaerobic fermentation leading to a variety of chemical reactions like reduction, dechlorination, dealkylation and oligomerisation, causing transformation of almost all type of pesticides containing besides carbon and hydrogen other elements like N, P and S. Since there is no conjugate formation, products may sometimes be more toxic. This is best illustrated from a few examples of microbial transformation products having possible ecotoxic significance.

Thus amongst the many transformation products of DDT, the universal pollutant, TDE formed by enzymatic reduction in the presence of reduced porphyrins produced by *E. coli* deserves special mention as this product is more toxic and has the same propensity of biomagnification as DDT.

The cyclodiene insecticides dieldrin and heptachlor produce on the other hand, the cytotoxic epoxides and by microbial oxidation. Microbial transformation of endrin yields a far more toxic product, ketoendrin, and has been one of the reasons for banning endrin.

Light Induced Transformations

Today no pesticide is allowed to be marketed without investigating the phototransformation products from toxicity considerations.

Cyclodiene insecticides

The older cyclodiene insecticides, aldrin, dieldrin, endrin and isodrin undergo a variety of photochemical changes on exposure to sunlight as thin films on foliage and soil surfaces even though they do not absorb light strongly. As such these changes must be sensitiser mediated. A large number of products, formed by dehalogenation from various sites and by rearrangement, are not only important as environmental pollutants but also in many cases identical with biotic metabolites formed particularly by sewage bacteria. The photoisomers of these cyclodienes are of special significance. Not only they are formed very easily in the presence of natural photosensitisers and therefore are major among transformation products but are also more toxic as well as more persistent than their precursor pesticides. Table 15.1 gives a list of these photo isomers and an idea about their toxic potential.

Table 15.1. Toxic cyclodiene photoproducts.

Cyclodiene (a)	Photoisomer (b)	Toxicity ratio (b/a)
Aldrin	Photoaldrin	4.0
Dieldrin	Photodieldrin	10.0
Isodrin	Photoisodrin	0.3
Heptachlor	Photoheptachlor	3.0
Cis-chlordane	Cis-photochlordane	1.5
Endrin	Ketoendrin	9.0

The recent decision of Government of India to ban/phase out the older cyclodienes is to be welcomed to reduce the environmental hazards posed by these transformation products. Amongst the cyclodienes, only endosulphan has retained its importance in Indian Agriculture as it is one of the environmentally safe insecticides. It gives mostly polar environmental metabolites like diol and lactone which are non-toxic. A major phototransformation product of endosulphan of some concern is the isomerisation of α-endosulphan to its more stable β-isomer under field conditions in sunlight. The β-isomer is known to be more toxic to bees.

The pesticides can translocate from the point of application to nontarget components of environment often creating problems. They also undergo a variety of transformations by biotic and abiotic agents mostly breaking down into small nontoxic products. However, a few of these transformation products are more toxic and their ecotoxic significance ought to be assessed seriously if safe use of pesticides is to be ensured. On the other hand some good pesticides undergo rapid breakdown and the consequent loss of efficacy is a matter of economic concern.

Removal of Pesticides from Air

The abatement of pesticidal contamination of the ambient air is a complex problem but it is being attempted. It appears that the contamination arising from the production processes can be controlled.

Although incidences of occupational poisonings have been reported, proper protective measures are available. Precautions similar to those used in general chemical industries are taken to prevent the dusts and fumes from leaving the production plant into the outside environment.

Bag packers, barrel fillers, blenders, mixing tanks, and grinding operations are generally completely enclosed or hooded and the air is vented through baghouses or cyclone separators. Similar control procedures are used when liquids are involved; liquid scrubbers are used however, instead of baghouses.

Although Tabor has monitored the air near a formulating plant and found air levels similar to that observed in earlier agricultural samplings in the same area, too little air monitoring data are available at the present time to properly evaluate the production air control measures.

The control of chemical drift as a source of pesticide air contamination has been studied extensively including research on drift control. It has been found that three factors affect the control of a given application: (i) the distribution equipment, (ii) the physical state of the pesticide, and (iii) the microclimatology of the area. Although the emphasis has been placed on control of aerial applications, applications made with ground equipment can also result in drift. However, greater operator control is possible with a ground unit, since it generally has a lower discharge rate than aerial equipment.

The physical state of the pesticide is quite important in drift. The drift potential from pesticide dust is very high because of particle size. Dust materials are generally screened to incorporate only particles ranging from 1 to 25 μ in size; on the average, 80 to 90 per cent of the particles in the formulation are under 25 μ. Spray droplets of 50 μ in size show less drift than dusts of smaller particle size.

In a study of a Vermont apple orchard where Tadlon dusting for apple insect control had been the standard practice for the previous years, found that drift could be a problem even under ideal weather conditions. It has been noted that under conditions of a windspeed of 1.3 mph, a temperature of 81°F, and relative humidity of 40 per cent, drift occurred up to 300 feet. He suggested that a buffer zone of at least 300 feet be used in future applications where forage areas are adjacent to the sprayed area.

It has been found that spray droplets ranging from 10 to 50 μ in size can drift several miles from the area of application, whereas 100 μ sized particles usually do not produce a drift hazard unless the winds are high.

Evaporating the spray droplet in spraying operations has been considered as a drift control measure. Water is the most frequently used diluent because of its availability, its low cost, and its non-phytotoxic effects on agricultural crops. However, water droplets may experience considerable evaporation and reduction in size while airborne in the spray, resulting in a sufficiently smaller droplet size and therefore more drift than had been anticipated. As a control measure, oils which are relatively non-evaporative are frequently used as the pesticide carrier in sprays. They are especially used as carriers for low volume (1 to 3 gal./acre) applications of very fine sprays for forests and range lands.

Improvement in delivery equipment and techniques is another partial solution to the abatement problem. Equipment has been developed for producing invert emulsions (water-in-oil emulsions) of which greater than 90 per cent is in the water phase. In contrast to conventional oil-in-water emulsions, preparations containing a high water content are quite viscous. When using the invert emulsions, sprays consisting of large droplets can be delivered aerially, minimising both drift and evaporation. Trials indicated that smaller quantities of pesticide could be used in the invert emulsion with equivalent results in terms of insect kill or herbicidal efficiency. At the same time, the accuracy of delivery should be improved so that the invert emulsions could be applied under more adverse meteorological conditions than conventional sprays. Various spray nozzles have been designed and used with varied pesticide formulations having different viscosity, density, and surface tension in attempts to control drift during application. Additional factors such as the angle of the nozzle with airstream or the use of screens or discs at the nozzle also contribute to the characteristic of the spray.

Meteorological conditions are extremely important parameters that are considered in the application of pesticides and the control of potential drifts. Wind direction and velocity, humidity and temperature at ground and higher levels, and the amount of sunshine or rain are all interrelated factors that are considered. Because of the importance of such meteorological information, the Indian Weather Bureau provides the specialised data as part of their service.

Dusting and spraying advisories are sent out on a tele-typewriter circuit 24 hours a day emphasising local weather conditions for aerial and ground applications for various agricultural chemicals. In addition, these advisories include information relating certain insect and other pest activities with weather conditions, so that pesticides can be applied at the proper time to produce maximum pest control.

Volatilisation of pesticides into the air from soil, water, plants, and other treated surfaces is known to occur. However, the control of this is complicated by the fact that the extent to which volatilisation occurs is not known. Some control over volatilisation from soil can be effected by the use of cover crops. It has been observed that two to three times more insecticidal residues were recovered from alfalfa covered plots than from fallow ones.

FERTILISERS

Fertilisers can be classified as nitrogenous fertilisers, phosphatic fertilisers and complex fertilisers. Plants may be producing nitrogenous fertilisers, like urea, ammonium sulphate, ammonium nitrate and ammonium chloride, or phosphatic fertilisers like super-phosphates; there are plants where complex fertilisers containing both nitrogen and phosphates, like ammonium phosphate and ammonium sulphate phosphate are produced. Also, some fertiliser units are only involved in mining and undertake no other activity.

Air Emissions

The major air emissions in fertiliser industry are given in Table 15.2.

Table 15.2. Air emissions in fertiliser industry.

Pollutants	Source
SO_2	Sulphuric acid plants, boilers using coal, fuel oil, LSHS, etc.
NO_x	Nitric acid plants, NP/NPK plants
	Particulate matter rock grinding in the phosphoric and NP/NPK plants. Prilling/granulation in urea, NP/NPK/SSP/TSP plants
Fluorine	Phosphoric acid, single super phosphate, triple super phosphate, NP/NPK plants

Control of air emissions

Over the years, the technology adopted has been upgraded on the basis of improved systems available in advanced countries. With the energy crisis, greater attention is being paid to this aspect along with stricter emission controls adopted due to increasing environmental consciousness. All the plants, therefore, have incorporated the most modern pollution control systems, minimising emissions to practicable limits with the best of technologies available to meet the standards. However, there are many old plants which have adopted older technologies where the emission levels are higher and so also is energy consumption.

Control of sulphur dioxide and acid mist emissions from sulphuric acid plants

Sulphuric acid is produced by the oxidation of sulphur to sulphur dioxide (SO_2) and conversion of this to sulphur trioxide (SO_3). This is absorbed to produce sulphuric acid. All plants utilise a vanadium catalyst to convert SO_2 to SO_3 according to the contact process.

In the past 20 years all new sulphuric acid plants that have come up are of DCDA type which gives a conversion efficiency of 99.5–99.7 per cent. Table 15.3 shows the amount of SO_2 emitted by sulphuric acid plants operating at several gas strengths and conversion levels.

Table 15.3. Sulphur dioxide emission from sulphuric acid plant.

% Conversion	*Exit SO_2, ppm*		
	6% SO_2 in feed	*8% SO_2*	*10% SO_2*
96	2626	3616	4672
97	1972	2716	3510
98	1316	1813	2344
99	658	907	1174
99.5	329	454	587
99.7	197	272	352

In a typical sulphuric acid plant, the conversion of SO_2 to SO_3 is achieved in a four bed vanadium pentoxide catalyst bed. After fourth bed, when SO_2 content has been reduced from 10 to 0.06 per cent corresponding to 600 ppm, the sulphur trioxide is absorbed in 98.5–99 per cent sulphuric acid in the final absorption tower (FAT). Before being vented to the atmosphere, the gases pass through a brink mist eliminator located after the initial absorption tower (IAT) and FAT where acid mist is removed. The catalyst bed temperature, catalyst activity, pressure drop and concentration of the circulating acid are vital parameters affecting the performance.

In spite of adopting double conversion double absorption (DCDA) system, during initial start up or start up after a shut down of more than 4 hours when catalyst beds are cold or relatively cold the required temperature for conversion of SO_2 to SO_3 is not present resulting in higher emissions of SO_2 during start up. Sulphuric acid plants are therefore equipped with either a start up heater to increase catalyst bed temperature to the required level or a start up scrubber (which is a packed tower using caustic soda to absorb the SO_2). In the case of start up scrubbers Na_2SO_3/$NaHSO_3$ is formed depending on the pH of the scrubbing solution.

Control of sulphuric acid mist

Acid mist comprises of liquid droplets which range in size from 10 microns down to 0.07 microns. It is a well known problem in sulphuric acid production. This not only produces a visible and persistent plume in the stack gases but also causes equipment corrosion. Of course, this is a pollution hazard. Mist is formed when water vapour in the air combines with SO_3 in the convertor and in the absorption circuit. By and large, this is controlled by drying the air thoroughly before feeding it to the convertor and by proper maintenance of temperatures and acid concentration in the absorption towers.

Efficient drying of air is achieved by circulating 93–99 per cent acid so that water vapour in the outlet gas is around 5–100 mg/nm^3. At higher concentration of acid higher operating temperatures can be tolerated while the reverse is true for lower acid concentrations. The type of mist eliminator used in the drying tower are either the impaction type fibre bed mist eliminator or the mesh pad type. Wire mesh pad require periodic replacement in drying towers. Although plastic mesh pad offer better resistance to corrosion there have been cases where back flow of hot gases during a shut down past a leaking shut off damper have melted the plastic mesh pad.

The absorption of SO_2 in acid gives rise to acid mist. The acid concentration should be maintained between 98–99 per cent to keep the $H_2SO_4 + H_2O + SO_3$ vapour pressure at a minimum. At lower concentrations, the partial pressure of water vapour is high enough to give rise to production of acid mist, by combination of this water vapour with the sulphur trioxide in the gas steam. Typical mist loadings in the IAT are 50–1500 mg/nm³ with an average particle size diameter of 1–2 μm. Brownian diffusion filter elements are hollow cylinders which look like candles containing hydrophobic glass mattress packed between two stainless steel cages. Most mist particles 1 μm in size and larger are removed upto 96 per cent of those smaller than 1 μm are collected. The comparative efficiencies of filters are given in Table 15.4.

Table 15.4. Comparative efficiencies of various type of filter.

Mist size (μ)	Single mesh pad	Double mesh pad	Single fibre glass pad	Candle filter
5	100%	100%	100%	100%
5 to 2	90–100%	100%	100%	100%
3 to 2	50%	95%	90%	99%
2 to 1	–	90%	–	98–99%
1	–	–	–	96%

Advances are being made in mist collection equipments. High efficiency fibre beds are being developed with different efficiencies for various submicron particle sizes and various companies offer mist eliminators for sulphuric acid plant service (Table 15.5)

Table 15.5. Efficiency of fibre beds.

Parameter	High efficiency	High velocity
Efficiency on particle > 3 μm	95–99%	90–98%
Efficiency on particles < 3 μm	100%	100%
Pressure drop (inches water)	5–15	6–8

Control of particulate matter

Urea prilling tower dust

Urea is produced by reacting NH_3 and CO_2 at elevated temperatures and pressures. Traditionally urea has been transformed into solid form by prilling techniques. The urea solution leaving the synthesis section has a concentration of 72–76 per cent and is treated in a vacuum concentration section to obtain a urea melt to feed to the prilling systems.

It is during this prilling that particulate matter is emitted. The conversion of urea to solid form is achieved in a prilling tower by cooling the solution with air to expel its water content. The quantity of air required has been estimated to be 10–1200 nm³/T of urea produced. In a typical 1000 tpd plant the total quantity of air is 5,00,000 nm³/hr and it contains around 200–500 mg/nm³ of urea dust which has to be controlled, otherwise it would cause air pollution and result in loss of the product. Most of the plants use wet scrubbers for particulate removal. Dust laden air rising through the prilling tower enters the annular duct. Air is drawn from this annular duct by a ring of liquid jets consisting of nozzles arranged in annular space which provides the energy required to overcome the pressure drop in the system. The liquid droplets act as a spherical collectors for the urea dust whose size ranges from 2–200 microns. The water is sprayed at a pressure of 4.5–6 kg/cm^2. It is finally discharged into the atmosphere after passing through demister packings. Fresh make up water is sprayed into the demisters for washing purposes while the jet nozzles are fed with the urea solution which collects in the annular basin. After the concentration of urea in the recirculating solution reaches around 10–15 per cent, it is drained and sent to a tank for further processing (Table 15.6).

Table 15.6. Absorption efficiency of wet scrubbers.

Urea concentration in recirculating solution %	Absorption efficiency %
0.0	99
5.0	95
20.0	92

The outlet dust concentration is 30–50 mg/nm^3.

Control of particulate matter in complex fertiliser plants

Dust is emitted during rock grinding, product cooling and granulation operations. Normally, fabric filters are used for rock grinding and wet scrubbers like venturi scrubbers for granulation (i.e. drying of slurry in an equipment called the 'Spherodiser'). The hot gases are sucked out by an exhauster and enter the Venturi scrubber whose essential feature is the presence of a constricted cross-section or 'throat', through which the gas is forced to flow at high velocity. Water is introduced ahead of the 'throat' and the atomised liquid drops act as collectors. The concentration of dust at the outlet is 60 mg/nm^3. Dust is emitted during the grinding of rock phosphate. Bag houses are commonly used for dust control. The dust concentration to the inlet of the bag house is normally in the range of 20–30 gms/nm^3 which is brought down to less than 150 mg/nm^3. The fabric usually used is polypropylene or polyester with a special coating. Pulse jet cleaning is adopted when the pressure drop is around 6'' (150 mm) of water.

Control of oxides of nitrogen

Nitric acid plants

Nitric acid is commercially produced by reacting ammonia with air to produce nitrogen oxides which are then absorbed in water to yield the acid. Despite the many variations in operating details among the plants producing nitric acid from ammonia, three basic steps are common to all:

1. Reaction of ammonia with air over a catalyst (platinum, radium) at a high temperature and moderate pressure to produce nitric oxide.
2. Oxidation of the nitric oxide by oxygen remaining in the gas stream to produce nitrogen dioxide.
3. Absorption of the nitrogen dioxide in water to produce nitric acid releasing additional nitric oxide which must be reoxidised.

After absorption, the tail gas containing, namely, nitrogen, water vapour and oxides of nitrogen (commonly known as NO) is normally preheated and expanded through a turbine to recover energy, which is used to power the nitric acid plant's compressor.

The NO_x content in the tail gas varies from 1000–3000 ppm of which approximately 60 per cent is NO_2 and the rest is NO. The NO_2 component imparts a yellowish colour to the tail gas. In general, there are three types of processes which can be used to reduce or control emissions:

1. Catalytic reduction (selective and nonselective): In selective catalytic reduction an improved process is adopted in some of the plants where selecting reduction of NO_x (i.e. NO and NO_2) in the gas system is done with slight excess of stoichiometric quantities of ammonia over a mixed catalyst. The catalyst is selective and the ammonia consumption is almost stoichiometric and virtually no ammonia reacts with oxygen present in the gas which results in much lower exit gas temperatures allowing use of simpler and cheaper equipment. Two catalysts are used in the reaction, the first is based on precious metals mixed with metallic oxides, noble or iron group or V_2O_5 on alumina which promotes the NO_x reduction and the second is platinum based which destroys or converts any excess ammonia fed to the system into harmless nitrogen. The exit temperature is raised by 20°–30°C due to exothermic reaction. The only care to be taken is to see that ammonia gas to the reactor is not fed when the tail gas temperature is less than 200°C, in order to avoid the formation of ammonium nitrite and nitrate. The NO_x could be brought down to less than 300 ppm to NO_x. This technique has become widely used in recent years.

   In nonselective catalytic reduction the oxides of nitrogen are formed during the combustion of ammonia are absorbed in water to form nitric acid. The unabsorbed portion of oxides of nitrogen and inerts are emitted to atmosphere which generally contain 0.2–0.4 per cent of NO_x and 2–3 per cent free oxygen. The tail gas leaving the absorber is mixed with a fuel gas like natural gas (or refinery gas or purge gas from the ammonia plant) and the mixture is passed over a catalyst (platinum vanadium, iron oxide or titanium based) bed. This converts the NO_2 to NO rendering the exit gas colourless. This step is called the decolourisation reaction. This alone may not be sufficient and complete destruction of NO by converting it to N_2 may be required. To achieve this the oxygen present in the tail gas (2–3 per cent) must first react with the fuel which must be present in slight excess of the stoichiometric amount. The reaction with oxygen results in a large amount of heat of combustion which results in temperatures of nearly of 650°–700°C.

2. Extended absorption: The extended absorption processes aims to continuing the process of absorption of the NO_x in water beyond the level at which it normally ends, i.e. 2000–3000 ppm NO_x. The extended absorption is made possible by provision of large absorption columns. The additional absorber produces nitric acid. Owing to the large size of the absorbers which have to be necessarily of stainless steel construction, the investment costs are the highest as compared to the other NO_x control processes. However, there is no additional operating cost involved.

3. Chemical absorption: The tail gas could be cleaned by scrubbing with a liquid containing caustic soda or urea. In case of the caustic, it results in the formation of $NaNO_2$ and $NaNO_3$. A means must be found to recover and reuse the $NaNO/NaNO_3$ from the solution otherwise it could be a source of pollution. Urea also reacts with the nitrogen oxides resulting in the liberation of N_2/CO_2 and formation of ammonium nitrate in solution. The scrubbing yields a steamy but colourless plume. Here again, the by-product ammonium nitrate solution must be utilised otherwise, it would result in the total use of urea as well as a source of water pollution. Liquid scrubbing systems are best adopted in combination with extended absorption. After cleaning the tail gas substantially in additional absorber, liquid scrubbing could be adopted to reduce the pollution levels. This would result in considerable reduction in the absorber volume while, at the same time, keeping chemical consumption reasonably low.

Control of NO$_x$ in the nitrophosphate plants

Rockphosphate [fluoropatite 3Ca$_3$ (PO$_4$) CaF$_2$], which is naturally available contains phosphorus in the form of tricalcium phosphate which is not soluble and hence not available to the plants. To convert this phosphorus to the soluble forms, i.e. dicalcium phosphate and monocalcium phosphate the CaO/P$_2$O$_5$ ratio has to be reduced. This is done either by removing part of the calcium nitrate formed (after addition of nitric acid to rock phosphate) by chilling and crystallisation or by the additional external P$_2$O$_5$ or by both. By this fertiliser of required grade containing nitrogen and phosphorous (NP) is produced.

The emission of NO$_x$ takes place during the treatment of rockphosphate with nitric acid. In theory no NO$_x$ should be formed as the nitric acid is being used purely as an acid to dissolve phosphate rock. In practice, however, the rock phosphate contains some oxidisable impurities such as sulphides and organic material and it is these which reacts with nitric acid to produce NO$_x$. The NO$_x$ emission can be minimised but not eliminated by maintaining proper temperature in the acidulation reactor. To combat the emission of NO$_x$, urea is added at the digestion stage to yield nitrogen, ammonium nitrate and water.

The operation and maintenance of this system is very simple provided urea addition is continuously done. This is achieved through an urea solution tank equipped with a pump to add urea at the rock digestion stage. Good results have been obtained with apparently no adverse effects on product quality.

Absorption

The absorption of NO$_x$ on a fixed bed of solids has not been extensively commercialised as other processes. Normal (dry) absorption on activated carbon or molecular sieves is very efficient but the absorbents must be regenerated either by thermal desorption or by pressure reduction.

Control of fluorine emissions

Fluoropatite ores are used for the production of fertilisers such as single superphosphates (SSP), triple superphosphate or converted to wet process phosphoric acid and there on to other fertilisers. Depending upon the source of origin the fluorine content in the rock varies between 1 and 5 per cent which is released as silicon tetrafluoride (SiF$_4$) and hydrogen fluoride (HF) in the presence of active silica contained in the rock when rock is acidulated and ammoniated for fertiliser manufacture. To circumvent this problem it has become common to fix volatile fluorine compounds as H$_2$SiF$_6$ (hydrofluorisilicic acid) by scrubbing with water which then forms a base for production of a variety of fluorine related compounds.

Fluorine is released during the attack on rock by sulphuric or phosphoric acids. In the manufacture of SSP 12–25 per cent of total fluorine in the rock is evolved as gaseous vapours in the form of hydrogen fluoride and silicon tetrafluoride. The hydrogen fluoride liberated during acidulation reacts with silica to produce SiF$_4$ which reacts with the water vapour generated during acidulation and with water in the scrubber towers to form hydrofluoride acid and finely divided silica. In the TSP manufacture 5–15 per cent of the total fluorine contained in the acid and rock is evolved during the acid attack and 20–40 per cent of the total fluorine during drying. The reaction is nearly the same with a part of fluorine getting converted to H$_2$SiF$_6$ and the rest remaining as HF and SiF$_4$ which is removed by further scrubbing. Phosphoric acid manufactured by action of sulphuric acids on rock at 70°–80°C evolves fluorine during the acidulation and concentration steps. Total volatalisation of flourine (F) during acidulation and concentration of phosphoric acid is approximately 40 per cent. The fluorine emissions are controlled by scrubbing these gases in either Venturi scrubbers, packed towers, crossflow scrubbers or a spray chamber. Water is used as scrubbing liquor which converts the HF and SiF$_4$ to H$_2$SiF$_6$. Efficient scrubbing system can operate upto 25 per cent H$_2$SiF$_6$ but above this point the SiF$_4$ vapour pressure rises steeply. Many towers operate at 15–20 per cent H$_2$SiF$_6$ or less.

One of the commonly encountered problems is deposition of silica when concentration of phosphoric acid is less than 50 per cent. This leads to choking of scrubber nozzles. Hence nozzles have to be periodically cleaned and specially designed for this service.

Single super phosphate is mainly manufactured by a continuous process. Ground rock phosphate and diluted sulphuric acid are mixed in a Broadfield mixer and passed through a closed slow moving conveyor (Den). Most Indian SSP manufacturers are using the above-mentioned process. During the acidulation process, some acidic fumes containing fluorine are emitted from the Den and the mixer. They are to be scrubbed to remove toxic gases. Hydrogen-silico-fluoride formed in the scrubber can be used to make sodium-silico-fluoride, a useful by-product.

To remove toxic gases, in older plants, these are channelised through zig-zag type scrubbing tower in series, and a diluted solution of hydrofluoro silicic acid is sprayed through different nozzles mounted on top of the towers. Suction in mixer and Den is maintained by a 6000 cfm capacity centrifugal fan (Fig. 15.5). Water is recirculated by a pump from a small storage sump of hydrofluoro silicic acid. Most SSP plants in India earlier used such technology to removes toxic gases, but due to certain earlier used problems, this scrubbing tower method failed. The problems were mainly:

1. Choking of nozzles.
2. Scrubbing efficiency was quite low as fluorine recovery was low.
3. Choking in tower and duct due to silica.
4. Temperature of scrubbing liquid getting increased.

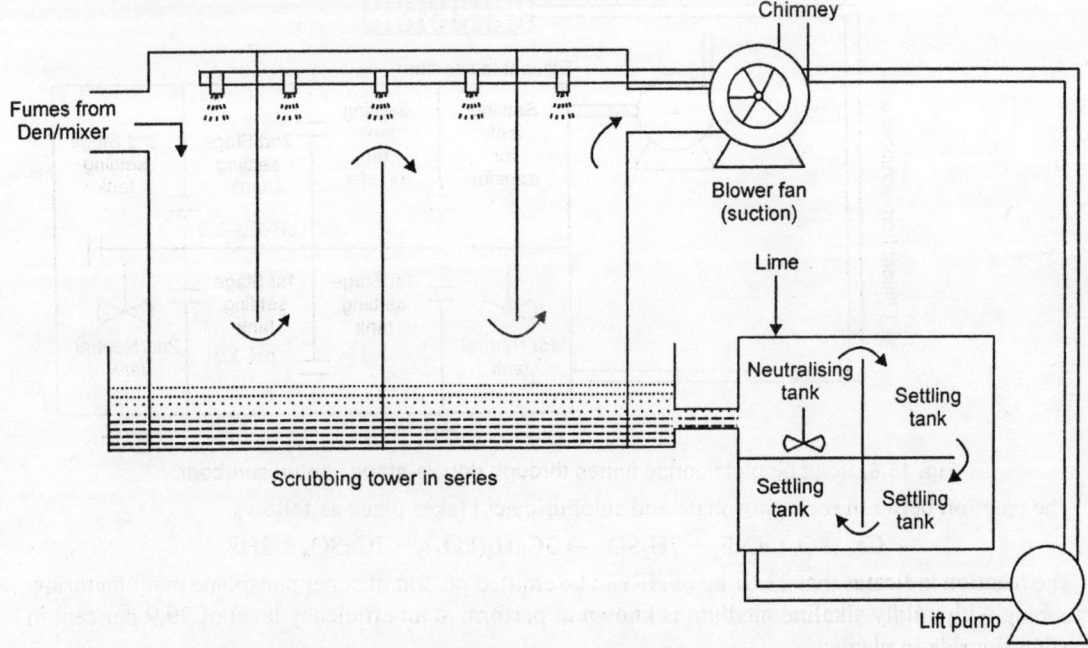

Fig. 15.5. Scrubbing of fluoride fumes in series of scrubber towers.

To overcome the above difficulties and increase scrubbing efficiency, the double-stage venturi scrubber method is used. In this, acidic fumes are scrubbed in a very advanced two-stage venturi scrubber, connected in series along with a cyclonic separator. In this case, water is used as an absorbent for the

acidic fumes and fluorides. The fresh water for absorption purpose is added in the final stage of scrubbing.
A schematic diagram of this is shown in Fig. 15.6.

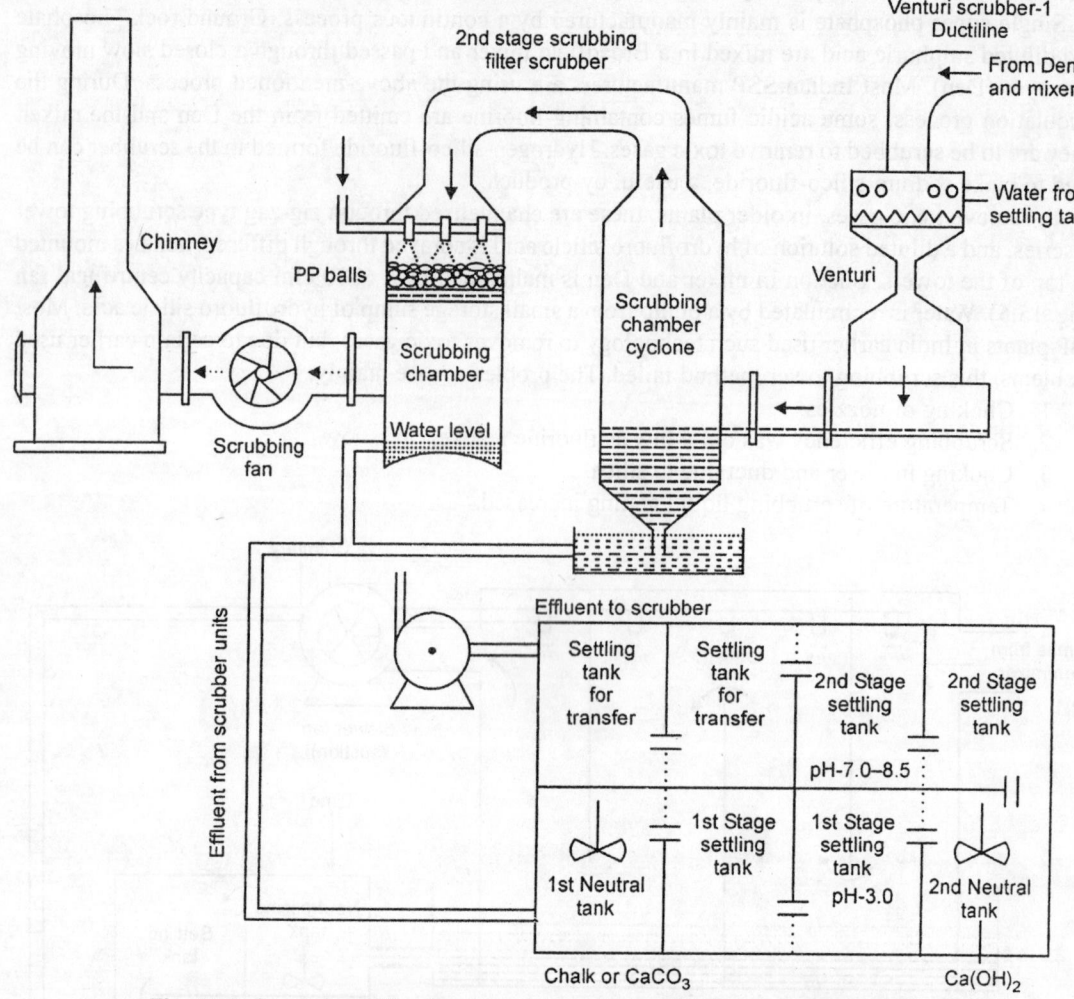

Fig. 15.6. Scrubbing of fluoride fumes through double-stage venturi scrubber.

The reaction between rock phosphate and sulphuric acid takes place as follows:

$$Ca_9 (PO_4)_6 \cdot CaF_2 + 7H_2SO_4 \rightarrow 3CaH_4(PO_4)_2 + 7CaSO_4 + 2HF$$

The reaction indicates that 23.61 kg of HF can be emitted per ton of super phosphate manufacturing.
Scrubbing with mildly alkaline medium is known to perform at an efficiency level of 99.9 per cent in
arresting fluoride in plants.

The controlled emission can be less than 0.04 kg of fluorine per ton of rock phosphate processed.
The various process units required for efficient control of HF emissions are:

1. Venturi scrubber-I.
2. Scrubber pit-I.

3. Scrubber pump-I.
4. Venturi scrubber-II.
5. Scrubber pit-II.
6. Scrubber pump-II.
7. Cyclonic separator.
8. Scrubber fan.
9. Chimney.

All the scrubbing system equipment are made of rubber-lined mild steel.

PAINT AND DYES

Paint is any liquid, liquefiable, or mastic composition which after application to a substrate in a thin layer is converted to an opaque solid film. A dye can generally be described as a coloured substance that has an affinity to the substrate to which it is being applied. The dye is generally applied in an aqueous solution, and may require a mordant to improve the fastness of the dye on the fibre.

Air Pollution

Emission fuel burning

1. Flue gases are emitted from boilers for raising steam. Pollutants produced are minimal in case of LPG and natural gases, somewhat higher for furnace oil, LSHS and light diesel oil (LDO), and maximum for coal, especially in form of particulates. Flue gases are generally discharged through 20–30 m high Stacks.
2. Thermopaes are often used for heating thermic fluids (e.g. oils) which in turn are used for heating in certain processes. Such units are operated on furnace oils, LSHS or LDO.

Emission process

Resin production

1. Phthalic condenser: Fumes from reaction kettles are passed through phthalic condensers. These condensers contain baffles to make the fumes follow a sinous path, forcing the phthalic particles to deposit on the walls. The efficiency of the condenser is further enhanced by providing a water jacket for cooling the fumes.
2. Special chimney: Here fine particles of phthalic are separated. The base of the chimney being conical, helps the fumes to slow down to give better deposition of phthalic particles. Weekly removal of phthalic particles is done to maintain the trap efficiency.

Varnish production

Decomposition products of resin and oils are released as fumes. These fumes are scrubbed by jets of water spray.

Bituminous paint production

During bitumen melting, fumes are given out and scrubbed by water.

Air extraction

Air extractors and exhaust fans are provided with suitably designed hoods at key positions to help improve ventilation in working areas. The air is dispersed through stacks. The flue gas emission from fuel sources should meet the standards prescribed in the emission regulation of the Central Pollution Control Board and Bureau of Indian standards.

In the dye industry, point—source generation of process emissions are from the reaction vessels. These emissions are scrubbed by water. The other emissions are of fugitive nature and should be controlled by proper modification of plant and maintenance of the equipment. The emission from steam generation plant and captive power plant should follow the standards prescribed in the Emission Regulations of the Central Pollution Control Board.

The air pollutants in production of dyes can be reaction gases, solvent vapours, dusts or liquid droplets. Gases include chlorine, phosgene, any of these toxic gases can produce health problems inside the plant itself. Large quantities of solvents are used in chemical synthesis and it is estimated that 1 to 2 per cent of the solvents used, escape into atmosphere. If the dilution by air is high and the toxicity of the solvent is low, then there is no serious problem. However, in the event of a breakdown of the cooling system, large quantities can also escape into the atmosphere. Some are not very toxic but have a bad odour, like mercaptans. In such cases there are persistent complaints from the surrounding community.

Air pollution may also result, when powders have to be filled into different containers or drums for further use. Unless this is done under a fume-chamber with an exhaust system, this can lead to atmospheric pollution. Storage of compressed gases normally posses no problem. However, if there is an accidental leakage or bursting of a cylinder, it may lead to a hazardous situation.

Generally reactions are carried out either in aqueous solution or in some solvent. Either it can be a batch process or a continuous process. Dyes and pigments are made by batch process in aqueous systems while drugs or pharmaceutical active ingredients are made by batch process in solvents. Large volume compounds like insecticides or herbicides are now produced by continuous solvent processes. Unless the reactions are carried out in a completely closed system, air pollution occurs during their production. In our country old plants and particularly some of the smaller factories have no arrangements to collect the outcoming gases right at the source. These gases should be passed through a wet scrubber or absorbed ion activated carbon or made to react with another chemical (e.g. off-phosgene with ammonia). However, care must be taken to ensure that the peak flow of pollutant does not exceed the absorption capacity of equipment, least a blow-through may occur. Problems may also arise when waste gases of several production buildings are treated together in a common multipurpose air treatment unit. As an example, ammonia or some amines may be given off in one building while hydrochloric acid gas may be coming out from another source. Ammonium chloride formed may result in a very fine mist, which is difficult to remove. Drying operations may be the source of particular problems. Volatile compounds like solvents, but in some special cases, also mercury vapour, when mercury is used as a catalyst, are given off in drying operations. Often, emissions of dust result, when dry substance is removed from the apparatus.

Grinding or milling of dry substances are quite often sources of dust. This also applies to final packaging steps for fine powders.

For filtration as well as for distillation of high boiling substances, for the replacement of air by an inert atmosphere, for evacuation of drying ovens vacuum systems are necessary. Gases and vapours pumped out of a vessel are thus transferred to the atmosphere by way of exhaust system of the vacuum pump. Depending on the type of vacuum pump used, the polluting compounds may be introduced and distributed differently between the water phase and air.

A mention may also be made of air pollution caused by some waste-treatment plants. The biological degradation of waste substances in an activated sludge system of open construction, especially using surface aerators, may be the cause of air pollution, again through stripping of volatile compounds of all sorts from the aqueous media. This stripping is promoted particularly by the large volumes of air blown into the mass by the aerators. Occasionally, this step may also give rise to air pollution problems, if the volume of oxygen introduced is not sufficient and if, as a consequence, part of the aeration tank changes to an anaerobic state, producing foul odours.

Incineration of chemical wastes (solid, semisolid or liquid) must be carried out with an appropriate scrubbing system for flue gases, because of toxic compounds resulting from sulphur, nitrogen or halogen content of these wastes. Incomplete incineration also can result in evolution of undesired products. It is important, that temperatures are above 1000°C to ensure complete incineration.

The attention must be paid to certain measures, which can reduce the quantities of substances emitted into the atmosphere. It is a generally accepted principle that the most efficient methods of environmental protection are those, which reduce or eliminate emissions right at their source. In almost all chemical reactions, there are by-products, some of which are potential air pollutants. Development work should be undertaken to reduce the quantity of such by-products. In the production of a certain pesticide, a bad-smelling mercaptan was used earlier as a starting material. This could be replaced by a different intermediate, reducing air pollution. Similarly, packaging of fine powders may lead to emissions of dust. The same product may be packed in the form of freely flowable granules, again reducing air pollution. Collection of vent gases at the source is very important. Hydrochloric acid coming out from various chlorination steps can be absorbed in water to give liquid hydrochloric acid. Certain highly volatile solvents, can be replaced by solvents of higher boiling point, resulting in lower emissions to the atmosphere. Finally, changing from a batch process to a continuous process can be of great help in this direction, though the latter cannot be applied to small scale productions.

Methods of Air Pollution Control

It is an important prerequisite to know the exact nature and quantities of pollutants to be removed. Unlike in water pollution, monitoring of air pollutants is more difficult, for the simple reason that, successive measurements will differ among themselves. Results are expressed in terms of deviations, in comparison with the mean value reported. Minimum average and maximum mass flow should be known. The sampling procedures used must be such that the small quantities and concentrations in question may be determined with sufficient accuracy. The methods of analysis are broadly the same as applied for organic compounds, but they have to be suitably adapted.

Technical equipment

The selection of individual methods and types of equipment to be installed depends on the following factors :

1. The absolute efficiency—does it meet emission limits?
2. Energy requirements.
3. Cost/benefit analysis.
4. Washing liquid/waste-water considerations.
5. Whether substances are to be removed and destroyed or regenerated.

Air pollutants from chemical industry, as mentioned earlier can be in the form of gases or dusts, though sometime liquid droplets also come out. Let us discuss control of dust pollution first, because it is somewhat simpler. Dust particles can be removed by gravity settling, filtration or electrostatic attraction. Though filtration may be quite efficient, still it cannot remove 100 per cent of all the particles, because the smaller ones pass through and go into the ambient air. The efficiency of a cyclone collector drops rapidly, when the particles are below 10 microns in diameter. It is most suitable when the dust is coarse, its concentration is high and where high efficiency is not critical, cyclone collectors, normally cost less both in investment and in maintenance and they are commonly used during grinding and mixing operations in dyestuffs and pesticides plants.

Scrubbers

The next most important equipment used in this industry is the scrubber. Dust particles can have an impact either with a wet surface or with individual droplets of the scrubbing liquid. Since the turbulance required for scrubbing can be produced in so many ways, there are various designs ranging from simple spray chambers to highly complex mechanical devices. Since they utilise liquid media, they can also be used as chemical mass transfer devices. They can also cool the exhaust gases. Operating costs are relatively high, because of higher energy input, though maintenance cost is low. If water is used as a scrubbing medium, an adequate supply of the same has to be ensured.

Fabric filters

The usual form comprises a number of cylindrical bags which gets inflated by the gases to be cleaned, gas passing through the fabric from inside. Particles adhere to the fabric and are thus removed from the gas stream. With the passage of time, the flow capacity is reduced or the bag ruptures. Hence the bag has to be cleaned periodically, which is done by mechanical shaking, collapsing or reverse flow etc. Such filters are normally used, where valuable material is to be collected dry and water availability and disposal are problems. Operating costs are low but maintenance cost is higher because of bag replacement.

Gaseous pollutants

Methods for removing obnoxious gaseous pollutants from air are adsorption, absorption and combustion.

Adsorption

In this process, molecules from a gas or liquid stream attach themselves on the surface of a solid. The process should be reversible, in order to be economic, so that the same absorbent can be used more than once. Activated carbon is the most important absorbent. It shows good performance with solvents and odour molecules and is relatively non-polar. Steam is normally used for regeneration of the used carbon at a temperature of approximately 65°F. The absorbate molecules are released and are ejected with steam, which is normally passed in a direction opposite to the flow. Afterwards, the bed is cooled and dried. Adsorption is a high initial cost process, which is justified only if the recovered product is valuable.

Absorption

This is a process, in which a soluble gas is transferred from a gas stream into a liquid. The gas may simply dissolve in or may react with the liquid. Usually, the gaseous components to be removed are present to an extent of 1 per cent or less. Water is the most suitable scrubbing medium for air pollution control application. Acid gases like HCl or SO_2 are absorbed in alkaline solutions to form nonvolatile salts. Hydrogen sulphide is absorbed in ethanolamine.

Combustion

When the fumes have no recovery value, but are combustible, one means of disposal is flame incineration, by which the organic matter (both gaseous and particulate) is converted into carbon dioxide and water. Hydrogen sulphide is also combustible, when it is converted into sulphur dioxide. Often auxiliary fuel is required to be burnt to heat the fumes to a sufficiently high temperature.

TEXTILE INDUSTRY

A textile or cloth is a flexible material consisting of a network of natural or artificial fibres often referred to as thread or yarn. Yarn is produced by spinning raw fibres of wool, flax, cotton, or other material to produce long strands. Textiles are formed by weaving, knitting, crocheting, knotting or pressing fibres together (felt).

The textile industry actually represents a range of industries with operations and processes as diverse as its products. It is almost impossible to describe a 'typical' textile effluent because of such diversity. After fabrics are manufactured, they are subjected to several wet processes collectively known as 'finishing' and it is in these finishing operations that the major waste effluents are produced. These finishing processes are complex and ever-changing. This is a fact of life that is reflected in the variety of chemicals that find their way into textile finishing waste-waters.

Air Emissions Characterisation

Possible emissions from textile processing include:

1. Oil mists and organic emissions produced when textile materials containing knitting and lubricating oils, plasticisers, and other materials that can volatilise or be thermally degraded into volatile substances, are subjected to heat. Processes that can be the sources of oil mists include tentering, calendaring, heat setting, drying, and curing.
2. Acid mists produced during the carbonising of wool.
3. Solvent vapours released during and after solvent processing operations such as dry cleaning.
4. Dust and lint produced by the processing of natural fibres and synthetic staple prior to and during spinning, as well as by napping and carpet shearing.

Air Pollution Control Measures

Emissions from finishing operations, including drying, curing, and heat setting, have been and continue to be one of the more significant air pollution problems in the textile industry. The blue haze and odour that are characteristic of these emissions have posed both technical and economic problems for large and small facilities alike. The amount of air pollution depends on the finishing processes used. Various kinds and amounts of oils and finishing resins can vapourise from the cloth in tenter frames (commonly used for drying, curing, and heat setting) operating at temperatures typically in the 300°–400°F range. Some exhausts also can carry a significant amount of lint.

There are process modifications that can reduce some of this air pollution. For example, most of the oils can be removed by effectively prescouring the cloth. Alternatively, the selection of 'less polluting' finishing chemicals may help, but cloth finishing quality and economics often limit the ability of such changes to correct the problem.

Over the longer term, new technology, such as solvent-free, radiation-curable coatings, may prove to be a factor in overcoming the air pollution problem. Such technology not only would eliminate the problem, but would result in significant energy savings, higher line speeds, and greatly reduced floor space. Unfortunately, lack of availability of radiation-curable coatings and high cost have been the major limitations to commercialising this technology.

Historically, the solution to the characteristic blue haze and odour problem has been the use of air pollution control equipment. Wet (two-stage) electrostatic precipitators, fabric-type (glass fibre-wool) mist eliminators, and fume incinerators have traditionally been the equipment of choice. Wet precipitators and mist eliminators have been plagued by continuous and costly operation and maintenance problems.

In general, their performance has been strongly dependent on the effectiveness of gas stream cooling prior to treatment in the control device.

Experience has shown that temperatures generally should not exceed 120°F. Fume incinerators operating at 1200°–1400°F with approximately a 0.3 second gas residence time have been effective in eliminating blue haze and odour problems; however, capital and operating costs may have a significant impact on the facility.

Air pollution problems resulting from greige processing of cotton consist predominantly of dust generation in the high-speed mechanical operations, such as spinning, drawing, carding, and twisting. Dry dust collectors of the vacuum type are well suited to maintain a dust-free working environment.

Finely divided lint is present in the air in the vicinity of various operations associated with spinning the yarn and weaving the cloth. Most modern mills have air tunnels built under the floor to effect a downward flow of air around the machines that produce the largest volumes of waste lint (carding, roving, spinning, twisting, weaving, etc.). In the opening and picking rooms in a cotton plant, stock is usually handled by overhead ductwork. Lint and dust are collected by the vacuum system as close as possible to their point of origin. Automatic travelling vacuum cleaning systems are frequently used in conjunction with the underfloor system.

The suspended lint and dust are conveyed by air movement through ducts to a centrally located filter system for removal. Automatic dry-type travelling disposable air filters are extensively used for this purpose. Since any filtering medium tends to hold the lint fibres tenaciously, a low-cost, disposable paper medium is often used. The fumes from wool carbonising pose another problem that must be dealt with. This process generates very fine carbon particles that appear as smoke, as well as some fumes and odours. These fumes probably include residual sulphur oxides (if sulphuric acid has been used for carbonising), in addition to organic decomposition products, and are generally very corrosive. Corrosion-resistant stainless steels and/or plastics have been used in the collection systems, which usually exhaust the fumes to the atmosphere. Since a significant quantity of particulate matter is in the submicrometre range, a visible emission is usually apparent unless there is a very efficient air pollution control system. The severity of the opacity problem will depend to a large extent on location, topography, and local meteorological conditions. Incinerators, wet scrubbers, and fabric filters have been used to resolve the problem. Wet scrubbing systems have the added advantage of reducing residual gases such as sulphur oxides, as well as particulate emissions.

Air Pollution Sources

Textile industries burn fuel in the boilers for generating steam, which is mainly used in wet processing departments. Average textile mills consume about 20 to 30 tons of coal per day and atmospheric pollution due to chimney gases ejected from the boilers can pose a serious threat to the population in cities like Ahmedabad and Mumbai where large number of mills are located. The flue gases ejected into the atmosphere through the chimney of the boilers can cause atmospheric contamination in two ways: (i) descending of particulate matters in the flue gases to the ground level, (ii) settling of sulphur dioxide gas formed due to reaction of sulphur of the fuels with oxygen during combustion.

Pollution due to particulate matter

Most of the boilers in textile mills which burn coal as fuel are normally manually fired. Only in the city of Mumbai, where there is a big concentration of textile industry in relation to other big and small industries, the mills are dissuaded from using coal as a fuel and instead, fuel oil is recommended despite its high cost because the pollution of air by the particulate matter of the chimney gases from the coal fired boilers seriously threaten the health of the residents. The particulate matter emitted through chimney is considerable in case of manually fired boilers, whereas it is not so high in case of boilers where coal is fired by means of mechanical stokers. In the city of Mumbai, where some of the mills have changed from oil firing to coal firing and many more want to change over to coal because of large savings in fuel bill due to comparatively much less price of coal. But because of pollution problem, the authorities in Mumbai have made the following two stipulations for the mills which want to resort to coal fired boilers:

1. The boilers should be provided with mechanical stokers.
2. The boilers should be provided with efficient soot collectors which reduce the particulate matter in the flue gases of the chimney.

The mills prefer simple manually fired boilers, such as Lancashire boilers, though they are less efficient because they are simple, rugged, need little maintenance and hardly have any serious break downs. It is possible to fit mechanical stokers on these boilers, but suitable ones are not available in India. Chain grate stokers suitable for Lancashire boilers have been developed in western countries and are known to be working satisfactorily in India. Arrangements, therefore, should be made for a large scale import of these stokers or for manufacturing indigenously with foreign collaboration if necessary.

All industrial stack emissions contain at least some particulates generated in operations like beating, grinding, pulverising, combustion, etc. or formed through condensation and chemical precipitation. Such particulates also include droplet clouds often called mists. Some times droplets or solid particulates may be formed through condensation or precipitation after the gaseous effluents have left the stacks. This especially happens when the temperature of the stack gases is high enough to keep the pollutants in gaseous forms, but after release in the atmosphere, they cool down to produce intense clouds. These particles ultimately settle down the larger and heavier ones fall early and close to the source sites and the finer spread over large areas-giving dirty look to land, buildings, roads, clothing and even plant leaves. Besides causing nuisance, particulates also adversely affect the health of the people breathing the dust laden air. Lung diseases like byssinosis, fibrosis and silicosis are directly caused due to large quantities of fibres or dust accumulating in the respiratory system. The adverse health effects are caused by the particles finer than 5 micron since any larger particles are effectively screened and removed by the nasal hair and mucous system. Particles of 1 to 2 micron size are the most hazardous from health point of view. Through proper stack design, it can be ensured that pollutants would disperse far and wide and thus keep the ground level concentration close to the stipulated limits. However, particulates are going to settle on the ground ultimately due to gravity unlike gaseous pollutants that can diffuse to the upper layers and ultimately get chemically converted to non-offensive forms. Thus, a proper stack design alone would be adequate for abatement of dust pollution on long-term, long range basis and other methods for pollution control of particulates as mentioned below have to be resorted to.

A very wide variety of duct cleaning equipments has been developed and is available off-the-shelf in the developed countries. Large number of dust cleaning equipments installed in India are of foreign make. Generally following dust cleaning equipments are used for pollution control due to the particulates.

1. Simple gravitational and impingement devices.
2. Cyclonic separators.
3. Bag filters.
4. Electrostatic precipitators.
5. Scrubbers and Washers.
6. Other miscellaneous devices.

Simple gravitational equipments like settling chambers have low initial costs, pressure drops and maintenance costs, but their removal efficiency is also low. Their performance is significantly impaired by presence of any turbulance or non-uniform distribution of gas velocities. Gas velocity through the chamber has to be kept less than 3 m/sec. to prevent re-entrainment.

They can remove the particles above 100 micron size. The impingement devices like grit arrestor can remove particles of about 30 micron size by suddenly changing the direction of flow 2 to 3 times.

Cyclonic separators are extremely versatile equipments which can handle gas streams at very high temperatures, can handle large dust loads, are very efficient in dust removal and their energy requirement is low. Generally, they can remove particles of 5 to 7 micron size with pressure losses as low as 10 to 15 cms of water and with removal efficiency ranging from 80 to 90 per cent. The diameter of such cyclones varies from 0.6 to 1.2 m and length from 2 to 4 metres. Several such units are operated in parallel each handling 100 m^3/min. of stack gases.

Bag filters made of such materials as cotton, jute, wool, nylon, terelyne, teflon fibres, glass fibres and even stainless steel mesh are commercially available and they offer high removal efficiencies for the particles below 5 micron size. It has been reported that a typical bag filter using processed wool felt as a filtering medium can separate 90 per cent of the particles of 2 micron size from the gas at a pressure loss of about 20 cms of water. A good bag filter of 45 cm diameter and 4 meter length can handle about 30 to 50 m^3/min of gas flow. Electrostatic precipitators operating on 30 to 50 KV voltage are known to have given 95 to 99 per cent or even higher removal efficiency for a variety of industrial dusts as fine as of one micron size or even less. Their functioning comprise of three stages: (i) inducing a small electrical charge on each particle by passing the gas stream through a high voltage corona discharge, (ii) collecting the particles so charged on earthed collector plates, and (iii) transferring the collected dust from the collector plates to a storage hopper by suitable rapping or other mechanism since a thick layer or collected dust on the collector plates can reduce the removal efficiency. The electrostatic precipitators can handle large gas flows, large dust concentrations and high temperatures without any problems. In fact, an electrostatic precipitator can be tailor made to suit the removal of specific particulates from a particular gas. The exact configurations and arrangements of the electrodes and other features not only affects the efficiency but the longevity, easy and quick repairability and maintenance costs. The most common failure of electrostatic precipitator is the electrode failure when wires are used. In case of horizontal layout, electrode failures may additionally pose danger of short circuits. A very important aspect of proper electrostatic precipitator operation is voltage control. It is extremely important to seek and operate at the optimal voltage for any particular operating conditions rather than operate at fixed voltage.

Pollution due to sulphur dioxide

Fortunately, Indian coals do not have high sulphur content. Generally they contain about 0.4 to 0.8 per cent sulphur. Even then sulphur dioxide concentration in the air can exceed the permissible limit when considerable amount of coal is burnt, so that its sulphur content is combusted to sulphur dioxide. It is reported that maximum concentrations of sulphur dioxide in the atmosphere upto 10 ppm at any particular time (not average over a day) is permissible. In the USA, it has been specified that the annual average of sulphur dioxide in the atmosphere should not exceed 0.02 ppm. A study conducted in Ahmedabad by National Institute of Occupational Health revealed that sulphur dioxide concentration was about 0.4 ppm in the mill area and 0.3 ppm in the residential area. Dispersion of the industry over large area would help in reducing the problem, but it is hardly a practical solution. Raising the chimney height, absorption of sulphur dioxide in the alkali solution or any other suitable solvent are the other two means by which this problem can be tackled, but they warrant substantial expenditure. In case of oil fired boilers, particulate matters are generally within the stipulated limits, but since the sulphur content of the fuel oil used in India is about 2 to 3 per cent, the sulphur dioxide concentration in the flue gases also will be much higher than that for coal firing. Sulphur dioxide in the atmosphere at high concentration can cause respiratory ailments to the inhabitants beside corrosion of the machinery and buildings and detrimental effect to the plants and trees.

Miscellaneous pollution

The atmosphere is generally unclean in spinning departments. It is contaminated with cotton and other fibres and dust particles. Persons exposed to this kind of atmosphere for prolonged periods tend to develop respiratory ailments and a specific ailment called Bissinosis. The problem is particularly serious in the ring frame section. With proper control of air movement and filtration, the internal pollution inside the building can be kept within safe limit. Certain wet processing operations in the textile mills give out some injurious or toxic fumes. For example, in the carbonising section in which cotton component of a fabric is dissolved in the sulphuric acid bath, acid fumes rise up and pollute the area. Likewise, when pigment printed fabric is dried in the hot air drier and polymerisers, kerosene fumes spread into the surroundings. Certain types of resin treatments also give out toxic fumes. These problems can be easily taken care of by provision of suitable hoods and sealing the leaking joints of the machines.

Accoustic pollution has not received much attention in India. Excessive noise level could impair the hearing ability of the persons exposed to it for a prolonged period. Some of the sections in the textile mill do create a high noise level which is above the permissible limit. Loomshed particularly is very bad in this regard not only in our country but in foreign countries also. The permissible safe limit of sound level for 8 hour exposure in a day is about 90 dBA whereas the actual sound level in the loomshed is higher than this. One way to reduce the sound level is to make suitable design modifications in the looms so that generation of noise is reduced. Some amount of research has been carried out in this regard in the advanced countries for large reduction in noise level. Besides, provision of sound absorbing surfaces at the walls and ceilings can also help reduce the noise level.

MAN-MADE FIBRE AND RAYON INDUSTRY

Fibre is a fundamental form of solid (usually crystalline), characterised by relatively high tenacity and an extremely high ratio of length to diameter (several hundred to one). Natural fibres are animal, such as wool and silk (proteins); vegetable, such as cotton (cellulose); and mineral (asbestos). Cotton fibre is called staple, and rarely exceeds 2 inches in length. Semisynthetic fibres include rayon and inorganic substances extruded in fibrous form, such as glass, boron, boron carbide, boron nitride, carbon, graphite, aluminium silicate, fused silica and some metals (steel). Synthetic fibres are made from high polymers (polyamides, polyesters, acrylics, and polyolefins) by extruding from spinnerets (nylon, Orlon, etc.). Some are being used in speciality papers, though the primary use is in textile fabrics.

Rayon is a generic name for a semisynthetic fibre composed of regenerated cellulose, as well as manufactured fibres composed of regenerated cellulose in which substituents have replaced not more than 15 per cent of the hydrogen, of the hydroxyl groups.

Gaseous Effluents

Gaseous effluents in this industry are in general not a health risk, potentially toxic gases are discharged, after treatment, through tall chimneys with considerable dilution. They can, however, present a considerable nuisance. There is a dilemma too in that good working conditions may require large volumes of air to be extracted from the work rooms. This dilutes the gas to be treated and so increases the size of the treatment plant.

Odours can arise from many textile processes; when the operation is a small one, for example stentering of fabrics, much can be done with simple scrubbers or small activated carbon units. A major problem arises however from the production of cellulose fibre or film from viscose solutions. Carbon disulphide needs to be used at a rate of about 30 per cent by weight of finished product. Since the product contains no sulphur all this material will appear in some form as effluent. During the process, some of the disulphide reacts to become hydrogen sulphide and very small quantities react to form more or less complex sulphur compounds. These latter substances have very low odour thresholds and are believed to be the main cause of the viscose 'reek' for which the industry was at one time notorious.

In a typical factory manufacturing viscose rayon staple the exhaust air would be extracted at a rate of $17000 \ m^3/min.$ containing about 600 ppm CS_2 and 300 ppm H_2S, together with detectable quantities of organic sulphur compounds, e.g. thioformaldehyde, carbonyl sulphide, furfural mercaptan, etc. From this whole stream it is necessary to remove as much hydrogen sulphide and related chemicals as possible. It is, however, possible to separate an enriched stream of carbon disulphide of about $8000 \ m^3/min.$

The method of removal of H_2S and related compounds is restricted by the high air flow and comparatively low concentration. The mass emission is such that dispersion through tall chimneys alone may in some units be sufficient. Absorption onto treated activated carbon with subsequent sulphur regain is feasible but not economic. Oxidation by combustion is uneconomic unless part of the stream can be used as the source of air for factory boiler plants though this is clearly limited by the demand of the furnaces and the great care needed to ensure that combustion is complete. Incomplete combustion can produce intermediate organic substances even more odourous than the original contaminants. Oxidation in this way also produces SO_2 which is undesirable although it does not give an odour nuisance.

The methods of choice are therefore limited to chemical oxidation in scrubbers. There are some 85 methods of chemical removal of H_2S listed in patent literature. The eventual choice must be governed by the efficiency and cost of the operation. The final choice can only be properly made after extensive pilot plant studies and when the process route has been selected, operating conditions optimised by further pilot studies. In the case instanced, the method chosen was the well-known 'Ferrox' process. The chemistry of the process is superficially simple, the sulphur gases are absorbed in alkaline liquor in the presence of ferric hydroxide acting as a catalyst. Air is blown into the liquor and the sodium hydrogen sulphide converted back to hydroxide with sulphur being liberated. The hydroxide restores the sodium carbonate/bicarbonate balance and the process is theoretically self supporting. In its simplest form the scrubber is fabricated as a long horizontal chamber, the liquor being sprayed through jets mounted in the side walls. The choice of a horizontal chamber in place of the more common vertical type was made because of the high pumping cost for some 800 litres/sec. of circulating liquor. Spray design and mounting are critical, it is advantageous to have a heavy curtain of liquor broken up by the impinging gases rather than a finely divided liquor spray.

The chamber floor is designed to collect and return the liquor to adjacent regeneration (aeration) tanks. After scrubbing the purified air stream passes through eliminator plates, any entrained liquor being released and returned to the circulating tank. The catalyst liquor is prepared by reacting ferrous sulphate and sodium carbonate giving ferrous hydroxide, oxidised to ferric hydroxide before use.

Regeneration is performed by oxidation with atmospheric air the liquor being stirred whilst diffused air is introduced below the stirrer. Alternatively a mixer-aerator of the helical type may be used. The flow of liquor through the tank carries the released sulphur into a trough where it is diluted before going forward as effluent. Sulphur recovery is possible. In operation the process requires better than average maintenance to keep high efficiency, the jets must be regularly checked for alignment and cleanliness and pressures maintained by attention to filter screens and regular cleaning of eliminator plates.

The chemistry of the process is simple but the liquor constituents must be optimised and kept in balance. Using the pilot plant referred to above a mathematical model has been constructed relating efficiency to air flow/liquor flow/concentration of individual chemicals. If this is observed and conditions kept at their optimum values, removal of greater than 95 per cent may be achieved.

The problem of removing the traces of sulphur gases from the outlet stream remains unsolved. Work is proceeding to investigate the oxidation of sulphides and related species by bacteriological means, initially in a packed tower using plastic media as a support for the biomass. At low gas flows very effective removal can be obtained even from very concentrated (2000 ppm) H_2S streams. The biological process also appears to be very effective in oxidising other sulphur gases (CS_2, etc.)

Carbon disulphide (CS_2) does not present so great an odour problem though its removal from air streams is desirable. In some instances it is possible to absorb this material onto beds of activated carbon and subsequently recover it for reuse. Both static and fluidised beds are in use for this purpose. In certain installations it is possible to obtain good removal efficiencies by condensation of rich streams of CS_2. Where mass emissions are smaller, other methods of odour control have been attempted. Masking agents are perhaps the least expensive method available. These aim to eliminate the perception of smell by superimposition of another odour, preferably pleasant. A wide range of masking chemicals are available and must be selected and compounded for each situation. Selection requires considerable knowledge and experience. Counteracting agents attempt to reduce odour intensity by the addition of second odour. Again the selection of a suitable agent or combination of agents requires a great deal of expertise.

Our experience with masking agents has been mixed. It has frequently happened that whereas under test conditions the agent is very effective when applied to a factory the effluent gas has a tendency to separate after dispersion and that two odour nuisances are found, the original contaminant and the agent itself. We are, therefore, convinced that only complete oxidation of the substance or its removal are sure remedies for odour complaint. As with other methods of effluent treatment it is important that the system chosen does not itself produce a further problem. Thus it remains common practice to treat sulphur gases in a scrubber using caustic soda to absorb them. The treatment of the resulting sodium hydrogen sulphide is however a major problem.

JUTE INDUSTRY

Jute processing is a major industry in India. Majority of jute mills are situated in a narrow belt along both banks of river Hugli in the State of West Bengal. The area being a highly industrialised urban sector, the environment is in a critical state. Jute processing industry is a major contributor to the pollution. In order to prevent any further deterioration of the environment it is necessary that urgent action be taken to check the contribution of the pollutants by the industry.

Air Pollution

Boiler

The major source of air pollution from a jute mill is the emission from the stack of the furnace used for the boiler. Coal is the principal fuel burnt in the furnaces used for raising steam. Jute waste/jute dust mixed with mineral oil collected during floor sweepings are also used as auxiliary fuel. The composition of the fuel and the performance efficiency of the furnace primarily determine the composition of stack emission (Table 15.7).

Table 15.7. Particulars of the stacks, stack emission and characteristics of flue gases.

Stack	Diameter of stack (internal)		2.4 m.
	Height of stack above ground level		37.8 m.
	Material of construction		
	Outer shell		Mild steel
	Inner shell		Brick
Flue gas	Discharge rate Nm³/sec.		14.2
	Velocity of flow m/sec.		8.9
Flue	SPM	(mg/Nm³)	398.0
Characteristics	SO_x	(mg/Nm³)	254.0
	CO	(mg/Nm³)	4.1
	NO_x	(mg/Nm³)	1.7

Diesel generator set

The other source of air pollution in jute processing industry is the DG set. Power shut downs are quite common in the industrial belt where the mills are situated. The emissions from the generators is comparatively low and intermittent in nature.

Miscellaneous sources

Fugitive emission

The third source of at pollution from jute mill is dust blown up in the ambient air by air currents. Dust and jute fibres present on floor, machine tops, beams and rafters they settle down on any resting place they find. Unless these are regularly removed and good housekeeping is observed, a strong wind is likely to lift them up and blow the dust into the atmosphere. Jute fibres are seen outside the mill shed as well. They too can be blown up by wind movement. Air pollution caused by such casual wind movement is of a minor nature. Good housekeeping, indoor and outdoor of the mill sheds, will minimise if not eliminate the pollution on this account.

In certain mills local exhaust ventilation is employed for control of dust and air contaminants in the working environment inside the mill shed. Exhaust air containing the dust and fumes is discharged outside the mill building. Unless the dust and the gaseous pollutants are separated from the exhaust air, the discharge of the exhaust air will add to the pollution of the ambient air. Suitable collectors, such as bag filters cyclones etc. need to be incorporated in the local exhaust system.

Dust fall

The dust discharged in the ambient air is carried by wind over long distances. Dust particles will settle down eventually. The distance and direction these dust particles are conveyed depend on several factors amongst which the meteorological data—such as humidity, precipitation, velocity and direction of wind movement are the major ones.

The following measures are recommended:
1. Use of coal of :
   (a) low sulphur content, (less than 0.4%).
   (b) low ash content, (less than 40%).
2. Jute waste/jute fines which are used as an auxiliary fuel in the furnace be sieved properly before use to remove inorganic dust.
3. Improve furnace performance by proper maintenance.
4. Improve the performance of the DG sets.

Standard for stack air emission

Stack height should satisfy the following equation:

$$H = 14(Q)^{0.3}$$

where, H = Stack height in metre.

Q = Sulphur dioxide emission in kg/hr.

In case the chimney height falls short of the requirement its height needs to be raised. The boiler of jute mill should meet the particulate matter emission standards as follows:

Capacity of the boiler	Particulate matter standard
2 ton/hr	1600 mg/Nm³
2 to 5 ton/hr	1200 mg/Nm³
More than 5 ton/hr	150 mg/Nm³

SOAP AND DETERGENT

In modern times, the soap and detergent industry, although a major one, produces relatively small volumes of liquid wastes directly. However, it causes great public concern when its products are discharged after use in homes, service establishments and factories.

Air Emissions

Slurry preparation

The formulation of slurry for detergent granules requires the intimate mixing of various liquids, powdered, and granulated materials. The soap crutcher is almost universally used for this mixing operation. Premixing of various minor ingredients is performed in a variety of equipment prior to charging to the crutcher or final mixer. The slurry, mixed in batch operations, is then held in surge vessels for continuous pumping to the spray dryer.

Air emissions characterisation

The receiving, storage, and hatching of the various dry ingredients create dust emissions. Pneumatic conveying of fine materials causes dust emissions when conveying air is separated from the bulk solids. Many detergent products require raw materials with a high percentage of fines. Typical specifications for some raw materials include the following percentage of fine materials passing a 200 mesh screen: sodium sulphate, 12 per cent; TSPP, 74 per cent; STP, 53 per cent. The storage and handling of the liquid ingredients, including the sulphonic acids, sulphonic salts, and sulphates, do not cause emission problems other than mild odours. In the batching and mixing of the fine dry ingredients to form slurry, dust emissions are generated at scale hoppers, mixers, and the crutcher. Liquid-ingredient addition to the slurry creates no visible emissions, but may cause odours.

Air pollution control measures

Control of dusts generated from pneumatic or mechanical conveying or from discharge of fine materials into bins or vessels can be controlled by fixing bag filters. No unique problems occur in hooding or exhaust systems for controlling dust emissions from conveying and slurry preparation. Baghouses are employed not only to reduce and eliminate the dust emissions, but also for the salvage of raw materials. None of the dusts causes any serious corrosion problems. Filter fabrics should be selected that have good resistance to alkalis. Filter ratios for baghouses with intermittent shaking cleaning mechanisms should be under 3 cfm/ft^2. Continuous process for fatty acids and soaps is shown in Fig. 15.7.

Spray drying

Process description

All spray-drying equipment designed for detergent granule production incorporates the following components: spray-drying tower, air heating and supply system, slurry atomising equipment, slurry pumping equipment, product cooling equipment, and conveying equipment. The towers are cylindrical with cone bottoms and range in size from 12 to 24 ft in diameter and 40 to 125 ft in height. Single towers may vary in diameter, being larger at the top and smaller at the bottom. Air is supplied to the towers from direct-heated furnaces fired with either natural gas or fuel oil. The products of combustion are tempered with outside air to lower temperatures and then blown to the dryer under forced draft. The towers are usually maintained under slightly negative pressure, between 0.05 and 1.5 inch of water column, with exhaust blowers adjusted to provide this balance.

Most towers designed for detergent production are of the countercurrent type, with the slurry introduced at the top and the heated air introduced at the bottom. A few towers of the concurrent type are used for detergent spray drying, with both hot air and slurry introduced at the top. Some towers are equipped for either mode of operation, as illustrated in Fig. 15.8. In most towers today, the slurry is atomised by spraying through a number of nozzles rather than by centrifugal action. The slurry is sprayed at pressures of 600–1000 psi in single-fluid nozzles and at pressures of 50–100 psi in two-fluid nozzles. Steam or air is used as the atomising fluid in the two-fluid nozzles.

Tower operations vary widely among manufacturers and among products. Heated air supplied to the tower varies from 350° to 750°F. Temperatures of air supplied to countercurrent towers are generally lower and most often range from 500° to 650°F. Concurrent tower temperatures are somewhat higher. Solids content of slurries for detergent spray drying varies from 50 to 65 per cent by weight, with some operations as high as 70 per cent. Moisture content of the dried product varies from 10 to 17 per cent. Towers are designed for specific airflow rates, and these rates are maintained throughout all phases of operation. Slurry temperatures may vary, but in most formulations they do not exceed 160°F and are frequently as low as 80°F. Exit gas temperatures range from 150° to 250°F, with wet-bulb temperatures of 120° to 150°F.

Air velocities in concurrent towers are usually higher than velocities in countercurrent towers. The concurrent towers produce granules that are mostly hollow beads of light specific gravity (0.05 to 0.20). Countercurrent towers produce multicellular, irregularly shaped granules that have higher specific gravities ranging from 0.25 to 0.45.

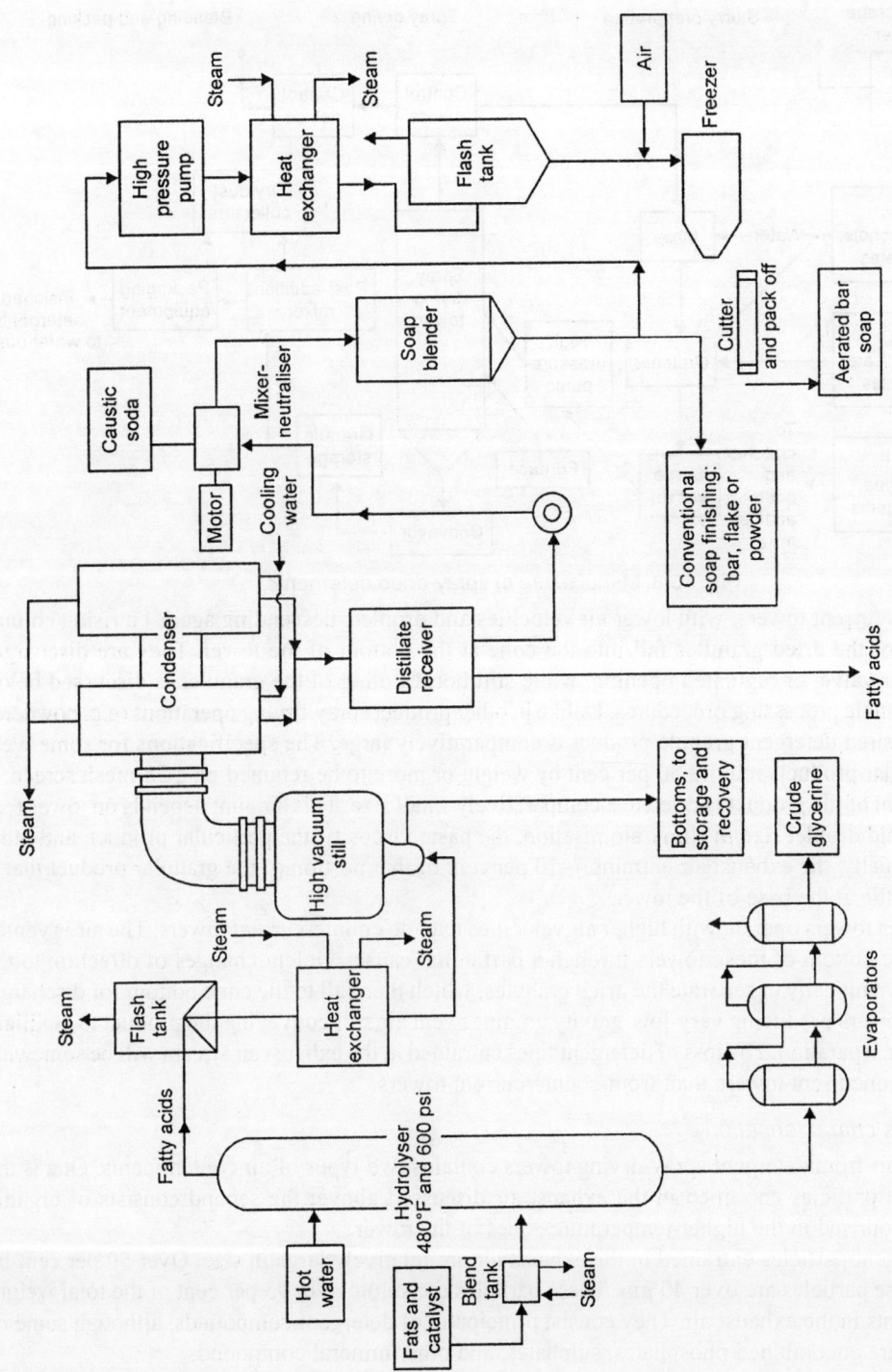

Fig. 15.7. Continuous process for fatty acids and soaps.

Receiving, storage and transfer   Slurry preparation   Spray drying   Blending and packing

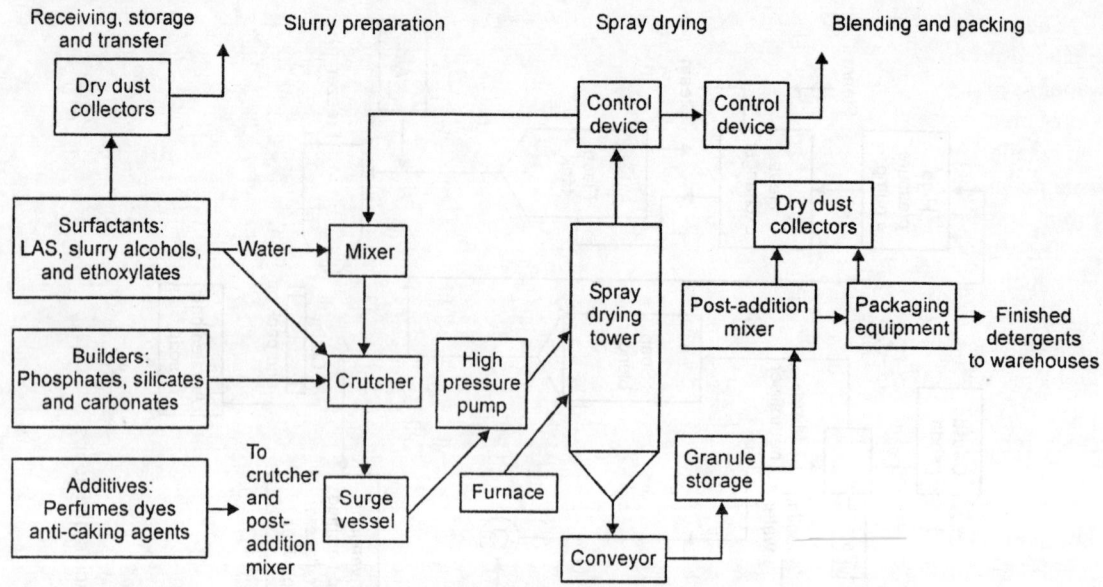

Fig. 15.8. Manufacture of spray-dried detergents.

In countercurrent towers, with lower air velocities and droplets descending against a rising column of air, most of the dried granules fall into the cone at the bottom of the tower. They are discharged through a star valve, or regulated opening, while still hot. Cooling of the granules is discussed below with other granule processing procedures. Unlike in other product spray-drying operations (e.g. powdered milk), the desired detergent granule product is comparatively large. The specifications for some well-known granular products require 50 per cent by weight or more to be retained on a 28-mesh screen. A certain amount of the product is dried to a comparatively small size. This amount depends on tower feed rates, the liquid droplet size in slurry atomisation, the paste viscosity, the particular product, and other variables. Usually, the exhaust air entrains 7–10 per cent of that portion of the granular product that is too fine to settle at the base of the tower.

Concurrent towers operate with higher air velocities than do countercurrent towers. The air is vented just above the bottom of these towers through a baffle that causes violent changes of direction to the exhaust air dynamically to separate the dried granules, which then fall to the cone bottom for discharge. Concurrent towers producing very-low-gravity granules vent air still conveying the product to auxiliary equipment for separation. The loss of detergent fines entrained in the exhaust air stream will be somewhat higher from concurrent towers than from countercurrent towers.

Air emissions characterisation

The exhaust air from detergent spray-drying towers contains two types of air contaminants. One is the fine detergent particles entrained in the exhaust air discussed above; the second consists of organic materials vapourised in the higher-temperature zones of the tower.

The detergent particles entrained in the exhaust air are relatively large in size. Over 50 per cent by weight of these particles are over 40 μm. These particles constitute over 95 per cent of the total weight of contaminants in the exhaust air. They consist principally of detergent compounds, although some of the particles are uncombined phosphates, sulphates, and other mineral compounds.

The second type of air contaminant, organic compounds, originates primarily from the surfactants included in the slurry. Various organic components in the slurry vapourise in the tower. The amount vapourised depends on many variables, such as tower temperatures and volatility of the organics in the slurry mixture. Volatility of the organics is a function of the chain length of surfactant compounds and the reaction completeness in the sulphonation and sulphation of anionic surfactants. The vapourised organic materials condense upon cooling in the tower exhaust air stream into micrometer- and submicrometer-sized droplets or particles.

The variety of possible detergent compounds is almost infinite, and manufacturers are continually introducing new formulations or reformulating older ones. It is not always possible to predict how certain organic compounds in a slurry mixture will affect stack emission. Work in the early 1970s indicated that if amides are present in the slurry in amounts greater than 0.5 per cent by weight, emission problems will occur. Source tests of exhaust air from an air-pollution control scrubber with amides present in the slurry being dried in the spray tower revealed 0.08 grain of organic particulate per standard cubic foot of exhaust gas. The presence of this relatively low concentration of submicrometer-sized aerosols causes water vapour plumes to persist for long distances. Following the break or end of the water vapour plume, a highly visible contaminant plume persists for even greater distances. The amide emission rate increases as tower operating temperatures rise. Many tower operating variables affect air contaminant emissions. The most significant variables are the formulation's final granule temperature and moisture, the dryer inlet air temperature, and temperature profiles in the dryer.

Slurry formulations containing alcohol ethoxylate surfactants or alcohol sulphates with high levels of unsulphated alcohols cause similar aerosol emissions. Source tests of the aerosol leaving the scrubber indicate a particle size range of 0.2 to 1 µm, where ethoxylated alcohol surfactants were added to the crutcher. It is believed that very small amounts of unreacted alcohols coat water droplets like amides, inhibiting the evaporation of the droplets and resulting in gray water vapour plumes that persist for long distances.

Air pollution control measures

The collection of air contaminants not only provides for the, economic return of detergent fines to the process, but it also provides for control of submicrometer particles and aerosols to ensure compliance with air pollution prohibitory rules.

Manufacturers producing detergent granules have developed several separate approaches for capturing the detergent fines in the spray-dryer effluent for return to process. Dry cyclones and cyclonic impingement scrubbers are the primary approaches used. Cyclonic impingement scrubbers return a slurry to the crutcher, while dry cyclones return detergent fines to the crutcher. The dry cyclone separators can remove 90 per cent or more by weight of the detergent product fines in the tower exhaust air. The detergent dust remaining in the effluent vented from the cyclones consists of particles 95 per cent by weight less than 10 µm. Particulate concentrations vary from 0.1 to 1.0 g/scf. The cyclones are designed for relatively high efficiencies and operate at pressure drops of from 8 to 10 inches of water column.

Secondary collection equipment is used to collect fine dust and organic aerosols that pass the primary collectors. Mist eliminators are used after cyclonic impingement scrubbers. Cyclones are generally followed by scrubbers, scrubber/precipitators, or fabric filters. Several types of scrubbers can be used following the cyclone collectors. Venturi scrubbers operating at 8 to 10 inches of water pressure drop (using water at 8 to 10 psig distributed through nozzles in the throat, velocities of 8500 fpm, and water supplied to the throat at a ratio of 4.5 to 5.0 gal per 1000 ft^3 of effluent) have been used following the cyclones. Venturi scrubbers have been replaced largely by packed bed scrubbers operating at 0.5 to 2.0 inches of water pressure drop and water rates of 1 to 3 gal per 1000 acfm. Packed-bed scrubbers are usually followed by wetpipe-type electrostatic precipitators constructed immediately above the packed bed in the same vessel. Fabric filters are sometimes used after cyclones, but their application is limited to spray dryers having low drying loads and drying products with low surfactant concentrations. On efficient spray dryers with exhaust conditions near the dew point drying products with more than 10–15 per cent surfactant, condensing water vapour and condensing organic aerosols will bind the filter fabrics.

Typical control efficiencies and emission factors for various control options are given in Table 15.8. As previously mentioned, small amounts of organic, especially paraffin alcohols and amides, can result in a highly visible plume that can persist after the condensed water vapour plume has dissipated. The precipitator/scrubber appears to provide the best control for these compounds, although no conclusive data exist.

Differences in formulation among manufacturers and differences in drying systems make comparison impossible. Fabric filters operating at higher temperatures will not collect organic aerosols, some of which will be caught on the filter.

Table 15.8. (Metric and English units) particulate emission factors for detergent spray drying[a] emission factor rating: E[b].

Control device	Efficiency (%)	Particulate	
		kg/mg of product	lb/ton of product
Uncontrolled	NA	–	–
(SCC 3–01–099–01)	–	45	90
Cyclone	85	7	14
Cyclone with			
Spray chamber	92	3.5	7
Packed scrubber	95	2.5	5
Venturi scrubber	97	1.5	3
Wet scrubber	99	0.544	1.09
Wet scrubber/ESP	99.9	0.023	0.046
Packed bed/ESP	99	0.47	0.94
Fabric filter	99	0.54	1.1

[a]Some type of primary collector, such as a cyclone, is considered integral to a spray-drying system. ESP, electrostatic precipitator; SCC, Source Classification Code.
[b]Emission factors are estimations and are not supported by current test data.

Opacity and the organics emitted are influenced by granule moisture and temperature at the end of drying, temperature profiles on the dryer, and formulation of the slurry. An alternative method for controlling visible emissions from the dryer caused by volatile organics in the slurry is to reformulate the slurry to eliminate these offending organic compounds. Sometimes the purity of raw materials such as alcohol ethoxylates can be improved by stripping short-chain molecules from the raw material. When amide compounds were identified as causing the emission problems, some manufacturers developed other formulations or methods for adding the amides to the spray-dried granules after the dryer to achieve a comparable product.

When reformulation is not possible, the tower production rate may be reduced, permitting operation at lower air inlet temperatures and lower exhaust gas temperatures. When tower temperatures are reduced, lower temperatures may result in lower granule temperature and higher granule moisture, thus reducing organic emissions.

DRUG AND PHARMACEUTICAL INDUSTRY

The pharmaceutical industry, although a strong and important entity in itself, is frequently considered to be part of the chemical industry. The pollution problems of the pharmaceutical industry, particularly those of the major companies, parallel those of the chemical industry. Because the problems of the industry are not unique, this section will be concerned with emphasising the problems, where they should be looked for, and some of the more interesting solutions to solve these. The pharmaceutical companies facing major problems in connection with both air and water pollution are the ones that operate fermentation plants, principally in the production of antibiotics (penicillin, streptomycin, etc.).

Both gaseous organic and inorganic compounds, as well as particulates, may be emitted during pharmaceutical manufacturing, separation, purification and formulation. Some of the volatile organic compounds (VOCs) and inorganic gases that are emitted may be hazardrous. The various sources of emission in the order of their priority are distillation units, storage and transfer of materials, filtration, extraction, centrifugation, and crystallisation.

During reaction: VOC emissions from reactor vents, material loading and unloading, acid gases (halogen acid, sulphur dioxide, nitrous oxides); fugitive emissions from pumps, sample collections, valves, tanks, etc.

During separation: VOC emissions from filtering systems which are not contained and fugitive emissions from valves, tanks and centrifuges.

During purification: Solvent vapours from purification tanks, fugitive emissions.

During formulation: Tablet dusts, other particulates.

Air Emissions Characterisation

This section describes the characteristics of emissions associated with bulk pharmaceutical chemicals (BPC) manufacturing processes. Subsections presented include: emissions sources, types of emissions, emission estimates, and estimation techniques. Guidance is provided on estimating emission rates.

Emission sources

A summary of typical material inputs and emissions for BPC manufacturing processes is presented in Table 15.9. Sources of emissions at BPC manufacturing plants vary widely from facility to facility, depending on the products manufactured. Sources of emissions associated with the major BPC manufacturing unit process operations are described below.

Reactors

Emissions from reactors may occur by several mechanisms, including: (i) displacement of air during the charging of reactants, (ii) inert gas flows during the reaction cycle, (iii) opening of the reactor during operation to collect samples or charge reactants, (iv) venting of emissions during refluxing, (v) solvent vapourisation during cleanup operations, and (vi) the relieving of pressure between cycles (where a), reactor is operated under pressure). Emissions are maximised when reactions involve highly volatile chemicals and elevated temperatures. In contrast, emissions are minimised with less volatile chemicals, low temperatures, and/or sometimes subatmospheric pressures and when water-based reactions are involved.

Table 15.9. Summary of typical material inputs and outputs in BPC processes.

Process	Inputs (examples of some commonly used chemicals provides)	Air emissions
Chemical synthesis reaction	Solvent, catalysts, reactants, e.g. benzene, chloroform, methylene chloride, toluene, methanol, ethylene glycol, methyl isobutyl ketone (MiBK), xylenes, hydrochloric acid, etc.	VOC emissions from reactor vents, manways, material loading and unloading, acid gases (halogen acids, sulphur dioxide, nitrous oxides); fugitive emissions, from pumps, sample collections, valves, tanks
Separation	Separation and extraction solvents, e.g. methanol, toluene, hexanes, etc.	VOC emissions from filtering systems that are not contained; fugitive emissions from valves, tanks and centrifuges
Purification	Purification solvent, e.g. methanol, toluene, hexanes, etc.	Solvent vapours from purification tanks; fugitive emissions
Drying	Finished active drug(s) or intermediates	VOC emissions from manual loading and unloading of dryers
Natural product extraction	Plants, roots, animal tissues, extraction solvents, e.g. ammonia, chloroform, phenol, toluene, etc.	Solvent vapours and VOCs from extraction chemicals
Fermentation	Inoculum, sugars, starches, nutrients, phosphates, fermentation solvents, e.g. ethanol, amyl alcohol, methanol, MiBK, acetone, etc.	Odouriferous gases, extraction solvent vapours, particulates

Distillation units

Emissions occur from the condensers used to recover the evaporated liquids and from the vents as a result of vapour displacement during startup.

Extractors

Emissions from extractors may occur from the: (i) displacement of air while charging, emptying, and cleaning the extractor, (ii) agitation of liquids during the extraction process, and (iii) filling and emptying of associated surge tanks.

Fermentors

Emissions occur during the fermentation process as a result of displacement of large volumes of solvent from one vessel to another, and recovery of product from the concentrated solvent by crystallisation, filtration, and drying of the solid product.

Centrifuges

A significant source of emissions is the vapourisation of organic compounds from the 'wet' solids as they are removed from the centrifuge and transported to the next operation. In addition, emissions can occur as the centrifuge is opened, as the filtrate is discharged into a holding tank as inert gases are vented. Some older facilities may have open-top centrifuges, where direct vapourisation to the ambient air can occur.

Filters

Emissions typically occur when a filter is opened to remove the collected solids or when the filter is purged with steam or inert gas prior to cleaning.

Crystallisers

Emissions from crystallisation operations are usually not significant, unless supersaturation is achieved through solvent evaporation. These emissions may be vented directly to the atmosphere or through a capture device (e.g. a condenser).

Dryers

In the pharmaceutical industry, drying is the unit operation with generally the highest VOC emissions. With a convective dryer, the exhaust gas is a high-flow, relatively dilute stream, which traditionally has often been vented directly to the atmosphere without emission control. With a conductive dryer, the vent gas is much more concentrated; so the flow rate is low. In the case of vacuum drying, VOC emissions occur downstream from the vacuum pump, where the air in-leakage is exhausted from the system.

Storage and transfer

The vapour space in a tank will in time become saturated with the stored organics. During tank filling, vapours are displaced causing an emission or a 'working loss'. Some vapours also are displaced as the temperature of the stored VOC rises, such as from solar radiation, or as atmospheric pressure drops; these are 'breathing losses'. The amount of loss depends on several factors: type of VOC stored, size of tank, type of tank, diurnal temperature changes, and tank throughput. In-plant transfer of VOC is done mainly by pipeline, but also may be done manually (e.g. loading or unloading 55-gal drums). Raw materials are delivered to the plant by tank truck, rail car, or in 55-gal drums. These storage devices are susceptible to breathing and working losses, the magnitude of which depends on the type of compound stored, the size and design of the tank, ambient temperature and diurnal temperature changes, and tank throughput. Emissions may also be associated with manual material transfer operations within the facility and with the transfer of liquids from tanker cars, trucks, and rail tank cars.

Types of emissions

Pharmaceutical manufacturing facilities are similar to many chemical-manufacturing operations in that both gaseous and particulate emissions can be significant.

VOCs

Usually of greatest concern at pharmaceutical facilities are emissions of VOCs. Many of these compounds are photochemically reactive and contribute to tropospheric (lower-level) ozone formation (a concern for US facilities located in ozone nonattainment areas), while others may deplete the stratospheric (upper-level) ozone layer. An added concern is that some compounds are air toxics or HAPs, categorised by the EPA as possible or probable human carcinogens, thereby making long-term human exposure undesirable. Finally, some compounds used in the industry may be acute toxicants, where short-term exposure, including emergency releases, can lead to adverse health effects.

Particulates

Emissions of particulates also may be of concern. One reason is that particulate emissions, especially those associated with formulation and packaging operations, may include biologically active ingredients. Additionally, some of these emissions may be in the respirable size range. Within this range, a significant fraction of the particulates may be inhaled directly into the lungs, thereby enhancing the likelihood of being absorbed into the body and damaging lung tissues. Fortunately, because of concern about worker health and product contamination, particulate emissions are often closely controlled, especially where toxic compounds are in use. Consequently, particulate emissions at most pharmaceutical manufacturing facilities are insignificant.

Odours

Additionally, odours can be a problem, especially where fermentation takes place. While odours can create a community nuisance, they generally do not represent a serious air pollution problem.

Other

Another class of gaseous emissions is acid gases; these are principally sulphur dioxide, nitrous oxides, and hydrochloric acid. The oxides of sulphur and nitrogen are typically associated with the large-scale combustion of fossil fuels are utilised for power generation or steam production by large chemical facilities. Because of the generally smaller-scale operations of the pharmaceutical industry, acid gases are usually not an issue. The exception may be halogen acids (e.g. hydrochloric), which are formed as by-products of chemical synthesis or as products of the combustion of halogenated compounds.

Air Pollution Control Measures

Information on air pollution control measures that apply to pharmaceutical processes has been divided into the following three subsections: emission control considerations, process design considerations, and operational practices. Emission control systems are commonly used to reduce emissions of VOCs (including odours) and particulates. Because some VOCs and particulates may contain toxic components (HAPs), these control systems are also effective in reducing emissions of air toxics. Indeed, air-toxics control practices have been followed in the industry for many years as a result of the biologically active and high-value materials handled by some operations. 'Air emissions characterisation' VOC emissions are of primary concern at pharmaceutical facilities and are thus our focus here.

The primary focus of the remainder of this discussion is the control of VOC emissions. A variety of control systems are available for use in the pharmaceutical industry, depending on the particular application and on process operating parameters. A listing of the types of controls that are in use or available for use at pharmaceutical facilities is presented below.

Control category	Available control systems
Add-on controls	Condensers
	Absorbers
	Adsorption units
	Thermal destruction
	Vapour containment
Process design	VOC minimisation
	Solvent substitution
Operational practices	Cleanup operations

Information is presented in the following on the use of these pollution control concepts: add-on controls, process design, and operational practices.

Emission Control Considerations

Because of the batch nature of many pharmaceutical process unit operations, it is common for critical emission parameters to vary throughout a process operating cycle. These parameters include temperature, pressure, flow rate, VOC composition and concentration, and particulate loading.

To control these variable emissions streams effectively, emission controls must be designed to handle both peak and nonpeak flow conditions. Additionally, the controls must be designed to be started up and stopped numerous times over a year and to be effective over short operational cycles. Consequently, it is very important that control equipment be sized correctly and designed to handle a range of operating parameters and conditions. Add-on controls available for use at pharmaceutical manufacturing facilities include condensation, absorption, carbon adsorption, thermal destruction, and vapour containment.

Condensation

The principle of operation of a condenser in emission control service is that a gas stream containing a (VOC can be cooled to below the saturation temperature ('dew point'), forming a second phase consisting of the VOC liquid. Separating the liquid from the gas (e.g. by gravity in the condenser combined with a demister pad to trap entrainment) removes a fraction of the VOC from the emission stream. The amount of VOC that remains in the gas stream as a vapour is a function of the temperature and the vapour–liquid equilibrium, of the VOC species. In general, the lower the temperature, the lower will be the VOC content in the exit gas stream. In theory, any desired level of VOC removal (control efficiency) approaching 100 per cent is possible with a condenser operating at a low enough temperature. There are some very real practical limitations, however. Low condensation temperatures are required to control high-volatility VOCs or low concentrations of VOCs with any significant degree of volatility. In the latter case, most of the cooling duty of the condenser is to cool the non-condensible gas to the saturation temperature. Because the heat-transfer coefficient is low under these circumstances, the required heat-transfer area is large, increasing the cost of the condenser.

Producing low temperatures requires energy to drive the refrigeration system. Lower temperatures have progressively higher energy consumption (and thus cost) per unit quantity of heat removed. One scheme to raise the required condensation temperature is to compress the vent stream to a higher pressure so that the dew-point temperature will be raised. This approach is particularly useful where high control efficiencies of low-boiling-point materials (e.g. methylene chloride) are required. It is sometimes possible to raise the dew-point temperature sufficiently so that the problem with water vapour to be described can be avoided.

The presence of water vapour in the emission stream along with the VOC is common in actual practice. If room air sweeps or air purges are used, as is often the case in pharmaceutical manufacture, normal humidity will be present. If the process contains water as well as organic solvents, then even isolated vent streams will contain water vapour. Even conditioned 'plant air' has traces of water, since the design operation of typical plant air dryers is to lower the water content to a dew point of only about –40°C (–40°F). If a gas stream containing water passes through a condenser that is operating below the freezing point, frost (rime) will form on the tubes, thereby effectively decreasing the capacity of the condenser by blocking efficient heat transfer.

Low-temperature condensation of volatiles from a gas, stream containing water can be accomplished with a series of two or more condensers with decreasing temperatures. The first condenser(s) removes the water vapour, while successive, lower-temperature units remove the organic compounds. A concern with this scheme is that the condensate containing the water may also have a substantial organic content (by condensation) and thus may pose a water pollution control problem. Because a vent condenser can be a modest cost-control device, it is often recommended. Based on the foregoing discussion, there are clearly some situations that are well suited to condensation. For example, when there is little or no water vapour in the stream and a reasonable concentration of a medium-or high-boiling-point VOC

(10 per cent or more), the use of a condenser is particularly appropriate. When the VOC is a low-boiling-point material, when there is water vapour present, or when the VOC concentration is low (less than 1 per cent) the applicability of condensers is marginal, and multiple units operated at low temperatures may be required. When the VOC concentration is quite low (at the fractional percent level), condensers are probably not appropriate, as the achievable level of control will be unacceptably low.

Absorption

Absorption is the selective transfer of one or more components of a gas mixture (solute) into a solvent liquid. For any given solvent, solute, and set of operating conditions (pressure and temperature), there is an equilibrium ratio of solute concentration in the gas mixture-to-solute concentration in the solvent. If the solute concentration is higher than the equilibrium value, there is a 'driving force' for mass transfer of that solute into the solvent, leading to its removal from the gas stream. If the equilibrium concentration in the gas is low compared with the concentration in the liquid solvent, the solute is said to have high solubility. Conversely, if a high concentration of the solute in the gas phase is in equilibrium with the liquid, the solute is said to have low solubility. The apparent solubility of a solute species may be enhanced if there is a chemical reaction between it and components of the solvent solution. This reaction has the practical effect of minimising the concentration of the actual solute species in the liquid by changing the solute species into another, more soluble species.

In emission control service for VOC removal, the solvents are chosen for their high organic solubility. These solvents often include water, high-boiling-point mineral oils, and sometimes-aqueous solutions of sodium carbonate and sodium hydroxide.

Emission control devices that are based on the principle of absorption are commonly called scrubbers and include spray towers, venturi scrubbers, packed columns, and trayed (plate) columns. The basic design concept of all these devices is to provide the most economical means of achieving a high degree of gas–liquid contact in order to promote the desired mass transfer. The degree of control achieved (i.e. the fraction of the solute that is transferred from the gas to the liquid) is a function of residence time, mass transfer area, and physical and thermodynamic properties of the VOC species involved.

Spray towers are inherently simple devices consisting of an open column up which the gas stream flows and a group of atomisation nozzles to distribute the solvent as a spray of fine drops. Spray towers are versatile in that particulate matter also can be removed, but they have low mass transfer coefficients and typically are used only with high-solubility gases.

Venturi scrubbers, which are slightly more complex mechanically than spray towers, have a high degree of gas-liquid mixing and high particulate removal efficiency, but have relatively short residence times that limit the mass transfer efficiency. They are typically used only with high-solubility gases. For controlling organic compound emissions, packed and trayed columns are generally preferred. Packed columns are usually more economical than trayed columns for small diameter (less than 0.6 metre) units and can be readily designed for low-gas-phase pressure drop and for handling corrosive materials and liquids that tend to foam or plug. Trayed columns are preferred for larger-diameter units, where internal cooling is required or where low liquid flow rates would be insufficient to wet the packing.

Scrubbers have been applied particularly to control water-soluble inorganic gases (e.g. sulphur dioxide, hydrogen chloride, hydrogen sulphide, and ammonia) in airstreams using water as the solvent. In the case of acid gases, using an alkaline scrubbing medium increases the effectiveness even more.

In the pharmaceutical industry, scrubbers are also used to remove VOCs. Solvent selection, equipment design parameters, and operating conditions all determine the control efficiency. To absorb organic compounds that have relatively high water solubility (e.g. most alcohols, organic acids, aldehydes, ketones, amines, and glycols), water is the preferred solvent. For organic compounds with low water solubility, another organic liquid (usually one with low vapour pressure) may be chosen. However, in the case of low solubility, water may still be selected, and the equipment will be designed to handle the high flow rate required.

Increasing the depth of the packing or the number of trays increases the control efficiency by increasing the amount of mass transfer area. Decreasing the operating temperature may result in a more favourable gas-liquid equilibrium and achieve higher control efficiency with a given piece of equipment.

Absorption can be used effectively with a wide range of concentrations, ranging from several per cent to as low as 200 or 300 ppm. Control efficiencies typically range from 60 to 95 per cent or better. However, because of the hydraulic behaviour of packing and trays, a relatively steady gas flow is required to maintain good efficiency. Turndown ratios of approximately 50 per cent are standard, thus limiting the ability of absorption to handle the intermittent flow of vents from batch processes.

Adsorption

The principle of operation for adsorption and the application of adsorption to VOC removal from gas streams are discussed below.

Principle of operation

Gas adsorption is the selective transfer of one or more components (solutes) of a gas mixture onto the surfaces of the pores of a microcrystalline solid sorbent. The selectivity is most pronounced in a monomolecular layer next to the solid surface, although selectivity may persist to a height of three or four molecules. Both natural and synthetic materials are suitable for use as sorbents if they have a microcrystalline structure. The large number of pores in these materials results in extremely high surface areas (e.g. more than a square kilometre per kilogram, or 0.2 square mile per pound, of solid in some instances). Adsorbents in large-scale commercial use include carbons, silicates, aluminas, and aluminosilicates (molecular sieves).

Generally, adsorption is reversible: The effective adsorption capacity (performance) of a sorbent for a particular solute increases with the concentration of the solute in the gas stream and decreases with increasing temperature. This means that the sorbent may be repeatedly 'regenerated'; that is to say, the solute can be desorbed by passing a lower-concentration or higher-temperature gas stream over the bed. Reversible adsorption is desired for VOC removal because it allows for reuse of the adsorbent. However, in some situations, the adsorption of a solute from a gas is irreversible. The capacity of a solid sorbent is nearly constant and independent of the gas stream composition, but because the solute is so tightly bonded to the sorbent, it cannot be reused.

Activated carbon is the preferred absorbent material for removing VOCs from gas streams. It has a high affinity for nonpolar compounds, with an even greater capacity for high-molecular-weight materials. The high-molecular-weight compounds can pose some problems, as they tend to be irreversibly adsorbed and effectively decrease the capacity of the carbon in a reversible system. A further problem is that high-molecular-weight compounds can displace lower molecular-weight materials already adsorbed.

The performance of a particular type of activated carbon for a particular solute species must be determined from equilibrium behaviour. Where only one species of solute is present in the gas stream, the equilibrium behaviour is conveniently represented by a simple curve that plots solute concentration in the solid phase as a function of solute concentration in the gas phase. These curves are usually only valid at a single temperature and are known as isotherms. Isotherms for common organic vapours being adsorbed on activated carbon are available in the open literature and from carbon suppliers. They may represent strictly experimental data or, more commonly, experimental data fitted to established algebraic formulas (e.g. Langmuir or freundlich isotherms). In some rare cases, they may represent purely theoretical calculations of equilibrium based on the molecular statistics of the solute–surface interactions.

Application to VOC removal from gas streams

In designing a carbon adsorption pollution control device in which the inlet VOC concentration is known and a certain outlet concentration is desired, it is possible to determine from the isotherm and material balance calculations how much carbon will be required. The process design is finalised by applying experience factors in terms of 'ageing' of the carbon beds (i.e. loss of capacity with time as a result of pore plugging or some other problem) and safety factors. The desired outlet concentration of the solute in the gas phase is used to determine the concentration of the solute on the activated carbon. When the entire mass of carbon reaches the 'saturation' level, the solute no longer will be adsorbed from the gas stream to the desired level, and 'breakthrough' is said to have occurred. The carbon then must be either regenerated (in the case of reversible adsorption) or disposed of (in the case of irreversible adsorption).

To allow vent streams or other gas streams containing VOCs to come into contact with the activated carbon, the carbon granules are usually arranged in a fixed-bed arrangement in either a vertical or horizontal cylindrical vessel. Small units are manufactured with the carbon already in place (e.g. a canister) with the intent that the entire unit will be replaced when the capacity limit of the carbon is reached. The supplier may regenerate the unit, or it may be disposed of. Larger units are constructed so that the carbon granules are loaded after installation of the vessel. In these units, the carbon may be periodically regenerated in place, or it can be removed and shipped off to a central location for regeneration on a contract basis.

Fixed-bed units are operated 'batchwise'; that is, the carbon is gradually saturated with the VOC and must be periodically regenerated, usually in place. This regeneration in place is done by having two or more beds installed and manifolded together so that one bed is in use while the other bed is undergoing regeneration. For very large adsorption operations, 'continuous' processes have been designed in which the carbon bed is continuously moved (by a fluidised bed, conveyor belt, or other mechanical means) from an adsorption zone through a regeneration zone and back to the adsorption zone. Continuous adsorption generally does not apply to the smaller scale operations at pharmaceutical plants.

Activated carbon beds in VOC removal service usually are regenerated using steam. The steam serves as a gas stream with zero concentration of the VOC solute (so that the solute tends to be desorbed) and acts to raise the temperature of the bed so that the equilibrium for desorption will be favoured. A much smaller volume of steam is required for regeneration compared with the volume of gas that is treated during the adsorption portion of the cycle. Following desorption, the steam is condensed (further reducing its volume) and sent to a waste-water treatment system to remove the organic content. In some cases, depending on the species and loading on the carbon bed, the VOC concentration in the water is sufficient such that two liquid layers form with the steam condensate. The second layer, with the higher organic compound concentration, can be readily decanted and either recycled or disposed of in a more efficient manner than waste-water.

In the pharmaceutical industry, because of the final product purity requirements, there are limited opportunities for in-plant recycling of recovered solvents. However, where practiced, recovery of a separate organic stream is desirable, since there may be opportunities to sell recovered solvents to another industry or to incinerate them for in plant heat recovery.

Activated carbon adsorption is currently in use in pharmaceutical plants to remove VOCs from vent streams, back up condensers, and control odours. Removal efficiencies of 95 per cent are common with values as high as 98–99 per cent seen in selected cases. In odour-control applications, the malodourous organic compounds are often fairly large molecules, and thus carbon adsorption may be substantially irreversible. In these cases, the used disposable carbon devices (such as canisters) may be favoured. A related issue is the problem of displacement. If the carbon bed is used to adsorb the larger molecules, it is unlikely that it also will be very effective at simultaneously removing the lighter-molecular-weight compounds. Thus, the removal efficiency of the lighter-weight compounds will be poor.

It should be noted that potential fire hazards are associated with using carbon adsorption on air vents or vent streams containing oxygen and flammable hydrocarbons. Whereas the gas streams may be sufficiently dilute in flammables as to be well below the lower flammability limit, the carbon bed concentrates the flammable materials, so that a potentially flammable mixture may be formed inside the bed. Furthermore, the carbon itself, with a high surface area, is combustible. For these reasons, special precautions for early fire detection or bed cooling may be required in some instances or an alternative to carbon adsorption may need to be used.

Thermal destruction

There are four main types of thermal destruction systems in general use: flares, boilers and process heaters, thermal incinerators, and catalytic (oxidisers) incinerators.

Flares

Flares burn flammable or combustible vapours with an open flame and can be either tower or ground mounted. Nozzle design, steam injection, and air injection are some of the design aspects used to enhance destruction efficiency. Flares are well suited for use with gas streams having a significant fuel value (more than 7.45 MJ/nm^3 or 200 Btu/scf) and are particularly well suited for handling a large, intermittent flow such as might occur from an emergency vent header. They have only limited capability for handling chlorine-containing compounds owing to corrosion of the burner elements and the lack of postcombustion emission control devices (for acid gas removal). The use of flares in the pharmaceutical industry for emission control is very limited; more typical applications are in petroleum refining or large chemical complexes.

Boilers and process heaters

Boilers and process heaters represent an opportunity to burn a waste stream while simultaneously recovering the stream's heating value. Typical applications are combusting a high-fuel-value liquid waste stream (uncommon in pharmaceutical operations) or using a pollutant-laden airstream as a source of combustion air. A fairly high-fuel-value vent gas stream is required for the boiler or process heater to function without added fuel. As with flares, boilers and process heaters operating with waste streams are more typically found in petroleum refining or large chemical complexes.

Thermal incinerators

Thermal incinerators are controlled combustion devices where fuel and air are added to a combustion chamber to maintain a high minimum operating temperature. Because of these high temperatures, thermal incinerators are able to handle dilute vent streams with high destruction efficiency. Gases with heating values below about 1.86 MJ/nm^3 (50 Btu/scf) can be handled with the addition of fuel. Combustion chamber temperatures range from 700° to 1300°C (1300°–2400°F). Packaged, single-unit thermal incinerators are available in a wide range of sizes and are able to handle flow rates ranging from 0.1 nm^3/sec (200 scfm) to about 24 nm^3/sec (50000 scfm).

The control efficiency achieved by thermal incineration is typically 98 per cent destruction or an exit gas concentration of 20 ppm by volume, whichever is less stringent. Thus, an inlet stream with a VOC concentration of 2000 ppm by volume or higher (which, with the typical 1:1 air dilution, becomes 1000 ppm by volume at the inlet to the combustion chamber) will have 98 per cent of the incinerator inlet VOC content removed. Inlet streams with lower VOC concentrations will result in outlet gas concentrations of 20 ppm by volume of unburned organics, but with lower destruction efficiencies. Thermal incinerators can handle chlorine and sulphur containing compounds but may require a scrubber on the outlet to control acid gases.

Due to the wide range of sizes available and the ability to handle a wide range of compounds in vent gas streams, thermal incineration is well suited for emission control in the pharmaceutical industry. It is, however, a capital- and energy-intensive approach.

Catalytic incinerators

Catalytic incineration is a variation of thermal incineration. A catalyst is used to promote oxidation of the inlet gas stream at lower temperatures than are required in standard thermal combustion. Catalytic units usually operate over a range of 320°–650°C (600°–1200°F), and these lower operating temperatures reduce energy requirements. Catalytic incineration units may be smaller than standard thermal incineration units, but the cost of the catalyst may tend to offset any potential savings in the capital investment. Two additional design constraints are that: (i) high organic concentrations may cause catalyst failure, and (ii) the catalyst may be poisoned by sulphur-containing compounds, heavy metals, and halogens.

The control efficiency of catalytic incineration is equivalent to that of thermal incineration, and a wide range of packaged unit sizes is available.

As with thermal incineration, catalytic incineration is well suited for emission control in the pharmaceutical industry.

Vapour containment

Transferring volatile organic liquids from delivery tank trucks and tank cars to storage tanks and from storage tanks to process vessels displaces gas from the headspaces that contain some fraction of VOC vapour. While these intermittent vent streams may be treated by one of the add-on control devices described above vapour containment may in some cases be the preferred option. In this technology, additional piping is provided so that as the liquid is transferred from the supply tank to the receiving tank, the displaced gas and vapour from the receiving tank are returned in a separate line to the supply tank. With a properly designed system, very little gas/vapour escapes into the atmosphere, and VOC emissions are minimised. This concept can be extended to filling reactors from drums by using a drum pump and vapour return line rather than picking up the drum and pouring.

Other unit operations that can be improved in terms of vapour containment are those designed to operate at atmospheric pressure. Because of the relatively unrestricted mixing of the process vapours with the atmosphere, reactor and condenser vents open to the atmosphere are difficult-to-control sources of emissions. An option is to operate them as closed systems, at a slight pressure or vacuum, so that the flow of non-condensible gas (air or inert gas) is restricted. This approach is likely to decrease emissions, and the restricted flow rate stream that occurs is easier to control.

Bulk quantities of volatile organic liquids typically are stored in atmospheric pressure tanks with vents to relieve excessive pressure and to break any vacuum that may form. However, diurnal temperature fluctuations will result in 'breathing losses' where the headspace vapour, approximately saturated with volatile organic vapours, is expelled when the temperature rises. When the temperature drops, fresh air (or an inert blanketing gas) is drawn into the tank, from which it may later be expelled with some degree of saturation. Installing 'conservation vents' raises, by a small amount, the pressure or vacuum threshold at which this event occurs, thereby reducing the emission stream volume. Isolated, especially high-pressure, rises or unusually low vacuums will still open the conservation vents, but normal daily fluctuation will not, so that the total volume vented to the atmosphere is substantially reduced.

Process design considerations

To use add-on controls, including vapour containment, a certain amount of process design must be undertaken. However, in the case of new products or revamps of existing processes, it may be possible to use a different pollution control concept—designing a process that has an inherently lower potential for air emissions. One strategy is to minimise the use of VOCs, to substitute lower-volatility (and/or less toxic) compounds, and to keep these compounds contained.

For example, in the case of drying emissions in pharmaceutical manufacture, a modified process design could use and transfer intermediates as solutions or slurries, rather than as dry products. Where solvents are required, a lower-vapour pressure solvent would tend to minimise evaporative losses and increase recovery efficiency when using condensers. With some processes, a reaction or formulation step traditionally conducted in a solvent possibly could be carried out in an aqueous medium. One ultimate goal may be the practice of a 'solventless' synthesis, where the reaction is conducted 'neat'. In this case, a customised reactor design may be required to handle potential heat transfer and rheological problems associated with the resulting more concentrated reaction mixture.

Although there are often physical or chemical reasons why a process design must incorporate a particular volatile compound, there are many other instances where, with ingenuity, VOC emissions can be minimised by eliminating the offending material altogether.

Another strategy in a modified process design is to minimise the transfer of materials from process vessels by having a particular unit perform more than one unit operation. An example is a filter-dryer combination. These units, increasingly being used in the pharmaceutical industry, are batch pressure filters that, instead of discharging a wet filter cake, remain closed and convectively dry the cake with a recirculating gas stream.

For maximum effectiveness with some add-on controls, process redesign may be required. An example is the use of closed-loop drying systems.

Acknowledging that a convective dryer is potentially a large source of emissions, a condenser, or a condenser combined with gas compression, is often a good choice. Instead of venting the drying gas, a further refinement of this approach is to recirculate the gas from the discharge of the condenser through the heater and back to the dryer. During the drying cycle there are, other than fugitives, essentially no emissions from the closed loop.

Operational practices

This last control concept deals with the potential for changing workplace practices in such a way as to minimise emissions. Much of the everyday operation of a pharmaceutical synthesis facility is governed by standard operating procedures (SOPs). With proper engineering and management review, the SOPs can sometimes be revised and modified to reduce emissions. One area where emission reductions are particularly possible is in the use of solvents in cleanup practices. Typical procedures may call for the filling and subsequent emptying and air drying of process vessels. The air-drying step results in significant emissions regardless of the vapour pressure of the solvent involved. As an improved SOP, it may be feasible to eliminate the drying step, leaving residual solvent present before the start of the next cycle. Alternatively, a low-VOC aqueous cleaner might be substituted. A similar emission source involves flushing and blowing transfer lines (piping). Again, it may be possible to change the SOPs so that the lines are left filled with solvent rather than blown dry to the atmosphere.

LEATHER AND TANNERY

Tanning is the chemical process that converts animal hides and skins into leather. The term hide is used for the skin of large animals (e.g. cows or horses), while skin is used for that of small animals (e.g. sheep). The hide is composed of three layers: epidermis, dermis and subcutaneous. The dermis consists of about 30 to 35 per cent protein, which is mostly collagen, with the remainder being water and fat. The dermis is used to make leather after the other layers have been removed using chemical and mechanical means. In the process of tanning, chemical reactions convert the semi-soluble protein 'collagen' present in the corium of animal skins and hides into tough, flexible and highly durable leather.

Gaseous Pollution

Tanneries discharge into the atmosphere odourous gases, smoke and dust. The main source of smells in a tannery are the compounds containing nitrogen and sulphur. The end products of anaerobic decomposition or putrefaction of proteins include indole, skatole, mercaptans and miscellaneous aldehydes, all of which are odourous.

The other odour producing compounds in tanneries include sulphide, fatty acids like butyric acid, valeric acid and caproic acids, solvents, lacquers, formalin and some of the chemicals used in finishing operations. Smells in tanneries intensify from unhygienic practices in skin and hide processing and delayed disposal of liquid and solid wastes. In many tanneries it is the foul odour which emanate from the putrescible solid and liquid wastes which account for much of the smell traditionally associated with the tanneries. Immediately after the hide or skin is removed from the animals, decay starts unless it is cured properly. The operations like soaking, liming, deliming, fleshing, etc. are the most disagreeable steps in leather manufacture. They involve use of bad smelling materials and production of putrescible organic matter like soak pit sludge, lime sludge, green fleshings, limed fleshings. During soaking, the removal of curing salt and dehydration of the skin introduces the possibility of bacterial growth and protein putrefaction. Many unhairing systems in practical use are based on balance between sodium sulphide, sulphohydrate, dimethylamine sulphate and sodium hydrosulphide. Some of these sulphides have potential to liberate hydrogen sulphide when mixed with pickling liquors. Hydrogen sulphide is an extremely bad smelling gas which can be detected in concentrations as low as 0.1 ppm. Hydrogen sulphide and organic sulphides cause odour nuisance when present in the air at concentrations of 10 to 150 times smaller than the lowest concentration of sulphur dioxide detectable by smell. There is a possibility of hydrogen sulphide generation during deliming from the sulphides sticking to hide if the limed hides are not properly washed.

Impact of gaseous pollutants on environment

The problems associated with gaseous pollutants from a tannery include noxious smell, hydrogen sulphide and dust. The effect of noxious smells on people is primarily a nuisance effect. The loss of property values near poorly operated tanneries is partly a consequence of offensive odours. Dust problems normally arise in tanneries from leather buffing operations. Leather dust of finer sizes are reported to be harmful to human health and comfort. Hydrogen sulphide which is liberated during some of the tanning operations is an irritant gas and has a very bad smell. The maximum allowable concentration for 8 hours exposure in working areas is 20 ppm.

Control of gaseous pollutants

Source control is the most effective means of abating odour. Good sanitation practices are usually cheaper than control measures. Source control of odours can be carried out by following methods given below:
1. Drawing the odourous air from the working atmosphere by exhaust fans and diluting with relatively clean air.
2. Removal of causative impurities from the tannery.
3. Masking the odour with objectionable additives.
4. Removal of odour bearing dusts by cyclone separators.
5. Sorption of odourous gases through a granular sorbents like activated carbon.

Chemical air scrubber can be used to eliminate the odour nuisance in the tannery waste-water treatment plant. Odour masking chemicals can be added. Masking agents can also be used to control odours outdoors in such places as waste lagoons.

The smell caused by putrefaction of solid and liquid wastes generated during tannery operations can be reduced by quickly disposing them off without allowing for putrefaction. Aerobic biological methods of effluent treatment and dewatering of sludge produced by methods like vacuum filtration, drying on sand drying bed will considerably reduce the smell caused by them.

PULP AND PAPER

Pulp is a dry fibrous material prepared by chemically or mechanically separating fibres from wood, fibre crops or waste paper. Wood pulp is the most common material used to make paper. The timber resources used to make wood pulp are referred to as pulpwood. Wood pulp comes from softwood trees such as spruce, pine, fir, larch and hemlock, and hardwoods such as eucalyptus, aspen and birch.

Gaseous Emissions and Air Pollution

Air pollution from big paper mills

Air pollution caused by the pulp and paper industry results from the use of mechanical processes, and the combustion equipment required for heating the fibre and pulp. The use of combustion equipment is common to both mechanical and chemical processes. Therefore, the industry is responsible for emissions of suspended particulates and pollutants, specifically related to fossil fuel combustion, carbon monoxide (CO), carbon dioxide (CO_2), sulphur dioxide (SO_2) and nitrogen oxides (NO_x). Pulp mill emissions contain very high levels of NO_x, which is responsible for the formation of acid rain and photochemical smog (an ozone-depleting gas). Pulp mills also emit large quantities of CO_2, a greenhouse gas. Mills using chemical pulping technology produce various sulphur oxides (SO_x), including SO_2, and a number of foul-smelling compounds, such as hydrogen sulphide (H_2S) and thiols. Sulphur dioxide (SO_2) is a major atmospheric pollutant that contributes to the formation of acid rain, precipitation and acid smog. This gas, which is almost completely absorbed between the nose and pharynx, is irritating to the respiratory tract. Sulphur dioxide is also fairly damaging to plants, with concentrations around 0.03 ppm causing acute lesions on foliage. Air pollution from pulp mills has not been well studied. Mills usually do not monitor the range of air emissions, such as particulate matter, carbon dioxide, sulphur dioxide, hydrogen sulphide, volatile organic compounds, chlorine, chloroform, and chlorine dioxide. The general gaseous wastes are digester relief gases, digester blow tank and water hood vent gases, chemical recovery furnace emissions, smelt dissolving tank gases and power/stream boiler emissions. The emissions of pulp and paper mills, using sulphate method of pulping, contain malodourous gases like mercaptans and hydrogen sulphide. Therefore, odour problem exists in all mills. An increase in temperature of the chlorination results in an increase in pollution loadings in chlorination and extraction effluent. An increase in the end pH and temperature of chlorination and chlorine charge, increases the formation of chloroform, potential mutagens and mutagens in the chloroform liquid. Big paper mills are those which produce above 10,000 tons of paper per annum.

Air pollution from small paper mills

Production capacity of small paper mills

Small paper mills are those which produce 10,000 tons per annum of paper without chemical recovery. Air pollution, due to release of gaseous emissions into atmosphere, occurs mainly from two sources in small paper mills, viz. (i) digesters, and (ii) steam boilers. The third source could be the captive power generation facilities provided in the mills.

In agricultural residue based mills, after the raw material digestion with caustic soda and/or lime is completed, the pressure in the digester is released after the digester attains a temperature of about 90°C. During the process about 1.4 tons of steam per ton of pulp escape containing volatile organics released during digestion process. The escaping gases have characteristic odour and cause aesthetic pollution problems. These mills unlike the conventional sulphate mills (kraft mills) do not generate hydrogen sulphide and mercaptans (organic sulphides) since sodium sulphide is not used in pulping chemicals. The pollution is mostly confined to the surroundings of the mill, intermittent due to batch process adopted and can be felt at the time of digester gas release.

Coal is the commonly used fuel for generating steam required in all the small paper mills. It is reported that about 3.4 tons of coal is required per ton of paper made in agricultural residue based mills and the corresponding coal requirement in waste paper based mills is 1.5 tons per ton of paper. In all these boilers, coal lumps of 2.5 to 5.0 cm diameter are fired. Very few boilers use powdered coal. Besides coal, paddy husk, diesel oil, etc. are also used in a very few mills. Indian coals are reported to contain 0.5 to 0.8 per cent sulphur.

Since the quantity of steam generation is small, the steam boilers are not provided with any air cleaning equipment in these mills. In a study conducted by NEERI on air pollution problems from steam boilers, it is observed that a boiler plant producing 57 tons steam per hour, and provided with an electrostatic precipitator (ESP) would release the following quantities of pollutants. The ESP is reported to be working with 80 per cent efficiency.

Suspended particulate matter	620–1030 mg/N cubic metre
Sulphur dioxide, SO_2	360–390 mg/N cubic metre
Oxides of nitrogen, NO_x	90–120 mg/N cubic metre

Captive power generation is reported in 5 mills among the 93 mills from which information were obtained. Four of the mills exclusively used diesel oil while the fifth one uses coal and diesel as fuels. The main purpose for providing captive power is to meet sudden power shut-offs and these units are put into operation for specific period whenever such exigency arises. The capacity of power generating units ranged from 110 KW to 2170 KW. Air pollution is also caused by the delignification of wood by alkalis producing alkyl sulphides, like dimethyl sulphide, and mercaptans, like methyl mercaptan. These products are volatile and, therefore, escape to the atmosphere. Such compounds are not only toxic, but also possess offensive odour.

However, as for as pulping is concerned, the air pollution problem is a relative minor one, being restricted to the areas in the immediate vicinity of the mills.

Treatment of air pollutants

A high concentration of contaminants are emitted from the stacks of significant sources like boilers, furnaces, incinerators, and electricity generating units. Particulates can be major air pollutants in many industries, and improved removal can reduce the impact of industrial pollution. Electrostatic precipitators, thermal precipitators and bioscrabbers are used for the removal of air pollutants from pulp and paper mills. Wet scrubbers are applied to provide a more effective removal of fine particles, in the size range of 0.1 to 2.0 m radius from an exhaust gas stream. The research is focused on ways to improve drop size and distribution in a venturi scrubber to optimise particle removal. Postcyclones, which use the energy in an exhaust stream leaving a cyclone to further remove fine particles, may provide a cost effective means of improving particle removal. The research involves both laboratory tests and computer modelling air and particulate movement through the cyclones.

CAUSTIC SODA/SODIUM HYDROXIDE

Caustic soda/Sodium hydroxide (NaOH), is a caustic metallic base. It is used in many industries, mostly as a strong chemical base in the manufacture of pulp and paper, textiles, drinking water, soaps and detergents and as a drain cleaner.

The caustic soda industry, among the other chemical industries, has of late attracted particular attention of pollution control authorities world over. The concern is due to the release of mercury into the environment from the industrial units using the mercury cell process for the manufacture of caustic soda. Caustic soda is manufactured either by electrolytic or chemical processes. In the electrolytic process either mercury or perforated steel plates are used as cathodes to which sodium ions migrate. The former is called mercury cell process and the latter diaphragm process because the plates are wrapped with either asbestos or synthetic membrane diaphragm.

The diaphragm allows sodium ions to pass through but hinder diffusion of products. In the chemical process, sodium carbonate and calcium hydroxide are reached to produce caustic soda and calcium carbonate.

Manufacturing Process

A flow diagram depicting the unit processes in the mercury cell electrolytic method of manufacture of caustic soda is presented in Fig. 15.9. Caustic soda is produced by the electrical decomposition of the solution of sodium chloride in water known as brine, the preparation of which is the first step in the manufacturing process. A saturated solution contains 310 grams of sodium chloride per litre.

Air Emissions Characterisation

The air emissions to be controlled from diaphragm-, mercury-, and membrane-cell chlorine plants include chlorine gas, mercury vapours, asbestos, and hydrogen.

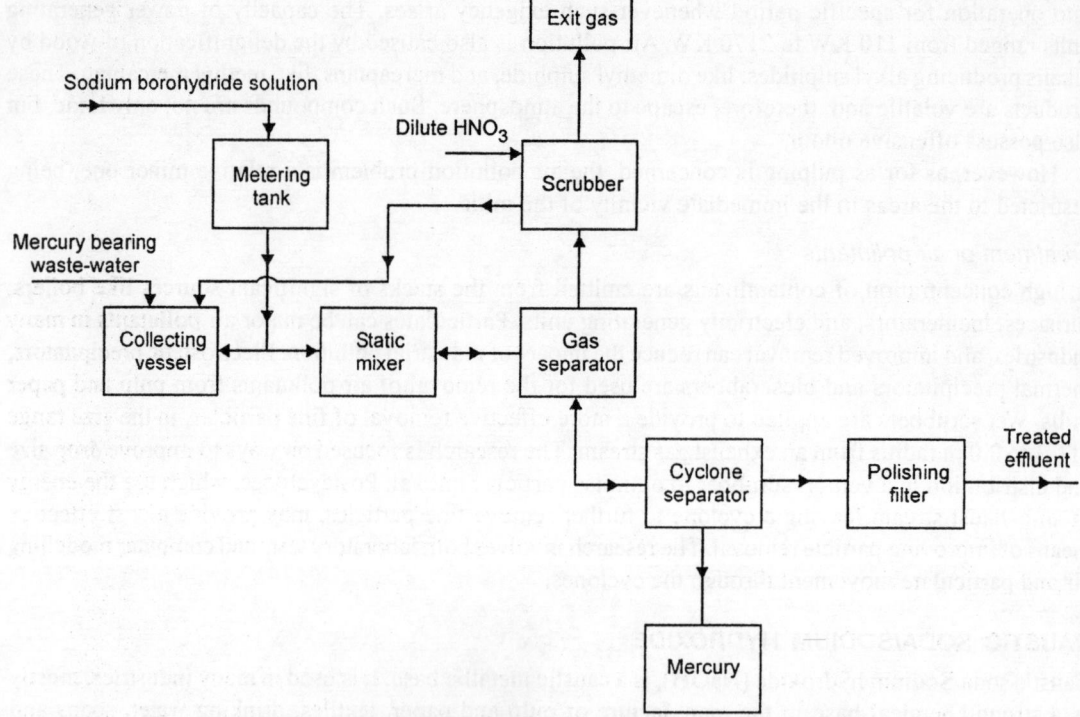

Fig. 15.9. Schematic diagram of reduction process using sodium borohydride.

Chlorine

Chlorine gas is listed as an extremely hazardous substance under, and any person handling in excess of 100 pounds of the gas must take certain emergency precautions. The potential sources of the various emissions are:

1. From the cell flanges, connections, and relief devices.
2. From the low-pressure headers during normal operations, start-ups and shutdowns, and emergency operations.
3. From chlorine dissolved in the sulphuric acid and then released to the atmosphere.
4. From chlorine compressor seals.
5. From high-pressure piping and liquefaction equipment.
6. Tail gas from recovery units and scrubbers.
7. From tanks and equipment being loaded.

Mercury

The potential sources of the various emissions are:

1. From the cells during operation and maintenance.
2. In the vent gas from the cells.
3. In the hydrogen from the cells.
4. From piping leaks.

Asbestos

The potential sources of the various emissions are:
1. Receiving and handling.
2. Diaphragm manufacture.
3. Disposal methods.

Hydrogen

This very light gas does not appear on the regulatory lists, and spill and release reporting is not required. Hydrogen is an extremely flammable gas, and extreme care and good equipment design are required in its handling to avoid fires and explosions. Among the potential sources of the various emissions are:
1. From the cells during operation.
2. From vent leak-through.

Air Pollution Control Measures

The control and elimination of fugitive and point-source emissions depend on plant design, operating philosophy, and the maintenance program.

Chlorine

Emissions from diaphragm cells and low-pressure headers are controlled or eliminated by the use of corrosion-resistant materials, the use of elastomer gaskets retained by bolts or heavy compression springs, and operation at atmospheric pressure or under a very slight vacuum. All headers and major equipment are vented through safety relief systems to caustic scrubbers during start-ups, shutdowns, or emergency conditions to eliminate releases to the atmosphere. These scrubbers must have sufficient capacity to neutralise full production capacity for the time necessary to start up with inerts or to shut down and evacuate the system. A control system capable of maintaining a consistent pressure throughout the system under varying rates and stream compositions is extremely important to successful operation. Operation of the low-pressure system above atmospheric pressure can result in the release of fugitive chlorine, and operation at deep vacuums can result in the intrusion of atmospheric air, which causes high noncondensible flows in the high-pressure system. Cells, piping headers, and major equipment are evacuated to a vacuum system and purged prior to opening them for repairs.

The design and operation of the sulphuric acid drying system are extremely important to long-term successful operation. The rate of corrosion of steel used in the high-pressure system is accelerated to unacceptable levels as the moisture level increases. The sulphuric acid used to dry the chlorine will contain dissolved chlorine, which must be recovered and recycled by vacuum stripping or neutralised with sulphides to prevent release to the atmosphere after discharge and neutralisation of the spent acid.

Most manufacturers now use multistage centrifugal compressors, although reciprocal and sulphuric acid wetted-lobe compressors have been used. Double seals with a dry air purge between the seals are used to prevent seal leakage. It is important to design all high-pressure piping and equipment with the minimum number of flanges, and no leakage is acceptable. Because of the rapid corrosion of steel when it is exposed to wet chlorine, any small leak will rapidly become a large leak. High-pressure equipment is also vented through safety relief valves to a caustic scrubber with sufficient capacity to handle full production rates for the period required to shut down and evacuate the equipment.

The high-pressure chlorine is condensed in liquefaction equipment at temperatures ranging from +75° to −50°F, pending on the operating pressure of the system and the chlorine removal desired. The tail gas from the system, which consists of the noncondensibles and residual chlorine, can be sent to chlorination processes that can use low-concentration streams or to recovery systems. Recovery systems are capable of complete recovery and recycling of the chlorine in the tail gas. The design of recovery systems must recognise the presence of low levels of hydrogen that can react violently with the chlorine in the system. The hydrogen level must be monitored closely and steps taken when required to ensure that it is kept below the explosive range. These systems consist of multiple stages of scrubbing, absorption/ reaction, and refrigeration of the stream, with the chlorine being recovered by desorption/reaction from the scrubbing medium. The capability to divert the tail gas to a caustic scrubber should be included in the process to handle upsets or outages. The liquid chlorine is transferred to scale tanks or spheres for storage prior to shipment. All liquid lines are equipped with expansion chambers between the valves to allow for the expansion of trapped liquid without pipeline rupture. The storage tank vapours are equalised to the drying tower system for recovery. Vapours from transportation equipment (i.e. tank cars, tank trucks, and barges) are recovered by a vacuum system. Pumps are equipped with double seals with a dry air pad to eliminate chlorine leakage at the seals. Valves and cocks are of a special leak-free design with bellows or multiseal construction.

Mercury

The control of mercury emissions is dependent on equipment and facility design, a good preventive maintenance program, and observation of appropriate operations procedures.

The primary cell is operated at a slight vacuum imposed by the chlorine compressor to eliminate emissions. The end boxes, which are the mercury inlet and outlet chambers, should have vapour-tight covers. In addition, water is added to the end boxes to wash, cool, and control vapour emissions from the mercury. A slight vacuum is pulled on the end boxes by a fume system further to reduce the possibility of mercury release. These vent streams are cooled, scrubbed, and/or passed through treated carbon to ensure a mercury-free discharge. Mercury pump tanks, pumps, and seal pots are treated in the same way. Cell maintenance procedures and program quality are extremely important to the prevention of the escape of mercury vapours. Leaks in pipes, flanges, and other equipment must be repaired immediately, and a preventive maintenance program should be in place to prevent their occurrence. Any time that the cell cover or top must be removed for maintenance, the cell should be cooled and vented and, when possible, the bottom kept covered with water or brine to reduce emissions. The time that the cell is open and the frequency of such openings should be held to a minimum. A potentially major source of mercury loss from the system is with the hydrogen generated in the decomposer or denuder cell. The hydrogen from the cells should be cooled at each individual cell and the pipes arranged to return condensed mercury directly to the cell, thus reducing handling of the mercury in later recovery steps. The hydrogen product is further cooled in refrigerated coolers and/or scrubbed with chemical scrubbers in series to recover the maximum quantity of mercury. Liquid mercury recovered is recycled directly to the cells, and the mercury recovered in scrubbers can be returned with the brine. The floors in the cell room area should be smooth with no cracks and all seams sealed, so that no mercury or mercury solutions can be held for later release. Trenches should be sloped to prevent mercury puddles and a layer of water kept on sumps. Where possible, waste-waters are recycled, and when this is not possible they are treated to remove mercury before they are discharged.

A good inspection program involving daily inspections and floor washdown can further limit emissions. Special vacuum cleaners equipped with carbon filters are available for the cleanup of mercury droplets discovered by the inspection program.

Operator training and compliance with operating standards are an important part of the maintenance of minimum emissions. They involve housekeeping, spill control, personal hygiene, and control of operating parameters that affect emission levels. Well-designed systems can reduce emissions to less than 100 g/d from end-box systems and to less than 10 g/d from the hydrogen system. Actual testing is required to meet the 1000 grams per day emission limit for point sources.

Asbestos

The control of asbestos (friable) emissions starts with the receipt of the asbestos, which arrives in airtight plastic bags, shrink wrapped with plastic, on pallets in boxcars or trailer trucks. The pallet loads are inspected before removal is started. If any tears or spills are noted, they are repaired with tape and the spilled material cleaned up using a vacuum cleaner. Workers wear special clothing and masks during all phases of unloading. The asbestos bags should be stored in a dedicated area with appropriate warning signs. If water is used for the cleanup of spills, care must be taken to limit pressure to ensure that fibres are not broadcast into the air.

Periodically, a pallet of asbestos bags is transferred to the cell repair area. Access to this area is limited, and appropriate warning signs are posted. When it is necessary to add asbestos to the liquid bath used to produce the asbestos diaphragms, a bag is placed inside a glove box through a hinged Plexiglas door. The box is equipped with rubber gloves hermetically sealed through openings into the box. These gloves are used for all operations inside the glove box. A vacuum is pulled on the box system through the vacuum mix tank, which ensures that no fibres escape the box system. The asbestos bags can be opened inside the glove box and the asbestos transferred to the mix tank through a transfer pipe. From this point on, the asbestos is handled in the wet (nonfriable) state, eliminating exposure. The empty asbestos bags are transferred to a clean disposal bag attached to an opening in the glove box. Before adding the next bag, the box can be cleaned with a vacuum wand located inside the box. All spills of asbestos slurry are hosed down as required to prevent any later generation of fibres.

Cell diaphragms eventually must be replaced because of the gradual plugging caused by trace impurities in the brine. At that time, the wet asbestos is washed from the cathode into a sump. It is then recovered in a filter operation and, while still in the wet form, is placed in plastic-lined boxes for secure landfill. Cleanliness is important, and any spilled material should be washed to the sump for recovery.

Hydrogen

Hydrogen is a very light flammable gas that does not appear on regulatory lists. Hydrogen from mercury cells can contain mercury, for which the relevant control measures are as discussed under 'mercury'. The major emphasis for hydrogen is on control systems to provide safety in the handling of this extremely flammable gas. Cell and suction header pressures are maintained near atmospheric pressure to reduce losses from flanges. Because of the difficulty in preventing hydrogen from leaking through isolation and vent valves, the use of water seals is recommended where pressure permits. Double-block and bleed arrangements are recommended for higher-pressure service. High-pressure valves can be checked for leak-through by venting through seal tanks that can be filled with water to determine valve tightness.

SODIUM CARBONATE

The basic method used in the manufacture of sodium carbonate or soda ash is the ammonia-soda process perfected by Solvay. The chemistry involved is fairly complex and is discussed in most elementary chemistry texts as a classic example of inorganic reaction. Raw materials for the process are salt, limestone and coke; products and by-products are sodium carbonate, calcium chloride and carbon dioxide. Carbon dioxide and ammonia are recycled, with the ammonia entering into a number of intermediate reactions.

Air Emissions Characterisation

There are a number of emission sources within the natural sodium carbonate industry. Those judged most significant include calciners, bleachers, dryers, and predryers. These are process emission sources that have the potential to emit large quantities of particulate matter.

Calciners

Calciners are the largest source of particulate emissions from plants using the monohydrate process. These particulates consist of sodium carbonate and inerts. The exit gas from coal-fired calciners also contains fly ash. Particulate emissions from calciners are affected by the gas velocity and the particle size distribution of the ore feed. As the gas velocity increases, the rate of increase in the particulate emissions steadily increases. Thus, coal-fired calciners may have higher particulate emissions than gas-fired calciners because of higher gas flow rates. Additionally, small particles are more easily entrained in a moving stream of gas than are larger particles.

Sulphur oxides are produced from fuel combustion. The quantities produced depend upon the sulphur content of the fuel. Organics are also emitted from calciners, some of which are present in the feed in the form of oil shale. At the calcination temperatures used, organics may vapourise or be partially combusted. In addition, some organics may result from partial or incomplete combustion of the fuel.

Nitrogen oxides are emitted from direct-fired process heating units such as calciners and soda ash dryers, but very little emission test data are available for these processes. Reported emission factors are 0.0008 kg/mg of feed (0.016 lb/T) for one calcimer with a cyclone and scrubber, and 0.056 kg/mg (0.111 lb/T) from another calcimer with a cyclone and ESP.

Dryers, predryers and bleachers

Three types of dryers are used for product drying in the monohydrate and direct carbonation processes: rotary steam tube, rotary gas fired, and fluid-bed steam tube. Sodium carbonate fines are emitted from each of these dryers. Particulate emissions from dryers are affected by the gas velocity of the feed. As the gas velocity increases, the rate of increase in the total emission rate of particulates steadily increases. Therefore, because of higher gas flow rates, and higher gas velocities, fluid-bed steam tube dryers and rotary gas-fired dryers have higher emission rates than rotary steam tube dryers.

In the direct carbonation process, rotary steam-heated predryers are used to lower the water content of wet sodium bicarbonate crystals before they are calcined. Particulates of sodium bicarbonate are the primary type of emissions from predryers. Emissions from bleachers consist mainly of particulates of sodium carbonate. Small amounts of compounds formed from the reactions of sodium nitrate may also be present in the particulates.

Air Pollution Control Measures

Particulate emission control techniques applicable to sources in sodium carbonate plants include:
1. Centrifugal separation.
2. Wet scrubbing.
3. Electrostatic precipitation.
4. Fabric filtration.

Centrifugal separators, or cyclones, rely on centrifugal forces to effect particulate separation from the gas stream. Cyclones are frequently used upstream of scrubbers or electrostatic precipitators (ESPs).

Scrubbers rely mainly on inertial impaction of particles with water droplets to effect particulate separation from the gas stream. Particles are contacted with wetted surfaces or atomised liquid droplets. As the gas stream diverges to pass such obstructions, the inertia of particles in the gas stream carries the particles into the water droplets or wetted surfaces. The particulate laden liquid is then separated from the gas stream, and either recycled to the production process or discharged as waste.

Electrostatic precipitators generate an electrical field by applying a high voltage to a discharge electrode system consisting of rows of vertical wires. The strength of the field depends in part on the gas composition. The subsequent migration of the charged particles to the collected plates depends on the particle size, resistivity, gas velocity and distribution, rapping, and field strength. The collecting electrodes are rigid plates that are baffled. Electromagnetic or pneumatic hammers are used to rap the electrodes, dislodging the collected particles, which then fall into hoppers. Baffling on the collecting electrodes provides shielded air pockets that reduce re-entrainment of particles after rapping.

Fabric filtration is a process where dust particles in a gas stream are filtered out and collected.

Ore and product handling

Particulate emissions from ore and product handling operations, such as conveyor transfer points, crushing, and product sizing, are typically controlled by either venturi scrubbers or fabric filters (baghouses). These control devices are an integral part of the manufacturing process, capturing raw materials and product for economic reasons.

Calciners and bleachers

Calciners and bleachers are typically controlled by cyclones in series with ESPs. Venturi scrubbers, also used to control emissions from calciners, typically achieve lower removal efficiencies than ESPs. Higher removal efficiencies may be achieved with sufficient pressure drop. It appears that a pressure drop of 154 cm (60 inches) of water may be required to achieve a removal efficiency comparable to that achieved with a four-stage ESP, based on the removal efficiency achieved from EPA test data from a gas-fired calciner with a scrubber pressure drop of about 85 cm (33.5 inches) of water.

Fabric filters (baghouses) have not been reported to be in use to control emissions from calciners or bleachers. The sticky, hygroscopic nature of sodium carbonate could lead to bag blinding and/or caking.

Dryers and predryers

Venturi scrubbers are the devices used to control emissions from rotary steam tube dryers. Cyclones in series with venturi scrubbers are used to control emissions from fluid-bed steam tube dryers and rotary steam-heated predryers. Both venturi scrubbers and ESPs have been used to control emissions from gas-fired dryers. The exhaust gas from both rotary and fluid-bed steam tube dryers and predryers is well suited to control by wet scrubbing. The removed sodium carbonate particles are quite soluble and hygroscopic. These characteristics enhance the removal of sodium carbonate particles in wet scrubbers. However, these characteristics coupled with the high water content of the dryer exit gas can result in operating problems for ESPs or baghouses. Moisture in the exit gas can condense in ESPs, or baghouses. Wet, sticky dust adheres to the electrodes and hoppers of ESPs or blinds and cakes the bags in baghouses. ESPs have been used to control emissions from rotary gas-fired dryers, but the exit gas from these dryers is at a higher temperature and lower relative humidity than gas from steam tube dryers.

Recent and Future Industry Trends

Sodium carbonate facilities are potential major sources of particulate emissions. These emissions can be effectively controlled through the proper use and maintenance of conventional add-on particulate control techniques. Tests conducted at sodium carbonate plants along with industry data contributed to the selection of ESPs as the best system of emission reduction for calciners and bleachers and venturi scrubbers as the best system for dryers and predryers. Small amounts of sulphur oxides and organics are also emitted from direct-fired calciners; however, source tests have indicated that these emissions are very low relative to uncontrolled particulate emissions. Environmental issues have contributed to the phase-out of the production of sodium carbonate by the solvay process in the United States; for example, substantial quantities of aqueous waste, containing high concentrations of calcium chloride, are produced in the solvay process. Another reason for the decline in the use of the solvay process has been high fuel costs, since the solvay process is more fuel intensive than any of the natural processes.

Most of the new sodium carbonate process lines constructed in Wyoming in the past decade use the monohydrate process. This process is likely to continue to be used in the future. Because of the proximity of the Wyoming trona mines to western coal mines, future plants are expected to make greater use of coal than older existing plants.

FOOD PROCESSING INDUSTRIES

Sugar Industry

Sugar is a term for a class of edible crystalline carbohydrates, mainly sucrose, lactose, and fructose characterised by a sweet flavour. In food, sugar almost exclusively refers to sucrose, which primarily comes from sugar cane and sugar beet. Other sugars are used in industrial food preparation, but are usually known by more specific names—glucose, fructose or fruit sugar, high fructose corn syrup, etc.

Bagasse combustion in sugar mills

Process description

Bagasse is the matted cellulose fibre residue from sugar cane that has been processed in a sugar mill. Previously, bagasse was burned as a means of solid waste disposal. However, as the cost of fuel oil, natural gas, and electricity has increased, bagasse has come to be regarded as a fuel rather than refuse. Bagasse is a fuel of varying composition, consistency, and heating value. These characteristics depend on the climate, type of soil upon which the cane is grown, variety of cane, harvesting method, amount of cane washing, and the efficiency of the milling plant. In general, bagasse has a heating value between 3000 and 4000 British thermal units per pound (Btu/lb) on a wet, as-fired basis. Most bagasse has a moisture content between 45 and 55 percent by weight.

Sugar cane is a large grass with a bamboo-like stalk that grows 8 to 15 feet tall. Only the stalk contains sufficient sucrose for processing into sugar. All other parts of the sugar cane (i.e. leaves, top growth, and roots) are termed 'trash'. The objective of harvesting is to deliver the sugar cane to the mill with a minimum of trash or other extraneous material.

The cane is normally burned in the field to remove a major portion of the trash and to control insects and rodents. The three most common methods of harvesting are hand cutting, machine cutting, and mechanical raking. The cane that is delivered to a particular sugar mill will vary in trash and dirt content depending on the harvesting method and weather conditions. Inside the mill, cane preparation for extraction usually involves washing the cane to remove trash and dirt, chopping, and then crushing.

Juice is extracted in the milling portion of the plant by passing the chopped and crushed cane through a series of grooved rolls. The cane remaining after milling is bagasse.

Firing practices

Fuel cells, horseshoe boilers, and spreader stoker boilers are used to burn bagasse. Horseshoe boilers and fuel cells differ in the shapes of their furnace area but in other respects are similar in design and operation. In these boilers (most common among older plants), bagasse is gravity-fed through chutes and piles onto a refractory hearth. Primary and overfire combustion air flows through ports in the furnace walls; burning begins on the surface pile. Many of these units have dumping hearths that permit ash removal while the unit is operating.

In more recently built sugar mills, bagasse is burned in spreader stoker boilers. Bagasse fed to these boilers enters the furnace through a fuel chute and is spread pneumatically or mechanically across the furnace, where part of the fuel burns while in suspension. Simultaneously, large pieces of fuel are spread in a thin, even bed on a stationary or moving grate. The flame over the grate radiates heat back to the fuel to aid combustion. The combustion area of the furnace is lined with heat exchange tubes (waterwalls).

Emissions

The most significant pollutant emitted by bagasse-fired boilers is particulate matter, caused by the turbulent movement of combustion gases with respect to the burning bagasse and resultant ash. Emissions of sulphur dioxide (SO_2) and nitrogen oxides (NO_x) are lower than conventional fossil fuels due to the characteristically low levels of sulphur and nitrogen associated with bagasse.

Auxiliary fuels (typically fuel oil or natural gas) may be used during startup of the boiler or when the moisture content of the bagasse is too high to support combustion; if fuel oil is used during these periods, SO_2 and NO_x emissions will increase.

Soil characteristics such as particle size can affect the magnitude of particulate matter (PM) emissions from the boiler. Cane that is improperly washed or incorrectly prepared can also influence the bagasse ash content. Upsets in combustion conditions can cause increased emissions of carbon monoxide (CO) and unburned organics, typically measured as volatile organic compounds (VOCs) and total organic compounds (TOCs).

Controls

Mechanical collectors and wet scrubbers are commonly used to control particulate emissions from bagasse-fired boilers. Mechanical collectors may be installed in single cyclone, double cyclone, or multiple cyclone (i.e. multiclone) arrangements. The reported PM collection efficiency for mechanical collectors is 20 to 60 per cent. Due to the abrasive nature of bagasse fly-ash, mechanical collector performance may deteriorate over time due to erosion if the system is not well maintained.

The most widely used wet scrubbers for bagasse-fired boilers are impingement and venturi scrubbers. Impingement scrubbers normally operate at gas-side pressure drops of 5 to 15 inches of water; typical pressure drops for venturi scrubbers are over 15 inches of water. Impingement scrubbers are in greater use due to their lower energy requirements and fewer operating and maintenance problems. Reported PM collection efficiencies for both scrubber types are 90 per cent or greater. Fabric filters and electrostatic precipitators have not been used to a significant extent for controlling PM from bagasse-fired boilers because both are relatively costly compared to other control options. Fabric filters also pose a potential fire hazard.

Gaseous emissions (e.g. SO_2, NO_x, CO, and organics) may also be absorbed to a significant extent in a wet scrubber. Alkali compounds are sometimes utilised in the scrubber to prevent low pH conditions. If carbon dioxide (CO_2)-generating compounds (such as sodium carbonate or calcium carbonate) are used, CO_2 emissions will increase. Fugitive dust may be generated by truck traffic and cane handling operations at the sugar mill.

Dairy Industry

The dairy industry involves processing raw milk into products such as consumer milk, butter, cheese, yogurt, condensed milk, dried milk (milk power), and ice cream, using processes such as chilling, pasteurisation, and homogenisation. Typical by-products include buttermilk, whey, and their derivatives.

Waste characteristics

Dairy effluents contain dissolved sugars and proteins, fats, and possible residues of additives. The key parameters are biochemical oxygen demand (BOD) (with an average ranging from 0.8 to 2.5 kilograms per metric ton (kg/T) of milk in the untreated effluent), chemical oxygen demand (COD) (normally about 1.5 times BOD level), total suspended solids (100 to 1000 milligrams per litre (mg/l), total dissolved solids, phosphorus (10 to 100 mg/l), and nitrogen (about 6 per cent of BOD level). Cream, butter, cheese, and whey production are major sources of BOD in waste-water.

The waste load equivalents of specific milk constituents are: 1 kg of milk fat = 3 kg COD; 1 kg of lactose = 1.13 kg COD; and 1 kg protein = 1.36 kg COD. The waste-water may contain pathogens from contaminated materials or production processes. A dairy often generates odour and in some cases, dust, which need to be controlled. Most of the solid wastes can be processed into other products and by-products.

Pollution prevention and control

Good pollution prevention practices in the dairy industry include:

1. Reduce product losses by better production control.
2. Use disposable packaging (or bulk dispensing of milk) instead of bottles where feasible.
3. Collect waste product for use in lower-grade products such as animal feed where feasible without exceeding cattle feed quality limits.
4. Optimise use of water and cleaning chemicals. Recirculate cooling waters.
5. Keep effluents from sanitary installations, process, and cooling (including condensation) systems segregated. This facilitates recycling of waste-water.
6. Use condensates instead of fresh water for cleaning.
7. Recover energy by using heat exchangers for cooling and condensing.
8. Use high pressure nozzles to minimise water usage.
9. Avoid the use of phosphorus-based cleaning agents.

Continuous sampling and measuring of key production parameters allow production losses to be identified and reduced, thus reducing the waste load. Table 15.10 presents product losses that are achieved in a well-run dairy.

Odour problems can usually be prevented with good hygiene and storage practices. Chlorinated fluorocarbons should not be used in the refrigeration system.

Table 15.10. Product losses in the dairy industry.

Operation	Product losses (%)[1]		
	Milk	Fat	Whey
Butter/transport skimmed milk	0.17	0.14	N/A
Butter + skimmed milk powder	0.60	0.20	N/A
Cheese	0.20	0.10	1.6
Cheese + whey evaporation	0.20	0.10	2.2
Cheese + whey powder	0.20	0.10	2.3
Consumer milk	1.9	0.7	N/A
Full cream milk powder	0.64	0.22	N/A

[1]Expressed as percentage of volume of milk, fat or whey processed.

N/A = Not applicable.

Air emissions

Odour controls (such as absorbents/biofilter on exhaust systems) should be implemented where necessary to achieve acceptable odour quality for nearby residents. Fabric filters should be used to control dust to below 50 milligrams per normal cubic meter (mg/Nm^3) from milk powder production.

Monitoring and reporting

Monitoring of the final effluent for the parameters listed above should be carried out at least once per month, or more frequently if the flows vary significantly. Monitoring data should be analysed and reviewed at regular intervals and compared with the operating standards so that any necessary corrective actions can be taken. Records of monitoring results should be kept in an acceptable format. These should be reported to the responsible authorities and relevant parties, as required, and provided to MIGA if requested.

Key issues

The following points summarises the key production and control practices that will lead to compliance with emission guidelines:

1. Monitor key production parameters to reduce product losses.
2. Use disposable packaging (or bulk dispensing of milk) instead of bottles where feasible.
3. Design and operate the production system to achieve recommended waste-water loads.
4. Recirculate cooling waters.
5. Collect wastes for use in low-grade products.

Meat Processing

The meat processing and rendering industry includes the slaughter of animals and fowl, processing of the carcasses into cured, canned, and other meat products, and the rendering of inedible and discarded remains into useful by-products such as lards and oils. In-plant measures that can be used to reduce the odour nuisance and the generation of solid and liquid wastes from the production processes. Biofilters, carbon filters, and scrubbers are used to control odours and air emissions from several processes, including ham processing and rendering. Recycling exhaust gases from smoking may be feasible in cases where operations are not carried out manually and smoke inhalation by workers is not of concern.

Emissions guidelines

Emissions levels for the design and operation of each project must be established through the environmental assessment (EA) process on the basis of country legislation and the Pollution Prevention and Abatement Handbook, as applied to local conditions. The guidelines given below present emissions levels normally acceptable to the World Bank Group in making decisions regarding provision of World Bank Group assistance. Any deviations from these levels must be described in the World Bank Group project documentation. The emissions levels given here can be consistently achieved by well-designed, well-operated, and well-maintained pollution control systems. The guidelines are expressed as concentrations to facilitate monitoring. Dilution of air emissions or effluents to achieve these guidelines is unacceptable. All of the maximum levels should be achieved at least 95 per cent of the time that the plant or unit is operating, to be calculated as a proportion of annual operating hours.

Air emissions

Odour controls should be implemented, where necessary, to minimise odour impacts on nearby residents. Particulate matter emissions of smokehouses should be kept below 150 milligrams per normal cubic meter (mg/Nm^3), with a carbon content of less than 50 mg/Nm^3.

Seafood Industry

The world seafood industry plays a significant role in the economic and social wellbeing of nations, as well as in the feeding of a significant part of the world's population. Fishing and fish farming has emerged as one of the major food processing occupations of mankind. In ancient times, economically and socially backward people were employed in this profession. The advent of modern mechanised fishing vessels has brought vast changes in the attitude of the public fishing and seafood processing. From low income and socially backward communities the profession has shifted to the hands of industrialists and technologists. Today fishing and processing activities provide employment to millions of people around the world. There are several product categories from seafood industry, based on raw material type (fresh/frozen) and value-addition (degree of processing and value content). The following figure depicts few seafood product categories in a seafood industry (Fig. 15.10).

Emission to air

Point source emission

These emissions are exhausted into a vent or stack and emitted through a single point source to the atmosphere. The major air pollution sources in a typical seafood industry are from combustion sources like boiler and generators for electric power. Boiler is used for steam supply during pre-cooking and sterilisation process. The examples of fuels used in the boilers are electricity, fuel oil, coal and LPG. The Table 15.11 highlights common air emissions and their sources from seafood processing, while in Table 15.12 are given emissions from boilers.

Table 15.11. Common air emissions from seafood processing.

Sources	Emission
Cooking	Volatile organic compound (VOC)
Fried dryer	VOC, particulate matter (PM10)
Pre-cooking and sterilisation	VOC, CO, NO_x, SO_2, CO_2
Refrigeration	NH_3
Disinfection/cleaning	Cl_2

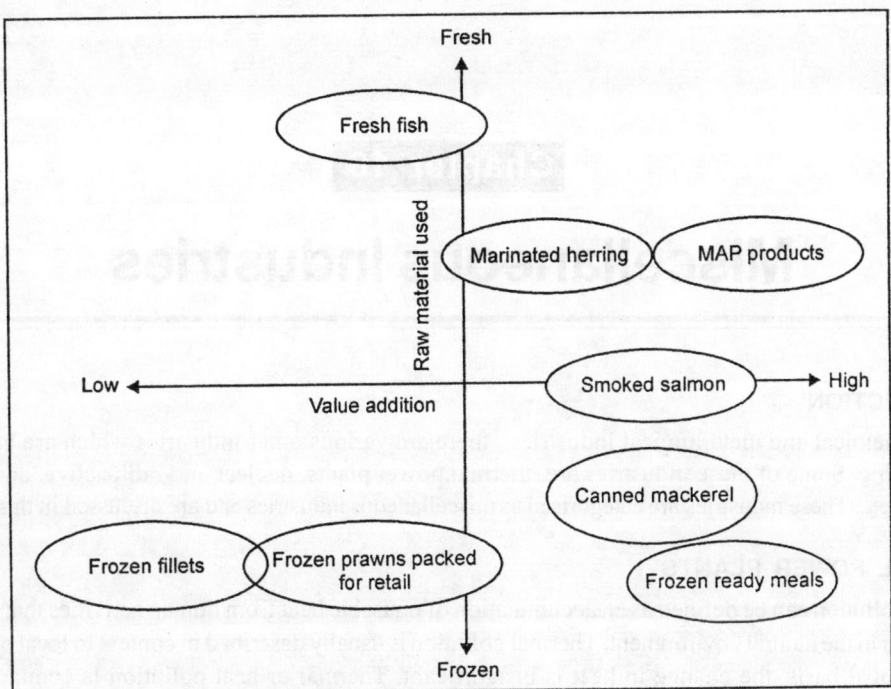

Fig. 15.10. Seafood products in a seafood industry.

Table 15.12. Typical emission from boilers using different fuels.

	PM	SO_2	CO	HC	NO_2
LPG-F	0.2 (g/l)	0.5 (g/l)	0.23 (g/l)	0.084 (g/l)	1.35 (g/l)
Coal	12.5 (g/kg)	12 (g/kg)	1 (g/kg)	0.5 (g/kg)	7.5 (g/kg)
Fuel oil	2.75 (g/l)	10 (g/l)	0.5 (g/l)	0.35 (g/l)	9.6 (g/l)

Odour

Odour is often the most significant form of air pollution in fish processing. Major sources include storage sites for processing waste, cooking by-products during fish meal production, fish drying processes, and odour emitted during filling and emptying of bulk tanks and silos. Fish quality may deteriorate under the anaerobic conditions found in onboard storage on fishing processing facilities. This deterioration causes the formation of odourous compounds such as ammonia, mercaptans, and hydrogen sulphide gas.

Chapter 16

Miscellaneous Industries

INTRODUCTION

Besides chemical and metallurgical industries, there are various other industries which are hazardous and polluting. Some of these industries are, thermal power plants, nuclear and radioactive, and cement industries, etc. These industries are categorised as miscellaneous industries and are discussed in this chapter.

THERMAL POWER PLANTS

Thermal pollution can be defined as an accumulation of unusable heat from human activities that distrupts ecosystems in the natural environment. Thermal pollution is usually described in context to local problems, as on a global basis, the change in heat is insignificant. Thermal or heat pollution is common in the vicinity of power plants generating electricity. The water used for cooling the working steam is led back to the source at an appreciable higher temperature, affecting aquatic organisms. The use of nuclear power for the industries opens up the possibility of radioactive wastes leaking out, which might cause enormous damage to the biosphere.

Fuels like coal and oil used in power plants prove very harmful to the environment as they release very heavy amounts of pollutants into the atmosphere causing air pollution. Thermal power stations though primarily encouraged for economic development have become a major source of pollution in the environment. Dust concentration is reduced in thermal power stations by using high efficiency electrostatic precipitators. Sulphur dioxide emissions can be reduced by removing sulphur from the fuel before combustion by removing sulphur dioxide from the flue gases (flue gas desulphurisation).

Thermal or heat pollution is common in the vicinity of power plants. The water used for cooling the working steam is led back to the source at an appreciably higher temperature, affecting the aquatic organisms. The use of nuclear power opens up the possibility of radioactive wastes leaking out, which might cause enormous damage to the biosphere around.

Air Pollution

Major air pollutants from fossil fuel fired power plants include particulate matter, sulphur dioxide, nitrogen oxides, carbon dioxide, carbon monoxide and hydrocarbon emissions are relatively small. Often the solution to air pollution is sought by dilution of pollutants in atmosphere. The control of particulate as well as acid gas emissions will become very essential. A summary of different control technologies for each of the major air pollutant is given below.

Particulate matter

Particles are emitted in the boiler stack gas at a rate of about 78 G/cu.m for the representative 210 MW plant. The use of 99.8 per cent efficient electrostatic precipitator reduces the particulate emission rate into the atmosphere to about 156 mg/cu.m which nearly satisfies the mandated limit of 150 mg/cu.m.

Techniques for controlling particulate emissions from a TPS are utilising control equipments which remove the particulate emissions released into the atmosphere. The basic techniques of particulate collection equipments are: (i) mechanical collector, (ii) wet collector, (iii) fabric filters, (iv) electrostatic precipitator.

Sulphur dioxide

The second most abundant air pollutant emitted by the coal fired power plant is sulphur dioxide. The sulphur content on an average in Indian coal (with the exception of Assam coal) is less compared to the coal found elsewhere in the world. Sulphur dioxide emissions are not a problem in India and however concentration of sulphur dioxide in ambient air can reach highlevels only due to clustering of thermal power plants. Besides using fuels low in sulphur content the possible methods for reducing sulphur dioxide emission from coal/fossil fuel combustion are:

1. Use of tall stacks to increase atmospheric dispersion and dilution.
2. Flue gas desulphurisation/flue gas treatment.
3. Desulphurisation of fuel itself.

Tall Stack Dispersion

This control method is based on natural dispersion at high elevation so that ground level concentrations are acceptable at all the times. This approach is also recommended by the Pollution Control Boards. The benefit of using tall stack for pollutant dispersion is that it does not clearly rely on sulphur dioxide removal process and/or fuel desulphurisation methods, which involve high operating costs. To certain extent, a tall stack also as a safeguard against removal equipment breakdowns and adverse meteorological conditions in the future.

Flue Gas Desulphurisation

Sulphur dioxide removal process may be grouped according to two classifications: (i) throwaway; and (ii) regenerative. While the throwaway process can be wet or dry, the regenerative system is only a dry process. Throwaway processes are those in which a solid waste product is formed which must be discarded. As a result, fresh chemicals also must be continuously added. In regenerative processes, the chemistry is such that the removal agents can be continuously regenerated in a closed loop system. Wet or dry processes are differentiated simply by whether or not the active removal agent is contained in a liquid solution. The removal system typically involves the use of absorption, adsorption or catalytical processes. A brief comparison of some of the major processes developed for SO_2 removal from stack gases appears in Table 16.1. Since SO_2 is an acidic gas almost all of the scrubbing processes use an aqueous solution or slurry and dispose the removed sulphur in the form of calcium type waste sludge. As a result, the alkaline make up is required except for sea water scrubbing where the natural alkalinity of sea water is utilised.

Throwaway processes frequently can be used to remove particulate matter as well, if the system is enlarged. However, in most regenerative systems a high efficiency particulate collector such as an ESP must precede the SO_2 removal equipment, because particulates are not acceptable in its operation. Dry nahcolite (a naturally occurring sodium-bicarbonate) injection processes is still under development, dry limestone injection process does not appear to offer a satisfactory solution to these SO_2 control problem due to its low sulphur removal efficiencies, poor utilisation of the reagent, tube fouling and adverse effect on ESP performance. Conventional (wet) lime and limestone processes continue to be installed world wide at thermal power stations.

Table 16.1. Comparison of sulphur dioxide removal processes.

Process	Process operations	Active material	Key sulphur product
Set throwaway processes			
Lime or limestone	Slurry scrubbing	CaO, $CaCO_3$	$CaSO_3/CaSO_4$
Double alkali	Na_2SO_3 solution regenerated by CaO or $CaCO_3$	$CaCO_3/Na_2SO_3$	$CaCO_3/CaSO_4$
Seawater	Scrubbing	Natural alkalinity	Sulphates
Wet regenerative processes			
Magnesium oxide	$MgSO_3$ slurry	MgO	H_2SO_4
Wellman lord	Na_2SO_3 solution	Na_2SO_3	Sulphur
Electrolyte	NaOH solution, acid, decomposition, electrolytic regeneration	NaOH	95% SO_2
Citrate	Sodium citrate solution, reaction with H_2S to form sulphur	H_2S	Sulphur
Sulphoxel	Na_2CO_3 solution, reaction with CO to form sulphur	Na_2CO_3/CO	Sulphur
Formate	KCOOH solution, reaction with CO to form H_2S	KCOOH/CO	H_2S
Dry processes			
Lime spray dryer	Reaction of lime with SO_2	$Ca(OH)_2$	$CaSO_3$
Dry furnace injection	High temperature absorption	$CaCO_3$	$CaSO_3/CaSO_4$
Catalytic oxidation	Oxidation at 725 K, scrubbing with H_2O	V_2O_5 Catalyst	80% H_2SO_4
Carbon absorption	Absorption at 400 K reaction with H_2S to S reaction with H_2 to H_2S	Activated carbon/H_2	Sulphur

Oxides of Nitrogen Control

Most of the advanced combustion concepts and modification techniques are directed towards preventing the formation of fuel nitrogen oxides. A fuel rich initial combustion zone is established in which the fuel nitrogen can be oxidised to molecular nitrogen, rather than to nitric oxide. Thus, all of the combustion modification techniques except flue gas recirculation (FGR) depend on reducing the availability of oxgyen in primary combustion zones.

Combustion modification techniques such as low excess air firing and stages combustion have been shown to be effective in reducing nitrogen oxide emission from coal combustion. In low excess air firing, the unit operate at a reduced level of total combustion air flow while maintaining acceptable flame and furnace conditions.

There are three methods of staged combustion:

1. Biased or off stoichiometric firing: The air/fuel mixture in the boiler is stratified, and part of the combustion air is diverted outside of the initial fuel/air mixing zone.
2. Burner out of service operations: The fuel flow to an individual burner is cut (while maintaining the air flow).
3. Over fire air ports: Part of the combustion air enters by ports above the burner to generate fuel-rich conditions at the burner. In FGR, gas is taken from the exhaust stream and reintroduced in the furnace. This technique has often been used for steam temperature control.

Both nitrogen oxide production and the effect of combustion modifications in reducing depends on how the furnace is fired. While modification techniques can significantly reduce nitrogen oxide emissions, more advanced concepts, such as new burner design (which can be retrofitted into existing units) or flue gas treatment, are required to achieve very low levels of nitrogen oxides. Presently control of pollutants from the flue gas are done in following manners:

1. Eighty per cent of ash produced in boiler comes out with the flue gas. This ash, called fly-ash, is major source of pollutants in coal-fired power station. It can be controlled by installing electrostatic precipitators (ESP) of high efficiency. It can be shown that with ESP of 99.78 per cent efficiency, even the emission from worst coal (ash content 45 per cent) can be kept within the limit of 150 mg/nm as set by CPCB.
2. Control of SO_2 and NO_x is taken care by the adequate dispersion of the pollutants in the atmosphere. Actual stack emission measurements have shown that NO_x emission from the coal fired power plant is quite low, remaining within 200 mg/nm. So its ground level concentration after dispersion remains insignificant.

NUCLEAR POLLUTION

The menace of nuclear radioactive pollution spreading into the environment has increased extensively as a result of the discovery of artificial radioactivity, particularly due to the development of the atom bomb, hydrogen bomb and of techniques of harnessing nuclear energy. Actually, this dangerous pollution enters into the environment in waste streams and stack gases from operations of power processing plants. From neutron bombardment of atomic fuel, heavy radio-nuclides are produced which are extremely toxic. Once these radio-elements find access into the environment, they enter the eco-cycling processes and ultimately into the food chain and metabolic pathways. Radioactive pollution poses a serious threat to the environment and future generation. Radioactive wastes from nuclear plants, reactors, etc. are, however, of a special kind in the sense that they do not smell bad or pollute the atmosphere like smoke, but are extremely lethal to living beings even in minute quantities. These wastes persist in the environment for a long time. Non-radiation pollutants and short half-life radio nuclides are being constantly released into the air and it is expected that they will disperse or degrade in a short span of time. Consequently, they are not regarded as a future pollution threat until their concentration exceeds the limit. They can not be disposed of into the environment like other industrial wastes. The nuclear establishments produce various types of solid and liquid radioactive wastes which contain different amounts of radioactivity and special care has to be taken for their safe disposal.

Radioactive wastes require royal disposal methods. During the phenomenon of radioactivity, some naturally unstable elements tend to become stable by emitting alpha (α), beta (β) and gamma (γ) rays. These rays can ionise the air and disarray the life activities in a cell when they pass through them.

Radioactive Waste Management

The radioactive wastes generated during the various nuclear fuel cycle operations are highly variable in nature, composition, volume, radioactivity levels and half-lives of radio-nuclides contained. They are classified into solid, liquid and gaseous wastes. Solid and liquid wastes are nominally labelled as low, medium and high-level wastes, depending upon the surface dose and or radio-nuclide content. An elaborate segregation and collection system is necessary before choosing the treatment and disposal methods. The basic scheme for management of radioactive wastes is given in Fig. 16.1. The widely accepted principles for management of radioactive wastes are: (i) dilute and disperse, (ii) delay and decay, and (iii) concentrate and contain.

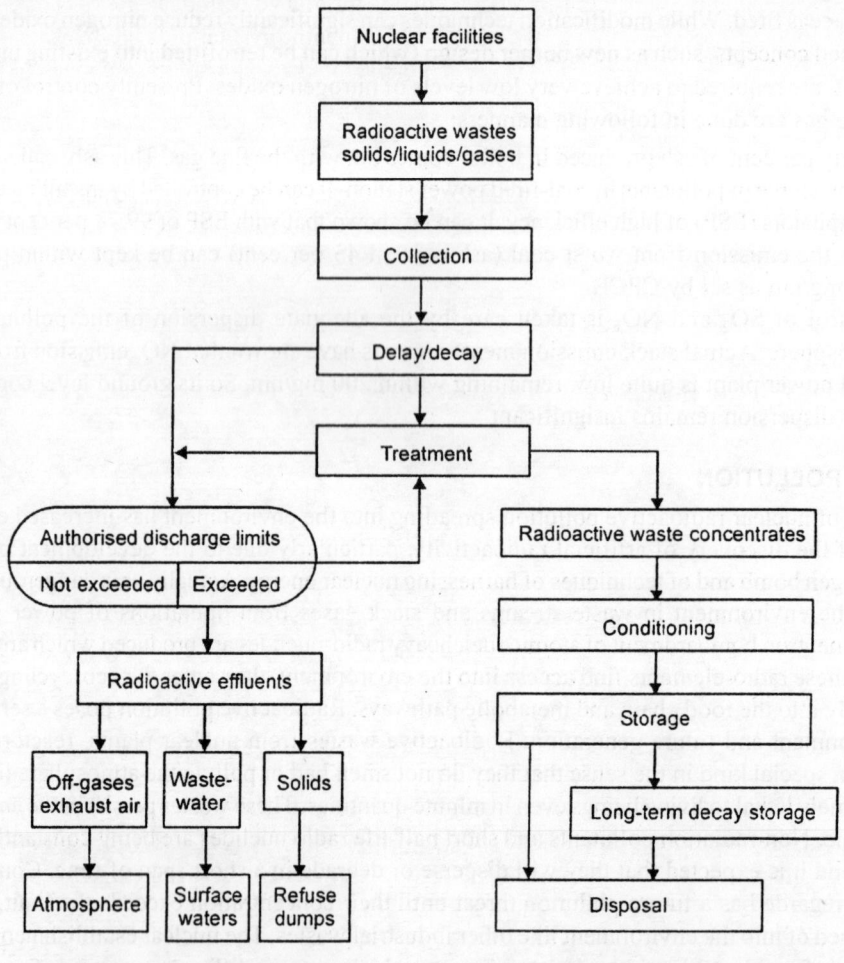

Fig. 16.1. Basic scheme for the management of radioactive wastes.

The first principle is generally applied for gaseous and low-level liquid effluents. It is ascertained that the dilution by air or water is to a level sufficiently low so that the resulting dose will be below the acceptable limits. These practices should conform to regulatory standards. 'Delay and decay' applies to those waste which contain only short life nuclides. 'Concentrate and contain' is generally applied for medium and high-level wastes, both solid and liquid. The wastes are processed, treated and or conditioned before containment. A variety of techniques have been developed for treatment, storage and disposal of radioactive wastes. The gaseous wastes generated are generally released into the atmosphere through stacks. On the basis of the meteorological data applicable to the site, the height of releases are optimised so that the resulting dose from atmospheric releases are maintained well within the stipulated limits.

Liquid effluents, especially low and medium categories, are treated by chemical coagulation, ion exchange or evaporation to remove a bulk the of radionuclides. The treated liquid effluents that conform to authorised limits are only allowed to be discharged. The residual solids are conditioned to immobilise the radionuclides by cementation, bituminisation or by polymerisation. The high-level liquid effluents generated in reprocessing operations, which contain over 99.9 per cent of the non-gaseous fission products and traces of urecovered actinides generated in power reactors, are converted into a non-leachable glass form. The solid wastes include a wide variety of materials including contaminated materials. filters, resins, chemical sludges, incineration ash, etc. The combustible items are incinerated in suitably designed incinerators with air cleaning facilities and the ash is conditioned in cement matrix for proper storage/ disposal. For low and medium-level solid wastes, underground disposal is the most viable option. Both reinforced concrete trenches as well as steel-lined concrete tile holes are used depending upon the activity of the solid matrix. In the case of vitrified high-level wastes, after an engineered storage for a period of 25 years, disposal in suitable deep underground geological formation will ensure isolation from groundwater sources. Thus, the main objective of waste management in nuclear operations is to contain (isolate from biosphere) as is much radioactive materials as is practicable so as to minimise releases into the environment.

Evaluation of Release Limits

For the purpose of evaluating authorised limits for limiting radioactive discharges from the site in the planning stage of the installation, an environmental impact analysis is carried out by the predictive methodology by the application of models. Both environmental models that attempt to describe the environmental processes and dosimetric models that relates the radioactivity levels in different environmental components to radiation exposure to the population are used. Essentially, there are three environments from which one may receive radiation exposure as a result of releases from nuclear installations. These components of environment–atmospheric (air), terrestrial (land including food from land) and aquatic (both fresh water and sea water, aquatic food and marine products)—provide for dispersion/dilution and concentration of radionuclides. Information on the source-term. behaviour of waste components in the receiving medium and transfer factors to environmental materials along with environmental habit survey data enable estimation of dose to the exposed subjects. There are various environmental models appropriate to each sector of the environment like atmosphere. land and water. The transport of radioactive materials released into the atmosphere is controlled by the normal atmospheric mixing processes. The major mixing processes incorporated into atmospheric transport models are diffusion and advection. The path for radiation exposure from atmospheric releases of gaseous effluents are direct irradiation (external exposure) and inhalation (internal exposure). For radioactive effluents, other than gases, both dry and wet deposition takes place on ground/vegetation. These deposited radionuclides cause external exposure or can result in internal exposure by entering the food chain or drinking water supplies or can cause exposure via inhalation through resuspensions in the air (Fig. 16.2).

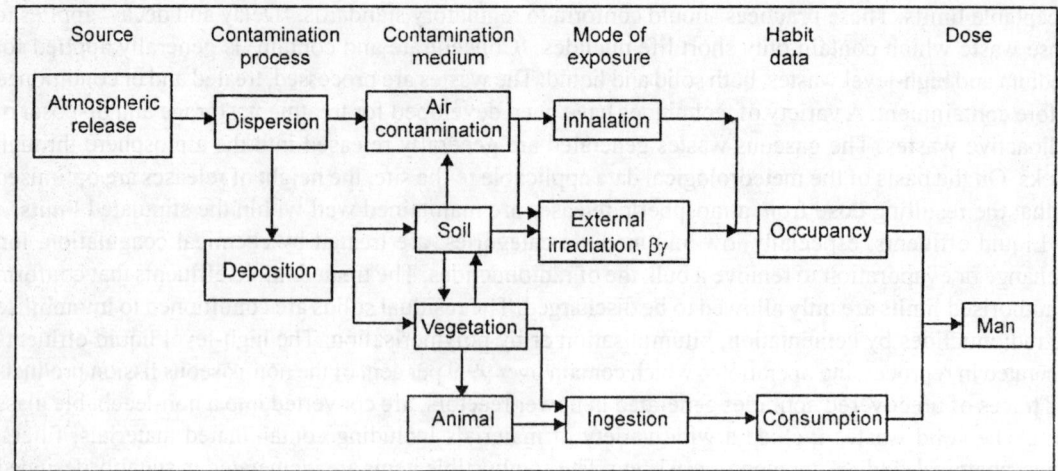

Fig. 16.2. Schematic representation of atmospheric pathways.

Terrestrial environmental models take into consideration possible external irradiation from deposited nuclides on land surfaces, concentration in vegetation and other food items (including foliar deposition and uptake from soil and irrigated water) concentration in animal feeds, milk, etc.

In the case of discharges of radioactive materials into the aquatic environment, the principal hydrodynamic mechanism of radionuclide transport are diffusion and advection. Different aquatic environments like semi-enclosed basins, near-shore waters of open sea, fresh water lakes and rivers are being used as recipients for liquid effluents. Aquatic models used take into consideration different pathways and hydrological factors (Fig. 16.3).

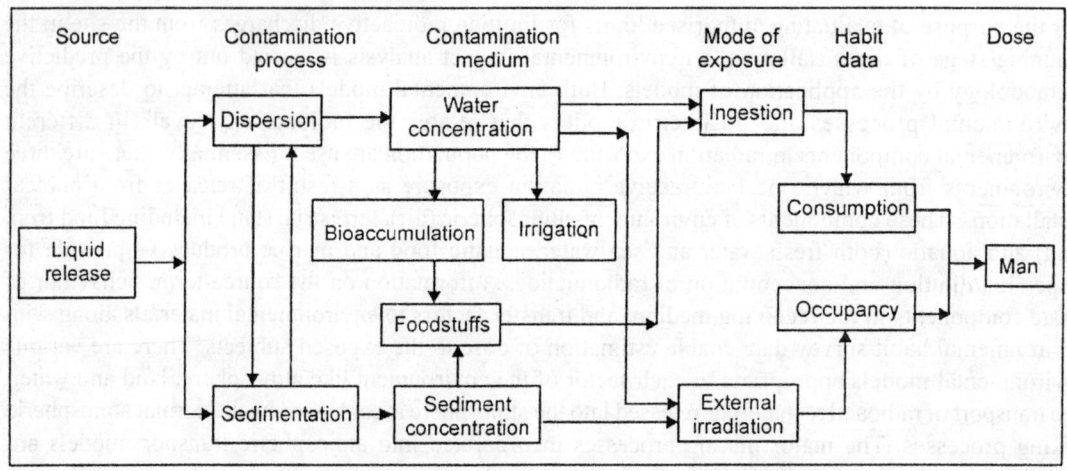

Fig. 16.3. Schematic representation of aquatic pathways.

In addition, geochemical and ecological factors that cause accumulation of radionuclides in sediments, aquatic food or marine products and utilisation of the aquatic environment greatly influence the recipient capacity of the medium. These environmental models are used for advance prediction of dose rates, for the most exposed (critical) group of population for the purpose of evaluating release limits for specific sites by the application of specific site characteristics and environmental parameters.

CEMENT INDUSTRIES

Cement is the general term given to the powdered materials which initially have plastic flow when mixed with water or other liquid, but have the property of setting to a hard solid structure in several hours, with varying degree of strength and bonding properties. Portland cement is chemically defined as finely ground mixture of calcium aluminates and silicates of varying compositions, which hydrate when mixed with water to form a rigid solid structure with good compressive strength. Portland cement is a mixture of compounds such as dicalcium silicate, tricalcium silicate, tricalcium aluminate, tetracalcium alumino ferrite, magnesium oxide and calcium oxide. The chief raw materials for the manufacture of portland cement are limestone and clay, and these are generally available in large amounts in the vicinity of cement factories. There are two methods of manufacturing portland cement—the wet process and the dry process. It should be noted that the two processes differ only in the treatment of the raw materials, (in the dry process, no water is added to the material in grinding and thus no slurry is made), otherwise the same equipment is used in both (Fig. 16.4).

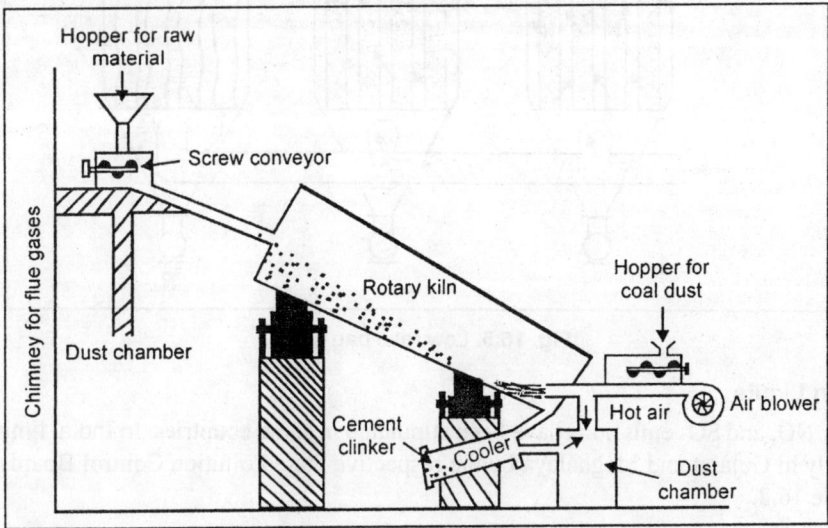

Fig. 16.4. Dry process for manufacturing cement.

Wet process is the most common and is almost universally employed for the manufacture of cement. Dry process is slow, costly and the cement produced is of low quality. Cement industry contributes to air pollution in the form of suspended particulate matter and gases such as oxides of nitrogen and sulphur dioxide. Particulate matter is the dominant environmental problem. The gaseous emissions are known to affect health and vegetation. NO_x and SO_2 emissions in cement plants in India have been monitored and found to be well within regulatory standards of developed countries.

Cement industry is known to contribute to atmospheric pollution, particularly from the stacks. Particulate emissions create nuisance in and around cement plants. Although gaseous emissions do not contribute significantly to atmospheric pollution, there are reported to have adverse effects on the surrounding environment, affecting human health, property and vegetation. The awareness to curb all possible types of pollution has created the need to assess the extent of gases generated in cement plants. Sulphur dioxides and oxides of nitrogen are the gases produced during pyro-processing. Discussed here are air pollution limits, mechanism of gaseous pollutants generation, monitoring techniques and analysis of data generated in dry process cement plants. Generally, low ratio bag filters (Fig. 16.5) are used for collection of dust.

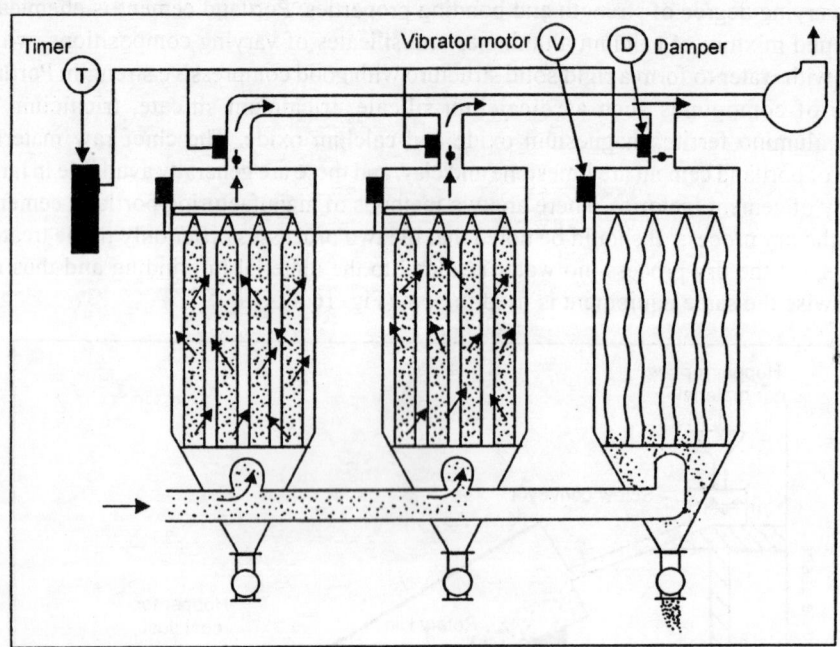

Fig. 16.5. Low ratio bag filters.

Air Pollution Limits

Standards for NO_x and SO_2 emissions have been stipulated in some countries. In India, limits have been stipulated only in Gujarat and Meghalaya by the respective State Pollution Control Boards (SPCB), as given in Table 16.2.

Table 16.2. Limits of NO_x and SO_2 emission prescribed by SPCB.

Name of board	Limit	
	NO_x	SO_2
Gujarat (ppm)		
Protected area	50	100
Other area	50	100
Meghalaya (mg/nm³)		
Protected area	100	100
Other area	500	500

Formation of nitrogen oxides (NO$_x$)

Nitrogen oxides mainly comprising of nitrogen monoxide (NO) and nitrogen dioxide (NO$_2$), are produced during combustion. The proportion of NO$_2$ is approximately 5–10 per cent of total nitrogen oxide emissions. The nitrogen oxides are generated in the kiln during pyro-processing mainly by two mechanisms:

1. Oxides of molecular nitrogen of combustion air, i.e. thermal NO$_x$.
2. Oxidation of nitrogen compounds in fuel, i.e. fuel NO$_x$.

Formation of thermal NO$_x$ is predominant at very high temperatures, i.e. oxidising flame with a temperature above 1200°C during sufficiently long residence time of the gases. The factors which determine the concentration of NO in kiln exit gases are flame temperature and shape, excess air rate, material temperature and retention time in burning zone. Fuel NO$_x$ is generated by oxidation of nitrogen compounds present in fuels at temperatures ranging between 820° and 1100°C. Fuel NO is predominantly found in the secondary firing zone of the pre-heater and precalciner kiln systems, where the combustion temperature is well below 1200°C. The NO formation is determined by nitrogen content in the fuel, oxygen level in firing zone, initial NO concentration in combustion gas, volatile content in the fuel and secondary firing zone temperature.

Formation of sulphur dioxide (SO$_2$)

SO$_2$ formation in kiln is mainly by two mechanisms:

1. Oxidation of sulphur compounds present in raw materials and fuel.
2. Sulphur cycle conditions in the kiln inlet zone.

SO$_2$ emissions depend more on the nature and quantity of various sulphur compounds with varying volatility than on the magnitude of sulphur intake from raw meal and fuel. SO$_2$ emitted from raw material and fuel gets absorbed in various zones of the kiln. The sulphur compounds, which exceed the amount that can be adsorbed, is emitted along with kiln exit gases. SO$_2$ generation from volatile sulphur compounds takes place in upper cyclone stages of the pre-heater systems at a temperature of 3000°–6600°C.

SO$_2$ emissions due to sulphur cycle occur in kilns without precalciner. When the sulphur and alkali inputs due to raw meal and fuel are excessive, formation and build-up of SO$_2$ occurs at the kiln inlet. The formation of SO$_2$ is influenced by sulphur alkali ratio, residence time of material, clinkerisation temperature, presence of carbon in kiln feed meal and sulphur content in fuel.

Measurement of Gaseous Emissions

Gaseous emissions are monitored by the wet chemical method and electrochemical sensors. In the wet chemical method, a measured volume of air is drawn by bubbling in gas absorbing solutions. The absorbing solutions for NO$_x$ and SO$_2$ are NaOH and H$_2$O$_2$ at pH 5, respectively. The resultant increase in acidity of solution is determined by simple titration using suitable indicators. For determination of NO$_x$ and SO$_2$ by electrochemical principle, a definite volume of air is drawn through the multigas analyser. The analyser consists of number of electrochemical sensors, which sense the particular gases, i.e. NO$_x$, SO$_2$, etc. The results of the gaseous emission are displayed on an alphanumeric LCD display. Measured values can be printed for record. NCB has used both the methods for measurement of NO$_x$ and SO$_2$.

Present Status of Emission

Monitoring of NO$_x$ and SO$_2$ emission was carried out in six dry process cement plants, having nine kilns of various capacities. Monitoring was done during stabilised conditions of the kilns. Cement plants were selected considering the various aspects, such as process, capacity, age, fuel, raw material, location, emission limit, dust collector, etc. The results of NO_x and SO_2 emissions are given in Table 16.3.

Table 16.3. NO_x and SO_2 emissions.

Plant code	Kiln capacity (TPD)	NO_x emission		SO_2 emission	
		mg/nm³	kg/T of clinker	mg/nm³	kg/T of clinker
A	4600	194	0.57	11	0.032
B	2800	196	0.37	9	0.017
	3300	105	0.31	7	0.020
C	1000	217	0.78	19	0.068
	1600	119	0.44	9	0.033
D	2200	350	0.61	7	0.013
	3500	319	0.49	6	0.019
E	1800	167	0.51	8	0.025
F	3300	475	1.20	0	0

It is observed that NO_x emission varies between 105–475 mg/nm³, which is well below the prescribed limits abroad. Higher values, i.e. 475 mg/nm³ is due to high burning zone temperature. In terms of kg NO_x per ton by clinker, it is in the range of 0.3–1.2, which is also less than the USEPA standards of 1.3 kg per ton of clinker.

SO_2 emission is in the range of 0–19 mg/nm³, which is very low because of the very low sulphur (average 0.5 per cent) content of Indian coal. In terms of kg/T of clinker, it varies between 0–0.068, which is much below the USEPA standard of 1.7 kg/T of clinker.

Thus, gaseous pollutants, i.e. NO_x and SO_2 are not the dominating environmental problems in the Indian cement industry. The extent of NO_x emissions are in the range of 105–475 mg/nm³. The SO_2 emission varies between 0–19 mg/nm³. The measured data reveals that gaseous emission in Indian cement plants are minimal and are well within the regulatory standards of the developed countries, like the USA, Japan and Germany.

SECTION IV

==

Special Topics

Chapter 17

Instrumental Techniques in Environmental Chemical Analysis

INTRODUCTION

Environmental science is one of the key concerns of the latter part of the twentieth century and will continue to be into the twenty-first. Concerns for environmental protection and public health worldwide have led to extensive legislation.

The investigation and modelling of environmental systems, together with the implementation of laws and regulations, has led to a demand for a large number of environmental measurements, many of which are made by techniques falling within the broad range of analytical chemistry. Many professionals make regular use of data obtained by techniques of analytical chemistry. Thus, although not primarily analytical chemists or even chemists, they need sufficient knowledge of the background of analytical chemistry to judge the quality and limitations of the environmental data obtained.

Both analytical chemistry and environmental science have an extensive literature at varying levels of sophistication. However, there have been few attempts to link the two.

In the final quarter of the twentieth century the health of our global environment has become of matter of wide concern. This concern has stimulated a wide ranging and intensive search for an understanding of the way in which the natural environment functions, and the way in which the human race is bringing about environmental changes. These studies are heavily dependent on observation and quantitative measurements. It is clearly impossible to discuss sensibly the greenhouse effect without measurement of atmospheric carbon dioxide level; eutrophication of a lake without measurement of dissolved oxygen and nutrients; or heavy metal pollution without measurement of metal concentrations in soils and water. Many different parameters are studied, and diverse types of measurements made. These measurements must be designed and executed so as to be both relevant to the problem being studied, and reliable in themselves, i.e. they must be valid. Only by a proper understanding of the capabilities and limitations of the measurement methods themselves, and the context within which they are being applied can such validity be ensured. Increasingly, measurements are also made to demonstrate compliance with a framework of environmental legislation and regulations.

The challenges of chemical analysis in the environmental context are immense, demanding high levels of expertise and skill across a wide range of analytical methodology.

Thus analytical in particular is considered as the backbone of chemistry of environment. A thorough knowledge of the environmental analytical chemistry greatly helps us to measure extent of environmental pollution and in turn adopt suitable devices to control pollution.

NEUTRON ACTIVATION ANALYSIS

In chemistry, neutron activation analysis (NAA) is a nuclear process used for determining the concentrations of elements in a vast amount of materials. NAA allows discrete sampling of elements as it disregards the chemical form of a sample, and focuses solely on its nucleus. The method is based on neutron activation and therefore requires a source of neutrons. The sample is bombarded with neutrons, causing the elements to form radioactive isotopes. The radioactive emissions and radioactive decay paths for each element are well known. Using this information, it is possible to study spectra of the emissions of the radioactive sample, and determine the concentrations of the elements within it. A particular advantage of this technique is that it does not destroy the sample, and thus has been used for analysis of works of art and historical artifacts. NAA can also be used to determine the activity of a radioactive sample.

Neutron activation analysis is a sensitive multi-element analytical technique used for both qualitative and quantitative analysis of major, minor, trace and rare elements. NAA was discovered in 1936 by Hevesy and Levi, who found that samples containing certain rare earth elements became highly radioactive after exposure to a source of neutrons. This observation led to the use of induced radioactivity for the identification of elements. NAA is significantly different from other spectroscopic analytical techniques in that it is based not on electronic transitions but on nuclear transitions. To carry out an NAA analysis the specimen is placed into a suitable irradiation facility and bombarded with neutrons, this creates artificial radioisotopes of the elements present. Following irradiation the artificial radioisotopes decay via the emission of particles or more importantly gamma rays, which are characteristic of the element from which they were emitted.

For the NAA procedure to be successful the specimen or sample must be selected carefully. In many cases small objects can be irradiated and analysed intact without the need of sampling. But more commonly a small sample is taken, usually by drilling in an inconspicuous place. About 50 mg (one-twentieth of a gram) is a sufficient sample, so damage to the object is minimised. It is often good practice to remove two samples using two different drill bits made of different materials. This will reveal any contamination of the sample from the drill bit material itself. The sample is then encapsulated in a vial made of either high purity linear polyethylene or quartz. These sample vials come in many shapes and sizes to accommodate many specimen types. The sample and a standard are then packaged and irradiated in a suitable reactor at a constant, known neutron flux. A typical reactor used for activation uses uranium fission, providing a high neutron flux and the highest available sensitivities for most elements. The neutron flux from such a reactor is in the order of 10^{12} neutrons cm^{-2} s^{-1}. The type of neutrons generated are of relatively low kinetic energy (KE), typically less than 0.5 eV. These neutrons are termed thermal neutrons. Upon irradiation a thermal neutron interacts with the target nucleus via a non-elastic collision, causing neutron capture. This collision forms a compound nucleus which is in an excited state. The excitation energy within the compound nucleus is formed from the binding energy of the thermal neutron with the target nucleus. This excited state is unfavourable and the compound nucleus will almost instantaneously de-excite (transmutate) into a more stable configuration through the emission of a prompt particle and one or more characteristic prompt gamma photons. In most cases this more stable configuration yields a radioactive nucleus. The newly formed radioactive nucleus now decays by the emission of both particles and one or more characteristic delayed gamma photons. This decay process is at a much slower rate than the initial de-excitation and is dependent on unique half-life of the radioactive nucleus. These unique half-lives are dependent upon the particular radioactive species and can range from fractions of a second to several years. Once irradiated the sample is left for a specific decay period then placed into a detector, which will measure the nuclear decay according to either the emitted particles, or more commonly the emitted gamma rays.

Variations

NAA can vary according to a number of experimental parameters. The kinetic energy of the neutrons used for irradiation will be a major experimental parameter. The above description is of activation by slow neutrons, slow neutrons are fully moderated within the reactor and have KE <0.5 eV. Medium KE neutrons may also be used for activation, these neutrons have been only partially moderated and have KE of 0.5 eV to 0.5 MeV, and are termed epithermal neutrons. Activation with epithermal neutrons is known as Epithermal NAA (ENAA). High KE neutrons are sometimes used for activation, these neutrons are unmoderated and consist of primary fission neutrons. High KE or fast neutrons have a KE >0.5 MeV. Activation with fast neutrons is termed Fast NAA (FNAA). Another major experimental parameter is whether nuclear decay products (gamma rays or particles) are measured during neutron irradiation (prompt gamma), or at some time after irradiation (delayed gamma). PGNAA is generally performed by using a neutron stream tapped off the nuclear reactor via a beam port. Neutron fluxes from beam ports are the order of 10^6 times weaker than inside a reactor. This is somewhat compensated for by placing the detector very close to the sample reducing the loss in sensitivity due to low flux. PGNAA is generally applied to elements with extremely high neutron capture cross-sections; elements which decay too rapidly to be measured by DGNAA; elements that produce only stable isotopes; or elements with weak decay gamma ray intensities. PGNAA is characterised by short irradiation times and short decay times, often in the order of seconds and minutes. DGNAA is applicable to the vast majority of elements that form artificial radioisotopes. DG analyses are often performed over days, weeks or even months. This improves sensitivity for long-lived radionuclides as it allows short-lived radionuclide to decay, effectively eliminating interference. DGNAA is characterised by long irradiation times and long decay times, often in the order of hours, weeks or longer (Fig. 17.1).

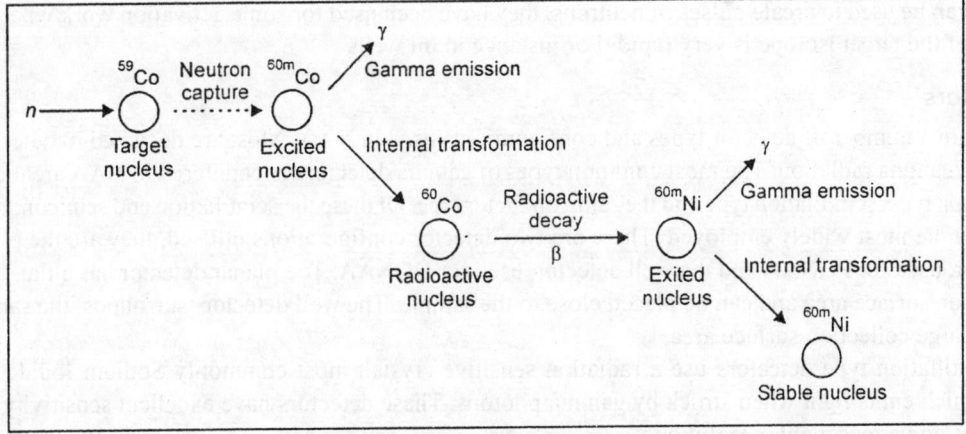

Fig. 17.1. Neutron activation analysis.

If NAA is conducted directly on irradiated samples it is termed instrumental neutron activation analysis (INAA). In some cases irradiated samples are subjected to chemical separation to remove interfering species or to concentrate the radioisotope of interest, this technique is known as radiochemical neutron activation analysis (RNAA).

Neutron Sources

A range of different sources can be used:
1. A nuclear reactor.
2. An actinoid such as californium which emits neutrons through spontaneous fission.
3. An alpha source such as radium or americium, mixed with beryllium; this generates neutrons by a $(\alpha, {}^{12}C + n)$ reaction.
4. A D-T fusion reaction in a gas discharge tube.

Reactors

Some reactors are used for the neutron irradiation of samples for radioisotope production for a range of purposes. The sample can be placed in an irradiation container which is then placed in the reactor; if epithermal neutrons are required for the irradiation then cadmium can be used to filter out the thermal neutrons.

Fusors

A relatively simple Farnsworth-Hirsch fusor can be used to generate neutrons for NAA experiments. The advantages of this kind of apparatus is that it is compact, often benchtop-sized, and that it can simply be turned off and on. A disadvantage is that this type of source will not produce the neutron flux that can be obtained using a reactor.

Isotope sources

For many workers in the field a reactor is an item which is too expensive, instead it is common to use a neutron source which uses a combination of an alpha emitter and berylium. These sources tend to be much weaker than reactors.

Gas discharge tubes

These can be used to create pulses of neutrons, they have been used for some activation work where the decay of the target isotope is very rapid. For instance in oil wells.

Detectors

There are a number of detector types and configurations used in NAA. Most are designed to detect the emitted gamma radiation. The most common types of gamma detectors encountered in NAA are the gas ionisation type, scintillation type and the semiconductor type. Of these the scintillation and semiconductor type are the most widely employed. There are two detector configurations utilised, they are the planar detector, used for PGNAA and the well detector, used for DGNAA. The planar detector has a flat, large collection surface area and can be placed close to the sample. The well detector 'surrounds' the sample with a large collection surface area.

Scintillation type detectors use a radiation sensitive crystal, most commonly Sodium Iodide NaI (TI), which emits light when struck by gamma photons. These detectors have excellent sensitivity and stability, and a reasonable resolution.

Semiconductor detectors utilise the semiconducting element germanium. The germanium is processed to form a p-i-n (positive-intrinsic-negative) diode, and when cooled to ~77 K by liquid nitrogen to reduce dark current and detector noise, produces a signal which is proportional to the photon energy of the incoming radiation. There are two types of germanium detector, the lithium-drifted germanium or Ge(Li) (pronounced 'jelly'), and the high-purity germanium or HPGe. The semiconducting element silicon may also be used but germanium is preferred, as its higher atomic number makes it more efficient at stopping and detecting high energy gamma rays. Both Ge(Li) and HPGe detectors have excellent sensitivity and resolution, but Ge(Li) detectors are unstable at room temperature, with the lithium drifting into the intrinsic region ruining the detector. The development of undrifted high purity germanium has overcome this problem.

Particle detectors can also be used to detect the emission of alpha (α) and beta (β) particles which often accompany the emission of a gamma photon but are less favourable, as these particles are only emitted from the surface of the sample and are often absorbed or attenuated by atmospheric gases requiring expensive vacuum conditions to be effectively detected. Whereas gamma rays are not absorbed or attenuated by atmospheric gases, and can also escape from deep within the sample with minimal absorption.

Analytical Capabilities

NAA can detect up to 74 elements depending upon the experimental procedure. With minimum detection limits ranging from 0.1 to 1×10^6 ng g^{-1} depending on element under investigation. Heavier elements have larger nuclei, therefore they have a larger neutron capture cross-section and are more likely to be activated. Some nuclei can capture a number of neutrons and remain relatively stable, not undergoing transmutation or decay for many months or even years. Other nuclei decay instantaneously or form only stable isotopes and can only be identified by PGNAA.

ANODIC STRIPPING VOLTAMMETRY

Anodic stripping voltammetry is a voltammetric method for quantitative determination of specific ionic species. The analyte of interest is electroplated on the working electrode during a deposition step, and oxidised from the electrode during the stripping step. The current is measured during the stripping step. The oxidation of species is registered as a peak in the current signal at the potential at which the species begins to be oxidised. The stripping step can be either linear, staircase, squarewave, or pulse (Fig. 17.2).

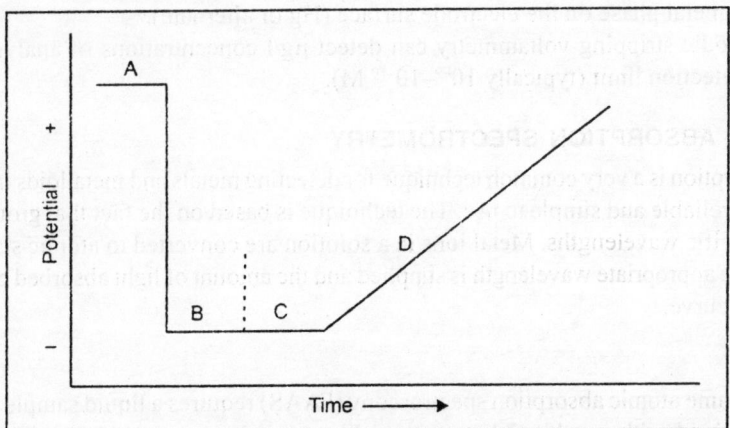

Fig. 17.2. Anodic stripping voltammetry.

Electrochemical Cell Set-up

Anodic stripping voltammetry usually incorporates three electrodes, a working electrode, auxiliary electrode (sometimes called the counter electrode), and reference electrode. The solution being analysed usually has an electrolyte added to it. For most standard tests, the working electrode is a mercury film electrode. The mercury film forms an amalgam with the analyte of interest, which upon oxidation results in a sharp peak, improving resolution between analytes. The mercury film is formed over a glassy carbon electrode. A mercury drop electrode has also been used for much the same reasons. In cases where the analyte of interest has an oxidising potential above that of mercury, or where a mercury electrode would be otherwise unsuitable, a solid, inert metal such as silver, gold, or platinum may also be used.

Steps

Anodic stripping voltammetry usually incorporates 4 steps if the working electrode is a mercury film or mercury drop electrode and the solution incorporates stirring. The solution is stirred during the first two steps at a repeatable rate. The first step is a cleaning step; in the cleaning step, the potential is held at a more oxidising potential than the analyte of interest for a period of time in order to fully remove it from the electrode. In the second step, the potential is held at a lower potential, low enough to reduce the analyte and deposit it on the electrode. After the second step, the stirring is stopped, and the electrode is kept at the lower potential. The purpose of this third step is to allow the deposited material to distribute more evenly in the mercury. If a solid inert electrode is used, this step is unnecessary. The last step involves raising the working electrode to a higher potential (anodic), and stripping (oxidising) the analyte. As the analyte is oxidised, it gives off electrons which are measured as a current.

Stripping analysis is an analytical technique that involves: (i) preconcentration of a metal phase onto a solid electrode surface or into Hg (liquid) at negative potentials, and (ii) selective oxidation of each metal phase species during an anodic potential sweep. Very sensitive and reproducible (RSD<5 per cent) method for trace metal ion analysis in aqueous media. Concentration limits of detection for many metals are in the low ppb to high ppt range (S/N = 3) and this compares favourably with AAS or ICP analysis. Field deployable instrumentation that is inexpensive. Approximately 12–15 metal ions can be analysed for by this method. The stripping peak currents and peak widths are a function of the size, coverage and distribution of the metal phase on the electrode surface (Hg or alternate).

Sensitivity: Anodic stripping voltammetry can detect $\mu g/l$ concentrations of analyte. This method has an excellent detection limit (typically 10^{-9}–10^{-10} M).

FLAME ATOMIC ABSORPTION SPECTROMETRY

Flame atomic absorption is a very common technique for detecting metals and metalloids in environmental samples. It is very reliable and simple to use. The technique is based on the fact that ground state metals absorb light at specific wavelengths. Metal ions in a solution are converted to atomic state by means of a flame. Light of the appropriate wavelength is supplied and the amount of light absorbed can be measured against a standard curve.

Basic Principle

The technique of flame atomic absorption spectroscopy (FAAS) requires a liquid sample to be aspirated, aerosolised, and mixed with combustible gases, such as acetylene and air or acetylene and nitrous oxide. The mixture is ignited in a flame whose temperature ranges from 2100° to 2800°C.

During combustion, atoms of the element of interest in the sample are reduced to free, unexcited ground state atoms, which absorb light at characteristic wavelengths, as shown in Fig. 17.3.

The characteristic wavelengths are element specific and accurate to 0.01–0.1 nm. To provide element specific wavelengths, a light beam from a lamp whose cathode is made of the element being determined is passed through the flame. A device such as photonmultiplier can detect the amount of reduction of the light intensity due to absorption by the analyte, and this can be directly related to the amount of the element in the sample.

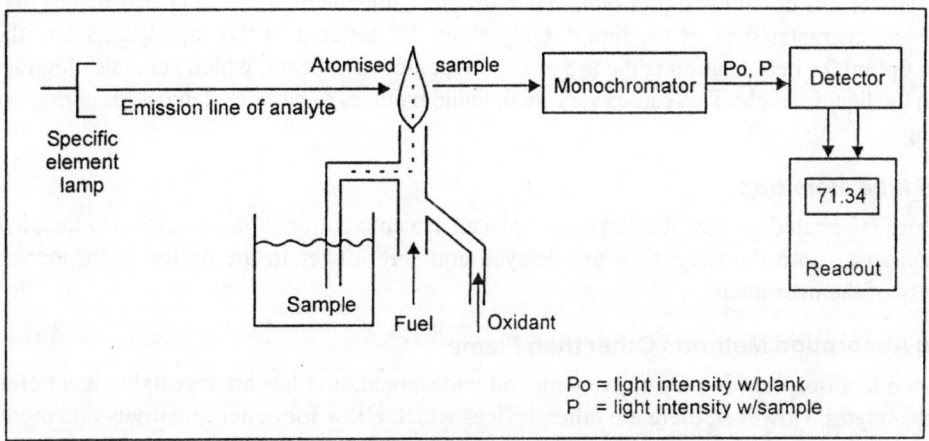

Fig. 17.3. Atomic spectroscopy with flames.

Flame atomic absorption hardware is divided into six fundamental groups that have two major functions: generating atomic signals and signal processing. Signal processing is a growing additional feature to be integrated or externally fitted to the instrument.

A cathode lamp is a stable light source, which is necessary to emit the sharp characteristic spectrum of the element to be determined. A different cathode lamp is needed for each element, although there are some lamps that can be used to determine three or four different elements if the cathode contains all of them. Each time a lamp is changed, proper alignment is needed in order to get as much light as possible through the flame, where the analyte is being atomised, and into the monochromator.

The atom cell is the part with two major functions: nebulisation of sample solution into a fine aerosol solution, and dissociation of the analyte elements into free gaseous ground state form. Not all the analyte goes through the flame, part of it is disposed.

As the sample passes through the flame, the beam of light passes through it into the monochromator. The monochromator isolates the specific spectrum line emitted by the light source through spectral dispersion, and focuses it upon a photomultiplier detector, whose function is to convert the light signal into an electrical signal. The processing of electrical signal is fulfilled by a signal amplifier. The signal could be displayed for readout, or further fed into a data station for printout by the requested format.

Operation of Flame for Atomic Absorption

Types of flame

Different flames can be achieved using different mixtures of gases, depending on the desired temperature and burning velocity. Some elements can only be converted to atoms at high temperatures. Even at high temperatures, if excess oxygen is present, some metals form oxides that do not redissociate into atoms. To inhibit their formation, conditions of the flame may be modified to achieve a reducing, nonoxidising flame.

Ultrasonic nebulisation

Proper nebulisation is required to break up an aqueous sample into a fine mist of uniform droplet size that can be readily burned in the flame. Most instruments utilise the direct aspiration. During aspiration, the gas flow breaks down the liquid sample into droplets, and the nebulisation performance depends on the physical characteristics of the liquid. Only about 10 per cent of the sample gets into the flame. Another option for nebulisation is the use of an ultrasonic wave beam, which generates high frequency waves in the liquid sample. This causes very small liquid particles to be ejected into a gas current forming a dense fog.

Slotted tube atom trap

This device is a heated quartz/tube that can be placed in a conventional flame. As the dissociated ground state atoms pass into the tube, they are delayed and stay longer in the optical path, increasing the sensitivity of the instrument.

Atomic Absorption Methods Other than Flame

Flame atomic absorption is very convenient and widespread, and has an acceptable level of accuracy for most analytes. However, there are other devices which allow for better sensitivity and more control over the chemical environment of the analyte.

Electrothermal atomisation

This type of atomisation requires a graphite furnace, where after thermal pre-treatment the sample is rapidly atomised. To maintain a dense fraction of free ground state elements in the optical path, an inert gas atmosphere is used. Since the dilution and expansion effects of flame cells are avoided, and the atoms have a longer residence time in the optical path, a higher peak concentration of atoms is obtained.

Carbon rod analyser

This device can be used to convert a powdered sample into atomic vapour. A current is applied to a very thin, heated carbon rod that contains the solid sample in order to vapourise it.

Tantalum boat analyser

This is another technique that produces an atomic vapour from a solid sample. A Tantalum boat is electrically heated in a manner similar to the carbon rod system, within an inert atmosphere.

Techniques of Measurement and EPA Methods Using FAAS

Atomic absorption spectrometry is a fairly universal analytical method for determination of metallic elements when present in both trace and major concentrations. The EPA employs this technique for determining the metal concentration in samples from a variety of matrices.

Sample preparation

Depending on the information required, total recoverable metals, dissolved metals, suspended metals, and total metals could be obtained from a certain environmental matrix.

Appropriate acid digestion is employed in these methods. Hydrochloric acid digestion is not suitable for samples which will be analysed by graphite furnace atomic absorption spectroscopy because it can cause interferences during furnace atomisation.

Calibration and standard curves

As with other analytical techniques, atomic absorption spectrometry requires careful calibration. EPA QA/QC demands calibration through several steps, including interference check sample, calibration verification, calibration standards, bland control, and linear dynamic range.

The idealised calibration or standard curve is stated by Beer's law that the absorbance of an absorbing analyte is proportional to its concentration. Unfortunately, deviations from linearity usually occur, especially as the concentration of metallic analytes increases due to various reasons, such as unabsorbed radiation, stray light, or disproportionate decomposition of molecules at high concentrations. Figure 17.4 shows an idealised and deviation of response curve. The curvature could be minimised, although it is impossible to be avoided completely. It is desirable to work in the linearity response range. The rule of thumb is that a minimum of five standards and a blank should be prepared in order to have sufficient information to fit the standard curve appropriately. Manufacturers should be consulted if a manual curvature correction function is available for a specific instrument.

If the sample concentration is too high to permit accurate analysis in linearity response range, there are three alternatives that may help bring the absorbance into the optimum working range:

1. Sample dilution.
2. Using an alternative wavelength having a lower absorptivity.
3. Reducing the path length by rotating the burner hand.

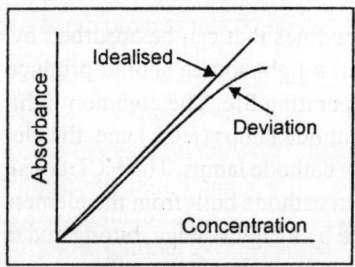

Fig. 17.4. Idealised/deviation response curve.

EPA method for metal analysis

Flame atomic absorption methods are referred to as direct aspiration determinations. They are normally completed as single element analyses and are relatively free of interelement spectral interferences. For some elements, the temperature or type of flame used is critical. If flame and analytical conditions are not properly used, chemical and ionisation interferences can occur. Graphite furnace atomic absorption spectrometry replaces the flame with an electrically heated graphite furnace. The major advantage of this technique is that the detection limit can be extremely low. It is applicable for relatively clean samples, however, interferences could be a real problem. It is important for the analyst to establish a set of analytical protocol which is appropriate for the sample to be analysed and for the information required.

Main Components of the Atomic Absorption Spectrophotometer

The basic components of any type or brand of atomic absorption include the following:

1. Light source.
2. Burner/nebuliser.

3. Monochromator.
4. Photomultiplier detector.
5. Output device.

The primary sources of radiation in atomic absorption are hollow cathode lamps. These lamps are composed of a cathode and an anode sealed in a tube with an inert gas (argon or neon). The cathode is made of the element to be determined. When a high voltage is applied the atoms of the inert gas are ionised and attracted by the cathode. These ions hit the cathode and excite the atoms of the elements used to make the cathode. Once the atoms are excited radiation is emitted at the characteristic wavelength of the element. The light from the hollow cathode lamp passes through the flame (Burner/nebuliser) where the sample is atomised. This fine mist of the sample is sprayed into the nebuliser. Atoms of the elements are formed from the sample mist and are able to absorb some of the light from the lamp at the wavelength set for that particular element.

The light passed through the flame is received by the monochromator, which is set to accept and transmit radiation at the specified wavelength. The light emerges from the monochromator exit slit and falls on the photomultiplier detector. At this point an output current, proportional to the incident light, is intensified, amplified, processed electronically and finally presented to a readout device (i.e. printer, digital display).

Light source

The light source generates resonance lines that can be absorbed by atoms of the element of unknown concentration. The ideal activity that a light source should produce is a steady, low noise signal with very little interference and a long operating life. The common light sources used in atomic absorption spectrophotometry are the hollow cathode lamps (HCL) and the electrode discharge lamps (EDL). The Perkin-Elmer model 460 uses hallow cathode lamps. The HCL consists basically of three parts: a sealed glass tube filled with argon or neon, a cathode built from the element of interest, and an anode. The gas ions are accelerant ahead the cathode by a high voltage, hitting and displacing atoms of the element that are then excited by collisions with gas ions. As atoms suffer excitation they radiate light of the proper wavelength for absorption by atoms of the same element in the flame. The HCL can be of two types depending on the material used to build the cathode: single or multi elements. HCL are commonly use as the light source for most elements, however with volatile elements short lamp life and low light intensity may be a problem.

Burner/nebuliser

The most common method to break chemical bonds to produce ground-state atoms is the flame atomiser. The flame system components include a burner and a nebuliser. The flame is generated in the burner where combustion occurs and an atomic vapour of the element to be analysed is produced. The nebuliser, which may be pneumatic or ultrasonic, converts sample solution into a fine mist or aerosol that is fed to the flame. The nebuliser is a device operating on the principle of a scent or paint spray. Today, most of the burners used are those of long-slot designed burning premixed fuel and oxidant gases and fitted with a pneumatic nebuliser. The selection of the flame is important for complete atomisation and to avoid ionisation. When the flame temperature is too low, atomisation will be incomplete since the flame cannot supply sufficient energy to dissociate the compounds in the sample. On the other hand, if the flame temperature is too high, the atoms formed may be ionised reducing the number of atoms present.

Monochromator

The monochromator is included as an important device of the optical system of an atomic absorption spectrophotometer. The function of this device in atomic absorption is to separate the spectral line of interest from others spectral lines with different wavelengths emitted by the hollow-cathode lamp. The desired spectral line is chosen with the preferred wavelength and bandwidth by an appropriate monochromator's setting named grating. A grating is a reflective surface, scored either mechanically or holographically with parallel grooves that can be designed for different wavelength regions. Generally, most of the instruments are equipped with two gratings with the goal to cover a wavelength range from 189 to 851 nm which is used in atomic absorption.

Photomultiplier detector

The monochromator receives light from the hollow-cathode lamp through the flame together with the light emitted from the flame. The signal arising at the detector from the flame emission will be rejected and only that from the hollow-cathode lamp will be accepted. The detector used almost universally is a photomiltiplier tube whose current output corresponds to the intensity of the light falling on its photocathode. This feeds the amplifier and output device, which displays the measured signals.

Output device

The simplest output devices consist of a moving–coil meter or a pen recorder displaying percentage transmission (%T). The model 460 has a digital display, which provides a direct readout of absorbance values. This, with a provision for curve linearisation, forms the basis for displaying outputs directly in concentration terms, using standard solutions for calibrations. At present, most aspects of instrument control, operation, standardisation and data processing or storage and carried out by a microcomputer or microprocessor built in into the atomic absorption or interfaced to it.

Cold Vapour Atomic Fluorescence Spectroscopy

Cold vapour atomic fluorescence spectroscopy, sometimes referred to by the acronym CVAFS, is a subset of the analytical technique known as atomic fluorescence spectroscopy (AFS). It is used in the measurement of trace amounts of volatile heavy metals such as mercury. cold vapour AFS makes use of the unique characteristic of mercury that allows vapour measurement at room temperature. Free mercury atoms in a carrier gas are excited by a collimated ultraviolet light source at a wavelength of 253.7 nanometres. The excited atoms reradiate their absorbed energy (fluoresce) at this same wavelength. Unlike the directional excitation source, the fluorescence is omnidirectional and may thus be detected using a photomultiplier tube or UV photodiode. The technique differs from the more conventional atomic absorption (AA) technique in that it is more sensitive, more selective, and is linear over a wide range of concentrations. However, any molecular species present in the carrier gas will quench the fluorescence signal and for this reason, the technique is most commonly used with an inert carrier gas such as argon. Gold coated traps may be used to collect mercury in ambient air or other media. The traps are then heated, releasing the mercury from the gold while passing argon through the cartridge. This preconcentrates the mercury, increasing sensitivity, and also transfers the mercury into an inert gas.

Inductively Coupled Plasma Optical Emission Spectroscopy

Inductively coupled plasma optical emission spectroscopy (ICP-OES) is a major technique for elemental analysis. The sample to be analysed, if solid, is normally first dissolved and then mixed with water before being fed into the plasma.

The first step in the procedure is the conversion of the molecules in the sample to individual atoms and ions using a high temperature radio frequency induced argon plasma. The sample is introduced into the plasma as a solution. Sample is pumped using a peristaltic pump to a nebuliser, where it is converted to a fine spray and mixed with argon in a spray chamber. The purpose of the spray chamber is to make sure that only droplets in a narrow size range make it through into the plasma. Most of the sample drains away from the chamber; the rest is carried into the plasma and instantly excited by the high temperatures (5000–10,000 K). Atoms become ionised with 99 per cent efficiency (arsenic and selenium are a couple of exceptions, ionising only at 52 per cent and 33 per cent). Either ICP-optical emission spectrometry (ICP-OES) or ICP mass spectrometry (ICP-MS) can be used to analyse samples. ICP-OES utilises UV and visible spectrometry to image the plasma at the exact wavelength of ionic excitation of the element of interest. This is a well-established technique. ICP-MS is a somewhat newer tool in the biological sciences and is described in detail below.

Sample throughput with atomic emission is very rapid when using automated systems capable of multi-elemental analysis. For example, sampling rates of 3000 determinations per hour have been achieved using an ICP with simultaneous analysis, and 300 determinations per hour with a sequential ICP. Flame emission is often accomplished using an atomic absorption spectrometer.

Plasma

A plasma consists of a hot, partially ionised gas, containing an abundant concentration of cations and electrons that make the plasma a conductor. The plasmas used in atomic emission are formed by ionising a flowing stream of argon, producing argon ions and electrons. The high temperatures in a plasma result from resistive heating that develops due to the movement of the electrons and argon ions. Because plasmas operate at much higher temperatures than flames, they provide better atomisation and more highly populated excited states. Besides neutral atoms, the higher temperatures of plasma also produce ions of the analyte. The ICP torch consists of three concentric quartz tubes, surrounded at the top by a radio-frequency induction coil. The sample is mixed with a stream of Ar using a spray chamber nebuliser similar to that used for flame emission and is carried to the plasma through the torch's central tube. Plasma formation is initiated by a spark from a Tesla coil.

An alternating radiofrequency current in the induction coils creates a fluctuating magnetic field that induces the argon ions and electrons to move in a circular path. The resulting collisions with the abundant unionised gas give rise to resistive heating, providing temperatures as high as 10,000 K at the base of the plasma, and between 6000 and 8000 K at a height of 15–20 mm above the coil, where emission is usually measured. At these high temperatures the outer quartz tube must be thermally isolated from the plasma. This is accomplished by the tangential flow of argon shown in Fig. 17.5.

Multi-elemental analysis atomic emission spectroscopy is ideally suited for multi-elemental analysis because all analytes in a sample are excited simultaneously. A scanning monochromator can be programmed to move rapidly to an analyte's desired wavelength, pausing to record its emission intensity before moving to the next analyte's wavelength. Proceeding in this fashion, it is possible to analyse three or four analytes per minute. Another approach to multi-elemental analysis is to use a multichannel instrument that allows for the simultaneous monitoring of many analytes. A simple design for a multichannel spectrometer consists of a standard diffraction grating and 48–60 separate exit-slits and detectors positioned in a semicircular array around the diffraction grating at positions corresponding to the desired wavelengths.

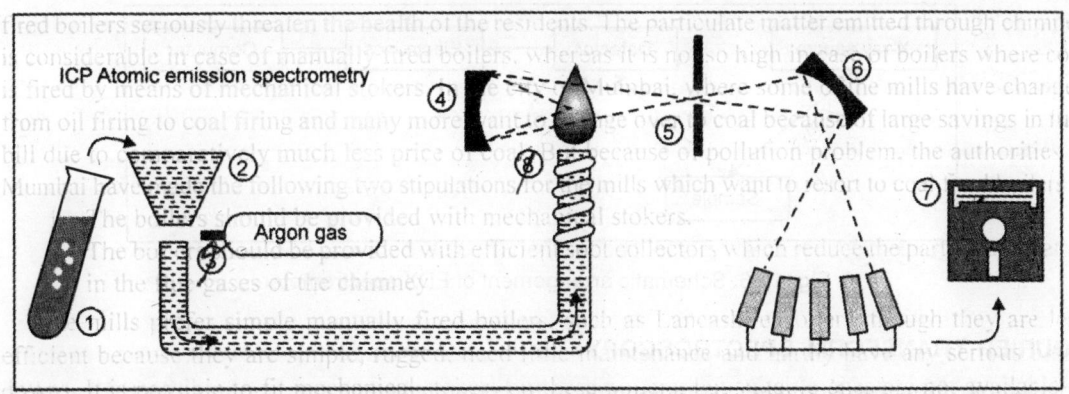

Fig. 17.5. Inductively coupled plasma-atomic emission spectrometery the (1) aqueous sample is pumped and (2) atomised with argon gas into the (3) hot plasma. The sample is excited, emitting light wavelengths characteristic of its elements. (4) A mirror reflects the light through the (5) entrance slit of the spectrometer onto a (6) grating that separates the element wavelengths onto (7) photomultiplier detectors.

X-ray Fluorescence

X-ray fluorescence (XRF) is the emission of characteristic 'secondary' (or fluorescent) X-rays from a material that has been excited by bombarding with high-energy X-rays or gamma rays. The phenomenon is widely used for elemental analysis and chemical analysis, particularly in the investigation of metals, glass, ceramics and building materials, and for research in geochemistry, forensic science and archaeology.

When materials are exposed to short-wavelength X-rays or to gamma rays, ionisation of their component atoms may take place. Ionisation consists of the ejection of one or more electrons from the atom, and may take place if the atom is exposed to radiation with an energy greater than its ionisation potential. X-rays and gamma rays can be energetic enough to expel tightly held electrons from the inner orbitals of the atom. The removal of an electron in this way renders the electronic structure of the atom unstable, and electrons in higher orbitals 'fall' into the lower orbital to fill the hole left behind. In falling, energy is released in the form of a photon, the energy of which is equal to the energy difference of the two orbitals involved. Thus, the material emits radiation, which has energy characteristic of the atoms present. The term fluorescence is applied to phenomena in which the absorption of radiation of a specific energy results in the re-emission of radiation of a different energy (generally lower).

Chemical Analysis

The use of a primary X-ray beam to excite fluorescent radiation from the sample was first proposed by Glocker and Schreiber in 1928. Today, the method is used as a non-destructive analytical technique, and as a process control tool in many extractive and processing industries. In principle, the lightest element that can be analysed is beryllium ($Z = 4$), but due to instrumental limitations and low X-ray yields for the light elements, it is often difficult to quantify elements lighter than sodium ($Z = 11$), unless background corrections and very comprehensive interelement corrections are made.

Energy dispersive spectrometry

In energy dispersive spectrometers (EDX or EDS), the detector allows the determination of the energy of the photon when it is detected. Detectors historically have been based on silicon semiconductors, in the form of lithium-drifted silicon crystals, or high-purity silicon wafers (Fig. 17.6).

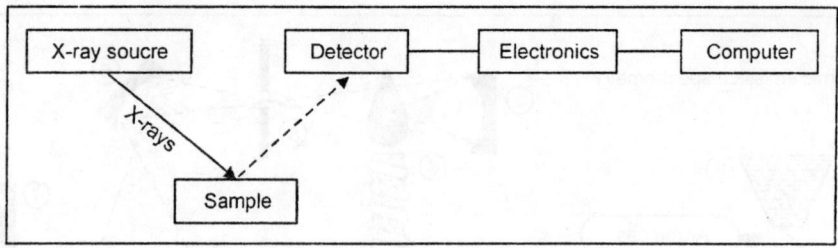

Fig. 17.6. Schematic arrangement of EDX spectrometer.

FOURIER TRANSFORM SPECTROSCOPY

Fourier transform spectroscopy is a measurement technique whereby spectra are collected based on measurements of the coherence of a radiative source, using time-domain or space-domain measurements of the electromagnetic radiation or other type of radiation. It can be applied to a variety of types of spectroscopy including optical spectroscopy, infrared spectroscopy (FTIR, FT-NIRS), nuclear magnetic resonance (NMR) and magnetic resonance spectroscopic imaging (MRSI), mass spectrometry and electron spin resonance spectroscopy. There are several methods for measuring the temporal coherence of the light, including the continuous wave Michelson or Fourier transform spectrometer and the pulsed Fourier transform spectrograph (which is more sensitive and has a much shorter sampling time than conventional spectroscopic techniques, but is only applicable in a laboratory environment).

The term Fourier transform spectroscopy reflects the fact that in all these techniques, a Fourier transform is required to turn the raw data into the actual spectrum, and in many of the cases in optics involving interferometers, is based on the Wiener–Khinchin theorem.

Measuring an Emission Spectrum

One of the most basic tasks in spectroscopy is to characterise the spectrum of a light source: How much light is emitted at each different wavelength. The most straightforward way to measure a spectrum is to pass the light through a monochromator, an instrument that blocks all of the light except the light at a certain wavelength (the un-blocked wavelength is set by a knob on the monochromator). Then the intensity of this remaining (single-wavelength) light is measured. The measured intensity directly indicates how much light is emitted at that wavelength. By varying the monochromator's wavelength setting, the full spectrum can be measured. This simple scheme in fact describes how some spectrometers work.

Fourier transform spectroscopy is a less intuitive way to get the same information. Rather than allowing only one wavelength at a time to pass through to the detector, this technique lets through a beam containing many different wavelengths of light at once, and measures the total beam intensity. Next, the beam is modified to contain a different combination of wavelengths, giving a second data point. This process is repeated many times. Afterwards, a computer takes all this data and works backwards to infer how much light there is at each wavelength (Fig. 17.7).

To be more specific, between the light source and the detector, there is a certain configuration of mirrors that allows some wavelengths to pass through but blocks others (due to wave interference). The beam is modified for each new data point by moving one of the mirrors; this changes the set of wavelengths that can pass through. Computer processing is required to turn the raw data (light intensity for each mirror position) into the desired result (light intensity for each wavelength). The processing required turns out to be a common algorithm called the Fourier transform (hence the name, 'Fourier transform spectroscopy'). The raw data is sometimes called an 'interferogram'.

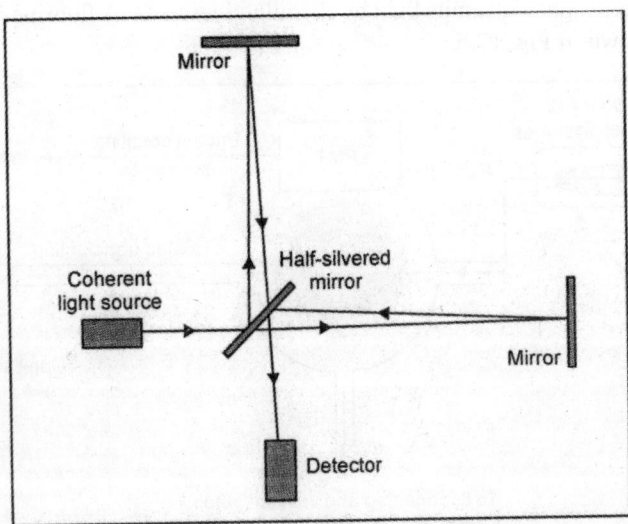

Fig. 17.7. The Fourier transform spectrometer is just a Michelson interferometer but one of the two fully-reflecting mirrors is movable, allowing a variable delay (in the travel-time of the light) to be included in one of the beams.

Measuring an Absorption Spectrum

The method of Fourier transform spectroscopy can also be used for absorption spectroscopy. The primary example is 'FTIR spectroscopy', a common technique in chemistry.

In general, the goal of absorption spectroscopy is to measure how well a sample absorbs or transmits light at each different wavelength. Although absorption spectroscopy and emission spectroscopy are different in principle, they are closely related in practice; any technique for emission spectroscopy can also be used for absorption spectroscopy. First, the emission spectrum of a broadband lamp is measured (this is called the background spectrum). Second, the emission spectrum of the same lamp shining through the sample is measured (this is called the 'sample spectrum'). The sample will absorb some of the light, causing the spectra to be different. The ratio of the 'sample spectrum' to the 'background spectrum' is directly related to the sample's absorption spectrum.

Accordingly, the technique of 'Fourier transform spectroscopy' can be used both for measuring emission spectra (for example, the emission spectrum of a star), and absorption spectra (for example, the absorption spectrum of a glass of liquid).

Chemiluminescence

Chemiluminescence (sometimes chemoluminescence) is the emission of light with limited emission of heat (luminescence), as the result of a chemical reaction. Chemiluminescence differs from fluorescence in that the electronic excited state is derived from the product of a chemical reaction rather than the more typical way of creating electronic excited states, namely absorption. It is the antithesis of a photochemical reaction, in which light is used to drive an endothermic chemical reaction. Here, light is generated from a chemically exothermic reaction. A standard example of chemiluminescence in the laboratory setting is found in the luminol test, where evidence of blood is taken when the sample glows upon contact with iron. When chemiluminescence takes place in living organisms, the phenomenon is called bioluminescence. A lightstick emits light by chemiluminescence. A fluorine-induced chemiluminescence detection is shown in Fig. 17.8.

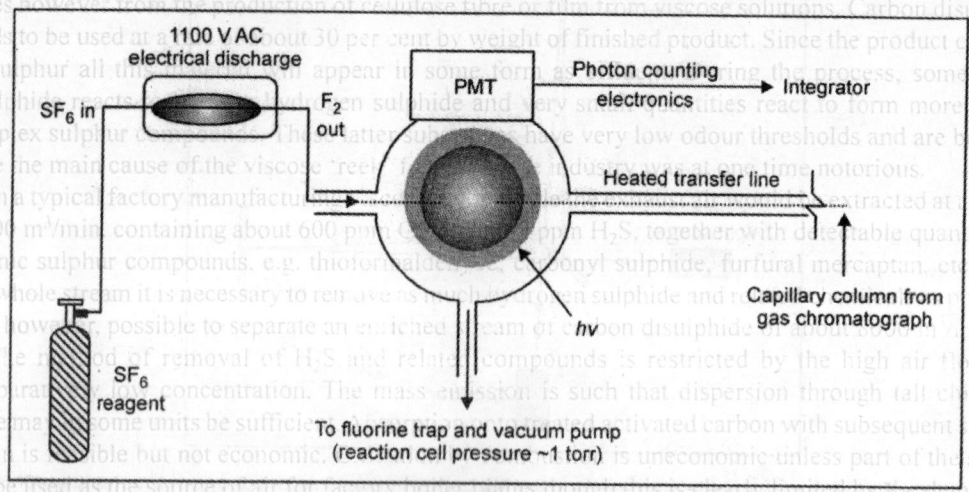

Fig. 17.8. Fluorine-induced chemiluminescence detection.

Gas-phase reactions

1. One of the oldest known chemoluminescent reactions is that of elemental white phosphorus oxidising in moist air, producing a green glow. This is a gas-phase reaction of phosphorus vapour, above the solid, with oxygen producing the excited states (PO_2) and HPO.

2. Another gas phase reaction is the basis of nitric oxide detection in commercial analytic instruments applied to environmental air quality testing. Ozone is combined with nitric oxide to form nitrogen dioxide in an activated state.

$$NO + O_3 \rightarrow NO_2 + O_2$$

The activated NO_2 luminesces broadband visible to infrared light as it reverts to a lower energy state. A photomultiplier and associated electronics counts the photons which are proportional to the amount of NO present. To determine the amount of nitrogen dioxide, NO_2, in a sample (containing no NO) it must first be converted to nitric oxide, NO, by passing the sample through a converter before the above ozone activation reaction is applied. The ozone reaction produces a photon count proportional to NO which is proportional to NO_2 before it was converted to NO. In the case of a mixed sample containing both NO and NO_2, the above reaction yields the amount of NO and NO_2 combined in the air sample, assuming that the sample is passed through the converter. If the mixed sample is not passed through the converter, the ozone reaction produces activated NO_2 only in proportion to the NO in the sample. The NO_2 in the sample is not activated by the ozone reaction. Though unactivated NO_2 is present with the activated NO_2, photons are only emitted by the activated species which is proportional to original NO. Final step, subtract NO from (NO + NO_2) to yield NO_2.

Ion Selective Electrode

An ion-selective electrode (ISE), also known as a specific ion electrode (SIE), is a transducer (or sensor) that converts the activity of a specific ion dissolved in a solution into an electrical potential, which can be measured by a voltmeter or pH meter. The voltage is theoretically dependent on the logarithm of the ionic activity, according to the Nernst equation. The sensing part of the electrode is usually made as an ion-specific membrane, along with a reference electrode. Ion-selective electrodes are used in biochemical and biophysical research, where measurements of ionic concentration in an aqueous solution are required, usually on a real time basis. Figure 17.9 shows the ion selective electrode.

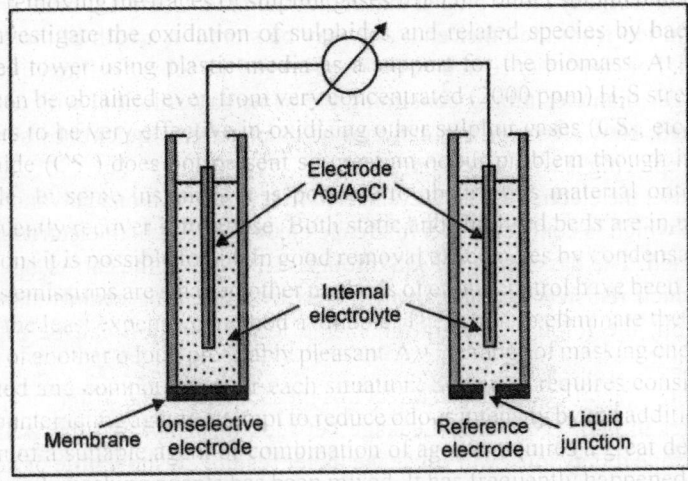

Fig. 17.9. Ion-selective electrode.

Types of ion-selective membrane

There are four main types of ion-selective membrane used in ion-selective electrodes: glass, solid state, liquid based, and compound electrode.

Glass membranes: Glass membranes are made from an ion-exchange type of glass (silicate or chalcogenide). This type of ISE has good selectivity, but only for several single-charged cations; mainly H^+, Na^+, and Ag^+. Chalcogenide glass also has selectivity for double-charged metal ions, such as Pb^{2-}, and Cd^{2+}. The glass membrane has excellent chemical durability and can work in very aggressive media. A very common example of this type of electrode is the pH glass electrode.

Crystalline membranes: Crystalline membranes are made from mono- or polycrystallites of a single substance. They have good selectivity, because only ions which can introduce themselves into the crystal structure can interfere with the electrode response. Selectivity of crystalline membranes can be for both cation and anion of the membrane-forming substance. An example is the fluoride selective electrode based on LaF_3 crystals.

Ion-exchange resin membranes: Ion-exchange resins are based on special organic polymer membranes which contain a specific ion-exchange substance (resin). This is the most widespread type of ion-specific electrode. Usage of specific resins allows preparation of selective electrodes for tens of different ions, both single-atom or multi-atom. They are also the most widespread electrodes with anionic selectivity.

However, such electrodes have low chemical and physical durability as well as 'survival time'. An example is the potassium selective electrode, based on valinomycin as an ion-exchange agent.

Construction: These electrodes are prepared from glass capillary tubing approximately 2 millimeters in diameter, a large batch at a time. Polyvinyl chloride is dissolved in a solvent and plasticisers (typically phthalates) added, in the standard fashion used when making something out of vinyl. In order to provide the ionic specificity, a specific ion channel or carrier is added to the solution; this allows the ion to pass through the vinyl, which prevents the passage of other ions and water.

One end of a piece of capillary tubing about an inch or two long is dipped into this solution and removed to let the vinyl solidify into a plug at that end of the tube. Using a syringe and needle, the tube is filled with salt solution from the other end, and may be stored in a bath of the salt solution for an indeterminate period. For convenience in use, the open end of the tubing is fitted through a tight o-ring into a somewhat larger diameter tubing containing the same salt solution, with a silver or platinum electrode wire inserted. New electrode tips can thus be changed very quickly by simply removing the older electrode and replacing it with a new one.

Applications: In use, the electrode wire is connected to one terminal of a galvanometer or pH meter, the other terminal of which is connected to a reference electrode, and both electrodes are immersed in the solution to be tested. The passage of the ion through the vinyl via the carrier or channel creates an electrical current, which registers on the galvanometer; by calibrating against standard solutions of varying concentration, the ionic concentration in the tested solution can be estimated from the galvanometer reading.

In practice there are several issues which affect this measurement, and different electrodes from the same batch will differ in their properties. Leakage between the vinyl and the wall of the capillary, thereby allowing passage of any ions, will cause the meter reading to show little or no change between the various calibration solutions, and requires that that electrode be discarded. Similarly, with use the ion-sensitive channels in the vinyl appear to gradually become blocked or otherwise inactivated, causing the electrode to lose sensitivity. The response of the electrode and galvanometer is temperature sensitive, and also 'drifts' over time, requiring recalibration frequently during a series of measurements, ideally at least one calibration sample before and after each test sample. On the other hand, after immersion in the solution there is a 'settling time' which can be five minutes or even longer, before the electrode and galvanometer equilibrate to a new reading; so that timing of the reading is critical in order to find the most accurate 'window' after the response has settled, but before it has drifted appreciably.

Chromatography

Chromatography is the collective term for a set of laboratory techniques for the separation of mixtures. It involves passing a mixture dissolved in a 'mobile phase' through a stationary phase, which separates the analyte to be measured from other molecules in the mixture based on differential partitioning between the mobile and stationary phases. Subtle differences in compounds partition coefficient results in differential retention on the stationary phase and thus separation. The basic equipment for chromatography is shown in Fig. 17.10.

Chromatography may be preparative or analytical. The purpose of preparative chromatography is to separate the components of a mixture for further use (and is thus a form of purification). Analytical chromatography is done normally with smaller amounts of material and is for measuring the relative proportions of analytes in a mixture. The two are not mutually exclusive.

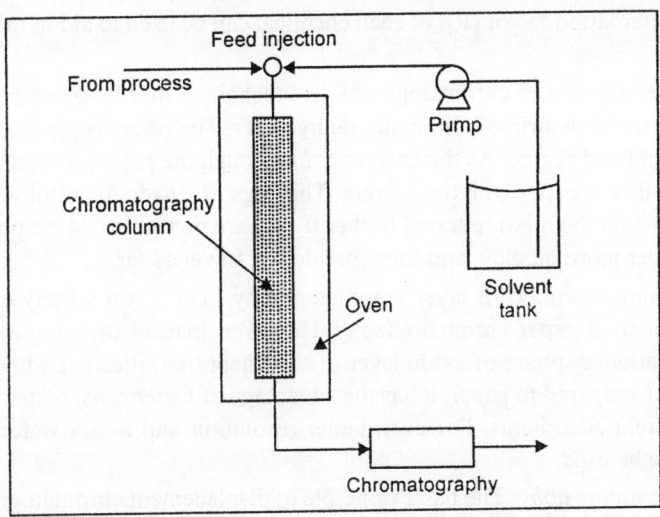

Fig. 17.10. Basic equipment for chromatography.

Techniques by chromatographic bed shape

Column chromatography: Column chromatography is a separation technique in which the stationary bed is within a tube. The particles of the solid stationary phase or the support coated with a liquid stationary phase may fill the whole inside volume of the tube (packed column) or be concentrated on or along the inside tube wall leaving an open, unrestricted path for the mobile phase in the middle part of the tube (open tubular column). Differences in rates of movement through the medium are calculated to different retention times of the sample.

The technique is very similar to the traditional column chromatography, except for that the solvent is driven through the column by applying positive pressure. This allowed most separations to be performed in less than 20 minutes, with improved separations compared to the old method. Modern flash chromatography systems are sold as pre-packed plastic cartridges, and the solvent is pumped through the cartridge. Systems may also be linked with detectors and fraction collectors providing automation. The introduction of gradient pumps resulted in quicker separations and less solvent usage.

A spreadsheet that assists in the successful development of flash columns has been developed. The spreadsheet estimates the retention volume and band volume of analytes, the fraction numbers expected to contain each analyte, and the resolution between adjacent peaks. This information allows users to select optimal parameters for preparative-scale separations before the flash column itself is attempted.

In expanded bed adsorption, a fluidised bed is used, rather than a solid phase made by a packed bed. This allows omission of initial clearing steps such as centrifugation and filtration, for culture broths or slurries of broken cells.

Planar chromatography: Planar chromatography is a separation technique in which the stationary phase is present as or on a plane. The plane can be a paper, serving as such or impregnated by a substance as the stationary bed (paper chromatography) or a layer of solid particles spread on a support such as a glass plate (thin layer chromatography). Different compounds in the sample mixture travel different distances according to how strongly they interact with the stationary phase as compared to the mobile phase. The specific retardation factor (R_f) of each chemical can be used to aid in the identification of an unknown substance.

Paper chromatography: Paper chromatography is a technique that involves placing a small dot or line of sample solution onto a strip of chromatography paper. The paper is placed in a jar containing a shallow layer of solvent and sealed. As the solvent rises through the paper, it meets the sample mixture which starts to travel up the paper with the solvent. This paper is made of cellulose, a polar substance, and the compounds within the mixture travel farther if they are non-polar. More polar substances bond with the cellulose paper more quickly, and therefore do not travel as far.

Thin layer chromatography: Thin layer chromatography (TLC) is a widely-employed laboratory technique and is similar to paper chromatography. However, instead of using a stationary phase of paper, it involves a stationary phase of a thin layer of adsorbent like silica gel, alumina, or cellulose on a flat, inert substrate. Compared to paper, it has the advantage of faster runs, better separations, and the choice between different adsorbents. For even better resolution and to allow for quantitation, high-performance TLC can be used.

Displacement chromatography: The basic principle of displacement chromatography is: A molecule with a high affinity for the chromatography matrix (the displacer) will compete effectively for binding sites, and thus displace all molecules with lesser affinities. There are distinct differences between displacement and elution chromatography. In elution mode, substances typically emerge from a column in narrow, Gaussian peaks. Wide separation of peaks, preferably to baseline, is desired in order to achieve maximum purification.

The speed at which any component of a mixture travels down the column in elution mode depends on many factors. But for two substances to travel at different speeds, and thereby be resolved, there must be substantial differences in some interaction between the biomolecules and the chromatography matrix. Operating parameters are adjusted to maximise the effect of this difference. In many cases, baseline separation of the peaks can be achieved only with gradient elution and low column loadings. Thus, two drawbacks to elution mode chromatography, especially at the preparative scale, are operational complexity, due to gradient solvent pumping, and low throughput, due to low column loadings. Displacement chromatography has advantages over elution chromatography in that components are resolved into consecutive zones of pure substances rather than 'peaks'. Because the process takes advantage of the nonlinearity of the isotherms, a larger column feed can be separated on a given column with the purified components recovered at significantly higher concentrations.

Techniques by physical state of mobile phase

Gas chromatography: Gas chromatography (GC), also sometimes known as gas-liquid chromatography, (GLC), is a separation technique in which the mobile phase is a gas. Gas chromatography is always carried out in a column, which is typically 'packed' or 'capillary'.

Gas chromatography (GC) is based on a partition equilibrium of analyte between a solid stationary phase (often a liquid silicone-based material) and a mobile gas (most often helium). The stationary phase is adhered to the inside of a small-diameter glass tube (a capillary column) or a solid matrix inside a larger metal tube (a packed column). It is widely used in analytical chemistry; though the high temperatures used in GC make it unsuitable for high molecular weight biopolymers or proteins (heat will denature them), frequently encountered in biochemistry, it is well suited for use in the petrochemical, environmental monitoring, and industrial chemical fields. It is also used extensively in chemistry research.

Liquid chromatography: Liquid chromatography (LC) is a separation technique in which the mobile phase is a liquid. Liquid chromatography can be carried out either in a column or a plane. Present day liquid chromatography that generally utilises very small packing particles and a relatively high pressure is referred to as high performance liquid chromatography (HPLC).

In the HPLC technique, the sample is forced through a column that is packed with irregularly or spherically shaped particles or a porous monolithic layer (stationary phase) by a liquid (mobile phase) at high pressure. HPLC is historically divided into two different sub-classes based on the polarity of the mobile and stationary phases. Technique in which the stationary phase is more polar than the mobile phase (e.g. toluene as the mobile phase, silica as the stationary phase) is called normal phase liquid chromatography (NPLC) and the opposite (e.g. water-methanol mixture as the mobile phase and C18 = octadecylsilyl as the stationary phase) is called reversed phase liquid chromatography (RPLC). Ironically the 'normal phase' has fewer applications and RPLC is therefore used considerably more.

Chapter 18

Fugitive Emissions

INTRODUCTION

Fugitive emissions are leaks or releases that occur whenever there are discontinuities in the solid barrier that maintains containment. Sources of fugitive emissions include pumps and compressors, storage and processing vessels, loading facilities, flow control and pressure relief valves, and leakage from pipelines carrying materials from one process to another. Fugitive emissions are usually small in quantities but are the origin of the continuous background exposure of workers. This chapter provides some examples of engineering measures and technological innovations to control and abate the common sources of fugitive emissions from chemical processing plants, petrochemical complex and refineries. Due to the extent and complexity of the chemical and petroleum industries, enumeration of sources of fugitive emissions would require detailed analysis of individual process unit operations. It is beyond the intent of this chapter to discuss in detail all sources of emissions and potential engineering control applicable to these emissions.

EMISSION SOURCES

Pump Emissions

Pump and compressor seals are a large source of emission leakage contributing to workplace exposures. In centrifugal pumps, the main source of leakage is the drive shaft that passes through the impeller casing, while in reciprocating pumps, the emissions are through the openings in the cylinder or fluid end, which actuates the piston. Proper sealing of the annular clearances between shafts and casings can reduce pump leakage. The packed seals can be replaced with more effective mechanical seals (Fig. 18.1).

The mechanical seals use springs to press the rotating element and stationary seal member in the stuffing box to reduce fugitive emissions. Several types of mechanical seals are available, viz. single, dual and double mechanical seals in the increasing order of superiority. Highly toxic or hazardous liquids can be transferred by sealless or canned pumps, which are magnetically driven. These encapsulated pumps have no shaft entry and thus eliminate seal leakage.

Loading Emissions

Loading of petrochemical or petroleum products is another major source of emissions. In conventional loading-by-loading racks or loading arm assemblies, the hydrocarbon vapours in the filling tank are displaced by the incoming liquid thus causing vapours emissions in the vicinity of the loading operators.

Splash or top filling generates turbulence and thus agitates the liquid being loaded and produces more hydrocarbon vapours.

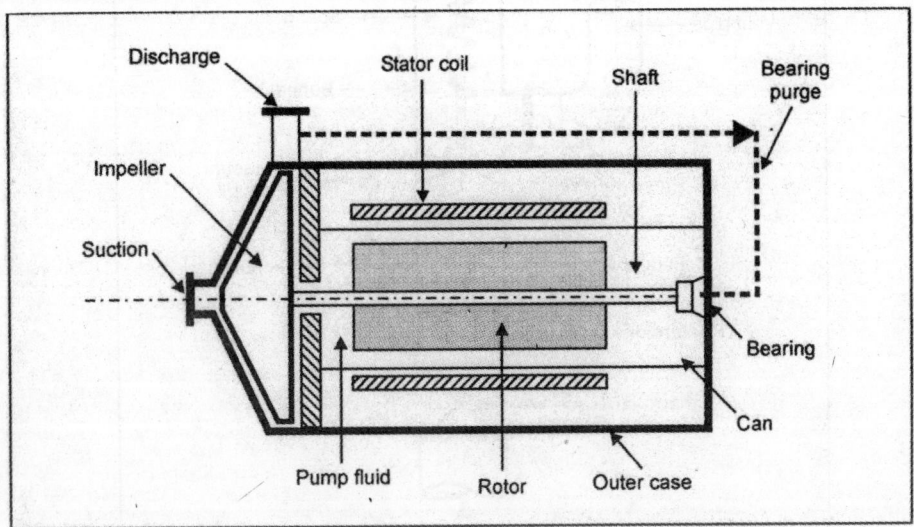

Fig. 18.1. A typical seal less pump.

A better solution is through bottom loading which introduces the incoming liquid under the surface of the liquid in the tank. Quick coupling valves and lines have facilitated the conversion from splash filling method to bottom loading techniques.

A further method of control fugitive emissions in loading operation is to provide a vapours return line to duct the displaced vapours to a suitable collection device such as vapours absorption or recovery system, or pollution control system such as an incinerator or a flare stack.

Pressure Relief Valve Emissions

In petrochemical and refinery operations, pressure relief or safety valves are used to protect process vessels from over-pressurisation. These valves are usually spring loaded to effect closure through a disk, which is held against process pressure. When the spring set pressure is exceeded by a high process pressure, the disk moves, releasing process fluid, thereby relieves the process pressure. The disk is reseated when the process pressure falls below the set pressure. However, disks do not always reseat properly, and corrosion of the valves often reduces the efficiency of re-seating which results in emission leakage (Fig. 18.2).

A rupture disk can be installed upstream of the pressure relief valve to minimise fugitive emission. The disk is a thin metal dish and is designed to burst at a specified pressure. It separates the process fluid from the safety relief valve and thereby prevents leakage through the valve. To check against backpressure on the rupture disk resulting from a 'pin-hole' leak, a pressure gauge could be provided between the disk and the relief valve. For highly toxic materials (e.g. substances with acute effects or chemical carcinogens) relief valves should discharge to a closed vent system either to flaring or through emission control equipment. Where this is not practicable, the discharge should be vented through an elevated vent.

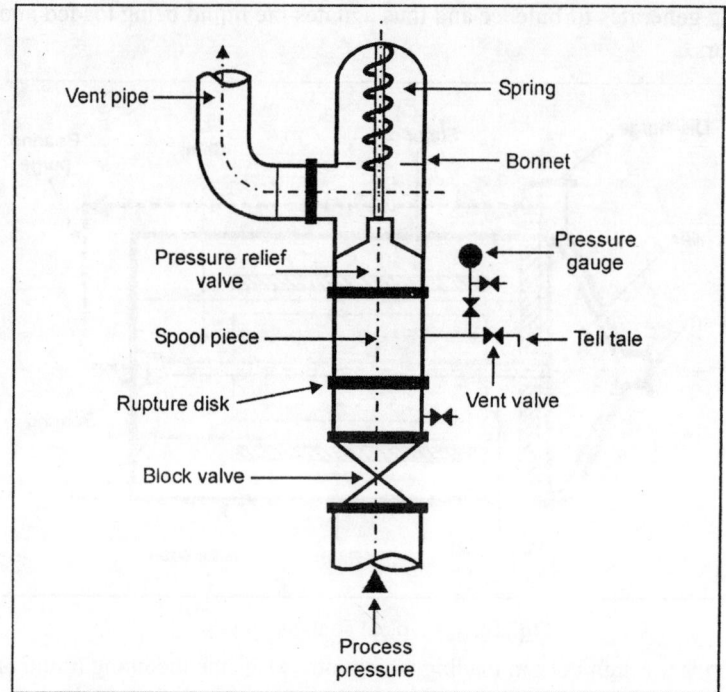

Fig. 18.2. Pressure relief valve with upstream rupture disk.

Release from Sampling Lines

A source of potential exposure to toxic hydrocarbons in petrochemical plants and refineries is during manual batch sampling operations. Conventional method of taking samples of streams for laboratory analysis requires draining sufficient liquid through a line before a representative sample is collected in a container.

A better method to control fugitive emission of the material being sampled is by using a closed loop sampling system. The loop is first purged or flushed with the material being conveyed. The valve drain to the container is then opened and the sample is collected. The drain valve connection to the sampling loop is short and the amount of 'dead' space is small, thereby reducing sample contamination.

A much better way of collecting toxic stream is by using a sample 'bomb' where the flushing takes place through the loop and the 'bomb' itself. The 'bomb' with the sample collected is disconnected from the loop for laboratory analysis.

The best sampling control is using automatic in-line automatic analyser sampling system. Although this automatic sampling is not entirely free of fugitive emissions, it greatly reduces potential toxic exposure problems posed by manual sampling.

Emissions from Various Tanks

Emissions from bulk storage tanks

Fugitive emissions from bulk storage tanks depend on the types and conditions of storage. The following are various types of storage tanks in increasing order of fugitive emission control.

Fixed roof tank with atmosphere vent

This type of storage tank is used for storing low toxicity liquid. It provides the least control over fugitive emissions. When the tank is being filled, the displaced vapour passes through the vent directly to the atmosphere. During emptying of liquid from the tank, air from the atmosphere passes through the vent into the tank.

Fixed roof tank with pressure-vacuum vent

A pressure-vacuum vent or 'breather' is used to compensate for diurnal effects, with pressure building during the heat of the day, and evening coolness or rain resulting in a reduced pressure which usually is slightly below atmospheric pressure. The latter condition opens the vacuum vent, admitting atmospheric air to balance the pressure.

External floating roof tank

A roof is constructed to float on the surface of the liquid. The roof is sealed against the walls of the tank to significantly reduce evaporative losses. However, bad weather conditions often create problems for this tankage control system.

Internal floating roof tank

An improvement on the external floating roof tanks is the installation of a fixed roof that provides protection against the sun and rain. Top and open side vents are provided on the fixed roof to allow dilution venting of evaporated vapour that may accumulate in the vapour space.

Closed floating roof tank

In a closed floating roof tank, the vents on the fixed roof are closed and a pressure-vacuum vent provides relief for any pressure variation in the vapour space. Injection of an inert gas, e.g. nitrogen provides blanketing in the vapour space to achieve an essentially zero emission tank (Fig. 18.3).

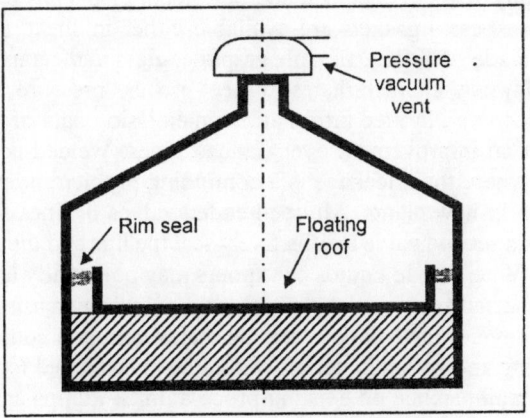

Fig. 18.3. Closed floating roof tank.

Valves, Flanges and Open-ended Pipes

Valves are the major contributors to fugitive emission losses. A variety of valves are available for use as block, manual and automatic control valves (Fig. 18.4).

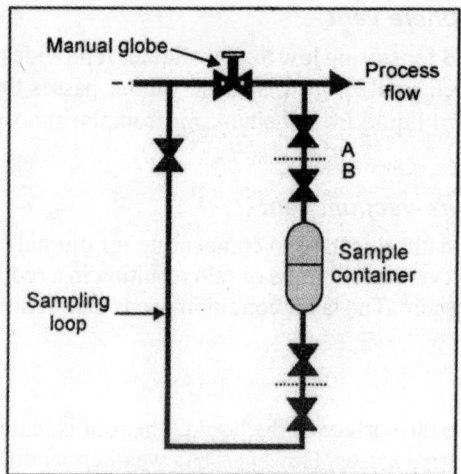

Fig. 18.4. Closed container.

Packing glands are used in these valves to reduce stem emissions. These include asbestos, graphite or carbon, glass fibre, plastic and metal packings. The performance of these packings varies and the choice depends on the process temperature, pressure and other factors. Where available, emission data on the control effectiveness of packing should be obtained from manufacturers before purchase or installation. Innovative types of valves have been developed to contain toxic or corrosive fluids and to eliminate fugitive emissions. These valves do not have packing glands or use packings as secondary protection. They are of the hermetically sealed types such as bellow, diaphragm and pinch valves.

Flange emissions are a relatively large fraction of total uncontrolled emissions in the process industry. The emission rates depend upon the gasket materials, surface roughness and the flange-bolting system. The standard gasket for many years in chemical and petroleum plants has been the asbestos gasket.

A wide variety of non-asbestos gaskets are available either in sheet, spiral wound, envelope or jacketed forms. These are made of Teflon, flexible graphite, glass and ceramic fibre. They are selected in accordance with their physical characteristics to meet process pressure, temperature and chemical resistance requirements. However, limited information on emission data of these gaskets is available.

Welded pipes represent an improvement over flanged pipes. Welded connections eliminate leaks and should be considered where toxic leakage is a continuing problem in existing plants or may be a potential exposure problem in new plants. All open-ended valves or lines should be equipped with a cap, plug or blind flange, or a second valve to effectively seal the line and thus prevent fugitive emission leakage to the atmosphere. Where toxic liquids or vapours may potentially leak from a bleed valve, the bleeder outlet should be connected to a closed drain with leakage indication using a level glass or rotameter.

Valves, pressure relief devices, and flanges are the major common sources of fugitive emissions. Newer and improved packing and gasket materials should be investigated for installation in new plants or as replacements during maintenance on existing process units. Pumps and compressors are a large source of fugitive emissions. Other sources of fugitive emissions include bulk loading facilities, storage tanks and manual sampling operations. Engineering innovations are the best solutions of eliminating or minimising these fugitive emissions. Engineering controls should be considered at the plant design stage, as these will be more economically installed than after the plant is operating. The largest contributor to in-plant emissions is lack of maintenance of plant equipment. Regular monitoring and maintenance is imperative for detecting and controlling fugitive emissions and leaks.

ENVIRONMENTAL GUIDELINES FROM PREVENTION AND CONTROL OF FUGITIVE EMISSIONS FROM CEMENT PLANTS

For achieving effective prevention and control of potential fugitive emission sources in cement manufacturing plants, specific requirements along with guidelines have been evolved. In order to establish proper management practices, requirements such as Operation and Maintenance aspects, trained manpower and documents and records to be maintained are also prescribed. In addition, general guidelines are also evolved for the sources otherwise not specified.

Requirements for Prevention and Control of Fugitive Emission for Various Potential Sources

For the purpose of effective prevention and control of fugitive emissions, the cement industry is required to implement the following for the sections mentioned in Tables 18.1 to 18.9.

Table 18.1. Unloading section (limestone, coal and other relevant material).

Control measures to be provided	Guidelines
Enclosure should be provided for all unloading operations, except wet materials like gypsum.	The enclosures for the unloading sides could be flexible curtain type material covering up to height of dumpers discharge from the roof.
Water shall be sprayed on the material prior and during unloading.	A dust suppression system should be provided to spray water. The amount of water sprayed should preferably be optimised by employing proper design of spray system. Suitable systems may be adopted to reduce the problems like choking, jamming of the moving parts.

Table 18.2. Material handling section (Including transfer points).

Control measures to be provided	Guidelines
All transfer point locations should be fully enclosed.	The enclosures from all sides with the provision for access doors, which shall be kept, closed during operation. Spillages should be periodically removed.
Airborne dust at all transfer operations/points should be controlled either by spraying water or by extracting to bag filter.	Either water spray system should be provided for suppressing the air borne dust or dry extraction cum bag filter with adequate extraction volume.
Belt conveyors should preferably be closed.	This will avoid wind blowing of fines.

Table 18.3. Coal storage section.

Control measures to be provided	Guidelines
Coal yard/storage area should be clearly earmarked. The pathways in coal yard for vehicle movement should be paved.	A board should be erected to display the area earmarked. Proper pathways with entry and exit point should be provided.
Accumulated dust shall be removed/swept regularly and water the area after sweeping.	Any deposits of dust on the concrete roads should be cleaned regularly by sweeping machines.
Coal other than coal stock pile should preferably be stored under covered shed.	Where ever blending activity is carried out by chaining in open ground, covered shed should be provided to reduce the fine coal dust getting airborne. The enclosure walls shall cover minimum three sides up to roof level.

(Contd...)

Control measures to be provided	Guidelines
The coal stock pile should preferably be under covered shed for new plants.	The enclosure should be from three sides and roof so as to contain the airborne emissions.
Instead of dust extraction cum bag filter system, If dust suppression measure is used, following additional control measures should be provided.	
Wetting before unloading.	Coal should be sufficiently moistened to suppress fines by spraying minimum quantity of water, if possible.
Spray water at crusher discharge and transfer points.	Water spray should also be applied at crusher discharge and transfer points.

Table 18.4. Clinker cooler section.

Control measures to be provided	Guidelines
Air borne fines extracted from clinker cooler shall be separated and sent to last possible destination directly, if possible.	The possibilities especially in new cement plant may be explored for the following:
	The unit may need to add on/install necessary provisions for separating fine particulates from the clinker cooler ESP collection. Fines separation may be achieved by passing collected dust through cyclone, the fines escaping cyclone to be separated, cyclone collection (coarse particles) could be recycled. The fines shall be recycled to the last possible destination (like clinker day silo) suitable or safely disposed.

Table 18.5. Clinker stock piles section.

Control measures to be provided	Guidelines
In new cement plant, clinker should be stored preferably in silo.	Bag filter may be provided before venting out the gases.
Clinker should be stored in closed enclosure covered from all sides and should have a venting arrangement along with a bag filter.	The enclosures should have a venting arrangement located at transfer point where clinker is dropped to the stockpile. The extraction/venting should be sufficient enough. Clinker stockpile access door should be covered by mechanical gate or by flexible rubber curtain. The access doors shall be kept closed at all possible times.
The dust extracted and captured in bag filter should be avoided to feed back/recycled to the clinker stockpile, if possible.	Extracted dust should be captured in bag filter and the collected dust should be avoided to feed back to the clinker stockpile, if layout permits. It may be recycled at last possible destination, i.e. cement mill section through suitable arrangement, if possible.
Generally open storage of clinker should be avoided. Only in case of emergency clinker should be stored in open with following control measures.	
Area for open storage of clinker should be clearly earmarked.	After earmarking the open storage area of clinker, a board should be erected to display the area earmarked.

(Contd...)

Control measures to be provided	Guidelines
Provide cover on openly stored clinker.	During the period when the openly stored clinker is inactive, it should be covered fully by HDPE or tarpaulin type sheets to prevent wind blowing of fugitive dust.
Provide windbreak walls or greenbelt on three sides of open stock piles	Install three sided enclosures, which extend to average height of the stockpile, wherever feasible.
Provide partial enclosure for retrieving area.	Flexible type wind breaking enclosure should be provided covering the clinker retrieval area as wind barrier to prevent dust carry over by wind. The enclosure could be of lightweight material like moulded plastic material or similar, which could be dismantled/assembled and shifted from one place to other.
The travel path of pay loaders should be paved and frequently swept.	Travel areas path used by the front-end pay loader shall be paved with concrete. It should be regularly swept by high efficiency vacuum sweeper to minimise the material build-up.
Provide loading of clinker by pay loaders into trucks/trailers be carried out in an enclosure vented to a bag filter.	The possibilities especially in new cement plant may be explored for the following: An enclosure fitted with bag filter could be located at the most central place adjacent to the clinker storage area. The pay loader moves to the fixed loading area from one end of the enclosure and the truck/trailer enters the enclosure from other end.

Table 18.6. Storage of limestone, gypsum, flyash and other additives.

Control measures to be provided	Guidelines
The storage should be done under covered shed.	The enclosure walls shall cover minimum two sides up to roof level.
Dry fly ash shall be transported by closed tankers. In case of wet fly ash trucks may be used for transportation.	Flyash shall be pumped directly from the tankers to silos pneumatically in closed loop or mechanically such that fugitive emissions do not occur.
Dry Fly ash shall be stored in silos only.	The silo vent be provided with a bag filter type system to vent out the air borne fines.
Flyash in the dry form should be encouraged and in wet form should be discouraged. In case wet flyash is to be used, it may be stored in open temporarily for the purpose of drying with necessary wind break arrangement to avoid wind carryover of fly ash. The flyash should be removed immediately after drying.	If possible, the dry flyash should be sent to closed silos. Otherwise, flyash should be transported through closed belt conveyors to avoid wind carryover of flyash.

Table 18.7. Cement packing section.

Control measures to be provided	Guidelines
Provide dust extraction arrangement for packing machines.	The packing machines should be equipped with dust extraction arrangement such that the packing operation is performed under negative pressure. The dust may be captured in bag filters.

(Contd...)

Control measures to be provided	Guidelines
Provide adequate ventilation for the packing hall.	Adequate ventilation for the packing hall should be provided for venting out suspended particulate thereby ensuring dust free work environment.
Spillage of cement on floor shall be minimised and cleared daily to prevent fugitive emissions.	The spilled cement from the packing machine should be collected properly and sent for recycling.
	The spilled cement on the shop floor should be swept by vacuum sweeping machines periodically.
	Proper engineering controls to prevent the fugitive emissions may include arrangements like providing guiding plate, scrapper brush for removing adhered dust on cement bag, etc.
Prevent emissions from the recycling screen by installing appropriate dust extraction system.	The vibratory screen provided for screening/recycling spilled cement should be provided with a dust extraction arrangement to prevent fugitive emission from that section.

Table 18.8. Silo section.

Control measures to be provided	Guidelines
The silo vent be provided with a bag filter type system to vent out the air borne fines.	The bag filter should be operated and maintained properly, especially the cleaning of bags to avoid pressurisation of silos thereby causing fugitive emissions from leakages, etc.

Table 18.9. Roads.

Control measures to be provided	Guidelines
All roads on which vehicle movement of raw materials or products take place should be paved.	The paved roads should be maintained as paved at all times and necessary repairs to be done immediately after damages to the road if any.
Limit the speed of vehicles.	Limit the speed of vehicle to 10 Km/hr for heavy vehicles with in the plant premises to prevent the road dust emissions.
Employ preventive measures to minimise dust build up on roads.	Preventive measures include covering of trucks and paving of access areas to unpaved areas.
Carry out regular sweeping of roads to minimise emissions.	Mitigative controls include vacuum sweeping, water flushing.

Requirement of Maintaining Documentation and Records

The industry shall maintain records to document the specific dust control actions taken and maintain such records for a period of not less than two years and make such records available to the regulatory authorities upon request. In addition documents of technical specifications of the control system and O&M guidelines should also be maintained. (Refer Appendix A1 for details of documents and records to be maintained).

Requirement of Trained Manpower

1. The industry shall employ or contract a 'dust control officer' who shall be available on site during working hours and should have authority to expeditiously employ sufficient dust mitigation measures to ensure control of fugitive emissions especially in abnormal circumstances. A suitably qualified person could be designated to operate as dust control officer. But, he should

be provided necessary training and should be aware of operational, maintenance aspects. He should be responsible for proper control of fugitive emissions. Environmental officer may act as a dust control officer.

2. Regular training should be given to the personnel operating and maintaining fugitive emissions control systems on the operational and maintenance aspects and record keeping responsibility.

Operation and Maintenance Requirement for all Dust Extraction cum Bag Filter Systems

1. A 'U'-tube manometer (of minimum 400 mm length) shall be fixed at all bag filters. It shall be connected with inlet and outlet side of the bag filter through flexible rubber tubes. Coloured water should be filled to zero level mark for proper visibility of the pressure drop across bag filter.
2. The minimum dust extraction volume should be based on the guidelines for ventilating various sources as per industrial ventilation hand book guidelines.
3. Un-interrupted supply of dry compressed air at desired pressure should be always ensured for pulsejet cleaning type bag filter.
4. The flow rate and static pressure at the bag filter inlet should be monitored at least quarterly and recorded to ensure appropriate functioning of the bag filter installed.
5. A sampling platform, portable and access ladder shall be provided at the final stack to carry out stack monitoring (in main stacks). Final emission should not exceed the prescribed standard.
6. In systems where water is also spread, it should be ensured that water does not get carried over/sucked to the bag filter. The details such as bag house specifications, layout drawing, operation and maintenance guidelines are to be maintained.
7. The details such as bag house specifications, layout drawing, operation and maintenance guidelines are to be maintained.

Operation and Maintenance Requirements for all Dust Suppression Systems

1. Basic details/specifications of the dust suppression systems installed at various locations should be maintained. The information should contain the quantity of water sprayed in LPH, number of nozzles, type of nozzles, desired water pressure, details of suppliers of spares, pipeline diagram, system layout, etc.
2. A fine mesh micro filter should be installed for filtering suspended solids from water prior to pumping to the nozzles to prevent choking of nozzles thereby ensuring proper sprays.
4. A pressure gauge and water flow meter shall be installed at major source for online measurements and a record be maintained for quantity of water sprayed.

SPM Concentration Standard for Assessing Effectiveness of Control Measures Adopted

1. The effectiveness of prevention cum control measures provided for controlling fugitive emissions from any source shall be said to be satisfactory, provided the SPM concentration, measured at 10 metre distance (from the enclosure wall housing the emission source or from the edge of the stockpiles/pavement area) in downwind direction shall not exceed 2000 microgram per cubic metre and 5000 microgram per cubic metre for coal yard /coal stock pile and rest other area respectively. These standards are for one year period and will be reviewed after one year. In cases where SPM concentrations exceed the prescribed limit, necessary corrective measures in terms of improving the controls shall be taken and action taken records of improvements carried out be maintained.

2. The measurement shall be carried out by high volume/respirable type samplers as per standard method prescribed by CPCB/BIS, covering at least 4 hours duration (240 minutes) during normal working hours with normal production rate of the operation/source being monitored on quarterly basis.

General Guidelines (For Areas Not Otherwise Specified)

Apart from the specific guidelines provided above for some specific sections/areas, for all other fugitive dust emitting areas, following general guidelines would apply:

1. The industry should prevent fugitive emission from all active operation and storage piles, such that the emissions are not visible in the atmosphere beyond the boundary line of the emission source.
2. The Industry shall conduct active operations by utilising the applicable best available control measures to minimise the fugitive dust emission from each fugitive dust source type within active operation.
3. Except for Gypsum and Clinker, all storage piles should be kept in moist condition by spraying water at regular intervals for controlling fugitive emission, wherever possible.
4. The operation of the pay loaders shall be slow down whenever the average wind speed is high exceeding 50 km/hr, which may cause fugitive emission.
5. All storage silos shall be vented to bag filters, which should have proper bag cleaning arrangement so as to avoid choking of filter bags, thereby to avoid pressurisation of silos.
6. Regular inspection at a pre-determined frequency be carried out of all fugitive dust control system and records be maintained of such inspection and corrective action taken if any.

Appendix A1. List of documents and records to be maintained for fugitive dust control.

Title of record to be maintained	Frequency of recording	Information to be recorded
Documents:		
List of fugitive emission management systems (FEMS) installed	To be up-dated once in a year	Location of FEMS, marked on process flow diagram, identity number, type of FEMS, year of installation, operating status
Technical specifications of FEMS installed		
Specification of dust suppression system	As and when installed/ modified	Locations of controlling emissions, identity number, supplier name, date of commissioning, pump HP, flow rate in LPM, pressure in kg/cm^2, nozzles type, numbers, LPM. O&M instruction from supplier
Specification of dust extraction cum APCD	As and when installed/modified	Location of system installed, identity number, name of system supplier, date of commissioning, flow rate in m^3/hr, time, flow m^3/hr static pressure mmWc, velocity m/sec, current drawn by ID fan motor, operation and maintenance instruction from supplier
Capacities of closed storages	Annually	For coal, limestone, clinker, gypsum, cement, additives, flyash, dimensions, bulk density, tons
Capacities of open storages	Annually	For coal, limestone, clinker, gypsum, additives, flyash, dimensions, bulk density, tons

(Contd...)

Title of record to be maintained	Frequency of recording	Information to be recorded
Records		
Replacement of damaged filter bags	As and when replaced	Number of bags replaced, date, bag filter identification number
Measurement of flow rate static pressure at bag filter inlet	Once a month	Bag filter number, date of monitoring, time, flow m³/hr, static pressure mmWc, velocity m/sec, current drawn by ID fan motor name of the person
Stack nonitoring of bag filters stack, wherever monitoring is feasible	Quarterly	Bag filter number, date of monitoring, time, measured data in m³/hr and mmWc, dust concentration in mg/Nm³
Operational details of dust suppression system	Once in a month	Quantity of material handled, quantity of water sprayed, number of operational nozzles, water pressure at filter inlet and outlet, details of damaged nozzles and replacements
Road sweeping record	Daily	Road location swept, date, running hours of sweeping machines
Quantity of coal in open storage, if any	Quarterly	Inventory of existing storage, add on, retrieved on quarterly basis, Date
Quantity of clinker in open storage, if any	Quarterly	Inventory of existing storage, add on, retrieved on quarterly basis, Date
Corrective actions taken for improving controls	As and when	Details of modifications carried out, level of reduction in SPM achieved

Chapter 19

Chemical Toxicology

INTRODUCTION

Toxicology is a branch of biology and medicine concerned with the study of the adverse effects of chemicals on living organisms. It is the study of symptoms, mechanisms, treatments and detection of poisoning, especially the poisoning of people.

Aquatic toxicology is the study of the effects of manufactured chemicals and other anthropogenic and natural materials and activities on aquatic organisms at various levels of organisation, from subcellular through individual organisms to communities and ecosystems.

In the United States aquatic toxicology plays an important role in the waste-water permit program. In additional to analytical testing for known pollutants, aquatic, whole effluent toxicity tests have been standardised and are performed routinely as a tool for evaluating the potential harmful effects of effluents discharged into surface waters.

Ecotoxicity, the subject of study of the field of ecotoxicology (a portmanteau of ecology and toxicology) refers to the potential for biological, chemical or physical stressors to affect ecosystems. Such stressors might occur in the natural environment at densities, concentrations or levels high enough to disrupt the natural biochemistry, physiology, behaviour and interactions of the living organisms that comprise the ecosystem.

The relationship between dose and its effects on the exposed organism is of high significance in toxicology. The chief criterion regarding the toxicity of a chemical is the dose, i.e. the amount of exposure to the substance. All substances are toxic under the right conditions. The term LD_{50} refers to the dose of a toxic substance that kills 50 per cent of a test population (typically rats or other surrogates when the test concerns human toxicity). LD_{50} estimations in animals are no longer required for regulatory submissions as a part of preclinical development package.

The conventional relationship (more exposure equals higher risk) has been challenged in the study of endocrine disruptors.

Many substances regarded as poisons are toxic only indirectly. An example is 'wood alcohol', or methanol, which is chemically converted to formaldehyde and formic acid in the liver. It is the formaldehyde and formic acid that cause the toxic effects of methanol exposure. As for drugs, many small molecules are made toxic in the liver, a good example being acetaminophen (paracetamol), especially in the presence of chronic alcohol use. The genetic variability of certain liver enzymes makes the toxicity of many compounds differ between one individual and the next. Because demands placed on one liver enzyme can induce activity in another, many molecules become toxic only in combination

498

with others. A family of activities that many toxicologists engage includes identifying which liver enzymes convert a molecule into a poison, what are the toxic products of the conversion and under what conditions and in which individuals this conversion takes place.

There are various specialised subdisciplines within the field of toxicology that concern diverse chemical and biological aspects of this area. For example, toxicogenomics involves applying molecular profiling approaches to the study of toxicology. Other areas include aquatic toxicology, chemical toxicology, ecotoxicology, environmental toxicology, forensic toxicology, and medical toxicology.

Chemical toxicology is a scientific discipline involving the study of structure and mechanism related to the toxic effects of chemical agents, and encompasses technology advances in research related to chemical aspects of toxicology. Research in this area is strongly multidisciplinary, spanning computational chemistry and synthetic chemistry, proteomics and metabolomics, drug discovery, drug metabolism and mechanisms of action, bioinformatics, bioanalytical chemistry, chemical biology, and molecular epidemiology.

The term ecotoxicology was coined by René Truhaut in 1969 who defined it as 'the branch of toxicology concerned with the study of toxic effects, caused by natural or synthetic pollutants, to the constituents of ecosystems, animal (including human), vegetable and microbial, in an integral context'.

Ecotoxicology is the integration of toxicology and ecology or, as Chapman suggested, 'ecology in the presence of toxicants'. It aims to quantify the effects of stressors upon natural populations, communities, or ecosystems. Ecotoxicology differs from environmental toxicology in that it integrates the effects of stressors across all levels of biological organisation from the molecular to whole communities and ecosystems, whereas environmental toxicology focuses upon effects at the level of the individual and below. This broader remit is distinct from the anthropocentric nature of classical toxicology and the legislative approach of environmental toxicology. Ecotoxicology incorporates aspects of ecology, toxicology, physiology, molecular biology, analytical chemistry and many other disciplines. The ultimate goal of this approach is to be able to predict the effects of pollution so that the most efficient and effective action to prevent or remediate any detrimental effect can be identified. In those ecosystems that are already impacted by pollution ecotoxicological studies can inform as to the best course of action to restore ecosystem services and functions efficiently and effectively.

Genetic toxicology, by definition, is the study of how chemical or physical agents affect the intricate process of heredity. Genotoxic chemicals are defined as compounds that are capable of modifying the hereditary material of living cells. The probability that a particular chemical will cause genetic damage inevitably depends on several variables, including the organism's level of exposure to the chemical, the distribution and retention of the chemical once it enters the body, the efficiency of metabolic activation and/or detoxification systems in target tissues, and the reactivity of the chemical or its metabolites with critical macromolecules within cells. The probability that genetic damage will cause disease ultimately depends on the nature of the damage, the cell's ability to repair or amplify genetic damage, the opportunity for expressing whatever alteration has been induced, and the ability of the body to recognise and suppress the multiplication of aberrant cells.

TOXIC METALS AND HUMAN HEALTH

Toxic metals comprise a group of minerals that have no known function in the body and, in fact, are harmful. Today mankind is exposed to the highest levels of these metals in recorded history. This is due to their industrial use, the unrestricted burning of coal, natural gas and petroleum, and incineration of waste materials worldwide. Toxic metals are now everywhere and affect everyone on planet earth.

They have become a major cause of illness, ageing and even genetic defects. The study of toxic metals is part of nutrition and toxicology, areas not emphasised in medical schools. For this reason, these important causes of disease are accorded little attention in conventional mainstream medicine. Today mankind is exposed to the highest levels in recorded history of lead, mercury, arsenic, aluminium, copper, nickel, tin, antimony, bromine, bismuth and vanadium. Levels are up to several thousand times higher than in primitive man. In my clinical experience, everyone has excessive amounts of some or all of the toxic metals. Toxic metals are also persistent and cumulative.

Toxic metals concentrate in the lipid tissues causing chromosome damages:

1. Physical agents: Inhalation of small fibre, particles of asbestos is responsible for causing diseases like aesbestosis and mesothelioma a rare form of cancer.
2. Oxides of nitrogen: Nitrogen oxides NO_2, N_2O_3 are extremely carcinogenic as they produce mutagenic nitrosoamines.
3. Organohalogens: A variety of these compounds are used in glues and nail polish removers, pesticides like Dieldrin, Aldrin and Heptachlor have been known to induce growth of malignant tumours.
4. Alkylating agents: Compounds like dialkyl sulphates, nitrosoamines, expoxides, lactones, ethyleneimine or aziridine have been known to cause chromosomal aberration in wheat and barley.
5. Aromatic hydrocarbons: Aromatic compounds like benzene, benzpyrine are well known for their carcinogenic effects.
6. Anaesthetics: Bromo-chloro trifluoro ethane (halotehene), N_2O, $CHCl_3$, are examples for anaesthetics.
7. Food additives: The common fast foods contain coal tar dyes and sugar substitutes. Chemical like sodium sulphite, benzoic acid and butylated organic compounds are used as food preservatives. These substances become toxic when they undergo certain reactions with other substances.
8. Naturally occurring toxicants: Fungal, sea food toxins, organic and some inorganic nitrogen containing compounds, excess of amino acids occurring in natural food are know to be not only toxic but also mutagenic and carcinogenic. Smoke preserved food contains toxic chemical 3,4-benzopyrene.

When toxic chemicals mentioned above are released into the environment, they enter into the human food chain. As soon as they enter our biological systems, they disturb the biological processes and so finally causes serious health hazards, leading to even death some times. The degree of toxicity depends upon the form in which the toxic chemical are present. Nevertheless, chemical toxicology refers to the science of study of toxic chemical and their modes of action.

Essential limit: The definite concentration of essential elements required for the normal growth and development of plants, animals and human beings is termed as 'essential limit'. It depends on the nature and also age of the living organism. Any chemical composition exceeds its essential limit, it becomes toxic, e.g. normal healthy young man required an average of 12 mg of iron per a day. It is the essential limit (12 mg of Fe) of iron for an young man.

ENVIRONMENTAL EFFECTS OF PESTICIDES

Use of pesticides can have unintended effects on the environment. Over 98 per cent of sprayed insecticides and 95 per cent of herbicides reach a destination other than their target species, including nontarget species, air, water, bottom sediments, and food. Pesticide contaminates land and water when it escapes from production sites and storage tanks, when it runs off from fields, when it is discarded, when it is sprayed aerially, and when it is sprayed into water to kill algae. The amount of pesticide that migrates from the intended application area is influenced by the particular chemical's properties: its propensity for binding to soil, its vapour pressure, its water solubility, and its resistance to being broken down over time. Factors in the soil, such as its texture, its ability to retain water, and the amount of organic matter contained in it, also affect the amount of pesticide that will leave the area. Some pesticides contribute to global warming and the depletion of the ozone layer.

Air

Pesticides can contribute to air pollution. Pesticide drift occurs when pesticides suspended in the air as particles are carried by wind to other areas, potentially contaminating them. Pesticides that are applied to crops can volatilise and may be blown by winds into nearby areas, potentially posing a threat to wildlife. Also, droplets of sprayed pesticides or particles from pesticides applied as dusts may travel on the wind to other areas, or pesticides may adhere to particles that blow in the wind, such as dust particles. Ground spraying produces less pesticide drift than aerial spraying does. Farmers can employ a buffer zone around their crop, consisting of empty land or non-crop plants such as evergreen trees to serve as windbreaks and absorb the pesticides, preventing drift into other areas. Such windbreaks are legally required in the Netherlands.

Pesticides that are sprayed onto fields and used to fumigate soil can give off chemicals called volatile organic compounds, which can react with other chemicals and form a pollutant called ozone, which accounts for a near 6 per cent loss of ozone layer.

Water

In the United States, pesticides were found to pollute every stream and over 90 per cent of wells sampled in a study by the US Geological Survey. Pesticide residues have also been found in rain and groundwater. Studies by the UK government showed that pesticide concentrations exceeded those allowable for drinking water in some samples of river water and groundwater.

Pesticide impacts on aquatic systems are often studied using a hydrology transport model to study movement and fate of chemicals in rivers and streams. As early as the 1970s quantitative analysis of pesticide runoff was conducted in order to predict amounts of pesticide that would reach surface waters.

There are four major routes through which pesticides reach the water: it may drift outside of the intended area when it is sprayed, it may percolate, or leach, through the soil, it may be carried to the water as runoff, or it may be spilled, for example accidentally or through neglect. They may also be carried to water by eroding soil. Factors that affect a pesticide's ability to contaminate water include its water solubility, the distance from an application site to a body of water, weather, soil type, presence of a growing crop, and the method used to apply the chemical.

Soil

Many of the chemicals used in pesticides are persistent soil contaminants, whose impact may endure for decades and adversely affect soil conservation.

The use of pesticides decreases the general biodiversity in the soil. Not using the chemicals results in higher soil quality, with the additional effect that more organic matter in the soil allows for higher water retention. This helps increase yields for farms in drought years, when organic farms have had yields 20–40 per cent higher than their conventional counterparts. A smaller content of organic matter in the soil increases the amount of pesticide that will leave the area of application, because organic matter binds to and helps break down pesticides.

Effects on Biota

Plants

Nitrogen fixation, which is required for the growth of higher plants, is hindered by pesticides in soil. The insecticides DDT, methyl parathion, and especially pentachlorophenol have been shown to interfere with legume-rhizobium chemical signalling. Reduction of this symbiotic chemical signalling results in reduced nitrogen fixation and thus reduced crop yields.

Pesticides can kill bees and are strongly implicated in pollinator decline, the loss of species that pollinate plants, including through the mechanism of Colony Collapse Disorder, in which worker bees from a beehive or Western honey bee colony abruptly disappear. Application of pesticides to crops that are in bloom can kill honeybees, which act as pollinators.

Animals

Pesticides inflict extremely widespread damage to biota, and many countries have acted to discourage pesticide usage through their Biodiversity action plans.

Animals may be poisoned by pesticide residues that remain on food after spraying, for example when wild animals enter sprayed fields or nearby areas shortly after spraying.

Widespread application of pesticides can eliminate food sources that certain types of animals need, causing the animals to relocate, change their diet, or starve. Poisoning from pesticides can travel up the food chain; for example, birds can be harmed when they eat insects and worms that have consumed pesticides. Some pesticides can bioaccumulate, or build up to toxic levels in the bodies of organisms that consume them over time, a phenomenon that impacts species high on the food chain especially hard.

Birds

Bald eagles are common examples of nontarget organisms that are impacted by pesticide use. Reductions in bird populations have been found to be associated with times and areas in which pesticides are used. In another example, some types of fungicides used in peanut farming are only slightly toxic to birds and mammals, but may kill off earthworms, which can in turn reduce populations of the birds and mammals that feed on them. Some pesticides come in granular form, and birds and other wildlife may eat the granules, mistaking them for grains of food. A few granules of a pesticide is enough to kill a small bird.

The herbicide paraquat, when sprayed onto bird eggs, causes growth abnormalities in embryos and reduces the number of chicks that hatch successfully, but most herbicides do not directly cause much harm to birds. Herbicides may endanger bird populations by reducing their habitat.

Aquatic life

Fish and other aquatic biota may be harmed by pesticide-contaminated water. Pesticide surface runoff into rivers and streams can be highly lethal to aquatic life, sometimes killing all the fish in a particular stream.

Application of herbicides to bodies of water can cause fish kills when the dead plants rot and use up the water's oxygen, suffocating the fish. Some herbicides, such as copper sulphite, that are applied to water to kill plants are toxic to fish and other water animals at concentrations similar to those used to kill the plants. Repeated exposure to sublethal doses of some pesticides can cause physiological and behavioural changes in fish that reduce populations, such as abandonment of nests and broods, decreased immunity to disease, and increased failure to avoid predators. Application of herbicides to bodies of water can kill off plants on which fish depend for their habitat. Pesticides can accumulate in bodies of water to levels that kill off zooplankton, the main source of food for young fish. Pesticides can kill off the insects on which some fish feed, causing the fish to travel farther in search of food and exposing them to greater risk from predators.

The faster a given pesticide breaks down in the environment, the less threat it poses to aquatic life. Insecticides are more toxic to aquatic life than herbicides and fungicides.

Amphibians

In the past several decades, decline in amphibian populations has been occurring all over the world, for unexplained reasons which are thought to be varied but of which pesticides may be a part. Mixtures of multiple pesticides appear to have a cumulative toxic effect on frogs. Tadpoles from ponds with multiple pesticides present in the water take longer to metamorphose into frogs and are smaller when they do, decreasing their ability to catch prey and avoid predators.

A Canadian study showed that exposing tadpoles to endosulphan, an organochloride pesticide at levels that are likely to be found in habitats near fields sprayed with the chemical kills the tadpoles and causes behavioural and growth abnormalities.

The herbicide atrazine has been shown to turn male frogs into hermaphrodites, decreasing their ability to reproduce.

Humans

Pesticides can enter the human body through inhalation of aerosols, dust and vapour that contain pesticides; through oral exposure by consuming food and water; and through dermal exposure by direct contact of pesticides with skin. Pesticides are sprayed onto food, especially fruits and vegetables, they secrete into soils and groundwater which can end up in drinking water, and pesticide spray can drift and pollute the air.

The effects of pesticides on human health are more harmful based on the toxicity of the chemical and the length and magnitude of exposure. Farm workers and their families experience the greatest exposure to agricultural pesticides through direct contact with the chemicals. But every human contains a percentage of pesticides found in fat samples in their body. Children are most susceptible and sensitive to pesticides due to their small size and underdevelopment. The chemicals can bioaccumulate in the body over time. Exposure to pesticides can range from mild skin irritation to birth defects, tumours, genetic changes, blood and nerve disorders, endocrine disruption, and even coma or death.

Persistent Organic Pollutants

Persistent organic pollutants (POPs) are compounds that resist degradation and thus remain in the environment for years. Some pesticides, including aldrin, chlordane, DDT, dieldrin, endrin, heptachlor, hexachlorobenzene, mirex, and toxaphene, are considered POPs. POPs have the ability to volatilise and travel great distances through the atmosphere to become deposited in remote regions. The chemicals also have the ability to bioaccumulate and biomagnify, and can bioconcentrate (i.e. become more concentrated) up to 70,000 times their original concentrations. POPs may continue to poison nontarget organisms in the environment and increase risk to humans by disruption in the endocrine, reproductive, and immune systems; cancer; neurobehavioural disorders, infertility and mutagenic effects, although very little is currently known about these chronic effects. Some POPs have been banned, while others continue to be used.

Pest Resistance

Pests may evolve to become resistant to pesticides. Many pests will initially be very susceptible to pesticides, but some with slight variations in their genetic make-up are resistant and therefore survive to reproduce. Through natural selection, the pests may eventually become very resistant to the pesticide. Pest resistance to a pesticide is commonly managed through pesticide rotation, which involves alternating among pesticide classes with different modes of action to delay the onset of or mitigate existing pest resistance. Tank mixing pesticides is the combination of two or more pesticides with different modes of action in order to improve individual pesticide application results and delay the onset of or mitigate existing pest resistance.

Pest Rebound and Secondary Pest Outbreaks

Nontarget organisms, organisms that the pesticides are not intended to kill, can be severely impacted by use of the chemicals. In some cases, where a pest insect has some controls from a beneficial predator or parasite, an insecticide application can kill both pest and beneficial populations. A study comparing biological pest control and use of pyrethroid insecticide for diamondback moths, a major cabbage family insect pest, showed that the insecticide application created a rebounded pest population due to loss of insect predators, whereas the biocontrol did not show the same effect. Likewise, pesticides sprayed in an effort to control adult mosquitoes, may temporarily depress mosquito populations, however they may result in a larger population in the long run by damaging the natural controlling factors. This phenomenon, wherein the population of a pest species rebounds to equal or greater numbers than it had before pesticide use, is called pest resurgence and can be linked to elimination of predators and other natural enemies of the pest. Loss of predator species can also lead to a related phenomenon called secondary pest outbreaks, an increase in problems from species which were not originally very damaging pests due to loss of their predators or parasites. An estimated third of the 300 most damaging insects in the US were originally secondary pests and only became a major problem after the use of pesticides. In both pest resurgence and secondary pest outbreaks, the natural enemies have been found to be more susceptible to the pesticides than the pests themselves, in some cases causing the pest population to be higher than it was before the use of pesticide.

Eliminating Pesticides

Many alternatives are available to reduce the effects pesticides have on the environment. There are a variety of alternative pesticides such as manually removing weeds and pests from plants, applying heat, covering weeds with plastic, and placing traps and lures to catch or move pests. Pests can be prevented by removing pest breeding sites, maintaining healthy soils which breed healthy plants that are resistant to pests, planting native species that are naturally more resistant to native pests, and use biocontrol agents such as birds and other pest eating organisms.

HEAVY METAL TOXICITY

There are 35 metals that concern us because of occupational or residential exposure; 23 of these are the heavy elements or 'heavy metals': antimony, arsenic, bismuth, cadmium, cerium, chromium, cobalt, copper, gallium, gold, iron, lead, manganese, mercury, nickel, platinum, silver, tellurium, thallium, tin, uranium, vanadium, and zinc. Interestingly, small amounts of these elements are common in our environment and diet and are actually necessary for good health, but large amounts of any of them may cause acute or chronic toxicity (poisoning). Heavy metal toxicity can result in damaged or reduced mental and central nervous function, lower energy levels, and damage to blood composition, lungs, kidneys, liver, and other vital organs. Long-term exposure may result in slowly progressing physical, muscular, and neurological degenerative processes that mimic Alzheimer's disease, Parkinson's disease, muscular dystrophy, and multiple sclerosis. Allergies are not uncommon and repeated long-term contact with some metals or their compounds may even cause cancer.

For some heavy metals, toxic levels can be just above the background concentrations naturally found in nature. Therefore, it is important for us to inform ourselves about the heavy metals and to take protective measures against excessive exposure. In most parts of the United States, heavy metal toxicity is an uncommon medical condition; however, it is a clinically significant condition when it does occur. If unrecognised or inappropriately treated, toxicity can result in significant illness and reduced quality of life. For persons who suspect that they or someone in their household might have heavy metal toxicity, testing is essential. Appropriate conventional and natural medical procedures may need to be pursued.

The association of symptoms indicative of acute toxicity is not difficult to recognise because the symptoms are usually severe, rapid in onset, and associated with a known exposure or ingestion: cramping, nausea, and vomiting; pain; sweating; headaches; difficulty breathing; impaired cognitive, and language skills; mania; and convulsions.

The symptoms of toxicity resulting from chronic exposure (impaired cognitive, motor, and language skills; learning difficulties; nervousness and emotional instability; and insomnia, nausea, lethargy, and feeling ill) are also easily recognised; however, they are much more difficult to associate with their cause. Symptoms of chronic exposure are very similar to symptoms of other health conditions and often develop slowly over months or even years. Sometimes the symptoms of chronic exposure actually abate from time to time, leading the person to postpone seeking treatment, thinking the symptoms are related to something else.

Heavy metals are chemical elements with a specific gravity that is at least 5 times the specific gravity of water. The specific gravity of water is 1 at 4°C (39°F). Simply stated, specific gravity is a measure of density of a given amount of a solid substance when it is compared to an equal amount of water. Some well-known toxic metallic elements with a specific gravity that is 5 or more times that of water are arsenic, 5.7; cadmium, 8.65; iron, 7.9; lead, 11.34; and mercury, 13.546.

Beneficial Heavy Metals

In small quantities, certain heavy metals are nutritionally essential for a healthy life. Some of these are referred to as the trace elements (e.g. iron, copper, manganese, and zinc). These elements, or some form of them, are commonly found naturally in foodstuffs, in fruits and vegetables, and in commercially available multivitamin products. Diagnostic medical applications include direct injection of gallium during radiological procedures, dosing with chromium in parenteral nutrition mixtures, and the use of lead as a radiation shield around X-ray equipment. Heavy metals are also common in industrial applications such as in the manufacture of pesticides, batteries, alloys, electroplated metal parts, textile dyes, steel, and so forth. Many of these products are in our homes and actually add to our quality of life when properly used.

Toxic Heavy Metals

Heavy metals become toxic when they are not metabolised by the body and accumulate in the soft tissues. Heavy metals may enter the human body through food, water, air, or absorption through the skin when they come in contact with humans in agriculture and in manufacturing, pharmaceutical, industrial, or residential settings. Industrial exposure accounts for a common route of exposure for adults. Ingestion is the most common route of exposure in children. Children may develop toxic levels from the normal hand-to-mouth activity of small children who come in contact with contaminated soil or by actually eating objects that are not food (dirt or paint chips). Less common routes of exposure are during a radiological procedure, from inappropriate dosing or monitoring during intravenous (parenteral) nutrition, from a broken thermometer, or from a suicide or homicide attempt.

As a rule, acute poisoning is more likely to result from inhalation or skin contact of dust, fumes or vapours, or materials in the workplace. However, lesser levels of contamination may occur in residential settings, particularly in older homes with lead paint or old plumbing. The Agency for Toxic Substances and Disease Registry (ATSDR) in Atlanta, Georgia, (a part of the US Department of Health and Human Services) was established by congressional mandate to perform specific functions concerning adverse human health effects and diminished quality of life associated with exposure to hazardous substances. The ATSDR is responsible for assessment of waste sites and providing health information concerning hazardous substances, response to emergency release situations, and education and training concerning hazardous substances.

Lead Toxicity

Lead has been mined and used in industry and in household products for centuries. The dangers of lead toxicity, the clinical manifestations of which are known as plumbism, have been known since ancient times. The twentieth century has seen both the greatest-ever exposure of the general population to lead and an extraordinary amount of new research on lead toxicity.

1. Populations are exposed to lead chiefly via paints, cans, plumbing fixtures, and leaded gasoline. The intensity of these exposures, while recently decreased by regulatory actions, remains high in some segments of the population because of the deterioration of lead paint used in the past and the entrainment of lead from paint and vehicle exhaust into soil and house dust.
2. Many other environmental sources of exposure exist, such as leafy vegetables grown in lead-contaminated soil, improperly glazed ceramics, lead crystal, and certain herbal folk remedies.
3. Many industries, such as battery manufacturing, demolition, painting and paint removal, and ceramics, continue to pose a significant risk of lead exposure to workers and surrounding communities.

Elemental lead and inorganic lead compounds are absorbed through ingestion or inhalation. Organic lead (e.g. tetraethyl lead, the lead additive to gasoline) is absorbed to a significant degree through the skin as well. Pulmonary absorption is efficient, particularly if particle diameters are <1 μm (as in fumes from burning lead paint). Children absorb up to 50 per cent of the amount of lead ingested, whereas adults absorb only about 10 to 20 per cent. Gastrointestinal absorption of lead is enhanced by fasting and by dietary deficiencies in calcium, iron, and zinc; such absorption is minimal, however, for lead in the form of lead sulphide, a common constituent of mining waste. Lead is absorbed into blood plasma, where it equilibrates rapidly with extracellular fluid, crosses membranes (such as the blood-brain barrier and the placenta), and accumulates in soft and hard tissues. In the blood, around 95 to 99 per cent of lead is sequestered in red cells, where it is bound to haemoglobin and other components. As a consequence, lead is usually measured in whole blood rather than in serum. The largest proportion of absorbed lead is incorporated into the skeleton, which contains more than 90 per cent of the body's total lead burden. Lead is excreted mainly in the urine (in a process that depends on glomerular filtration and tubular secretion) and in the feces. Lead also appears in hair, nails, sweat, saliva, and breast milk. The half-life of lead in blood is approximately 25 days; in soft tissue, about 40 days; and in the nonlabile portion of bone, more than 25 years. Thus, blood lead levels may decline significantly while the body's total burden of lead remains heavy.

The toxicity of lead is probably related to its affinity for cell membranes and mitochondria, as a result of which it interferes with mitochondrial oxidative phosphorylation and sodium, potassium, and calcium ATPases. Lead impairs the activity of calcium-dependent intracellular messengers and of brain protein kinase C. In addition, lead stimulates the formation of inclusion bodies that may translocate the metal into cell nuclei and alter gene expression.

Toxicity of Mercury

Mercury is a highly toxic element that is found both naturally and as an introduced contaminant in the environment. The risk is determined by the likelihood of exposure, the form of mercury present (some forms are more toxic than others), and the geochemical and ecological factors that influence how mercury moves and changes form in the environment.

The toxic effects of mercury depend on its chemical form and the route of exposure. Methylmercury [CH_3Hg] is the most toxic form. It affects the immune system, alters genetic and enzyme systems, and damages the nervous system, including coordination and the senses of touch, taste, and sight. Methylmercury is particularly damaging to developing embryos, which are five to ten times more sensitive than adults. Exposure to methylmercury is usually by ingestion, and it is absorbed more readily and excreted more slowly than other forms of mercury. Elemental mercury, $Hg(0)$, the form released from broken thermometers, causes tremors, gingivitis, and excitability when vapours are inhaled over a long period of time. Although it is less toxic than methylmercury, elemental mercury may be found in higher concentrations in environments such as gold mine sites, where it has been used to extract gold. If elemental mercury is ingested, it is absorbed relatively slowly and may pass through the digestive system without causing damage. Ingestion of other common forms of mercury, such as the salt $HgCl_2$, which damages the gastrointestinal tract and causes kidney failure, is unlikely from environmental sources. People are exposed to methylmercury almost entirely by eating contaminated fish and wildlife that are at the top of aquatic foodchains.

HARMFUL CHEMICALS IN OUR ENVIRONMENT

Over the past century humans have introduced a large number of chemical substances into the environment. Some are the waste from industrial and agricultural processes. Some have been designed as structural materials and others have been designed to perform various functions such as healing the sick or killing pests and weeds. Obviously some chemicals are useful but many are toxic and their harm to the environment and our health far outweighs their benefit to society. We need to manage the risks better by only using chemicals, which are safe.

Chemicals enter air as emissions and water as effluent. Industrial and motor vehicle emissions of nitrogen and sulphur oxides cause acid rain, which poisons fish and other aquatic organisms in rivers and lakes and affects the ability of soil to support plants. Carbon dioxide causes the greenhouse effect and climate change. Chlorofluorocarbons (CFCs) cause the destruction of ozone in the stratosphere and create the possibility of serious environmental damage from ultraviolet radiation. Chemical fertilisers and nutrients runoff from farms and gardens cause the build up of toxic algae in rivers, making them uninhabitable to aquatic organisms and unpleasant for humans. Some toxic chemicals find their way from landfill waste sites into our groundwater, rivers and oceans and induce genetic changes that compromise the ability of life to reproduce and survive. The impact of human activities on the environment is complex and affects a chain of interconnecting ecosystems. The extinction of species all along the chain may mean the loss of useful genetic material or life saving cancer drugs or safer alternatives to the dangerous chemicals in use at the moment.

Organochlorines: Organochlorine compounds such as polychlorinated biphenyls or PCBs were developed originally for use in electric equipment as cooling agents and are very dangerous chemicals. During the manufacture and disposal of products containing PCBs, and as a result of accidents, millions of gallons of PCB oil have leaked out. Although their manufacture in the United States was halted in the 1970s and they are being phased out, they are difficult to detect, are nearly indestructible and large quantities remain in existence and they will remain in the environment for a long time. They accumulate in the food chain and significant levels of them have been found in marine species, particularly mammals and sea birds, decades after their production was discontinued. They are carcinogenic and capable of damaging the liver, nervous system and the reproductive system in adults. When PCBs are burned, even more toxic dioxins are formed.

Dioxins: Dioxins, are a class of super-toxic chemicals formed as a by-product of the manufacture, moulding, or burning of organic chemicals and plastics that contain chlorine. They are the most toxic man-made organic chemicals known. They cause serious health effects even at levels as low as a few parts per trillion. Only radioactive waste is more toxic. They are virtually indestructible and are excreted by the body extremely slowly. Dioxins became known when Vietnam War veterans and Vietnamese civilians, exposed to dioxin-contaminated Agent Orange, became ill.

Dioxins enter the body in food and accumulate in body fat. They bind to cell receptors and disrupt hormone functions in the body and they also affect gene functions. Our bodies have no defence against dioxins which may cause a wide range of problems, from cancer to reduced immunity to nervous system disorders to miscarriages and birth deformity. The effects can be very obvious or subtle. Because they change gene functions, they can cause genetic diseases to appear and they can interfere with child development. Attention Deficit Disorder, diabetes, endometriosis, chronic fatigue syndrome, rare nervous and blood disorders have been linked to exposure to dioxins and PCBs.

Over the past 40 years there has been a dramatic increase in the manufacture and use of chlorinated organic chemicals in plastics, insecticides and herbicides. Dioxins have been found in high concentrations ar to the sites where these chemicals have been produced and where insecticides and herbicides have been heavily used, such as on farms, orchards, or along electric and railway lines. They have also been a found downstream from paper mills where chlorine chemicals have been used to bleach wood pulp.

In the last few years we have begun to discard our unfashionable household plastic products, together with industrial and medical waste by burning them in incinerators. Dioxins formed during the combustion process have been carried for hundreds of miles on tiny specks of ash and contaminated the countryside. They settle on pastures and crops and get eaten by cows, pigs and chickens. They get into lakes, streams, and ocean and are taken up by fish. They go through the food chain and appear in meat and milk and accumulate in the fat cells of our bodies.

Chapter 20

Environmental Implications of Nanotechnology

INTRODUCTION

The environmental implications of nanotechnology are the possible effects that the use of nanotechnological materials and devices will have on the environment. As nanotechnology is an emerging field, there is great debate regarding to what extent industrial and commercial use of nanomaterials will affect organisms and ecosystems.

Nanotechnology's environmental implications can be split into two aspects: the potential for nanotechnologcal innovations to help improve the environment, and the possibly novel type of pollution that nanotechnological materials might cause if released into the environment.

APPLICATIONS OF NANOTECHNOLOGY IN ENVIRONMENT

In industrialised nations the air is filled with numerous pollutants caused by human activity or industrial processes, such as carbon monoxide (CO), chlorofluorocarbons (CFC), heavy metals (arsenic, chromium, lead, cadmium, mercury, zinc), hydrocarbons, nitrogen oxides, organic chemicals (volatile organic compounds, known as VOCs, and dioxins), sulphur dioxide and particulates. The presence of nitrogen and sulphur oxide in the air generates acid rain that infiltrates and contaminates the soil. The elevated levels of nitrogen and sulphur oxide in the atmosphere are mainly due to human activities, particularly burning of oil, coal and gas. Only a small portion comes from natural processes such as volcanic action and decay of soil bacteria. Water pollution is caused by numerous factors, including sewage, oil spills, leaking of fertilisers, herbicides and pesticides from land, by-products from manufacturing and extracted or burned fossil fuels.

Contaminants are most often measured in parts per million (ppm) or parts per billion (ppb) and their toxicity defined by a 'toxic level'. The toxic level for arsenic, for instance, is 10 ppm in soil whereas for mercury is 0.002 ppm in water. Therefore, very low concentrations of a specific contaminant can be toxic. In addition, contaminants are mostly found as mixtures. Consequently, there is a need for technologies that are capable of monitoring, recognising and, ideally, treating such small amount of contaminants in air, water and soil. In this context, nanotechnology offers numerous opportunities to prevent, reduce, sense and treat environment contamination. Nanoscience allows designing and manipulating materials at the atomic and molecular level. Nanomaterials can be fabricated with specific properties that can 'recognise' a particular pollutant within a mixture. The small size of nanomaterials, together with their high surface-to-volume ratio, can lead to very sensitive detection. These properties will allow developing highly miniaturise, accurate and sensitive pollution-monitoring devices (nano-sensors).

Nanomaterials can also be engineered to actively interact with a pollutant and decompose it in less toxic species. Thus, in the future nanotechnology could be used not only for detecting contaminated sites but also treating them. Finally, this technology can be used to reduce the production of harmful wastes in manufacturing processes by reducing the amount of material used, and by employing less toxic compounds.

NANOPOLLUTION

Nanopollution is a generic name for all waste generated by nanodevices or during the nanomaterials manufacturing process. This kind of waste may be very dangerous because of its size. It can float in the air and might easily penetrate animal and plant cells causing unknown effects. Most human-made nanoparticles do not appear in nature, so living organisms may not have appropriate means to deal with nanowaste. It is probably one great challenge to nanotechnology: how to deal with its nanopollutants and nanowaste.

Environmental assessment is justified as nanoparticles present novel (new) environmental impacts. Scrinis raises concerns about nano-pollution, and argues that it is not currently possible to 'precisely predict or control the ecological impacts of the release of these nano-products into the environment'. Ecotoxicological impacts of nanoparticles and the potential for bioaccumulation in plants and micro-organisms remain under-researched. The capacity for nanoparticles to function as a transport mechanism also raises concern about the transport of heavy metals and other environmental contaminants. A May 2007 Report to the UK Department for Environment, Food and Rural Affairs noted concerns about the toxicological impacts of nanoparticles in relation to both hazard and exposure. The report recommended comprehensive toxicological testing and independent performance tests of fuel additives.

Not enough data exists to know for sure if nanoparticles could have undesirable effects on the environment. Two areas are relevant here: (i) in free form nanoparticles can be released in the air or water during production (or production accidents) or as waste by-product of production, and ultimately accumulate in the soil, water or plant life, and (ii) in fixed form, where they are part of a manufactured substance or product, they will ultimately have to be recycled or disposed of as waste. It is not known yet whether certain nanoparticles will constitute a completely new class of non-biodegradable pollutant. In case they do, it is not known how such pollutants could be removed from air or water because most traditional filters are not suitable for such tasks (their pores are too big to catch nanoparticles).

Of the US $950 million spent in 2006 by the US government on nanotechnology research, only $8,00,000 was spent on environmental impact assessments. Risks identified by Uskokovic include: self-replicating nanobots aggressively or through slowly rising supremacy wiping out the whole biosphere; further destabilising the already endangered diversity of the biosphere or extending the existing gap between the rich and poor.

Concerns have been raised about Silver Nanotechnology used by Samsung in a range of appliances such as washing machines and air purifiers.

POLLUTION REMEDIATION AND TREATMENT

Soil and groundwater contamination arising from manufacturing processes are a matter of great complexity and concern. Affected sites include contaminated industrial sites (including lakes and rivers in their vicinity), underground storage tank leakages; landfills; and abandoned mines. Pollutants in these areas include heavy metals (e.g. mercury, lead, cadmium) and organic compounds (e.g. benzene, chlorinated solvents, creosote). Nanotechnology can develop techniques that will allow for more specific and cost-effective remediation tools. Currently, many of the methods employed to remove toxic contaminants involve laborious, time-consuming and expensive techniques. A pretreatment process and removal of the contaminated area is often required, with a consequent disturbance of the ecosystem. Nanotechnology allows developing technologies that can perform *in situ* remediation and reach inaccessible areas such as crevices and aquifers, thus eliminating the necessity for costly 'pump-and-treat' operations. In addition, thanks to its ability to manipulate matter at a molecular level, nanoscience can be used to develop remediation tools that are specific for a certain pollutant (e.g. metal), therefore increasing affinity and selectivity, as well as improving the sensitivity of the technique.

Drinking water quality and its contamination from pollutants is another matter of concern. Mercury and arsenic are in particular two extremely toxic metals that pose very high health risks. Remediation methods that allow fast, economic and effective treatment of water polluted with such contaminants is highly needed. Nanotechnology can introduce new methods for the treatment and purification of water from pollutants, as well as new techniques for waste-water management and water desalinisation.

Life Cycle Responsibility

To properly assess the health hazards of engineered nanoparticles the whole life cycle of these particles needs to be evaluated, including their fabrication, storage and distribution, application and potential abuse, and disposal. The impact on humans or the environment may vary at different stages of the life cycle.

The Royal Society report identified a risk of nanoparticles or nanotubes being released during disposal, destruction and recycling, and recommended that 'manufacturers of products that fall under extended producer responsibility regimes such as end-of-life regulations publish procedures outlining how these materials will be managed to minimise possible human and environmental exposure'. Reflecting the challenges for ensuring responsible life cycle regulation, the Institute for Food and Agricultural Standards has proposed standards for nanotechnology research and development should be integrated across consumer, worker and environmental standards. They also propose that NGOs and other citizen groups play a meaningful role in the development of these standards.

Remediation Using Nanoparticles

The use of zero-valent (Fe^0) iron nanoparticles for the remediation of contaminated groundwater and soil is a good example of how environmental remediation can be improved with nanotechnology. When exposed to air, iron oxidises easily to rust; however, when it oxidises around contaminants such as trichloroethylene (TCE), carbon tetrachloride, dioxins, or PCBs, these organic molecules are broken down into simple, far less toxic carbon compounds. Since iron is non-toxic and is abundant in the natural environment (in rocks, soil, water, etc.), some industries have started using an 'iron powder' to clean up their new industrial wastes. However, the 'iron powder' (that is, granular zero-valent iron with dimensions in the micron range) is not effective for decontaminating old wastes that have already soaked into the soil and water. Moreover, bioremediation using granular iron powder is often incomplete: some chlorinated compounds, such as PCE or TCE, are only partially treated and toxic by-products (such as DCE) are still found after treatment. This effect is due to the low reactivity of iron powders. Another matter of concern is the decrease of reactivity of iron powders over time, possibly due to the formation of passivation layers over the surface of the iron granules.

Nanotechnology has offered a solution to this remediation technology in the form of iron nanoparticles. These nanoparticles are 10 to 1000 times more reactive then commonly used iron powders). They have a larger surface area available for reacting with the organic contaminant and their small size (1–100 nm) allow them to be much more mobile, so they can be transported effectively by the flow of groundwater. A nanoparticle—water slurry can be injected to the contaminated plume where treatment is needed (Fig. 20.1). The nanoparticles do not change by soil acidity, temperature or nutrient levels, so they can remain in suspension maintaining their properties for extended periods of time to establish an *in situ* treatment zone. Experimental results collected both in laboratory and in the field have shown that nanoscale iron particles are very effective for the complete transformation and detoxification of a wide variety of common environmental contaminants, such as chlorinated organic solvents, organochlorine pesticides, and PCBs. When nano-sized iron powders are used, no toxic by-products are formed, a result of the increased reactivity and stability of the nanoparticles compared to the granular iron powder. Contaminant levels around the injection level is considerably reduced in a day or two and nearly eliminated within a few days. Thanks to their stability, nano-iron particles remain active in a site for six to eight weeks before they become dispersed completely in the groundwater and become less concentrated than naturally occurring iron. Researchers are assessing whether the technique could also be used for the remediation of dense nonaqueous phase liquid (DNAPL) sources within aquifers, as well as for the immobilisation of heavy metals and radionucleotides.

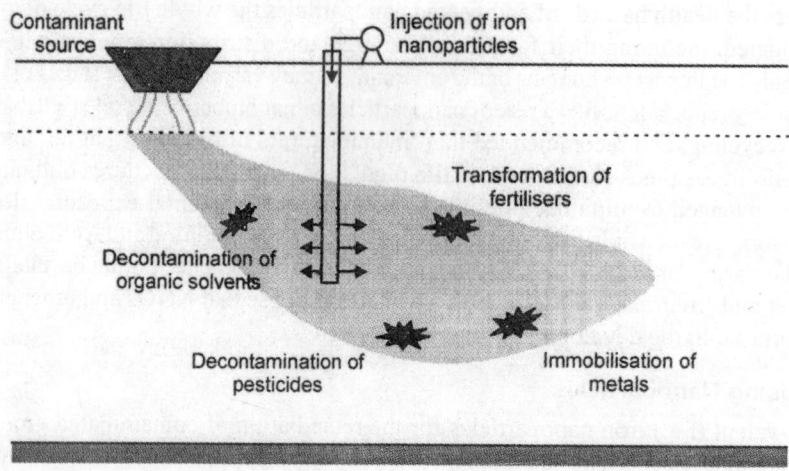

Fig. 20.1. Nanoscale iron particles for *in situ* remediation.

Bimetallic iron nanoparticles, such as iron/palladium, have been shown to be even more active and stable then zero-valent iron nanoparticles, therefore further improving this remediation technology. Finally, iron or bimetallic nanoparticles could be anchored on solid supports such as activated carbon or silica for the *ex situ* treatment of contaminated water and industrial wastes.

Nanoparticles that are activated by light are also investigated for their capability of removing contaminants from various media, especially water. The semiconductors TiO_2 and ZnO are of particular interest, since these are readily available and inexpensive, so their use for remediation has been studied for many years. Recently, nanosized TiO_2 and ZnO have been considered as these have more active surface given the same volume of material. The vision is to create some solar photocatalysis remediation systems, where TiO_2 or ZnO are used to convert toxic contaminants, such as chlorinated detergents, into benign products using sun radiation. There is evidence that those semiconductors can photo degrade numerous toxic compounds, but the technology requires improvements in term of efficiency, since TiO_2

or ZnO only adsorb UV light which represents only 5 per cent of the solar spectrum. In this context, nanotechnology could bring an improvement in the form of nanoparticles with surfaces modified with organic or inorganic dyes to increase the photoresponse window of TiO_2 and ZnO from UV to visible light. Gold nanoparticles are also investigated as a material to increase the activity of TiO_2. Finally, a recent work has shown that ZnO nanoparticles can act both as a sensor of chlorinated phenols and as a photocatalytic degradation tool. Nanomaterials have also been found able to remove metal contaminants from air. For instance silica-titania nanocomposites are investigated for the removal of elementary mercury (Hg) from vapours such as those coming from combustion sources. In these nanocomposites, silica acts as a support material and titania transforms mercury to a less volatile form (mercury oxide).

Chelating agents in the form of dendrimers are also studied for the removal of metal contaminants. Dendrimers are highly branched polymers with controlled composition and nanoscale dimensions. These can be designed so to able to act as 'cages' and trap metal ions and zero-valent metals, making them soluble in appropriate media or able to bind to certain surfaces. The vision is to use dendrimers as nanoscale chelating agents for polymers supported ultrafiltration systems.

Another class of nanoparticles that have environmental applications is magnetic nanoparticles. For instance, researchers from Rice University's Centre for Biological and Environmental Nanotechnology (CBEN) have recently found that nanoparticles of rust can be used to remove arsenic from water using a magnet. The concept is simple: arsenic sticks to rust which, being essentially iron oxide tends to be magnetic so it can be removed from water using a magnet. Nanosized rust, about 10 nm in diameter, with its high surface area, was found to improve removal efficiency while reducing the amount of material used. Compared to other techniques currently used to remove arsenic from contaminated water, such as centrifuges and filtration systems, this one has the advantage of being simple, and most importantly, not requiring electricity. This is very important, given that arsenic-contaminated sites are often found in remote areas with limited access to power. Magnetic nanoparticles modified with specific functional groups are also used for the detection of bacteria in water samples (Fig. 20.2).

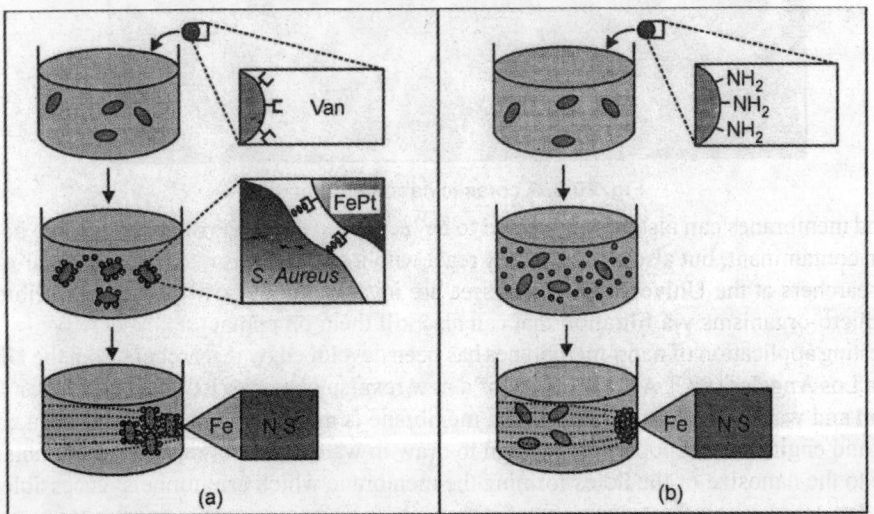

Fig. 20.2. Illustration of the capture of bacteria by modified magnetic nanoparticles via a plausible multivalent interaction and the corresponding control experiment.

Arsenic and arsenate may be also precipitated using nano-scale zero-valent iron (Fe^0) as indicated by recent studies. The removal mechanism in this case involves the spontaneous adsorption and coprecipitation of arsenic with the oxidised forms of Fe^0. As already noted, zero-valent iron is extremely reactive when it is nanosized, so it is currently considered a suitable candidate for both *in situ* and *ex situ* groundwater treatment.

Nano-Membranes and Nano-Filters

Nanotechnology can also be employed for the fabrication of nano-filters, nanoadsorbents and nano-membranes with specific properties to be used for decontaminating water and air. As with other applications, it is the ability to manipulate matter at a molecular level that makes nanotechnology so promising in this field, together with the small size and high surface-to-volume ratio of nanomaterials that are employed for the fabrication of these products. In principle, 'nanotraps' designed for a certain contaminant can be produced, for instance having a specific pore size and surface reactivity. An example is given by the work carried out at Rice's CBEN, where researchers are developing reactive iron oxide ceramic membranes (ferroxane membranes) that are capable of remediating organic waste in water (Fig. 20.3).

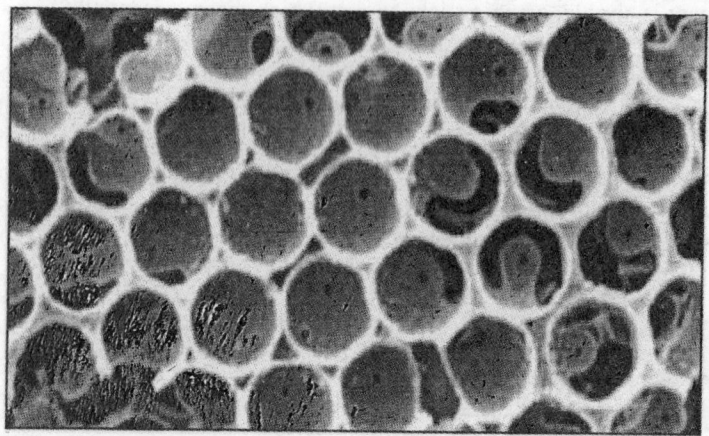

Fig. 20.3. A ceramic nanomembrane.

Filters and membranes can also be engineered to be 'active' in the sense of being capable not only to trap a certain contaminant, but also to chemically react with it and convert it to a non-toxic product. For instance, researchers at the University of Tennessee are investigating a new type of nano-fibre for the removal of micro-organisms via filtration that can also kill them on contact.

An interesting application of nano-membranes has been developed by researchers from the University of California Los Angeles (UCLA) in the form of a new reverse osmosis (RO) membrane for seawater desalinisation and waste-water remediation. The membrane is made of a uniquely cross-linked matrix of polymers and engineered nanoparticles design to draw in water ions but repel contaminants. This is possible due to the nanosize of the holes forming the membrane which are 'tunnels' accessible only to the water molecules.

Another distinctive feature of this nano-membrane is its ability to repel organics and bacteria, thanks to the chemical composition of the nanoparticles embedded in the membrane. Compared with conventional RO membrane, these ones are thus less prone to clogging, which increases the membrane lifetime with an obvious economic benefit.

POLLUTION PREVENTION/REDUCTION

Reduction of waste in manufacturing processes; reduction in the use of harmful chemicals; reduction in the emission of 'greenhouse' effect gases during fuel combustions; use of biodegradable plastics: these are only few of the many approaches that can be taken to reduce the pollution of the environment. Nanotechnology is already actively involved in this sector, either as a technology to produce advanced materials that pollute less, or as a method to increase the efficiency of certain industrial processes.

Materials

Materials that are more environment-friendly fabricated using nanotechnology include biodegradable plastics made of polymers that have a molecular structure optimal for degradation; non-toxic nanocrystalline composite materials to replace lithium-graphite electrodes in rechargeable batteries; and self-cleaning glasses, such as Activ™ Glass, a commercial product available worldwide from Pilkington. The glass is composed of a special coating made of nanocrystals of TiO_2 which, once exposed to daylight, reacts in two ways. First, it breaks down any organic dirt deposits on the glass and second, when exposed to water, it allows rain to 'sheet' down the glass easily and washes the loosened dirt away.

Another area where nanotechnology is making a contribution is the development of fertilisers and wood treatment products that are more stable and leach less into the environment. For instance, researchers at the Michigan State University have incorporated biocides for wood treatment inside polymeric nanoparticles.

Their small size allows them to efficiently travel inside the very fine, sieve-like structure of wood. At the same time, the biocide, being safely trapped inside a 'nanoshell', is protected from leach and random degradative processes.

Nanocatalysis

A catalyst is a substance that increases a chemical reaction rate without being consumed or chemically altered. One of the most important properties of a catalyst is its 'active surface' where the reaction takes place. The 'active surface' increases when the size of the catalysts is decreased (Fig. 20.4). The higher is the catalysts active surface, the greater is the reaction efficiency. Also, research has shown that the spatial organisation of the active sites in a catalyst is important as well. Both properties (nanoparticle size and molecular structure/distribution) can be controlled using nanotechnology. In the environmental field, nanocatalysis is being investigated for desulphurising fuels, with the aim of developing 'clean' fuels containing very low sulphur products (produced in the fuel during its refining process and responsible for generating sulphuric acid upon fuel combustion).

A commercial example is Oxonica's Envirox fuel which uses nanosized cerium oxide as a catalyst to enhance efficiency.

This enhanced-fuel has been tested in 2005 and 2006 in 1000 buses in the UK (another 500 buses were tracked as control). It was found that the test buses used 5 per cent less fuel than the controls and that the fuel savings more than paid for the additive. Nanoscale catalysts are also promising for improving air quality and for treating particularly challenging contaminants in water that must be reduced to a very low level.

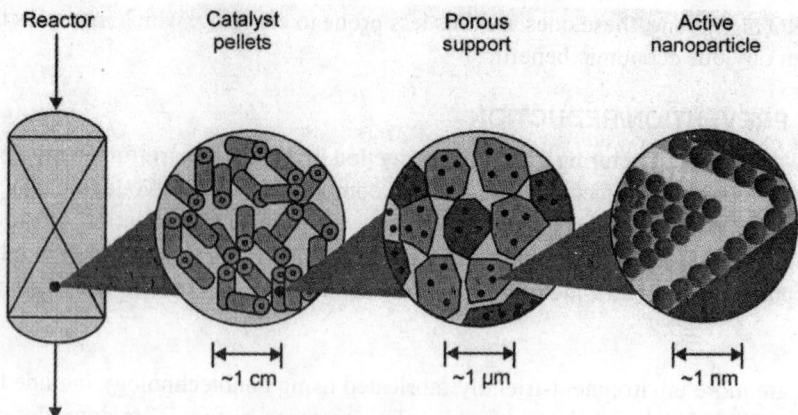

Fig. 20.4. Schematic representation showing how nanoparticles could be included in the catalyst material of a desulphurising fuel reactor. The image on the far right is a real Scanning Tunnelling Microscope (STM) image of a MoS_2 nanocrystal on Au(111) showing peculiar atomic distribution at the edges of the crystal.

Green Manufacturing

Manufacturing processes are always accompanied by the production of diverse waste products, many of which pose a threat to the environment and thus need to be removed and treated. Ideally, manufacturing processes should be designed to minimise material usage and waste production, while ensuring the use of the least amount of energy possible. 'Green manufacturing' is a generic name to broadly cover methods and technologies that are directed towards achieving this goal. It includes the development of new chemical and industrial procedures (for instance water-based rather the solvent-based processes); reduction in the use of unsafe compounds (such as metals); development of 'green' chemicals that are more environment-compatible; and efficient use of energy. In terms of its application to the reduction of manufacturing waste, nanotechnology can contribute in two ways: by directing the manufacturing to be more controlled and efficient, and by using nanomaterials (such as catalysts) that can raise the manufacturing efficiency while reducing or eliminating the use of toxic materials. Overall, nanotechnology has the potential of making industrial processes more efficient in terms of energy usage and material usage, while minimising the production of toxic wastes. The application of 'green nanotechnology' to manufacturing includes bottom-up, atomic-level synthesis for developing improved catalysts; inserting information into molecules to build new materials (such as DNA) through highly specific synthetic routes; scaling down material usage during chemical reaction by using nanoscale reactors; and improving manufacturing to require less energy and less toxic materials.

An example of 'green nanotechnology' is the development of aqueous-based microemulsions to be used in alternative to volatile organic compounds (VOCs) in the cleaning industry. These toxic and potentially carcinogenic compounds, such as chloroform, hexane, percholoroethylene, are conventionally used in the cleaning industry (like the textile industry) as well as in the oil extraction industry. Microemulsions contain nanosized aggregates that can be used as 'receptors' for extracting specific molecules at a nanoscale level. Researchers from the University of Oklahoma have synthesised microemulsions having water attractive and water-repellent 'linkers' inserted between the head and tail parts of a surfactant molecule. The result is a surfactant that has a very low interfacial tension with a wide range of oils. When tested for cleaning textiles from motor oil residues, as well as for extracting edible oil from oilseeds, the microemulsions were found to be very competitive with conventionally used VOCs, both in terms of extraction yield and simplicity of the process.

ENVIRONMENT SENSING

Protection of the human health and of the environment requires the rapid, sensitive detection of pollutants and pathogens with molecular precision. Accurate sensors are needed for *in situ* detection, as miniaturised portable devices, and as remote sensors, for the real-time monitoring of large areas in the field. Generally speaking, a sensor is a device built to detect a specific biological or chemical compound, usually producing a digital electronic signal upon detection. Sensors are now used for the identification of toxic chemical compounds at ultra low levels (ppm and ppb) in industrial products, chemical substances, water, air and soil samples or in biological systems. Nanotechnology can improve current sensing technology in various ways. First, by using nanomaterials with specific chemical and biological properties, the sensor selectivity can be improved, thus allowing isolating a specific chemical or biological compound with little interference. Hence, the accuracy of the sensors is improved. As with other nanoengineered products discussed in this chapter, the high surface-to-volume ratio of nanomaterials increases the surface area available for detection, which in turn has a positive effect on the limit of detection of the sensor, therefore improving the sensitivity of the device. Scaling down using nanomaterials allows packing more detection sites in the same device, thus allowing the detection of multiple analytes. This scaling-down capability, together with the high specificity of the detection sites obtainable using nanotechnology, will allow the fabrication of super-small 'multiplex' sensors, this way lowering the cost of the analysis and reduce the number of devices needed to perform the analysis with an economic benefit. Advancements in the field of nanoelectronics will also allow the fabrication of nanosensors capable of continuous, real time monitoring.

Research in the field of nanosensors includes various areas, like synthesising new nanomaterials with specific detection sites able to recognise a certain pollutant; developing new detection methods, to increase the limit of detection of the sensors while ensuring a 'readable' electrical signal; and miniaturising the size of the sensor elements while integrating these with larger parts of the device. An example of how nanoscience can be applied to the sensing technology is shown in Fig. 20.5, which schematises the operational principle of a heavy-metal nanosensor developed for monitoring heavy metals in drinking water. The sensor is made of an array of electrode pairs fabricated on a silicon chip and separated by few nanometres. When the electrodes are exposed to a solution of water containing metal ions, these deposit inside the nano-gap in between the electrodes.

Once the deposited metal bridges the gap a 'jump' in conductance between the electrodes is registered. The size of the gap, being only few nanometres, allows the detection of a very low concentration of metal ions. This type of sensor is called 'nanocontact sensor'.

Some nanomaterials in the form of nanowires or nanotubes offer outstanding opportunities as sensor elements in chemical and biological sensors. Individual single-walled carbon nanotubes (SWNTs) have been demonstrated to exhibit a faster response and a substantially higher sensitivity for instance towards gaseous molecules (such as NO_2 and NH_3) than that of existing solid-state sensors. In this case, direct binding of the gaseous molecule to the surface of the SWNT is the mechanism involved in sensing, upon which the electrical resistance of the SWNT dramatically increases or decreases. Moreover, this sensitivity was registered at room temperature, whereas conventional solid-state sensors operate at very high (200° to 600°C) temperatures in order to achieve enhanced chemical reactivity between molecules and the sensor material.

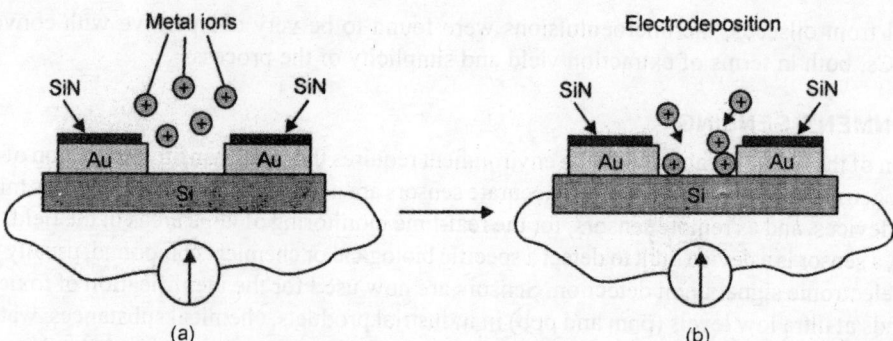

Fig. 20.5. Schematic diagram of a nanocontact sensor. (a) A drop of sample solution containing metal ion is placed onto a pair of nanoelectrodes separated with an atomic scale gap on a silicon chip. (b) Holding the nanoelectrodes at a negative potential, electrochemical deposition of a single or a few metal atoms into the gap can form a nanocontact between the two nanoelectrodes and result in a quantum jump in the conductance

Although SWNTs are promising candidates as nanosensors, they also have some limitations that could limit their development. First, existing synthetic methods produce a mixture of metallic and semiconducting NTs, only the latter being useful as sensors. Second, in order to be able to sense a variety of chemical and biological species, the surface of NTs needs to be modified to have specific functionalities to bind those species. Flexible methods to modify the surface of NTs to bind a large variety of analytes are not well established yet. Conversely, nanowires of semiconductors such as Si don't have these limitations: they are always semiconductors and there is established knowledge for the chemical modification of their surface. Boron-doped silicon nanowires (SiNWs) have been used for the sensitive real-time electrical detection of proteins, antibodies the metabolic indicator calcium and glucose in water. The small size and the capability of these semiconductor nanowires to detect in real-time a wide range of analytes could be used for developing sensors for detecting pathogens, chemical and biological agents in water, air and food.

ENVIRONMENTAL BENEFITS OF NANOTECHNOLOGY

Energy

Nanotechnology could potentially have a great impact on clean energy production. Research is underway to use nanomaterials for purposes including more efficient solar cells, practical fuel cells, and environmentally-friendly batteries. The most advanced nanotechnology projects related to energy are: storage, conversion, manufacturing improvements by reducing materials and process rates, energy saving (by better thermal insulation for example), and enhanced renewable energy sources. Current commercially available solar cells have low efficiencies of 15–20 per cent. Research is ongoing to use nanowires and other nanostructured materials with the hope of to create cheaper and more efficient solar cells than are possible with conventional planar silicon solar cells. It is believed that these nanoelectronics-based devices will enable more efficient solar cells, and would have a great effect on satisfying global energy needs.

Another example for an environmentally-friendly form of energy is the use of fuel cells powered by hydrogen. Probably the most prominent nanostructured material in fuel cells is the catalyst consisting of carbon supported noble metal particles with diameters of 1–5 nm. Suitable materials for hydrogen storage contain a large number of small nanosized pores.

Nanotechnology may also find applications in batteries. Because of the relatively low energy density of conventional batteries the operating time is limited and a replacement or recharging is needed, and the huge number of spent batteries represent a disposal problem. The use of nanomaterials may enable batteries with higher energy content or supercapacitors with a higher rate of recharging, which could be helpful for the battery disposal problem.

POLLUTION PREVENTION AND TREATMENT USING NANOTECHNOLOGY

This section is intended to give an overview of the various aspects of nanotechnology and the environment, mainly looking at it from the side of applications rather than from the risk side. It should have become clear that nanotechnology in general and nanoparticles in particular will have important impacts on various fields of environmental technology and engineering. However, we should always keep in mind that nanotechnology has a Janus face and that each positive and desired property of nanomaterials could be problematic under certain conditions and pose a risk to the environment. A careful weighing up of the opportunities and risks of nanotechnology with respect to their effects on the environment is therefore needed. Environmental nanotechnology is considered to play a key role in the shaping of current environmental engineering and science. Looking at the nanoscale has stimulated the development and use of novel and cost-effective technologies for remediation, pollution detection, catalysis and others. However, there is also a wide debate about the safety of nanoparticles and their potential impact on environment and biota, not only among scientists but also the public. Especially the new field of nanotoxicology has received a lot of attention in recent years. Nanotechnology and the environment— is it therefore a Janus-faced relationship.

There is the huge hope that nanotechnological applications and products will lead to a cleaner and healthier environment. Maintaining and re-improving the quality of water, air and soil, so that the earth will be able to support human and other life sustainably, are one of the great challenges of our time. The scarcity of water, in terms of both quantity and quality, poses a significant threat to the well-being of people, especially in developing countries. Great hope is placed on the role that nanotechnology can play in providing clean water to these countries in an efficient and cheap way. On the other hand, the discussion about the potential adverse effects of nanoparticles has increased steadily in recent years and is a top priority in agencies all over the world. Figure 20.6 shows the hits for a search for risk related to nanotechnology in the Web of Science. Publications that deal in one-way or other with risk have skyrocketed in the last few years since 2002.

The same properties that can be deleterious for the environment can be advantageous for technical applications and are exploited for treatment and remediation. Table 20.1 shows a few examples of nanotechnology: engineered particles with high mobility are needed for efficient groundwater remediation, but at the same time this property will render a particle more difficult to remove during water treatment. The toxicity of some nanoparticles can be used for water disinfection where killing of micro-organisms is intended, whereas the same property is unwanted when nanoparticles eventually enter the environment.

Table 20.1. Examples of nanotechnology highlighting opportunities and risks of nanomaterial.

Opportunities	Nature	Risks
Replacement of toxic materials	Materials efficiency	Unforeseen effects
Less energy use	Energy efficiency	Unforeseen effects
Mobile NP for groundwater remediation	Mobility	Not removed in enviornment

(Contd ...)

Opportunities	Nature	Risks
Water sterilisation	Toxicity	Ecotoxicity
Degradation of pollutants	Catalytic activity	Ecotoxicity
Removal of pollutants	Sorption capacity	Mobilisation and transport of pollutants
Medical uses	Cell uptake	Ecotixicity
Tailored uses	Functionalisation	Higher mobility, ecotoxicity

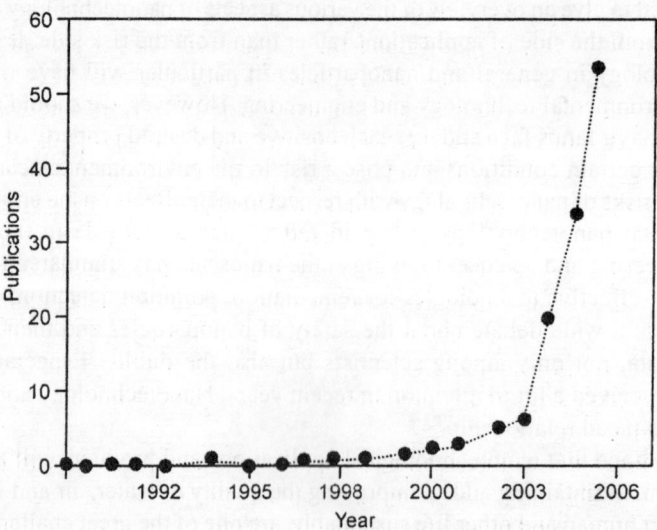

Fig. 20.6. Hits in the web of science for the search terms '(nanotechnol or nanopart or nanotub) and risk' for the years 1990–2006.

The catalytic activity of a nanoparticle can be advantageous when used for the degradation of pollutants, but can induce a toxic response when taken up by a cell. The high sorption capacity of certain nanoparticles is exploited for the removal of organic and inorganic pollutants while this property may also mobilise sequestered pollutants in the environment.

The engineering of nanoparticles that are easily taken up by cells will have a huge impact on medicine and pharmacological research, but the dispersion of such particles in the environment can lead to unwanted nd unexpected effects.

Also the fact that many engineered nanoparticles are functionalised and therefore have a different surface activity from pristine particles is pivotal for many applications where a tailored property is needed, but such particles may behave in a completely different way from standard particles in the environment and may, for example, be much more mobile or show an increased (or decreased, as the case may be) toxicity.

This shortlist of properties exemplifies the fact that engineered nanoparticles or nanotechnological applications make use of the same properties that are looked for by environmental scientists. This section will give a general overview of potential environmental applications of nanotechnology and nanoparticles and will also give a short overview of the current knowledge about possible risks for the environment.

MORE EFFICIENT RESOURCE AND ENERGY CONSUMPTION

Pollution prevention by nanotechnology refers on the one hand to a reduction in the use of raw materials, water or other resources and the elimination or reduction of waste and on the other hand to more efficient use of energy or involvement in energy production. The implementation of green chemistry principles for the production of nanoparticles and for nanotechnological applications in standard chemical engineering will lead to a great reduction in waste generation, less hazardous chemical syntheses, improved catalysis and finally an inherently safer chemistry. However, there are very few data that actually show quantitatively that these claims are true and that replacing traditional materials with nanoparticles really does result in less energy and materials consumption and that unwanted or unanticipated side effects do not occur.

Nanomaterials can be substituted for conventional materials that require more raw material, are more energy intensive to produce or are known to be environmentally harmful. Some new nanocatalysts can be used at much lower temperatures than conventional catalysts and therefore require less energy input. The capacity of nanocatalysts to function at room temperature opens the way for broad applications of nanomaterials in many consumer products. Another example of how nanotechnology can reduce energy costs is nanomaterial coatings on ships, which are expected to realise fuel savings on the order of $460 million per year for commercial shipping in the USA. Nanodiamonds are expected to increase the life expectancy of automotive paints and therefore to reduce material costs and expenditure. Nanotechnology may also transform energy production and storage by providing alternatives to current practices. One example is nanoparticulate catalysts for fossil fuels, which will lead to reduced emissions or better energy efficiency, higher storage capacity for hydrogen, biohydrogen production and more effective and cheaper solar cells or coatings on windows that reduce heat loss. Nanoparticles can increase the storage capacity of batteries and rechargeable batteries or are used in flat screens where they reduce the amount of heavy metals.

POLLUTION DETECTION AND SENSING

Various nanostructured materials have been explored for their use in sensors for the detection of different compounds. An example is silver nanoparticle array membranes that can be used as flow-through Raman scattering sensors for water quality monitoring. The particular properties of carbon nanotubes (CNTs) make them very attractive for the fabrication of nanoscale chemical sensors and especially for electrochemical sensors. A majority of sensors described so far use CNTs as a building block. Upon exposure to gases such as NO_2, NH_3 or O_3, the electrical resistance of CNTs changes dramatically, induced by charge transfer with the gas molecules or due to physical adsorption. The possibility of a bottom-up approach makes the fabrication compatible with silicon microfabrication processes. The connection of CNTs with enzymes establishes a fast electron transfer from the active site of the enzyme through the CNT to an electrode, in many cases enhancing the electrochemical activity of the biomolecules. In order to take advantage of the properties of CNTs, they need to be properly functionalised and immobilised. CNT sensors have been developed for glucose, ethanol, sulphide and sequence-specific DNA analysis. Trace analysis of organic compounds, e.g. for the drug fluphenazine, has also been reported. Nanoimmunomagnetic labelling using magnetic nanoparticles coated with antibodies specific to a target bacterium have been shown to be useful for the rapid detection of bacteria in complex matrices.

ENVIRONMENTAL RISKS

Behaviour in the Environment

The use of nanoparticles in environmental applications will inevitably lead to the release of nanoparticles into the environment. Assessing their risks in the environment requires an understanding of their mobility, bioavailability, toxicity and persistence. Whereas airborne particles and inhalation of nanoparticles have attracted a lot of attention, much less is known about the possible exposure of aquatic and terrestrial life to nanoparticles in water and soils. Nanoparticles agglomerate rapidly into larger aggregates or are contained within other materials (e.g. polymers). Cations, for example, are able to coagulate acid-treated CNTs with critical coagulation concentrations of 37 mM for Na, 0.2 mM for Ca and 0.05 mM for trivalent metals (e.g. La^{3+}).

Aggregation of CNTs added as a suspension to filtered pond water has been reported. Sedimentation and therefore removal from water can be expected under such conditions. The coagulation and interception by surfaces also determine the fate of nanoparticles in porous media and rapid removal has been observed in many, but not all, cases. However, a recent study shows that humic and fulvic acids are able to solubilise CNTs under natural conditions and that stable suspensions are obtained. Most nanoparticles in technical applications are functionalised and therefore, studies using pristine nanoparticles may not be relevant for assessing the behaviour of the actually used particles. As mentioned already on groundwater remediation, functionalisation is often used to decrease agglomeration and therefore, increase mobility of particles.

Very little is known to date about the influence of functionalisation on the behaviour of nanoparticles in the environment.

Ecotoxicology

A consistent body of evidence shows that nanosized particles can be taken up by a wide variety of mammalian cell types, are able to cross the cell membrane and become internalised. The uptake of nanoparticles is size dependent. Most of the toxicological studies have been carried out with mammalian cells and therefore were carried out in a cell culture medium containing a mixture of proteins and other biological compounds. In this medium, nanoparticles are coated with proteins and have a negative surface charge irrespective of the charge of the pristine particles. Results from such studies therefore cannot be directly transferred to environmental conditions.

Ecotoxicological studies show that nanoparticles are also toxic to aquatic organisms, both unicellular (e.g. bacteria or protozoa) and animals (e.g. daphnia or fish). Whereas bulk TiO_2 is considered to have no health effects on aquatic organisms, this is clearly not the case for nanosized TiO_2.

This was found both for inorganic nanoparticles such as TiO_2, CeO_2 and ZnO and for carboncontaining particles such as fullerenes and CNTs. The observed effects ranged from higher activity of certain stress-related genes, lipid peroxidation and glutathione depletion and antibacterial activity (growth inhibition) for micro-organisms to increased mortality and reduced fertility at high particle concentrations. Inorganic nanoparticular TiO_2 had a toxic effect on bacteria and the presence of light was a significant factor increasing the toxicity. In copepods purified CNTs did not show any effect whereas unpurified CNTs with all their by-products increased mortality.

Organisms are able to use a lipid coating of CNTs as a food source and therefore alter the solubility and toxicity of the CNT in the organism. Nanosized CeO_2 particles were adsorbed on the cell wall of *E. coli* but the microscopic methods were not sensitive enough to discern whether internalisation had taken place. Nanosized ZnO was internalised by bacteria. Nanoparticles that damage bacterial cell walls have been found to be internalised, whereas those without this activity were not taken up. CNTs have been shown to be taken up by a unicellular protozoan and they induced a dose-dependent growth inhibition. The CNTs were localised with the mitochondria of the cells.

These results from ecotoxicological studies show that certain nanoparticles will have effects on organisms on the environment, at least at elevated concentrations. The next step towards an assessment of the risks of nanoparticles in the environment will therefore be to estimate the exposure to the different nanoparticles.

SECTION V

Case Studies Related to Air Pollution

Chapter 21

Case Studies

INTRODUCTION

This chapter discusses case studies of Bhopal gas tragedy — India, Chernobyl Explosion — Russia, London smog (UK), Assessment of ambient air quality with special reference to SO_x in the Rourkela Industrial Complex, Environmental management at Tata Iron and Steel Company (TISCO), Bangkok's, strategy to tackle air pollution, Air quality and air pollution control in Hongkong, and Hurricane Katrina costliest natural disaster in US.

CASE STUDY-1: BHOPAL GAS TRAGEDY—INDIA

In the early morning hours of December 3,1984, a poisonous grey cloud (forty tons of toxic gases) from Union Carbide India Limited (UCIL's) pesticide plant at Bhopal spread throughout the city. Water carrying catalytic material had entered Methyl Isocyanate (MIC) storage tank No. 610. What followed was a nightmare. The killer gas spread through the city, sending residents scurrying through the dark streets. No alarm ever sounded a warning and no evacuation plan was prepared. When victims arrived at hospitals breathless and blind, doctors did not know how to treat them, as UCIL had not provided emergency information. It was only when the sun rose the next morning that the magnitude of the devastation was clear. Dead bodies of humans and animals blocked the streets. leaves turned black, the smell of burning chilli peppers lingered in the air. Estimates suggested that as many as 10,000 may have died immediately and 30,000 to 50,000 were too ill to ever return to their jobs. The catastrophe raised some serious ethical issues. The pesticide factory was built in the midst of densely populated settlements. UCIL chose to store and produce MIC, one of the most deadly chemicals (permitted exposure levels in USA and Britain are 0.02 parts per million), in an area where nearly 1,20,000 people lived. The MIC plant was not designed to handle a runaway reaction. When the uncontrolled reaction started, MIC was flowing through the scrubber (meant to neutralise MIC emissions) at more than 200 times its designed capacity. MIC in the tank was filled to 87 per cent of its capacity while the maximum permissible was 50 per cent. MIC was not stored at zero degree centigrade as prescribed and the refrigeration and cooling systems had been shut down five months before the disaster, as part of Union Carbide Chemicals (UCC) global economy drive. Vital gauges and indicators in the MIC tank were defective. The flare tower meant to burn off MIC emissions was under repair at the time of the disaster and the scrubber contained no caustic soda.

As part of UCC's drive to cut costs, the work force in the Bhopal factory was brought down by half from 1980 to 1984. This had serious consequences on safety and maintenance. The size of the work crew for the MIC plant was cut in half from twelve to six workers. The maintenance supervisor position had been eliminated and there was no maintenance supervisor. The period of safety-training to workers in the MIC plant was brought down from 6 months to 15 days.

In addition to causing the Bhopal disaster, UCC was also guilty of prolonging the misery and suffering of the survivors. By withholding medical information on the chemicals, it deprived victims of proper medical care. By denying interim relief, as directed by two Indian courts, it caused a lot hardship to the survivors. In February 1989, the Supreme Court of India ruled that UCC should pay US $470 million as compensation in full and final settlement. UCC said it would accept the ruling provided Government of India (GoI) did not pursue any further legal proceedings against the company and its officials. GoI accepted the offer without consulting with the victims.

Journey from Virginia to Bhopal

In the beginning of the 20th century, UCC was born of a merger of four US companies producing batteries and arc lamps for street lighting and headlamps for cars. By the second half of the 20th century, UCC had 130 subsidiaries in 40 countries, approximately 500 production sites and 1,20,000 employees. UCC manufactured industrial gases, such as nitrogen, oxygen, methane, ethylene and propane, used in petroleum industry as well as chemical substances like ammonia and urea used in the manufacture of fertilisers. It also produced sophisticated metallurgical specialities based on alloys of cobalt, chrome and tungsten, used in airplane turbines. In addition to all these, it produced a whole range of plastic goods for general use. In the 1950s, parasites were creating havoc in the United States, as well as Mexico, Central America and several South American countries, destroying fodder crop, and plantations.

These parasites also found in Malaysia, Japan, and southern Europe attacked potato crops as well as fruit trees and vegetables. The red vine spider was another threat to food crops. The chemical industry had to come up with something to eradicate this. A number of companies went into action. One of them was UCC. In 1954, UCC embarked on a mission of devising a product to exterminate a wide range of parasites, while at the same time respecting the prevailing standards for the protection and safety of human beings and their environment. Thus was born the 'Experimental Insecticide Seven Seven' which soon came to be known as 'Sevin'. To manufacture Sevin phosgene gas was made to react with another gas called monomethylamine.

The reaction of these two gases produced a new molecule, MIC. MIC was one of the most dangerous compounds ever invented in the chemical history. UCC's toxicologists had tested it on rats and the results had been so terrifying that the company banned production of their work. Other experiments had shown that animals exposed to MIC vapours would face instantaneous death. MIC was so volatile that as soon as it came into contact with a few drops of water or a few ounces of metal dust, it got off an uncontrollably violent reaction. No safety system, no matter how sophisticated, would then be able to stop it emitting a fatal cloud into the atmosphere. To prevent explosion, MIC had to be kept permanently at a temperature near zero. Therefore, provision had to be made for the refrigeration of any drums or tanks that were to hold it.

UCC's operations in India started in the beginning of the twentieth century. In 1924, an assembly plant for batteries was opened in Kolkata. By 1983 UCC had 14 plants in India manufacturing chemicals pesticides, batteries and other products. UCC held a 50.9 per cent stake in the Indian subsidiary. The balance of 49.1 per cent was owned by various Indian investors. Normally foreign investors were limited to 40 per cent ownership of equity in Indian companies, but GoI waived this requirement in the case of UCC because of the sophistication of its technology and the company's potential for export. In 1966, an agreement was signed between GoI and UCIL. Under the agreement, UCIL would import 1200 tons of Sevin from the parent company in the United States. UCC would build a factory in India to produce Sevin within five years. The location of the factory would be Kali Grounds in Bhopal (Madhya Pradesh). In 1969, UCC set up its pesticide unit in Bhopal. The GoI granted a license to UCIL to manufacture 5000 tons of Sevin a year. UCIL would produce Sevin and all the chemical ingredients required in India itself. Eduardo Munoz, the Argentinean agronomic engineer, who was with UCC, was entrusted the responsibility of making the project a success.

Eduardo Munoz felt that manufacturing 5000 tons of Sevin would require considerable quantities of MIC to be manufactured and stored. He was not in favour of storing huge quantity of MIC and suggested an alternative like batch production of MIC to meet production line requirements as they rose. This would eliminate the need to store large quantity of MIC on site. However, this production philosophy was against the American industrial culture and UCC officials turned down the suggestion saying, 'You have absolutely no need to worry, dear Eduardo Munoz. Your Bhopal plant will be as inoffensive as a chocolate factory'. Eduardo Munoz was also against the proposed site of the factory as it was too close to areas where people lived, such as the slums in Oriya Bustee, Jayprakash Nagar and Chola.

However, UCC officials thought Kali Grounds was the right place to build the plant. These officials submitted their request for a sixty hectare plot of land on Kali Grounds. According to municipal planning regulations, no industry likely to give off toxic emissions could be set up on a site where the prevailing wind might carry effluents into densely populated areas.

At the Kali Grounds the wind usually blew from north to south, toward the slums, the railway station and finally toward the overpopulated parts of the old town. Under such circumstances, the application should have been rejected. But the UCC officials did not mention that their proposed factory would be making pesticides out of the most toxic gases available in the chemical industry. At the beginning of the summer of 1972, UCC dispatched to UCIL all the plans for the factory's construction and development. In 1979, the Bhopal plant was inaugurated and work started. Initially, when the factory was not ready to make the MIC needed to produce Sevin, the UCIL management decided to import several hundred barrels from the parent company's factory in the United States. In May 1980, the chemical reactors of the Bhopal plant produced their first gallons of MIC and dispatched them into three huge tanks. The new CEO of UCC, Warren Anderson, came over specially from the United States for the event.

All's Not Well with the Bhopal Plant

Since 1980, the Bhopal plant had caused death and injury to many. In December 1981, plant operator Mohammed Ashraf was killed by a phosgene gas leak. Two other workers were injured. In May 1982, three American engineers from the chemical products and household plastics division of UCC came to Bhopal. Their task was to appraise the running of the plant and confirm that everything was functioning according to the standards laid down by UCC.

The report presented to the UCC officials revealed that all was not well with the Bhopal plant. The report described the surroundings of the site as being 'strewn with oily old drums, used piping, pools of used oil and chemical waste likely to cause fire'. It condemned the shoddy workmanship on certain connections, the warping of equipment, the corrosion of several circuits, the absence of automatic sprinklers in the MIC and phosgene production zones, and the risk of explosion in the gas evacuation flares.

It also reported leaks of phosgene, MIC and chloroform, ruptures in pipework and sealed joints, absence of any earth wire on one of the three MIC tanks and poor adjustment of certain devises where excessive pressure could lead to water entering the circuits.

At the same time, the report expressed concern at the inadequately trained staff, unsatisfactory instruction methods and sloppy maintenance reports. Local newspapers in Bhopal published articles criticising the poor management of the Bhopal plant. One newspaper said, 'the day is not far off when Bhopal will be a dead city, when only scattered stones and debris will bear witness to its tragic end'. In October 1982, MIC escaped from a broken valve, seriously affecting four workers and causing eye irritation and breathlessness among people in the nearby communities. This incident was a clear indication of the potential risk to public life.

In the early 1980s, UCC appointed Warren Woomer as the managing director of its pesticide plant in Bhopal. Analysts felt that this signaled the degree of control UCC wanted to exercise over UCIL. In 1982, Woomer retired and Jagannathan Mukund (Mukund) became the managing director. In 1983, under pressure from the parent company, Mukund devoted all his energies to cost cutting. Two hundred skilled workers and technicians were asked to resign. In the MIC unit alone, the manpower in each shift was cut by half. In the control room, only one man was left to oversee some seventy dials, counters and gauges, which relayed, among other things, the temperature and pressure of the three tanks containing the MIC.

The issue of the danger posed by the pesticide plant to Bhopal was raised in the Madhya Pradesh Assembly in December 1982. However, T. S. Viyogi, labour minister in the Arjun Singh government allayed all fears saying, 'A sum of Rs. 250 million has been invested in this unit. The factory is not a small stone, which can be shifted elsewhere. There is no danger to Bhopal, nor will there ever be'. Equally confident was Mukund: 'The gas leak just can't be from my plant. The plant is shut down. Our technology just can't go wrong, we just can't have such leaks', he said.

In the autumn of 1983, Mukund ordered the shutting down of the principal safety systems in the plant. He felt that because the factory was no longer active, these systems were no longer needed. According to analysts Mukund did not pay heed to the fact that sixty tons of MIC were stored in the tanks. Interrupting the refrigeration of these tanks might possibly save a few hundred rupees worth of electricity a day, but it violated a fundamental rule laid down by UCC's chemists, which stipulated that MIC must in all circumstances be kept at a temperature close to zero degree celsius.

In order to save coal, the flames which burnt off any toxic gases emitted into the atmosphere in the event of an accident that burned day and night at the top of the flare, was also extinguished. Other essential equipment, such as the scrubber cylinder used to decontaminate any gas leaks, were subsequently deactivated. All this served as a signal for many well-trained and experienced engineers and operators to leave the Bhopal factory in search of more secure and satisfactory employment.

Between one-half and two-thirds of the skilled engineers who had worked with the plant right from the project stage had left the plant by 1983. Analysts felt that the top officials at UCC were neglecting the Bhopal plant because they were no longer interested in it.

The Bhopal plant was licensed to manufacture 5000 tons of MIC based pesticides per year. However, peak production was only 2704 tons in 1981, which fell to 1657 tons in 1983. Thus the quantity of pesticides manufactured in 1983 was only 33.14 per cent of its licensed capacity. In the first ten months of 1984, UCIL's losses amounted to Rs. 50 million. UCC planned to close the plant and put it up for sale. When no buyer came forward in India, plans were made to dismantle the factory and ship it to another country. Negotiations to this end were completed by the end of November 1984. Financial losses and plans to dismantle the plant exacerbated UCIL's already negligent management practices.

Tragedy

On the night of December 2, 1984, during routine maintenance operations at the MIC plant, at about 9.30 pm a large quantity of water entered storage Tank No. 610 containing over 40 tons of MIC. This triggered off a reaction, resulting in a tremendous increase of temperature and pressure in the tank. 40 tons of MIC, along with Hydrogen Cyanide and other reaction products burst past the ruptured disc into the night air of Bhopal at around 12.30 am. Safety systems were grossly under-designed and inoperative. Senior factory officials knew of the lethal build-up in the tank at least one hour before the leakage, yet the siren to warn neighbourhood communities was sounded more than one hour after the leak started.

By then, the poisonous gases had covered an area of 40 sq/kms. killing thousands of people. Over 500 thousand experienced acute breathlessness, pain in the eyes, and vomiting as they inhaled the deadly vapours. They ran in panic to get away from the poisonous cloud that hung close to the ground for more than four hours. When people poured into hospitals by thousands, their eyes and lungs in burning, choking agony, the doctors called up the plant medical officer to find out what they ought to do. Dr. Loya, UCIL's official doctor in Bhopal replied, 'It is not a deadly gas, just irritating, a sort of tear gas'. Unofficial estimates put the death toll at over 16,000. A study carried out by a non-governmental organisation in March, 1985 showed that between 50–70 per cent of the non-hospitalised population in exposed areas of Bhopal had one or more symptoms of MIC poisoning.

According to an epidemiological study sponsored by Jawaharlal Nehru University, New Delhi, in October 1989, 70 to 80 per cent of the people in the severely affected communities and 40 to 50 per cent in the mildly affected communities continued to suffer from MIC exposure related illnesses five years after the disaster. A house to house symptom survey in one community, conducted as part of a doctoral dissertation in Delhi University in early 1993, showed 65.7 per cent people suffering from respiratory symptoms, 68.4 per cent with neurological problems and 49 per cent with ophthalmic symptoms. Among the women in the reproductive age, 43.2 per cent suffered from reproductive disorders.

Union Carbide Takes the Offensive

Following the accident, the GoI filed a compensation lawsuit against the UCC for an estimated US $3 billion. However, UCC felt that the GoI was to blame for the disaster. In December 1986, UCC filed a countersuit against the GoI and the State of Madhya Pradesh. The company charged the governments with 'contributory' responsibility for the leak of poisonous gases, saying both governments knew of the toxicity of MIC but failed to take adequate precautions to prevent a disaster.

Under the two sections 'First Steps at Control' and 'Contingency Planning and Experience Help', UCC listed all the things that it did immediately following the first call it got about the tragedy. The document said that vital decisions were made the UCC facility making MIC in the US was shut down; a task force led by Warren Anderson was set up; and medical and technical teams were dispatched to the site of the tragedy 'within 24 hours'. The document also said that 'Union Carbide had a contingency plan for emergencies'.

However analysts felt that contrary to what was said in UCC's document, UCC did not have any kind of emergency plans in place at its Indian subsidiary. So much so, that when the accident occurred and people started pouring into the hospitals in Bhopal complaining about the various ailments, the hospital staff had no idea of what had happened or what to do.

UCC tried to defend its position by saying that it had only a 50.9 per cent stake in UCIL. The company also said that all the employees in the company were Indians and that 'the last American employee at the site had left two years before'. UCC maintained that it did not have any hold over its Indian affiliate. UCC further argued that the day-to-day working of UCIL was independent of the parent company and therefore it could not to be held responsible for the gas leak.

However investigations revealed that this was not really true. In spite of denials, it appeared that UCC had substantial authority over its affiliate. Many of the day to day details, such as staffing and maintenance, were left to Indian officials, but every major decision, such as the annual budget, had to be cleared with the parent company.

Settlement

Within months after the disaster, the GoI issued an ordinance appointing itself as the sole representative of the victims for any legal dealings with UCC as regards compensation. The ordinance was later replaced by the Bhopal Gas Leak (Processing of Claims) Act, 1985. Armed with this power, the GoI filed its suit for compensation and damages against UCC in the United States District Court for the Southern District of New York.

Besides filing the suit, one of its prime responsibilities was to register the claims of each and every gas victim in Bhopal. Analysts felt that this job was never done, or rather, not with any seriousness for the next ten years. The government set up various inquiry commissions to investigate the causes of the disaster; they remained half-hearted initiatives at best. UCC, on the other hand, moved more quickly with its 'investigations': it announced by March 1985 that the disaster was due to 'an act of sabotage' by a Sikh terrorist. Then they shifted blame to a disgruntled worker.

In May 1986, Judge J.F. Keenan ruled that India and not the US was the appropriate forum for the Bhopal compensation litigation. In the first pre-trial hearing in the consolidated Bhopal litigation in US federal courts, John F Keenan, asked UCC as 'a matter of fundamental human decency' to provide an interim relief payment of $5-10 million. UCC agreed to provide $5 million, provided a satisfactory plan of distribution and accounting of the funds was devised. For 8 months, the UCC and the GoI haggled over terms of reference and conditions for using the $5 million interim relief.

Finally, in November 1985, the parties agreed to channel the money through the American Red Cross to the Indian Red Cross. Even after one year of the tragedy, no one—not even the official of the MP Government in charge of relief for the victims—had any idea what the Red Cross would do with the money. On December 17, 1987, a Bhopal District Court Judge passed an order directing UCC to pay Rs. 3.5 billion as interim relief.

UCC challenged this order in the MP high court (at Jabalpur) on the grounds that the trial judge was not authorised to pass the order under any provisions of the Indian Civil Penal Code. On April 4, Justice S. K. Seth of the High Court upheld the liability of UCC for the Bhopal disaster, but reduced the interim compensation to Rs. 2.5 billion. UCC appealed to the Supreme Court of India against the High Court order saying 'No court that we know of in India or elsewhere in the world has previously ordered interim compensation where there is no proof of damages or where liability is strongly contested'.

On February 14, 1989, the Supreme Court directed UCC to pay up US $470 million in 'full and final settlement' of all claims, rights, and liabilities arising out of the disaster. The Supreme Court of India ruled that the $470 million settlement was 'just, equitable and reasonable'. UCC described the court's decision as fair and reasonable, and the company's stock soared in the London market. Analysts felt that the Bhopal Gas disaster, which left thousands of people dead and injured, was settled for a mere US $470 million-which worked out to around Rs. 10,000 per victim (if it was divided equally).

In the same year, a leading national daily stated that approximately US $40,000 was spent on the rehabilitation of every sea otter affected by the Alaska oil spill. Each sea otter was given rations of lobsters costing US $500 per day. Thus the life of an Indian citizen in Bhopal was clearly much cheaper than that of a sea otter in America.

In 1991, the Bhopal court summoned Warren Anderson to appear on a charge of 'homicide in a criminal case'. However, he did not turn up. On September 9, 1993, UCC sold its entire 50.9 per cent stake in UCIL to the Calcutta based Mc Leod Russell India Ltd., a company of the B. M. Khaitan Group. Till 2000, attempts to serve a summon on Warren Anderson by victims' organisations in the Federal Court on Southern district of New York have been unsuccessful. Kenneth McCallion, who was the lawyer for some of the victims and their family members, said a private investigator also hired to deliver the summons at Anderson's residences in Vero Beach, Florida, and Manhattan and Long Island in New York was unable to locate him. Asked if he believed Warren Anderson had gone into hiding to avoid the summons, McCallion said, 'We are just surprised we have been unable to find him, a former CEO of a major corporation'. He observed, 'And there is also a legal process which has been issued by the courts in India for him to appear in Bhopal district court to answer criminal charges and those attempts to serve him... have been unsuccessful as well'. In 2001, in their book. It was five past midnight in Bhopal, Dominique Lapierre and Javier Moro wrote that bringing UCC to justice was unlikely because UCC had been sold out. In August 1999, Dow Chemical purchased UCC for US $9.3 billion.

CASE STUDY-2: CHERNOBYL EXPLOSION—RUSSIA

At 1:24 am on April 26, 1986, there was an explosion at the Soviet nuclear power plant at Chernobyl. One of the reactors overheated, igniting a pocket of hydrogen gas. The explosion blew the top off the containment building, and exposed the molten reactor to the air. Thirty-one power plant workers were killed in the initial explosion, and radioactive dust and debris spewed into the air.

It took several days to put out the fire. Helicopters dropped sand and chemicals on the reactor rubble, finally extinguishing the blaze. Then the Soviets hastily buried the reactor in a sarcophagus of concrete. Estimates of deaths among the clean-up workers vary widely. Four thousand clean-up workers may have died in the following weeks from the radiation.

The countries now known as Belarus and Ukraine were hit the hardest by the radioactive fallout. Winds quickly blew the toxic cloud from Eastern Europe into Sweden and Norway. Within a week, radioactive levels had jumped over all of Europe, Asia, and Canada. It is estimated that seventy-thousand Ukrainians have been disabled, and five million people were exposed to radiation. Estimates of total deaths due to radioactive contamination range from 15,000 to 45,000 or more. To give you an idea of the amount of radioactive material that escaped, the atomic bomb dropped on Hiroshima had a radioactive mass of four and a half tons. The exposed radioactive mass at Chernobyl was fifty tons. In the months and years following, birth defects were common for animals and humans. Even the leaves on the trees became deformed.

Today, in Belarus and Ukraine, thyroid cancer and leukemia are still higher than normal. The towns of Pripyat and Chernobyl in the Ukraine are ghost towns. They will be uninhabitable due to radioactive contamination for several hundred years. The worst of the contaminated area is called 'The Zone', and it is fenced off. Plants, meat, milk, and water in the area are still unsafe. Despite the contamination, millions of people live in and near The Zone, too poor to move to safer surroundings. Further, human genetic mutations created by the radiation exposure have been found in children who have only recently been born. This suggests that there may be another whole generation of Chernobyl victims. Recent reports say that there are some indications that the concrete sarcophagus at Chernobyl is breaking down.

How a Nuclear Power Plant Works?

The reactor at Chernobyl was composed of almost 200 tons of uranium. This giant block of uranium generated heat and radiation. Water ran through the hot reactor, turning to steam. The steam ran the turbines, thereby generating electricity. The hotter the reactor, the more electricity would be generated. Left to itself, the reactor would become too *reactive* — it would become hotter and hotter and more and more radioactive. If the reactor had nothing to cool it down, it would quickly *meltdown* — a process where the reactor gets so hot that it melts — melting through the floor. So, engineers needed a way to control the temperature of the reactor, to keep it from the catastrophic meltdown. Further, the engineers needed to be able to regulate the temperature of the reactor — so that it ran hotter when more electricity was needed, and could run colder when less electricity was desired.

The method they used to regulate the temperature of the reactor was to insert heat-absorbing rods, called control rods. These control rods absorb heat and radiation. The rods hang above the reactor, and can be lowered into the reactor, which will cool the reactor. When more electricity is needed, the rods can be removed from the reactor, which will allow the reactor to heat up. The reactor has hollow tubes, and the control rods are lowered into these reactor tubes, or raised up out of the reactor tubes. At the Chernobyl-type reactors, there are 211 control rods. The more control rods that are inserted, the colder the reactor runs. The more control rods that are removed, the hotter the reactor becomes (Fig. 21.1).

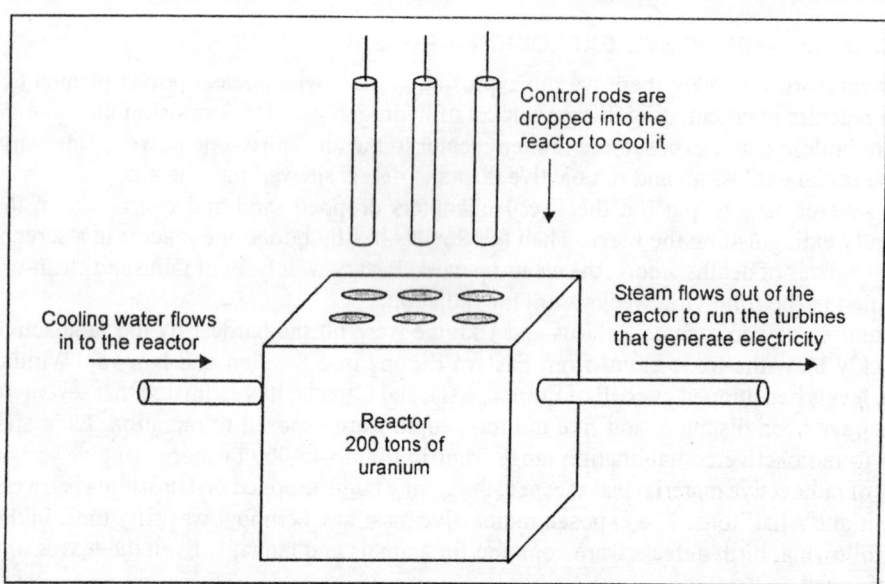

Fig. 21.1. Working of nuclear power plant.

Soviet safety procedures demanded that at least 28 rods were inserted into the Chernobyl reactor at all times. This was a way to make sure that the reactor wouldn't overheat. Water was another method to moderate the temperature of the reactor. When more water ran through the reactor, the reactor cooled faster. When less water ran through the reactor, the reactor stayed hot.

Chernobyl Background

The list of senior engineers at Chernobyl was as follows: Viktor Bryukhanov, the plant director, was a pure physicist, with no nuclear experience.

Anatoly Dyatlov, the deputy chief engineer, served as the day-to-day supervisor. He had worked with reactor cores but had never before worked in a nuclear power plant. When he accepted the job as deputy chief engineer, he exclaimed, 'you don't have to be a genius to figure out a nuclear reactor'. The engineers were Aleksandr Akimov, serving his first position in this role; Nikolai Fomin, an electrical engineer with little nuclear experience; Gennady Metlenko, an electrical engineer; and Leonid Toptunov, a 26 year-old reactor control engineer. The engineers were heavy in their experience of electric technology, but had less experience with the uniqueness of neutron physics.

The confidence of these engineers was exaggerated. They believed they had decades of problem-free nuclear work, so they believed that nuclear power was very safe. The engineers believed that they could figure out any problem. In reality, there had been many problems in the Soviet nuclear power industry. The Soviet state tried to keep problems a secret because problems are bad PR.

The Soviets had a number of nuclear accidents (this is a partial list of Soviet accidents before Chernobyl). In 1957 in Chelyabinsk, there was a substantial release of radioactivity caused by a spontaneous reaction in spent fuel; in 1966 in Melekess the nuclear power plant experienced a spontaneous surge in power, releasing radiation. In 1974, there was an explosion at the nuclear power plant in Leningrad. Later in 1974, at the same nuclear power plant, three people were killed and radiation was released into the environment. In 1977, there was a partial meltdown of nuclear fuel at Byeloyarsk. In 1978 at Byeloyarsk, the reactor went out of control after a roof panel fell onto it..In 1982 at Chernobyl, radioactivity was released into the environment. In 1982, there was there was a fire at Armyanskaya. In 1985, fourteen people were killed when a relief valve burst in Balakovo. Had the engineers at Chernobyl had the information of the previous nuclear accidents, perhaps they would have known to be more careful. It is often from mistakes that we learn, and the engineers at Chernobyl had no opportunity to learn.

As a footnote, don't think that the problems were just those mistake-laden Soviets. Here is a partial list of American accidents before Chernobyl: In 1951, the Detroit reactor overheated, and air was contaminated with radioactive gases. In 1959, there was a partial meltdown in Santa Susanna, California. In 1961, three people were killed in an explosion at the nuclear power plant at Idaho Falls, Idaho. In 1966, there was a partial meltdown at a reactor near Detroit. In 1971, 53,000 gallons of radioactive water were released into the Mississippi River from the Monticello plant in Minnesota. In 1979, there was population evacuation and a discharge of radioactive gas and water in a partial meltdown at Three Mile Island. In 1979 there was a discharge of radiation in Irving Tennessee. In 1982, there was a release of radioactive gas into the environment in Rochester, New York. In 1982, there was a leak of radioactive gases into the atmosphere at Ontario, New York. In 1985, there was a leak of radioactive water near New York City. In 1986, one person was killed in an explosion of a tank of radioactive gas in Webbers Falls, Oklahoma.

The engineers at Chernobyl didn't know about these nuclear accidents. These were secrets that the Soviets kept from the nuclear engineers. Consequently, no one was able to learn from the mistakes of the past. The nuclear plant staff believed that their experience with nuclear power was pretty much error-free, so they developed an overconfidence about their working style. So, according to Gregori Medvedev (the Soviet investigator of Chernobyl), their practice became lazy and their safety practices slipshod. Further, the heavy bureaucracy and hierarchy of the Soviet system created an atmosphere where every decision had to be approved at a variety of higher levels. Consequently, the hierarchical system had quelled the operators' creativity and motivation for problem-solving.

April 25th, 1:00 pm

The engineers at Chernobyl had volunteered to do a safety test proposed by the Soviet government. In the event of a reactor shutdown, a back-up system of diesel generators would crank up, taking over the electricity generation. However, the diesel engines took a few minutes to start producing electricity.

The reactor had a turbine that was meant to generate electricity for a minute or two until the diesel generators would start operating. The experiment at Chernobyl was meant to see exactly how long that turbine would generate the electricity.

The experiment required that the reactor be operating at 50 per cent of capacity. On April 25th, 1986, at 1:00 pm, the engineers began to reduce the operating power of the reactor, by inserting the control rods into the reactor.

This had the effect, you may recall, of cooling off the reactor—making it less reactive. They also shut down the emergency cooling system. They were afraid that the cooling system might kick in during the test, thereby interfering with the experiment. They had no authorisation to deactivate the cooling system, but they went ahead and deactivated it. The experiment called for running the reactor at 50 per cent capacity, thereby generating only half the electricity. At 2:00 pm, a dispatcher at Kiev called and asked them to delay the test because of the higher-than-expected energy usage. They delayed the test, but did not reactivate the emergency cooling system.

April 25th, 11:00 pm

At 11:00 pm, they began the test again. Toptunov, the senior reactor control engineer, began to manually lower the reactor to 50 per cent of its capacity so that they could begin the turbine safety experiment.

Lowering the power generation of a nuclear reactor is a tricky thing. It is not like lowering the thermostat in a house. When you lower the thermostat in the house from 72° to 68°C, the temperature in the house will drop to 68°C and stay there. But in a nuclear reactor, the dropping of the temperature is not only the result of lowering the reactivity, but it is also a cause of lowering the reactivity. In other words, the coldness of the reactor will make the reactor colder. This is called the self-damping effect. Conversely, when the reactor heats up, the heat of the reactor will make itself hotter (the self-amplifying effect).

So, when the control rods are dropped into the reactor, the reactivity goes down. And the water running through the reactor also lessens reactivity. But the lower reactivity also makes the reactor itself less reactive. So, the Chernobyl reactor damped itself, even as the water and the control rods damped its reactivity. It is typically hard for people to think in terms of exponential reduction or exponential increase. We naturally think of a linear (straight-line) reduction or a linear increase. We have trouble with self-damping and self-amplifying effects, because they are nonlinear by definition. So, the engineers oversteered the process, and hit the 50 per cent mark, but they were unable to keep it there. By 12:30 am, the power generation had dropped to 1 per cent of capacity.

Chernobyl-type reactors are not meant to drop that low in their capacity. There are two problems with the nuclear reactor running at 1 per cent of capacity. When reactivity drops that low, the reactor runs unevenly and unstably, like a bad diesel engine. Small pockets of reactivity can begin that can spread hot reactivity through the reactor. Secondly, the low running of the reactor creates unwanted gasses and by-products (xenon and iodine) that poison the reactor. Because of this, they were strictly forbidden to run the reactor below 20 per cent of capacity.

In the Chernobyl control room, Dyatlov (the chief engineer in charge of the experiment), upon hearing the reactor was at 1 per cent, flew into a rage. With the reactor capacity was so low, he would not be able to conduct his safety experiment. With the reactor at 1 per cent capacity, Dyatlov had two options:

1. One option was to let the reactor go cold, which would have ended the experiment, and then they would have to wait for two days for the poisonous by-products to dissipate before starting the reactor again. With this option, Dyatlov would no doubt have been reprimanded, and possibly lost his job.

2. The other option was to immediately increase the power. Safety rules prohibited increasing the power if the reactor had fallen from 80 per cent capacity. In this case, the power had fallen from 50 per cent capacity—so they were not technically governed by the safety protocols.

Dyatlov ordered the engineers to raise power. Today, we know the horrible outcome of this Chernobyl chronology. It is easy for us to sit back in our armchairs, with the added benefit of hindsight, and say Dyatlov made the wrong choice. Of course, he could have followed the spirit of the protocols and shut the reactor down. However, Dyatlov did not have the benefit of hindsight. He was faced with the choice of the surety of reprimand and the harming of his career vs. the possibility of safety problems. And, we know from engineers and technical operators everywhere, safety protocols are routinely breached when faced with this kind of choice. Experts tend to believe that they are experts, and that the safety rules are for amateurs.

Further, safety rules are not designed so that people are killed instantly when the safety standard is broken. On a 55 mile per hour limit on a highway, cars do not suddenly burst into flames at 56 miles per hour. In fact, there is an advantage to going 56 miles an hour as opposed to 55 (you get to your destination faster). In the same way, engineers frequently view safety rules as troublesome, and there is an advantage to have the freedom to disregard them.

In fact, we experience this psychologic every day, usually without thinking about it. When you come toward an intersection, and the light turns yellow, you reach a point where you either have to go through on a yellow light, or come to a stop. Many people go through on the yellow, even though there is a greater risk. So, in a split second, we decide between the surety of sitting at a red light or the possibility, albeit slight, of a safety problem to go through the yellow light. There is a clear advantage to take the risk (as long as you aren't in an accident). While the stakes were higher at Chernobyl, the same psychologic applies. At this point in the Chernobyl process, there were 28 control rods in the reactor—the minimum required. Increasing power would mean that even more control rods would have to be removed from the reactor. This would be a breach of protocol—the minimum number of rods was 28. Dyatlov gave the order to remove more control rods.

Toptunov, the reactor control engineer, refused to remove any more rods. He believed it would be unsafe to increase the power. With the reactor operating at 1 per cent, and the minimum number of control rods in the reactor, he believed it would be unsafe to remove more rods. He was abiding by a strict interpretation of the safety protocols of 28 rods. But Dyatlov continued to rage, swearing at the engineers and demanding they increase power. Dyatlov threatened to fire Toptunov immediately if he didn't increase the power.

The 26 year-old Toptunov was faced with a choice. He believed he had two options:

1. He could refuse to increase power—but then Dyatlov would fire him immediately, and his career would be over.

2. His other choice was to increase power, recognising that something bad might happen.

Toptunov looked around. All the other engineers—including his supervisors—were willing to increase power. Toptunov knew he was young and didn't have much experience with reactors. Perhaps this kind of protocol breach was normal. Toptunov was faced with that choice of the surety of his career ending, vs. the possibility of safety problems. Toptunov decided to agree and increase the power. Tragically, it would be the last decision Toptunov would ever make.

April 26th, 1:00 am

By 1:00 am, the power of the reactor was stable at 7 per cent of capacity. Only 18 control rods were in the reactor (safety protocols demanded that no less than 28 control rods should always be in the reactor).

At 1:07 am, the engineers wanted to make sure the reactor wouldn't overheat, so they turned on more water to ensure proper cooling (they were now pumping five times the normal rate of water through the reactor). The extra water cooled the reactor, and the power dropped again. The engineers responded by withdrawing even more control rods. Now, only 3 control rods were inserted in the reactor.

The reactor stabilised again. The engineers, satisfied with the amount of steam they were getting (they needed steam for their experiment) shut off the pumps for the extra water. They shut off the water, apparently only considering the effect that the water would have on the experiment—and did not consider the effect that the water was having on the reactor. At this point, with only 3 control rods in the reactor, the water was only thing keeping the reactor cool. Without the extra cool water, the reactor began to get hot. Power increased slowly at first. As the reactor got hotter, the reactor itself made the reactor hotter—the self-amplifying effect. The heat and reactivity of the reactor increased exponentially.

The engineers were trying to watch multiple variables simultaneously. The water, the steam, the control rods, and the current temperature of the reactor all were intertwined to affect the reactivity of the reactor. People can easily think in cause and effect terms. Had their only been one variable that controlled the reactivity, the results would probably have been different. However, people have difficulty thinking through the process when there are a multitude of variables, all interacting in different ways.

People are not processors of unlimited information. There is a limited amount of information with which a person can work. With the safety of hindsight, we can sit back and make a judgment saying, 'they didn't think through all their information'. However, this kind of linear judgment does not tell us why they didn't see what is obvious to our hindsight.

At 1:22 am (90 seconds before the explosion), the engineers were still relaxed and confident. Dyatlov, in fact, was seeing his turbine safety experiment coming to a successful conclusion. In what turned out to be a tragic irony, he encouraged his engineers by suggesting, 'in two or three minutes it will all be over'. Thirty seconds before the explosion, the engineers realised the reactor was heating up too fast. With only 3 control rods in the reactor, and then shutting off the water, the reactor was superheating. In a panic, they desperately tried to drop control rods into the reactor, but the heat of the reactor had already melted the tubes into which the control rods slid.

The floor of the building began to shake, and loud banging started to echo through the control room. The coolant water began to boil violently, causing the pipes to burst. The super-heating reactor was creating hydrogen and oxygen gases. This explosive mixture of gases accumulated above the reactor. The heat of the reactor was building fast, and the temperature of the flammable gases was rising.

April 26th, 1:24 am

Finally, the gases detonated, destroying the reactor and the protective containment building. The control room was far enough away from the containment building to escape destruction, but the explosion shook the entire plant. Debris caved in around the control room members, and Dyatlov, Akimov, Toptunov, and the others were knocked to the floor. Dust and chalk filled the air. While they knew there had been an explosion, they hoped and prayed the explosion had not come from the reactor. Toptunov and Akimov ran over the broken glass and ceiling debris to the open door, and ran across the compound toward the containment building. There, they saw the horrifying, unspeakable sight. There was rubble where the reactor had been. They saw flames shooting up 40 feet high, burning oil squirting from pipes onto the ground, black ash falling to the ground, and a bright purple light emanating from the rubble.

Within a few minutes, fire fighters had arrived. The fire fighters, mostly with no protective equipment, heroically worked to extinguish the fire, hoping to prevent further damage to the three other reactors at the plant. Most of the fire fighters died from the radiation exposure. Bryukhanov (the plant director), who was not at the plant at the time, had been contacted and told about an explosion. In the chaos, those informing Bryukhanov of the explosion still did not know the total amount of devastation. Bryukhavov, still desperately hoping that the reactor was intact, called Moscow to inform them that while there had been an explosion, the reactor had not sustained any damage.

Again, with the benefit of hindsight, we can say that Bryukhanov should have acted quicker. It's true that many lives could have been saved if he had acted differently. However, his actions are not uncommon in these kinds of situations. A common reaction is called 'horizontal flight', where people retreat from the worst-case scenario, convincing themselves to believe the best-case scenario. Bryukhanov had convinced himself that the reactor was not in danger. And after all, someone from the plant had called and given an ambiguous message. Surely they would have known if the reactor had been destroyed.

April 26th, 4:00 am

At 4:00 am, the command from Moscow came back: *Keep the reactor cool.* The authorities in Moscow had no idea that the damage was so catastrophic. Akimov, Dyatlov, and Toptunov, their skin brown from the radiation, and their bodies wrenched from internal damage, had already been taken away to the medical center. At 10:00 am, Bryukhanov, the plant director, was informed that the reactor had been destroyed. Bryukhanov rejected the information, preferring to believe that the reactor was still intact. He informed Moscow that the reactor was intact and radiation was within normal limits.

Later that day, experts from around the Soviet Union came to Chernobyl, and found the horrifying truth. The reactor had indeed been destroyed, and fifty tons of radioactive fuel had instantly evaporated. The wind blew the radioactive plume in a northwesterly direction. Belarus and Finland were going to be in the path of the radioactive cloud.

Days Afterward

The secretive Soviet state was slow to act. Soviet bureaucracy debated whether to evacuate nearby cities, and how much land should be evacuated. They were slow in their response, slow to evacuate, and slow to inform the world of the disaster. It took over 36 hours before authorities began to evacuate nearby residents. Two days later, the nightly news (the fourth story) reported that one of the reactors was 'damaged.' Within a few days, radiation detectors were going off all over the world. The Soviets continued to try to hide the issue from the world and their own residents.

Several months later, Bryukhanov was arrested, still believing that he did everything right. Dyatlov survived the radiation sickness, and was arrested in December of that year. He believed he was a scapegoat for the accident. Akimov died a few weeks after the disaster, but till the very end continued to say, 'I did everything right. I don't know how it happened'.

Radiation Cloud on April 27, 1986

Problem-solving errors

Wooden-headedness

Barbara Tuchman, in her book, March of Folly (1984), describes a characteristic she calls 'wooden-headedness'. In the book, she chronicles people in history that made monumental errors — and continued to make the errors after there was evidence that the policy was folly. Wooden-headedness is the 'I've made up my mind, don't confuse me with the facts' mentality.

Wooden-headedness is not an ignorance of the facts, but a refusal to see the facts. It's a tunnel vision, a preoccupation with one course of action even when there is evidence that the course of action is either not working, or even backfiring. Wooden-headedness begins with the assumption that 'I'm correct'. The wooden-header believes that all their assumptions are correct, their view of reality is untarnished, and their strategies are perfect. They continue to pursue the course of action without ever reevaluating their assumptions or rethinking their strategy.

Wooden-headedness and overconfidence often go hand in hand. Part of the overconfidence at Chernobyl came from no one speaking up about the dangerous path they were pursuing (only Toptunov spoke up against the process, but he quickly caved under Dyatlov's pressure to reheat the reactor). Part of the overconfidence came from the lack of independent thinking caused by the out-of-control hierarchical Soviet bureaucracy — the engineers didn't feel a need to question the path they were going down. Part of the overconfidence came from a lack of knowledge about past nuclear accidents.

Preoccupation and overfocus

The engineers at Chernobyl were grossly preoccupied with their safety experiment. They were so obsessed with completing the experiment that they lost sight of the larger picture of nuclear safety. Dyatlov was so obsessed with finishing the experiment that he re-fired the reactor when it had dropped to 1 per cent of capacity. He was so obsessed with finishing the experiment that he threatened to fire Toptunov — who had refused to re-heat the already unstable reactor. Even after the disaster, Dyatlov (and many others) still believed everything they did was right — and showed no capacity to rethink their assumptions.

Preoccupation is a common problem among leaders. People can become so overfocused on one aspect of the whole picture, that they don't give new incoming information the importance it deserves. Psychologists call this a 'loss of situation awareness,' although the term does not explain why they became situationally unaware. Dorner found that when a decision-maker gets preoccupied with a detail, they will frequently ignore incoming information that tells them that their detail has become irrelevant.

Horizontal flight

Horizontal flight is the tendency to retreat into an idea or strategy that is safe when everything around the person is going to hell. Bryukhanov showed horizontal flight by his absolute refusal to believe that the reactor had exploded, even when he was being told by his own staff. Dyatlov showed horizontal flight by believing that he could do anything to the reactor and still be safe. The Soviet bureaucracy showed horizontal flight in their failure to act quickly.

Again, this is a common response to difficult problems. At the end of World War II, Joseph Goebbels was designing new medals for the soldiers — only a few days before the end of the Third Reich. When situations become intolerable, we frequently move into areas where we have control — whether or not it makes sense.

Thus, while there were a few design errors that contributed to the disaster, the Chernobyl case study illustrates that people's problem-solving errors can be catastrophic. The grim chronology of Chernobyl outlines the unfolding of a logic of human error. It illustrates the development of human error — how it is not a decision, but a process. Human error, ubiquitous in all its forms and modes, unfolds with a peculiar logic that is all its own.

CASE STUDY-3: LONDON SMOG

A fog so thick and polluted it left thousands dead wreaked havoc on London in 1952. The smoke-like pollution was so toxic it was even reported to have choked cows to death in the fields. It was so thick it brought road, air and rail transport to a virtual standstill. This was certainly an event to remember, but not the first smog of its kind to hit the capital.

Smog had become a frequent part of London life, but nothing quite compared to the smoke-laden fog that shrouded the capital from Friday 5 December to Tuesday 9 December 1952. While it heavily affected the population of London, causing a huge death toll and inconveniencing millions of people, the people it affected were also partly to blame for the smog.

During the day on 5 December, the fog was not especially dense and generally possessed a dry, smoky character. When nightfall came, however, the fog thickened. Visibility dropped to a few metres. The following day, the sun was too low in the sky to burn the fog away. That night and on the Sunday and Monday nights, the fog again thickened. In many parts of London, it was impossible at night for pedestrians to find their way, even in familiar districts. In The Isle of Dogs area, the fog there was so thick people could not see their feet.

History of Smog

Britain has long been affected by mists and fogs, but these became much more severe after the onset of the Industrial Revolution in the late 1700s. Factories belched gases and huge numbers of particles into the atmosphere, which in themselves could be poisonous. The pollutants in the air, however, could also act as catalysts for fog, as water clings to the tiny particles to create polluted fog or smog.

When some of the chemicals mix with water and air, they can turn into acid which can cause skin irritations, breathing problems, and even corrode buildings. Smog can be identified easily by its thick, foul-smelling, dirty-yellow or brown characteristics, totally different to the clean white fog in country areas.

There are reports of thick smog, smelling of coal tar, which blanketed London in December 1813. Lasting for several days, people claimed you could not see from one side of the street to the other. A similar fog in December 1873 saw the death rate across London rise 40 per cent above normal. Marked increases in death rate occurred, too, after the notable fogs of January 1880, February 1882, December 1891, December 1892 and November 1948. The worst affected area of London was usually the East End, where the density of factories and homes was greater than almost anywhere else in the capital. The area was also low-lying, making it hard for fog to disperse.

How the Smog of 1952 Formed?

The weather in November and early December 1952 had been very cold, with heavy snowfalls across the region. To keep warm, the people of London were burning large quantities of coal in their homes. Smoke was pouring from the chimneys of their houses.

Under normal conditions, smoke would rise into the atmosphere and disperse, but an anticyclone was hanging over the region. This pushes air downwards, warming it as it descends. This creates an inversion, where air close to the ground is warmer than the air higher above it. So when the warm smoke comes out of the chimney, it is trapped. The inversion of 1952 also trapped particles and gases emitted from factory chimneys in the London area, along with pollution which the winds from the east had brought from industrial areas on the continent.

Early on 5 December, in the London area, the sky was clear, winds were light and the air near the ground was moist. Accordingly, conditions were ideal for the formation of radiation fog. The sky was clear, so a net loss of long-wave radiation occurred and the ground cooled. When the moist air came into contact with the ground it cooled to its dew-point temperature and condensation occurred. Beneath the inversion of the anticyclone, the very light wind stirred the saturated air upwards to form a layer of fog 100–200 metres deep. Along with the water droplets of the fog, the atmosphere beneath the inversion contained the smoke from innumerable chimneys in the London area.

During the period of the fog, huge amounts of impurities were released into the atmosphere. On each day during the foggy period, the following pollutants were emitted: 1000 tons of smoke particles, 2000 tons of carbon dioxide, 140 tons of hydrochloric acid and 14 tons of fluorine compounds. In addition, and perhaps most dangerously, 370 tons of sulphur dioxide were converted into 800 tons of sulphuric acid.

Impact of the Smog

The fog finally cleared on December 9, but it had already taken a heavy toll:
1. About 4000 people were known to have died as a result of the fog, but it could be many more.
2. Many people suffered from breathing problems.
3. Press reports claimed cattle at Smithfield had been asphyxiated by the smog.
4. Travel was disrupted for days.

Response to the Smog

A series of laws brought in to avoid a repeat of the situation. This included the Clean Air Acts of 1956 and 1968. These acts banned emissions of black smoke and decreed residents of urban areas and operators of factories must convert to smokeless fuels.

People were given time to adapt to the new rules, however, and fogs continued to be smoky for some time after the Act of 1956 was passed. In 1962, for example, 750 Londoners died as a result of a fog, but nothing on the scale of the 1952 Great Smog has ever occurred again. This kind of smog has now become a thing of the past, tanks partly to pollution legislation and also to modern developments, such as the widespred use of central heating.

CASE STUDY-4: DONORA SMOG

The towns of Donora and Webster, Pennsylvania, along the Monongahela River southwest of Pittsburgh, were the site of a lethal air pollution disaster in late October 1948 that convinced members of the scientific and medical communities, as well as the public, that air pollution could kill people, as well as cause serious damage to health. The disaster took place over the course of five days, when weather conditions known as a temperature inversion trapped cooled coal smoke and pollution from a zinc smelter and steel mill beneath a layer of warm air over the river valley that enclosed the two towns and the surrounding farmland. Almost half of the area's 14000 residents reported becoming ill and about two dozen deaths were attributed to the badly polluted air.

After the disaster, fact-finding studies conducted by federal, state, and local government, as well as the steel industry and private investigators, never definitively identified the exact mix of pollutants that caused the deaths and illnesses. It is believed that a thick blanket of sulphur oxides, carbon monoxide, and particulate literally smothered the towns. Donora is remembered as a key event that inspired federal air pollution legislation in the 1960s and 1970s and contributed indirectly to the establishment of the US Environmental Protection Agency in 1970. It helped mobilise public sentiment in favour of federal regulation rather than continued state and local jurisdiction over polluters.

Incident

Sulphur dioxide emissions from US Steel's Donora Zinc Works and its American Steel and Wire plant were frequent occurrences in Donora. What made the 1948 event more severe was a temperature inversion, in which a mass of warm, stagnant air was trapped in the valley, the pollutants in the air mixing with fog to form a thick, yellowish, acrid smog that hung over Donora for five days. The sulphuric acid, nitrogen dioxide, fluorine and other poisonous gases that usually dispersed into the atmosphere were caught in the inversion and accumulated until the rain ended the weather pattern.

One of the heroes to emerge during the four-day smog was Chief John Volk of the Donora Fire Department and his assistant Russell Davis. Volk and Davis responded to calls from Friday night until Sunday night, depleting their supply of 800 cubic feet (23 m^3) of Oxygen, borrowing more from all nearby municipalities including, McKeesport, Monessen, and Charleroi.

The eight doctors in the town, who belonged to the Donora Medical Association, made house calls much like the firefighters during the period of intense smog, often visiting the houses of patients who were treated by the other doctors in town. This was a result of patients calling every doctor in town in the hope of getting treatment faster. It was not until mid-day Saturday that Mrs. Vernon had it set up so that all calls going to the doctors' offices, would be switched to the emergency center being established in the town hall. The smog was so intense that driving was nearly abandoned; those who chose to continue driving were risky. 'I drove on the left side of the street with my head out the window. Steering by scraping the curb'. recalls Davis.

It was not until Sunday morning the 31st of October, that a meeting occurred between the operators of the plants, and the town officials. Burgess Chambon requested the plants temporarily cease operations. The superintendent of the plants, L.J. Westhaver, said the plants already began to shut down operation at around 6 am that morning. With the rain alleviating the smog, the plants resumed normal operation the following morning.

Researchers analysing the event have focused likely blame on pollutants from the zinc plant, whose emissions had killed almost all vegetation within a half-mile radius of the plant. Dr. Devra L. Davis, director of the Center for Environmental Oncology at the University of Pittsburgh Cancer Institute, has pointed to autopsy results showing fluorine levels in victims in the lethal range, as much as 20 times higher than normal. Fluorine gas generated in the zinc smelting process became trapped by the stagnant air and was the primary cause of the deaths.

Aftermath

Preliminary results of a study performed by Dr. Clarence A. Mills of the University of Cincinnati and released in December 1948 showed that thousands more Donora residents could have been killed if the smog had lasted any longer than it had, in addition to the 20 humans and nearly 800 animals killed during the incident.

Lawsuits were filed against US Steel, which never acknowledged responsibility for the incident, calling it 'an act of God'. While the steel company did not accept blame, it reached a settlement in 1951 in which it paid about $235,000, which was stretched over the 80 victims who had participated in the lawsuit, leaving them little after legal expenses were factored in. Representatives of American Steel and Wire settled the more than $4.6 million claimed in 130 damage suits at about 5 per cent of what had been sought, noting that the company was prepared to show at trial that the smog had been caused by a 'freak weather condition' that trapped over Donora 'all of the smog coming from the homes, railroads, the steamboats, and the exhaust from automobiles, as well as the effluents from its plants'. US Steel closed both plants by 1966. By 1949, a year after the disaster, the total value of the predominantly residential property in Donora had declined by nearly 10 per cent.

The Donora Smog marked one of the incidents where Americans recognised that exposure to large amounts of pollution in a short period of time can result in injuries and fatalities. The event is often credited for helping to trigger the clean-air movement in the United States, whose crowning achievement was the Clean Air Act of 1970, which required the United States Environmental Protection Agency to develop and enforce regulations to protect the general public from exposure to hazardous airborne contaminants. The incident was little spoken of in Donora until a historical marker was placed in the town in 1998 to mark the 50th anniversary of the incident. The 60th anniversary, in 2008, was commemorated with memorials for the families of the victims and other educational programs. The Donora Smog Museum was opened on October 20, 2008, located in an old storefront at 595 McKean Avenue near Sixth Street, with the slogan 'Clean Air Started Here'. Fewer than 6000 people still live in Donora.

CASE STUDY-5: ASSESSMENT OF AMBIENT AIR QUALITY (SO_x) IN THE ROURKELA INDUSTRIAL COMPLEX

Air pollution is woven throughout the fabric of our modern life. Pollution problems began with the growth of cities and population, and their related water, industrial and disposal needs. With increasing urbanisation, man began to intrude on the beautiful, balanced mechanism of the atmosphere. Today, in our homes, in our factories and through our automobiles and planes, we are recklessly injecting pollutants into our atmosphere. By the use of various fuels in factories and homes, cars and planes, we are reducing the available oxygen in the atmosphere by nearly 10 per cent every year. Sulphur dioxide is the most dangerous smoke pollutant for vegetation, and it is most dangerous when shoots and blossoms are just forming, and when moisture is present. In the atmosphere, the SO_2 comes mainly from coal and petroleum combustion. When rain falls, it dissolves sulphur dioxide and becomes very acidic and harms the forest, soil and vegetables. Therefore, control of SO_x pollutant is essential, and this can be achieved by regular monitoring of the level of environmental pollution.

In this regard, a study was conducted of the ambient air quality in the Rourkela Industrial Complex, with special reference to SO_x pollutants. The study was conducted from September 2007 to April 2008. The Rourkela region was chosen for the study as very little work had earlier been done on the quality of ambient air in the region.

Rourkela Industrial Complex

Rourkela is the biggest industrial city in Orissa. The area comes under the Rourkela Development Authority Master Plan Area. Industrialisation started in the area more than three decades ago and has made rapid progress. The entire Rourkela region comes under tropical monsoon climate and is more like that of Deccan Plateau. Being in the north-eastern corner of the Deccan Plateau, the climate here is milder than that in the main Deccan region. The climate is hot and dry in summer with high humidity. Normally, there is heavy rainfall during the south-west monsoon and light rain during the pre-monsoon season. Environmental quality of the area is determined as a result of the increasing industrial activity. In order to find out the current status of the air pollution in the area, it is highly essential to identify the various sources of air pollution. The major sources to air pollution in the area are the Rourkela Steel Plant and a fertiliser plant.

Besides these two heavy industries, there are several medium-scale industries like cement, refractories, explosives, etc. and nearly 300 small-scale industries, which are contributing to the pollution load in the area. In addition, the other major source of air pollution in the region are automobiles. We present here the results of a survey on the concentration of SO_x in six areas of Rourkela.

Survey

The major sources of SO_x pollution in the Rourkela Industrial Complex are the Rourkela Steel Plant (RSP), the fertiliser plant and two captive power plants. The burning of fossil fuels (coal, etc.) in the captive power plants, manufacturer of sulphuric acid and fertilisers, smelting units of RSP and automobiles are the major contributors to the SO_x emission. Six sampling studies stations were selected for the study, of which two were residential areas and four industrial areas. Fertiliser Township (FT) and Indo-German Club (IGC) belong to residential areas while the Environmental Engineering Laboratory (EEL) of RSP, the Fertiliser Plant Laboratory (FPL), Research and Development Centre for Iron and Steel (RDCIS) and the Captive Power Plant-II (CPP-II) were located in industrial areas. The survey was conducted at these sampling stations on a weekly basis using high-volume samplers. At each sampling station, the samplers were allowed to run for 8–10 hours. The specifications of the samplers used in the survey (Envirotech APM 410–411) are presented in Table 21.1. The analysis of SO_x was done by the well-known Well-Gaeke spectrophotometric method.

Table 21.1. Specifications of the high-volume samplers used.

Item	Description/specification
Model	Envirotech APM 410–411
Sampling rate	1.0–1.3 m³/min. for particulates and 0.2–3.0 l/min. for gaseous pollutants.
Particulate size	Down to 0.5 µm (depending upon the filter medium used)
Particulate concentration	Reliable results down to 1.0 µg/m³
Gaseous concentration	Reliable results down to 5.0 µg/m³ for SO_x, NO_x, NH_3, H_2S, Cl_2, etc.
Sampling time	30 min. to 24 hour
Power requirement	220v, 2 amp. single-phase, AC (with built-in voltage stabiliser)
Recommended filter	Whatman's GF/A grade for normal use; Whatman's EPM 2000 for quantitative analysis of dust collected
Accessories	Gaseous sampling attachment APM 411, impingers, filter paper and flow recording system.

Results and Discussion

The results of investigations are presented in Table 21.2. Since the monitoring was done on a weekly basis, the results shown in Table 21.2 are mean values averaged over four weeks of the month and expressed for the sampling duration of 24 hours. The recommended national standards for ambient air quality are presented in Table 21.3 for comparison.

From Table 21.3, we find that the recommended upper limit for SO_x concentration in residential and rural areas is 80 µg/m³ and that in industrial and mixed-use areas is 120 µg/m³. From the comparison of the results of the survey (Table 21.2) with the national air quality standards, it is clear that there is no SO_x pollution in two areas, viz. Fertiliser Plant Laboratory (FPL) and the Research and Development Centre for Iron and Steel (RDCIS). However, the other four sampling stations show high values of SO_x

concentration during some winter months and are free from SO_x pollution during the summer and rainy months. From the Table 21.2, it is clear that there is higher concentration of SO_x in January in the Fertiliser Township, in January and March in the Indo-German Club, in January and February in the Environmental Engineering Laboratory and in November and March in the Captive Power Plant-II.

Table 21.2. Mean concentration of SO_x in six localities of the Rourkela Industrial Complex.

Month	Mean SO_x concentration ($\mu g/m^3$)					
	FT	IGC	EEL	FPL	RDCIS	CPP-II
2007	10.04	61.32	84.07	56.16	58.85	112.64
2007	25.44	25.44	92.17	70.25	66.10	112.75
2007	47.36	76.71	98.31	69.17	72.19	132.32
2007	73.61	76.29	115.79	71.12	77.16	40.54
2008	83.55	83.31	121.54	71.03	74.63	62.83
2008	47.58	64.26	136.09	56.72	75.64	76.43
2008	69.85	88.57	94.14	87.56	65.60	132.95
2008	55.59	77.74	112.06	71.88	77.80	44.00

Table 21.3. National standards for ambient air quality.

Type of area	Pollutant concentration ($\mu g/m^3$)			
	SPM	SO_x	NO_x	CO
Industrial and mixed use	500	120	120	5000
Residential and rural	200	80	80	2000
Sensitive	100	30	30	1000

The formation of smog in winter months at Rourkela may be attributed to the higher concentration of SO_x. Also, a greater percentage of people at Rourkela complain of acute and chronic bronchitis, with emphysema and other lung diseases, eye and respiratory track irritation, and many other allergic diseases in the winter season, which may be due to SO_x pollution. Hence, the Rourkela Steel Plant and other industries around Rourkela should adopt effective pollution control measures for the control of air pollution. Also, restrictions should be imposed on burning of fossil fuels within the city limit.

CASE STUDY-6: ENVIRONMENTAL MANAGEMENT AT TATA IRON AND STEEL COMPANY (TISCO)

As a good corporate citizen, deeply conscious of its obligations to the community, Tata Steel has for a number of years striven to minimise pollution incidental to any steel industry, through a series of effective measures both in the steel works and outside it. A new, greatly intensified action plan has now been drawn up, which promises to further bring down air and water pollution in the Steel City and conform not only with the stringent limits laid down by the Water and Air Pollution Control Acts as well as the Environmental Protection Bill of 1986, but also to ensure development without destruction.

Despite the fact that TISCO was set up at a time when 'pollution control' was not in the industrial vocabulary, for several decades an ambitious and continuing tree plantation program has proved extremely effective in countering the effects of heat, dust and air pollution, and in ensuring soil conservation, thus making Jamshedpur India's greenest industrial city. In 1986, tree plantation programs have been intensified and over two lakh trees have been planted, both inside the works and in the township, to greatly enhance protection to the environment.

In recent times, TISCO has made huge investments in securing sophisticated pollution control technologies. Today, many high capacity electrostatic precipitators in boiler houses and other locales substantially inhibit pollutants, such as dust, dirt, flyash and toxic materials. The steel plant is rapidly modernising and all units are being equipped with fool proof closed circuit pollution control systems. A substantial portion of the outlay of Rs. 1000 crores to be spent over the next five years on the ongoing modernisation program in the steel plant will be on acquiring sophisticated pollution control equipment and using it for the best results. Several other schemes now underway in the plant are proper flyash control from power houses by converting flyash at the boiler houses into slurry for quenching molten slag, the mixing of dry ash with slag, installing wet circuit electrostatic precipitators at boiler houses, gas cleaning at the L.D. shop and lime calcining plants and dedusting units at several other places.

Tata Steel is an old plant containing a large number of antiquated machines and equipment which do not easily conform to pollution control activities. But the new plan envisages substantial capital investment in retrofitting pollution control equipment at all old installations to achieve desired results of environmental management. Other exciting new directions envisaged are a large biological oxidation treatment plant for bacteriological control of coke oven effluents; the complete conversion into synthetic oxides of pickled liquor from the sheet mills and the tubes division, a substantial part of which today runs into the many canals connecting the plant to the river, by Tata Pigments; the construction of a number of catchment ponds for treatment of overflow slurry from the slag pits in order to prevent particulate matter from polluting natural water resources; and better operational discipline to make full use of fuel in order to control chimney emanations of toxic gases.

The company also intends to install the latest monitoring devices, such as electronic stack monitoring equipment, calorimeters, etc. in an effort to check pollution of ambient air. Sophisticated laboratory facilities will also be installed and the company intends to acquire in the near future a self-contained mobile van for monitoring pollution at various sites around the steel city.

As a prelude to the new action plan, the steel company is about to embark on an 'environmental impact assessment program' with the help of a specialised consultant in order to assess measures that are needed to be done on the short — and long-term basis.

CASE STUDY-7: BANGKOK'S STRATEGY TO TACKLE AIR POLLUTION

As a consequence of population increase, city development and a growing number of motor vehicles on its roads, Bangkok, the capital of Thailand, has experienced serious air pollution problems over the past several decades. Measures recently adopted by the Thai government, however, have helped the growing city manage its air quality, putting Bangkok on the path to cleaner air and better quality of life for its residents.

Transport is the greatest source of air pollutants in Bangkok. Street-level concentrations of air pollutants along the city's major roads can reach hazardous levels, owing to increased numbers of high emission motor vehicles coupled with long distances travelled and extreme traffic congestion. The number of motor vehicles registered in Bangkok soared from 6,00,000 in 1980 to 4,163,000 at the end of 1999 — a seven-fold increase. Between 1999 and 2007, vehicle registrations continued to rise. By the end of 2007, there were 5,614,294 vehicles choking Bangkok's inadequate street and roadway networks, comprising 3,208,462 passenger cars; 2,261,545 motorcycles; 110,571 trucks; and 33,716 buses. Results of ambient air quality monitoring indicate that the air pollutants of concern in Bangkok are particulate matter (PM), ozone (O_3), carbon monoxide (CO), sulphur dioxide (SO_2), and nitrogen dioxide (NO_2). The ambient air quality in the city and in general background and roadside areas is shown in Fig. 21.2. This illustrates that PM (PM_{10} and total suspended particles) is the pollutant of greatest concern, and the pollution near roads is more serious than elsewhere in the city.

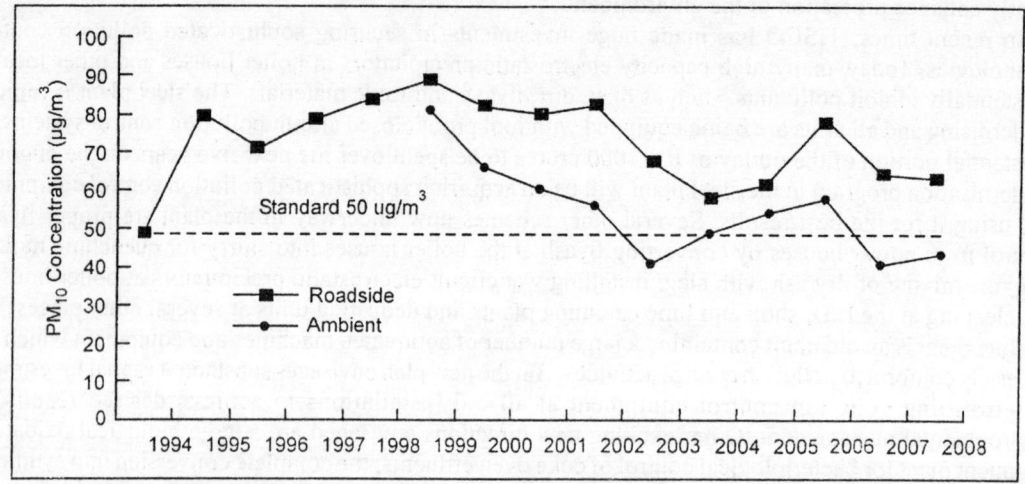

Fig. 21.2. Annual average of PM_{10} in Bangkok during 1994–2008.

The Royal Thai Government has adopted a number of measures to mitigate Bangkok's air pollution problems, focusing on maintaining a good quality of life for the general public. The government's ultimate goal is to bring emissions and ambient air quality in line with the National Air Quality Standards or better. One important milestone was the elimination of lead from gasoline in 1996. Now, the ambient air lead concentration in Bangkok is near zero. Since the 1990s, the government has facilitated ongoing collaboration among the municipality of Bangkok, various sectors impacting air pollution, and the public, resulting in the adoption of air pollution control strategies for transport-related sources, such as improving fuel quality; enforcing emission standards for new and in-use vehicles; implementing an inspection and maintenance program; reducing vehicle kilometers travelled; and performing roadside inspection, traffic management and gasoline vapour recovery. Air pollution control strategies for stationary sources have also evolved, including requiring environmental impact assessments, enforcing emission standards and fuel oil standards, and implementing monitoring requirements. To provide transport alternatives and decrease the number of vehicles on the roads, Bangkok developed a new public transport system, featuring a subway line and an above-ground Skytrain, in 2004. The mass-rapid-transit system has helped improve air quality somewhat, but the limited area covered by the system does little to alleviate traffic and curb the city's overall pollution. Bangkok is now working to expand the distance reachable by Skytrain, which will help ensure good air quality even while the population increases.

Thailand has succeeded in mitigating air pollution in Bangkok, but the government continues to conduct research to make use of new knowledge and keep up with rapidly advancing technology. Thailand is disseminating and sharing its experiences in air pollution control with other countries in Asia.

Air pollution: Levels of carbon monoxide and oxidants in the Bangkok air are twice the amount recommended by WHO. Also nitrogen dioxide levels are too high. Both of these particles are harmful for human health, they affect visibility, irrigate lung tissue causing long-term disorders and cancer. Air pollution is highest in the early mornings, when the traffic conditions are heavy and the air is calm. Noise levels in the city are also all above the internationally accepted level of 70 dBa.

CASE STUDY-8: AIR QUALITY AND AIR POLLUTION CONTROL IN HONGKONG

This case study gives an overview on air pollution control in Hongkong. Air pollution in Hongkong is considered a serious problem. Visibility is currently less than eight kilometres for 30 per cent of the year. Cases of asthma and bronchial infections have soared in recent years due to reduced air quality. Hongkong has been facing two air pollution issues. One is local street-level pollution. The other is the regional smog problem. Diesel vehicles are the main source of street-level pollution. Smog, however, is caused by a combination of pollutants from motor vehicles, industry and power plants both in Hongkong and in the Pearl River Delta region.

The Hongkong special administrative region government gives high priority to controlling both street-level air pollution and smog. The main strategies include:

1. Implementing a wide range of measures to control emissions from motor vehicles, power plants, and industrial and commercial processes locally.
2. Working with Guangdong Provincial Authorities to implement a joint plan to tackle the regional smog problem.

Reducing Emissions from Vehicles

Motor vehicles, especially diesel vehicles, are the main causes of high concentrations of respirable suspended particulates (RSPs) and nitrogen oxides (NO_x) at street level in Hongkong.

To tackle this problem, the Government introduced a comprehensive program in 2000 with targets to reduce RSPs and NO_x emissions from motor vehicles by 80 per cent and 30 per cent respectively by the end of 2005. The key measures include:

1. Adopt tighter fuel and vehicle emission standards.
2. Adopt cleaner alternatives to diesel where practicable.
3. Control emissions from remaining diesels with devices that trap pollutants.
4. Strengthen vehicle emission inspections and enforcement against smoky vehicles.
5. Promote better vehicle maintenance and eco-driving habits.

The measures had borne fruit — air quality in districts with heavy traffic has already improved. Compared with 1999, the roadside concentrations of the major air pollutant emissions from vehicles, namely respirable suspended particulates (RSP) and nitrogen oxides (NO_x), had been reduced by 22 per cent and 23 per cent respectively in 2008, and the number of smoky vehicles spotted has also been reduced by about 80 per cent. The Government introduced additional measures in 2007 and 2008 to strengthen its effort in reducing vehicle emissions. The measures include:

1. To incentivise early replacement of old diesel commercial vehicles with vehicles that comply with the prevailing statutory emission standard for newly registered vehicles, which is now Euro IV standard.
2. To encourage the use of environment-friendly petrol private cars through tax concession;
3. To introduce a concessionary duty of $0.56 for Euro V diesel. Since then, all petrol filling stations in Hongkong are exclusively offering this fuel. Starting from 14 July 2008, the duty rate for Euro V diesel has been waived to further encourage drivers to use this more environment-friendly fuel.
4. To complete a public consultation on whether to introduce legislation to ban idling vehicles from running their engines.
5. To encourage the use of environment-friendly commercial vehicles through tax concession.

Figure 21.3 shows measures and effectiveness in reducing number of smoky vehicles spotted.

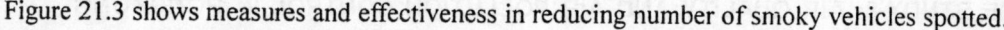

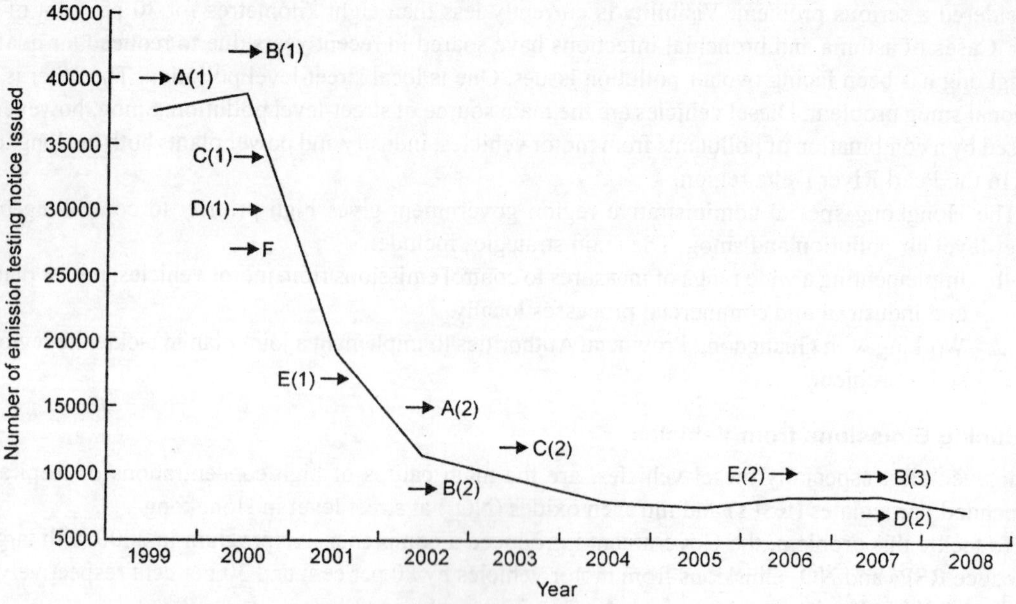

Measures to reduce vehicle emissions

Advancement of smoke testing method:
→ A(1)–Dynamometer smoke test for light duty vehicles (Sep. 1999)
→ A (2)–Dynamometer smoke test for heavy duty vehicles (Jan. 2002)

One-off grant for vehicle replacement:
→ B(1)–Diesel to LPG taxi (Aug. 2000)
→ B(2)–Diesel to LPG light bus (Aug. 2002)
→ B(3)–Replacement of pre-euro and euro I commercial vehicles (Apr. 2007)

Retrofitting emission reduction device:
→ C(1)–Trap/DOC retrofitting for pre-euro LDV (Sep. 2000)
→ C(2)–DOC retrofitting for per-euro HDV (Jan. 2003)

Cleaner diesel:
→ D(1)–ULSD (Jul. 2000)
→ D(2)–Euro V diesel (Dec. 2007)

Siringent vehicle import standard:
→ E(1)–Euro III standard (Oct. 2001)
→ E(2)–Euro IV standard (Oct. 2006)

Punishment for smoky vehicle:
→ F–The fine or fixed penalty ticket raised to $1000 (Dec. 2000)

Fig. 21.3. Measures and effectiveness in reducing number of smoky vehicles spotted.

The number of smoky vehicles spotted has reduced substantially as a result of measures taken to reduce vehicular emissions in recent years.

Reducing Emissions from Industrial Sources and Power Plants

The air pollution control ordinance and its subsidiary regulations provide the control of emissions from power plants, industrial and commercial sources, construction activities, open burning, asbestos, petrol filling stations and dry-cleaning machines. A regulation introduced in 1990 limiting the sulphur content of industrial fuel has reduced sulphur dioxide pollution to very low levels (Fig. 21.4).

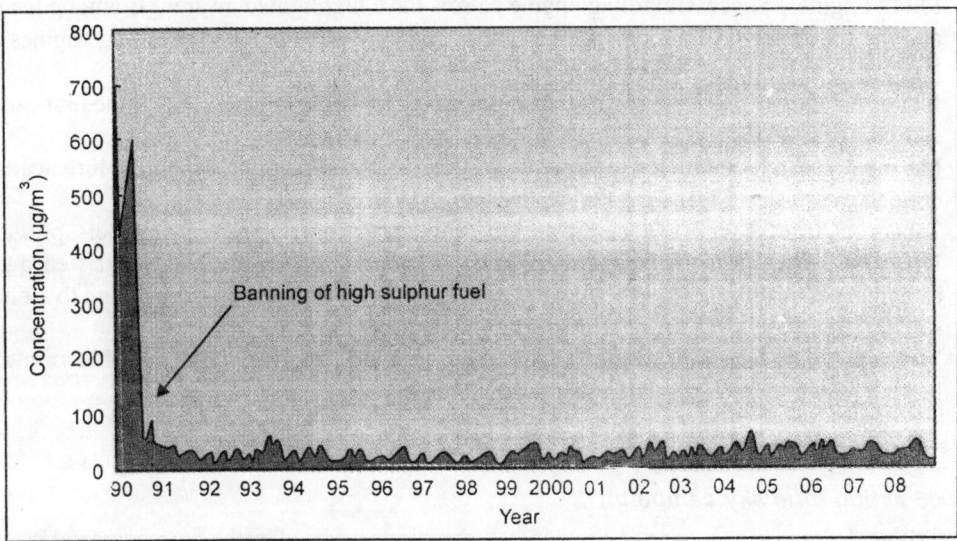

Fig. 21.4. Sulphur dioxide levels hit by fuel restriction regulations (sulphur dioxide at previously badly affected area).

Emissions from power plants have been substantially reduced over the years even though demand for power has increased.

Actions Implemented

Switch to cleaner motor fuels

All HK taxis and many LPB now run on LPG.

2006 indoor smoking ban

The Smoking (Public Health) (Amendment) Ordinance 2006 bans smoking indoors in restaurants, workplaces, schools, karaoke lounges, as well as beaches, swimming pools, sports grounds and public parks with effect from 1st January, 2007.

Organisations Working Against Air Pollution

Clear the air Hongkong

Clear the air is a charity organisation committed to improving air quality in Hongkong. Current Projects include:

1. Diesel: Recent government policies centred on voluntary schemes to phase out old polluting diesel vehicles (pre Euro 3) have proven ineffective. Further campaigns have to mandate a compulsory and scheduled phasing out of old polluting vehicles.

2. Energy: Aim to reduce harmful pollutants (PM 2.5) from power station emissions through more stringent Air Quality Objectives (AQO). Work to improve energy savings and efficiency in all public and private buildings across Hongkong.
3. Events and Education: Participating in and/or driving various territory wide environmental events and educational programs on clean air.
4. Idling Engines: Successful idling engine patrols have highlighted awareness among the public and the government of increased road side pollution leading to a ban on idling engines across Hongkong to be tabled at Legco.
5. Indoor Air Quality (IAQ): Free IAQ assessments for schools to be extended; the first campaign started April, 2008.
6. Marine: Research conducted to implement a 'smoky vessel spotter' scheme. More needs to be done to establish a Clean Port Policy scheme.
7. Tobacco: Clear the air is advocating the government to license all tobacco retailers, pressing the Financial Secretary to increase tobacco tax, pressing Legco to rescind the flawed qualified establishment exemptions and educating the public that the true cost of smoking to Hongkong society is in excess of $73.32 billion per year.
8. Town planning: Clear the air defends the benchmark of 'canyon effect' and Electronic Road Pricing (ERP) to auto-regulate urban traffic density.

Actions Discussed or Forums Created

July 2006 action blue sky campaign

The action blue sky campaign was an environmental campaign organised by the Environmental Protection Department, and launched by chief executive Donald Tsang in July 2006. Its campaign slogan in English is 'Clean air for a cool Hongkong'! The campaign hoped to win support from the public as well as the business community, including those businesses investing in the Pearl River Delta Region.

November 2007 vehicle idling ban

In November 2007, the government launched a public consultation on the proposal which would impose a fixed penalty of HK$320 on drivers who would violate a ban on idling, with taxi and minibus drivers likely to bear the brunt of the ban. The government said its action is due to the failure of motorists to heed many past campaigns switch off engines while waiting. Taxi and minibus drivers were opposed to the proposal.

Currently it is illegal for any driver to leave their engine running if they get out of their vehicle. The courts have been awarding fines of HK$700. It is also illegal for taxis to loiter and minibuses to stop longer than necessary to pick up or put down passengers. It is also illegal to park anywhere except in a designated parking place. This means that the vast majority of drivers who idle their engines are already in violation of at least one existing traffic safety law.

However, traffic wardens are under strict policy guidelines not to give out any tickets unless there has already developed a 'serious' obstruction of the roadway or there have been multiple complaints made by the public; this is the 'Selective Traffic Enforcement Policy' (STEP).

Traffic safety policing of idling vehicles, therefore, falls to private organisations like 'mini spotters' who act as volunteer traffic wardens, making statements to police that can be prosecuted without traffic wardens having to issue tickets directly to the transport trade.

CASE STUDY-9: HURRICANE KATRINA

Hurricane Katrina of the 2005 Atlantic hurricane season was the costliest natural disaster, as well as one of the five deadliest hurricanes, in the history of the United States. Among recorded Atlantic hurricanes, it was the sixth strongest overall. At least 1836 people lost their lives in the actual hurricane and in the subsequent floods, making it the deadliest US hurricane since the 1928 Okeechobee hurricane; total property damage was estimated at $81 billion, nearly triple the damage wrought by Hurricane Andrew in 1992.

Hurricane Katrina formed over the Bahamas on August 23, 2005 and crossed southern Florida as a moderate Category 1 hurricane, causing some deaths and flooding there before strengthening rapidly in the Gulf of Mexico. The storm weakened before making its second landfall as a Category 3 storm on the morning of Monday, August 29 in southeast Louisiana. It caused severe destruction along the Gulf coast from central Florida to Texas, much of it due to the storm surge. The most severe loss of life occurred in New Orleans, Louisiana, which flooded as the levee system catastrophically failed, in many cases hours after the storm had moved inland. Eventually 80 per cent of the city and large tracts of neighbouring parishes became flooded, and the floodwaters lingered for weeks. However, the worst property damage occurred in coastal areas, such as all Mississippi beachfront towns, which were flooded over 90 per cent in hours, as boats and casino barges rammed buildings, pushing cars and houses inland, with waters reaching 6–12 miles (10–19 km) from the beach.

The hurricane protection failures in New Orleans prompted a lawsuit against the US Army Corps of Engineers (USACE), the builders of the levee system as mandated in the Flood Control Act of 1965. Responsibility for the failures and flooding was laid squarely on the Army Corps in January 2008, but the federal agency could not be held financially liable due to sovereign immunity in the Flood Control Act of 1928. There was also an investigation of the responses from federal, state and local governments, resulting in the resignation of Federal Emergency Management Agency (FEMA) director Michael D. Brown, and of New Orleans Police Department (NOPD) Superintendent Eddie Compass. Conversely, the United States Coast Guard (USCG), National Hurricane Center (NHC) and National Weather Service (NWS) were widely commended for their actions, accurate forecasts and abundant lead time.

Five years later, thousands of displaced residents in Mississippi and Louisiana are still living in temporary accommodation. Reconstruction of each section of the southern portion of Louisiana has been addressed in the Army Corps of Engineers LACPR. Final Technical Report which identifies areas not to be rebuilt and areas and buildings that need to be elevated.

METEOROLOGICAL HISTORY

Hurricane Katrina formed as Tropical Depression Twelve over the southeastern Bahamas on August 23, 2005 as the result of an interaction of a tropical wave and the remains of Tropical Depression Ten. The system was upgraded to tropical storm status on the morning of August 24 and at this point, the storm was given the name Katrina. The tropical storm continued to move towards Florida, and became a hurricane only two hours before it made landfall between Hallandale Beach and Aventura on the morning of August 25. The storm weakened over land, but it regained hurricane status about one hour after entering the Gulf of Mexico.

The storm rapidly intensified after entering the Gulf, growing from a Category 3 hurricane to a Category 5 hurricane in just nine hours. This rapid growth was due to the storm's movement over the 'unusually warm' waters of the Loop Current, which increased wind speeds. On Saturday, August 27, the storm reached Category 3 intensity on the Saffir-Simpson Hurricane Scale, becoming the third major hurricane of the season. An eyewall replacement cycle disrupted the intensification, but caused the storm to nearly double in size. Katrina again rapidly intensified, attaining Category 5 status on the morning of August 28 and reached its peak strength at 1:00 pm CDT that day, with maximum sustained winds of 175 mph (280 km/hr) and a minimum central pressure of 902 mbar. The pressure measurement made Katrina the fourth most intense Atlantic hurricane on record at the time, only to be surpassed by Hurricanes Rita and Wilma later in the season; it was also the strongest hurricane ever recorded in the Gulf of Mexico at the time (a record also later broken by Rita).

Katrina made its second landfall at 6:10 am CDT on Monday, August 29 as a Category 3 hurricane with sustained winds of 125 mph (205 km/hr) near Buras-Triumph, Louisiana. At landfall, hurricane-force winds extended outward 120 miles (190 km) from the center and the storm's central pressure was 920 mbar. After moving over southeastern Louisiana and Breton Sound, it made its third landfall near the Louisiana/Mississippi border with 120 mph (195 km/hr) sustained winds, still at Category 3 intensity.

Katrina maintained strength well into Mississippi, finally losing hurricane strength more than 150 miles (240 km) inland near Meridian, Mississippi. It was downgraded to a tropical depression near Clarksville, Tennessee, but its remnants were last distinguishable in the eastern Great Lakes region on August 31, when it was absorbed by a frontal boundary. The resulting extratropical storm moved rapidly to the northeast and affected eastern Canada.

Preparations

Federal government

On the morning of Friday, August 26, at 10 am CDT (1500 UTC), Katrina had strengthened to a Category 3 storm in the Gulf of Mexico. Later that afternoon, the NHC realised that Katrina had yet to make the turn toward the Florida Panhandle and ended up revising the predicted track of the storm from the panhandle to the Mississippi coast. The NHC issued a hurricane watch for southeastern Louisiana, including the New Orleans area at 10 am CDT Saturday, August 27. That afternoon the NHC extended the watch to cover the Mississippi and Alabama coastlines as well as the Louisiana coast to Intracoastal City.

The United States Coast Guard began prepositioning resources in a ring around the expected impact zone and activated more than 400 reservists. On August 27, it moved its personnel out of the New Orleans region prior to the mandatory evacuation. Aircrews from the Aviation Training Center, in Mobile, staged rescue aircraft from Texas to Florida. All aircraft were returning back towards the Gulf of Mexico by the afternoon of August 29. Air crews, many of whom lost their homes during the hurricane, began a round-the-clock rescue effort in New Orleans, and along the Mississippi and Alabama coastlines.

President of the United States George W. Bush declared a state of emergency in selected regions of Louisiana, Alabama, and Mississippi on Saturday, the 27th, two days before the hurricane made landfall. That same evening, the NHC upgraded the storm alert status from hurricane watch to hurricane warning over the stretch of coastline between Morgan City, Louisiana to the Alabama-Florida border, 12 hours after the watch alert had been issued, and also issued a tropical storm warning for the westernmost Florida Panhandle. During video conferences involving the president on August 28 and 29, the director of the National Hurricane Center, Max Mayfield, expressed concern that Katrina might push its storm surge over the city's levees and flood walls. In one conference, he stated, 'I do not think anyone can tell you with confidence right now whether the levees will be topped or not, but that's obviously a very, very great concern'.

On Sunday, August 28, as the sheer size of Katrina became clear, the NHC extended the tropical storm warning zone to cover most of the Louisiana coastline and a larger portion of the Florida Panhandle. The National Weather Service's New Orleans/Baton Rouge office issued a vividly worded bulletin predicting that the area would be 'uninhabitable for weeks' after 'devastating damage' caused by Katrina, which at that time rivaled the intensity of Hurricane Camille. 'On Sunday, August 28, President Bush spoke with Governor Blanco to encourage her to order a mandatory evacuation of New Orleans'.

Voluntary and mandatory evacuations were issued for large areas of southeast Louisiana as well as coastal Mississippi and Alabama. About 1.2 million residents of the Gulf Coast were covered under a voluntary or mandatory evacuation order.

Investigation of state of emergency declaration

In a September 26, 2005 hearing, former FEMA chief Michael Brown testified before a US House subcommittee about FEMA's response. During that hearing, Representative Stephen Buyer (R-IN) inquired as to why President Bush's declaration of state of emergency of August 27 had not included the coastal parishes of Orleans, Jefferson, and Plaquemines. (In fact, the declaration did not include any of Louisiana's coastal parishes, whereas the coastal counties were included in the declarations for Mississippi and Alabama.) Brown testified that this was because Louisiana Governor Blanco had not included those parishes in her initial request for aid, a decision that he found 'shocking'. After the hearing, Blanco released a copy of her letter, which showed she had requested assistance for 'all the southeastern parishes including the New Orleans Metropolitan area and the mid state Interstate I-49 corridor and northern parishes along the I-20 corridor that are accepting'.

Gulf coast

On August 26, the state of Mississippi activated its National Guard in preparation for the storm's landfall. Additionally, the state government activated its Emergency Operations Center the next day, and local governments began issuing evacuation orders. By 7:00 pm EDT on August 28, 11 counties and eleven cities issued evacuation orders, a number which increased to 41 counties and 61 cities by the following morning. Moreover, 57 emergency shelters were established on coastal communities, with 31 additional shelters available to open if needed. Louisiana's hurricane evacuation plan calls for local governments in areas along and near the coast to evacuate in three phases, starting with the immediate coast 50 hours before the start of tropical storm force winds. Persons in areas designated Phase II begin evacuating 40 hours before the onset of tropical storm winds and those in Phase III areas (including New Orleans) evacuate 30 hours before the start of such winds.

Many private caregiving facilities that relied on bus companies and ambulance services for evacuation were unable to evacuate their charges because they waited too long. Louisiana's Emergency Operations Plan Supplement 1C calls for use of school and other public buses in evacuations. Although buses that later flooded were available to transport those dependent upon public transportation, not enough bus drivers were available to drive them as Governor Blanco did not sign an emergency waiver to allow any licensed driver to transport evacuees on school buses. However, 20 year old Jabbar Gibson armed with only a standard operator's permit took it upon himself to take a school bus and drive it to Houston with 50 to 70 evacuees. Rental cars were scarce and most public transportation had shut down well before Katrina arrived. Some estimates claimed that 80 per cent of the 1.3 million residents of the greater New Orleans metropolitan area evacuated, leaving behind substantially fewer people than remained in the city during the Hurricane Ivan evacuation.

By Sunday, August 28, most infrastructure along the Gulf Coast had been shut down, including all Canadian National Railway and Amtrak rail traffic into the evacuation areas as well as the Waterford Nuclear Generating Station. The NHC maintained the coastal warnings until late on August 29, by which time Hurricane Katrina was over central Mississippi.

City of new orleans

By August 26, the possibility of unprecedented cataclysm was already being considered. Many of the computer models had shifted the potential path of Katrina 150 miles (240 km) westward from the Florida Panhandle, putting the city of New Orleans directly in the center of their track probabilities; the chances of a direct hit were forecast at 17 per cent, with strike probability rising to 29 per cent by August 28. This scenario was considered a potential catastrophe because some parts of New Orleans and the metro area are below sea level. Since the storm surge produced by the hurricane's right-front quadrant (containing the strongest winds) was forecast to be 28 feet (8.5 m), emergency management officials in New Orleans feared that the storm surge could go over the tops of levees protecting the city, causing major flooding.

At a news conference at 10 am on August 28, shortly after Katrina was upgraded to a Category 5 storm, New Orleans mayor Ray Nagin ordered the first-ever mandatory evacuation of the city, calling Katrina 'a storm that most of us have long feared'. The city government also established several 'refuges of last resort' for citizens who could not leave the city, including the massive Louisiana Superdome, which sheltered approximately 26,000 people and provided them with food and water for several days as the storm came ashore.

Florida

Many people living in the South Florida area were unaware when Katrina strengthened from a tropical storm to a hurricane in one day and struck southern Florida near the Miami-Dade–Broward county line. The hurricane struck between the cities of Aventura, in Miami-Dade County, and Hallandale, in Broward County, on Thursday, August 25, 2005. However, National Hurricane Center (NHC) forecasts had correctly predicted that Katrina would intensify to hurricane strength before landfall, and hurricane watches and warnings were issued 31.5 hours and 19.5 hours before landfall, respectively — only slightly less than the target thresholds of 36 and 24 hours.

Florida Governor Jeb Bush declared a state of emergency on August 24 in advance of Hurricane Katrina's landfall in Florida. Shelters were opened and schools closed in several counties in the southern part of the state. A number of evacuation orders were also issued, mostly voluntary, although a mandatory evacuation was ordered for vulnerable housing in Martin County.

Impact

On August 29, Katrina's storm surge caused 53 different levee breaches in greater New Orleans submerging eighty per cent of the city. A June 2007 report by the American Society of Civil Engineers indicated that two-thirds of the flooding were caused by the multiple failures of the city's floodwalls. Not mentioned were the flood gates that were not closed. The storm surge also devastated the coasts of Mississippi and Alabama, making Katrina the most destructive and costliest natural disaster in the history of the United States, and the deadliest hurricane since the 1928 Okeechobee Hurricane. The total damage from Katrina is estimated at $81.2 billion, nearly double the cost of the previously most expensive storm, Hurricane Andrew, when adjusted for inflation.

The confirmed death toll (total of direct and indirect deaths) is 1,836, mainly from Louisiana (1577) and Mississippi (238). However, 135 people remain categorised as missing in Louisiana, and many of the deaths are indirect, but it is almost impossible to determine the exact cause of some of the fatalities. The relative lack of status, power, and resources put many women at risk of being sexually assaulted during Hurricane Katrina.

Federal disaster declarations covered 90,000 square miles (2,33,000 km^2) of the United States, an area almost as large as the United Kingdom. The hurricane left an estimated three million people without electricity. On September 3, 2005, Homeland Security Secretary Michael Chertoff described the aftermath of Hurricane Katrina as 'probably the worst catastrophe, or set of catastrophes,' in the country's history, referring to the hurricane itself plus the flooding of New Orleans. Even in 2010, debris remained in some coastal communities.

South Florida and Cuba

Hurricane Katrina first made landfall on August 25, 2005 in South Florida where it hit as a Category 1 hurricane, with 80 mph (130 km/hr) winds. Rainfall was heavy in places and exceeded 14 inches (350 mm) in Homestead, Florida, and a storm surge of 3–5 feet (1.5 m) was measured in parts of Monroe County. More than 1 million customers were left without electricity, and damage in Florida was estimated from $1–$2 billion, with most of the damage coming from flooding and overturned trees. There were 14 fatalities reported in Florida as a result of Hurricane Katrina.

Most of the Florida Keys experienced tropical-storm force winds from Katrina as the storm's center passed to the north, with hurricane force winds reported in the Dry Tortugas. Rainfall was also high in the islands, with 10 inches (250 mm) falling on Key West. On August 26, a strong F1 tornado formed from an outer rain band of Katrina and struck Marathon. The tornado damaged a hangar at the airport there and caused an estimated $5 million in damage.

Although Hurricane Katrina stayed well to the north of Cuba, on August 29 it brought tropical-storm force winds and rainfall of over 8 inches (200 mm) to western regions of the island. Telephone and power lines were damaged and around 8000 people were evacuated in the Pinar del Río Province. According to Cuban television reports the coastal city of Surgidero de Batabano was 90 per cent underwater.

Louisiana

On August 29, Hurricane Katrina made landfall near Buras-Triumph, Louisiana with 125 mph (205 km/hr) winds, as a strong Category 3 storm. However, as it had only just weakened from Category 4 strength and the radius of maximum winds was large, it is possible that sustained winds of Category 4 strength briefly impacted extreme southeastern Louisiana. Although the storm surge to the east of the path of the eye in Mississippi was higher, a very significant surge affected the Louisiana coast. The height of the surge is uncertain because of a lack of data, although a tide gauge in Plaquemines Parish indicated a storm tide in excess of 14 feet (4.3 m) and a 12-foot (3 m) storm surge was recorded in Grand Isle. Hurricane Katrina made final landfall near the mouth of the Pearl River, with the eye straddling St. Tammany Parish, Louisiana and Hancock County, Mississippi, on the morning of August 29 at about 9:45 am CST.

Hurricane Katrina also brought heavy rain to Louisiana, with 8–10 inches (200–250 mm) falling on a wide swath of the eastern part of the state. In the area around Slidell, the rainfall was even higher, and the highest rainfall recorded in the state was approximately 15 inches (380 mm). As a result of the rainfall and storm surge the level of Lake Pontchartrain rose and caused significant flooding along its northeastern shore, affecting communities from Slidell to Mandeville. Several bridges were destroyed, including the I-10 Twin Span Bridge connecting Slidell to New Orleans. Almost 9,00,000 people in Louisiana lost power as a result of Hurricane Katrina.

Katrina's storm surge inundated all parishes surrounding Lake Pontchartrain, including St. Tammany, Tangipahoa, St. John the Baptist and St. Charles Parishes. St. Tammany Parish received a two-part storm surge: First, as Lake Pontchartrain rose and the storm blew water from the Gulf of Mexico into the lake. Second, as the eye of Katrina passed, westerly winds pushed water into a bottleneck at the Rigolets Pass, forcing it farther inland. The range of surge levels in eastern St. Tammany Parish is estimated at 13 to 16 feet (4.9 m), not including wave action.

Hard-hit St. Bernard Parish was flooded due to breaching of the levees that contained a navigation channel called the Mississippi River Gulf Outlet (MR-GO) and the breach of the Levee Board designed and built 40 Arpent canal levee. The search for the missing was undertaken by the St. Bernard Fire Department due to the assets of the United States Coast Guard being diverted to New Orleans. Many of the missing in the months after the storm were tracked down by searching flooded homes, tracking credit card records, and visiting homes of family and relatives.

According to the US Dept. of Housing and Urban Development, in St. Bernard Parish, 81 per cent (20,229) of the housing units were damaged. In St. Tammany Parish, 70 per cent (48,792) were damaged and in Placquemines Parish 80 per cent (7212) were damaged.

New orleans

As the eye of Hurricane Katrina swept to the northeast, it subjected the city to hurricane conditions for hours. Although power failures prevented accurate measurement of wind speeds in New Orleans, there were a few measurements of hurricane-force winds. From this the NHC concluded that it is likely that much of the city experienced sustained winds of Category 1 or Category 2 strength.

Katrina's storm surge led to 53 levee breaches in the federally built levee system protecting metro New Orleans and the failure of the 40 Arpent Canal levee. Nearly every levee in metro New Orleans was breached as Hurricane Katrina passed just east of the city limits. Failures occurred in New Orleans and surrounding communities, especially St. Bernard Parish. The Mississippi River Gulf Outlet (MRGO) breached its levees in approximately 20 places, flooding much of east New Orleans. most of Saint Bernard Parish and the East Bank of Plaquemines Parish. The major levee breaches in the city included breaches at the 17th Street Canal levee, the London Avenue Canal, and the wide, navigable Industrial Canal, which left approximately 80 per cent of the city flooded.

Most of the major roads travelling into and out of the city were damaged. The only routes out of the city were the westbound Crescent City Connection and the Huey P. Long Bridge, as large portions of the I-10 Twin Span Bridge travelling eastbound towards Slidell, Louisiana had collapsed. Both the Lake Pontchartrain Causeway and the Crescent City Connection only carried emergency traffic.

On August 29, at 7:40 am CST, it was reported that most of the windows on the north side of the Hyatt Regency New Orleans had been blown out, and many other high rise buildings had extensive window damage. The Hyatt was the most severely damaged hotel in the city, with beds reported to be flying out of the windows. Insulation tubes were exposed as the hotel's glass exterior was completely sheared off.

The Superdome, which was sheltering many people who had not evacuated, sustained significant damage. Two sections of the Superdome's roof were compromised and the dome's waterproof membrane had essentially been peeled off. Louis Armstrong New Orleans International Airport was closed before the storm but did not flood. On August 30, it was reopened to humanitarian and rescue operations. Limited commercial passenger service resumed at the airport on September 13 and regular carrier operations resumed in early October.

Levee breaches in New Orleans also caused widespread loss of life, with over 700 bodies recovered in New Orleans by October 23, 2005. Some survivors and evacuees reported seeing dead bodies lying in city streets and floating in still-flooded sections, especially in the east of the city. The advanced state of decomposition of many corpses, some of which were left in the water or sun for days before being collected, hindered efforts by coroners to identify many of the dead.

The first deaths reported from the city were reported shortly before midnight on August 28, as three nursing home patients died during an evacuation to Baton Rouge, most likely from dehydration. While there were also early reports of fatalities amid mayhem at the Superdome, only six deaths were confirmed there, with four of these originating from natural causes, one from a drug overdose, and one a suicide. At the Convention Center, four bodies were recovered. One of the four is believed to be the result of a homicide.

Mississippi

The Gulf coast of Mississippi suffered massive damage from the impact of Hurricane Katrina on August 29, leaving 238 people dead, 67 missing, and billions of dollars in damage: bridges, barges, boats, piers, houses and cars were washed inland. Katrina travelled up the entire state, and afterwards, all 82 counties in Mississippi were declared disaster areas for federal assistance, 47 for full assistance.

After making a brief initial landfall in Louisiana, Katrina had made its final landfall near the state line, and the eyewall passed over the cities of Bay St. Louis and Waveland as a Category 3 hurricane with sustained winds of 120 mph (195 km/hr). Katrina's powerful right-front quadrant passed over the west and central Mississippi coast, causing a powerful 27-foot (8.2 m) storm surge, which penetrated 6 miles (10 km) inland in many areas and up to 12 miles (20 km) inland along bays and rivers; in some areas, the surge crossed Interstate 10 for several miles. Hurricane Katrina brought strong winds to Mississippi, which caused significant tree damage throughout the state. The highest unofficial reported wind gust recorded from Katrina was one of 135 mph (217 km/hr) in Poplarville, in Pearl River County.

The storm also brought heavy rains with 8–10 inches (200–250 mm) falling in southwestern Mississippi and rain in excess of 4 inches (100 mm) falling throughout the majority of the state. Katrina caused eleven tornadoes in Mississippi on August 29, some of which damaged trees and power lines.

Battered by wind, rain and storm surge, some beachfront neighbourhoods were completely levelled. Preliminary estimates by Mississippi officials calculated that 90 per cent of the structures within half a mile of the coastline were completely destroyed, and that storm surges traveled as much as six miles (10 km) inland in portions of the state's coast. One apartment complex with approximately thirty residents seeking shelter inside collapsed. More than half of the 13 casinos in the state, which were floated on barges to comply with Mississippi land-based gambling laws, were washed hundreds of yards inland by waves.

A number of streets and bridges were washed away. On US Highway 90 along the Mississippi Gulf Coast, two major bridges were completely destroyed: the Bay St. Louis—Pass Christian bridge, and the Biloxi-Ocean Springs bridge. In addition, the eastbound span of the I-10 bridge over the Pascagoula River estuary was damaged. In the weeks after the storm, with the connectivity of the coastal US Highway 90 shattered, traffic travelling parallel to the coast was reduced first to State Road 11 (parallel to I-10) then to two lanes on the remaining I-10 span when it was opened.

All three coastal counties of the state were severely affected by the storm. Katrina's surge was the most extensive, as well as the highest, in the documented history of the United States; large portions of both Hancock, Harrison, and Jackson Counties were inundated by the storm surge, in all three cases affecting most of the populated areas. Surge covered almost the entire lower half of Hancock County, destroying the coastal communities of Clermont Harbour and Waveland, much of Bay St. Louis, and flowed up the Jourdan River, flooding Diamondhead and Kiln. In Harrison County, Pass Christian was completely inundated, along with a narrow strip of land to the east along the coast, which includes the cities of Long Beach and Gulfport; the flooding was more extensive in communities such as D'Iberville, which borders Back Bay. Biloxi, on a peninsula between the Back Bay and the coast, was particularly hard hit, especially the low-lying Point Cadet area. In Jackson County, storm surge flowed up the wide river estuary, with the combined surge and freshwater flooding cutting the county in half. Remarkably, over 90 per cent of Pascagoula, the easternmost coastal city in Mississippi, and about 75 miles (121 km) east of Katrina's landfall near the Louisiana-Mississippi border, was flooded from surge at the height of the storm. Other large Jackson County neighborhoods such as Porteaux Bay and Gulf Hills were severely damaged with large portions being completely destroyed, and St. Martin was hard hit; Ocean Springs, Moss Point, Gautier, and Escatawpa also suffered major surge damage.

Mississippi Emergency Management Agency officials also recorded deaths in Forrest, Hinds, Warren, and Leake counties. Over 9,00,000 people throughout the state experienced power outages.

Southeast United States

Although Hurricane Katrina made landfall well to the west, Alabama and the Florida Panhandle were both affected by tropical-storm force winds and a storm surge varying from 12 to 16 feet (3–5 m) around Mobile Bay, with higher waves on top. Sustained winds of 67 mph (107 km/hr) were recorded in Mobile, Alabama, and the storm surge there was approximately 12 feet (3.7 m). The surge caused significant flooding several miles inland along Mobile Bay. Four tornadoes were also reported in Alabama. Ships, oil rigs, boats and fishing piers were washed ashore along Mobile Bay: the cargo ship M/V *Caribbean Clipper* and many fishing boats were grounded at Bayou La Batre.

An oil rig under construction along the Mobile River broke its moorings and floated 1.5 miles (2 km) northwards before striking the Cochrane Bridge just outside Mobile. No significant damage resulted to the bridge and it was soon reopened. The damage on Dauphin Island was severe, with the surge destroying many houses and cutting a new canal through the western portion of the island. An offshore oil rig also became grounded on the island. As in Mississippi, the storm surge caused significant beach erosion along the Alabama coastline. More than 6,00,000 people lost power in Alabama as a result of Hurricane Katrina and two people died in a traffic accident in the state. Residents in some areas, such as Selma, were without power for several days.

Along the Florida Panhandle the storm surge was typically about five feet (1.5 m) and along the west-central Florida coast there was a minor surge of 1–2 feet (0.3–0.6 m). In Pensacola, Florida 56 mph (90 km/hr) winds were recorded on August 29. The winds caused damage to some trees and structures and there was some minor flooding in the Panhandle. There were two indirect fatalities from Katrina in Walton County as a result of a traffic accident. In the Florida Panhandle, 77,000 customers lost power.

Northern and central Georgia were affected by heavy rains and strong winds from Hurricane Katrina as the storm moved inland, with more than 3 inches (75 mm) of rain falling in several areas. At least 18 tornadoes formed in Georgia on August 29, the most on record in that state for one day in August.

The most serious of these tornadoes was an F2 tornado which affected Heard County and Carroll County. This tornado caused 3 injuries and one fatality and damaged several houses. In addition this tornado destroyed several poultry barns, killing over 1,40,000 chicks. The other tornadoes caused significant damages to buildings and agricultural facilities. In addition to the fatality caused by the F2 tornado, there was another fatality in a traffic accident.

Other US States and Canada

Hurricane Katrina weakened as it moved inland, but tropical-storm force gusts were recorded as far north as Fort Campbell, Kentucky on August 30, and the winds damaged trees in New York. The remnants of the storm brought high levels of rainfall to a wide swath of the eastern United States, and rain in excess of 2 inches (50 mm) fell in parts of 20 states. A number of tornadoes associated with Katrina formed on August 30 and August 31, which caused minor damages in several regions. In total, 62 tornadoes formed in eight states as a result of Katrina.

Eastern Arkansas received light rain from the passage of Katrina. Gusty winds downed some trees and power lines, though damage was minimal. In Kentucky, a storm that had moved through the weekend before had already produced flooding and the rainfall from Katrina added to this. As a result of the flooding, Kentucky Governor Ernie Fletcher declared three counties disaster areas and a statewide state of emergency. One person was killed in Hopkinsville, Kentucky and part of a high school collapsed. Flooding also prompted a number of evacuations in West Virginia and Ohio, the rainfall in Ohio leading to two indirect deaths. Katrina also caused a number of power outages in many areas, with over 1,00,000 customers affected in Tennessee, primarily in the Memphis and Nashville areas.

The remnants of Katrina were absorbed by a new cyclone to its east across Pennsylvania. This second cyclone continued north and affected Canada on August 31. In Ontario there were a few isolated reports of rain in excess of 100 mm (4 inches) and there were a few reports of damage from fallen trees. Flooding also occurred in both Ontario and Quebec, cutting off a number of isolated villages in Quebec, particularly in the Côte-Nord region.

Aftermath

Economic effects

The economic effects of the storm were far-reaching. The Bush Administration sought $105 billion for repairs and reconstruction in the region, which did not account for damage to the economy caused by potential interruption of the oil supply, destruction of the Gulf Coast's highway infrastructure, and exports of commodities such as grain. Katrina damaged or destroyed 30 oil platforms and caused the closure of nine refineries; the total shut-in oil production from the Gulf of Mexico in the six-month period following Katrina was approximately 24 per cent of the annual production and the shut-in gas production for the same period was about 18 per cent. The forestry industry in Mississippi was also affected, as 1.3 million acres (5,300 km^2) of forest lands were destroyed. The total loss to the forestry industry from Katrina is calculated to rise to about $5 billion. Furthermore, hundreds of thousands of local residents were left unemployed, which will have a trickle-down effect as fewer taxes are paid to local governments. Before the hurricane, the region supported approximately one million non-farm jobs, with 6,00,000 of them in New Orleans. It is estimated that the total economic impact in Louisiana and Mississippi may exceed $150 billion.

Katrina redistributed over one million people from the central Gulf coast elsewhere across the United States, which became the largest diaspora in the history of the United States. Houston, Texas, had an increase of 35,000 people; Mobile, Alabama, gained over 24,000; Baton Rouge, Louisiana, over 15,000; and Hammond, Louisiana received over 10,000, nearly doubling its size. Chicago received over 6000 people, the most of any non-southern city. By late January, 2006, about 2,00,000 people were once again living in New Orleans, less than half of the pre-storm population. By July 1, 2006, when new population estimates were calculated by the US Census Bureau, the state of Louisiana showed a population decline of 2,19,563, or 4.87 per cent. Additionally, some insurance companies have stopped insuring homeowners in the area because of the high costs from Hurricanes Katrina and Rita, or have raised homeowners' insurance premiums to cover their risk.

Environmental effects

Katrina also had a profound impact on the environment. The storm surge caused substantial beach erosion, in some cases completely devastating coastal areas. In Dauphin Island, approximately 90 miles (150 km) to the east of the point where the hurricane made landfall, the sand that comprised the barrier island was transported across the island into the Mississippi Sound, pushing the island towards land. The storm surge and waves from Katrina also obliterated the Chandeleur Islands, which had been affected by Hurricane Ivan the previous year. The US Geological Survey has estimated 217 square miles (560 km^2) of land was transformed to water by the hurricanes Katrina and Rita.

The lands that were lost were breeding grounds for marine mammals, brown pelicans, turtles, and fish, as well as migratory species such as redhead ducks. Overall, about 20 per cent of the local marshes were permanently overrun by water as a result of the storm.

The damage from Katrina forced the closure of 16 National Wildlife Refuges. Breton National Wildlife Refuge lost half its area in the storm. As a result, the hurricane affected the habitats of sea turtles, Mississippi sandhill cranes, Red-cockaded woodpeckers and Alabama Beach mice.

The storm caused oil spills from 44 facilities throughout southeastern Louisiana, which resulted in over 7 million US gallons (26 million L) of oil being leaked. Some spills were as small as a few hundred gallons; the largest are tabulated to the right. While most of the spills were contained on-site, some oil entered the ecosystem, and the town of Meraux was flooded with a blend of water and oil. Unlike Hurricane Ivan no offshore oil spills were officially reported after Hurricane Katrina. However, Skytruth reported some signs of surface oil in the Gulf of Mexico.

Finally, as part of the cleanup effort, the flood waters that covered New Orleans were pumped into Lake Pontchartrain, a process that took 43 days to complete. These residual waters contained a mix of raw sewage, bacteria, heavy metals, pesticides, toxic chemicals, and oil, which sparked fears in the scientific community of massive numbers of fish dying.

Prior to the storm, subsidence and erosion caused erosion in the Louisiana wetlands and bayous. This, along with the canals built in the area, allowed for Katrina to maintain more of its intensity when it struck.

Looting and violence

Shortly after the hurricane moved away on August 30, 2005, some residents of New Orleans who remained in the city began looting stores. Many were in search of food and water that were not available to them through any other means, as well as non-essential items.

Reports of carjacking, murders, thefts, and rapes in New Orleans flooded the news. Some sources later determined that many of the reports were inaccurate, because of the confusion. Thousands of National Guard and federal troops were mobilised (the total went from 7841 in the area the day Katrina hit to a maximum of 46,838 on September 10) and sent to Louisiana along with numbers of local law enforcement agents from across the country who were temporarily deputised by the state. 'They have M16s and are locked and loaded. These troops know how to shoot and kill and I expect they will', Louisiana Governor Kathleen Blanco said. Congressman Bill Jefferson (D-LA) told ABC News: 'There was shooting going on. There was sniping going on. Over the first week of September, law and order were gradually restored to the city'. Several shootings were between police and New Orleans residents, including a fatal incident at Danziger Bridge. A number of arrests were made throughout the affected area, including some near the New Orleans Convention Center. A temporary jail was constructed of chain link cages in the city train station.

In Texas, where more than 3,00,000 refugees were located, local officials ran 20,000 criminal background checks on the refugees, as well as on the relief workers helping them and people who opened up their homes. The background checks found that 45 per cent of the refugees had a criminal record of some nature, and that 22 per cent had a violent criminal record. The number of homicides in Houston from September 2005 through February 22, 2006 went up by 23 per cent relative to the same period a year before; 29 of the 170 murders involved displaced Louisianans as victims or suspects.

Government response

Within the United States and as delineated in the National Response Plan, disaster response and planning is first and foremost a local government responsibility. When local government exhausts its resources, it then requests specific additional resources from the county level. The request process proceeds similarly from the county to the state to the federal government as additional resource needs are identified. Many of the problems that arose developed from inadequate planning and back-up communications systems at various levels.

Some disaster recovery response to Katrina began before the storm, with Federal Emergency Management Agency (FEMA) preparations that ranged from logistical supply deployments to a mortuary team with refrigerated trucks. A network of volunteers began rendering assistance to local residents and residents emerging from New Orleans and surrounding parishes as soon as the storm made landfall (even though many were directed to not enter the area), and continued for more than six months after the storm. Of the 60,000 people stranded in New Orleans, the Coast Guard rescued more than 33,500. Congress recognised the Coast Guard's response with an official entry in the Congressional Record, and the Armed Service was awarded the Presidential Unit Citation.

The United States Northern Command established Joint Task Force (JTF) Katrina based out of Camp Shelby, Mississippi, to act as the military's on-scene response on Sunday, August 28, with US Army Lieutenant General Russel L. Honoré as commander. Approximately 58,000 National Guard personnel were activated to deal with the storm's aftermath, with troops coming from all 50 states. The Department of Defense also activated volunteer members of the Civil Air Patrol.

Michael Chertoff, Secretary of the Department of Homeland Security, decided to take over the federal, state, and local operations officially on August 30, 2005, citing the National Response Plan. This was refused by Governor Blanco who indicated that her National Guard could manage. Early in September, Congress authorised a total of $62.3 billion in aid for victims. Additionally, President Bush enlisted the help of former presidents Bill Clinton and George H.W. Bush to raise additional voluntary contributions, much as they did after the 2004 Indian Ocean earthquake and tsunami. American flags were also ordered to be half-staff from September 2, 2005 to September 20, 2005 in honour of the victims.

FEMA provided housing assistance (rental assistance, trailers, etc.) to more than 7,00,000 applicants — families and individuals. However, only one-fifth of the trailers requested in Orleans Parish were supplied, resulting in an enormous housing shortage in the city of New Orleans. Many local areas voted to not allow the trailers, and many areas had no utilities, a requirement prior to placing the trailers. To provide for additional housing, FEMA has also paid for the hotel costs of 12,000 individuals and families displaced by Katrina through February 7, 2006, when a final deadline was set for the end of hotel cost coverage. After this deadline, evacuees were still eligible to receive federal assistance, which could be used towards either apartment rent, additional hotel stays, or fixing their ruined homes, although FEMA no longer paid for hotels directly. As of March 30, 2010, there were still 260 families living in FEMA-provided trailers in Louisiana and Mississippi.

Law enforcement and public safety agencies, from across the United States, provided a 'mutual aid' response to Louisiana and New Orleans in the weeks following the disaster. Many agencies responded with manpower and equipment from as far away as California, Michigan, Nevada, New York, and Texas. This response was welcomed by local Louisiana authorities as their staff were either becoming fatigued, stretched too thin, or even quitting from the job. Two weeks after the storm, more than half of the states were involved in providing shelter for evacuees. By four weeks after the storm, evacuees had been registered in all 50 states and in 18,700 zip codes—half of the nation's residential postal zones. Most evacuees had stayed within 250 miles (400 km), but 2,40,000 households went to Houston and other cities over 250 miles (400 km) away and another 60,000 households went over 750 miles (1200 km) away.

Criti ism of government response

The criticisms of the government's response to Hurricane Katrina primarily consisted of criticism of mismanagement and lack of leadership in the relief efforts in response to the storm and its aftermath. More specifically, the criticism focused on the delayed response to the flooding of New Orleans, and the subsequent state of chaos in the Crescent City. The neologism Katrinagate was coined to refer to this controversy, and was a runner-up for '2005 word of the year'.

Within days of Katrina's August 29, 2005 landfall, public debate arose about the local, state and federal governments' role in the preparations for and response to the hurricane. Criticism was initially prompted by televised images of visibly shaken and frustrated political leaders, and of residents who remained stranded by flood waters without water, food or shelter. Deaths from thirst, exhaustion, and violence, days after the storm had passed, fueled the criticism, as did the dilemma of the evacuees at facilities such as the Louisiana Superdome (designed to handle 800, yet 30,000 arrived) and the New Orleans Civic Center (not designed as an evacuation center, yet 25,000 arrived). Some alleged that race, class, and other factors could have contributed to delays in government response. The percentage of black victims among storm-related deaths (49 per cent) was below their proportion in the area's population (approx. 60 per cent).

In accordance with federal law, President George W. Bush directed the Secretary of the Department of Homeland Security, Michael Chertoff, to coordinate the Federal response. Chertoff designated Michael D. Brown, head of the Federal Emergency Management Agency, as the Principal Federal Official to lead the deployment and coordination of all federal response resources and forces in the Gulf Coast region. However, the President and Secretary Chertoff initially came under harsh criticism for what some perceived as a lack of planning and coordination. Brown claimed that Governor Blanco resisted their efforts and was unhelpful. Governor Blanco and her staff disputed this. Eight days later, Brown was recalled to Washington and Coast Guard Vice Admiral Thad W. Allen replaced him as chief of hurricane relief operations. Three days after the recall, Michael D. Brown resigned as director of FEMA in spite of having received recent praise from President Bush.

During *A Concert for Hurricane Relief*, a benefit concert for victims of the hurricane, rapper Kanye West veered off script and harshly criticised the government's response to the crisis, stating that 'George Bush doesn't care about black people'. Although the camera quickly cut away, and the scene was deleted from delayed broadcasts, West's comments still reached the East Coast broadcasts, and were replayed and discussed afterwards.

Criticism from politicians, activists, pundits and journalists of all stripes was directed at the local and state and governments headed by Mayor Ray Nagin of New Orleans and Louisiana Governor Kathleen Blanco. Nagin and Blanco were criticised for failing to implement New Orleans' evacuation plan and for ordering residents to a shelter of last resort without any provisions for food, water, security, or sanitary conditions. Perhaps the most important criticism of Nagin was that he delayed his emergency evacuation order until 19 hours before landfall, which led to hundreds of deaths of people who (by that time) could not find any way out of the city.

The destruction wrought by Hurricane Katrina raised other, more general public policy issues about emergency management, environmental policy, poverty, and unemployment. The discussion of both the immediate response and of the broader public policy issues may have affected elections and legislation enacted at various levels of government. The storm's devastation also prompted a Congressional investigation, which found that FEMA and the Red Cross 'did not have a logistics capacity sophisticated enough to fully support the massive number of Gulf coast victims'. Additionally, it placed responsibility for the disaster on all three levels of government.

An ABC News Poll conducted on September 2, 2005, showed more blame was being directed at state and local governments (75 per cent) than at the Federal government (67 per cent), with 44 per cent blaming Bush's leadership directly. A later CNN/USAToday/Gallup poll showed that respondents disagreed widely on who was to blame for the problems in the city following the hurricane — 13 per cent said Bush, 18 per cent said federal agencies, 25 per cent blamed state or local officials and 38 per cent said no one was to blame.

Five former police officers have pleaded guilty to charges connected to the Danziger Bridge shootings in the aftermath of the hurricane. Six other former or current officers will appear in court in June, 2011. Two unarmed civilians were killed and four others seriously wounded when police opened fire on people attempting to cross the bridge.

International response

Over seventy countries pledged monetary donations or other assistance. Notably, Cuba and Venezuela (both hostile to US government themselves) were the first countries to offer assistance, pledging over $1 million, several mobile hospitals, water treatment plants, canned food, bottled water, heating oil, 1100 doctors and 26.4 metric tons of medicine, though this aid was rejected by the US government. Kuwait made the largest single pledge, $500 million; other large donations were made by Qatar and United Arab Emirates (each $100 million), South Korea ($30 million), Australia ($10 million), India, China (both $5 million), New Zealand ($2 million), Pakistan ($1.5 million), and Bangladesh ($1 million).

India sent tarps, blankets and hygiene kits. An Indian Air Force IL-76 aircraft delivered 25 tons of relief supplies for the Hurricane Katrina victims at the Little Rock Air Force Base, Arkansas on September 13, 2005. Israel sent an IDF delegation to New Orleans to transport aid equipment including 80 tons of food, disposable diapers, beds, blankets, generators and additional equipment which were donated from different governmental institutions, civilian institutions and the IDF. The Bush Administration announced in mid-September that it did not need Israeli divers and physicians to come to the United States for search and rescue missions, but a small team landed in New Orleans on September 10 to give assistance to operations already under way. The team administered first aid to survivors, rescued abandoned pets and discovered hurricane victims.

Countries like Sri Lanka, which was still recovering from the Indian Ocean Tsunami, also offered to help. Countries including Canada, Mexico, Singapore, and Germany sent supplies, relief personnel, troops, ships and water pumps to aid in the disaster recovery. Belgium sent in a team of relief personnel. Britain's donation of 3,50,000 emergency meals did not reach victims because of laws regarding mad cow disease. Russia's initial offer of two jets was declined by the US State Department but accepted later. The French offer was also declined and requested later.

In addition to receiving aid from around the world, there was criticism to go a long with it, including accusations of racism. Quoted from the UK Mirror, 'Many things about the United States are wonderful, but it has a vile underbelly which is usually kept well out of sight. Now in New Orleans it has been exposed to the world'.

Non-governmental organisation response

The American Red Cross, America's Second Harvest (now known as Feeding America), Southern Baptist Convention, Salvation Army, Oxfam, Common Ground Collective, Emergency Communities, Habitat for Humanity, Catholic Charities, Service International, 'A River of Hope', The Church of Jesus Christ of Latter-day Saints (Mormons), and many other charitable organisations provided help to the victims of the storm. They were not allowed into New Orleans proper by the National Guard for several days after the storm because of safety concerns. These organisations raised US $4.25 billion in donations by the public, with the Red Cross receiving over half of the donations.

Volunteers from amateur radio's emergency service wing, the Amateur Radio Emergency Service, provided communications in areas where the communications infrastructure had been damaged or totally destroyed, relaying everything from 911 traffic to messages home. In Hancock County, Mississippi, ham radio operators provided the only communications into or out of the area, and even served as 911 dispatchers. Many corporations also contributed to relief efforts. On September 13, 2005, it was reported that corporate donations to the relief effort were $409 million, and were expected to exceed $1 billion.

During and after the Hurricanes Katrina, Wilma and Rita, the American Red Cross had opened 1470 different shelters across and registered 3.8 million overnight stays. None were allowed in New Orleans however. A total of 2,44,000 Red Cross workers (95 per cent of which were non-paid volunteers) were utilised throughout these three hurricanes. In addition, 3,46,980 comfort kits (such as toothpaste, soap, washcloths and toys for children) and 205,360 cleanup kits (containing brooms, mops and bleach) were distributed. For mass care, the organisation served 68 million snacks and meals to victims of the disasters and to rescue workers. The Red Cross also had its Disaster Health services meet 5,96,810 contacts, and Disaster Mental Health services met 8,26,590 contacts. Red Cross emergency financial assistance was provided to 1.4 million families. Hurricane Katrina was the first natural disaster in the United States in which the American Red Cross utilised its 'Safe and Well' family location website.

In the year following Katrina's strike on the Gulf Coast, The Salvation Army allocated donations of more than $365 million to serve more than 1.7 million people in nearly every state. The organisation's immediate response to Hurricane Katrina included more than 5.7 million hot meals served in and around New Orleans, 8.3 million sandwiches, snacks and drinks. Its SATERN network of amateur radio operators picked up where modern communications left off to help locate more than 25,000 survivors. Salvation Army pastoral care counselors were on hand to comfort the emotional and spiritual needs of 2,77,000 individuals. As part of the overall effort, Salvation Army officers, employees and volunteers contributed more than 9,00,000 hours of service.

Analysis of New Orleans levee failures

A June 2007 report released by the American Society of Civil Engineers states that the failures of the locally built and federally funded levees in New Orleans were found to be primarily the result of system design flaws. The US Army Corps of Engineers who by federal mandate is responsible for the conception, design and construction of the region's flood-control system failed to pay sufficient attention to public safety. According to new modelling and field observations by a team from Louisiana State University, the Mississippi River Gulf Outlet (MRGO), a 200 meter-wide (660 foot-wide) canal designed to provide a shortcut from New Orleans to the Gulf of Mexico, helped provide a funnel for the storm surge, making it 20 per cent higher and 100–200 per cent faster as it crashed into the city. St. Bernard Parish, one of the more devastated areas, lies just south of the MRGO. The Army Corps of Engineers disputes this causality and maintains Katrina would have overwhelmed the levees with or without the contributing effect of the MRGO. The water flowing west from the storm surge was perpendicular to MRGO, and thus the canal had a negligible effect.

There is also the ongoing argument made by residents concerning a possible planned levee breach. This would not be the first time the Army Corps of Engineers has breached a levee. Many references are made to the 1927 flood in which the levee was breached south of New Orleans in order to divert floodwater to the Gulf of Mexico. Recently, the US Fish and Wildlife Service and the Nature Conservancy have developed a floodplain reconnection project in which the Ouachita River would be connected to its floodplain and the Gulf of Mexico. A breach in the levee caused the water level downstream to drop six inches (152 mm) in a previous event in the early 1990s. Both cases show the many benefits of allowing the river to run its course.

On April 5, 2006, months after independent investigators had demonstrated that levee failures were not caused by natural forces beyond intended design strength, Lieutenant General Carl Strock testified before the United States Senate Subcommittee on Energy and Water that 'We have now concluded we had problems with the design of the structure'. He also testified that the US Army Corps of Engineers did not know of this mechanism of failure prior to August 29, 2005. The claim of ignorance is refuted, however, by the National Science Foundation investigators hired by the Army Corps of Engineers, who point to a 1986 study by the Corps itself that such separations were possible in the I-wall design.

Many of the levees have been reconstructed since the time of Katrina. In reconstructing them, precautions were taken to bring the levees up to modern building code standards and to ensure their safety. For example, in every situation possible, the Corps of Engineers replaced I-walls with T-walls. T-walls have a horizontal concrete base that protects against soil erosion underneath the floodwalls.

However, there are funding battles over the remaining levee improvements. In February 2008, the Bush administration requested that the state of Louisiana pay about $1.5 billion of an estimated $7.2 billion for Army Corps of Engineers levee work, a proposal which angered many Louisiana leaders.

On May 2, 2008, Louisiana Gov. Bobby Jindal used a speech to The National Press Club to request that President Bush free up money to complete work on Louisiana's levees. Bush promised to include the levee funding in his 2009 budget, but rejected the idea of including the funding in a war bill, which would pass sooner.

Media Involvement

Many representatives of the news media reporting on the aftermath of Hurricane Katrina became directly involved in the unfolding events, instead of simply reporting. Because of the loss of most means of communication, such as land-based and cellular telephone systems, field reporters in many cases became conduits for information between victims and authorities.

The authorities, who monitored local and network news broadcasts, as well as internet sites, would then attempt to coordinate rescue efforts based on the reports. One illustration was when Geraldo Rivera of Fox News tearfully pleaded for authorities to either send help or evacuate the thousands of evacuees stranded at the Ernest N. Morial Convention Center.

The storm also brought a dramatic rise in the role of Internet sites—especially blogging and community journalism. One example was the effort of NOLA.com, the web affiliate of New Orleans' Times-Picayune, which was awarded the Breaking News Pulitzer Prize, and shared the Public Service Pulitzer with the Biloxi-based Sun Herald. The newspaper's coverage was carried for days only on NOLA's blogs, as the newspaper lost its presses and evacuated its building as water rose around it on August 30. The site became an international focal point for news by local media, and also became a vital link for rescue operations and later for reuniting scattered residents, as it accepted and posted thousands of individual pleas for rescue on its blogs and forums. NOLA was monitored constantly by an array of rescue teams—from individuals to the Coast Guard—which used information in rescue efforts. Much of this information was relayed from trapped victims via the SMS functions of their cell phones, to friends and relatives outside the area, who then relayed the information back to NOLA.com. The aggregation of community journalism, user photos and the use of the internet site as a collaborative response to the storm attracted international attention, and was called a watershed moment in journalism. In the wake of these online-only efforts, the Pulitzer Committee for the first time opened all its categories to online entries.

The role of AM radio was of importance to the hundreds of thousands of persons with no other ties to news. AM radio provided emergency information regarding access to assistance for hurricane victims. Immediately after Hurricane Katrina, radio station WWL-AM (New Orleans) was one of the few area radio stations in the area remaining on the air. The 870 kHz frequency has a clear channel high power designation and the on-going nighttime broadcasts continued to be available up to 500 miles (800 km) away. Announcers continued to broadcast from improvised studio facilities after the storm damaged their transmitter tower. During the period of several weeks when most area radio stations were off the air, WWL-AM's emergency coverage was simulcast on the frequencies of other area radio stations. This emergency service was named 'The United Radio Broadcasters of New Orleans'. To reach emergency radio operators in storm-ravaged areas, many of whom made their volunteer services available to the Red Cross and government entities, WWL-AM was simulcast on shortwave outlet WHRI, owned by World Harvest Radio International. The cellular phone antenna network was severely damaged and completely inoperable for several months.

As the US military and rescue services regained control over the city, there were restrictions on the activity of the media. On September 9, the military leader of the relief effort announced that reporters would have 'zero access' to efforts to recover bodies in New Orleans. Immediately following this announcement, CNN filed a lawsuit and obtained a temporary restraining order against the ban. The next day the government backed down and reversed the ban.

Hurricane Katrina has also been the centerpiece of several documentary films, including Spike Lee's film, *When the Levees Broke*, and Darren Martinez's film, *Hellp*. An episode of the Fox TV series House first broadcast on May 16, 2006, featured a teenage victim of Hurricane Katrina at the center of the main medical storyline. An episode of the BBC show Top Gear was praised by some for being one of the first to show the total scale of the destruction once the waters had receded.

Retirement of Katrina name

Because of the large loss of life and property along the Gulf Coast, the name Katrina was officially retired on April 6, 2006 by the World Meteorological Organisation at the request of the US government. It was replaced by Katia on List III of the Atlantic hurricane naming lists, which will next be used in the 2011 Atlantic hurricane season.

Reconstruction

Reconstruction of each section of the southern portion of Louisiana has been addressed in the Army Corps LACPR (Louisiana Coastal Protection and Restoration) Final Technical Report which identifies areas to not be rebuilt and areas buildings need to be elevated.

The Technical Report includes:

1. Locations of possible new levees to be built.
2. Suggested existing levee modifications.
3. 'Inundation zones', 'Water depths less than 14 feet, Raise-In-Place of Structures', 'Water depths greater than 14 feet, Buyout of Structures', 'Velocity Zones' and 'Buyout of Structures' areas for five different scenarios.

The Corps of Engineers submitted the report to Congress for consideration, planning, and response in mid-2009.

In Popular Culture

In August 2006, HBO and 40 Acres and A Mule Filmworks released When the Levees Broke: A Requiem in Four Acts. A four-hour documentary directed by Spike Lee, When the Levees Broke chronicles the events leading up to Katrina, as well as its immediate aftermath. In August 2010, a sequel documentary by Lee was released. Entitled If God Is Willing and Da Creek Don't Rise, the film marks the fifth anniversary of Katrina's arrival and the status of New Orleans and the Gulf region five years after the storm. Katrina has inspired numerous novels, plays and films, of which the HBO television series Tremé is perhaps the best known.

Glossary

Absorption	:	A process whereby a material extracts one or more substances present in the atmosphere or a mixture of gases or liquids; accompanied by physical or chemical change, or both of the material.
Accretion	:	A phenomenon consisting of the increase in size of particles by the process of external addition; for example, water to salt particles.
Acid rain	:	Rain made artificially acidic by pollutants, particularly by oxides of sulphur and nitrogen. (Natural rainwater is slightly acidic owing to the effect of carbon dioxide dissolved in the water).
Active solar energy systems	:	Direct use of solar energy that requires mechanical power; usually consists of pumps and other machinery to circulate air, water, or other fluids from solar collectors to a heat sink where the heat may be stored.
Adsorption	:	A physical process in which the molecules of either a gas or a liquid are condensed on the surface of a solid material, such as activated carbon, silica gel, etc. Commercial absorbent materials have enormous internal surfaces.
Aerobic	:	Characterised by the presence of free oxygen.
Aerosols	:	Finely divided solid or liquid particles suspended in a gas. Usually considered to range from 50μ to sub-micron. This includes dust, mist, fume, fog, haze and smoke.
Afterburner	:	A device, which includes an auxiliary fuel burner and a combustion chamber, in which combustible air contaminants are incinerated.
Agglomeration	:	The clustering or adhering together of a number of a small particles to form a body or structure (an agglomerate), which then acts as a larger single particle.
Agroecosystem	:	An ecosystem created by agriculture. It has low genetic, species and habitat diversity.
Air contaminants	:	Aerosols or gases that are discharged into the atmosphere, where they might exhibit an adverse effect.
Air monitoring	:	The continuous sampling and measuring of the quantity of air pollutants present in the atmosphere.
Air pollution	:	Presence of abnormal quantities of certain matter in the atmosphere that are detrimental to the health and/or welfare of humankind.
Air quality standards	:	Levels of air pollutants that delineate acceptable levels of pollution over a particular time period. Valuable because they are often tied to emission standards that attempt to control air pollution.
Air, primary	:	The air supplied to fuel in its early stages of combustion.
Air, secondary	:	Air introduced above or beyond the bed of burning fuel to promote complete combustion of volatile material produced in the first stage of combustion.

Ambient air	:	The surrounding local air.
Ambient	:	Surrounding, it is used to describe the physical properties of air (temperature, humidity, pressure, etc.) or air pollution concentration in the open air, as against, that at the point of emission or that indoors, for example, ambient temperature, ambient air quality.
Atmosphere	:	Layer of gases surrounding earth.
Atmospheric inversion	:	A condition in which warmer air is found above cooler air, restricting air circulation; often associated in urban areas with a pollution event.
Autotroph	:	An organism that produces its own food from inorganic compounds and a source of energy. There are photoautotrophs (photosynthetic plants) and chemical autotrophs.
Average residence time	:	A measure of the time it takes for a given part of the total pool or reservoir of a particular material in a system to be cycled through the system. When the size of the pool and rate of throughput are constant, average residence time is the ratio of the total size of the pool or reservoir to the average rate of transfer through the pool.
Baghouse/Bag filter	:	Fabric filter used for dust removal. Usually of tubular or envelope shape. The entire structure housing such bags is called a baghouse.
Biochemical oxygen demand (BOD)	:	A measure of the amount of oxygen necessary to decompose organic material in a unit volume of water. As the amount of organic waste in water increases, more oxygen is used, resulting in a higher BOD.
Biogeochemical cycle	:	The cycling of a chemical element through the biosphere; its pathways, storage locations and chemical forms in the atmosphere, oceans, sediments and lithosphere.
Biological control	:	A set of methods to control pest organisms by using natural ecological interactions, including predation, parasitism and competition.
Biological diversity	:	Used loosely to mean the variety of life on the earth, but technically this concept consists of three components: (i) genetic diversity—the total number of genetic characteristics, (ii) species diversity, and (iii) habitat or ecosystem diversity—the number of kinds of habitats or ecosystems in a given unit area. Species diversity in turn includes three concepts: species richness, evenness and dominance.
Biological evolution	:	The change in inherited characteristics of a population from generation to generation, which can result in new species.
Biomagnification	:	Also called biological concentration. The tendency for some substances to concentrate with each trophic level. Organisms preferentially store certain chemicals and excrete others. When this occurs consistently among organisms, the stored chemicals increase as a percentage of the body weight as the material is transferred along a food chain or trophic level. For example, the concentration of DDT is greater in herbivores than in plants and greater in plants than in the non-living environment.
Biomass	:	The amount of living material, or the amount of organic material contained in living organisms, both as live and dead material, as in the leaves (live) and stem wood (dead) of trees.
Biosphere	:	That part of a planet where life exists. On earth it extends from the depths of the oceans to the summit of mountains, but most life exists within a few meters of the surface.

Biota	:	A general term for all the organisms of all species living in an area or region up to and including the biosphere, as in 'the biota of the Mojave Desert' or 'the biota in that aquarium'.
Carbon cycle	:	Combined biochemical cycles of carbon, oxygen and hydrogen. Carbon combines with and is chemically and biologically linked with the cycles of oxygen and hydrogen that form the major compounds of life.
Carcinogen	:	Any material that is known to produce cancer in human or other animals.
Carnivores	:	Organisms that feed on other live organisms; usually applied to animals that eat other animals.
Chemoautotrophs	:	Autotrophic bacteria that can derive energy from chemical reactions of simple inorganic compounds. The most common use for inorganic sulphur compounds.
Climate	:	The representative or characteristic conditions of the atmosphere at particular places on earth. Climate refers to the average or expected conditions over long periods; weather refers to the particular conditions at one time in one place.
Climatic change	:	Change in mean annual temperature and other aspects of climate over periods of time ranging from decades to hundreds of years to several million years.
Coarse solid particles	:	Solid particles having a size equal to or greater than 50 microns and solid particles contained in or on liquid particles.
Cogeneration	:	The capture and use of waste heat; for example, using waste heat from a power plant to heat adjacent factories and other buildings.
Combustion	:	The chemical combination of oxygen and combustible matter, resulting in the rapid release of energy and products of combustion (incompletely burned organic species, CO, CO_2, H_2O, NO_x, etc.).
Controlled burning	:	Using prescribed fire to reduce the risk from wildfires, control tree diseases, increase food and habitat for wildlife and manage forests for greater production of desirable tree species.
Convection	:	The transfer of heat involving the movement of particles; for example, the boiling water in which hot water rises to the surface and displaces cooler water, which moves toward the bottom.
Cyclone collectors	:	A mechanical system employing centrifugal force for removal of aerosols (liquids or wet or dry solids) from gas streams by imposing a rapid whirling motion to the entering gas stream.
Desertification	:	The process of creating a desert where there was not one before. Farming in marginal grasslands, which destroys the soil and prevents the future recovery of natural vegetation, is an example of desertification.
Dispersion	:	A general term for a system consisting or particulate matter suspended in air or other gases.
Diversity, species	:	Used loosely to mean the variety of species in an area or on the earth. Technically, it is composed of three components: species richness — the total number of species; species evenness — the relative abundance of species; and species dominance — the most abundant species.
Dust	:	Finely divided particles formed by mechanical disintegration (size = 100 to several hundred microns). Dust particles usually are irregular in shape.
Ecology	:	The science of the study of the relationships between living things and their environment.

Ecosystem	:	An ecological community and its local, nonbiological community. An ecosystem is the minimum system that includes and sustains life. It must include at least an autotroph, a decomposer, a liquid medium, a source and sink of energy and all the chemical elements required by the autotroph and the decomposer.
Ecosystem effect	:	Effects that result from interactions among different species, effects of species on chemical elements in their environment and conditions of the environment.
El Niño	:	Natural perturbation of the physical earth system that effects global climate. Characterised by development of warm oceanic water in the eastern part of the tropical Pacific Ocean, a weakening or reversal of the trade winds and a weakening or even reversal of the equatorial ocean currents. Reoccurs periodically and effects the atmosphere and global temperature by pumping heat into the atmosphere.
Electrostatic precipitator	:	An electrical system for the removal of aerosols from a gas stream by giving them an electrical charge and then collecting them on a plate imparted with an opposite charge.
Environment	:	Literally surroundings; may apply to the indoor or working conditions but commonly refers to man's total surroundings, natural and man-made. Parts of the environment, are often specified. For example, biological environment, water environment, air environment, marine environment, etc.
Environmental ethics	:	A school, or theory, in philosophy that deals with the ethical value of the environment, including especially the rights of nonhuman objects and systems in the environment, for example, trees and ecosystems.
Environmental impact	:	The effects of some action on the environment, particularly action by human beings.
Environmental risk	:	Used in discussions of endangered species to mean variation in the physical or biological environment, including variations in predator, prey, symbiotic, or competitor species that can threaten a species with extinction.
Fission	:	The splitting of an atom into smaller fragments with the release of energy.
Flue gas	:	Gaseous product of combustion, for example, a boiler furnace or a kiln.
Fluidised-bed combustion	:	A process used during the combustion of coal to eliminate sulphur oxides. Involves mixing finely ground limestone with coal and burning it in suspension.
Fly ash	:	Finely divided particles of ash entrained in flue gases arising from the combustion of fossil fuels (predominantly coal). The particles of ash may contain incompletely burned fuel.
Fossil fuels	:	Fossil fuels are forms of stored solar energy created from incomplete biological decomposition of dead organic matter. Include coal, crude oil and natural gas.
Fugitive sources	:	Type of stationary air pollution sources that generate pollutants from open areas exposed to wind processes.
Geochemical cycles	:	The pathways of chemical elements in geologic processes, including the chemistry of the lithosphere, atmosphere and hydrosphere.
Geologic cycle	:	The formation and destruction of earth materials and the processes responsible for these events. The geologic cycle includes the following subcycles: hydrologic, tectonic, rock and geochemical.
Global forecasting	:	Process of predicting or forecasting future change in environmental areas such as world population, natural resource utilisation and environmental degradation.
Global warming	:	Natural or human-induced increase in the average global temperature of the atmosphere near the earth's surface.

Greenhouse effect	:	Process of trapping heat in the atmosphere. Water vapour and several other gases warm the earth's atmosphere because they absorb and remit radiation, that is, they trap some of the heat radiating from the earth's atmospheric system.
Growth rate	:	The net increase in some factor per unit time. In ecology, the growth rate of a population is sometimes measured as the increase in numbers of individuals or biomass per unit time and sometimes as a percentage increase in numbers or biomass per unit time.
Hazardous air pollutant	:	An air pollutant to which no ambient air quality standard is applicable and which in the judgement of the control authority causes, or contributes to air pollution that may reasonably be anticipated to result in an increase in mortality, or in serious irreversible, or reversible incapacitating illness.
Herbivore	:	An organism that feeds on an autotroph.
Heterotrophs	:	Organisms that cannot make their own food from inorganic chemicals and a source of energy and therefore live by feeding on other organisms.
Hydrocarbons	:	Compounds containing only hydrogen and carbon.
Incineration	:	The process of burning solid, semi-solid or gaseous combustible wastes to an inoffensive gas and a sterile residue, containing little or no combustible material.
Industrial ecology	:	Process of designing industrial systems to behave more like ecosystems where waste from one part of the system is a resource for another part.
Lithosphere	:	Outer layer of earth, approximately 100 km thick, of which the plates that contain the ocean basins and the continents are composed.
Microclimate	:	The climate of a very small local area. For example, the climate under a tree, near the ground within a forest, or near the surface of streets in a city.
Migration	:	The movement of an individual, population or species from one habitat to another or more simply from one geographic area to another.
Mitigation	:	Process that identifies actions to avoid, lessen, or compensate for anticipated adverse environmental impacts.
Monitoring	:	Process of collecting data on a regular basis at specific sites to provide a database from which to evaluate change. For example, collection of water samples from beneath a landfill to provide early warning should a pollution problem arise.
Nitrogen cycle	:	A complex biogeochemical cycle responsible for moving important nitrogen components through the biosphere and other earth systems. This is an extremely important cycle because nitrogen is required by all living things.
Noise pollution	:	A type of pollution characterised by unwanted or potentially damaging sound.
Non-renewable energy	:	Alternative energy sources, including nuclear and geothermal, that are dependent on fuels or a resource that may be used up much faster than it is replenished by natural processes.
Ocean thermal conversion	:	Direct utilisation of solar energy using part of a natural oceanic environment as a gigantic solar collector.
Oxidants	:	Those substances in the air (such as ozone, PAN and nitrogen dioxide) that are capable of oxidising other chemicals or elements in oxidation-reduction type chemical reactions. Oxidants are made up mostly of ozone and only small amounts of the other materials.

Ozone (O₃)	:	Form of oxygen in which three atoms of oxygen occur together. Is chemically active and has a short average lifetime in the atmosphere? Forms a natural layer high in the atmosphere (stratosphere) that protects us from harmful ultraviolet radiation from the sun. Is an air pollutant when present in the lower atmosphere above the National Air Quality Standards?
Ozone shield	:	Stratospheric ozone layer that absorbs ultraviolet radiation.
Particulate matter	:	Any liquid or solid matter in the atmosphere. Size may range from 0.0002μ to 500μ in diameter.
Phosphorus cycle	:	Major biogeochemical cycle involving the movement of phosphorus throughout the biosphere and lithosphere. This cycle is important because phosphorus is an essential element for life and often is a limiting nutrient for plant growth.
Photochemical oxidants	:	Result from atmospheric interactions of nitrogen dioxide and sunlight. Most common is ozone (O_3).
Photochemical reactions	:	Chemical reactions occuring in the atmosphere under the influence of sunlight. The energy for these reactions is obtained from sunlight, particularly ultraviolet rays. The results of these reactions are photochemical smogs.
Photochemical smog	:	Sometimes called L.A.-type smog or brown air. Directly related to automobile use and solar radiation. Reactions that occur in the development of the smog are complex and involve both nitrogen oxides and hydrocarbons in the presence of sunlight.
Photosynthesis	:	Synthesis of sugars from carbon dioxide and water by living organisms using light as energy. Oxygen is given off as a by-product.
Point sources	:	Sources of pollution such as smokestacks, pipes, or accidental spills that are readily identified and stationary. They are often thought to be easier to recognise and control area sources. This is true only in a general sense, as some very large point sources emit tremendous amounts of pollutants into the environment.
Pollutant	:	In general terms, any factor that has a harmful effect on living things or their environment.
Primary pollutants	:	Air pollutants emitted directly into the atmosphere. Included particulates are, sulphur oxides, carbon monoxide, nitrogen oxides and hydrocarbons.
Pyrolysis	:	Breaking down burnable waste by combustion in the absence of oxygen. Usually, high heat is applied to the waste in a closed chamber evaporating all moisture and breaking down materials into gaseous hydrocarbons and carbon-like residue.
Recycle	:	Integral part of waste management that attempts to identify resources in the waste stream that may be collected and reused.
Rule of climatic similarity	:	Similar environment lead to the evolution of organisms similar in form and function (but not necessarily in genetic heritage or internal makeup) and to similar ecosystems.
Scrubbing/scrubber	:	An absorption operation in which gaseous or fine particulate pollutants are removed from a stream of air or gas by contact with a liquid spray or a bath or wetted packing/wet surfaces in a tower. The apparatus/equipment used is known as scrubber.
Secondary pollutants	:	Air pollutants produced through reactions between primary pollutants and normal atmospheric compounds. An example is ozone that forms over urban areas through reactions of primary pollutants, sunlight and natural atmospheric gases.

Sediment pollution	:	By volume and mass, sediment is our greatest water pollutant. It may choke streams, fill reservoirs, bury vegetation and generally create a nuisance that is difficult to remove.
Solar energy	:	Collecting and using energy from the sun directly.
Sulphur dioxide (SO_2)	:	Colourless and odourless gas normally present at the earth's surface in low concentrations. An important precursor to acid rain. Major anthropogenic source is burning fossil fuels.
Sulphurous smog	:	Produced primarily by burning coal or oil at large power plants. Sulphur oxides and particulates combine under certain meteorological conditions to produce a concentrated form of this smog.
Sustainable ecosystem	:	An ecosystem that is subject to some human use, but at a level that leads to no loss of species or of necessary ecosystem functions.
Thermal pollution	:	A type of pollution that occurs when heat is released into water or air and produces undesirable effects on the environment.
Threshold odour	:	The lowest concentration of an odour bearing gas at which only half of a panel of sniffers can detect its presence.
Toxic pollutant	:	Those pollutants, or combinations of pollutants, including disease causing agents, which after discharge and upon exposure, ingestion, inhalation or assimilation into any organism, either directly from the environment or indirectly through food chains, will cause death, disease, behavioural abnormalities, cancer, genetic mutations, physiological malfunction (including malfunctions in reproduction) or physical deformations, in such organisms or their offspring.
Toxicology	:	The science concerned with study of poisons (or toxins) and their effects on living organisms. The subject also includes the clinical, industrial, economic and legal problems associated with toxic materials.

References

Allen, E., *Environment and Man*, Blackwell Scientific Publications, London.

Alabaster, M., *Man Made Climate Change Acid Precipitation*, Marcel Dekker Inc., New York.

Andrew, K.W. and Lloyd, R., *Chemistry of Air Pollution*, Butterworths, London.

Baker, J.M., *Automotive Pollution*, Applied Science Publishers, London.

Brown, R.L., *Pesticides in Clinical Practice*, Academic Press, London.

Bryan, G.W., *Effects of Pollutants on Aquatic Organisms*, Cambridge University Press, Cambridge.

Bryan, G.W., *Elements of Air Pollution*, Academic Press, London.

Budyko, M.I., *Global Ecology*, Progress Publishers, Moscow.

Cavallaro, A. and Galli, G., *Air Pollution Analysis*, Academic Press, London.

Chapman, S.B., *Air Pollution Sources*, Blackwell Scientific Publications, London.

Connell, D.W. and Miller, G.J., *Chemistry and Ecotoxicology of Pollution*, John Wiley & Sons, New York.

Coolingwood, R.W., *Biological Aspects of Air Pollution*, John Wiley & Sons, New York.

Curds, C.R. and Hawkes, H.A., *Ecological Aspects of Air Pollution*, Academic Press, London.

Curtis, J.T., *Pollution Prevention*, University of Wisconsin Press, Madison.

Dix, H.M., *Environmental Pollution*, John Wiley & Sons, New York.

Dugan, P.R., *Environmental Chemistry*, Plenum Publishing Corporation, London.

Fruh, G.E., Gloyna, E.F. and Eckenfelder, W.W., *Global Environment*, University of Texas Press, Austin.

Goldberg, E.D., *Atmosphere and Pollution*, Gordon and Breach, Science Publishers, New York.

Golterman, L., *Meteorological Aspects of Air Pollution*, John Wiley & Sons, New York.

James, A. and Evison L., *Air Pollution Technology*, John Wiley & Sons, New York.

Kathern, R.L., *Radioactivity in the Environment: Sources, Distribution and Surveillance*, Harwood Academic Publishers, New York.

Ledbetter, J.O., *Air Pollution: Analysis*, Marcel Dekker, New York.

Lehr, J.H., Tyler, E.G., Wayne, A.P. and Jack, D., *Industrial Health Engineering*, McGraw-Hill, New York.

Lenihan, J., *Aquatic Chemistry*, Blackie, Glasgow and London.

McCaull, J. and Crossland, J., *Air Pollution*, Harcourt Brace Jovanovich, New York.

Palmer, C.M., *Fundamentals of Air Pollution*, Castle Housing Publications Ltd., London.

Smith, K.M., *Methods for Air Sampling and Analysis*, 2nd ed. American Public Health Association, USA.

Stumm, W. and Stumm-Zollineger, E., *Air Pollution Microbiology*, Wiley Interscience, New York.

Teal, J.M., *Meteorological Aspects of Air Pollution*, Pergamon Press, New York.

Tebbutt, T.H.Y., *Air Quality*, Pergamon Press, Oxford, London.

Vinogradov, A.P., *Prediction and Dispersion of Airborne Effects*, Consultants Bureau, New York.

Index